Modeling and Optimization in Science and Technologies

Volume 1

Series Editors

For further volumes:
http://www.springer.com/series/10577

About This Series

The book series *Modeling and Optimization in Science and Technologies (MOST)* publishes basic principles as well as novel theories and methods in the fast-evolving field of modeling and optimization. Topics of interest include, but are not limited to: methods for analysis, design and control of complex systems, networks and machines; methods for analysis, visualization and management of large data sets; use of supercomputers for modeling complex systems; digital signal processing; molecular modeling; and tools and software solutions for different scientific and technological purposes. Special emphasis is given to publications discussing novel theories and practical solutions that, by overcoming the limitations of traditional methods, may successfully address modern scientific challenges, thus promoting scientific and technological progress. The series publishes monographs, contributed volumes and conference proceedings, as well as advanced textbooks. The main targets of the series are graduate students, researchers and professionals working at the forefront of their fields.

Edward Y.L. Gu

A Journey from Robot to Digital Human

Mathematical Principles and Applications with MATLAB Programming

Springer

Edward Y.L. Gu
Dept. of Electrical and Computer
Engineering
Oakland University
Rochester, Michigan
USA

ISSN 2196-7326 ISSN 2196-7334 (electronic)
ISBN 978-3-642-39046-3 ISBN 978-3-642-39047-0 (eBook)
DOI 10.1007/978-3-642-39047-0
Springer Heidelberg New York Dordrecht London

Library of Congress Control Number: 2013942012

Printed on acid-free paper

Springer is part of Springer Science+Business Media (www.springer.com)

To my family Sabrina,
Heather and Jacob

Preface

This book is intended to be a robotics textbook with an extension to digital human modeling and MATLAB^{TM} programming for both senior undergraduate and graduate engineering students. It can also be a research book for researchers, scientists, and engineers to learn and review the fundamentals of robotic systems as well as the basic methods of digital human modeling and motion generation. In the past decade, I wrote and annually updated two lecture notes: Robotic Kinematics, Dynamics and Control, and Modern Theories of Nonlinear Systems and Control. Those lecture notes were successfully adopted by myself as the official textbooks for my dual-level robotics course and graduate-level nonlinear control systems course in the School of Engineering and Computer Science, Oakland University. Now, the major subjects of those two lecture notes are systematically mixed together and further extended by adding more topics, theories and applications, as well as more examples and MATLAB^{TM} programs to form the first part of the book.

I had also been invited and worked for the Advance Manufacturing Engineering (AME) of Chrysler Corporation as a summer professor intern for the past 12 consecutive summers during the 2000's. The opportunity of working with the automotive industry brought to me tremendous real-world knowledge and experience that was almost impossible to acquire from the classroom. In more than ten years of the internship program and consulting work, I was personally involved in their virtual assembly and product design innovation and development, and soon became an expert in major simulation software tools, from IGRIP robotic models, the early product of Deneb Robotics (now Dassault/Delmia) to the Safework mannequins in CATIA. Because of this unique opportunity, I have already been on my real journey from robot to digital human.

Therefore, it has been my long-term intention to merge both the robot analysis and digital human modeling into one single book in order to share my enjoyable journey with the readers. On the other hand, it is, indeed, not an easy job to integrate these two rapidly and dynamically growing research areas together, even though the latter often borrows the modeling theories and motion generation algorithms from the former.

Almost every chapter in the book has a section of exercise problems and/or computer projects, which will be beneficial for students to reinforce their understanding of every concept and algorithm. It is the instructor's discretion to select sections and chapters to be covered in a single-semester robotics course. In addition, I highly recommend that the instructor teach students to write a program and draw a robot or a mannequin in MATLAB^{TM} with realistic motion by following the basic approaches and illustrations from the book.

I hereby acknowledge my indebtedness to the people who helped me with different aspects of collecting knowledge, experience, data and programming skills towards the book completion. First, I wish to express my grateful appreciations to Dr. Leo Oriet who was the former senior manager when I worked for the AME of Chrysler Corporation, and Yu Teng who was/is a manager and leader of the virtual assembly and product design group in the AME of Chrysler. They both not only provided me with a unique opportunity to work on the digital robotic systems and human modeling for their ergonomics and product design verification and validation in the past, but also gave me every support and encouragement in recent years. I also wish to thank Michael Hicks who is an engineer working for General Dynamics Land Systems, and Ashley Liening who is a graduate student majoring in English at Oakland University for helping me polish my writing.

Furthermore, the author is under obligation to Fanuc Robotics, Inc., Robotics Research Corporation, and Aldebaran Robotics, Paris, France for their courtesies and permissions to include their photographs into the book.

Edward Y.L. Gu, *Rochester, Michigan*
guy@oakland.edu
April, 2013

Contents

List of Figures

Chapter 1
Introduction to Robotics and Digital Human Modeling

1.1 Robotics Evolution: The Past, Today and Tomorrow

Robotics research and technology development have been on the road to grow and advance for almost half a century. The history of expedition can be divided into three major periods: the early era, the middle age and the recent years. The official definition of robot by the Robot Institute of America (RIA) early on was:

"A robot is a reprogrammable multi-functional manipulator designed to move material, parts, tools, or specialized devices through variable programmed motions for the performance of a variety of tasks."

Today, as commonly recognized, beyond such a professional definition from history, the general perception of a robot is a manipulatable system to mimic a human with not only the physical structure, but also the intelligence and even personality. In the early era, people often remotely manipulated material via a so-called teleoperator as well as to do many simple tasks in industrial applications. The teleoperator was soon "married" with the computer numerically controlled (CNC) milling machine to "deliver" a new-born baby that was the **robot**, as depicted in Figure 1.1.

Since then, the robots were getting more and more popular in both industry and research laboratories. A chronological overview of the major historical events in robotics evolution during the early era is given as follows:

1947- The 1st servoed electric powered teleoperator was developed;
1948- A teleoperator was developed to incorporate force feedback;
1949- Research on numerically controlled milling machines was initiated;
1954- George Devol designed the first programmable robot;
1956- J. Engelberger bought the rights to found Unimation Co. and produce the Unimate robots;
1961- The 1st Unimate robot was installed in a GM plant for die casting;
1961- The 1st robot incorporating force feedback was developed;
1963- The 1st robot vision system was developed;

E.Y.L. Gu, *A Journey from Robot to Digital Human*,
Modeling and Optimization in Science and Technologies 1,
DOI: 10.1007/978-3-642-39047-0_1, © Springer-Verlag Berlin Heidelberg 2013

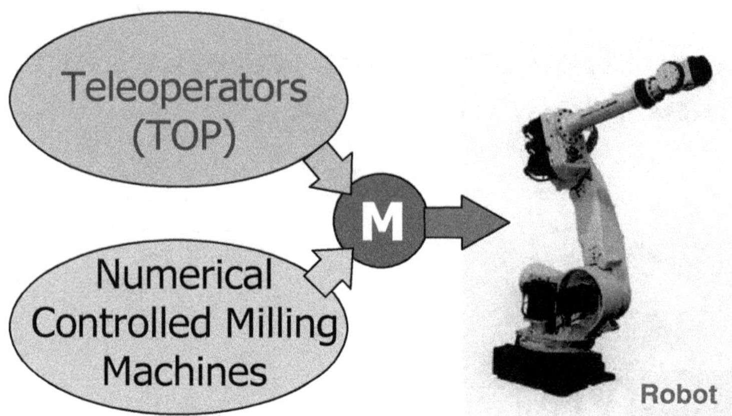

Fig. 1.1 Married with a child

1971- The Stanford arm was developed at Stanford University;
1973- The 1st robot programming language (WAVE) was developed at Stanford University;
1974- Cincinnati Milacron introduced the T3 robot with Computer Control;
1975- Unimation Inc. registered its first financial profit;
1976- The RCC (Remote Center Compliance) device for part insertion was developed at Draper Labs;
1978- Unimation introduced the PUMA robot based on a GM study;
1979- The SCARA robot design was introduced in Japan;
1981- The 1st direct-drive robot was developed at Carnegie-Mellon University.

Those historical and revolutionary initiations are unforgettable, and almost every robotics textbook acknowledges and refers to the glorious childhood of industrial robots [1, 2, 3]. Following the early era of robotics, from 1982 to 1996 at the middle age of robotics, a variety of new robotic systems and their kinematics, dynamics, and control algorithms were invented and extensively developed, and the pace of growth was almost exponential. The most significant findings and achievements in robotics research can be outlined in the following representative aspects:

- The Newton-Euler inverse-dynamics algorithm;
- Extensive studies on redundant robots and applications;
- Study on multi-robot coordinated systems and global control of robotic groups;
- Control of robots with flexible links and/or flexible joints;
- Research on under-actuated and floating base robotic systems;
- Study on parallel-chain robots versus serial-chain robots;
- Intelligent and learning control of robotic systems;

- Development of advanced force control algorithms and sensory devices;
- Sensory-based control and sensor fusion in robotic systems;
- Robotic real-time vision and pattern recognition;
- Development of walking, hopping, mobile, and climbing robots;
- Study on hyper-redundant (snake-type) robots and applications;
- Multi-arm manipulators, reconfigurable robots and robotic hands with dexterous fingers;
- Wired and Wireless networking communications for remote control of robotic groups;
- Mobile robots and field robots with sensor networks;
- Digital realistic simulations and animations of robotic systems;
- The study of bio-mimic robots and micro-/nano-robots;
- Research and development of humanoid robots;
- Development and intelligent control of android robots, etc.

After 1996, robotics research has advanced into its maturity. The robotic applications were making even larger strides than the early era to continuously grow and rapidly deploy the robotic technologies from industry to many different fields, such as the military applications, space exploration, underground and underwater operations, medical surgeries as well as the personal services and homeland security applications. In order to meet such a large variety of challenges from the applications, robotic systems design and control have been further advanced to a new horizon in the recent decades in terms of their structural flexibility, dexterity, maneuverability, reconfigurability, scalability, manipulability, control accuracy, environmental adaptability as well as the degree of intelligence [4]–[8]. One can witness the rapid progress and great momentum of this non-stop development in the large volume of internet website reports. Figure 1.2 shows a new Fanuc M-900iB/700 super heavy industrial robot that offers 700 Kg. payload capacity with built-in iRVision and force sensing integrated systems.

Parallel to the robotics research and technology development, virtual robotic simulation also has a long history of expedition. In the mid-1980's, Deneb Robotics, known as Dassault/Delmia today, released their early version of a robot graphic simulation software package, called IGRIP. Nearly as the same time, Technomatix (now UGS/Technomatix) introduced a ROBOCAD product, which kicked off a competition. While both the major robotic simulation packages impressed the users with their 3D colorful visualizations and realistic motions, the internal simulation algorithms could not accurately predict the reaching positions and cycle times, mainly due to the parameter uncertainty. As a result of the joint effort between the software firms and robotic manufacturers, a Realistic Robot Simulation (RRS) specification was created to improve the accuracy of prediction.

In the mid-1990's, robotic simulation technology was maturing. The capability of robotic simulation had also been extended to Product Lifecycle Management (PLM) [13, 14]. The robot arms, fixtures and workcells in a graphic simulation study were not only getting larger in scale, but also be-

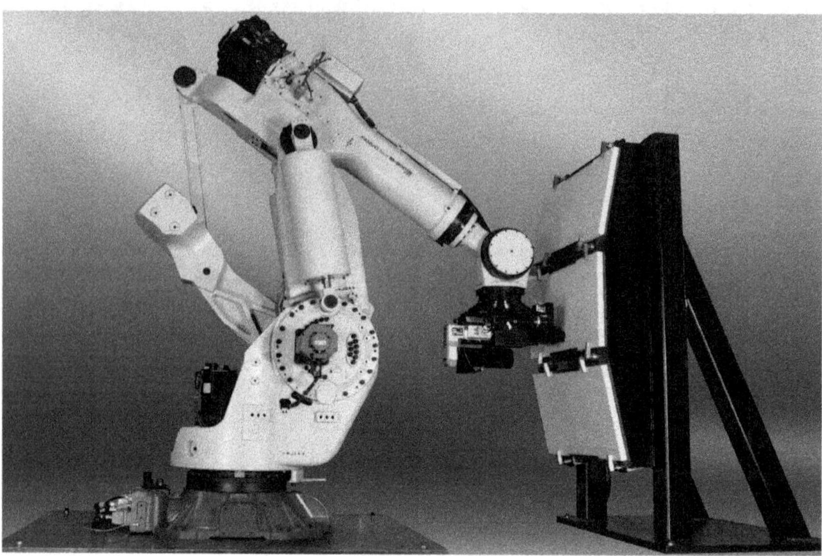

Fig. 1.2 A Fanuc M-900iB/700 industrial robot in drilling operation. Photo courtesy of Fanuc Robotics, Inc.

came more capable of managing product design in association with the manufacturing processes from concept to prototyping, to production. Today, the status of robotic simulation has further advanced to a more sophisticated and comprehensive new stage. It has become a common language to communicate design issues between the design teams and customers, and also an indispensable tool for product and process design engineers and managers as well as researchers to verify and validate their new concepts and findings.

The new trends of robotics research, robotic technology development and applications today and tomorrow will possibly grow even faster and be more flexible and dexterous in mechanism and more powerful in intelligence. Due to the potentially huge market and social demand, robotic systems design, performance, and technology have already jumped into a new transitional era from industrial and professional applications to social and personal services. Facing the pressing competitions and challenges from the transition, robotics research will never be running behind. Instead, by keeping up the great momentum of growth, it will rapidly move forward to create better solutions, make more innovations and achieve new findings to speed up the robotic technology development in the years to come [9]–[12].

Figure 1.3 depicts a robotics research and robotic systems evolution tree. The innovation and continuous development of industrial robots in the early era are the main stem of the tree. The robotics research that was initiated, motivated and challenged by industrial robot development becomes the top of the tree stem before it branches. As the robotics research was rapidly

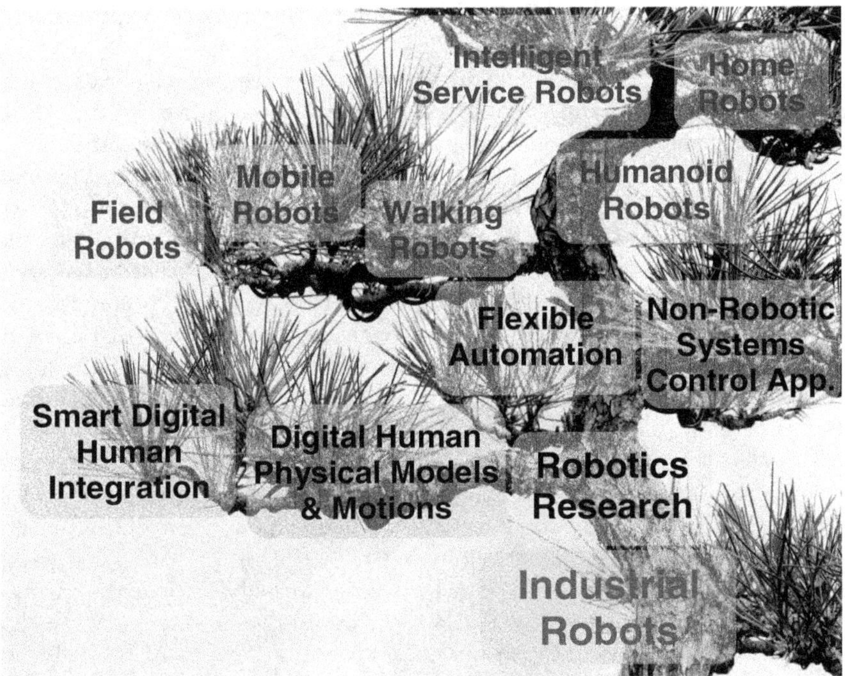

Fig. 1.3 Robotics research and evolutions

growing and getting mature, it became more capable of helping new robotic systems creation and fueling new research branches to sprout and grow. In addition to creating and developing a variety of service robots, a number of new research and application branches have also been created and fed by the robotic systematic modeling approaches and control theories, which benefited their developments. One of those beneficiaries is digital human modeling and applications. The others may include many non-robotic systems dynamic modeling and control strategy design, such as a gun-turret control system for military vehicles, helicopters and platforms, and a ball-board control system that will be discussed in Chapter 8.

While a large number of new service robots and humanoid robots are taking over today's performing stage of robotics, the development of industrial robotic technologies has never slowed down. Instead, they are gaining even more momentum to continuously innovate new robot models and systems to enhance their flexible automation in better serving manufacturing and production lines. A task that used to be operated by a single robot arm is now automated by two-robot coordination, or even by a large number of robots in a group. A typical example of recent applications is to employ a group of more than 20 industrial robots to be globally controlled by an Ethernet/wireless communication-based PLC (Programmable Logic Controller) to weld and fabricate car bodies in an automotive body-in-white assembly station.

One of the most remarkable achievements that deserves celebration is the development of humanoid robots, which underlies an infrastructure of various service robots and home robots. The history of the humanoid robot development is even longer than the industrial robots [15]. An Italian mathematician/engineer Leonardo da Vinci designed a humanoid automaton that looks like an armored knight, known as Leonardo's robot in 1495. The more contemporary human-like machine Wabot-1 was built at Waseda University in Tokyo, Japan in 1973. Wabot-1 was able to walk, to communicate with a person in Japanese by an artificial mouth, and to measure distances and directions to an object using external receptors, such as artificial ears and eyes. Ten years later, they created a new Wabot-2 as a musician humanoid robot that was able to communicate with a person, read a normal musical score by his eyes and play tones of average difficulty on an electronic organ. In 1986, Honda developed seven biped robots, called E0 (Experimental Model 0) through E6. Model E0 was created in 1986, E1-E3 were built between 1987 and 1991, and E4-E6 were done between 1991 and 1993. Then, Honda upgraded the biped robots to P1 (Prototype Model 1) through P3, as an evolutionary model series of the E series, by adding upper limbs. In 2000, Honda completed its 11th biped humanoid robot, known as ASIMO that was not only able to walk, but also to run.

Since then, many companies and research institutes followed to introduce their respective models of humanoid robots. A humanoid robot, called Actroid, which was covered by silicone "skin" to make it look like a real human, was developed by Osaka University in conjunction with Kokoro Company, Ltd. in 2003. Two years later, Osaka University and Kokoro developed a new series of ultra-realistic humanoid robots in Tokyo. The series initial model was Geminoid HI-1, followed by Geminoid-F in 2010 and Geminoid-DK in 2011.

It is also worth noting that in 2006, NASA and GM collaborated to develop a very advanced humanoid robot, called Robonaut 2. It was originally intended to assist astronauts in carrying out scientific experiments in a space shuttle or in the space station. Therefore, Robonaut 2 has only an upper body without legs for use in a gravity-free environment to perform advanced manipulations using its dexterous hands and arms [16, 17].

Almost every year, a large number of new humanoid robots are reported to show up worldwide. While the degree of intelligence and the realistic dynamic motion may still be two major challenges to the humanoid robot research and development, their appearance and motion speed have made a revolutionary breakthrough and climbed to a new height. We are quite optimistic that sooner rather than later, a smart humanoid robot would come to reality, and a true intelligent home robot would be a family addition to serve and assist in daily housework and to entertain family members and guests, and even replace a desktop or laptop computer to do every computation and documentation work in the home. However, to achieve this goal, only making a technological development effort is not enough. Instead, it must also rely on more new findings and solutions in theoretical development and basic research to overcome every challenging hurdle.

As a summary, in the recent status of basic research in robotics, there is a number of topics that still remain open:

1. Adaptive control of under-actuated robots or robotic systems under non-holonomic constraints;
2. Dynamic control of flexible-joint and/or flexible-link robots;
3. The dual relationship between open serial and closed parallel-chain robots;
4. Real-time image processing and intelligent pattern recognition;
5. Stability of robotic learning and intelligent control;
6. Robotic interactions and adaptations to complex environments;
7. Perceptional performance in a closed feedback loop between robot and environment;
8. Cognitive interactions with robotic physical motions;
9. More open topics in robot dynamic control and human-machine interactions.

In conclusion, the robot analysis part of this book is intended to motivate and encourage the reader to accept all the new challenges and make every effort and contribution to the current and future robotics research, systems design and applications. The robotics part of the book will cover and focus primarily on the three major fundamental topics: kinematics, dynamics, and control, along with the related MATLAB^{TM} programming. Specifically, Figure 1.4 illustrates the formal definitions in the covered topics of robotics. However, the book does not intend to include discussions on robotic force control, learning and intelligent control, robotic vision and recognition, sensory-feedback control, and programmable logic controller (PLC) and human-machine interface (HMI) based networking control of robotic groups. The reader can refer to the literature or application documents to learn more about those application-oriented topics.

1.2 Digital Human Modeling: History, Achievements and New Challenges

Dr. Don Chaffin from HUMOSIM Research Laboratory in the University of Michigan has made a comprehensive review in 2008 [19]. At the beginning of this review, he emphasized that many human factors/ergonomics specialists have long desired to have a robust, analytic model that would be capable of simulating the physical and cognitive performance capabilities of specific, demographically defined groups of people. He also referred to a 1990 report from the U.S. National Research Council on human performance modeling that highlights the following benefits of such models:

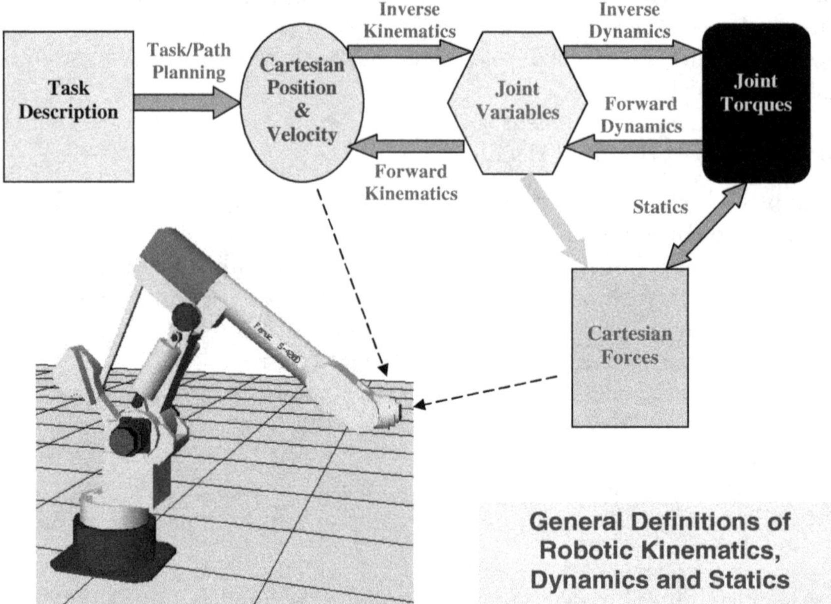

Fig. 1.4 Important definitions in robotics

1. Experts in ergonomics can simulate and test various underlying human behavior theories with these models, thus better prioritizing areas of new research;
2. Experts can use the models to gain confidence about their own knowledge regarding people's performance under a variety of circumstances;
3. The models provide a means to better communicate human performance attributes and capabilities to others who want to consider ergonomics in proposed designs.

Due to the limitation of computer power, the early attempt of digital human physical modeling was undertaken only conceptually until the late 1970's. With the exponential growth of computational speed, memory and graphic performance, a mannequin and its motion could be realistically visualized in a digital environment to allow the ergonomics specialists, engineers, designers and managers to more effectively assess, evaluate and verify their theoretical concepts, product designs, job analysis and human-involved pilot operations.

One of the earliest efforts of computerized human performance models in history, according to Chaffin's review, was done by K. Kilpatrick in 1970. He made a 3D human graphic model to demonstrate how the model reaches and moves in a seated posture. After the 1970's, a number of sophisticated digital human models emerged. SAMMIE (System for Aiding Man-Machine

Interaction Evaluation) was developed in the United Kingdom at that time and is now one of the leading packages in the world to run digital human simulations. During the late 1980's, Safework and Jack were showing their new mannequins with real-time motions as well as their unique features and functions. In the early 1990's, a human musculoskeletal model was developed in a digital environment by AnyBody Technology in Denmark to simulate a variety of work activities for automotive industry applications [18].

One of the most remarkable achievements in recent digital human modeling history was the research and development of a virtual soldier model: Santos in Center for Computer Aided Design at the University of Iowa, led by Dr. Karim Abdel-Malek during the 2000's [22, 23, 24]. It is now under continuous development in a spin-off company SantosHuman, Inc. Not only has the Santos mannequin demonstrated its unique high-fidelity of appearance with deformable muscle and skin in a digital environment, but it has also made a pioneering leap and contribution to the digital human research community in borrowing and applying robotic modeling theories and approaches. Their multi-disciplinary research has integrated many major areas in digital human modeling and simulation, such as:

- Human performance and human systems integration;
- Posture and motion prediction;
- Task simulation and analysis;
- Muscle and physiological modeling;
- Dynamic strength and fatigue analysis;
- Whole body vibrations;
- Body armor design and analysis;
- Warfighter fightability and survivability;
- Clothing and fabric modeling;
- Hand modeling;
- Intuitive interfaces.

To model and simulate dynamics, one of the most representative software tools is MADYMO (Mathematical Dynamic Models) [20]. MADYMO was developed as a digital dummy for car crash simulation studies by the Netherlands Organization for Applied Scientific Research (TNO) Automotive Safety Solutions division (TASS) in the early 1990's. It offers several digital dummy models that can be visualized in real-time dynamic responses to a collision. It also possesses a powerful post-processing capability to make a detailed analysis and check the results against the safety criteria and legal requirements. In addition, MADYMO provides a useful simulation tool of airbag and seat-belt design as well as the reconstruction and analysis of real accidents.

While all the achievements after three decades of extensive investigations in digital human modeling for design and engineering applications are quite encouraging [20, 21], there are still many big challenges ahead, and they can be summarized as follows:

1. Although the realism of digital human appearance has made a breakthrough, the high-fidelity of digital human motion may need more improvements, especially in a sequential motion, high-speed motion and motion in complex restricted environments;
2. Further efforts need to be made for modeling human-environment interactions in a more effective and adaptive fashion;
3. More work must be done to enhance the digital human physical models in adapting to the complex anthropometry, physiology and biomechanics, as well as taking digital human vision and sound responses into modeling consideration;
4. Develop a true integration between the digital human physical and non-physical models in terms of psychology, feeling, cognition and emotion.

1.3 A Journey from Robot Analysis to Digital Human Modeling

After screening the history and evolution of research and technology development in both robotics and digital human modeling, it is foreseeable that all progresses and cutting-edge innovations can always be mirrored in leading commercial simulation software products. However, most of such graphic simulation packages render a small "window" as a feature of open architecture to allow the user to write his/her own application program for research, testing or verification. When the user's program is ready to communicate the product, it often requires a special API (Application Program Interface) in order to acknowledge and run the user's application program. Thus, it becomes very limited and may not be suitable for academic research and education. Therefore, it is ideal to place the modeling, programming, modification, refinement and graphic animation all in one, such as MATLAB^{TM}, to create a flexible, user-friendly and true open-architectural digital environment for future robotics and digital human graphic simulation studies.

This book aims to take a journey from robot to digital human by providing the reader with a means to build a theoretical foundation at the beginning. Then, the reader will be able to mock up a desired 3D solid robot model or a mannequin in MATLAB^{TM} and drive it for motion. It will soon be realized that writing a MATLAB^{TM} code may not be difficult, because it is the highest-level computer language. The most challenging issue is the necessary mathematical transformations behind the robot or mannequin drawing. This is the sole reason why the theoretical foundation must be built up before writing a MATLAB^{TM} program to create a desired digital model for animation. Since MATLAB^{TM} has recently added a Robotics toolbox into the family, it will certainly reinforce the conceptual understanding of robotic theories and help for learning numerical solutions to robotic modeling procedures and motion algorithms.

Therefore, to make the journey more successful and exciting, this book will specifically focus on the basic digital modeling procedures, motion algorithms and optimization methodologies in addition to the theoretical fundamentals in robotic kinematics, statics, dynamics, and control. Making a realistic appearance, adapting various anthropometric data and digital human cognitive modeling will not be the emphasis in this book. Instead, once a number of surfaces are created to be further assembled together, more time can always be spent to sculpture each surface more carefully and microscopically to make it look like a real muscle/skin as long as the surface has a sufficient enough resolution. Moreover, one can also concatenate the data between the adjacent surfaces to generate a certain effect of deformation. For this reason, this book will introduce a few examples of basic mathematical sculpturing and deforming algorithms as a typical illustration, and leave to the reader to extend the basic algorithms to more advanced and sophisticated programs.

Furthermore, in the digital human modeling part of the book, each set of kinematic parameters, such as joint offsets and link lengths for a digital mannequin is part of the anthropometric data. They can be easily set or reset from one to another in a modeling program, and the parameter exchange will never alter the kinematic structure. For example, when evaluating the joint torque distribution by statics for a digital human in operating a material-handling task, it is obvious that the result will be different from a different set of kinematic parameters. However, once entering a desired set of parameters, the resulting joint torque distribution should exactly reflect the person's performance under the particular anthropometric data. There is a large number of anthropometry databases available now [20], such as CAESAR, DINED, A-CADRE, U.S. Army Natick, NASA STD3000, MIL-STD-1472D, etc. The reader can refer to those documents and literature to find appropriate data sets for high-credibility digital assessment and evaluation.

It is quite recognizable that in terms of real human musculoskeletal structure, the current rigid body-based digital human physical model would hardly be considered an accurate and satisfactory model until every muscle contraction and joint structure of real human are taken into account. Nevertheless, the current digital human modeling underlies a framework of the future targeting model. With continuous research and development, such an ideal digital human model with realistic motion and true smart interaction to complex environments would not be far away from today.

On the other hand, due to the maturity of robotics research, developing a digital human model and motion can be harvested by borrowing the systematic robotic modeling theories and motion algorithms. Therefore, this book is organized to trace the journey from robot analysis to digital human modeling. Chapters 2 and 3 introduce all the useful and relevant mathematical fundamentals. Chapter 4 starts a robotic modeling procedure and kinematic formulation. Chapter 5 will study the robots with redundancy, as well as the forward and inverse kinematics for serial/parallel hybrid-chain robotic systems. Once the foundations of robotics are built up, Chapter 6 will describe and illustrate the major steps to create parts and assemble them to mock up a complete robotic system with 3D solid drawing in MATLAB^{TM}. The robotic dynamics, such as modeling, formulation, analysis and algorithms, will then be introduced and further discussed in Chapter 7. It will be followed by an introductory presentation and an advanced lecture on robotic control: from independent joint-servo control to global dynamic control in Chapter 8. Some useful control schemes for both robotic systems and digital humans, such as the adaptive control and backstepping control design procedure, will be discussed in detail as well.

Starting from Chapter 9, the subject will turn to digital human modeling: local and global kinematics and statics of a digital human in Chapter 9, and creating parts and then assembling them together to build a 3D mannequin in MATLAB^{TM} as well as to drive the mannequin for basic and advanced motions in Chapter 10. The hand modeling and digital sensing will also be included in Chapter 10. The last chapter, Chapter 11, will introduce digital human dynamic models in a global sense, and explore how to generate a realistic motion using the global dynamics algorithm. At the end of Chapter 11, two typical digital human dynamic motion cases will be modeled, studied and simulated, and finally, it will be followed by a general strategy of interactive control of human-machine dynamic interaction systems that can be modeled as a k-cascaded large-scale system with backstepping control design.

References

1. Asada, H., Slotine, J.: Robot Analysis and Control. John Wiley and Sons, New York (1986)
2. Fu, K., Gonzalez, R., Lee, C.: Robotics: Control, Sensing, Vision and Intelligence. McGraw-Hill, New York (1987)
3. Spong, M., Vidyasagar, M.: Robot Dynamics and Control. John Wiley & Sons, New York (1989)
4. Murray, R., Li, Z., Sastry, S.: A Mathematical Introduction to Robotic Manipulation. CRC Press, Boca Raton (1994)
5. Craig, J.: Introduction to Robotics: Mechanics and Control, 3rd edn. Pearson Prentice Hall, New Jersey (2005)
6. Bekey, G.: Autonomous Robots, From Biological Inspiration to Implementation and Control. The MIT Press, Cambridge (2005)
7. Choset, H., Lynch, K., Hutchinson, S., Kantor, G., Burgard, W., Kavraki, L., Thrun, S.: Principles of Robot Motion, Theory, Algorithms, and Implementation. The MIT Press, Cambridge (2005)
8. Siciliano, B., Khatib, O. (eds.): Springer Handbook of Robotics. Springer (2008)
9. Sciavicco, L., Siciliano, B.: Modeling and Control of Robot Manipulators. McGraw-Hill (1996)
10. Lenari, J., Husty, M. (eds.): Advances in Robot Kinematics: Analysis and Control. Kluwer Academic Publishers, the Netherlands (1998)
11. Cubero, S. (ed.): Industrial Robotics: Theory, Modelling and Control. Pro Literatur Verlag, Germany/ARS, Austria (2006)

12. Siciliano, B., Sciavicco, L., Villani, L., Oriolo, G.: Robotics, Modeling, Planning and Control. Springer (2009)
13. Bangsow, S.: Manufacturing Simulation with Plant Simulation and SimTalk Usage and Programming with Examples and Solutions. Springer, Heidelberg (2009)
14. Wikipedia, Plant Simulation (2012), http://en.wikipedia.org/wiki/Plant_Simulation
15. Wikipedia, Humanoid Robot (2011), http://en.wikipedia.org/wiki/Humanoid_robot
16. Ambrose, R., et al.: ROBONAUT: NASAs Space Humanoid. IEEE Intelligent Systems Journal 15(4), 5763 (2000)
17. Wikipedia, Robonaut (2012), http://en.wikipedia.org/wiki/Robonaut
18. Chaffin, D.: On Simulating Human Reach Motions for Ergonomics Analysis. Human Factors and Ergonomics in Manufacturing 12(3), 235–247 (2002)
19. Chaffin, D.: Digital Human Modeling for Workspace Design. In: Reviews of Human Factors and Ergonomics, vol. 4, p. 41. Sage (2008), doi:10.1518/155723408X342844
20. Moes, N.: Digital Human Models: An Overview of Development and Applications in Product and Workplace Design. In: Proceedings of Tools and Methods of Competitive Engineering (TMCE) 2010 Symposium, Ancona, Italy, April 12-16, pp. 73–84 (2010)
21. Duffy, V. (ed.): Handbook of Digital Human Modeling: Research for Applied Ergonomics and Human factors Engineering. CRC Press (2008)
22. Abdel-Malek, K., Yang, J., et al.: Towards a New Generation of Virtual Humans. International Journal of Human Factors Modelling and Simulation 1(1), 2–39 (2006)
23. Abdel-Malek, K., et al.: Santos: a Physics-Based Digital Human Simulation Environment. In: The 50th Annual Meeting of the Human Factors and Ergonomics Society, San Francisco, CA (October 2006)
24. Abdel-Malek, K., et al.: Santos: A Digital Human In the Making. In: IASTED International Conference on Applied Simulation and Modeling, Corfu, Greece, ADA542025 (June 2008)

Chapter 2
Mathematical Preliminaries

2.1 Vectors, Transformations and Spaces

In general, a vector can have the following two different types of definition:

1. *Point Vector* – A vector depends only on its length and direction, and is independent of where its tail-point is. Under this definition, any two parallel vectors with the same sign and length are equal to each other, no matter where the tail of each vector is located. To represent such a type of vector, we conventionally let its tail be placed at the origin of a reference coordinate frame and its arrow directs to the point, the coordinates of which are augmented to form a point vector.
2. *Line Vector* – A vector depends on, in addition to its length and direction, also the location of the straight line it lies on. Therefore, two line vectors that lie on two parallel but distinct straight lines even with the same sign and length are treated as different vectors. Intuitively, to uniquely represent such a line vector, one has to define two independent point vectors, one of which determines its direction and length, and the other one determines its tail position, or the "moment" of the resided straight line. In other words, a line vector should be 6-dimensional in 3D (3-dimensional) space. A typical and also effective approach to the mathematical representation of line vector is the so-called Dual-Number Algebra [1] that will be introduced later in this chapter.

Figure 2.1 depicts two parallel vectors v_1 and v_2 with the same direction and length. Thus, $v_1 = v_2$ under the point vector definition, while $v_1 \neq v_2$ if the two vectors are considered as line vectors, because they are separated with a nonzero distance d. As a default, we say a vector is a point vector, unless it is specifically indicated as a line vector. For example, in the later analysis and applications of robotics and digital human modeling, to uniquely determine an axis for a robot link coordinate system with respect to the common base, two point vectors have to be defined: one is a 3D unit vector indicating

E.Y.L. Gu, *A Journey from Robot to Digital Human*,
Modeling and Optimization in Science and Technologies 1,
DOI: 10.1007/978-3-642-39047-0_2, © Springer-Verlag Berlin Heidelberg 2013

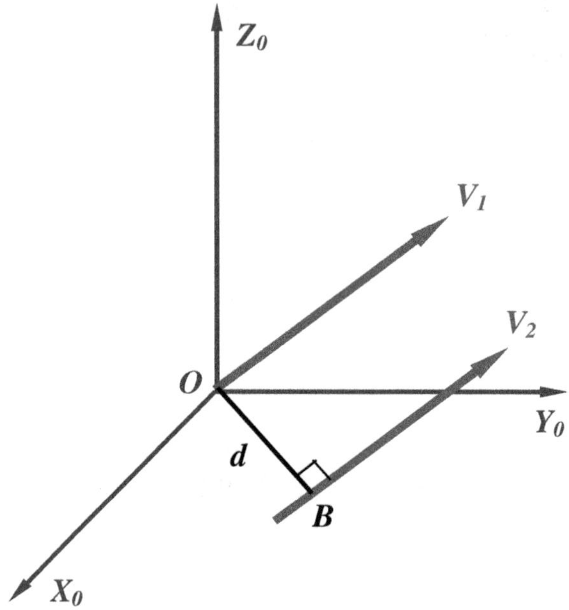

Fig. 2.1 Two parallel vectors have a common length

its direction and the other one is a 3D position vector that determines the location of the origin of that link coordinate frame with respect to the base.

A 3D vector (of course, this is a point vector as default) is often denoted by a 3 by 1 column. The mathematical operations between two vectors include their addition, subtraction and multiplication. The vector multiplication has two different categories: the dot (or inner) product and the cross (or vector) product [2, 3]. Their definitions are given as follows:

1. Dot Product – For two 3D vectors

$$v_1 = \begin{pmatrix} a_1 \\ b_1 \\ c_1 \end{pmatrix} \quad \text{and} \quad v_2 = \begin{pmatrix} a_2 \\ b_2 \\ c_2 \end{pmatrix},$$

$$v_1 \cdot v_2 = v_2 \cdot v_1 = v_1^T v_2 = v_2^T v_1 = a_1 a_2 + b_1 b_2 + c_1 c_2,$$

   which is a scalar.

2. Cross Product – To determine $v_1 \times v_2$, first, let a 3 by 3 skew-symmetric matrix for v_1 be defined as

$$S(v_1) = v_1 \times = \begin{pmatrix} 0 & -c_1 & b_1 \\ c_1 & 0 & -a_1 \\ -b_1 & a_1 & 0 \end{pmatrix}.$$

Then, $v_1 \times v_2 = S(v_1) \cdot v_2$ which is still a 3 by 1 column vector.

The major properties for the two categories of vector multiplication are outlined as follows:

1. The dot product $v_1 \cdot v_2 = \|v_1\|\|v_2\| \cos\theta$, where θ is the angle between the positive directions of the two vectors, while the cross product $\|v_1 \times v_2\| = \|v_1\|\|v_2\| \sin\theta$, and its direction is perpendicular to both v_1 and v_2 and determined by the Right-Hand Rule.
2. For two non-zero vectors, $v_1 \cdot v_2 = 0$ if and only if $v_1 \perp v_2$, and $v_1 \times v_2 = 0$ if and only if $v_1 \parallel v_2$.
3. Dot product is commutable, as shown above, while the cross product is not but is skew-symmetric, i.e., $v_1 \times v_2 = -v_2 \times v_1$.
4. Both dot and cross products have no inverse. In other words, there is no identity vector in any of the two vector multiplication operations.
5. A successive cross product operation, in general, is not associative, i.e., for three arbitrary non-zero vectors, $(v_1 \times v_2) \times v_3 \neq v_1 \times (v_2 \times v_3)$. For instance, if $v_2 \parallel v_3$ but they both are not parallel to v_1, and all the three vectors are non-zero, then, $(v_1 \times v_2) \times v_3 \neq 0$, while $v_1 \times (v_2 \times v_3) = 0$.
6. A *Triple Scalar Product* is defined by $(v_1\, v_2\, v_3) = v_1 \cdot (v_2 \times v_3)$. By using the skew-symmetric matrix for v_2, it can be re-written as $(v_1\, v_2\, v_3) = v_1^T S(v_2) v_3$. If taking the transpose, then,

$$(v_1\, v_2\, v_3) = -v_3^T S(v_2) v_1 = v_3^T S(v_1) v_2 = v_3 \cdot (v_1 \times v_2) = (v_3\, v_1\, v_2).$$

Similarly,

$$\begin{aligned}(v_1\, v_2\, v_3) &= (v_3\, v_1\, v_2) = (v_2\, v_3\, v_1) \\ &= -(v_1\, v_3\, v_2) = -(v_2\, v_1\, v_3) = -(v_3\, v_2\, v_1).\end{aligned}$$

Furthermore, if each $v_i = (a_i\ b_i\ c_i)^T$ for $i = 1, 2, 3$, then

$$(v_1\, v_2\, v_3) = \det \begin{pmatrix} a_1 & b_1 & c_1 \\ a_2 & b_2 & c_2 \\ a_3 & b_3 & c_3 \end{pmatrix}.$$

Therefore, the three vectors lie on a common plane if and only if

$$(v_1\, v_2\, v_3) = 0.$$

7. A *Triple Vector Product* is defined by $v_1 \times (v_2 \times v_3)$. Using a direct derivation, it can be proven that

$$v_1 \times (v_2 \times v_3) = (v_1 \cdot v_3) v_2 - (v_1 \cdot v_2) v_3 = (v_1^T v_3) v_2 - (v_1^T v_2) v_3. \quad (2.1)$$

This triple vector product formula will be very useful in many later discussions. Furthermore, based on this property, the following identity can be easily shown:

$$v_1 \times (v_2 \times v_3) + v_2 \times (v_3 \times v_1) + v_3 \times (v_1 \times v_2) \equiv 0.$$

As a numerical example, let

$$v_1 = \begin{pmatrix} 2 \\ -1 \\ 0 \end{pmatrix}, \quad v_2 = \begin{pmatrix} 0 \\ 1 \\ 3 \end{pmatrix} \quad \text{and} \quad v_3 = \begin{pmatrix} -4 \\ 0 \\ 2 \end{pmatrix}.$$

Then,

$$v_1 \times v_2 = S(v_1)v_2 = \begin{pmatrix} 0 & 0 & -1 \\ 0 & 0 & -2 \\ 1 & 2 & 0 \end{pmatrix} \begin{pmatrix} 0 \\ 1 \\ 3 \end{pmatrix} = \begin{pmatrix} -3 \\ -6 \\ 2 \end{pmatrix}.$$

The triple scalar product becomes

$$(v_3\, v_1\, v_2) = v_3 \cdot (v_1 \times v_2) = (-4 \;\; 0 \;\; 2) \begin{pmatrix} -3 \\ -6 \\ 2 \end{pmatrix} = 16.$$

On the other hand,

$$\det \begin{pmatrix} v_3^T \\ v_1^T \\ v_2^T \end{pmatrix} = \det \begin{pmatrix} -4 & 0 & 2 \\ 2 & -1 & 0 \\ 0 & 1 & 3 \end{pmatrix} = 16.$$

Thus, both approaches to evaluating the triple scalar product agree with each other.

Furthermore, the triple vector product can be calculated by

$$v_3 \times (v_1 \times v_2) = S(v_3)(v_1 \times v_2) = \begin{pmatrix} 0 & -2 & 0 \\ 2 & 0 & 4 \\ 0 & -4 & 0 \end{pmatrix} \begin{pmatrix} -3 \\ -6 \\ 2 \end{pmatrix} = \begin{pmatrix} 12 \\ 2 \\ 24 \end{pmatrix}.$$

It can also be evaluated by

$$v_3 \times (v_1 \times v_2) = (v_3^T v_2)v_1 - (v_3^T v_1)v_2 = 6 \begin{pmatrix} 2 \\ -1 \\ 0 \end{pmatrix} + 8 \begin{pmatrix} 0 \\ 1 \\ 3 \end{pmatrix} = \begin{pmatrix} 12 \\ 2 \\ 24 \end{pmatrix}.$$

Therefore, the two different ways for calculating the triple vector product match their answers with each other, too.

Both the two vector multiplication categories are very useful, especially in robotics and digital human modeling. In physics, the work W done by a force vector f that drives an object along a vector s can be evaluated by the dot-product $W = f \cdot s = f^T s$. If a force vector f is acting on an object for rotation and a radial vector r is tailed at the rotating center with its arrow at the force acting point, then the torque vector τ is defined by $\tau = r \times f$.

After the vector definitions and operations are reviewed, we now turn our attention to 3D transformations. A linear transformation in vector space is a linear mapping $T: \mathbb{R}^n \to \mathbb{R}^m$, which can be represented by an m by n matrix. In robotics, since rotation is one of the most frequently used operations, let us take a 3 by 3 rotation matrix as a typical example to introduce the 3D linear transformation.

Because a vector to be rotated should keep its length unchanged, a rotation matrix must be a length-preserved transformation [4, 5]. A 3 by 3 orthogonal matrix that is a member of the following *Special Orthogonal Group* can perfectly play such a unique role in performing a rotation as well as in representing the orientation of a coordinate frame:

$$SO(3) = \left\{R \in \mathbb{R}^{3\times 3} \mid RR^T = I \text{ and } \det(R) = +1\right\}.$$

Based on the definition, a 3 by 3 rotation matrix $R \in SO(3)$ has the following major properties and applications:

1. $R^{-1} = R^T$.
2. If a rotation matrix R is also symmetric, then $R^2 = RR^T = I$. This means that a vector rotated by such a symmetric R successively twice will return to itself. This also implies that a symmetric rotation matrix acts either a 0^0 rotation or a 180^0 rotation, and nothing but these two cases.
3. One of the three eigenvalues for a rotation matrix $R \in SO(3)$ must be $+1$, and the other two are either both real or complex-conjugate such that their product must be $+1$. This means that there exists an eigenvector $x \in \mathbb{R}^3$ such that $Rx = x$.
4. A 3 by 3 rotation matrix R_b^i can be used to uniquely represent, in addition to vector or frame rotations, the orientation of a coordinate frame i with respect to a reference frame b. Namely, each column of the rotation matrix R_b^i is a unit vector, and they represent, respectively in order, the x-axis, the y-axis and the z-axis of frame i with respect to frame b.
5. If a vector v_i is currently projected on frame i and we want to know the same vector but projected on frame b, then $v_b = R_b^i v_i$ if R_b^i is the orientation of frame i with respect to frame b.

The Lie group $SO(3)$ is also a 3D topological space, which can be spanned by the basis of three elementary rotations: $R(x, \alpha)$ that rotates a frame about its x-axis by an angle α, $R(y, \beta)$ that rotates about the y-axis by β, and $R(z, \gamma)$ to rotate about the z-axis by γ. The detailed equations of the basis are given as follows:

$$R(x, \alpha) = \begin{pmatrix} 1 & 0 & 0 \\ 0 & c\alpha & -s\alpha \\ 0 & s\alpha & c\alpha \end{pmatrix}, \quad R(y, \beta) = \begin{pmatrix} c\beta & 0 & s\beta \\ 0 & 1 & 0 \\ -s\beta & 0 & c\beta \end{pmatrix},$$

and

$$R(z, \gamma) = \begin{pmatrix} c\gamma & -s\gamma & 0 \\ s\gamma & c\gamma & 0 \\ 0 & 0 & 1 \end{pmatrix}, \tag{2.2}$$

where each $c\alpha = \cos\alpha$ and $s\alpha = \sin\alpha$ as well as for $c\beta = \cos\beta$ and $c\gamma = \cos\gamma$ as a conventional abbreviation in robotics.

It will be seen later that any orientation change for a frame can always be composed by a certain combination of the above three elementary rotations successively, such as the well-known Euler angles [7, 8] that will be discussed in the next chapter.

In contrast to rotation, a translation of either a vector tail-point or a coordinate frame is very straightforward without any complication, and we can simply use the addition of 3D vectors to perform any successive translations. In general, for a 6 degrees of freedom (d.o.f.) motion in 3D space, one may desire to define a unified transformation to perform both translation and rotation simultaneously for a coordinate frame. Such unified 6-dimensional transformations will span the following *Special Euclidean Space*:

$$SE(3) = \{(p, R) \mid p \in \mathbb{R}^3,\ R \in SO(3)\} = \mathbb{R}^3 \times SO(3),$$

which is also a Lie group [8].

2.2 Lie Group and Lie Algebra

In mathematics [4, 5], a collection, or a set of elements is often studied by associating it with a certain operation. This is the primary notion of group theory that distinguishes it from set theory.

Definition 1. A group is defined as a set G along with a binary operation $\circ$ such that all the following conditions hold:

1. Closure: For any $a,\ b \in G$, $a \circ b = c \in G$;
2. Associativity: For every $a,\ b,\ c \in G$, $(a \circ b) \circ c = a \circ (b \circ c)$;
3. Identity: There is an identity element $\iota \in G$ such that $\iota \circ g = g \circ \iota = g$ for all $g \in G$;
4. Inverse: For each $g \in G$, there exists an element $h \in G$ such that $g \circ h = h \circ g = \iota$.

All the real integers associated with addition form an additive group, but not under multiplication because of violating the inverse condition. All the real (complex) numbers under either addition or multiplication constitute an additive or a multiplicative group, respectively, and they are qualified to form a field, called the real field (complex field). If a collection along with a certain operation satisfies every condition but the inverse, even though the identity condition can still hold, then, the collection just forms a semigroup. In contrast, the set of all 3D real vectors in $\mathbb{R}^3$ under the cross product is neither a group, nor a semigroup, because it violates the associativity, identity and inverse conditions.

A group can be classified into finite and infinite groups, and discrete and continuous groups, depending on the number of elements and the property of element variations. A Lie group is a typical infinite and continuous group. As examples, all the non-singular n by n real matrices with multiplication form a general linear Lie group $GL(n)$. All the n by n real orthogonal matrices with multiplication form an orthogonal Lie group $O(n)$, and by further imposing the determinant of each member of $O(n)$ to be $+1$, it becomes a special orthogonal group $SO(n)$. Every rotation matrix belongs to the $SO(3)$ group.

However, many useful sets under certain binary operations violate either one or more conditions of the group definition, and they cannot be a group, though they are quite useful. In order to keep them for further study and application, we have to relax the conditions. Lie algebra is one of the most typical and important approaches to rescuing the useful sets that have been ruled out by the group definition.

Definition 2. A Lie algebra over the real field $\mathbb{R}$ or the complex field $\mathbb{C}$ is a vector space $\mathcal{G}$, in which a bilinear map $(X, Y) \to [X, Y]$ is defined from $\mathcal{G} \times \mathcal{G} \to \mathcal{G}$ such that

1. $[X, Y] = -[Y, X]$, for all $X, Y \in \mathcal{G}$, and
2. $[X, [Y, Z]] + [Y, [Z, X]] + [Z, [X, Y]] = 0$ for all $X, Y, Z \in \mathcal{G}$.

The second equation in the above definition is called the Jacobi Identity. Now, all the 3-dimensional real vectors along with the cross product operation form a Lie algebra, although they cannot be a Lie group. Since for $a \times b = [a, b] = c$, one can rewrite it as $S(a)b = c$, where $S(a)$ is the 3 by 3 skew-symmetric matrix of the vector $a \in \mathbb{R}^3$. In other words, $[a, \cdot] = S(a)\cdot = a\times$ is treated as an operator. Such a Lie algebra is often denoted as $so(3)$ in lower case. For example, if two vectors

$$a = \begin{pmatrix} 3 \\ 2 \\ -1 \end{pmatrix} \quad \text{and} \quad b = \begin{pmatrix} -1 \\ 0 \\ 2 \end{pmatrix}$$

are given, to find $c = [a, b] = a \times b$, let us first construct a skew-symmetric matrix $S(a)$ for the vector a by

$$S(a) = a\times = \begin{pmatrix} 0 & 1 & 2 \\ -1 & 0 & -3 \\ -2 & 3 & 0 \end{pmatrix} \in so(3). \tag{2.3}$$

Then, it is easily seen that

$$c = [a, b] = a \times b = S(a)b = \begin{pmatrix} 4 \\ -5 \\ 2 \end{pmatrix},$$

which agrees with the result using the conventional determinant method in calculus and physics as follows:

$$c = a \times b = \begin{vmatrix} i & j & k \\ 3 & 2 & -1 \\ -1 & 0 & 2 \end{vmatrix} = 4i - 5j + 2k.$$

More typical examples of Lie algebra include the matrix commutations and vector field derivatives. All the n by n square real matrices along with the commutator $[A, B] = AB - BA$ constitute a Lie algebra. One can also verify that all the n-dimensional real vector fields, each of which is a smooth function of point $x \in \mathbb{R}^n$ under the following Lie bracket

$$[f, g] = \frac{\partial g}{\partial x} f - \frac{\partial f}{\partial x} g \tag{2.4}$$

also constitute a Lie algebra.

As an interesting property, let us look at two vectors a and b and their skew-symmetric matrices $S(a) = a\times$ and $S(b) = b\times$ as the cross-product operators of a and b, respectively. We wish to know what the commutator $[S(a), S(b)] = S(a)S(b) - S(b)S(a)$ is supposed to be. First, by taking transpose on both sides, we find that the result of the skew-symmetric matrices commutator $[S(a), S(b)]$ is still a skew-symmetric matrix, and thus, satisfies the closure condition. Let $v \in \mathbb{R}^3$ be an arbitrary vector. Then,

$$[S(a), S(b)]v = S(a)S(b)v - S(b)S(a)v = a \times (b \times v) - b \times (a \times v),$$

i.e., the difference between two triple vector products. By applying the triple vector product formula in (2.1), we obtain

$$\begin{aligned}[S(a), S(b)]v &= (a^T v)b - (a^T b)v - (b^T v)a + (b^T a)v \\ &= (a^T v)b - (b^T v)a = (a \times b) \times v.\end{aligned}$$

Since v is arbitrary, the above equation implies that

$$[S(a), S(b)] = (a \times b)\times = S(a \times b), \tag{2.5}$$

i.e. the cross-product operator of $a \times b$.

Let us now continue the previous numerical example, i.e., $S(a)$ has been found in (2.3), while

$$S(b) = b\times = \begin{pmatrix} 0 & -2 & 0 \\ 2 & 0 & 1 \\ 0 & -1 & 0 \end{pmatrix}.$$

Thus,

$$[S(a), S(b)] = S(a)S(b) - S(b)S(a)$$

$$= \begin{pmatrix} 2 & -2 & 1 \\ 0 & 5 & 0 \\ 6 & 4 & 3 \end{pmatrix} - \begin{pmatrix} 2 & 0 & 6 \\ -2 & 5 & 4 \\ 1 & 0 & 3 \end{pmatrix} = \begin{pmatrix} 0 & -2 & -5 \\ 2 & 0 & -4 \\ 5 & 4 & 0 \end{pmatrix},$$

which exactly meets $S(c) = c\times = S(a \times b)$.

In mathematical history, one of the most significant discoveries in the theory of Lie group and Lie algebra is the following exponential mapping:

$$EXP: \; so(3) \longrightarrow SO(3). \tag{2.6}$$

This mapping means that for each 3 by 3 skew-symmetric matrix $S \in so(3)$, its exponential function $\exp(S) = R \in SO(3)$ is always a rotation matrix. In more general words, the exponential mapping can convert a Lie algebra to a Lie group in any finite dimension. This mapping property will be very useful and underlie the robotic kinematics to seek a new representation for rotation and orientation of coordinate frames [4, 6, 8].

2.3 The Exponential Mapping and k–ϕ Procedure

As discussed previously, to represent a 3D rotation, one often uses a rotation matrix that is a member of the Lie group $SO(3)$. This has been recognized to be the most consistent, secure and unique approach. However, in many application cases, we wish to perform a rotation for a given frame by rotating itself with a certain angle ϕ about an instantaneously fixed unit vector k, instead of using a successive rotation sequence that is a product of up to three elementary rotation matrices from the basis (2.2). Such a rotation approach is intuitively more natural, effective and easier for sampling in robotic path planning applications because the multiplication of matrices in a successive rotation process is not commutable in general.

Let a unit vector $k = \begin{pmatrix} k_1 \\ k_2 \\ k_3 \end{pmatrix}$ be referred and projected onto a given frame to be performed for a rotation. Clearly, $\|k\|^2 = k_1^2 + k_2^2 + k_3^2 = 1$. Its corresponding skew-symmetric matrix is given by

$$S(k) = K = \begin{pmatrix} 0 & -k_3 & k_2 \\ k_3 & 0 & -k_1 \\ -k_2 & k_1 & 0 \end{pmatrix}.$$

It can be shown without difficulty that for this 3 by 3 unity skew-symmetric matrix,

1. K is skew-symmetric with $\mathrm{tr}(K) = 0$;
2. K^2 is symmetric with $\mathrm{tr}(K^2) = -2$;
3. $K^3 = -K$.

With the above interesting properties owned by the unity skew-symmetric matrix K, we now develop a realistic rotation process, called the k–ϕ procedure. First, let $\phi k \in \mathbb{R}^3$ be defined as a 3 by 1 vector. Substituting its skew-symmetric matrix $\phi K \in so(3)$ into the exponential mapping (2.6) yields

$$\exp(\phi K) = R \in SO(3). \tag{2.7}$$

The result of the exponential mapping is a rotation matrix that should be equivalent to the rotation by ϕ about the k-axis.

Expanding the left-hand side of (2.7) into Taylor series and noticing the above properties of K, we obtain

$$\begin{aligned}\exp(\phi K) &= I + \phi K + \frac{\phi^2}{2!}K^2 + \frac{\phi^3}{3!}K^3 + \frac{\phi^4}{4!}K^4 \cdots \\ &= I + \left(\phi - \frac{\phi^3}{3!} + \cdots\right)K + \left(\frac{\phi^2}{2!} - \frac{\phi^4}{4!} + \cdots\right)K^2 \\ &= I + \sin\phi K + (1 - \cos\phi)K^2 = R,\end{aligned} \tag{2.8}$$

where the well-known Taylor expansions have been applied:

$$\sin\phi = \phi - \frac{\phi^3}{3!} + \frac{\phi^5}{5!} - \cdots \quad \text{and} \quad \cos\phi = 1 - \frac{\phi^2}{2!} + \frac{\phi^4}{4!} - \cdots.$$

Equation (2.8), commonly called the Rodrigues formula, provides a means as a forward direction to determine the equivalent rotation matrix R if a frame is to be rotated by a given angle ϕ about a given unit vector k.

In order to find its inverse solution, first, let us take the trace of (2.8), i.e.,

$$\text{tr}(R) = 3 - 2(1 - \cos\phi) = 1 + 2\cos\phi$$

so that

$$\phi = \arccos\left(\frac{\text{tr}(R) - 1}{2}\right). \tag{2.9}$$

Since

$$R^T = I - \sin\phi K + (1 - \cos\phi)K^2, \tag{2.10}$$

subtracting (2.10) from (2.8) yields

$$R - R^T = 2\sin\phi K,$$

and finally we have

$$K = \frac{R - R^T}{2\sin\phi}. \tag{2.11}$$

Both equations (2.9) and (2.11) offer an inverse formula in a recursive order to determine ϕ and k in terms of a given rotation matrix R. The formula, at first glance, does not uniquely set the sign. However, since for each cosine value, there can be $\pm$ angle solutions, if you choose a $+\phi$ (or $-\phi$) from equation (2.9), then, equation (2.11) will give you a corresponding k (or $-k$). Therefore, you may pick any one of the two possible pairs ϕ and k, and $-\phi$ and $-k$, and either one is correct.

As the first example, if we wish a coordinate frame to be rotated by 60^0 or $\pi/3$ about an axis $(1 \;\; 2 \;\; 1)^T$ that is referred to the frame itself, then the unit axis and its skew-symmetric matrix become

$$k = \frac{1}{\sqrt{6}}\begin{pmatrix} 1 \\ 2 \\ 1 \end{pmatrix}, \quad \text{and} \quad K = \frac{1}{\sqrt{6}}\begin{pmatrix} 0 & -1 & 2 \\ 1 & 0 & -1 \\ -2 & 1 & 0 \end{pmatrix}.$$

Therefore, the equivalent rotation matrix turns out to be

$$R = I + \sin(60^0)K + (1-\cos(60^0))K^2 = \begin{pmatrix} 0.5833 & -0.1869 & 0.7904 \\ 0.5202 & 0.8333 & -0.1869 \\ -0.6238 & 0.5202 & 0.5833 \end{pmatrix}.$$

If you want to reverse the direction, you may apply the inverse formula to determine both the angle ϕ and the axis k in terms of the above rotation matrix R. Based on equation (2.9), since $\mathrm{tr}(R) = 2$, $\phi = \arccos\left(\frac{1}{2}\right) = 60^0$ so that $\sin\phi = \frac{\sqrt{3}}{2}$. Then, based on (2.11), we have

$$K = \frac{R - R^T}{2\sin\phi} = \begin{pmatrix} 0 & -0.4082 & 0.8165 \\ 0.4082 & 0 & -0.4082 \\ -0.8165 & 0.4082 & 0 \end{pmatrix}. \tag{2.12}$$

Hence, the unit vector is $k = (0.4082 \;\; 0.8165 \;\; 0.4082)^T$. Clearly, because cosine is an even function, you can also pick -60^0 for the angle ϕ. In this case, your axis k becomes negative accordingly. The two picks are obviously equivalent, because rotating a frame about an axis by an angle will always have the same destination as rotating the original frame about the opposite axis by the negative angle in 3D space.

The second practical example is to find a new frame 1 such that its z_1 axis is along a vector $z_0^1 = (1 \;\; -1 \;\; -1)^T$ that is referred to the base frame 0. While it may have a few different approaches to finding the orientation R_0^1 of frame 1, we just try the $k - \phi$ procedure to do so. Let the common normal vector to both the base z_0-axis $z_0 = (0 \;\; 0 \;\; 1)^T$ and the new z_1-axis be determined through the cross-product by

$$b = z_0^1 \times z_0 = \begin{pmatrix} 0 & 1 & -1 \\ -1 & 0 & -1 \\ 1 & 1 & 0 \end{pmatrix}\begin{pmatrix} 0 \\ 0 \\ 1 \end{pmatrix} = \begin{pmatrix} -1 \\ -1 \\ 0 \end{pmatrix}.$$

Then, we pick the unit vector of b to be the rotating axis k:

$$k = \frac{b}{\|b\|} = \frac{1}{\sqrt{2}}\begin{pmatrix} -1 \\ -1 \\ 0 \end{pmatrix} \quad \text{such that} \quad K = \frac{1}{\sqrt{2}}\begin{pmatrix} 0 & 0 & -1 \\ 0 & 0 & 1 \\ 1 & -1 & 0 \end{pmatrix}.$$

To find the rotating angle ϕ from the base z_0-axis to the new z_1-axis about the above k, we use the dot-product:

$$z_0^T z_0^1 = -1 = \|z_0\|\|z_0^1\| \cos\phi = \sqrt{3}\cos\phi,$$

so that

$$\cos\phi = -\frac{1}{\sqrt{3}} \quad \text{and} \quad \sin\phi = \sqrt{1-\cos^2\phi} = \sqrt{\frac{2}{3}}.$$

Note that since $\cos\phi < 0$, it can be predicted that ϕ is in the second quadrant and thus $\sin\phi > 0$. Therefore, the new frame orientation can be found by

$$R_0^1 = I + \sin\phi K + (1-\cos\phi)K^2 = I + \sqrt{\frac{2}{3}}K + (1+\frac{1}{\sqrt{3}})K^2$$

$$= \begin{pmatrix} 0.2113 & 0.7887 & -0.5774 \\ 0.7887 & 0.2113 & 0.5774 \\ 0.5774 & -0.5774 & -0.5774 \end{pmatrix}.$$

Moreover, if R is symmetric, equation (2.11) will result in a $\frac{0}{0}$ uncertainty. This means that the frame is to be rotated either by nothing (0^0) if $R = I$, or by 180^0 if $R \neq I$ but $R^T = R$. In the first case $R = I$, equation (2.9) leads to $\phi = \arccos(1) = 0$, and you may just skip over (2.11) to avoid its singularity and do nothing. However, in the second case, $\phi = \arccos(-1) = 180^0$, and you cannot determine K directly from (2.11). Nevertheless, in this special case, $\sin\phi = 0$ and $\cos\phi = -1$. Based on the forward equation (2.8), $R = I + \sin\phi K + (1-\cos\phi)K^2 = I + 2K^2$, or

$$K^2 = \frac{1}{2}(R - I). \tag{2.13}$$

Therefore, you have to use this equation to determine K if $R = R^T$ but $R \neq I$.

For example, if

$$R = \begin{pmatrix} 0 & -1 & 0 \\ -1 & 0 & 0 \\ 0 & 0 & -1 \end{pmatrix},$$

which is symmetric but $R \neq I$. Thus, $\text{tr}(R) = -1$ and $\phi = 180^0$. Now, let

$$K = \begin{pmatrix} 0 & -k_3 & k_2 \\ k_3 & 0 & -k_1 \\ -k_2 & k_1 & 0 \end{pmatrix}$$

such that

$$K^2 = \begin{pmatrix} -k_3^2 - k_2^2 & k_1 k_2 & k_1 k_3 \\ k_1 k_2 & -k_3^2 - k_1^2 & k_2 k_3 \\ k_1 k_3 & k_2 k_3 & -k_1^2 - k_2^2 \end{pmatrix}.$$

Because

$$R - I = \begin{pmatrix} -1 & -1 & 0 \\ -1 & -1 & 0 \\ 0 & 0 & -2 \end{pmatrix},$$

by comparing it with K^2, based on (2.13), we first have the off-diagonal elements

$$k_1 k_2 = -\frac{1}{2}, \quad k_1 k_3 = 0 \quad \text{and} \quad k_2 k_3 = 0.$$

This implies that $k_3 = 0$. Then, with the diagonal elements,

$$k_3^2 + k_2^2 = k_2^2 = \frac{1}{2} \quad \text{and} \quad k_3^2 + k_1^2 = k_1^2 = \frac{1}{2}$$

such that either $k_1 = \sqrt{2}/2$ and $k_2 = -\sqrt{2}/2$, or $k_1 = -\sqrt{2}/2$ and $k_2 = \sqrt{2}/2$ in corresponding to $\phi = \pm 180^0$, which results in the same rotation. Let us just pick the first pair so that the unit vector

$$k = \begin{pmatrix} \frac{\sqrt{2}}{2} \\ -\frac{\sqrt{2}}{2} \\ 0 \end{pmatrix} \quad \text{and} \quad \phi = 180^0.$$

Therefore, only in the case of $R^T = R$ but $R \neq I$, we can follow the above procedure to determine k and just set $\phi = 180^0$. This example will be revisited again in Chapter 3.

If now postmultiplying the unit vector k to both sides of equation (2.8), we immediately obtain

$$Rk = k + \sin\phi K k + (1 - \cos\phi) K^2 k = k$$

because $Kk = k \times k \equiv 0$. This result shows that *the unit vector k in the k–ϕ procedure is actually the eigenvector of the rotation matrix R corresponding to its eigenvalue* $+1$. Therefore, if you program numerically the k–ϕ procedure for inverse solution in MATLAB^{TM}, you may take advantage of the internal function eig(R) for a given rotation matrix R without going through equation (2.11) to determine K to avoid the possible singularity. In the previous first numerical example,

```
>> R=[0.5833  -0.1869  0.7904
      0.5202  0.8333  -0.1869
     -0.6238  0.5202  0.5833];

>> [vec, val] = eig(R)

vec =

  -0.1291 - 0.6325i  -0.1291 + 0.6325i   0.4082
  -0.2582 + 0.3162i  -0.2582 - 0.3162i   0.8165
   0.6455             0.6455             0.4082

val =

   0.5000 + 0.8660i        0                  0
        0             0.5000 - 0.8660i        0
        0                  0             1.0000

>>
```

As you can see clearly, once the internal function eig(R) is called, the element "val" prints the three eigenvalues along the diagonal of the output matrix, while "vec" prints the three column eigenvectors for the input matrix R. Obviously, since the last diagonal element of "val" is 1.0000, its corresponding eigenvector on the thrid column of "vec" is $(0.4082 \quad 0.8165 \quad 0.4082)^T$, which agrees with the result in (2.12). If we try $R = I$, the identity matrix, then the working window of MATLAB^{TM} will show the following answer:

```
>> [vec, val]=eig(eye(3))

vec =

     1     0     0
     0     1     0
     0     0     1

val =

     1     0     0
     0     1     0
     0     0     1

>>
```

This shows that all the three eigenvalues of the 3 by 3 identity I are $+1$ and their corresponding eigenvectors are given by the unit vectors of the x, y and z axes of the reference frame. Finally, if we try the previous third numerical example in the case of a symmetric but non-identity rotation matrix in MATLAB^{TM}, then,

```
>> [vec, val]=eig([0 -1 0; -1 0 0; 0 0 -1])

vec =

   -0.7071         0   -0.7071
   -0.7071         0    0.7071
         0    1.0000         0

val =

    -1     0     0
     0    -1     0
     0     0     1

>>
```

This result shows that the unit eigenvector corresponding to the eigenvalue $+1$ is $k = (-0.7071\ \ 0.7071\ \ 0)^T$. This is just the negative of the vector from the above example, and it does not affect the final rotation at all because the rotation angle ϕ can be chosen by $\pm 180^0$ in this special case.

In conclusion, both the forward and inverse compact equations for the rotation matrix $R \in SO(3)$ in (2.8) and the k–ϕ procedure in (2.9) and (2.11) are developed based on the beautiful exponential mapping in mathematical history between $so(3)$ and $SO(3)$ in (2.7). One will find later that the k–ϕ procedure is extremely useful in robotic kinematics and digital human realistic motion-planning, especially for the orientation path-planning.

2.4 The Dual Number, Dual Vector and Their Algebras

In mathematics, two real numbers a and b can be combined under a certain unit quantity j to create a new number $c = a + jb$. In general, the unit quantity j can be defined independently only in the following three different cases:

$$j^2 = +1, \quad j^2 = 0 \quad \text{and} \quad j^2 = -1.$$

In history, the three types of new combined numbers have already been studied [9, 10, 12], and they can briefly be described in Table 2.1.

Table 2.1 Three types of combined numbers and their mathematical properties

$j^2 =$	**Name of** $c = a + jb$	**Magnitude**	**Algebra**	**Geometry**
$+1$	Double Number	$\sqrt{\lvert a^2 - b^2 \rvert}$	Ring	Hyperbolic
0	Dual Number	$\lvert a \rvert$	Ring	Parabolic
-1	Complex Number	$\sqrt{a^2 + b^2}$	Field	Elliptic

In the above table, each type of the combined numbers constitutes an additive group, but only the set of complex numbers is also qualified to be a multiplicative group and thus, forms an algebraic field. The other two: double number and dual number can only form a multiplicative semigroup each because of the inverse loss if $a = b$ for a double number and $a = 0$ for a dual number. Therefore, each of them just constitutes an algebraic ring, instead of a field, and is called the double ring and dual ring, respectively.

It is noticeable that the third type is the well-known complex field that has been widely used as one of the most popular mathematical tools. The double number is least useful because its algebraic properties are not quite simple nor meaningful. The dual number, however, could be overlooked in history. In fact, it is quite unique and attractive with a relatively simpler algebra, especially for the application of 3D space transformations.

The above three definitions of combined numbers can also be linearly mapped to a 2 by 2 matrix with all the operations of an algebraic field (or ring) preserved [4]. Namely, let

$$\Gamma(a + jb) = \begin{pmatrix} a & b \\ j^2 b & a \end{pmatrix} = C \tag{2.14}$$

such that $\Gamma(a + jb) = \begin{pmatrix} a & b \\ b & a \end{pmatrix}$ for a double number, $\Gamma(a + jb) = \begin{pmatrix} a & b \\ 0 & a \end{pmatrix}$ for a dual number and $\Gamma(a + jb) = \begin{pmatrix} a & b \\ -b & a \end{pmatrix}$ for a complex number. For instance, if $c_1 = a_1 + jb_1$ and $c_2 = a_2 + jb_2$ are two complex numbers with $j^2 = -1$, then,

$$\Gamma(c_1 + c_2) = \Gamma(a_1 + a_2 + j(b_1 + b_2)) = \begin{pmatrix} a_1 + a_2 & b_1 + b_2 \\ -(b_1 + b_2) & a_1 + a_2 \end{pmatrix}$$

and

$$\Gamma(c_1 \cdot c_2) = \Gamma(a_1 a_2 - b_1 b_2 + j(a_1 b_2 + a_2 b_1))$$

that is equal to

$$\begin{pmatrix} a_1 & b_1 \\ -b_1 & a_1 \end{pmatrix} \begin{pmatrix} a_2 & b_2 \\ -b_2 & a_2 \end{pmatrix} = \begin{pmatrix} a_1 a_2 - b_1 b_2 & a_1 b_2 + a_2 b_1 \\ -(a_1 b_2 + a_2 b_1) & a_1 a_2 - b_1 b_2 \end{pmatrix}.$$

Likewise, if both c_1 and c_2 are dual numbers with $j^2 = 0$, then, the addition is obvious, while the multiplication becomes

$$\Gamma(c_1 \cdot c_2) = \Gamma(a_1 a_2 + j(a_1 b_2 + a_2 b_1))$$

that is also equal to

$$\begin{pmatrix} a_1 & b_1 \\ 0 & a_1 \end{pmatrix} \begin{pmatrix} a_2 & b_2 \\ 0 & a_2 \end{pmatrix} = \begin{pmatrix} a_1 a_2 & a_1 b_2 + a_2 b_1 \\ 0 & a_1 a_2 \end{pmatrix}.$$

One can easily verify that the above mapping is also valid for two double numbers. In summary, the mapping Γ satisfies

$$\Gamma(c_1 \circ c_2) = \Gamma(c_1) \circ \Gamma(c_2),$$

where $\circ$ can be either addition or multiplication. Thus, the mapping Γ is a homomorphism that preserves the algebraic structure. The mapping is also applicable to finding the inverse for any one of the three combined numbers.

By inverting the 2 by 2 matrix in (2.14), we obtain

$$C^{-1} = \begin{pmatrix} a & b \\ j^2 b & a \end{pmatrix}^{-1} = \frac{\text{adj}\, C}{\det C} = \frac{\begin{pmatrix} a & -b \\ -j^2 b & a \end{pmatrix}}{a^2 - j^2 b^2},$$

where adj C is the adjugate matrix of C. For a complex number, $j^2 = -1$ such that

$$C^{-1} = \frac{\begin{pmatrix} a & -b \\ b & a \end{pmatrix}}{a^2 + b^2},$$

which is clearly equivalent to the inversion of a complex number via its complex conjugate:

$$(a + jb)^{-1} = \frac{a - jb}{(a + jb)(a - jb)} = \frac{a - jb}{a^2 + b^2}.$$

In contrast, the inverse of a dual number with $j^2 = 0$ becomes

$$C^{-1} = \frac{\begin{pmatrix} a & -b \\ 0 & a \end{pmatrix}}{a^2},$$

or

$$(a + jb)^{-1} = \frac{a - jb}{(a + jb)(a - jb)} = \frac{a - jb}{a^2}.$$

This shows that a dual number will lose its inverse if the real part $a = 0$ while the dual part b remains nonzero. Therefore, the set of dual numbers can only constitute a multiplicative semigroup, and they just form an algebraic ring, referred to as the dual ring.

Likewise, for a double number with $j^2 = +1$,

$$C^{-1} = \frac{\begin{pmatrix} a & -b \\ -b & a \end{pmatrix}}{a^2 - b^2},$$

or

$$(a + jb)^{-1} = \frac{a - jb}{(a + jb)(a - jb)} = \frac{a - jb}{a^2 - b^2}.$$

This also reveals that a double number will lose its inverse if $a = b$, and thus the set of double numbers can only form a ring as well, instead of a field.

Nevertheless, many unique and excellent algebraic properties from the dual ring will make it very attractive in the application of space transformations, especially for the representation of a line vector. In order to clearly distinguish a dual number from the others in the next discussions, let j be re-denoted by ϵ as $a + \epsilon b$ under $\epsilon^2 = 0$.

2.4.1 Calculus of the Dual Ring

The dual number, first introduced by Clifford in 1873 [9], was further developed by Study in 1901 [10] to represent a dual angle in spatial geometry [11]–[18]. Suppose that the shortest distance between two straight lines in 3D space is d and the angle between their directions is α, then a *dual angle* was defined by

$$\hat{\alpha} = \alpha + \epsilon d$$

under $\epsilon^2 = 0$. The geometric meaning for this so-defined dual angle is like a screw motion. Namely, using a screw driver to twist a screw by an angle α while pressing and moving the screw down for a vertical distance d.

Like most common functions in the complex field, a smooth function of a dual number $f(a + \epsilon b)$ can be expanded into a formal Taylor series at the point a, and due to $\epsilon^2 = 0$, the expansion will soon become finite:

$$\begin{aligned} f(a + \epsilon b) &= f(a) + \epsilon b f'(a) + \epsilon^2 \frac{b^2}{2!} f''(a) + \cdots \\ &= f(a) + \epsilon b f'(a), \end{aligned} \tag{2.15}$$

where $f'(a)$ and $f''(a)$ are the first and second derivatives, respectively, and evaluated at the point a. It is interesting that the above general Taylor expansion over the dual ring generates a new dual function. Based on such a compact expansion, it can be easily seen that

$$e^{a+\epsilon b} = e^a \cdot e^{\epsilon b} = e^a + \epsilon b e^a, \tag{2.16}$$

and also

$$\sin(\alpha + \epsilon d) = \sin\alpha + \epsilon d \cos\alpha \tag{2.17}$$

and

$$\cos(\alpha + \epsilon d) = \cos\alpha - \epsilon d \sin\alpha. \tag{2.18}$$

More significantly, the general expansion in (2.15) can be directly used through the dual-number calculus to determine the first derivative of any smooth function numerically without error and without a symbolical derivation needed by hand. As we have all experienced in numerical computation, to integrate a function numerically in a computer can easily reach at least the forth-order accuracy, such as using the well-known Runge-Kutta algorithm. However, to calculate the derivative for a function numerically, it is often difficult to reach a satisfactory high-order approximation without a memory of previous sampling data. Now, if a computational software package, such as MATLAB^{TM}, can implement an internal subroutine to operate the dual-number calculus, like the complex number calculation that has already been built in MATLAB^{TM}, we can calculate numerically any kind of derivatives of a function exactly and instantaneously at a point of interest in a computer via equation (2.15) without going through a symbolical derivation.

As an illustrative example, let

$$f(x) = x\exp(-x^2).$$

We wish to calculate the first derivative of $f(x)$ evaluated at $x = 3$, i.e., $f'(3)$ numerically. Let $x = 3 + \epsilon 1 = 3 + \epsilon$. Then,

$$f(x) = (3+\epsilon)\exp(-(3+\epsilon)^2) = (3+\epsilon)e^{-9-\epsilon 6} = (3+\epsilon)e^{-9}(1-\epsilon 6) = (3-\epsilon 17)e^{-9}.$$

Clearly, taking the dual part of the above result yields

$$\mathfrak{Du}\{f(3+\epsilon)\} = -17e^{-9} = -2.098 \times 10^{-3},$$

which is just the value of $f'(3)$ because we purposely set $b = 1$ in x. To verify the numerical result, we may symbolically calculate

$$f'(x) = (1 - 2x^2)\exp(-x^2).$$

Then, substituting $x = 3$ into the above derivative, we reach the exactly same answer $-17e^{-9} = -2.098 \times 10^{-3}$.

In the later chapters, we will have to calculate derivatives for much more complicated functions, or high-dimensional vector fields or matrix fields. It can be difficult or even unmanageable to find their symbolical derivatives by hand before programming them for numerical computation. In this case, the above property owned by the dual ring will overwhelmingly demonstrate its unique advantage over any other algorithms for a numerical derivative solution [1].

Furthermore, suppose that we replace $a + \epsilon b$ in (2.15) by $q + \epsilon\dot{q}$, where $q \in \mathbb{R}^n$ could be an n-dimensional vector for $n = 1, 2, \cdots$ and $\dot{q}$ is its time-derivative, while the function itself $f(\cdot)$ could also be a high-dimensional vector or even a matrix (note that in this case, the dimensions in each matrix multiplication must be compatible). Then,

$$f(q + \epsilon\dot{q}) = f(q) + \epsilon\frac{\partial f(q)}{\partial q}\dot{q} = f(q) + \epsilon\dot{f}(q), \tag{2.19}$$

where $\dot{f}(q) = \frac{d}{dt}f(q) = \frac{\partial f(q)}{\partial q}\dot{q}$, the time-derivative of the function $f(q)$ based on the derivative chain-rule. This new property in (2.19) will definitely extend the unique advantage of (2.15) to directly finding a time-derivative numerically for a large-scale function, which will be more often required in the later robotics and digital human modeling applications. Of course, the derivative may also be generalized with respect to any parameter other than time, and this will be exhibited more in later developments.

Let us now look at a real example. Suppose that the orientation for a robot link frame is given by the following rotation matrix with respect to the fixed base:

$$R = \begin{pmatrix} c_1c_{23} & s_1 & c_1c_{23} \\ s_1c_{23} & -c_1 & s_1s_{23} \\ s_{23} & 0 & -c_{23} \end{pmatrix},$$

where $s_1 = \sin\theta_1$, $c_1 = \cos\theta_1$, $s_{23} = \sin(\theta_2 + \theta_3)$ and $c_{23} = \cos(\theta_2 + \theta_3)$ as a conventional short notation. At a certain time instant, the three joint angles of the robot are $\theta_1 = \pi/6$, $\theta_2 = \pi/2$ and $\theta_3 = 0$ in radians, and the joint velocities are given by $\dot{\theta}_1 = 0.5$, $\dot{\theta}_2 = -0.4$ and $\dot{\theta}_3 = 0.8$ in radians/second. Then, programming it into MATLAB^{TM}, the rotation matrix R will have the following numerical result at the time instant:

```
>> R =

    0.0000    0.5000    0.8660
    0.0000   -0.8660    0.5000
    1.0000         0   -0.0000

>>
```

Now, let each joint angle θ_i be replaced by $\theta_i + \epsilon\dot{\theta}_i$ for $i = 1, 2, 3$, and taking the dual part out from the final numerical result $\hat{R} = R + \epsilon S$, we will immediately see the output in the MATLAB^{TM} working window:

```
>> S =

   -0.3464    0.4330   -0.2500
   -0.2000    0.2500    0.4330
    0.0000         0    0.4000

>>
```

Based on (2.19), this matrix S value is just the instantaneous time-derivative of R, i.e., $S = \dot{R}$.

Furthermore, if we wish to determine numerically the partial derivative of R with respect to θ_2 at the same time instant, i.e., $S = \frac{\partial R}{\partial \theta_2}$, then only θ_2 in R is replaced by $\theta_2 + \epsilon 1$ and the other two joint angles remain unchanged. Clearly, this is equivalent to setting the joint velocity $\dot{\theta}_2 = 1$ and the others $\dot{\theta}_1 = 0$ and $\dot{\theta}_3 = 0$ in the above MATLAB^{TM} program. The final result is popped out as follows:

```
>> S =

   -0.8660         0    0.0000
   -0.5000         0    0.0000
    0.0000         0    1.0000

>>
```

which is just the instantaneous partial derivative of R with respect to θ_2. Using the same method, we can also calculate the numerical partial derivatives $\frac{\partial R}{\partial \theta_1}$ and $\frac{\partial R}{\partial \theta_3}$.

2.4.2 Dual Vector and Dual Matrix

The dual number definition can be extended to combining two vectors or even two matrices to form either a dual vector or a dual matrix as follows:

$$\hat{a} = \begin{pmatrix} a_1 + \epsilon b_1 \\ a_2 + \epsilon b_2 \\ a_3 + \epsilon b_3 \end{pmatrix} = \begin{pmatrix} a_1 \\ a_2 \\ a_3 \end{pmatrix} + \epsilon \begin{pmatrix} b_1 \\ b_2 \\ b_3 \end{pmatrix} = a + \epsilon b,$$

or

$$\hat{A} = A + \epsilon B$$

for two same dimensional matrices A and B. Since it will soon be discovered that the high-dimensional extensions of dual numbers possess very unique and elegant properties over the complex numbers and double numbers, it deserves to be further discussed in detail.

First, the inner product between two dual vectors is similar to the multiplication of two dual numbers. Let $\hat{a} = a + \epsilon b$ and $\hat{c} = c + \epsilon d$ with each of a, b, c and d to be a 3 by 1 vector. Then, the inner (dot) product between $\hat{a}$ and $\hat{c}$ becomes

$$\hat{a}^T \hat{c} = a^T c + \epsilon (b^T c + a^T d).$$

To perform the cross (vector) product $\hat{a} \times \hat{c}$, let the skew-symmetric matrix of $\hat{a}$ be defined as

$$S(\hat{a}) = \begin{pmatrix} 0 & -a_3 - \epsilon b_3 & a_2 + \epsilon b_2 \\ a_3 + \epsilon b_3 & 0 & -a_1 - \epsilon b_1 \\ -a_2 - \epsilon b_2 & a_1 + \epsilon b_1 & 0 \end{pmatrix} = S(a) + \epsilon S(b).$$

Then,

$$\hat{a} \times \hat{c} = S(\hat{a})\hat{c} = S(a)c + \epsilon(S(b)c + S(a)d) = a \times c + \epsilon(b \times c + a \times d).$$

Likewise, the product between two dual matrices $\hat{A} = A + \epsilon B$ and $\hat{C} = C + \epsilon D$ for four n by n square matrices A, B, C and D is similar to the dual vector products:

$$\hat{A}\hat{C} = AC + \epsilon(BC + AD),$$

and also

$$\hat{A}\hat{a} = Aa + \epsilon(Ab + Ba)$$

without exception.

Furthermore, to determine the inverse of a dual square matrix $\hat{A} = A + \epsilon B$, we have to first assume that the real part A must be nonsingular. Then, the inverse of $\hat{A}$ can be determined by

$$\hat{A}^{-1} = (A + \epsilon B)^{-1} = A^{-1} - \epsilon A^{-1}BA^{-1},$$

which may easily be verified by testing

$$\hat{A}^{-1}\hat{A} = (A^{-1} - \epsilon A^{-1}BA^{-1})(A + \epsilon B) = I + \epsilon O = I,$$

where I and O are the n by n real identity and zero matrix, respectively, and so is its commutation $\hat{A}\hat{A}^{-1} = I$.

As we can recall, there is a well-known equation to link the determinant and trace of a real square matrix A together in linear algebra:

$$\det(e^A) = e^{\mathrm{tr}A}. \tag{2.20}$$

In fact, this equation can be justified without difficulty through some major properties for a square matrix in linear algebra. First, a square matrix A can always be diagonalized or Jordan-decomposed via a similarity transformation P as follows:

$$P^{-1}AP = \begin{pmatrix} \lambda_1 & & X \\ & \ddots & \\ O & & \lambda_n \end{pmatrix},$$

where λ_1 through λ_n are the eigenvalues of A and the upper-right triangle block X may have some 1's based on the Jordan decomposition, but the elements at the lower-left triangle corner are all zeros. Thus,

$$P^{-1}e^{A}P = \exp(P^{-1}AP) = \begin{pmatrix} e^{\lambda_1} & & \bar{X} \\ & \ddots & \\ O & & e^{\lambda_n} \end{pmatrix}$$

for some upper-right corner $\bar{X}$. No matter what $\bar{X}$ is, since all the elements at the lower-left corner are always zeros, this ensures that the determinant

$$\det(e^{A}) = \det(P^{-1}e^{A}P) = \prod_{i=1}^{n} e^{\lambda_i} = \exp(\lambda_1 + \cdots + \lambda_n) = e^{\mathrm{tr}A}.$$

Now, applying the Taylor expansion on the exponential function of any dual square matrix ϵM yields

$$\exp(\epsilon M) = I + \epsilon M.$$

This implies that if A is nonsingular, $\exp(\epsilon A^{-1}B) = I + \epsilon A^{-1}B$ so that

$$\det(\hat{A}) = \det(A + \epsilon B) = \det(A) \cdot \det(I + \epsilon A^{-1}B) = \det(A) \cdot \det(e^{\epsilon A^{-1}B}).$$

Using equation (2.20), we finally reach a new property for the determinant of a dual square matrix $\hat{A}$ as follows:

$$\det(\hat{A}) = \det(A + \epsilon B) = \det(A) \cdot e^{\epsilon \mathrm{tr}(A^{-1}B)} = \det(A) \cdot (1 + \epsilon \mathrm{tr}(A^{-1}B)). \tag{2.21}$$

This means that the determinant of a dual matrix $\hat{A}$ is a dual scalar number, whose real part is just the determinant of the real part of $\hat{A}$ and the dual part is $\det(A)\mathrm{tr}(A^{-1}B)$. Obviously, it must be operated under $\det(A) \neq 0$.

For example, let a 3 by 3 dual matrix be

$$\hat{A} = \begin{pmatrix} 3+\epsilon 2 & \epsilon 3 & -\epsilon 4 \\ 1 & -2+\epsilon & 0 \\ -1+\epsilon 4 & 3-\epsilon 2 & -4+\epsilon 3 \end{pmatrix}$$

$$= \begin{pmatrix} 3 & 0 & 0 \\ 1 & -2 & 0 \\ -1 & 3 & -4 \end{pmatrix} + \epsilon \begin{pmatrix} 2 & 3 & -4 \\ 0 & 1 & 0 \\ 4 & -2 & 3 \end{pmatrix} = A + \epsilon B.$$

Since $\det(A) = 24$, and

$$A^{-1} = \begin{pmatrix} 0.3333 & 0 & 0 \\ 0.1667 & -0.5000 & 0 \\ 0.0417 & -0.3750 & -0.2500 \end{pmatrix},$$

we have

$$A^{-1}B = \begin{pmatrix} 0.6667 & 1.0000 & -1.3333 \\ 0.3333 & 0 & -0.6667 \\ -0.9167 & 0.2500 & -0.9167 \end{pmatrix}.$$

Thus, $\text{tr}(A^{-1}B) = -0.25$ and $\det(\hat{A}) = 24(1 - \epsilon 0.25) = 24 - \epsilon 6$.

One can also directly calculate the determinant of $\hat{A}$ by dual number addition and multiplication:

$$\det(\hat{A}) = -\epsilon 4 \begin{vmatrix} 1 & -2+\epsilon \\ -1+\epsilon 4 & 3-\epsilon 2 \end{vmatrix} + (-4+\epsilon 3) \begin{vmatrix} 3+\epsilon 2 & \epsilon 3 \\ 1 & -2+\epsilon \end{vmatrix} = 24 - \epsilon 6.$$

Therefore, the results between two different methods are exactly same.

2.4.3 Unit Screw and Special Orthogonal Dual Matrix

Definition 3. A dual vector $\hat{a}_0 = a_0 + \epsilon b_0$ is said to be a *unit screw* if $\|\hat{a}_0\|^2 = \hat{a}_0^T \hat{a}_0 = 1$.

Definition 4. A dual square matrix $\hat{R} = R + \epsilon S$ is called a *special orthogonal dual matrix* if $\hat{R}^T \hat{R} = \hat{R}\hat{R}^T = I$, the real identity, and $\det(R) = +1$. In this case, $\hat{R} \in SO(3, d)$, the special orthogonal group over the dual ring.

Based on Definition 3, since $\hat{a}_0^T \hat{a}_0 = a_0^T a_0 + \epsilon 2 a_0^T b_0$, a unit screw must have both $a_0^T a_0 = 1$ and $a_0^T b_0 = 0$. Actually, the definition suggests to us how to create a dual part for a dual vector such that it can be the best to represent a line vector. As we all know, if a force f is applied to rotate a rigid body, the torque τ is commonly defined as $\tau = p \times f$, where p is a radial vector tailed at the pivoting point and its arrow touches the force line. Clearly, this implies that the force f and the torque τ are always perpendicular to each other. Such a torque definition inspires us to adopt a similar concept to define a dual vector for uniquely representing both the position and orientation of a frame with respect to the reference.

Conventionally in robotics we often use a rotation matrix R to represent the orientation of a frame and a 3 by 1 (point) vector p to determine the position of the origin of the frame with respect to the reference. Although each of the three coordinate axes of the frame can be described by the corresponding column r_i of the rotation matrix R for $i = 1, 2, 3$, each of which is just a point vector, it still needs a position vector p as an addition to uniquely and sufficiently cover both the orientation and location of the frame. Now, let us define a "moment" vector s_i, like the torque τ, by $s_i = p \times r_i$, where each of the three unit column vectors r_i in R is interpreted as a force f. Then, a complete dual vector for each axis of the frame becomes

$$\hat{r}_i = r_i + \epsilon s_i = r_i + \epsilon(p \times r_i). \tag{2.22}$$

By augmenting all such three dual vectors together in order, we finally reach a 3 by 3 dual rotation matrix:

$$\hat{R} = R + \epsilon S \tag{2.23}$$

to uniquely represent both the orientation and position of the frame with respect to the reference. In other words, each axis of the frame is now treated as a line vector, instead of a point vector, and is represented by the dual vector in (2.22).

It is now quite clear that since each r_i is a unit vector and the moment vector $s_i = p \times r_i$ is perpendicular to r_i, i.e., $r_i^T r_i = 1$ and $r_i^T s_i = 0$ for $i = 1, 2, 3$, each dual column vector $\hat{r}_i$ in $\hat{R}$ is obviously a unit screw and the dual rotation matrix $\hat{R}$ in (2.23) is a special orthogonal dual matrix, i.e., $\hat{R}^T \hat{R} = \hat{R}\hat{R}^T = I$, which implies that $R^T R = I$ and $R^T S + S^T R = O$ or $RS^T + SR^T = O$. In other words, $R^T S$ must be a 3 by 3 skew-symmetric matrix so that the trace $\mathrm{tr}(R^T S) = 0$. Substituting this zero trace into the determinant equation in (2.21) and noticing that the real part A in (2.21) is now the real rotation matrix R and the dual part B is now the matrix S, we reach $\det(\hat{R}) = \det(R) = +1$. This has sufficiently justified the following two lemmas:

Lemma 1. *If a 3 by 3 dual matrix $\hat{R} = R + \epsilon S \in SO(3, d)$, then each column or each row of $\hat{R}$ is a unit screw.*

Lemma 2. *A 3 by 3 dual matrix $\hat{R} = R + \epsilon S$ is a special orthogonal dual matrix, i.e., $\hat{R} \in SO(3, d)$, if and only if $R \in SO(3)$ and SR^T is a skew-symmetric matrix.*

Based on Lemma 2, for a $\hat{R} = R + \epsilon S \in SO(3, d)$, SR^T must be a skew-symmetric matrix so that $RS^T = -SR^T$ or $RS^T + SR^T = O$. By recalling how to augment all the three dual vectors, each of which is defined by (2.22) to form a dual rotation matrix $\hat{R}$ in (2.23), we conclude that

$$S = PR \quad \text{or} \quad SR^T = P, \tag{2.24}$$

and the matrix $P = SR^T$ is just the skew-symmetric matrix of the position vector p, i.e., $P = p\times$. This shows that if both the orientation and position for a frame are represented by a 3 by 3 special orthogonal dual matrix $\hat{R}$ defined by (2.23), then the orientation of the frame is uniquely described by the real part R and the position of the frame origin is determined by the cross-product operator $P = SR^T$.

For example, given a 3 by 3 dual matrix

$$\hat{R} = \begin{pmatrix} 1 & \epsilon 5 & \epsilon 7 \\ \epsilon 7 & -\epsilon 2 & -1 \\ -\epsilon 5 & 1 & -\epsilon 2 \end{pmatrix} = R + \epsilon S.$$

According to Lemma 2, one cannot say at this time that it is a special orthogonal dual matrix until testing its real part R and SR^T to see if $R \in SO(3)$ and SR^T is a skew-symmetric matrix. In fact,

$$RR^T = \begin{pmatrix} 1 & 0 & 0 \\ 0 & 0 & -1 \\ 0 & 1 & 0 \end{pmatrix} \begin{pmatrix} 1 & 0 & 0 \\ 0 & 0 & 1 \\ 0 & -1 & 0 \end{pmatrix} = I \quad \text{with} \quad \det(R) = +1,$$

while

$$SR^T = \begin{pmatrix} 0 & 5 & 7 \\ 7 & -2 & 0 \\ -5 & 0 & -2 \end{pmatrix} \begin{pmatrix} 1 & 0 & 0 \\ 0 & 0 & 1 \\ 0 & -1 & 0 \end{pmatrix} = \begin{pmatrix} 0 & -7 & 5 \\ 7 & 0 & -2 \\ -5 & 2 & 0 \end{pmatrix}$$

is obviously a skew-symmetric matrix. Thus, both the conditions hold, and this $\hat{R}$ is a special orthogonal dual matrix, i.e., $\hat{R} \in SO(3, d)$. Also, based on (2.24), the position vector at the origin of the frame of interest can be found directly from the above skew-symmetric matrix result $SR^T = P$, i.e., $p = (2\ 5\ 7)^T$.

The special orthogonal dual matrices in $SO(3, d)$ Lie group over the dual ring and their transformations will be revisited and further developed in Chapter 4 towards an important and useful theorem for robot kinematic modeling and path-planning applications.

2.5 Introduction to Exterior Algebra

The exterior algebra was first introduced by Hermann Grassmann in 1844, and was further developed and extended later by many great mathematicians, including Poincare, Cartan and Darboux [19, 20, 21]. It offers a useful computational and conceptual tool for many applications from the multivariable calculus to differential geometry, because it reveals a great fact that many mathematical relations, especially in differentiation and integration, are skew-symmetric in nature.

Definition 5. Let $x_1, \cdots, x_m$ be m vectors in $\mathbb{R}^n$ space ($m \leq n$). For any $i, j = 1, \cdots, m$, $x_i \wedge x_j$ is called a wedge or exterior product if the following conditions hold:

1. Skew-symmetric: $x_i \wedge x_j = -x_j \wedge x_i$;
2. Distributive: $(a_1 x_i + a_2 x_j) \wedge x_k = a_1 x_i \wedge x_k + a_2 x_j \wedge x_k$, for two arbitrary scalar constants a_1 and a_2 ; and
3. Associative: $x_i \wedge (x_j \wedge x_k) = (x_i \wedge x_j) \wedge x_k$.

The first condition of skew-symmetry in the definition implies that for the same vector, $x_i \wedge x_i = 0$ always. Without loss of generality, let x_1, x_2 and x_3 be three independent 3D vectors in $\mathbb{R}^3$. We now pick any two of them for wedge product and all the combinations form a basis of the so-called second-order multivector space $\Lambda^2(\mathbb{R}^3)$:

$$\{x_1 \wedge x_2,\ x_2 \wedge x_3,\ x_1 \wedge x_3\}.$$

If we pick only one of the three vectors at a time, then the basis becomes $\{x_1, x_2, x_3\}$ without any wedge product for $\Lambda^1(\mathbb{R}^3)$. Whereas if all three vectors are picked, then the basis $x_1 \wedge x_2 \wedge x_3$ of $\Lambda^3(\mathbb{R}^3)$ is only one-dimensional. Thus, in an n-dimensional linear space $\mathbb{R}^n$, the dimension of the k-th order multivector space $\Lambda^k(\mathbb{R}^n)$ is the total number of combinations:

$$\binom{n}{k} = \frac{n!}{k!(n-k)!}.$$

Therefore, the wedge or exterior product is fundamentally different from the cross product, though they both are skew-symmetric.

The applications of such an exterior product are quite significant. First, let three non-parallel vectors in 3D space with the basis $\{e^1, e^2, e^3\}$ be given by $u = u_1e^1 + u_2e^2 + u_3e^3$, $v = v_1e^1 + v_2e^2 + v_3e^3$ and $w = w_1e^1 + w_2e^2 + w_3e^3$. Here, we intentionally denote each vector of u, v and w in a covariant way to match with the traditional notation in engineering applications. Then,

$$u \wedge v \wedge w = \sum_{i,j,k=1}^{3} u_i v_j w_k (e^i \wedge e^j \wedge e^k).$$

Based on the rules of wedge product, expanding the above summation and noticing that $e^1 \wedge e^3 \wedge e^2 = -e^1 \wedge e^2 \wedge e^3$, $e^3 \wedge e^1 \wedge e^2 = -e^1 \wedge e^3 \wedge e^2 = e^1 \wedge e^2 \wedge e^3$ and so on, we will immediately realize that there is finally only one single common factor $e^1 \wedge e^2 \wedge e^3$ left in the summation, which is just the basis of $\Lambda^3(\mathbb{R}^3)$ along with the coefficient given by

$$u_1v_2w_3 + u_2v_3w_1 + u_3v_1w_2 - u_1v_3w_2 - u_2v_1w_3 - u_3v_2w_1.$$

This coefficient is exactly same as the determinant of the following 3 by 3 matrix V augmented by the three vectors u, v and w:

$$V = \begin{pmatrix} u_1 & u_2 & u_3 \\ v_1 & v_2 & v_3 \\ w_1 & w_2 & w_3 \end{pmatrix}.$$

Furthermore, the determinant $\det V = u \cdot (v \times w)$ is just the triple scalar product of the three vectors and is thus equal to the signed volume of a 3D parallelepiped spanned by the three non-parallel vectors.

Secondly, let us look at a case of coordinate transformation. Suppose that $z = f(x)$ is a smooth coordinate system mapping from $x = (x^1 \ \cdots \ x^n)^T \in \mathbb{R}^n$ to $z = (z^1 \ \cdots \ z^n)^T \in \mathbb{R}^n$ under the same dimension and expressed now in a contravariant way. We want to know what is the differential volume change after the mapping? It will soon be seen that the exterior algebra can provide us with a unified and automatic approach to the answer. Differentiating both sides of the mapping yields

$$dz^i = \sum_{j=1}^{n} \frac{\partial f^i}{\partial x^j} dx^j = \frac{\partial f^i}{\partial x^j} dx^j$$

for each $i = 1, \cdots, n$, where the last equal term that is omitting the summation symbol is to exhibit here the widely-adopted Einstein notation in physics and mathematics, but the sum must be operated as default for every pair of common upper and lower indices.

In order to see more explicitly the wedge product derivation, let us begin with the 2D case by setting $n = 2$. Since

$$dz^1 = \frac{\partial f^1}{\partial x^1} dx^1 + \frac{\partial f^1}{\partial x^2} dx^2$$

and

$$dz^2 = \frac{\partial f^2}{\partial x^1} dx^1 + \frac{\partial f^2}{\partial x^2} dx^2,$$

by wedging them together and noticing that $x^1 \wedge x^2 = -x^2 \wedge x^1$ and $x^i \wedge x^i = 0$ for each $i = 1, 2$, we obtain

$$dz^1 \wedge dz^2 = \left(\frac{\partial f^1}{\partial x^1} \frac{\partial f^2}{\partial x^2} - \frac{\partial f^2}{\partial x^1} \frac{\partial f^1}{\partial x^2} \right) dx^1 \wedge dx^2.$$

This 2D case evidently manifests that the coefficient of the basis vector $dx^1 \wedge dx^2$ of the space $\Lambda^2(\mathbb{R}^2)$ is just the determinant of the Jacobian matrix of the mapping, i.e., $\det\left(\frac{\partial f}{\partial x}\right)$. By inducing it from the 2D case to n-dimensional space, we conclude that

$$dz^1 \wedge \cdots \wedge dz^n = \det\left(\frac{\partial f}{\partial x}\right) dx^1 \wedge \cdots \wedge dx^n. \tag{2.25}$$

In other wards, the signed differential volume change after a smooth coordinate transformation $z = f(x)$ is to multiply the original differential volume by the determinant of the Jacobian matrix of that mapping, which exactly agrees with the result from the multi-variable calculus for high-dimensional integration.

The next application is to justify the well-known Stokes formula in calculus, i.e.,

$$\oint_C F_1(x,y,z)dx + F_2(x,y,z)dy + F_3(x,y,z)dz =$$

$$\iint_S \left(\frac{\partial F_2}{\partial x} - \frac{\partial F_1}{\partial y} \right) dxdy + \left(\frac{\partial F_3}{\partial y} - \frac{\partial F_2}{\partial z} \right) dydz + \left(\frac{\partial F_1}{\partial z} - \frac{\partial F_3}{\partial x} \right) dzdx,$$

for a surface S and its contour closed-loop C. Let a differential 1-form be defined as

$$\sigma = F_1(x,y,z)dx + F_2(x,y,z)dy + F_3(x,y,z)dz. \tag{2.26}$$

Now taking the so-called exterior derivative on (2.26), the one higher order differential of the 1-form σ, called a differential 2-form, can be manipulated through the exterior calculus as follows:

$$d\sigma = \left(\frac{\partial F_2}{\partial x} - \frac{\partial F_1}{\partial y}\right) dx \wedge dy + \left(\frac{\partial F_3}{\partial y} - \frac{\partial F_2}{\partial z}\right) dy \wedge dz + \left(\frac{\partial F_1}{\partial z} - \frac{\partial F_3}{\partial x}\right) dz \wedge dx. \tag{2.27}$$

Therefore, the Stokes formula can be written in a following extremely compact unified form:

$$\oint_C \sigma = \iint_S d\sigma.$$

This once again showcases the great advantage of the exterior calculus.

As a final discussion on the exterior calculus, for a given differential 1-form, such as the 3D case σ in (2.26), an interesting conceptual question that is frequently asked is, "does there exist a smooth function or vector field $\mu(x,y,z) \in \mathbb{R}^3$ such that its differential $d\mu = \sigma$?" In other words, can any of such 1-form σ be exactly integrated to reach a vector field $\mu(x,y,z)$? Note that the condition for $d\mu = \sigma$ is referred to as *exact*. It is conceivable that if the 1-form σ is exact, then each F_i in (2.26) must be $F_i = \frac{\partial \mu}{\partial x^i}$ for each $x^i = x$ or y or z. Its partial derivative with respect to x^j becomes

$$\frac{\partial F_i}{\partial x^j} = \frac{\partial^2 \mu}{\partial x^i \partial x^j}.$$

It is well-known in calculus that for a smooth function or vector field $\mu(x^1, x^2, x^3)$,

$$\frac{\partial^2 \mu}{\partial x^i \partial x^j} = \frac{\partial^2 \mu}{\partial x^j \partial x^i}$$

for every $i, j = 1, 2, 3$. By more closely inspecting the 2-form equation in (2.27), we conclude that a differential 1-form σ is exact only if it is *closed*, i.e., $d\sigma = 0$ or

$$\frac{\partial F_i}{\partial x^j} = \frac{\partial F_j}{\partial x^i} \tag{2.28}$$

for every $i, j = 1, 2, 3$. This means that a closed 1-form, $d\sigma = 0$, is a necessary condition for this 1-form σ to be exact. Furthermore, the great French mathematician Poincare showed in his lemma that it is also sufficient if the vector field is in a contractible topological space. However, it is at least true that *if a differential 1-form is not closed, then it is not exact.*

We will revisit in the next chapters these important concepts discussed here to study the differential properties of the "Jacobian matrix" that is commonly named in the robotics research community and is also extremely useful in applications to robotics and digital human kinematics, dynamics and control. In other words, a conventional "Jacobian matrix" J is widely used as a tangent mapping between joint space and Cartesian space in robotics. However, in many cases, such a differential 1-form is not exact, or the matrix J is not always a genuine Jacobian matrix. Will it affect or interfere with our kinematics, dynamics and control of robotic systems and digital human models? The question is fundamental and will be further addressed in the next chapter when the velocities and tangent space transformations will be formally defined and investigated.

2.6 Exercises of the Chapter

1. Given three 3D vectors a, b and c with $a \neq 0$.

   a. If $a \cdot b = a \cdot c$, does it imply $b = c$?
   b. If $a \times b = a \times c$, does it imply $b = c$?
   c. If both $a \cdot b = a \cdot c$ and $a \times b = a \times c$, do they imply $b = c$?

   If the answer is no in each of the cases, give your counterexamples.
2. If a force vector f with its magnitude $\|f\| = 6$ in Newtons is acting at a point P of a rigid body that is pivoted at the origin O, with a radial vector defined as $r = \overrightarrow{OP}$, then the torque $\tau = r \times f$. Using this definition, find the torque shown in Figure 2.2 when $\theta = 20^0$ and $\theta = -20^0$, where the rigid body is an equilateral triangle.

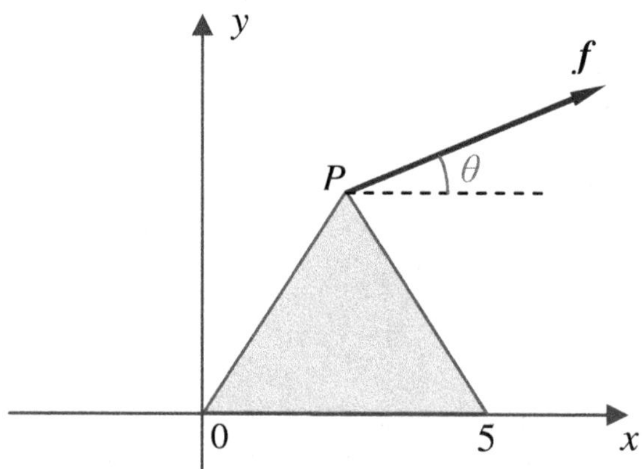

Fig. 2.2 Problem 2

3. Given two straight lines in 3D space. Line 1 is along a vector $v_1 = (1 \;\; 1 \;\; 1)^T$ in direction and passes a point A=$\{0, 0, 2\}$. Line 2 is along a vector $v_2 = (2 \;\; 0 \;\; -1)^T$ and passes another point B=$\{1, 0, 0\}$.

   a. Find the common normal vector of the two lines;
   b. Find the minimum distance between the two 3D lines.

4. There are three vectors $a = (1\ \ 0\ \ 0)^T$, $b = (1\ \ 3\ \ 3)^T$ and $c = (2\ \ 2\ \ 6)^T$ in 3D space.

   a. Determine the parallelepiped volume that is spanned by the three vectors;
   b. Find the height h of the arrow point of vector c with respect to the plane spanned by vectors a and b.

5. Given the same three vectors a, b and c from Problem 4.

   a. Find the skew-symmetric matrix $S(\cdot)$ for each of the three vectors;
   b. What is $[S(a), S(b)]\,c$?
   c. What is the result of $[[S(a), S(b)], S(c)]$?

6. If a is a 3D unit vector, i.e., $a^T a = \|a\|^2 = 1$, and b is an arbitrary 3D vector, show that
$$a \times (a \times (a \times b)) = b \times a.$$
If a is not unity, what will the above equation change?
7. Let $v_1 = (2\ \ 0\ \ 1)^T$ and $v_2 = (0\ \ 1\ \ 2)^T$. Find a rotation matrix R such that $v_2 = Rv_1$.
8. A plane in 3D space is defined by $x - y + 2z = 0$ and also a 3D vector is given by $v = (2\ \ 1\ \ 0)^T$.

   a. Determine the orthogonal projection v_p of the vector v on the plane;
   b. Find a rotation matrix R to perform the projection $\bar{v}_p = R\bar{v}$, where $\bar{v}_p = v_p/\|v_p\|$ and $\bar{v} = v/\|v\|$ are the unity vectors of v_p and v, respectively.

9. Given two rotation matrices:
$$R_1 = \begin{pmatrix} -0.3329 & -0.6675 & 0.6661 \\ 0.1343 & -0.7327 & -0.6671 \\ 0.9334 & -0.1326 & 0.3336 \end{pmatrix},$$
and
$$R_2 = \begin{pmatrix} -0.4286 & -0.2857 & 0.8571 \\ -0.2857 & -0.8571 & -0.4286 \\ 0.8571 & -0.4286 & 0.2857 \end{pmatrix}.$$
Find the corresponding angle ϕ and unit vector k in each of the two cases.
10. In the $k - \phi$ procedure, show that

   a.
$$K^2 + I = kk^T,$$
   where $S(k) = k\times = K$ and I is the 3 by 3 identity;
   b.
$$K^2 = \frac{1}{1 - \cos\phi}\left(\frac{R + R^T}{2} - I\right);$$

c. If the unity axis k is considered to be time-varying and $\dot{K}$ is the time-derivative of K, then,

$$K\dot{K}K = O,$$

where O is the 3 by 3 zero matrix.

11. A dual square matrix $\hat{A}$ can also have its dual eigenvalues $\hat{\lambda}$ and dual eigenvectors $\hat{x}$, and both satisfy $\hat{A}\hat{x} = \hat{\lambda}\hat{x}$. Given

$$\hat{A} = \begin{pmatrix} 3 - \epsilon 2 & 1 + \epsilon \\ -1 + \epsilon 3 & 2 + \epsilon \end{pmatrix}.$$

Find its two dual eigenvalues.

12. For two 3 by 3 dual matrices:

$$\hat{A} = \begin{pmatrix} 0.5403 - \epsilon 1.6830 & 0 & 0.8415 + \epsilon 1.0806 \\ 0 & 1.0000 + \epsilon 0.2000 & 0 \\ -0.8415 - \epsilon 1.0806 & 0 & 0.5403 - \epsilon 1.6830 \end{pmatrix},$$

and

$$\hat{B} = \begin{pmatrix} 0.5403 - \epsilon 1.6830 & 0 & 0.8415 + \epsilon 1.0806 \\ 0 & 1.0000 & 0 \\ -0.8415 - \epsilon 1.0806 & 0 & 0.5403 - \epsilon 1.6830 \end{pmatrix},$$

which one is a special orthogonal dual matrix $\in SO(3, d)$? If yes, find the position vector p.

13. Let a trigonometric function be defined by

$$f(x) = \frac{8 \sin 2x}{3 \cos x + 5 \sin 3x}.$$

Suppose that $x = 1.5$ and $\dot{x} = -1.2$ at a time instant.

a. Using the dual-number calculus, find the time-derivative of the function $\dot{f}(x)$ at this time instant;
b. Determine the value of $df(x)/dx$ at the same time instant.

14. Suppose that a 2D coordinate system is transformed from $\{x, y\}$ to $\{\xi, \eta\}$ by

$$\begin{cases} \xi = x^2 + y^2 \\ \eta = 3xe^{-y}. \end{cases}$$

Using the exterior calculus, determine their differential area change between $d\xi d\eta$ and $dxdy$.

15. Determine the following differential 1-forms to see if anyone is exact. If either one of them is exact, find a scalar function $\mu(\cdot)$ such that $\sigma = d\mu$.

a. $\sigma = ydx + xdy$;
b. $\sigma = y^2dx + x^2dy$;
c. $\sigma = (3x^2 + 3yz + 2y^2)dx + (3xz + 4xy + z^2)dy + (3xy + 2yz + 3z^2)dz$.

References

1. Gu, E., Luh, J.: Dual-Number Transformation and Its Applications to Robotics. IEEE Journal of Robotics and Automation RA-3(6), 615–623 (1987)
2. Fomenko, A., Manturov, O., Trofimov, V.: Tensor and Vector Analysis: Geometry, Mechanics and Physics. Mathematics (1998)
3. Danielson, D.: Vectors and Tensors in Engineering and Physics, 2nd edn. Westview, Perseus (2003)
4. Gilmore, R.: Lie Groups, Lie Algebras, and Some of Their Applications. John Wiley, New York (1974)
5. Karger, A., Novak, J.: Space kinematics and Lie Groups. Gordon and Breach Science, New York (1985)
6. Brockett, R.: Robotic Manipulators and the Product of Exponentials Formula. In: Fuhrman, P. (ed.) Mathematical Theory of Networks and Systems. LNCIS, vol. 58, pp. 120–129. Springer, Heidelberg (1984)
7. Spong, M., Vidyasagar, M.: Robot Dynamics and Control. John Wiley & Sons, New York (1989)
8. Murray, R., Li, Z., Sastry, S.: A Mathematical Introduction to Robotic Manipulation. CRC Press, Boca Raton (1994)
9. Clifford, W.: Preliminary Sketch of Bi-Quaternions. Proc. London Math. Soc. 4, 381–395 (1873)
10. Study, E.: Geometrie der Dynamen, Leipzig, Germany (1903)
11. Yang, A.: Calculus of Screws. In: Spillers, W. (ed.) Basic of Design Theory, pp. 266–281. North-Holland/American Elsevier, New York (1974)
12. Yaglom, I.: Complex Numbers in Geometry. Academic, New York (1968)
13. Rooney, J.: A Comparison of Representations of General Spatial Screw Displacement. Envision, Planning (England), Series B 5, 45–88 (1978)
14. Rooney, J.: On the Three Types of Complex Number and Planar Transformation. Envision, Planning (England), Series B 5, 89–99 (1978)
15. Rooney, J.: On the Principle of Transference. In: Inst. Mechanical Engineering Proceedings of 4th World Congress, Newcastle-upon-Tyne, England, pp. 1089–1094 (September 1975)
16. Pennock, G., Yang, A.: Application of Dual-Number Matrices to the Inverse Kinematics Problem of Robot Manipulators. ASME Trans. Journal of Mechanisms, Transmissions, Automation in Design 107, 201–208 (1985)
17. McCarthy, J.: Dual Orthogonal Matrices in Manipulator Kinematics. International Journal of Robotics Research 5, 45–51 (1986)
18. Duffy, J.: Analysis of Mechanisms and Robot Manipulators. Halstead Press (1980)
19. Flanders, H.: Differential Forms with Applications to the Physical Sciences. Dover Publications, New York (1989)
20. Ramanan, S.: Global Calculus. American Mathematical Society, Providence (2005)
21. Dubrovin, B., Fomenko, A., Novikov, S.: Modern Geometry – Methods and Applications, Part I - The Geometry of Surfaces, Transformation Groups, and Fields. Springer, New York (1992)

Chapter 3
Representations of Rigid Motion

3.1 Translation and Rotation

A rigid body is a theoretical model where any two distinct points on the body (or object) have a fixed time-invariant distance. Otherwise, the object is called a deformable body. In the areas of robotics and digital human modeling research, we often assume that each link of a robotic manipulator or a human body segment (link) is a rigid body, unless it indicates otherwise. A rigid body is relatively easy to be studied in both kinematics and dynamics.

The motion of a rigid body is called a rigid motion. Based on the Chasles Theorem in classical mechanics [1, 2], any rigid motion can be completely decomposed without coupling into a translation of the body mass center (or the center of gravity CG) and a rotation of the body about the mass center. Note that the mass center of a rigid body would be same as the centroid that is the geometric center of the body, if the mass of the rigid body could be uniformly distributed, or its mass density could be a constant. To represent both the translation and rotation for a rigid body with respect to a fixed base coordinate frame, we wish to have a unified formula. Due to the complication of rotation, it is not easy to find such a unique unification.

One may augment a 3 by 1 position vector p that can uniquely represent a position and perform a translation along with a 3 by 1 vector $\rho = \phi k$ that can represent an orientation for a frame, as introduced and discussed around equation (2.8) in Chapter 2, to form a 6 by 1 vector $\begin{pmatrix} p \\ \rho \end{pmatrix}$. At first glance, it seems to be a unification in representing every 6 d.o.f. (degrees of freedom) rigid motion. However, it will soon be seen that such a 3 by 1 so-called conformal vector $\rho = \phi k$ cannot offer a unique description for orientation and rotation of a rigid body in motion. Therefore, we may have to stay in the $SO(3)$ group and try to unify it with the 3 by 1 position vector to uniquely and safely represent any 6 d.o.f. rigid motion.

It has been shown that a vector v_1 can be both translated and rotated simultaneously to v_2 by an affine transformation:

E.Y.L. Gu, *A Journey from Robot to Digital Human*,
Modeling and Optimization in Science and Technologies 1,
DOI: 10.1007/978-3-642-39047-0_3,

$$v_2 = Rv_1 + p, \tag{3.1}$$

where $p \in \mathbb{R}^3$ is a 3 by 1 position vector and $R \in SO(3)$ is a 3 by 3 rotation matrix. Obviously, such a transformation given by (3.1) is not a unified form due to its non-homogeneous appearance. We are therefore motivated to seek a homogeneous transformation that combines both the translation and rotation in one form.

Let a rotation matrix be

$$R = \begin{pmatrix} r_{11} & r_{12} & r_{13} \\ r_{21} & r_{22} & r_{23} \\ r_{31} & r_{32} & r_{33} \end{pmatrix} \in SO(3)$$

and a position vector be

$$p = \begin{pmatrix} p_1 \\ p_2 \\ p_3 \end{pmatrix} \in \mathbb{R}^3.$$

A homogeneous transformation is defined by the following 4 by 4 square matrix:

$$H = \begin{pmatrix} r_{11} & r_{12} & r_{13} & p_1 \\ r_{21} & r_{22} & r_{23} & p_2 \\ r_{31} & r_{32} & r_{33} & p_3 \\ 0 & 0 & 0 & 1 \end{pmatrix}. \tag{3.2}$$

Then, each vector to be transformed has to be temporarily redefined as a 4 by 1 vector in the form:

$$v = \begin{pmatrix} p_1 \\ p_2 \\ p_3 \\ 1 \end{pmatrix},$$

such that

$$v_2 = Hv_1 \tag{3.3}$$

performs a unified operation of both translation and rotation simultaneously and homogeneously.

Furthermore, an indexed homogeneous transformation matrix H_j^i can also be used to uniquely represent both the relative position and orientation of frame i with respect to frame j. The unification with uniqueness, consistency and convenience has already made such a homogeneous transformation be the most popular and the best approach thus far to representing every 6 d.o.f. rigid motion. Clearly, all such 4 by 4 homogeneous transformation matrices constitute a multiplicative special Euclidean group $SE(3) = \mathbb{R}^3 \times SO(3)$ [3, 4, 5].

To invert a homogeneous transformation that is based on (3.2), given by

$$H_j^i = \begin{pmatrix} R_j^i & p_j^i \\ 0_3 & 1 \end{pmatrix}, \tag{3.4}$$

where $0_3 = (0 \ \ 0 \ \ 0)$ is the 1 by 3 zero covector, it may take advantage of the orthogonality of the rotation matrix $R^i_j \in SO(3)$ at the upper-left 3 by 3 corner of the above H^i_j. In fact, it can be easily verified that

$$(H^i_j)^{-1} = H^j_i = \begin{pmatrix} R^{iT}_j & -R^{iT}_j p^i_j \\ 0_3 & 1 \end{pmatrix} = \begin{pmatrix} R^j_i & -R^j_i p^i_j \\ 0_3 & 1 \end{pmatrix}. \tag{3.5}$$

Furthermore, to perform a series of successive 6 d.o.f. transformations, it is also convenient to operate multiplications between the homogeneous transformation matrices in certain order. Suppose that both the position and orientation of frame #1 with respect to the base is given by a 4 by 4 homogeneous transformation matrix H^1_0. If both the position and orientation of frame #2 referred to frame #1 is H^2_1, then $H^2_0 = H^1_0 H^2_1$ determines the position and orientation of frame #2 with respect to the base. On the other hand, if H^2_0 is given, then we can find $H^1_0 = H^2_0 H^1_2 = H^2_0 (H^2_1)^{-1}$. Therefore, each occurrence of multiplication must follow a *common diagonal-index cancellation rule*, such as $H^2_0 H^1_2$, to cancel the common indices 2 and 2 on the diagonal and result in H^1_0.

In many relatively simple cases, one can determine both the orientation and position for frame #2 with respect to frame #1 directly through 3D geometrical inspection. A general procedure for the inspection is outlined as follows:

Given two coordinate frames drawn in a common picture,

1. Project the unit vector of the x-axis of frame #2 onto frame #1 and then place it as the first column of a rotation matrix R^2_1 to be determined;
2. Repeat step 1 for the y-axis and z-axis of frame #2 and fill the normalized (unity) projections into the second and third columns of R^2_1, respectively, to complete the determination of R^2_1;
3. Connect the origins of the two frames to make a vector whose arrow is pointing to the origin of frame #2, and its tail is sitting at the origin of frame #1. Then, project the vector onto frame #1 to determine the 3 by 1 position vector p^2_1;
4. Finally, the homogeneous transformation H^2_1 is constructed through (3.2) or (3.4).

Note that each column of the rotation matrix must be normalized and the order of the three columns must follow exactly the x, y and z-axes and cannot be arbitrary.

Let us now look at a realistic example to illustrate both the determination and operation of homogeneous transformations. Figure 3.1 shows a webcam mounted on the top of a post, Q, looking down at the floor and ready to acquire a video image. Before taking a picture, the camera is required to calibrate itself with reference to the base coordinate frame #0. Given the height of the post $h = 2$, the distance of the bottom B of the post from the base origin along the x_0-axis is $d_0 = 0.8$ and the distance of B from the origin

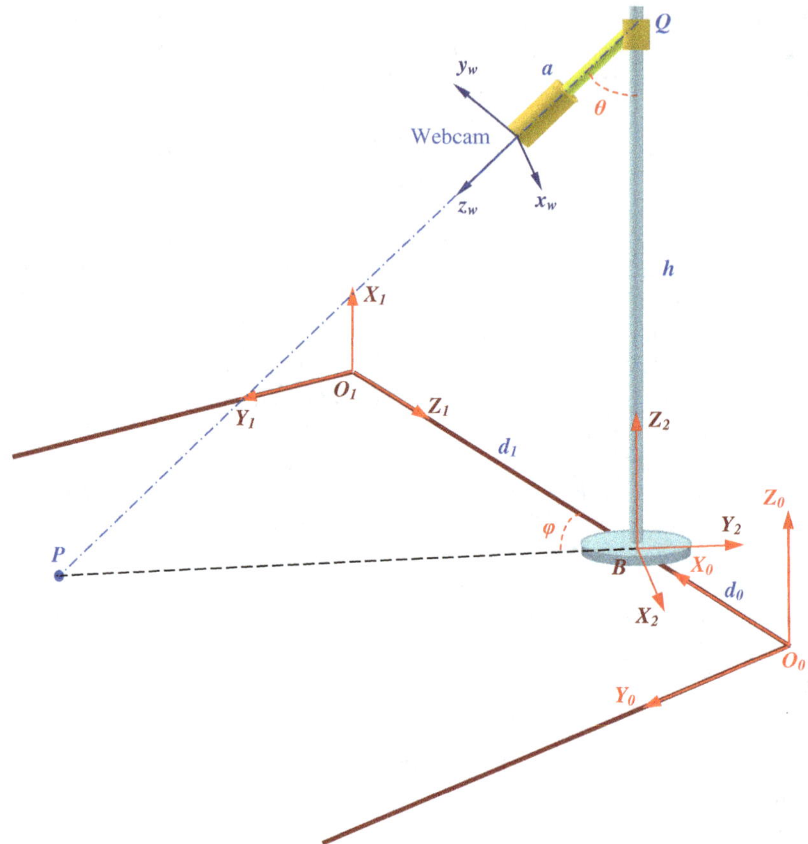

Fig. 3.1 The webcam position and orientation

of frame #1 along the z_1-axis is $d_1 = 2.2$, and the length between the post top point Q and the camera lens is $a = 0.6$, all in meters. Suppose that the webcam is shooting at a point P on the floor with the coordinates (1.8, 2.0, 0) referred to the base. We ask what are the position and orientation of the webcam lens frame w with respect to frame #1?

First, let the base frame be shifted to the bottom B of the post and also rotate it to form a new frame #2 such that its y_2-axis extends the connecting line from P to B, as depicted in Figure 3.1. Obviously, the vector tailed at the base origin O_0 with the arrow pointing to B with respect to the base should be $p_0^2 = (d_0 \;\; 0 \;\; 0)^T = (0.8 \;\; 0 \;\; 0)^T$. Similarly, the vector tailed at the point B with the arrow pointing to P and projected on the base is

$$p_0^p = \begin{pmatrix} 1.8 - d_0 \\ 2.0 \\ 0 \end{pmatrix} = \begin{pmatrix} 1.0 \\ 2.0 \\ 0 \end{pmatrix}.$$

Thus, the angle $\phi = \tan^{-1}(p^p_{0y}/p^p_{0x}) = \tan^{-1}(2.0/1.0) = 63.43^0$. The webcam angle θ can thus be determined by $\tan\theta =$ the length of p^p_0 over h so that $\theta = \tan^{-1}(\sqrt{1.0^2 + 2.0^2}/h) = 48.19^0$.

It can now be seen that frame #2 is formed by rotating the base about the z_0-axis by an angle $\psi = \phi + 90^0 = 153.43^0$ in counterclockwise. Therefore, according to the rotation basis given in equation (2.2), we obtain

$$H_0^2 = \begin{pmatrix} \cos\psi & -\sin\psi & 0 & d_0 \\ \sin\psi & \cos\psi & 0 & 0 \\ 0 & 0 & 1 & 0 \\ 0 & 0 & 0 & 1 \end{pmatrix} = \begin{pmatrix} -0.8944 & -0.4473 & 0 & 0.8 \\ 0.4473 & -0.8944 & 0 & 0 \\ 0 & 0 & 1 & 0 \\ 0 & 0 & 0 & 1 \end{pmatrix}.$$

Based on the geometrical definition of frame w, the axis x_w is the same as the x_2-axis of frame #2, and the y_w-axis has the angle θ above the negative y_2-axis while the z_w-axis has the angle θ to the left of the negative z_2-axis. Therefore, the first column of the rotation matrix R_2^w is the projection of the unit vector of x_w-axis onto frame #2, which is $(1\ \ 0\ \ 0)^T$. By the same rule, the second column of R_2^w is the projection of the unit vector of the y_w-axis onto frame #2, which should be $(0\ \ -\cos\theta\ \ \sin\theta)^T$. The third column of R_2^w is thus equal to $(0\ \ -\sin\theta\ \ -\cos\theta)^T$.

We can also see that the position vector from B to the origin of frame w projected onto frame #2 is the sum of the two vectors, i.e., $\overrightarrow{BQ} + \overrightarrow{QO_w}$. Since $\overrightarrow{BQ} = (0\ \ 0\ \ h)^T$ and $\overrightarrow{QO_w} = (0\ \ -a\sin\theta\ \ -a\cos\theta)^T$, we obtain

$$H_2^w = \begin{pmatrix} 1 & 0 & 0 & 0 \\ 0 & -\cos\theta & -\sin\theta & -a\sin\theta \\ 0 & \sin\theta & -\cos\theta & h - a\cos\theta \\ 0 & 0 & 0 & 1 \end{pmatrix}$$

$$= \begin{pmatrix} 1 & 0 & 0 & 0 \\ 0 & -0.6667 & -0.7454 & -0.4472 \\ 0 & 0.7454 & -0.6667 & 1.6000 \\ 0 & 0 & 0 & 1 \end{pmatrix}.$$

Moreover, the homogeneous transformation between frame #1 and the base is relatively straightforward. Since $d_0 + d_1 = 0.8 + 2.2 = 3.0$, it becomes

$$H_0^1 = \begin{pmatrix} 0 & 0 & -1 & 3.0 \\ 0 & 1 & 0 & 0 \\ 1 & 0 & 0 & 0 \\ 0 & 0 & 0 & 1 \end{pmatrix}.$$

Finally,

$$H_1^w = H_1^0 H_0^2 H_2^w = (H_0^1)^{-1} H_0^2 H_2^w = \begin{pmatrix} 0 & 0.7454 & -0.6667 & 1.6 \\ 0.4473 & 0.5963 & 0.6667 & 0.4 \\ 0.8944 & -0.2982 & -0.3334 & 2.0 \\ 0 & 0 & 0 & 1 \end{pmatrix}.$$

This shows that the orientation of frame w with respect to frame #1 is given by the 3 by 3 upper-left corner of the resultant homogeneous matrix H_1^w, and the position of frame w referred to frame #1 is determined by the last column of H_1^w.

In robotic task/path planning, **Euler angles** are also a frequently useful approach to describing rotation or orientation [4, 5]. Euler angles have many different definitions, but they all commonly use three successive elementary rotations, each of which rotates about one of the three coordinate frame axes, to compose a resultant rotation. The most popular ones include

1. **The Z-Y-Z Rotation:** First rotating a frame about the z-axis by ϕ. Then, rotating the new frame about the new y-axis by θ and followed by rotating the new frame about the new z-axis by ψ. The orientation of the resultant frame after the three successive rotations with respect to the initial one is given by

$$R = \begin{pmatrix} c\phi & -s\phi & 0 \\ s\phi & c\phi & 0 \\ 0 & 0 & 1 \end{pmatrix} \begin{pmatrix} c\theta & 0 & s\theta \\ 0 & 1 & 0 \\ -s\theta & 0 & c\theta \end{pmatrix} \begin{pmatrix} c\psi & -s\psi & 0 \\ s\psi & c\psi & 0 \\ 0 & 0 & 1 \end{pmatrix}, \tag{3.6}$$

   where each $c\phi$ or $s\phi$ is a short notation of $\cos\phi$ or $\sin\phi$.

2. **The Roll-Pitch-Yaw Rotation:** First, rotate the initial frame about x_0-axis by a Yaw angle ψ. Then, rotate the new frame about the initial y_0-axis by a Pitch angle θ. Finally, rotate the new frame about the initial z_0-axis by a Roll angle ϕ. The final orientation of the resultant frame with respect to the initial one is given by

$$R = \begin{pmatrix} c\phi & -s\phi & 0 \\ s\phi & c\phi & 0 \\ 0 & 0 & 1 \end{pmatrix} \begin{pmatrix} c\theta & 0 & s\theta \\ 0 & 1 & 0 \\ -s\theta & 0 & c\theta \end{pmatrix} \begin{pmatrix} 1 & 0 & 0 \\ 0 & c\psi & -s\psi \\ 0 & s\psi & c\psi \end{pmatrix}. \tag{3.7}$$

   In fact, this Roll-Pitch-Yaw definition is also equivalent to the **Z-Y-X Rotation**. Namely, first, rotate the initial frame about z-axis by ϕ. Then, rotate the new frame about the new y-axis by θ. Finally, rotate the new frame about the new x-axis by ψ.

It is quite noticeable that since a multiplication between matrices is not commutable in general, the order of multiplication in each of the above Euler-angle definitions is sensitive. We can show, based on the similarity transformation in linear algebra, the following **rule of multiplication order** in any successive rotation case:

Suppose that a frame k currently has an orientation R_n^k that is referred to frame n. If frame k is to be rotated about one of its own current axes, to determine the new orientation, the multiplication order is from the left to the right. In contrast, if frame k is rotated about an axis of frame n, the multiplication order is from the right to the left.

Clearly, the above Z-Y-Z rotation procedure belongs to the first case, i.e., each rotation is about the new axis, while the Roll-Pitch-Yaw procedure follows the second case, i.e., each rotation is about the old axis.

As we all know, for any successive transformation, the order of matrix multiplication commonly proceeds from the left to the right, as a mathematical convention. Why does the above rule require a reversed order in the second case? To answer this question and justify the rule, without loss of generality, let an original frame be rotated about its z_0-axis by an angle α. The orientation of the new frame, called frame #1, can be determined by the rotation matrix $R(z_0, \alpha)$. If frame #1 is to be rotated again about the new x-axis, i.e., the x_1-axis of frame #1, by an angle β, the new rotation matrix should be $R(x_1, \beta)$ and the final destination can be found by

$$R_0^2 = R(z_0, \alpha) R(x_1, \beta),$$

whose multiplication order is from the left to the right. Now, if the second rotation is about the old x_0-axis of the original frame by the same angle, then the orientation of frame #1, $R(z_0, \alpha)$, is supposed to be post-multiplied by a new rotation matrix, too. However, since the old x_0-axis has been a "history" behind the previous rotation $R(z_0, \alpha)$, $R(x_0, \beta)$ cannot directly be such a new rotation matrix. Instead, it has to go through a similarity transformation (actually an orthogonal transformation in this special case) based on linear algebra to find an equivalence $\tilde{R}(x_1, \beta)$ before following the mathematical convention of post-multiplication [6, 7]. More explicitly, this equivalence is determined by the following similarity (orthogonal) transformation:

$$\tilde{R}(x_1, \beta) = R(z_0, \alpha)^T R(x_0, \beta) R(z_0, \alpha),$$

and the final orientation turns out to be

$$\begin{aligned} R_0^2 = R(z_0, \alpha) \tilde{R}(x_1, \beta) &= R(z_0, \alpha) R(z_0, \alpha)^T R(x_0, \beta) R(z_0, \alpha) \\ &= R(x_0, \beta) R(z_0, \alpha). \end{aligned}$$

Apparently, this destination of rotating about the old axis is calculated by reversing the multiplication order, and is obviously different from the first case of rotating about the new axis by the same angle. In fact, it is not surprising that according to (2.2), $R(x_0, \beta)$ and $R(x_1, \beta)$ have exactly the same matrix form:

$$R(x_0, \beta) = R(x_1, \beta) = \begin{pmatrix} 1 & 0 & 0 \\ 0 & \cos\beta & -\sin\beta \\ 0 & \sin\beta & \cos\beta \end{pmatrix},$$

no matter which frame the x-axis lies in, as long as the rotation is about the x-axis of some frame by the same angle. The different destinations between the above two cases are solely due to the commutation in the product of two rotation matrices. Therefore, the *rule of multiplication order* becomes a key to determining a correct destination in any successive rotation process.

For example, let a successive rotation be planned to rotate an original frame, called frame #0, about its z_0-axis by an angle α. Then, rotate the new frame, called frame #1, about the new x_1-axis by β. It is further followed by rotating the latest one, called frame #2, about the original x_0-axis by γ to reach frame #3. Clearly, the rotation matrix for the first two rotations should be $R_0^2 = R(z_0, \alpha)R(x_1, \beta)$, i.e., their multiplication is in the order from the left to the right. Since the last rotation is about the original x_0-axis, the final orientation should be

$$R_0^3 = R(x_0, \gamma)R_0^2 = R(x_0\gamma)R(z_0, \alpha)R(x_1, \beta) =$$

$$\begin{pmatrix} 1 & 0 & 0 \\ 0 & c\gamma & -s\gamma \\ 0 & s\gamma & c\gamma \end{pmatrix} \begin{pmatrix} c\alpha & -s\alpha & 0 \\ s\alpha & c\alpha & 0 \\ 0 & 0 & 1 \end{pmatrix} \begin{pmatrix} 1 & 0 & 0 \\ 0 & c\beta & -s\beta \\ 0 & s\beta & c\beta \end{pmatrix}.$$

Moreover, by manipulating and solving a number of trigonometric equations, we can find their inverse solutions for the above Euler-angle representations. The inverse direction for the Z-Y-Z Euler angles is shown as follows:

Given $R = \begin{pmatrix} r_{11} & r_{12} & r_{13} \\ r_{21} & r_{22} & r_{23} \\ r_{31} & r_{32} & r_{33} \end{pmatrix}$. First, $\theta = \text{atan2}(\sqrt{r_{31}^2 + r_{32}^2},\ r_{33})$.

If $\sin\theta \neq 0$, then

$$\phi = \text{atan2}(r_{23}/\sin\theta,\ r_{13}/\sin\theta),$$
$$\psi = \text{atan2}(r_{32}/\sin\theta,\ -r_{31}/\sin\theta).$$

If $\theta = 0$, then

$$\phi = 0,$$
$$\psi = \text{atan2}(-r_{12},\ r_{11}).$$

If $\theta = 180^0$, then

$$\phi = 0,$$
$$\psi = \text{atan2}(r_{12},\ -r_{11}).$$

The inverse solution for the Roll-Pitch-Yaw Euler angles is given by the following procedure:

Given the same R as the above. First, $\theta = \text{atan2}(-r_{31},\ \sqrt{r_{11}^2 + r_{21}^2})$.

If $\cos\theta \neq 0$, then

$$\phi = \text{atan2}(r_{21}/\cos\theta,\ r_{11}/\cos\theta),$$
$$\psi = \text{atan2}(r_{32}/\cos\theta,\ r_{33}/\cos\theta).$$

If $\theta = 90^0$, then

$$\phi = 0,$$
$$\psi = \mathrm{atan2}(r_{12},\ r_{22}).$$

If $\theta = -90^0$, then

$$\phi = 0,$$
$$\psi = -\mathrm{atan2}(r_{12},\ r_{22}).$$

Note that all the above inverse solutions use the four-quadrant arc-tangent function atan2(· , ·) that is an internal function in MATLAB^{TM}. The first dummy element is y-component and the second one is x-component. In any angle determination, for better uniqueness, we have to avoid using the two-quadrant arc-tangent function atan(·), or the arc-sine function asin(·) or arc-cosine function acos(·) in the current and future inverse algorithms.

Let us now look at a numerical example. Consider a frame to be rotated by a Roll angle $\phi = 60^0$, a Pitch angle $\theta = -20^0$ and a Yaw angle $\psi = 80^0$ successively. Then, the final rotation matrix R for this Roll-Pitch-Yaw procedure becomes

$$R = \begin{pmatrix} \cos(60^0) & -\sin(60^0) & 0 \\ \sin(60^0) & \cos(60^0) & 0 \\ 0 & 0 & 1 \end{pmatrix} \times \begin{pmatrix} \cos(-20^0) & 0 & \sin(-20^0) \\ 0 & 1 & 0 \\ -\sin(-20^0) & 0 & \cos(-20^0) \end{pmatrix} \times$$

$$\begin{pmatrix} 1 & 0 & 0 \\ 0 & \cos(80^0) & -\sin(80^0) \\ 0 & \sin(80^0) & \cos(80^0) \end{pmatrix} = \begin{pmatrix} 0.4698 & -0.3188 & 0.8232 \\ 0.8138 & -0.2049 & -0.5438 \\ 0.3420 & 0.9254 & 0.1632 \end{pmatrix}.$$

To illustrate its reversed procedure, let us substitute the elements of the above R into the Roll-Pitch-Yaw inverse algorithm. We first obtain $\theta = \mathrm{atan2}(-r_{31},\ \sqrt{r_{11}^2 + r_{21}^2}) = -20^0$. Since $\cos\theta \neq 0$ in this case, $\phi = \mathrm{atan2}(r_{21}/\cos\theta,\ r_{11}/\cos\theta) = 60^0$ and $\psi = \mathrm{atan2}(r_{32}/\cos\theta,\ r_{33}/\cos\theta) = 80^0$, as expected.

Now, let us modify the Pitch angle to $\theta = -90^0$ and the other two remain unchanged. Then, we have the following new rotation matrix:

$$R = \begin{pmatrix} 0.0000 & -0.6428 & 0.7660 \\ 0.0000 & -0.7660 & -0.6428 \\ 1.0000 & 0.0000 & 0.0000 \end{pmatrix}.$$

To perform its inverse algorithm, first, $\theta = \mathrm{atan2}(-r_{31},\ \sqrt{r_{11}^2 + r_{21}^2}) = -90^0$. At this time, we cannot continue to use the same equations as we did in the first case before the Pitch angle modification, otherwise the algorithm will diverge. Thus, we have to apply the inverse equations under the condition $\theta = -90^0$, which yield $\phi = 0$ and $\psi = -\mathrm{atan2}(r_{12},\ r_{22}) = 140^0$.

This demonstrates that at the two special angles $\theta = \pm 90^0$, the solution of the Roll-Pitch-Yaw angles is no longer unique, although they all reach the same resultant rotation matrix. The other types of Euler angles inevitably pose the same complication on the uniqueness issue. Therefore, to deal with a rotation in 3D space is always much more complicated than a translation, and requires carefully to watch every condition or possibility in both computation and programming.

3.2 Linear Velocity versus Angular Velocity

In order to find an instantaneous velocity for a given translation represented by a 3 by 1 position vector $p \in \mathbb{R}^3$, or commonly called a linear velocity v in rigid motion, we can simply take time-derivative on the position vector p, i.e., $v = \dot{p}$. However, it should be done with the precaution that before taking the time-derivative, the position vector p must be projected on a fixed base, instead of any moving frame. Namely, if p_i is currently projected on frame i, which is not a fixed frame, one has to find R_b^i, the orientation of frame i with respect to the fixed base and then to determine $p_b = R_b^i p_i$ before calculating its linear velocity $v_b = \dot{p}_b$. The reason is obvious, $v_b = \dot{p}_b = \dot{R}_b^i p_i + R_b^i \dot{p}_i \neq R_b^i \dot{p}_i$, unless R_b^i is a constant matrix, which, however, is trivial and just means that frame i is motionless.

In contrast, finding an angular velocity ω for frame i with respect to the base is far more complicated than that for a linear velocity, because there is no 3 by 1 vector available to uniquely represent a frame rotation. The only unique and secure representation for both orientation and rotation is recognized to be the $SO(3)$ group in mathematics. In general, a 3 by 1 angular velocity ω is not an exact time-derivative of some 3 by 1 vector. In other words, there is no angular position vector $\rho \in \mathbb{R}^3$ such that $\omega = \dot{\rho}$ in general, unless the rotation is about a permanently fixed axis, such as a 2D spin. Using the exterior calculus language, the differential 1-form $\sigma = \omega dt$ is not exact, nor is it closed in general.

In order to provide a better understanding and insight into the 3D rotation and orientation, let us start seeking a relationship between the conventionally defined angular velocity ω and the time-derivative of a rotation matrix $R \in SO(3)$.

Based on the orthogonality of a rotation matrix $R \in SO(3)$, we have $RR^T = I$. Taking time-derivative on both sides yields

$$\dot{R}R^T + R\dot{R}^T = O,$$

where O is the 3 by 3 zero matrix. This result immediately tells us that $\dot{R}R^T$ is a skew-symmetric matrix. On the other hand, if we recall the notion of the $k - \phi$ procedure, as introduced and discussed in Chapter 2, the conventional definition for an angular velocity should be interpreted as

$$\omega = \dot{\phi} k \tag{3.8}$$

for a fixed unity axis k, or its corresponding skew-symmetric matrix

$$S(\omega) = \omega \times = \Omega = \dot{\phi} K.$$

In other words, an angular velocity is conventionally defined as a rotation rate of a frame about some fixed axis that is projected on that frame.

Now, applying equation (2.8) and taking time-derivative on (2.8) under the fact that K is fixed during a certain rotation period, we have

$$\dot{R} = \cos\phi \dot{\phi} K + \sin\phi \dot{\phi} K^2. \tag{3.9}$$

Post-multiplying $\dot{R}$ by R^T that can be found in equation (2.10), and noticing all the special properties owned by the unity skew-symmetric matrix K in Chapter 2, we finally obtain

$$\dot{R}R^T = (\cos\phi\dot{\phi}K + \sin\phi\dot{\phi}K^2)(I - \sin\phi K + (1-\cos\phi)K^2) = \dot{\phi}K = S(\omega) = \Omega. \tag{3.10}$$

This compact equation, $\Omega = \dot{R}R^T$, provides us with a direct relationship between a frame orientation change rate in terms of $\dot{R}$ and the conventional angular velocity ω for the frame rotation.

To better understand the geometrical meaning of the equation $\Omega = \dot{R}R^T$ derived in (3.10), let the orientation of a given frame #1 be represented by R_0^1 that is referred to the base frame #0. We wish to change its orientation to a new one R_0^2, as the orientation of frame #2 with respect to the base. Thus, the "difference" for the change between the two orientations is $R_1^2 = R_1^0 R_0^2 = (R_0^1)^T R_0^2$. Applying equations (2.9) and (2.11) in the $k - \phi$ procedure on R_1^2, we can find both ϕ and k to perform such an orientation change. Intuitively, this orientation change can be represented either in terms of the time-rate of R_1^2, or by the conventional angular velocity $\omega = \dot{\phi}k$, as defined in (3.8).

Now, equation (3.10) offers a desired close relationship. Since the unit vector k is referred to frame #1, the resulting $\Omega = \omega\times$ should also be projected onto frame #1, instead of the base. Noticing that when ϕ and k are determined from R_1^2 through the $k - \phi$ procedure, the unit vector k is a fixed axis until the change R_1^2 is completed. If one wants to make a new orientation change, say R_2^3 from frame #2 towards frame #3, then a new pair of ϕ and k should be calculated again, and this new k may not necessarily be the same as the first one in general. This means that the unit vector k in the conventional angular velocity definition (3.8) is tentatively fixed, but not permanently.

As a simple example, consider a frame to be rotated about its z-axis at a constant speed ρ radian/second. According to equation (2.2), the matrix to describe such a rotation can be written as

$$R = \begin{pmatrix} \cos\rho t & -\sin\rho t & 0 \\ \sin\rho t & \cos\rho t & 0 \\ 0 & 0 & 1 \end{pmatrix}.$$

Its time-derivative becomes

$$\dot{R} = \begin{pmatrix} -\rho\sin\rho t & -\rho\cos\rho t & 0 \\ \rho\cos\rho t & -\rho\sin\rho t & 0 \\ 0 & 0 & 0 \end{pmatrix}.$$

Then, one can easily verify that the product

$$\dot{R}R^T = \Omega = \begin{pmatrix} 0 & -\rho & 0 \\ \rho & 0 & 0 \\ 0 & 0 & 0 \end{pmatrix}$$

is clearly a skew-symmetric matrix and its corresponding vector is just the 3-dimensional angular velocity $\omega = (0 \quad 0 \quad \rho)^T$ with the only nonzero z-component for this particular frame rotation about the z-axis.

Another example is to calculate time-derivatives for the Euler angles and to demonstrate how useful the compact equation $\Omega = \dot{R}R^T$ from (3.10) will be. Let a rotation matrix R be formed by the Roll-Pitch-Yaw Euler angles. Based on equation (3.7),

$$R = R(z,\phi)R(y,\theta)R(x,\psi).$$

Taking time-derivative yields

$$\dot{R} = \dot{R}(z,\phi)R(y,\theta)R(x,\psi) + R(z,\phi)\dot{R}(y,\theta)R(x,\psi) + R(z,\phi)R(y,\theta)\dot{R}(x,\psi).$$

Thus, the corresponding angular velocity in skew-symmetric matrix form becomes

$$\Omega = \dot{R}R^T = \underline{\dot{R}(z,\phi)R(z,\phi)^T} + R(z,\phi)\underline{\dot{R}(y,\theta)R(y,\theta)^T}R(z,\phi)^T +$$

$$+ R(z,\phi)R(y,\theta)\underline{\dot{R}(x,\psi)R(x,\psi)^T}R(y,\theta)^T R(z,\phi)^T. \tag{3.11}$$

Since each term on the right-hand side of the above equation (3.11) contains a $\dot{R}R^T$ factor underlined with the rotation about each individual axis in the middle of each term, let us simplify the equation by first noticing that

$$\dot{R}(z,\phi)R(z,\phi)^T = \begin{pmatrix} 0 & -1 & 0 \\ 1 & 0 & 0 \\ 0 & 0 & 0 \end{pmatrix}\dot{\phi}, \quad \dot{R}(y,\theta)R(y,\theta)^T = \begin{pmatrix} 0 & 0 & 1 \\ 0 & 0 & 0 \\ -1 & 0 & 0 \end{pmatrix}\dot{\theta},$$

and

$$\dot{R}(x,\psi)R(x,\psi)^T = \begin{pmatrix} 0 & 0 & 0 \\ 0 & 0 & -1 \\ 0 & 1 & 0 \end{pmatrix} \dot{\psi}.$$

Then, substituting them into (3.11), with some multiplications, we reach that the first term is same as the above first equation for $\dot{\phi}$, and the second term for $\dot{\theta}$ becomes

$$\begin{pmatrix} c\phi & -s\phi & 0 \\ s\phi & c\phi & 0 \\ 0 & 0 & 1 \end{pmatrix} \begin{pmatrix} 0 & 0 & 1 \\ 0 & 0 & 0 \\ -1 & 0 & 0 \end{pmatrix} \begin{pmatrix} c\phi & s\phi & 0 \\ -s\phi & c\phi & 0 \\ 0 & 0 & 1 \end{pmatrix} \dot{\theta}$$

$$= \begin{pmatrix} 0 & 0 & c\phi \\ 0 & 0 & s\phi \\ -c\phi & -s\phi & 0 \end{pmatrix} \dot{\theta},$$

and the third term for $\dot{\psi}$ turns out to be

$$\begin{pmatrix} c\phi c\theta & -s\phi & c\phi s\theta \\ s\phi c\theta & c\phi & s\phi s\theta \\ -s\theta & 0 & c\theta \end{pmatrix} \begin{pmatrix} 0 & 0 & 0 \\ 0 & 0 & -1 \\ 0 & 1 & 0 \end{pmatrix} \begin{pmatrix} c\phi c\theta & s\phi c\theta & -s\theta \\ -s\phi & c\phi & 0 \\ c\phi s\theta & s\phi s\theta & c\theta \end{pmatrix} \dot{\psi}$$

$$= \begin{pmatrix} 0 & s\theta & s\phi c\theta \\ -s\theta & 0 & -c\phi c\theta \\ -s\phi c\theta & c\phi c\theta & 0 \end{pmatrix} \dot{\psi}.$$

Now, adding them together and converting the sum from the skew-symmetric matrix back to the vector form, we obtain

$$\omega = \begin{pmatrix} -s\phi\dot{\theta} + c\phi c\theta\dot{\psi} \\ c\phi\dot{\theta} + s\phi c\theta\dot{\psi} \\ \dot{\phi} - s\theta\dot{\psi} \end{pmatrix} = \begin{pmatrix} 0 & -s\phi & c\phi c\theta \\ 0 & c\phi & s\phi c\theta \\ 1 & 0 & -s\theta \end{pmatrix} \begin{pmatrix} \dot{\phi} \\ \dot{\theta} \\ \dot{\psi} \end{pmatrix}. \tag{3.12}$$

This result shows the relation between the time-derivatives of the Roll-Pitch-Yaw Euler angles and the angular velocity ω. Since the 3 by 3 coefficient matrix of the time-derivative column $(\dot{\phi}\ \dot{\theta}\ \dot{\psi})^T$ on the right-hand side of the above equation (3.12) will never be the identity or a constant matrix, the exactness of the differential 1-form $\sigma = \omega dt$ is almost hopeless. In other words, the angular velocity ω cannot be the time-derivatives of the three Roll-Pitch-Yaw Euler angles.

One can follow the same procedure to find the relation between the time-derivatives of the Z-Y-Z Euler angles and the angular velocity ω, too. We will leave it as one of the exercise problems at the end of this chapter.

In the future computations, a frequently asked question is how will a skew-symmetric matrix $S(a) = a\times$ change if the corresponding vector a is re-projected onto a new frame? Similarly, in equation (3.10), how will the skew-symmetric matrix Ω_1 of an angular velocity ω_1 that is currently referred to frame #1 be converted to Ω_0 that is referred to the base? Suppose that the re-projection is performed by a rotation matrix R. Then, we wish to know what $S(Ra)$ will be? Actually, it can be proven just by using the expansion of element-by-element multiplications for both $S(Ra)R$ and $RS(a)$ that

$$S(Ra) = RS(a)R^T. \tag{3.13}$$

This reveals a fact that after a vector is re-projected, its new skew-symmetric matrix will be related to the original one by an orthogonal transformation. Obviously, both sides of (3.13) are still skew-symmetric. Therefore, based on (3.13), since $\omega_0 = R_0^1\omega_1$,

$$\Omega_0 = S(\omega_0) = R_0^1\Omega_1(R_0^1)^T = R_0^1\Omega_1R_1^0.$$

Clearly, since $Ra \times Rb = S(Ra)Rb = RS(a)R^TRb = RS(a)b$, equation (3.13) can also imply that

$$R(a \times b) = Ra \times Rb, \tag{3.14}$$

for any two vectors $a, b \in \mathbb{R}^3$ to be re-projected by a rotation matrix $R \in SO(3)$.

It is further observable that based on the definition in (3.8), an angular velocity ω can be an exact time-derivative of some vector $\rho \in \mathbb{R}^3$ if the rotation axis is fixed permanently. This is due to the fact that if $\rho = \phi k$, then

$$\dot{\rho} = \dot{\phi}k + \phi\dot{k} = \omega + \phi\dot{k},$$

which implies that $\omega = \dot{\rho}$ if $\dot{k} = 0$. However, in the $k - \phi$ procedure, k is only tentatively fixed for a certain rotation process given by a specified orientation change R, but is not a permanently fixed axis in general. In fact, we will see in the next chapter that for an angular velocity ω, even if it is projected onto a fixed base, the differential 1-form $\omega dt = \sigma$ is not closed, i.e., its 2-form $d\sigma \neq 0$ in many general real examples.

Moreover, if we turn to utilize the dual-number calculus that was introduced in Chapter 2, it will be more convenient to calculate $\dot{R}$ required for finding an angular velocity ω numerically without going through any symbolical manipulation. Consider a symbolical rotation matrix that is a function of two joint angles θ_1 and θ_2 for a robot and given by

$$R = \begin{pmatrix} c_1c_2 & s_1 & c_1s_2 \\ s_1c_2 & -c_1 & s_1s_2 \\ s_2 & 0 & -c_2 \end{pmatrix},$$

where, once again, $s_i = \sin\theta_i$ and $c_i = \cos\theta_i$ for $i = 1, 2$. Suppose that we specify at a time instant $\theta_1 = \pi/6$ and $\theta_2 = \pi/3$ in radians, and $\dot{\theta}_1 = 0.4$ and $\dot{\theta}_2 = -0.6$ in radians/second. Then, replacing each θ_i by $\theta_i + \epsilon\dot{\theta}_i$ in the given matrix R, and splitting the real part and dual part for the dual rotation matrix $\hat{R} = R + \epsilon\dot{R}$, we immediately obtain R and $\dot{R}$ numerically in MATLAB^{TM} and further get $\Omega = \dot{R}R^T$:

```
>> R =

    0.4330    0.5000    0.7500
    0.2500   -0.8660    0.4330
    0.8660         0   -0.5000


>> dot R =

    0.3500    0.3464   -0.4330
    0.4330    0.2000    0.1500
   -0.3000         0   -0.5196


>> Omega = dot R * R'

   -0.0000   -0.4000    0.5196
    0.4000         0    0.3000
   -0.5196   -0.3000         0

>>
```

Thus, the angular velocity is resolved as $\omega = (-0.3000 \;\; 0.5196 \;\; 0.4000)^T$.

3.3 Unified Representations between Position and Orientation

As discussed earlier in this chapter, to uniquely determine a position and perform a translation, a 3 by 1 position vector $p \in \mathbb{R}^3$ is the best way to do so without any complication, and it is a member of the real additive group in $\mathbb{R}^3$. To unify the representation for both position and orientation, it now only depends on how to determine an orientation and perform a rotation. In general, to define a way to unify the representation for both position and orientation, and also for both translation and rotation simultaneously, the following conditions must hold:

1. It is unique and one-to-one correspondence at every position and orientation without any condition or constraint imposed;
2. It can uniquely represent and perform both translation and rotation simultaneously and form either an additive or a multiplicative group over a number field or ring;
3. With a certain step of differentiation, it can also uniquely and simultaneously determine velocities of both translation and rotation;
4. The major computations involved in every transformation and differentiation are manageable without complication.

The first condition indicates that a single unified definition is unique and cannot correspond to two distinct positions or orientations, and in turn, one single position or orientation cannot have two different values. The second one implies that when performing a simultaneous translation/rotation, the representing form must be closed and invertible, and should also satisfy distributivity and associativity for successive translations/rotations. The third one further requires a unique way of finding a unified tangent space to evaluate the rate of 6 d.o.f. rigid motion. Of course, we wish the unified 6 d.o.f. velocity could be an exact differential 1-form, which would be more ideal for our future robotic control systems design, especially in the robotic trajectory-tracking control applications. However, we are not imposing such a restrictive condition on the unification here, because at the level of modeling robotic and digital human kinematics, the exact 1-form may not necessarily be required.

The homogeneous transformation, defined in equation (3.2) or (3.4), clearly meets all the requirements of unification, because it is a member of the 6-dimensional multiplicative Lie group $SE(3)$ over the real field. The 3 by 3 dual rotation matrix, as a special orthogonal dual matrix given in (2.23), can also be an ideal unification to represent both position and orientation and also uniquely and simultaneously perform both translation and rotation without exception, because all such dual rotation matrices constitutes a Lie group $SO(3, d)$ over the dual ring. In particular, its exclusive advantage in numerical computations promotes itself to be one of the most promising approaches to uniquely representing 6 d.o.f. rigid motion, provided that every dual-number operation, including all the basic expansions, such as the exponential, sine and cosine given in equations (2.16), (2.17) and (2.18), can be implemented in a computer program as built-in functions, like the built-in complex number operations in MATLAB^{TM}. More advantages inherent in the dual rotation matrices will be further discussed in the next chapter.

As mentioned earlier in this chapter, we wish to directly define a 6 by 1 vector $\xi \in \mathbb{R}^6$ that augments a 3 by 1 position vector $p \in \mathbb{R}^3$ and the aforementioned 3 by 1 "orientation vector" $\rho = \phi k \in \mathbb{R}^3$ to represent 6 d.o.f. rigid motion. Namely, let $\xi = \begin{pmatrix} p \\ \rho \end{pmatrix}$, and we wish its time-derivative $\dot{\xi} = \begin{pmatrix} \dot{p} \\ \dot{\rho} \end{pmatrix}$ would also be able to represent a total velocity of 6 d.o.f. rigid motion. In the position part, $\dot{p} = v$ is obviously a linear velocity without complication. However, the time-derivative of ρ has to be determined by

$$\dot{\rho} = \dot{\phi}k + \phi\dot{k} = \omega + \phi\dot{k}.$$

This indicates that the unit vector k is no longer a fixed rotation axis so that $\dot{k} \neq 0$. On the other hand, the uniqueness in definition and the closure in successive rotation are the other two major concerns on the orientation part defined by $\rho = \phi k$. In other words, such a conformal orientation vector definition $\rho = \phi k$ may be able to represent an orientation, but may not perform a rotation correctly.

Recall that each pair of ϕ and k in the orientation vector ρ is determined by a corresponding rotation matrix R that uniquely represents a frame orientation, i.e., $R_0^1 = \exp(\phi_1 K_0^1) = \exp(S(\rho_1))$. If we perform a rotation for this frame #1 by R_1^2 in order to change the original orientation R_0^1 of frame #1 to a new one $R_0^2 = R_0^1 R_1^2$ for frame #2, then the corresponding ϕ_2 and k_2 for this orientation change can be found from R_1^2 through the $k - \phi$ procedure. Note that the new unit vector k_2 is now referred to frame #1, instead of the base frame #0. Thus, we have to use $R_0^1 k_2 = k_{2(0)}$ to find the new unit axis projected onto the base, which is equivalent to its skew-symmetric matrix $K_{1(0)}^2 = R_0^1 K_1^2 (R_0^1)^T = R_0^1 K_1^2 R_1^0$ based on (3.13), where the index inside the parenthesis (0) at each subscript denotes the frame of projection. If the rotation from frame #1 to frame #2 is represented by the orientation vector $\rho_2 = \phi_2 k_{2(0)}$, in order to unify it with its counterpart, the position vector that uses addition to perform a translation, then the resulting orientation of frame #2 is supposed to be $\rho = \rho_1 + \rho_2$, or its skew-symmetric matrix $S(\rho) = \phi_1 K_0^1 + \phi_2 K_{1(0)}^2$, where the two rotation axes have to be projected onto the common reference before being added together. A fundamental question now arises and must be justified before continuing:

$$R_0^2 = \exp(S(\rho)) = \exp(\phi_1 K_0^1 + \phi_2 K_{1(0)}^2) = R_0^1 R_1^2 = \exp(\phi_1 K_0^1) \cdot \exp(\phi_2 K_1^2)?$$

To simplify the question, let us just test the truth of the general form:

$$\exp(K_1 + K_2) = \exp(K_1) \cdot \exp(K_2)\ ? \tag{3.15}$$

where each of K_i for $i = 1, 2$ is a skew-symmetric matrix.

According to the well-known Baker-Campbell-Hausdorff (BCH) formula in mathematics [8, 9], for any two square matrices A and B,

$$\exp(A) \cdot \exp(B) = \exp(A + B + \frac{1}{2}[A, B] + \frac{1}{12}[A, [A, B]] - \frac{1}{12}[B, [A, B]] - \cdots),$$

where each $[A, B] = AB - BA$ is the Lie bracket defined as a commutator in Lie algebra, see Chapter 2. Therefore, equation (3.15) is true if and only if K_1 and K_2 are commutable in multiplication, i.e., $K_1 K_2 = K_2 K_1$. Since each K_i is skew-symmetric, we have $(K_1 K_2)^T = K_2^T K_1^T = K_2 K_1$, but not $K_1 K_2 = K_2 K_1$, unless the product $K_1 K_2$ is symmetric. In fact, based on equation (2.5) in Chapter 2,

$$[K_1, K_2] = K_1 K_2 - K_2 K_1 = (k_1 \times k_2)\times = S(k_1 \times k_2). \tag{3.16}$$

This means that $[K_1, K_2] = 0$ if and only if the two vectors $k_1 \parallel k_2$, which becomes a trivial case, because the two successive rotations are performed about the same axis. Therefore, the definition of an orientation vector using $\rho = \phi k$ under regular vector addition to perform a rotation is fundamentally incorrect.

As a numerical example, consider an orientation of frame #1 referred to frame #0 is given by R_0^1 and a rotation from frame #1 to frame #2 is given by R_1^2, and the resultant orientation is $R_0^2 = R_0^1 R_1^2$. The corresponding orientation vectors are ρ_1, ρ_2 and ρ_3, respectively, in the following MATLAB^{TM} working window:

```
>> R01 =

    0.0000    0.5000    0.8660
    0.0000   -0.8660    0.5000
    1.0000         0   -0.0000


R12 =

    0.1294    0.8660    0.4830
    0.2241   -0.5000    0.8365
    0.9659         0   -0.2588


R02 =

    0.9486   -0.2500    0.1941
    0.2888    0.4330   -0.8539
    0.1294    0.8660    0.4830


rho_1 =

   -1.9269
   -0.5163
   -1.9269


rho_2 =

   -1.8198
   -1.0507
   -1.3964


rho_3 =

    1.0717
    0.0403
    0.3358

>> rho_1 + rho_2
```

```
ans =

    -3.7467
    -1.5670
    -3.3233

>> rho_1 + R01*rho_2

ans =

    -3.6615
    -0.3046
    -3.7467

>>
```

It can be clearly seen that $\rho_1+\rho_2 \neq \rho_3$. Even if the projection of $\rho_2 = \phi_2 k_2$ is taken care by $R_0^1 \rho_2$ before adding it to ρ_1, the sum $\rho_1 + R_0^1 \rho_2 \neq \rho_3$. Therefore, the 3 by 1 conformal orientation vector can only be used to represent an instantaneous orientation for a frame given by a rotation matrix, but it cannot be utilized to perform a rotation. The following chart summarizes this fact:

$$\begin{array}{ccccc} R_0^2 & = & R_0^1 & \times & R_1^2 \\ \downarrow & & \downarrow & & \downarrow \\ \rho_3 & \neq & \rho_1 & + & \rho_2 \end{array}$$

In history, there were other similar definitions attempted for orientation vectors than the conformal definition $\rho = \phi k$ [10, 11]. All such attempts thus far may just improve their computational complexity in their conversions and may also uniquely represent an instantaneous orientation, but none of them could fix the infrastructural defect in performing a rotation. In general, we may define them in such a way that

$$\rho = \gamma(\phi) k$$

with a periodic and odd scalar function $\gamma(\phi)$. Typical definitions are listed below:

$$\rho^{(1)} = \sin\phi k; \quad \rho^{(2)} = \sin\frac{\phi}{2} k; \quad \text{and} \quad \rho^{(3)} = \tan\frac{\phi}{2} k.$$

Based on all the equations related to both forward and inverse conversions between the rotation matrix $R \in SO(3)$ and $k - \phi$ in (2.8), (2.9), (2.11) and (2.13), each of the skew-symmetric matrices $S(\rho^{(i)}) = \rho^{(i)} \times$ of the above orientation vector definitions can be determined in terms of R as follows:

$$S(\rho^{(1)}) = \frac{1}{2}(R - R^T); \quad S(\rho^{(2)}) = \frac{R - R^T}{2\sqrt{1 + \text{tr}(R)}}; \quad \text{and} \quad S(\rho^{(3)}) = \frac{R - R^T}{1 + \text{tr}(R)}.$$

In comparison with the conformal definition:

$$S(\rho) = \phi K = \frac{\phi}{2\sin\phi}(R - R^T) \quad \text{under} \quad \phi = \cos^{-1}\left(\frac{\mathrm{tr}(R) - 1}{2}\right),$$

it becomes quite interesting that the first three definitions can be determined directly in terms of only a given rotation matrix R without need to go through any inverse conversion in (2.9) and (2.11). Whereas the conformal definition $\rho = \phi k$ requires ϕ, and it has to rely upon equation (2.9). However, conversely to find the corresponding rotation matrix R in terms of a given $\rho^{(i)}$, no one can directly solve the above equation for each $S(\rho^{(i)})$ to find the equivalent R. Instead, we have to rely on both the inverse and forward conversions given in (2.9), (2.11) and (2.8). For instance, let us look at the third one $\rho^{(3)} = \tan\frac{\phi}{2}k$. If the numerical value of $\rho^{(3)}$ is given, first find its norm to determine $\tan\frac{\phi}{2} = \|\rho^{(3)}\|$ so that $\phi = 2\arctan(\|\rho^{(3)}\|)$. Second, the unit vector k is just its normalized vector, i.e., $k = \frac{\rho^{(3)}}{\|\rho^{(3)}\|}$. Once we have ϕ and k, apply the forward conversion (2.8) to reach the corresponding rotation matrix R.

All the attempts to define an orientation vector form have either a singularity issue or a limited range of ϕ. When R is symmetric but not the identity, $\phi = \pm\pi$ and $\mathrm{tr}(R) = -1$ so that the denominators of both $S(\rho^{(2)})$ and $S(\rho^{(3)})$ vanish. However, we can directly determine $\rho^{(2)} = \sin\frac{\phi}{2}k = \pm k$ and use equation (2.13) to find k, but $\rho^{(3)}$ will go to infinity. The first one $\rho^{(1)}$ has the simplest conversion over the others, but it only offers half of the range for the rotation angle ϕ.

In summary, unlike the rotation matrix $R \in SO(3)$, all such attempts to produce an orientation vector are conditional and will bring inconvenience in computer programming, plus none of them can uniquely perform rotations. The only advantage is that the velocity of each definition is obviously using its time-derivative $\dot{\rho}$ so that such an orientation velocity guarantees to be an exact differential 1-form, which may benefit to the future robotic control systems design.

Any kind of Euler angles may also possibly be adopted to define an orientation vector:

$$\rho = \begin{pmatrix} \phi \\ \theta \\ \psi \end{pmatrix}$$

for either the Z-Y-Z or Roll-Pitch-Yaw Euler angle definitions, which may perform a rotation by adding another Euler angles-based vector $\Delta\rho = (\Delta\phi\ \Delta\theta\ \Delta\psi)^T$. However, the resultant orientation vector, for instance, in the first Z-Y-Z case,

$$\rho_1 = \rho + \Delta\rho = \begin{pmatrix} \phi + \Delta\phi \\ \theta + \Delta\theta \\ \psi + \Delta\psi \end{pmatrix}$$

may not be consistent to the resultant rotation matrix defined in (3.6). Namely,

$$R_1 = R(z, \phi + \Delta\phi)R(y, \theta + \Delta\theta)R(z, \psi + \Delta\psi)$$

is not equal to the product

$$R(z, \phi)R(y, \theta)R(z, \psi)R(z, \Delta\phi)R(y, \Delta\theta)R(z, \Delta\psi)$$

at all due to matrix multiplication not being commutable. Although you may not need to care if the results are consistent and just treat the sum ρ_1 as a new orientation after a rotation $\Delta\rho$, it still has a non-unique issue in their inverse solutions, see the preceding discussion on the Euler angles in this chapter.

The well-known *quaternion* method can also be employed to uniquely represent orientation of a frame as well as to perform rotation effectively [12]. According to group theory, a quaternion q is defined as a "complex number" of two complex numbers. Let $c_1 = q_0 + j_3 q_3$ and $c_2 = q_2 + j_3 q_1$ be two complex numbers, where $j_3^2 = -1$. Then, a quaternion q is created by

$$q = c_1 + j_2 c_2 = q_0 + j_3 q_3 + j_2 q_2 + j_2 j_3 q_1 = q_0 + j_1 q_1 + j_2 q_2 + j_3 q_3,$$

where j_1, j_2 and j_3 are three independent complex units with the following properties:

$$j_1^2 = j_2^2 = j_3^2 = j_1 j_2 j_3 = -1,$$

and

$$j_1 j_2 = -j_2 j_1 = j_3, \quad j_2 j_3 = -j_3 j_2 = j_1, \quad j_3 j_1 = -j_1 j_3 = j_2.$$

It has been shown that each quaternion consists of four independent components $(q_0 \; q_1 \; q_2 \; q_3)$, and all such quaternions associated with the above multiplication rules constitute a non-commutative field, called the quaternion field. In other words, they are the members of both additive and multiplicative groups.

Regarding the addition between two quaternions, let $q^a = q_0^a + j_1 q_1^a + j_2 q_2^a + j_3 q_3^a$ and $q^b = q_0^b + j_1 q_1^b + j_2 q_2^b + j_3 q_3^b$ be two quaternions. Then,

$$q^a + q^b = (q_0^a + q_0^b) + j_1(q_1^a + a_1^b) + j_2(q_2^a + q_2^b) + j_3(q_3^a + q_3^b).$$

The multiplication between q^a and q^b follows the distributive law and the result is called the Hamilton product:

$$q^a \cdot q^b = (q_0^a q_0^b - q_1^a q_1^b - q_2^a q_2^b - q_3^a q_3^b) + j_1(q_0^a q_1^b + q_1^a q_0^b + q_2^a q_3^b - q_3^a q_2^b) +$$
$$+ j_2(q_0^a q_2^b - q_1^a q_3^b + q_2^a q_0^b + q_3^a q_1^b) + j_3(q_0^a q_3^b + q_1^a q_2^b - q_2^a q_1^b + q_3^a q_0^b).$$

Similar to the complex number, the conjugate of q is $q^* = q_0 - j_1 q_1 - j_2 q_2 - j_3 q_3$ such that

$$q \cdot q^* = \|q\|^2 = q_0^2 + q_1^2 + q_2^2 + q_3^2.$$

Actually, a quaternion can be linearly mapped to a 2 by 2 complex matrix with all the operations of algebraic field preserved, i.e.,

$$\Gamma(q) = \Gamma(q_0 + j_1 q_1 + j_2 q_2 + j_3 q_3) = \begin{pmatrix} q_0 + jq_1 & q_2 + jq_3 \\ -q_2 + jq_3 & q_0 - jq_1 \end{pmatrix} \quad \text{with } j^2 = -1.$$

One can easily verify that the above Hamilton product can also be found by

$$\Gamma(q^a \cdot q^b) = \begin{pmatrix} q_0^a + jq_1^a & q_2^a + jq_3^a \\ -q_2^a + jq_3^a & q_0^a - jq_1^a \end{pmatrix} \begin{pmatrix} q_0^b + jq_1^b & q_2^b + jq_3^b \\ -q_2^b + jq_3^b & q_0^b - jq_1^b \end{pmatrix}.$$

With the perfect algebraic structure owned by the quaternions, it is common to interpret a quaternion as a combination of a scalar and a 3-dimensional vector: $q = (q_0, \; v)$ with the vector $v = (q_1 \; q_2 \; q_3)^T$. Based on the rules of complex units multiplication, for two quaternions: $q^a = (q_0^a, \; v^a)$ and $q^b = (q_0^b, \; v^b)$,

$$q^a \cdot q^b = (q_0^a q_0^b - v^{aT} v^b, \; q_0^a v^b + q_0^b v^a + v^a \times v^b). \tag{3.17}$$

Now, let the four components of a quaternion be further specified by the four Euler's parameters:

$$q_0 = \cos\frac{\phi}{2}, \quad q_i = k_i \sin\frac{\phi}{2}, \quad \text{for } i = 1, 2, 3,$$

where $(k_1 \; k_2 \; k_3)^T = k$ is a unit vector, about which a frame is rotated by ϕ. In other words, the quaternion is specified to be

$$q = \cos\frac{\phi}{2} + (j_1 k_1 + j_2 k_2 + j_3 k_3)\sin\frac{\phi}{2} = \cos\frac{\phi}{2} + \hat{k}\sin\frac{\phi}{2}, \tag{3.18}$$

where $\hat{k} = j_1 k_1 + j_2 k_2 + j_3 k_3$ is the unit vector k in the form of a pure imaginary quaternion. Equation (3.18) can be rewritten as

$$q = \left(\cos\frac{\phi}{2}, \; \sin\frac{\phi}{2} k\right).$$

Since $\hat{k}^2 = (j_1 k_1 + j_2 k_2 + j_3 k_3)^2 = -1$, equation (3.18) can also be interpreted as a generalized Euler formula in complex analysis:

$$e^{\hat{k}\frac{\phi}{2}} = \cos\frac{\phi}{2} + \hat{k}\sin\frac{\phi}{2}.$$

Such a specification makes the quaternion be able to uniquely perform a rotation as well as to uniquely represent a frame orientation. We can easily prove that (3.18) is a unity quaternion if k is a unit vector. Namely, since $q^* = \cos\frac{\phi}{2} - \hat{k}\sin\frac{\phi}{2} = \left(\cos\frac{\phi}{2}, \; -\sin\frac{\phi}{2} k\right)$,

$$q \cdot q^* = \left(\cos^2 \frac{\phi}{2} + \sin^2 \frac{\phi}{2}, \ 0 \right) = 1.$$

Under the unity quaternion definition in (3.18), for two successive rotations, based on (3.17), the resultant orientation is

$$q = q^a q^b = \left(\cos \frac{\phi_1}{2}, \ \sin \frac{\phi_1}{2} k_1 \right) \left(\cos \frac{\phi_2}{2}, \ \sin \frac{\phi_2}{2} k_2 \right)$$

$$= \left(\cos \frac{\phi_1}{2} \cos \frac{\phi_2}{2} - \sin \frac{\phi_1}{2} \sin \frac{\phi_2}{2} k_1^T k_2, \right.$$

$$\left. \cos \frac{\phi_1}{2} \sin \frac{\phi_2}{2} k_2 + \cos \frac{\phi_2}{2} \sin \frac{\phi_1}{2} k_1 + \sin \frac{\phi_1}{2} \sin \frac{\phi_2}{2} (k_1 \times k_2) \right).$$

If $k_1 = k_2 = k$, i.e., the two rotations are about the same unit axis, then,

$$q = q^a q^b = \left(\cos \frac{\phi_1 + \phi_2}{2}, \ \sin \frac{\phi_1 + \phi_2}{2} k \right) = \cos \frac{\phi_1 + \phi_2}{2} + \hat{k} \sin \frac{\phi_1 + \phi_2}{2}.$$

For a 3D vector v, we may consider it as a special quaternion without the scalar part: $q_v = (0, \ v)$. Conventionally, to perform a rotation on this vector, the new vector $v_{new} = Rv$ for some rotation matrix $R \in SO(3)$. If we ask the quaternion field to perform the same rotation, using the $k - \phi$ procedure to find the corresponding rotating angle ϕ and the unit vector k, and then to define a quaternion $q = \left(\cos \frac{\phi}{2}, \ \sin \frac{\phi}{2} k \right)$. However, the resulting new vector v_{new} is the vector part, i.e., the imaginary part of the following new quaternion:

$$q_{new} = (0, \ v_{new}) = q \cdot q_v \cdot q^{-1} = q \cdot q_v \cdot q^*,$$

instead of just $q \cdot q_v$, where q^* is the conjugate of q and $q^{-1} = q^* / \|q\| = q^*$ for the unity quaternion q.

Based on (3.17), the product

$$q \cdot q_v = \left(\cos \frac{\phi}{2}, \ \sin \frac{\phi}{2} k \right) (0, \ v) = \left(- \sin \frac{\phi}{2} k^T v, \ \cos \frac{\phi}{2} v + \sin \frac{\phi}{2} (k \times v) \right).$$

Then, the scalar (real) part of $q \cdot q_v \cdot q^* = q \cdot q_v \cdot \left(\cos \frac{\phi}{2}, \ - \sin \frac{\phi}{2} k \right)$ is, indeed, equal to 0, while its imaginary part, based on (3.17) again, becomes

$$v_{new} = \sin^2 \frac{\phi}{2} (k^T v) k + \cos^2 \frac{\phi}{2} v + 2 \sin \frac{\phi}{2} \cos \frac{\phi}{2} (k \times v) + \sin^2 \frac{\phi}{2} k \times (k \times v).$$

Now, since $2 \sin \frac{\phi}{2} \cos \frac{\phi}{2} = \sin \phi$, and according to the triple cross-product equation (2.1) in Chapter 2, $k \times (k \times v) = (k^T v) k - (k^T k) v = (k^T v) k - v$ so that $(k^T v) k = k \times (k \times v) + v$. Substituting the results into the above imaginary part of the new quaternion q_{new} and noticing that $k \times v = Kv$ and

$k \times (k \times v) = K^2 v$ with $K = S(k)$, the skew-symmetric matrix of the unit vector k, we obtain

$$v_{new} = 2\sin^2\frac{\phi}{2}K^2 v + v + \sin\phi K v = v + \sin\phi K v + (1 - \cos\phi)K^2 v.$$

In contrast, based on the rotation matrix expansion in equation (2.8) of Chapter 2, the new vector after a rotation about the unit axis k by the angle ϕ for a vector v should be

$$v_{new} = Rv = (I + \sin\phi K + (1 - \cos\phi)K^2)v.$$

The above answer evidently exposes how many more computations are incurred by a quaternion to handle a vector rotation in comparison with the rotation matrix.

Therefore, the quaternion method is not quite as easy as the conventional rotation matrix approach for performing a 3D rotation and representing a frame orientation, although its infrastructure is a perfect algebraic field. In addition to the computational inconvenience, as evidently seen from the above introduction, a position vector has three coordinate elements, while a quaternion consists of four components in nature, which lacks advantage for the unification of representing 6 d.o.f. rigid motion.

3.4 Tangent Space and Jacobian Transformations

In robotic kinematics, one often intends to find a velocity for both position and orientation of the robotic end-effector frame referred to the base. In general robotic applications, a TCP (Tool Center Point) is usually the origin of the end-effector frame. If we have a unified vector $\xi = \begin{pmatrix} p \\ \rho \end{pmatrix} \in \mathbb{R}^6$ to represent the 6 d.o.f. end-effector rigid motion, then the velocity is its time-derivative: $\dot{\xi} = \begin{pmatrix} \dot{p} \\ \dot{\rho} \end{pmatrix}$. Suppose that a robot has n joints from the base to the end-effector. This 6-dimensional Cartesian position/orientation vector ξ is obviously a function of all the joint positions that form an n-dimensional vector $q = (q_1 \ \cdots \ q_n)^T \in \mathbb{R}^n$, and its velocity

$$\dot{\xi} = \frac{\partial \xi}{\partial q}\dot{q} = J\dot{q},$$

where

$$J = \frac{\partial \xi}{\partial q} = (g_1 \ \cdots \ g_n)$$

is called the Jacobian matrix of the vector ξ and each column $g_i = \partial\xi/\partial q^i$ is a 6-dimensional vector field for $i = 1, \cdots, n$. Using a differential form, we can rewrite it as

$$d\xi = \sum_{i=1}^{n} g_i dq^i = \sum_{i=1}^{n} \frac{\partial \xi}{\partial q^i} dq^i.$$

Based on the exterior calculus in Chapter 2, the above differential 1-form is clearly exact because its 2-form $d^2\xi = 0$. Namely,

$$\frac{\partial g_i}{\partial q^j} = \frac{\partial^2 \xi}{\partial q^j \partial q^i} = \frac{\partial^2 \xi}{\partial q^i \partial q^j} = \frac{\partial g_j}{\partial q^i} \quad \text{for every } i, j = 1, \cdots, n. \tag{3.19}$$

However, since all the definitions of the 3 by 1 orientation vector ρ so far have a common fundamental defect in performing rotations, such an ideal exact 1-form may not be a reality.

Of course, we cannot take a time-derivative directly on a homogeneous transformation matrix

$$H = \begin{pmatrix} R & p \\ 0_3 & 1 \end{pmatrix}$$

to directly find its velocity, where 0_3 is the 1 by 3 zero vector. Instead, inspired by the angular velocity $\Omega = \dot{R}R^T$, as given in equation (3.10), let us try to post-multiply the time-derivative of the above 4 by 4 homogeneous matrix by its inverse. Note that taking the transpose of a homogeneous transformation matrix will destroy its group membership, and thus it can only take the inverse. Namely, according to (3.5),

$$\dot{H}H^{-1} = \begin{pmatrix} \dot{R} & \dot{p} \\ 0_3 & 0 \end{pmatrix} \begin{pmatrix} R^T & -R^T p \\ 0_3 & 1 \end{pmatrix} = \begin{pmatrix} \Omega & \dot{p} - \Omega p \\ 0_3 & 0 \end{pmatrix}. \tag{3.20}$$

As we can see, the fourth column of the result is $\dot{p} - \omega \times p$. In mechanics, the absolute linear velocity of a rigid body motion at a point on the body is generally equal to $v_{ab} = v_{re} + \omega \times p$, where v_{ab} is the absolute velocity, v_{re} is the relative velocity, and ω is the angular velocity of the body. Therefore, the last column of the resulting matrix represents a relative velocity of the origin of the robot end-effector frame. In other words, this velocity is just the sliding velocity along the direction of the position vector p.

Coincidentally, the dual number representation will reach exactly the same result. Let $\hat{R} = R + \epsilon S$ be a 3 by 3 special orthogonal dual matrix to represent both the position and orientation of the robotic end-effector frame with respect to the base, where, based on equation (2.24), $SR^T = P = p\times$ is the skew-symmetric matrix of the position vector p, as introduced in Chapter 2. Likewise, we try

$$\dot{\hat{R}}\hat{R}^T = (\dot{R} + \epsilon\dot{S})(R^T + \epsilon S^T) = \dot{R}R^T + \epsilon(\dot{R}S^T + \dot{S}R^T).$$

The real part of the above dual matrix is obviously $\dot{R}R^T = \Omega$. For its dual part, let us first compare it with the time-derivative of $P = SR^T$, i.e., $\dot{P} = S\dot{R}^T + \dot{S}R^T$. We find that the second term $\dot{S}R^T$ of the dual part is exactly the same as the second term of $\dot{P}$ so that the dual part can be $\dot{R}S^T + \dot{S}R^T = \dot{R}S^T - S\dot{R}^T + \dot{P}$. Now, replacing each S by PR and noticing $\Omega = \dot{R}R^T$ as well as that both P and Ω are skew-symmetric, we obtain the dual part

$$\dot{R}S^T + \dot{S}R^T = \dot{R}R^T P^T - PR\dot{R}^T + \dot{P} = \dot{P} - \Omega P + P\Omega = \dot{P} - [\Omega, P],$$

where $[\Omega, P] = \Omega P - P\Omega$ is the Lie commutator and $[\Omega, P] = (\omega \times p)\times$ according to (3.16). Finally,

$$\dot{\hat{R}}\hat{R}^T = \Omega + \epsilon(\dot{P} - [\Omega, P]) = \Omega + \epsilon(\dot{P} - (\omega \times p)\times). \tag{3.21}$$

Comparing it with (3.20) in the above study on homogeneous transformation, it turns out to be so consistent that both the two most promising approaches to unifying the representation for every 6 d.o.f. rigid motion, the 4 by 4 homogeneous matrix and the 3 by 3 special orthogonal dual matrix, all suggest to augment a linear velocity $v = \dot{p}$ and an angular velocity ω as a unification of the 6-dimensional Cartesian velocity representation. Namely, $V = \begin{pmatrix} v \\ \omega \end{pmatrix}$, even though the differential 1-form $\sigma = \omega dt$ is not exact in general.

Under this suggestion, a 6-dimensional Cartesian velocity is simply defined as

$$V = \begin{pmatrix} v \\ \omega \end{pmatrix} = \sum_{i=1}^{n} \bar{g}_i dq^i = J\dot{q}, \tag{3.22}$$

where the top 3 by 1 part of each 6 by 1 column $\bar{g}_i$ in the matrix $J = (\bar{g}_1 \ \cdots \ \bar{g}_n)$ is $\partial p/\partial q^i$ while the bottom 3 by 1 part may not be a partial derivative of some vector. Therefore, more precisely, the above matrix J is not a genuine mathematical Jacobian matrix in general. However, since traditionally such a transformation matrix from joint velocity to Cartesian velocity is named as a Jacobian matrix in the robotics community, let us just keep this tradition in all of our later discussions. Once again, it must be emphasized that for the linear velocity $v = \dot{p}$, the position vector p must be referred to and projected onto the fixed base before taking a time-derivative on it.

Actually, equation (3.19) can also be adopted as a criterion to test an m by n matrix J to see if it is a genuine Jacobian matrix. Namely, by picking up any two columns g_i and g_j from J, if

$$\frac{\partial g_i}{\partial q^j} = \frac{\partial g_j}{\partial q^i} \quad \text{for every } i, j = 1, \cdots, n, \tag{3.23}$$

then, J is a genuine Jacobian matrix, i.e., there is an m-dimensional smooth function $\xi(q)$ for $q \in \mathbb{R}^n$ such that

$$J = \frac{\partial \xi}{\partial q}.$$

This criterion implies that if one single pair of columns g_k and g_l from the matrix J violates the criterion, the genuine Jacobian conclusion is blown up.

As a typical example, let us look at a 6-joint Stanford robot arm with five revolute joints and one prismatic joint. The joint position vector is given by $q = (\theta_1\ \theta_2\ d_3\ \theta_4\ \theta_5\ \theta_6)^T$. Its detailed kinematic modeling will be discussed in the next chapter. Here, we just show its Jacobian matrix $J_{(3)}$ that is projected onto frame 3, the frame of link 3, as follows:

$$J_{(3)} = \begin{pmatrix} d_2c_2 & d_3 & 0 & 0 & 0 & 0 \\ -d_3s_2 & 0 & 0 & 0 & 0 & 0 \\ d_2s_2 & 0 & 1 & 0 & 0 & 0 \\ s_2 & 0 & 0 & 0 & -s_4 & c_4s_5 \\ 0 & 1 & 0 & 0 & c_4 & s_4s_5 \\ -c_2 & 0 & 0 & 1 & 0 & c_5 \end{pmatrix},$$

where each $c_i = \cos\theta_i$ and $s_i = \sin\theta_i$ for a short notation.

According to the criterion given in (3.23) to test if a differential 1-form is closed, this 6 by 6 matrix $J_{(3)}$ is not a genuine Jacobian matrix at all, because by just picking up the second and third columns g_2 and g_3, we have

$$\frac{\partial g_3}{\partial \theta_2} = \begin{pmatrix} 0 \\ 0 \\ 0 \\ 0 \\ 0 \\ 0 \end{pmatrix}, \quad \text{while} \quad \frac{\partial g_2}{\partial d_3} = \begin{pmatrix} 1 \\ 0 \\ 0 \\ 0 \\ 0 \\ 0 \end{pmatrix},$$

and thus,

$$\frac{\partial g_3}{\partial \theta_2} \neq \frac{\partial g_2}{\partial d_3},$$

which has already violated the criterion (3.23).

Now, let the matrix $J_{(3)}$ be converted to $J_{(0)}$ by pre-multiplying a 6 by 6 augmented rotation matrix:

$$\begin{pmatrix} R_0^3 & O \\ O & R_0^3 \end{pmatrix}$$

with

$$R_0^3 = \begin{pmatrix} c_1c_2 & s_1 & c_1s_2 \\ s_1c_2 & -c_1 & s_1s_2 \\ s_2 & 0 & -c_2 \end{pmatrix}$$

for re-projecting the matrix onto the base. Then, the top half of $J_{(0)}$ becomes

$$J_{(0)}(1\!:\!3,:) = \begin{pmatrix} d_2c_1 - d_3s_1s_2 & d_3c_1c_2 & c_1s_2 & 0 & 0 & 0 \\ d_2s_1 + d_3c_1s_2 & d_3s_1c_2 & s_1s_2 & 0 & 0 & 0 \\ 0 & d_3s_2 & -c_2 & 0 & 0 & 0 \end{pmatrix}.$$

This is, indeed, a genuine Jacobian matrix, because the criterion (3.23) holds for every $i, j = 1, \cdots, 6$. For instance,

$$\frac{\partial g_1}{\partial \theta_2} = \frac{\partial g_2}{\partial \theta_1} = \begin{pmatrix} -d_3 s_1 c_2 \\ d_3 c_1 c_2 \\ 0 \end{pmatrix}, \quad \frac{\partial g_1}{\partial d_3} = \frac{\partial g_3}{\partial \theta_1} = \begin{pmatrix} -s_1 s_2 \\ c_1 s_2 \\ 0 \end{pmatrix},$$

and

$$\frac{\partial g_2}{\partial d_3} = \frac{\partial g_3}{\partial \theta_2} = \begin{pmatrix} c_1 c_2 \\ s_1 c_2 \\ s_2 \end{pmatrix}.$$

In fact, the integral of $J_{(0)}(1:3,:)$ is just the position vector of frame 6 under $d_6 = 0$, i.e.,

$$p_0^6 = \begin{pmatrix} d_3 c_1 s_2 \\ d_3 s_1 s_2 \\ d_1 - d_3 c_2 \end{pmatrix}$$

such that the top half of $J_{(0)}$ is exactly equal to

$$J_{(0)}(1{:}3,:) = \frac{\partial p_0^6}{\partial q}.$$

This also means that the differential 1-form $\sigma_v = v_0 dt = dp_0^6$ for the linear velocity is exact.

In contrast, the bottom half of either $J_{(3)}$ or $J_{(0)}$ directly violate the criterion (3.23) and thus, it will never be a genuine Jacobian matrix. For instance, since

$$J_{(0)}(4{:}6,1{:}3) = R_0^3 J_{(3)}(4{:}6,1{:}3) = \begin{pmatrix} 0 & s_1 & 0 \\ 0 & -c_1 & 0 \\ 1 & 0 & 0 \end{pmatrix},$$

for the first and second columns g_1 and g_2 in $J_{(0)}(4{:}6,:)$,

$$\frac{\partial g_1}{\partial \theta_2} = 0, \quad \text{while} \quad \frac{\partial g_2}{\partial \theta_1} = \begin{pmatrix} c_1 \\ s_1 \\ 0 \end{pmatrix} \neq 0.$$

Therefore, this Stanford robot example demonstrates that the linear velocity part can achieve a genuine Jacobian matrix, provided that the linear velocity and the Jacobian matrix are all projected onto the base. Whereas the angular velocity part is not, or the differential 1-form $\sigma_\omega = \omega dt$ for the angular velocity is neither closed, nor exact. Reality once again exposes the complication of rotation.

In future studies on robotic system control models and applications, since each robot is a nonlinear dynamic system, we will have to exactly linearize it before developing its control strategy. Often, in a feedback linearization procedure, a complete velocity to cover all the d.o.f (degrees of freedom) of motion is required to be the time-derivative of a Cartesian position/orientation, so that the Jacobian matrix has to be genuine. If each velocity could be an exact differential 1-form, developing an effective control law for the nonlinear robotic system would be much easier. The detailed discussion and extensive development of robotic systems control as well as the human-machine dynamic interactive control will be addressed and covered in later chapters.

3.5 Exercises of the Chapter

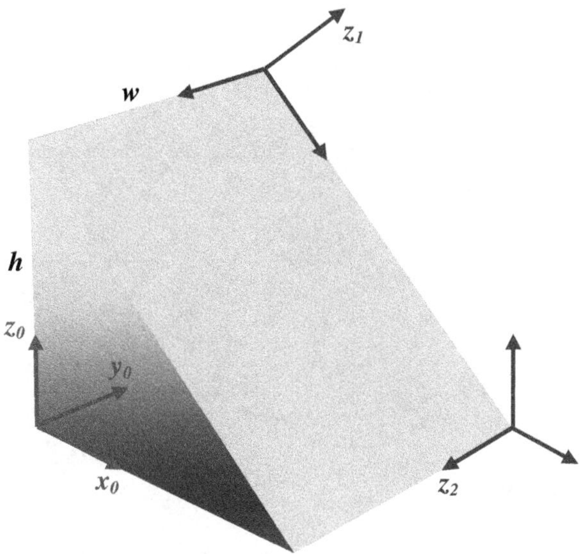

Fig. 3.2 Problem 1

1. A wedge block sits on the floor, as shown in Figure 3.2. Its side view is a right isosceles triangle with the bottom angle of 45^0. The height $h = 6$ and width $w = 4$ meters. There are three coordinate frames attached.
   a. Find the position vector p_0^1 and the rotation matrix R_0^1 for frame 1 w.r.t. frame 0;
   b. Find the position vector p_0^2 and the rotation matrix R_0^2 for frame 2 w.r.t. frame 0;
   c. Determine the position and orientation differences $p_{1(0)}^2$ and R_1^2 between frames 1 and 2;
   d. What is the position p_1^2 projected on frame 1?

2. Suppose that frame 1 is formed by rotating the base frame about the x_0-axis by 45^0.

   a. If frame 2 is formed by rotating frame 1 by 90^0 about a vector $v = (0\ 1\ 1)^T$ that is referred to frame 1, find the orientation of frame 2 w.r.t. the base;
   b. If frame 3 is formed by rotating frame 1 by 60^0 about the z_1-axis of frame 1, what is the orientation of frame 3 w.r.t. the base?
   c. What is the orientation of frame 3 with respect to frame 2?

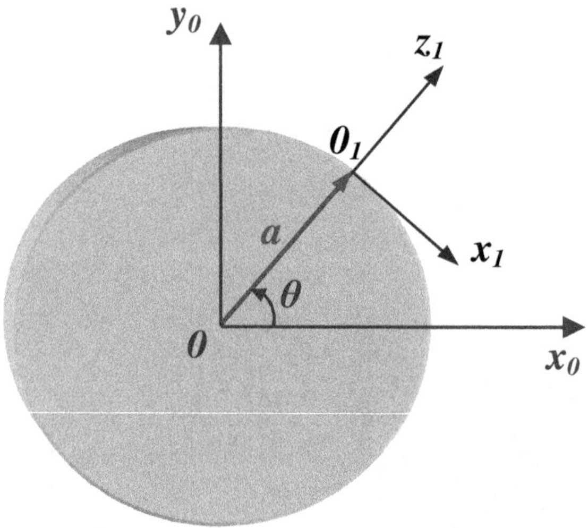

Fig. 3.3 Problem 3

3. A disc can be rotated about z_0-axis of the base. Frame 1 is fixed on the edge of the disc, as shown in Figure 3.3.

   a. Find the orientation R_0^1 and position p_0^1 of frame 1 with respect to the base in terms of the disc rotation angle θ ;
   b. As the angular velocity ω of the disc is defined by $\dot{\theta}\begin{pmatrix} 0 \\ 0 \\ 1 \end{pmatrix}$, show that the skew-symmetric matrix $S(\omega) = \dot{R}_0^1 R_1^0$;
   c. Also, show that $\dot{p}_0^1 = \omega \times a$, where a is the radial vector of the disc.

4. A new frame is generated by a rotation of the base frame in terms of Roll (ϕ), Pitch (θ) and Yaw (ψ) Euler angles. Suppose that $\phi = -90^0$, $\theta = -30^0$ and $\psi = 30^0$.

   a. Determine the orientation of the new frame w.r.t. the base;
   b. Now, using the $k - \phi$ procedure, find the unit vector k and angle ϕ for the above new orientation in (a) w.r.t. the base;

c. If the rotation from the base to the new frame is now broken into two even successive rotations, each of which rotates the frame about the vector k determined in (b) by $\frac{1}{2}\phi$, find the rotation matrix that represents the orientation of the midway between the two even successive rotations.

5. Suppose that frame 0 is rotated about its x_0-axis by $\alpha = 30^0$ to create frame 1. Then, frame 1 is rotated about its y_1-axis by $\beta = -45^0$ to form frame 2. If frame 2 is now rotated about the z_0-axis of frame 0 by $\gamma = 60^0$ to reach frame 3, determine the orientations R_0^1, R_0^2, R_0^3 and R_2^3.
6. A vector v is oriented w.r.t. the base frame by the angles α between v and x_0-axis, β between v and y_0-axis and γ between v and z_0-axis. If frame 1 is formed by sliding the base frame with $d = 20$ along the vector v and rotating the base about the vector v by 60^0.

   a. Find the homogeneous transformation matrix H_0^1 to represent both the position and orientation of frame 1 w.r.t. the base in terms of α, β and γ;
   b. If frame 2 is obtained by rotating the base frame by 135^0 about a vector $(0 \;\; -4 \;\; 0)^T$, what is H_0^2 of frame 2 w.r.t. the base?
   c. Determine the homogeneous transformation matrix H_1^2 of frame 2 w.r.t. frame 1 if $\alpha = \beta = \gamma$.

7. A vector v and a coordinate frame a are given as follows:

$$v_0 = \begin{pmatrix} 10 \\ 20 \\ 30 \end{pmatrix}, \qquad H_0^a = \begin{pmatrix} 0.866 & -0.50 & 0 & 11.0 \\ 0.50 & 0.866 & 0 & -3.0 \\ 0 & 0 & 1 & 9.0 \\ 0 & 0 & 0 & 1 \end{pmatrix},$$

   all of which are referred to the base. Write the vector v w.r.t. frame a.
8. Suppose that a base frame is given. Frame 1 is made by (1) rotating the base frame by α about the base z_0-axis, then (2) rotating the new frame by -90^0 about the new y-axis, and (3) the origin of frame 1 is located at (1, 1, 1) on the base. Frame 2 is made by (1) rotating the base frame by 90^0 about the base x_0-axis, then (2) rotating the new frame by β about the base z_0-axis, and (3) the origin of frame 2 is fixed at $(0, \; -1, \; 0)$ on the base. Determine the representation (both the position and orientation) of frame 2 w.r.t. Frame 1 in terms of α and β.
9. A rotation matrix is given in terms of two variable angles α and β by

$$R = \begin{pmatrix} c\alpha c\beta & -s\alpha & c\alpha s\beta \\ s\alpha c\beta & c\alpha & s\alpha s\beta \\ -s\beta & 0 & c\beta \end{pmatrix}.$$

   a. Determine the value of R if $\alpha = -45^0$ and $\beta = 60^0$ at a time instant;

b. With the same instantaneous angles, find the angular velocity ω if $\dot{\alpha} = 0.8$ and $\dot{\beta} = -0.6$ in radians/sec.

10. Using the same procedure as we did for the Roll-Pitch-Yaw Euler angles in equations (3.11) and (3.12), derive a relationship between the time-derivatives of the Z-Y-Z Euler angles and the angular velocity ω.
11. In equation (3.12), the coefficient matrix of the time-derivative vector for the Roll-Pitch-Yaw Euler angles is given by

$$D = \begin{pmatrix} 0 & -s\phi & c\phi c\theta \\ 0 & c\phi & s\phi c\theta \\ 1 & 0 & -s\theta \end{pmatrix}.$$

Determine the singularity of this matrix D and discuss its connection to the non-uniqueness between the Roll-Pitch-Yaw Euler angles $\{\phi, \theta, \psi\}$ and rotation matrix R.
12. If the three columns of a rotation matrix R are r_1, r_2 and r_3, show that the angular velocity can be written as

$$\omega = \frac{1}{2} \sum_{i=1}^{3} r_i \times \dot{r}_i.$$

13. Let k be a unit vector and let its skew-symmetric matrix be $K = S(k) = k\times$. If $\dot{k}$ and $\dot{K}$ are the time-derivatives of k and K, respectively, show that

a. $\dot{K}K\dot{K} = O$, the 3 by 3 zero matrix;
b. $K^2\dot{k} = -\dot{k}$;
c. $[K, [K, \dot{K}]] = -\dot{K}$.

References

1. Arnold, V.: Mathematical Methods of Classical Mechanics. Springer, New York (1978)
2. Marsden, J., Ratiu, T.: Introduction to Mechanics and Symmetry. Springer, New York (1994)
3. Spong, M., Vidyasagar, M.: Robot Dynamics and Control. John Wiley & Sons, New York (1989)
4. Murray, R., Li, Z., Sastry, S.: A Mathematical Introduction to Robotic Manipulation. CRC Press, Boca Raton (1994)
5. Craig, J.: Introduction to Robotics: Mechanics and Control, 3rd edn. Pearson Prentice Hall, New Jersey (2005)
6. Horn, R., Johnson, C.: Matrix Analysis. Cambridge University Press, New York (1985)
7. Bretscher, O.: Linear Algebra with Applications, 3rd edn. Prentice Hall (2004)
8. Jacobson, N.: Lie Algebras. John Wiley and Sons (1966)

9. Bakhturin, Y.: CampbellHausdorff Formula. In: Hazewinkel, M. (ed.) Encyclopedia of Mathematics. Springer (2001)
10. Koren, Y.: Robotics for Engineers. McGraw-Hill, New York (1985)
11. Gu, E.: An Exploration of Orientation Representation by Lie Algebra for Robotic Applications. IEEE Transactions on Systems, Man, and Cybernetics 20(1), 243–248 (1990)
12. Wehage, R.: Quaternions and Euler Parameters – A Brief Exposition. In: Hang, E. (ed.) Computer-Aided Analysis and Optimization of Mechanical System Dynamics, pp. 147–180. Springer, New York (1984)

Chapter 4
Robotic Kinematics and Statics

4.1 The Denavit-Hartenberg (D-H) Convention

In order to model an open serial-chain robot arm kinematically, we need not only to represent a 6 d.o.f. rigid motion in **Cartesian space** for the purpose of task/path description, but also to determine a complete transformation along the entire serial-chain robot body in **joint space** for the purpose of joint values determination. During the 1950's, Denavit and Hartenberg had independently developed a wonderful procedure to uniquely determine all the joint parameters plus variables for an open serial-chain robotic manipulator. Today, almost everyone recognizes that the D-H procedure is the best approach to uniquely modeling robotic kinematics, which is commonly called the **D-H Convention** [1, 2, 4].

Based on the D-H Convention, each joint and its host link possess in total four independent parameters: a joint angle θ, a joint offset d, a link twist angle α, and a link length a. Only one of them is selected as a joint variable. The detailed definitions are shown in Figure 4.1 and can be procedurized as follows:

1. First, determine a base coordinate frame for the robot and let the z_0-axis of the base be the axis for the first joint rotation if it is revolute, or translation if it is prismatic;
2. Then, assign z_{i-1} axis for the i-th joint as its rotation (or translation) axis for each $i = 2, \cdots, n$ if the total number of robotic joints is n;
3. After all the n z-axes are assigned along the robot body, determine the x_i axis for each $i = 1, \cdots, n$, which must be the common normal vector between z_{i-1} and z_i and must also be attached on both the two z_{i-1} and z_i axes;
4. After all the x-axes are so determined, then

   a. The joint angle θ_i is defined exactly to rotate link i about z_{i-1} and evaluated from x_{i-1} to x_i by looking into the z_{i-1} axis;
   b. The joint off-set d_i is defined as a distance along z_{i-1} from x_{i-1} to x_i;

E.Y.L. Gu, *A Journey from Robot to Digital Human*,
Modeling and Optimization in Science and Technologies 1,
DOI: 10.1007/978-3-642-39047-0_4, © Springer-Verlag Berlin Heidelberg 2013

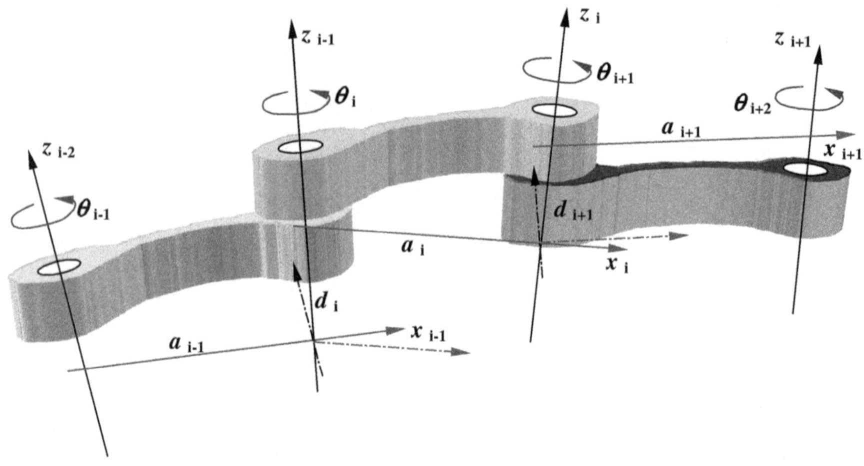

Fig. 4.1 Definition of the Denavit-Hartenberg (D-H) Convention

c. The link twist angle α_i is defined as an angle about x_i from z_{i-1} to z_i by looking into the x_i axis; and
d. The link length a_i is defined as a distance along x_i from z_{i-1} to z_i.

Through the above procedure, each link of the robot is guaranteed to have one and only one coordinate frame attached on it forever, no matter where the link moves. Namely, frame i is fixed on link i as the reference frame of link i, and this will also be the key important and convenient reference frame to determine all the dynamic parameters of the link, such as the mass center and the moments of inertia, in the subsequent dynamic modeling. The n-joint robot consists of n links in total and therefore has $n + 1$ frames assigned, including frame 0 on the base and from frame 1 through frame n with the n-th frame as the robotic hand or the end-effector frame. The above procedure of the D-H Convention allows us to determine the z_i and x_i axes for each frame, and the corresponding y_i axis can be determined automatically by the Right-Hand Rule.

Let us now look at a 6-joint robot arm as an example to illustrate how to follow the procedure and further to determine a so-called **D-H Table** for the robot. This robot arm, called a Stanford-Type robot, has a spherical structure over the first three links, i.e., the first three joints: Waist, Shoulder and Elbow are revolute, revolute and prismatic (RRP), respectively, as shown in Figure 4.2. Following the D-H Convention, one can determine without difficulty all the six frames as well as the four kinematic parameters for each link/joint. The D-H table can accordingly be created and is given in Table 4.1.

Table 4.1 The D-H table for a Stanford-type robot

Joint Angle	**Joint Offset**	**Twist Angle**	**Link Length**	$\cos\alpha$	$\sin\alpha$
θ_i	d_i	α_i	a_i	$c\alpha_i$	$s\alpha_i$
θ_1	d_1	90^0	0	0	1
θ_2	0	90^0	0	0	1
$\theta_3 = 0$	d_3	0	0	1	0
θ_4	0	-90^0	0	0	-1
θ_5	0	90^0	0	0	1
θ_6	d_6	0	0	1	0

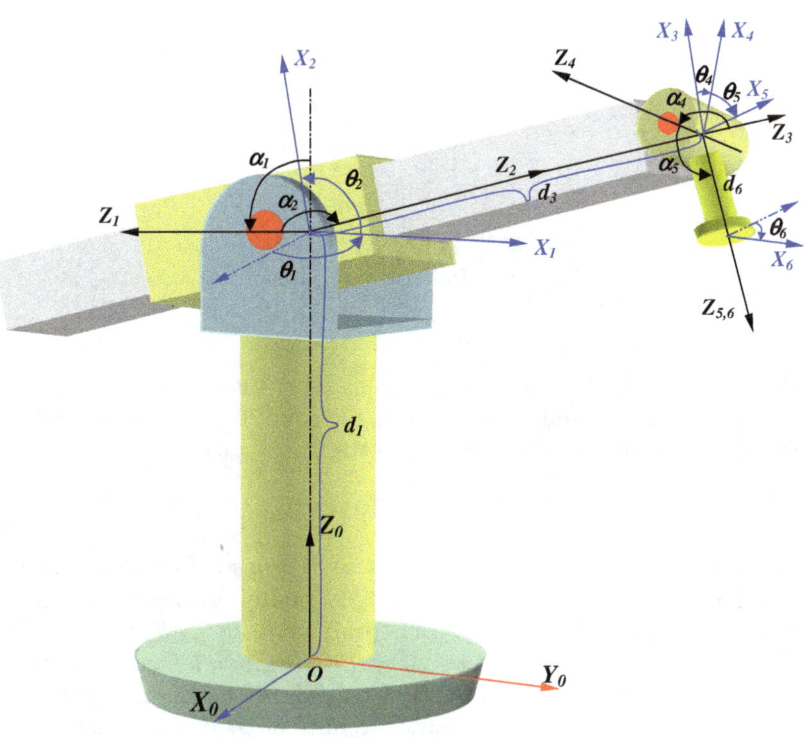

Fig. 4.2 A 6-joint Stanford-type robot arm

Since each set of the four parameters in the D-H table represents the corresponding coordinate frame position and orientation referred to the previous adjacent frame, and each of the joint angles θ_i and offsets d_i is about/along the z_{i-1} axis and each of the twist angles α_i and link lengths a_i is about/along the x_i axis, we can readily find a homogeneous transformation to describe such an adjacent one-step relationship between frame $i-1$ and frame i as follows:

$$A_{i-1}^{i} = \begin{pmatrix} c_i & -s_i & 0 & 0 \\ s_i & c_i & 0 & 0 \\ 0 & 0 & 1 & d_i \\ 0 & 0 & 0 & 1 \end{pmatrix} \begin{pmatrix} 1 & 0 & 0 & a_i \\ 0 & c\alpha_i & -s\alpha_i & 0 \\ 0 & s\alpha_i & c\alpha_i & 0 \\ 0 & 0 & 0 & 1 \end{pmatrix}, \tag{4.1}$$

where once again $c_i = \cos\theta_i$, $s_i = \sin\theta_i$, and $c\alpha_i = \cos\alpha_i$ and $s\alpha_i = \sin\alpha_i$ as a traditional short notation in robotics community.

As a result, by substituting the parameters/variables from the D-H table for the Stanford-Type robot into (4.1), we obtain all the **One-Step Homogeneous Transformations** for the robot.

$$A_0^1 = \begin{pmatrix} c_1 & 0 & s_1 & 0 \\ s_1 & 0 & -c_1 & 0 \\ 0 & 1 & 0 & d_1 \\ 0 & 0 & 0 & 1 \end{pmatrix}, \quad A_1^2 = \begin{pmatrix} c_2 & 0 & s_2 & 0 \\ s_2 & 0 & -c_2 & 0 \\ 0 & 1 & 0 & 0 \\ 0 & 0 & 0 & 1 \end{pmatrix},$$

$$A_2^3 = \begin{pmatrix} 1 & 0 & 0 & 0 \\ 0 & 1 & 0 & 0 \\ 0 & 0 & 1 & d_3 \\ 0 & 0 & 0 & 1 \end{pmatrix}, \quad A_3^4 = \begin{pmatrix} c_4 & 0 & -s_4 & 0 \\ s_4 & 0 & c_4 & 0 \\ 0 & -1 & 0 & 0 \\ 0 & 0 & 0 & 1 \end{pmatrix},$$

$$A_4^5 = \begin{pmatrix} c_5 & 0 & s_5 & 0 \\ s_5 & 0 & -c_5 & 0 \\ 0 & 1 & 0 & 0 \\ 0 & 0 & 0 & 1 \end{pmatrix}, \quad A_5^6 = \begin{pmatrix} c_6 & -s_6 & 0 & 0 \\ s_6 & c_6 & 0 & 0 \\ 0 & 0 & 1 & d_6 \\ 0 & 0 & 0 & 1 \end{pmatrix}. \tag{4.2}$$

Then, we may further determine the homogeneous transformation between frame i and frame j, and in particular, the robotic hand frame (#6) homogeneous transformation with respect to the base (#0):

$$A_0^6 = A_0^1 A_1^2 A_2^3 A_3^4 A_4^5 A_5^6. \tag{4.3}$$

The above successive transformation establishes a foundation of robotic kinematics. The right-hand side of (4.3) is clearly a function of all the six joint positions (variables) of the robot in $q = (\theta_1\ \theta_2\ d_3\ \theta_4\ \theta_5\ \theta_6)^T$. Let the hand frame that is represented by $A_0^6 = F_{4\times4}(q)$ match with a given desired task frame H_0^{task} with respect to the base. Then, the **inverse kinematics**, or I-K, is to solve all the trigonometric equations involved in the following 4 by 4 matrix equation to find all the joint values θ_1 through θ_6, including d_3:

$$A_0^6 = F_{4\times4}(q) = H_0^{task}. \tag{4.4}$$

Conversely, if all the joint positions in q are given, substituting them into A_0^6 to determine the task frame H_0^{task} is called the **forward kinematics**, or F-K.

Indeed, to solve each trigonometric equation involved in (4.4) is often difficult and tedious, and in some cases of 6-joint industrial robotic manipulators, they may not always have a closed form of the I-K solution. However, each robotic manipulator is required to derive its symbolical inverse kinematic solution to be programmed into its controller, but only needs to do **once**, and it will then be used forever in all future applications. On the other hand, a number of internal functions, such as fsolve($\cdots$) or fmincon($\cdots$) with constraints or fgoalattain($\cdots$) carrying objective functions in MATLAB^{TM}, can directly be called to solve the I-K numerically in a real-time fashion for a robot without need of a symbolical I-K algorithm. This will bring a lot of conveniences to robotic simulation studies performed in MATLAB^{TM}.

4.2 Homogeneous Transformations for Rigid Motion

After the D-H Convention with the D-H parameter table for an open serial-chain robotic system is introduced, all the possible homogeneous transformations will form a closed loop along both the inside and outside robot bodies. Each link has also been assigned with an attached coordinate frame and in total $n+1$ frames are assigned inside the robot body, including the base frame, if the robot has n joints. We often use A and H to denote a homogeneous transformation inside the body and outside the body, respectively, to distinguish them as different representations. However, either A_j^i or H_j^i describes both the relative position and orientation of frame i with respect to frame j. Therefore, they are all unique and collectively form the $SE(3) = \mathbb{R}^3 \times SO(3)$ multiplicative group to represent every 6 d.o.f. rigid motion.

In a robotic task/path planning phase, it is often required to determine a trajectory, along which the robotic hand or end-effector frame travels from a starting point to an ending or destination point. To uniquely represent either the starting or the ending point of the trajectory, we can now easily adopt a 4 by 4 homogeneous transformation matrix to do so. If H_0^1 is the position and orientation at the starting point with respect to the base and H_0^2 is the position and orientation at the ending point with respect to the base, too, then,

$$H_1^2 = H_1^0 H_0^2 = (H_0^1)^{-1} H_0^2$$

represents the "difference" of both the positions and orientations between frame #1 and frame #2.

In order to sample the trajectory into N uniform sampling points, we have to take position and orientation into separate consideration due to the uncommon natures between them, i.e., the former is additive in nature and the latter is multiplicative in nature. For **the position part**, if the trajectory is a straight line, we may first find the difference between the last columns of H_0^2 and H_0^1, i.e., $p_{1(0)}^2 = p_0^2 - p_0^1$, and then to determine the sampling interval $\Delta p_0 = \frac{p_{1(0)}^2}{N}$. Finally, at the i-th sampling step ($0 \le i \le N$), the position vector of the robotic hand with respect to the base is determined by

$$p_0^{(i)} = p_0^1 + i \cdot \Delta p_0 = p_0^1 + \frac{i}{N}(p_0^2 - p_0^1). \tag{4.5}$$

Obviously, $p_0^{(i)}$ is started at p_0^1 when $i = 0$ and ended at p_0^2 when $i = N$.

If the trajectory is a curve, then it is ideal to describe the path by a 3D parametric equation in the form

$$\begin{cases} x = x(t) \\ y = y(t) \\ z = z(t) \end{cases}$$

with each coordinate referred to the base as a function of time t or the other parameters, such as an arc length, etc. The robotic hand position vector at the i-th sampling point will become straightforward:

$$p_0^{(i)} = \begin{pmatrix} x(t_i) \\ y(t_i) \\ z(t_i) \end{pmatrix}.$$

For instance, if the path is specified as an ellipse on the y-z plane of the base with its center at $(x_c\ y_c\ z_c)$ and semi-axes a and b, then,

$$p_0^{(i)} = \begin{pmatrix} x_c \\ y_c + a\cos(\omega t_i) \\ z_c + b\sin(\omega t_i) \end{pmatrix},$$

where ω is the angular frequency to determine how fast the robotic hand travels along the elliptic path.

If the desired path is difficult to describe with an explicit function form, we may use a numerical solution. First, find a few desired discrete points on the path, and then apply the internal function spline$(\cdots)$ or pchip$(\cdots)$ in MATLAB^{TM} to make a more detailed and smoother new curve by certain interpolations for each of the three coordinates versus time t. The new curve is better to have the same number of the sampling points N so that the i-th position $p_0^{(i)}$ will just be the i-th x, y and z values of the new curve. For example,

```
t = 0:5;              % The Original Scale for 6 Known Points
x = [4 5 7 5 3 2]; % The X-Coordinates of the 6 Known Points
y = [10 9 7 4 2 3];    % The Y-Coordinates
z = [8 6 4 2 4 6];     % The Z-Coordinates

tt = 0:0.1:5;      % Make As Detail As N=51 Sampling Points

x1 = spline(t,x,tt);   % Call the spline(.) Internal Function
y1 = spline(t,y,tt);
z1 = spline(t,z,tt);

x2 = pchip(t,x,tt);    % Call the pchip(.) Internal Function
y2 = pchip(t,y,tt);
```

```
z2 = pchip(t,z,tt);

plot3(x,y,z,x1,y1,z1,x2,y2,z2), grid
                        % Make a 3D Plot for Comparison
```

In MATLAB^{TM}, the spline function provides a piecewise polynomial form of the cubic spline interpolation to the input data values at the point t, while the pchip function offers a piecewise polynomial form of a certain shape-preserving piecewise cubic Hermite interpolation to the input values at the point t. Thus, the former has more overshooting and oscillation than the latter. In the above program, six original points with their 3D coordinates are known. Then, in each of the three coordinate versus time profiles, we make a 10 times more detailed scale towards $N = 51$. The six known points are interpolated by spline and by pchip functions respectively, and the results are plotted in Figure 4.3 to compare each other among the three curves. If the sampling is also required to have $N = 51$ points along the entire desired path, then the position vector $p_0^{(i)}$ at t_i just takes the three coordinate values at the i-th point of either $(x_1\ y_1\ z_1)^T$ by the spline or $(x_2\ y_2\ z_2)^T$ by the pchip in the above program.

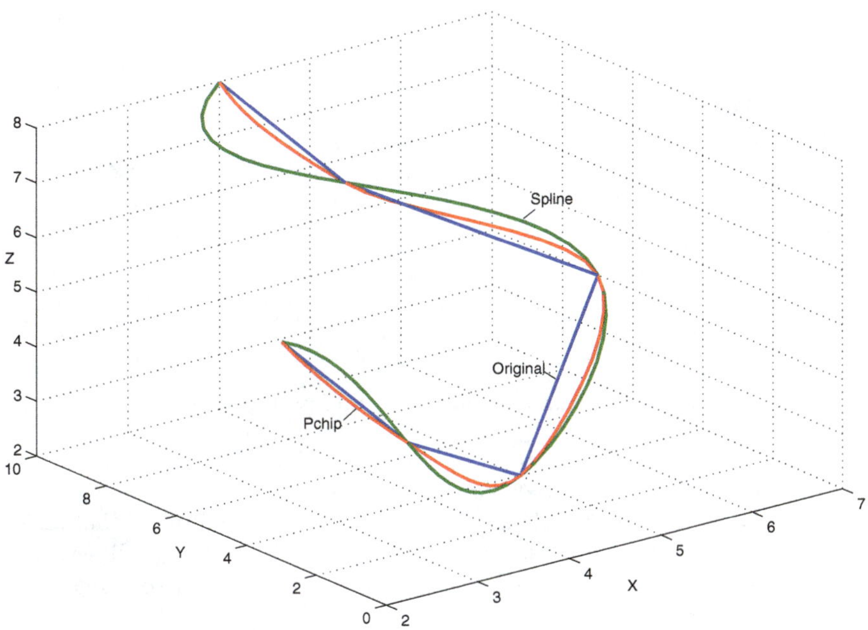

Fig. 4.3 A curved path before and after the spline and pchip interpolations

In contrast, for **the orientation part**, we have to first find the "difference" of the orientations between frame 1 and frame 2, i.e. $R_1^2 = R_1^0 R_0^2$. Then, using the $k - \phi$ procedure, as introduced in Chapter 2, to convert the "difference" R_1^2 into the equivalent k and ϕ using equations (2.9) and (2.11), and further to sample the angle ϕ into N intervals. Namely, $\Delta\phi = \frac{\phi}{N}$, and then using its forward conversion to determine $R_1^{(i)}$ by substituting the same unit vector k but a different angle $i\Delta\phi = \frac{i}{N}\phi$ into equation (2.8). Therefore, the orientation of the hand frame with respect to the base at the i-th sampling point on the trajectory is given by

$$R_0^{(i)} = R_0^1 R_1^{(i)}. \tag{4.6}$$

It can also be verified that $R_0^{(i)}$ matches R_0^1 when $i = 0$ and reaches R_0^2 when $i = N$.

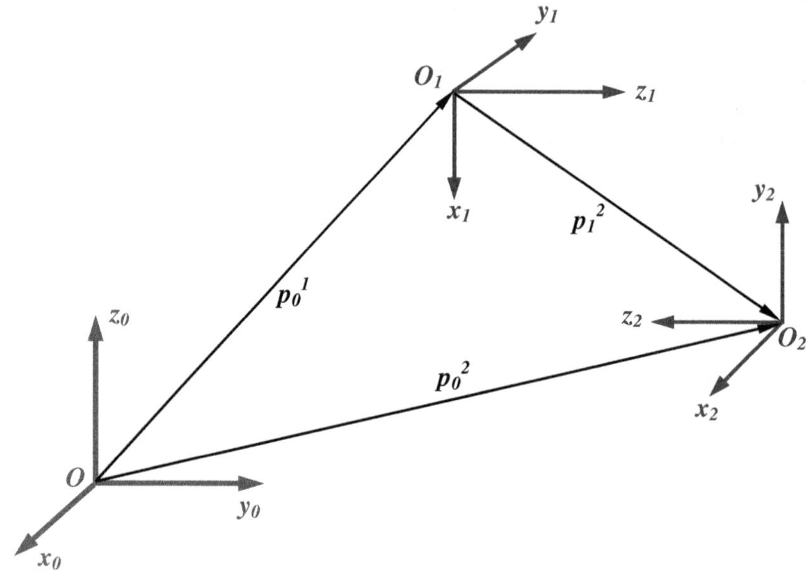

Fig. 4.4 Example of the position and orientation path planning

As a numerical example, let two coordinate frames #1 and #2 be placed near the base, as shown in Figure 4.4. Using the procedure of finding both position and orientation between two frames, as discussed and illustrated in the last chapter, we can readily determine the following two homogeneous transformations:

$$H_0^1 = \begin{pmatrix} 0 & -1 & 0 & 1.0 \\ 0 & 0 & 1 & 2.0 \\ -1 & 0 & 0 & 2.0 \\ 0 & 0 & 0 & 1 \end{pmatrix} \quad \text{and} \quad H_0^2 = \begin{pmatrix} 1 & 0 & 0 & -1.0 \\ 0 & 0 & -1 & 3.0 \\ 0 & 1 & 0 & 1.0 \\ 0 & 0 & 0 & 1 \end{pmatrix}$$

for frames #1 and #2 with respect to the base frame #0, respectively, where the position vectors are defined in meters. Now we split the path-planning from frame #1 to frame #2 into two portions. For the position portion, $p^2_{1(0)} = p^2_0 - p^1_0 = (-2.0 \ \ 1.0 \ \ -1.0)^T$. If we want the path to be a straight line and sampled into 10 even intervals, then $N = 10$, and $\Delta p_0 = p^2_{1(0)}/N = (-0.2 \ \ 0.1 \ \ -0.1)^T$. Thus, at the 4th sampling point, i.e. $i = 4$ for instance, $p^{(4)}_0 = p^1_0 + 4\Delta p_0 = (0.2 \ \ 2.4 \ \ 1.6)^T$ that is referred to the base.

For the orientation portion, let us first find the "difference" of the orientations between frames #2 and #1, i.e.,

$$R^2_1 = R^0_1 R^2_0 = \begin{pmatrix} 0 & 0 & -1 \\ -1 & 0 & 0 \\ 0 & 1 & 0 \end{pmatrix} \begin{pmatrix} 1 & 0 & 0 \\ 0 & 0 & -1 \\ 0 & 1 & 0 \end{pmatrix} = \begin{pmatrix} 0 & -1 & 0 \\ -1 & 0 & 0 \\ 0 & 0 & -1 \end{pmatrix}.$$

This rotation matrix happens to be symmetric, but is not equal to the identity I, and is exactly the same as the example in Chapter 2. We have found that $\phi = 180^0$ and $k = (\frac{\sqrt{2}}{2} \ \ -\frac{\sqrt{2}}{2} \ \ 0)^T$. Thus, for each step of rotation, $\Delta\phi = \phi/N = 18^0$ about the same axis k. At the 4th sampling point, $i = 4$,

$$R^{(4)}_1 = I + \sin(4 \times 18^0)K + (1 - \cos(4 \times 18^0))K^2$$

$$= \begin{pmatrix} 0.6545 & -0.3455 & -0.6725 \\ -0.3455 & 0.6545 & -0.6725 \\ 0.6725 & 0.6725 & 0.3090 \end{pmatrix}$$

such that

$$R^{(4)}_0 = R^1_0 R^{(4)}_1 = \begin{pmatrix} 0.3455 & -0.6545 & 0.6725 \\ 0.6725 & 0.6725 & 0.3090 \\ -0.6545 & 0.3455 & 0.6725 \end{pmatrix}.$$

This is the orientation at the 4th sampling point as frame #1 is traveling towards frame #2, which is referred to the base. Finally, the homogeneous transformation matrix $H^{(4)}_0$ at the 4th sampling point is augmenting $R^{(4)}_0$ and $p^{(4)}_0$ together, i.e.,

$$H^{(4)}_0 = \begin{pmatrix} 0.3455 & -0.6545 & 0.6725 & 0.2 \\ 0.6725 & 0.6725 & 0.3090 & 2.4 \\ -0.6545 & 0.3455 & 0.6725 & 1.6 \\ 0 & 0 & 0 & 1 \end{pmatrix}.$$

The above trajectory sampling procedure is often referred to as the **Cartesian Motion** planning, because the path sampling is taken in Cartesian space. The counterpart of the Cartesian motion is the **Joint Motion** planning.

In the joint motion path-planning, we only need to apply the I-K (inverse kinematics) algorithm for a given robot arm at the starting point with H^1_0

and the ending point with H_0^2 of the desired trajectory. If the I-K solution for all joint positions based on H_0^1 is q^I and that based on H_0^2 is q^{II}, the difference between the ending and starting joint positions becomes

$$q^{(2-1)} = q^{II} - q^I.$$

Then, we determine how much each joint should move in each sampling interval by

$$\Delta q = \frac{q^{(2-1)}}{N} = \frac{1}{N}(q^{II} - q^I).$$

Finally, all the joint positions of the robot at the i-th sampling point ($0 \leq i \leq N$) is given as

$$q^{(i)} = q^I + i \cdot \Delta q = q^I + \frac{i}{N}(q^{II} - q^I). \tag{4.7}$$

While the joint motion planning looks simple, when you implement the above algorithm, you have to make sure that each joint angle difference is the minimum in absolute value. For example, if one joint angle is currently at $\theta_i = 90^0$, and to move to $\theta_i = 220^0$. Often, in most computer programs, either the starting or the destination joint angle is solved by the two-dummy elements arc-tangent, i.e., θ_i =atan2(y, x) for some y and x, which gives the angle solution in the range of $(-180^0,\ 180^0]$. Therefore, the above destination after I-K calculation in a computer program will give $\theta_i = -140^0$, instead of 220^0, though they are considered as the same angles. Now, we need an if-then or a subroutine in the program to determine the minimum difference in absolute value. Namely, $220^0 - 90^0 = 130^0$, while $-140^0 - 90^0 = -230^0$. Clearly, the first one is smaller than the second one in absolute value and the computer program should make sure to move θ_i from 90^0 to -140^0 or 220^0 counterclockwise, instead of clockwise, in order to ensure that each joint of the robot will have the shortest joint motion. The following short MATLAB^{TM} program is a typical way to minimize the difference.

```
dq = qd - qs;   % Find the Difference Between the Starting qs
                % and the Destination qd Joint Positions
for j = 1:6        % 6 Joints Totally
    if dq(j) > pi
        dq(j) = dq(j)-2*pi;
    end
    if dq(j) < -pi
        dq(j) = dq(j)+2*pi;
    end
end
```

Since any robotic system can understand only the "joint language", the above joint motion planning appears to be very simple and straightforward in terms of formulation and computation. However, the drawback is unnatural-looking when the robot moves and is also unable to follow a desired trajectory, because the joint motion planning cannot meet or satisfy any requirement that is specified in Cartesian space.

As a comparative study, we summarize the advantages and disadvantages between the Cartesian motion and joint motion into Table 4.2.

Table 4.2 A comparison between Cartesian and joint motions

Type of Motion	Cartesian Motion	Joint Motion
No. of Inverse Kinematics	$N+1$	2
Traj. Tracking Accuracy	Very High	Unable to Offer
Computational Complexity	High	Very Low
Singularity Chance	Possible	Never
Body Move Appearance	Excellent	Uncontrollable

In this comparison table, the number of inverse kinematics means how many times the I-K algorithm of a given robot is required to be repeatedly computed. In general application cases, people often apply the Cartesian motion for a main task that requires a desired trajectory-tracking while using the joint motion just for the Leaving-Home trip before being ready to start the main task as well as the Back-Home trip after the main task is completed. Traditionally, almost every robotic system has its **Home** position defined as a long idle configuration for the robot. In addition, the singularity in Table 4.2 means a special configuration, where the robot loses one or more degree(s) of freedom. The singularity case often makes the I-K algorithm diverge, for instance, some denominator in the I-K algorithm vanishes. In the next sections, we will give more detailed analysis of the singularity issue and also a discussion on its avoidance approaches through an introduction to the third type of motion planning, called the **Differential Motion** for a robotic system.

4.3 Solutions of Inverse Kinematics

In this section, we introduce a general procedure of how to solve inverse kinematics (I-K) for a relatively simple robot with fewer joints. In the cases of real industrial robots with at least six joints, to find a closed form of I-K solution requires a well-trained robotics professional with solid mathematical skills. In many six-joint robotic systems, the number of distinct sets of I-K solutions is not just one, and it could have much more than one, depending on the mechanical structure and the number of revolute joints for each particular robot. A 6-joint PUMA robot arm can have up to 8 distinct sets of its I-K solutions due to all revolute joints plus a shoulder offset, while a 6-joint Stanford robot arm has only 4 distinct sets because one of the six joints is prismatic. In robotics, this phenomenon of more distinct sets of I-K solutions is referred to as a **multi-configuration** factor. This means that the mapping from Cartesian space to joint space is not one-to-one.

The simplest example of multi-configuration is a 2-link planar robot arm, as depicted in Figure 4.5. Fortunately, any two of the distinct sets of I-K solutions are not connected to each other, and they are often a non-trivial "distance" away. In general speaking, the "boundary" between two adjacent distinct configurations under the common hand frame position/orientation is either a singular configuration or a mechanical limit case for a robot. Therefore, the multi-configuration factor, in general, makes no trouble to a robotic path planning.

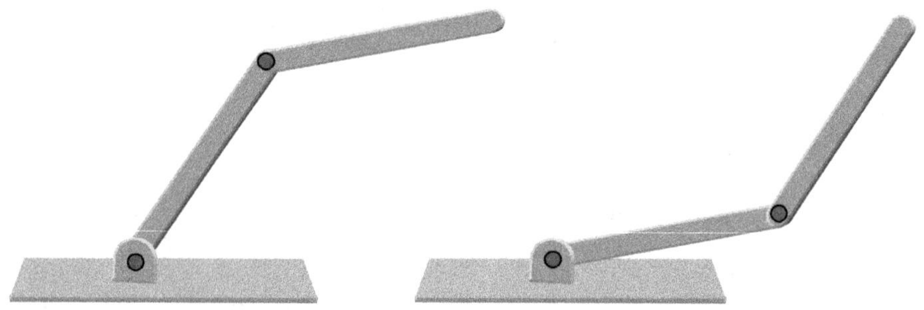

Fig. 4.5 Multi-configuration for a two-link arm

Since a symbolical algorithm of I-K for a robot is developed by solving a number of trigonometric equations involved in equation (4.4), to ensure the algorithm is more robust, each joint angle must be solved by using atan2$(y,\ x)$, i.e., the two-dummy-element four-quadrant arc tangent function, instead of the two-quadrant atan$(\frac{y}{x})$. Also, avoid using either arcsin or arccos functions as much as possible to determine each joint angle.

Let us look back at the Stanford-Type robot arm in Figure 4.2 to illustrate how to develop an I-K algorithm. This time we simplify the illustration by considering the position p_0^3 of frame 3 as the only Cartesian output to solve for the first three joint values: θ_1, θ_2 and d_3. According to the first three one-step homogeneous transformation matrices A_0^1, A_1^2 and A_2^3 of the Stanford-Type robot in the previous section, we can readily compute

$$A_0^3 = A_0^1 A_1^2 A_2^3 = \begin{pmatrix} c_1c_2 & s_1 & c_1s_2 & d_3c_1s_2 \\ s_1c_2 & -c_1 & s_1s_2 & d_3s_1s_2 \\ s_2 & 0 & -c_2 & d_1 - d_3c_2 \\ 0 & 0 & 0 & 1 \end{pmatrix}. \quad (4.8)$$

Now, the last column of A_0^3 is obviously the Cartesian position output p_0^3 pointing to the origin of frame 3 with respect to the base. Let

$$\begin{pmatrix} d_3c_1s_2 \\ d_3s_1s_2 \\ d_1 - d_3c_2 \end{pmatrix} = p_0^3 = \begin{pmatrix} p_1 \\ p_2 \\ p_3 \end{pmatrix}.$$

We can immediately solve that

$$d_3 = \sqrt{p_1^2 + p_2^2 + (p_3 - d_1)^2}, \tag{4.9}$$

where we only pick the positive solution for d_3, because the negative one or zero is physically meaningless for a prismatic joint. Since by inspecting Figure 4.2, and based on the working envelope of the Stanford-Type robot, $0 < \theta_2 < 180^0$, $\sin\theta_2$ should always be positive, and never vanish. Therefore,

$$\theta_1 = \text{atan2}(p_2,\ p_1). \tag{4.10}$$

In order to solve θ_2 more robustly, we notice that $p_1^2 + p_2^2 = d_3^2 s_2^2$ so that $d_3s_2 = \sqrt{p_1^2 + p_2^2}$, and also $d_3c_2 = d_1 - p_3$. Hence,

$$\theta_2 = \text{atan2}\left(\sqrt{p_1^2 + p_2^2},\ d_1 - p_3\right). \tag{4.11}$$

The shoulder angle θ_2 can alternatively be determined by using $p_1 = d_3c_1s_2$ or $p_2 = d_3s_1s_2$ so that $d_3s_2 = p_1/c_1$ or p_2/s_1, and then placing it into the atan2 function with $d_1 - p_3 = d_3c_2$. However, the waist angle θ_1 must have a wide range of rotation and it cannot be guaranteed that either $c_1 \neq 0$ or $s_1 \neq 0$. Therefore, equation (4.11) is clearly more robust than the alternative way.

In general cases, to develop an I-K algorithm for a robotic system, there are often many alternatives for each joint angle solution. Therefore, we have to check all possible constraints based on both the mathematical and physical restrictions of the robot to narrow them down towards the best and most robust solutions. Indeed, this requires a high skill, especially for the robots with six or more joints.

As a demonstration, we provide a complete symbolical I-K algorithm, but without giving detailed derivations, for each of the following two typical 6-joint robot arms: the Stanford arm and an educational model of industrial manipulator, as shown in Figure 4.6. Note that each of their I-K algorithms is subject to the link coordinate frame assignment under the D-H convention. In other words, **a different definition of x-axis or z-axis for each link/joint of the robot may result in different I-K solution**. Due to this fact, we show the frame assignment that corresponds to the given I-K algorithm for the Stanford robot in part (a) of Figure 4.6.

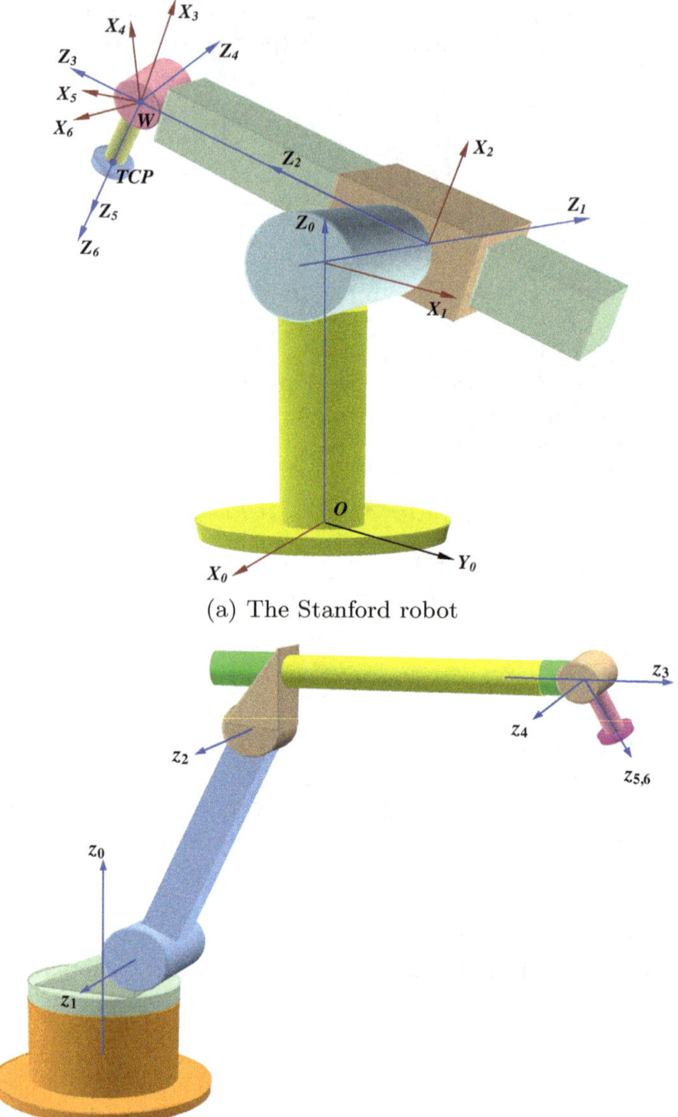

(a) The Stanford robot

(b) The industrial robot model

Fig. 4.6 Two robot arms with their z-axes

Algorithm 1 – Inverse kinematics for the Stanford robot, as seen in part (a) of Figure 4.6:

Given $\begin{pmatrix} r_{11} & r_{12} & r_{13} \\ r_{21} & r_{22} & r_{23} \\ r_{31} & r_{32} & r_{33} \end{pmatrix} = R_0^6$ and $\begin{pmatrix} x_6 \\ y_6 \\ z_6 \end{pmatrix} = p_0^6$. Since

$$p_0^6 = p_0^5 + d_6 \begin{pmatrix} r_{13} \\ r_{23} \\ r_{33} \end{pmatrix},$$

let p_0^5 be prepared as follows:

$$p_0^5 = \begin{pmatrix} x \\ y \\ z \end{pmatrix} = p_0^6 - d_6 \begin{pmatrix} r_{13} \\ r_{23} \\ r_{33} \end{pmatrix} = \begin{pmatrix} x_6 \\ y_6 \\ z_6 \end{pmatrix} - d_6 \begin{pmatrix} r_{13} \\ r_{23} \\ r_{33} \end{pmatrix}.$$

Then,

$$\begin{aligned}
\theta_1 &= \text{atan2}\left(\pm y\sqrt{x^2 + y^2 - d_2^2} - xd_2,\ yd_2 \pm x\sqrt{x^2 + y^2 - d_2^2} \right), \\
\theta_2 &= \text{atan2}\left(\pm\sqrt{x^2 + y^2 - d_2^2},\ z - d_1 \right), \\
d_3 &= \sqrt{x^2 + y^2 + (z - d_1)^2 - d_2^2}, \\
\theta_4 &= \text{atan2}(r_{23}c_1 - r_{13}s_1,\ r_{13}c_1c_2 + r_{23}s_1c_2 - r_{33}s_2) \ \text{ or } \ \theta_4 \pm 180^0, \\
\theta_5 &= \text{atan2}(\sqrt{(r_{23}c_1 - r_{13}s_1)^2 + (r_{13}c_1c_2 + r_{23}s_1c_2 - r_{33}s_2)^2}, \\
&\qquad r_{13}c_1s_2 + r_{23}s_1s_2 + r_{33}c_2) \text{ or } -\theta_5, \\
\theta_6 &= \text{atan2}(r_{12}c_1s_2 + r_{22}s_1s_2 + r_{32}c_2,\ -r_{11}c_1s_2 - r_{21}s_1s_2 - r_{31}c_2) \\
&\quad \text{or } \ \theta_6 \pm 180^0.
\end{aligned}$$

Algorithm 2 – Inverse kinematics for the industrial robot model, as seen in part (b) of Figure 4.6:

Given $p_0^6 = \begin{pmatrix} x_6 \\ y_6 \\ z_6 \end{pmatrix}$ and $R_0^6 = \begin{pmatrix} r_{11} & r_{12} & r_{13} \\ r_{21} & r_{22} & r_{23} \\ r_{31} & r_{32} & r_{33} \end{pmatrix}$. Then,

$$p_0^5 = \begin{pmatrix} x \\ y \\ z \end{pmatrix} = p_0^6 - d_6 \begin{pmatrix} r_{13} \\ r_{23} \\ r_{33} \end{pmatrix} = \begin{pmatrix} x_6 \\ y_6 \\ z_6 \end{pmatrix} - d_6 \begin{pmatrix} r_{13} \\ r_{23} \\ r_{33} \end{pmatrix}.$$

Using the following steps to determine all the six joint angles:

$$\begin{aligned}
\theta_1 &= \text{atan2}(y,\ x), \quad \text{and defining} \\
&\qquad \eta_x = \sqrt{x^2 + y^2} - a_1,\ \eta_z = z - d_1, \text{ and } \ \tilde{d}_4 = \sqrt{a_3^2 + d_4^2}, \\
\tilde{s}_3 &= \frac{\eta_x^2 + \eta_z^2 - \tilde{d}_4^2 - a_2^2}{2a_2\tilde{d}_4}, \\
\tilde{c}_3 &= \pm\sqrt{1 - \tilde{s}_3^2}, \\
\tilde{\theta}_3 &= \text{atan2}(\tilde{s}_3,\ \tilde{c}_3), \quad \text{and}
\end{aligned}$$

$$\theta_3 = \tilde{\theta}_3 - atan\left(\frac{a_3}{d_4}\right),$$
$$\theta_2 = \text{atan2}(\eta_x \tilde{d}_4 \tilde{c}_3 + \eta_z(\tilde{d}_4 \tilde{s}_3 + a_2),\ \eta_x(\tilde{d}_4 \tilde{s}_3 + a_2) - \eta_z \tilde{d}_4 \tilde{c}_3),$$
$$\theta_4 = \text{atan}\left(\frac{\rho_{23}}{\rho_{13}}\right), \quad \text{or} \quad \theta_4 = \theta_4 \pm 180^0,$$
$$\theta_5 = \text{atan2}(c_4\rho_{13} + s_4\rho_{23},\ -\rho_{33}),$$
$$\theta_6 = \text{atan2}(s_4\rho_{11} - c_4\rho_{21},\ s_4\rho_{12} - c_4\rho_{22}),$$

where

$$\begin{pmatrix} \rho_{11} & \rho_{12} & \rho_{13} \\ \rho_{21} & \rho_{22} & \rho_{23} \\ \rho_{31} & \rho_{32} & \rho_{33} \end{pmatrix} = R_3^6 = R_3^0 R_0^6 = \begin{pmatrix} c_1c_{23} & s_1c_{23} & s_{23} \\ s_1 & -c_1 & 0 \\ c_1s_{23} & s_1s_{23} & -c_{23} \end{pmatrix} R_0^6.$$

Note that in both the above symbolical I-K algorithms, the last joint #6 has an offset d_6 that is the length along the z_6-axis so that $p_0^6 = p_0^5 + d_6z_6$, where z_6 is the unit vector of the z_6-axis. In Algorithm 1, the Stanford robot arm has totally **four** different sets of I-K solutions for each given Cartesian position/orientation and they are listed in Table 4.3.

Table 4.3 Four different I-K solutions for the Stanford robot

Joint Angles	Solution 1	Solution 2	Solution 3	Solution 4
θ_1	pick +	pick −	pick +	pick −
θ_2	pick +	pick −	pick +	pick −
θ_4	θ_4	θ_4	$\theta_4 \pm 180^0$	$\theta_4 \pm 180^0$
θ_5	θ_5	θ_5	$-\theta_5$	$-\theta_5$
θ_6	θ_6	θ_6	$\theta_6 \pm 180^0$	$\theta_6 \pm 180^0$

Figures 4.7 and 4.8 realistically exhibit the four different I-K solutions that are animated in MATLAB^{TM} under the exact same position/orientation of frame #6. The four sets of joint values are also printed below with each angle in degrees and the third prismatic joint value in meters:

```
Q =

  -173.0324     46.1625   -173.0324     46.1625
    64.6057    -64.6057     64.6057    -64.6057
     0.4664      0.4664      0.4664      0.4664
   -15.9062    112.3812   -195.9062    -67.6188
    26.2715     51.2669    -26.2715    -51.2669
  -165.6655   -123.3500   -345.6655   -303.3500

>>
```

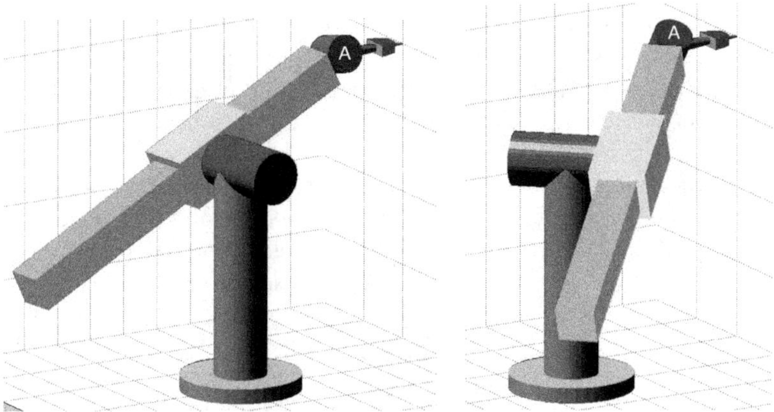

Fig. 4.7 The first and second I-K solutions for the Stanford arm

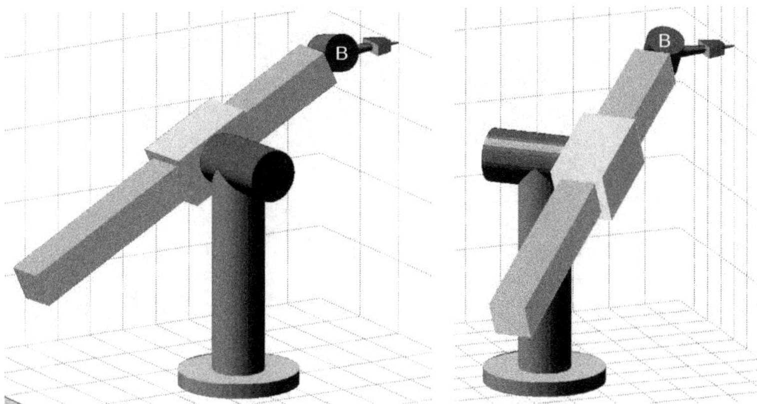

Fig. 4.8 The third and fourth I-K solutions for the Stanford arm

Likewise, in Algorithm 2, the industrial manipulator model also has **four** sets of inverse kinematics solutions, and each one of the four is a combination between each $\pm$ choice of $\tilde{c}_3$ and the choice of either θ_4 or $\theta_4 \pm 180^0$.

In fact, MATLAB^{TM} has an internal function in its optimization toolbox: fsolve(@fun, q0, option) that can numerically solve a vector or matrix equation $F(q) = 0$ described in a function subroutine fun.m for a vector argument q with its initial value q0 and some options. Using this internal function, the inverse kinematics (I-K) can be resolved numerically for any robot arm without need of symbolical I-K derivation. Of course, sometimes a symbolical form of the I-K algorithm is often more stable than the numerical solution, as long as the initial value $q(0)$ is not far away from the solution q. Otherwise, it may be stuck with a solution in some unwanted configuration. Unlike the symbolical way, where you can always make a decision on which configuration the robot arm should go for before executing the I-K algorithm, once the solution q reaches an unwanted configuration, the numerical I-K algorithm is unable to recover itself back to the right configuration.

To illustrate the numerical way to solve an I-K problem, let us look back at the Stanford-type robot arm, as shown in Figure 4.2. According to the six one-step homogeneous transformation matrices in equation (4.2), a MATLAB^{TM} main program plus a function subroutine are given as follows:

```
d1 = 1;  d6 = 0.2;  % In Meters

q = [90 110 0 0 -70 0]'*pi/180;
           % The Initial Joint Angles
q(3) = 1;  % The Prismatic Joint in Meters

option = optimset('Algorithm', ...
      'levenberg-marquardt','Display','off');

q = fsolve(@f_stanford,q,option);

  % The Main Program Above %

  % ---------------------- %

  % The Function F = 0 Subroutine Below %

function F = f_stanford(q)

   for i = 1:6
      s(i) = sin(q(i));  c(i) = cos(q(i));
   end
   d3 = q(3);

   A1=[c(1) 0 s(1) 0; s(1) 0 -c(1) 0; 0 1 0 d1; 0 0 0 1];
   A2=[c(2) 0 s(2) 0; s(2) 0 -c(2) 0; 0 1 0 0; 0 0 0 1];
   A3=[1 0 0 0; 0 1 0 0; 0 0 1 d3; 0 0 0 1];
   A4=[c(4) 0 -s(4) 0; s(4) 0 c(4) 0; 0 -1 0 0; 0 0 0 1];
   A5=[c(5) 0 s(5) 0; s(5) 0 -c(5) 0; 0 1 0 0; 0 0 0 1];
   A6=[c(6) -s(6) 0 0; s(6) c(6) 0 0; 0 0 1 d6; 0 0 0 1];

   A06 = A1*A2*A3*A4*A5*A6;  % The Hand Position/Orientation

   H06=[-1 0 0 0.1; 0 0 1 1.1; 0 1 0 1.3; 0 0 0 1];
                              % The Destination

   F = A06(1:3,:)-H06(1:3,:);  % To Solve F = 0
```

```
end

  % ---------------------- %
  %  The Output Joint Values %

>> q =

    83.66  108.33  0.95  19.46  -19.36  -108.43

    % All in Degrees Except q(3) in Meters
```

Before the joint values in q are updated by calling the internal function fsolve($\cdots$), the set of initial joint positions yields the following homogeneous transformation:

$$H_0^6(0) = \begin{pmatrix} 0.00 & 1.00 & 0.00 & 0.00 \\ 0.77 & 0.00 & 0.64 & 1.07 \\ 0.64 & 0 & -0.77 & 1.19 \\ 0 & 0 & 0 & 1 \end{pmatrix}.$$

After calling the internal function, the new set of joint values is given in the last output data line of the above program, and the destination by substituting the new joint values into the symbolical A_0^6 in MATLAB^{TM} exactly matches the numerical homogeneous transformation matrix H_0^6 specified in the above function subroutine. Therefore, this numerical way can directly be implemented in MATLAB^{TM} to replace the symbolical I-K solution as an alternative to updating all the joint values at each sampling point for a robot in Cartesian motion.

Note that in the above MATLAB^{TM} code, the internal function fsolve($\cdots$) uses three different algorithms: trust-region-dogleg, trust-region-reflective and Levenberg-Marquardt with the first one as default. We pick the third option to be more suitable for solving the 3 by 4 matrix equation $F = 0$.

If the function-call is repeated inside a for-loop or a while-loop and in each sampling interval, H_0^6 is varying in order to follow a desired Cartesian motion planning, then, the desired destination H_0^6 at each sampling point can be treated as a parameter to pass between the main program and function subroutine by an anonymous function, as illustrated below. The advantage of such an anonymous function approach is to avoid placing each destination H_0^6 as a global variable in both the main program and function subroutine.

```
for i = 0:N

    % ................ %

    H06 = ......;         % From a Cartesian Motion Planning
```

```
    option = optimset('Algorithm','levenberg-marquardt', ...
                      'Display','off');

    fs = @(q)f_stanford(q,H06);  % Define an Anonymous Function

    q = fsolve(fs,q,option);

    % ................. %

end

    % Then, the Function Should Be Modified To %

function F = f_stanford(q,H06)

    % Same as the Previous Subroutine %

end
```

4.4 Jacobian Matrix and Differential Motion

As discussed in the last chapter, a Cartesian velocity $V = \begin{pmatrix} v \\ \omega \end{pmatrix}$ is linearly related to the joint velocity $\dot{q}$ by a Jacobian matrix J, as expressed in (3.22). Since the Cartesian velocity V contains an angular velocity vector ω, which, according to equation (3.8), is not a time-derivative of some vector, the Jacobian matrix J cannot be determined directly by taking the partial derivative of a 6-dimensional position/orientation vector $\xi = \begin{pmatrix} p \\ \rho \end{pmatrix}$ with respect to the joint position vector q, i.e., $J \neq \partial\xi/\partial q$ in general. Although the upper half J_v of the Jacobian matrix $J = \begin{pmatrix} J_v \\ J_\omega \end{pmatrix}$ is the partial derivative of a 3D position vector $p(q)$, i.e., $J_v = \partial p/\partial q$, provided that p is projected on the fixed base frame, the bottom half J_ω of the Jacobian matrix is not. Nevertheless, we can still find a symbolical form for the traditional Jacobian matrix J that is 6 by n and playing a role of linear transformation in tangent space from joint to Cartesian velocities if a given robot has n joints such that $q \in \mathbb{R}^n$.

In order to develop an explicit symbolical form for the Jacobian matrix in terms of all the joint positions in $q \in \mathbb{R}^n$ of a given robot, we now start with a detailed analysis of the Cartesian velocity $V = \begin{pmatrix} v \\ \omega \end{pmatrix}$ that contains a linear velocity v and an angular velocity ω of frame n attached on the hand or end-effector link n. It can be observed that the motion of each link, say link i, will contribute its linear velocity v_i of frame i as part to the total amount v of the hand frame and also contribute its angular velocity ω_i as part to the total amount ω of the hand. In other words, the robotic hand, i.e.

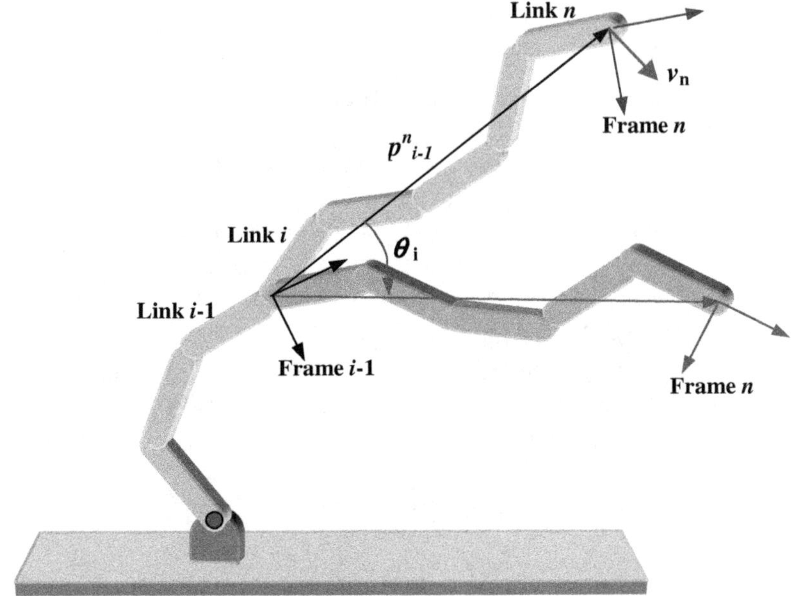

Fig. 4.9 The motion of link n superimposed by the motion of link i

link n, is kinematically the "busiest" link of the robot, because every other link movement will impose on the last link, as shown in Figure 4.9.

With the above interpretation, we can reasonably write

$$v = v_1 + \cdots + v_n, \quad \text{and} \quad \omega = \omega_1 + \cdots + \omega_n,$$

where each term of the equation must refer to a common projection frame. After a further derivation, each term of the robotic hand linear velocity $v_{(n)}$ and angular velocity $\omega_{(n)}$, both of which are projected on frame n of the hand, can be written as

$$\omega_{i(n)} = r_n^{i-1}\dot{\theta}_i, \quad \text{and} \quad v_{i(n)} = \omega_{i(n)} \times p^n_{i-1(n)} = (p_n^{i-1} \times r_n^{i-1})\dot{\theta}_i \tag{4.12}$$

for $i = 1, \cdots, n$ if all the n joints are **revolute**, where r_n^{i-1} is the last (third) column of the rotation matrix R_n^{i-1} and represents the z-axis of frame $i-1$ with respect to frame n. This shows that

$$V_{(n)} = \begin{pmatrix} v_{(n)} \\ \omega_{(n)} \end{pmatrix} = \begin{pmatrix} \sum_{i=1}^n (p_n^{i-1} \times r_n^{i-1})\dot{\theta}_i \\ \sum_{i=1}^n r_n^{i-1}\dot{\theta}_i \end{pmatrix}. \tag{4.13}$$

Comparing this to $V = J\dot{q}$ with $\dot{q} = (\dot{\theta}_1 \ \cdots \ \dot{\theta}_n)^T$, we achieve a basic equation that describes the formation of a 6 by n Jacobian matrix:

$$J_{(n)} = \begin{pmatrix} p_n^0 \times r_n^0 & \cdots & p_n^{n-1} \times r_n^{n-1} \\ r_n^0 & \cdots & r_n^{n-1} \end{pmatrix}. \tag{4.14}$$

If the j-th joint of the robot is **prismatic**, since equation (4.12) has to be changed to

$$\omega_{j(n)} = 0, \quad \text{and} \quad v_{j(n)} = r_n^{j-1} \dot{d}_j,$$

the j-th column $\begin{pmatrix} p_n^{j-1} \times r_n^{j-1} \\ r_n^{j-1} \end{pmatrix}$ in equation (4.14) must be replaced by $\begin{pmatrix} r_n^{j-1} \\ 0_3 \end{pmatrix}$ with the 3 by 1 zero column vector 0_3 at the bottom due to no rotation contributed by a prismatic joint.

Once we have $J_{(n)}$, it becomes easier to find a Jacobian matrix projected on any one of the link frames other than the n-th hand frame. For instance, if we wish to find $J_{(k)}$ that is projected on frame k for $0 \le k < n$, then

$$J_{(k)} = \begin{pmatrix} R_k^n & O_{3\times3} \\ O_{3\times3} & R_k^n \end{pmatrix} J_{(n)}, \tag{4.15}$$

where R_k^n is the 3 by 3 rotation matrix representing the orientation of frame n with respect to frame k, and $O_{3\times3}$ is the 3 by 3 zero matrix.

To calculate the Jacobian matrix using (4.14) for an n-joint robot, it can be seen that each p_n^j and each r_n^j for $j = 0, \cdots, n-1$ are the last (fourth) column and the third column of the homogeneous transformation $A_n^j = (A_j^n)^{-1}$, respectively. It is also observable that the farther the "distance" between frame j and n is, the more lengthy each element of A_n^j will be. Therefore, we often try to first develop a relatively shorter form of Jacobian matrix by projecting it on a middle frame between the base and the hand. For instance, if $n = 6$, then $J_{(3)}$ may be the Jacobian matrix with the simplest symbolical form for the robot. Afterwards, through equation (4.15), it can be converted to the Jacobian matrix projected onto any desired frame.

As an illustrative example, let us look back at the Stanford-Type robot in Figure 4.2 along with its D-H table and all the one-step homogeneous transformations in (4.2). Let us also assume $d_6 = 0$ so that there are no more length parameters beyond frame 3 in this particular robot. This assumption is also very common and is equivalent to merging the origins of frames 5 and 6. It will offer a huge simplification for the next Jacobian matrix derivation.

Under this assumption, we now develop the shortest symbolical form of Jacobian matrix $J_{(3)}$ that is projected on frame 3 for the Stanford-Type robot arm. According to equation (4.14), let us first invert the first three one-step homogeneous transformation matrices given by (4.2) as follows:

$$A_1^0 = \begin{pmatrix} c_1 & s_1 & 0 & 0 \\ 0 & 0 & 1 & -d_1 \\ s_1 & -c_1 & 0 & 0 \\ 0 & 0 & 0 & 1 \end{pmatrix}, \quad A_2^1 = \begin{pmatrix} c_2 & s_2 & 0 & 0 \\ 0 & 0 & 1 & 0 \\ s_2 & -c_2 & 0 & 0 \\ 0 & 0 & 0 & 1 \end{pmatrix},$$

and

$$A_3^2 = \begin{pmatrix} 1 & 0 & 0 & 0 \\ 0 & 1 & 0 & 0 \\ 0 & 0 & 1 & -d_3 \\ 0 & 0 & 0 & 1 \end{pmatrix}.$$

Then, we calculate

$$A_3^1 = A_3^2 A_2^1 = \begin{pmatrix} c_2 & s_2 & 0 & 0 \\ 0 & 0 & 1 & 0 \\ s_2 & -c_2 & 0 & -d_3 \\ 0 & 0 & 0 & 1 \end{pmatrix},$$

and

$$A_3^0 = A_3^1 A_1^0 = \begin{pmatrix} c_1c_2 & s_1c_2 & s_2 & -d_1s_2 \\ s_1 & -c_1 & 0 & 0 \\ c_1s_2 & s_1s_2 & -c_2 & d_1c_2 - d_3 \\ 0 & 0 & 0 & 1 \end{pmatrix}.$$

By noticing that joint 3 of the Stanford-Type robot is prismatic and all the others are revolute, it is now ready to determine the Jacobian matrix $J_{(3)}$ in the following form:

$$J_{(3)} = \begin{pmatrix} p_3^0 \times r_3^0 & p_3^1 \times r_3^1 & r_3^2 & 0_3 & 0_3 & 0_3 \\ r_3^0 & r_3^1 & 0_3 & r_3^3 & r_3^4 & r_3^5 \end{pmatrix}.$$

With a further cross-product calculation, we finally obtain

$$J_{(3)} = \begin{pmatrix} 0 & d_3 & 0 & 0 & 0 & 0 \\ -d_3s_2 & 0 & 0 & 0 & 0 & 0 \\ 0 & 0 & 1 & 0 & 0 & 0 \\ s_2 & 0 & 0 & 0 & -s_4 & c_4s_5 \\ 0 & 1 & 0 & 0 & c_4 & s_4s_5 \\ -c_2 & 0 & 0 & 1 & 0 & c_5 \end{pmatrix}. \tag{4.16}$$

Note that the above procedure to find a shortest symbolical Jacobian matrix $J_{(k)}$ at a middle frame k is valid if and only if there is no more length beyond frame k. In other words, if $d_6 \neq 0$ for the Stanford-Type robot, we have no other choice than finding $J_{(6)}$ first before re-projecting it to any other frame by (4.15).

Since the top 3 by 6 block of $J_{(3)}$ in (4.16) corresponds to the Cartesian linear velocity part $J_{v(3)}$, after it is re-projected on the base via $R_0^3 J_{v(3)} = J_{v(0)}$, one can alternatively determine it and match it by the mathematical Jacobian definition:

$$J_{v(0)} = \frac{\partial p_0^6}{\partial q}.$$

Due to the assumption $d_6 = 0$, the origins of the last three frames are merged to the common wrist center point so that $p_0^6 = p_0^3$ which is the last (fourth) column of A_0^3 given in (4.8), i.e.,

$$p_0^6 = p_0^3 = \begin{pmatrix} d_3 c_1 s_2 \\ d_3 s_1 s_2 \\ d_1 - d_3 c_2 \end{pmatrix}. \tag{4.17}$$

After taking partial derivative of p_0^3, the non-zero leftmost 3 by 3 block of $J_{v(0)}$ becomes

$$\frac{\partial p_0^3}{\partial q} = \begin{pmatrix} -d_3 s_1 s_2 & d_3 c_1 c_2 & c_1 s_2 \\ d_3 c_1 s_2 & d_3 s_1 c_2 & s_1 s_2 \\ 0 & d_3 s_2 & -c_2 \end{pmatrix}.$$

On the other hand, if we pre-multiply $J_{v(3)}$, the top 3 by 6 block of (4.16), by the rotation matrix

$$R_0^3 = \begin{pmatrix} c_1 c_2 & s_1 & c_1 s_2 \\ s_1 c_2 & -c_1 & s_1 s_2 \\ s_2 & 0 & -c_2 \end{pmatrix}, \tag{4.18}$$

we can immediately verify that the two results are identical. However, the above partial derivative approach is not applied to the bottom 3 by 6 block of (4.16), i.e. $J_{\omega(3)}$, which corresponds to the angular velocity ω.

In general, for a robot with n joints, the first half of its Jacobian matrix, $J_{v(k)}$ that is 3 by n and corresponding to the linear velocity v, determined by (4.14) through (4.15) and projected onto frame k for $0 \le k \le n$ is always related to the partial derivative of the position vector by

$$\frac{\partial p_0^n}{\partial q} = R_0^k J_{v(k)}. \tag{4.19}$$

Therefore, in addition to (4.14) and (4.15), directly taking partial derivative on the position vector p_0^n that must be referred to and projected onto the fixed base with respect to the joint position vector q is an alternative way to find J_v, but not for J_ω at all.

If we add an offset d_2 between the centers of joint 1 and joint 2 for the above Stanford-Type arm, then, it immediately returns back to the original Stanford robot. Using the same algorithm, the Jacobian matrix of the Stanford robot can be found as

$$J_{(3)} = \begin{pmatrix} d_2 c_2 & d_3 & 0 & 0 & 0 & 0 \\ -d_3 s_2 & 0 & 0 & 0 & 0 & 0 \\ d_2 s_2 & 0 & 1 & 0 & 0 & 0 \\ s_2 & 0 & 0 & 0 & -s_4 & c_4 s_5 \\ 0 & 1 & 0 & 0 & c_4 & s_4 s_5 \\ -c_2 & 0 & 0 & 1 & 0 & c_5 \end{pmatrix}, \tag{4.20}$$

which is also projected to frame 3, and needs to be pre-multiplied by

$$\begin{pmatrix} R_0^3 & O \\ O & R_0^3 \end{pmatrix}$$

in order to get $J_{(0)}$ that is projected onto the base.

Once a Jacobian matrix projected on any desired frame is developed in symbolical form for a given robot, it will be used forever in future numerical computations. One of the most typical applications for a Jacobian matrix is to play a pivotal role in robotic **differential motion** planning.

A robot arm is often digitally controlled by updating every joint position at each sampling point during the course of trajectory-tracking. In addition to the Cartesian motion and joint motion planning methods that have been introduced in the previous section, we can also rely on directly calculating differential increments of all the joint values to control the robotic hand to follow a given desired trajectory. Then, the desired trajectory has to be represented in tangent space, such as in the form of Cartesian velocity $V = \begin{pmatrix} v \\ \omega \end{pmatrix}$.

Let $q(t) = (\theta_1(t) \ \cdots \ \theta_n(t))^T$ be the instantaneous joint position vector for an n-joint robot arm during the motion. Obviously, the updated joint position vector after a Δt time interval is given by the following Taylor series:

$$q(t + \Delta t) = q(t) + \dot{q}(t)\Delta t + \frac{1}{2!}\ddot{q}(t)\Delta t^2 + \cdots.$$

If we consider the first-order approximation and notice that $\dot{q} = J^{-1}V$, then,

$$q(t + \Delta t) \approx q(t) + J^{-1}V\Delta t, \tag{4.21}$$

where the Jacobian matrix J evaluated at $q(t)$ is assumed to be invertible and both J and the Cartesian velocity vector V must be projected on the same reference frame. For an acceptable tracking accuracy, the sampling time interval Δt in (4.21) must be very small. Otherwise, we have to consider the following second-order approximation:

$$q(t + \Delta t) \approx q(t) + J^{-1}V\Delta t + \frac{1}{2}J^{-1}(\dot{V} - \dot{J}J^{-1}V)\Delta t^2. \tag{4.22}$$

The second-order approximation does offer much better tracking accuracy than the first-order one under the same Δt, but it requires to calculate $\dot{J}$ and to know the Cartesian acceleration vector $\dot{V}$. It is always possible to calculate $\dot{J}$ as long as J is available symbolically, even though it is tedious. However, to know both v and $\dot{v}$ of the translation part for a given trajectory, it is better to have the trajectory represented in a parametric equation.

For example, if the desired path is a planar circle centered at $(x_0,\ y_0)$ with a radius r on the $z = 0$ plane, instead of using the Cartesian circular equation $(x - x_0)^2 + (y - y_0)^2 = r^2$, we may employ the following equivalence:

$$\begin{cases} x(t) = x_0 + r\cos(\beta t) \\ y(t) = y_0 + r\sin(\beta t), \end{cases}$$

where β is a parameter to be specified as the tracking speed. In this case, the Cartesian linear velocity turns out to be

$$v = \begin{pmatrix} \dot{x} \\ \dot{y} \\ \dot{z} \end{pmatrix} = \begin{pmatrix} -r\beta\sin(\beta t) \\ r\beta\cos(\beta t) \\ 0 \end{pmatrix},$$

and further we can have the Cartesian linear acceleration by directly taking time-derivative of v.

However, to determine a parametric equation for the rotation part, we have to utilize the $k - \phi$ procedure, as discussed in Chapter 2, to determine the non-integrable angular velocity ω. Namely, for any desired orientation change R, first, find the corresponding k and ϕ by equations (2.9) and (2.11), and second, uniformly sample the angle ϕ to determine an increment $\Delta\phi = \phi/N$ in each sampling interval Δt, where N is the total number of sampling points. Then, the angular velocity for the desired orientation change is

$$\omega = \dot{\phi}k \approx \frac{\Delta\phi}{\Delta t}k,$$

which, of course, is the first-order approximation, too.

A symbolical form of Jacobian matrix will also be quite useful for a robot arm to determine its kinematic singularity. For a 6-joint robot, its Jacobian matrix J is 6 by 6, and all possible singular configurations are the solutions of the following equation:

$$\det(J) = 0, \tag{4.23}$$

and based on (4.15), all the solutions of (4.23) are independent of the projection frame of the Jacobian matrix J, because the determinant of any rotation matrix is always $+1$.

A singular configuration of the robot means that some of the 6 d.o.f. is (are) lost. Since solving the inverse kinematics in differential motion always requires inverting the Jacobian matrix, a vanishing determinant of the Jacobian matrix is clearly the sufficient and necessary condition of singularity. In the case of non-square Jacobian matrices, such as a redundant robotic system, where the number of joints exceeds the number of the Cartesian d.o.f., equation (4.23) has to be modified to

$$\det(JJ^T) = 0. \tag{4.24}$$

For example, we have just derived a symbolical form of the Jacobian matrix $J_{(3)}$ given in (4.16) for the 6-joint Stanford-Type robot. By taking advantage of the zero upper-right 3 by 3 corner of $J_{(3)}$, we can readily calculate $\det(J_{(3)}) = -d_3^2 s_2 s_5$. Since the prismatic joint value d_3 is never zero in general cases, either $\sin\theta_2 = 0$ or $\sin\theta_5 = 0$ is the singular point of this particular robot. We can further observe that $\theta_2 = 0$ or 180^0 is a "hanging hand down" or "raising hand up" special configuration, where the robot cannot move forward in any horizontal radial direction from the base origin. Moreover, if $\theta_5 = 0$, the z_3 and z_5 axes are lined up so that both the θ_4 and θ_6 rotations will be about the same axis and thus, will no longer be independent to each other. In other words, one of the three rotational d.o.f. is lost.

Intuitively, a redundant robotic system may directly reduce the possibility of singularity, but still be hard to completely eliminate the singularity, in particular the rotational singularity, such as in the case of $\theta_5 = 0$ for the Stanford-Type robot. To indirectly avoid a singular point, often one has to sacrifice the accuracy of path tracking before the singular point is reached. Unfortunately, most singular cases are unpredictable, and often, it would be too late to escape from the dead configuration once it occurs. In the next chapter, we will introduce and study this in more detail on redundant robotic systems. By taking advantage of the kinematic redundancy, we may have more room to control the robot configurations to always stay away from the singularity.

4.5 Dual-Number Transformations

In addition to the homogeneous transformation, a special orthogonal dual matrix can also play a vital role in representing robot kinematic models and motions. As introduced in Chapter 2, a dual matrix $\hat{R} = R + \epsilon S$ is said to be a special orthogonal dual matrix if $\hat{R}^T\hat{R} = I$ and $\det(R) = +1$. This implies that its real part $R \in SO(3)$ and its dual part obeys that SR^T must be skew-symmetric. We also recall that to make this happen, the dual part S can be constructed by augmenting three moment vectors, each of which is $s_i = p \times r_i$, where r_i is the unit vector of the i-th axis of a coordinate frame for $i = 1, 2, 3$ and p is the radial position vector for the origin of the frame. If the skew-symmetric matrix of this radial position vector p is $P = p\times$, then it is clear that $SR^T = P$ or $S = PR$.

However, during the robot kinematic modeling phase as a D-H table is determined, we want a special orthogonal dual matrix $\hat{R} = R + \epsilon S$ to be automatically constructed to uniquely represent both the position and orientation of each robotic link frame with respect to a reference frame, as an alternative but independent approach. If one just picks the matrix R from the homogeneous matrix A as the real part of $\hat{R}$ while using the last column p of A to find the dual part $S = PR$, then, the dual-number representation becomes meaningless.

In fact, it can be further observed from a D-H table that among four parameters at each link/joint level, the joint angle θ_i and the joint offset d_i can form a screw motion, because they both are acting about/along the common axis z_{i-1} of the $(i-1)$-st link frame. Likewise, both the twist angle

α_i and the link length a_i also act about/along the same axis x_i of the i-th link frame, and thus form another screw motion. Therefore, using the dual-number notation, each link/joint kinematic relation in a D-H table can be determined by two screw motions:

$$\hat{\theta}_i = \theta_i + \epsilon d_i \quad \text{and} \quad \hat{\alpha}_i = \alpha_i + \epsilon a_i.$$

This can be considered as another major credit owned by the D-H convention to add as an additional advantage [6].

Based on this notion, let us now state and show the following important theorem to demonstrate how a dual-number transformation is created once a D-H table is determined for a given robotic system.

Theorem 1. *Let R_k^n be a rotation matrix of frame n referred to frame k, which is a function of all related joint angles θ_i and twist angles α_i between frames k and n under the D-H convention. If each θ_i and each α_i are replaced by their corresponding dual screw motion angles $\hat{\theta}_i = \theta_i + \epsilon d_i$ and $\hat{\alpha}_i = \alpha_i + \epsilon a_i$, respectively, then the resultant dual rotation matrix $\hat{R}_k^n = R_k^n + \epsilon S_k^n$ has the real part equal to the given rotation matrix R_k^n and its dual part $S_k^n = P_k^n R_k^n$, where $P_k^n = p_k^n \times$ is the skew-symmetric matrix of the position vector from the origin of frame k to the origin of frame n.*

Proof. First, let each angle θ_i and α_i in R_k^n be replaced by $\theta_i + \epsilon\dot{\theta}_i$ and $\alpha_i + \epsilon\dot{\alpha}_i$, respectively. Based on equation (2.19), the resulting dual rotation matrix will become $\hat{R}_k^n = R_k^n + \epsilon\dot{R}_k^n$. Since $\dot{R}_k^n R_n^k = \Omega_k^n$, the skew-symmetric matrix of the angular velocity ω_k^n for frame n with respect to frame k, clearly, the dual part is $\dot{R}_k^n = \Omega_k^n R_k^n$.

According to the angular velocity formula in equation (4.12) or (4.13) for a robot kinematic model under the D-H convention, a more general equation of determining an angular velocity due to both joint angle and twist angle variations can be written as follows:

$$\omega_k^n = \sum_{i=k+1}^{n} \dot{\theta}_i r_{zk}^{i-1} + \dot{\alpha}_i r_{xk}^i, \tag{4.25}$$

where r_{zk}^{i-1} is the 3rd column of the rotation matrix R_k^{i-1} to represent the z-axis of frame $i-1$ referred to frame k, and r_{xk}^i is the first column of the rotation matrix R_k^i to represent the x-axis of frame i referred to frame k. If we replace each of $\dot{\theta}_i$ and $\dot{\alpha}_i$ in (4.25) by the corresponding joint offset d_i and link length a_i, respectively, then based on the D-H convention, it becomes exactly the position vector for the robot,

$$p_k^n = \sum_{i=k+1}^{n} d_i r_{zk}^{i-1} + a_i r_{xk}^i, \tag{4.26}$$

Thus, under each $\dot{\theta}_i = d_i$ and $\dot{\alpha}_i = a_i$, the skew-symmetric matrix for the position vector: $P_k^n = \Omega_k^n$.

Now, returning back to the first step of the proof, the above condition is equivalent to replacing each θ_i and α_i in the original rotation matrix R_k^n by the two screw motion angles $\hat{\theta}_i = \theta_i + \epsilon d_i$ and $\hat{\alpha}_i = \alpha_i + \epsilon a_i$, and $S_k^n R_n^k = \dot{R}_k^n R_n^k = \Omega_k^n = P_k^n$ under the conditions $\dot{\theta}_i = d_i$ and $\dot{\alpha}_i = a_i$ for each $i = k+1, \cdots, n$. Therefore, $S_k^n = P_k^n R_k^n$. □

To implement the replacement suggested by Theorem 1, we need to modify and reduce the one-step homogeneous matrix in (4.1) by taking its upper-left 3 by 3 corner:

$$R_{i-1}^i = \begin{pmatrix} c_i & -s_i & 0 \\ s_i & c_i & 0 \\ 0 & 0 & 1 \end{pmatrix} \begin{pmatrix} 1 & 0 & 0 \\ 0 & c\alpha_i & -s\alpha_i \\ 0 & s\alpha_i & c\alpha_i \end{pmatrix} = \begin{pmatrix} c_i & -s_i c\alpha_i & s_i s\alpha_i \\ s_i & c_i c\alpha_i & -c_i s\alpha_i \\ 0 & s\alpha_i & c\alpha_i \end{pmatrix}. \tag{4.27}$$

Then, replacing θ_i by $\hat{\theta}_i = \theta_i + \epsilon d_i$ and α_i by $\hat{\alpha}_i = \alpha_i + \epsilon a_i$, the one-step dual rotation matrix $\hat{R} = R + \epsilon S$ is immediately achieved.

To illustrate finding a dual rotation matrix via the replacement of angles, let us look back at the Stanford-Type robot arm again. Recalling the rotation matrix of frame 3 with respect to the base that was given in (4.18) by

$$R_0^3 = \begin{pmatrix} c_1 c_2 & s_1 & c_1 s_2 \\ s_1 c_2 & -c_1 & s_1 s_2 \\ s_2 & 0 & -c_2 \end{pmatrix},$$

we have to recover all three joint and twist angles from this already simplified version. First, according to its D-H table, the three pairs of dual joint angles and dual twist angles for the first three links/joints of the robot can be given as

$$\hat{\theta}_1 = \theta_1 + \epsilon d_1, \quad \hat{\alpha}_1 = 90^0 + \epsilon 0 = 90^0,$$
$$\hat{\theta}_2 = \theta_2 + \epsilon 0 = \theta_2, \quad \hat{\alpha}_1 = 90^0 + \epsilon 0 = 90^0,$$
$$\hat{\theta}_3 = \theta_3 + \epsilon d_3 = \epsilon d_3, \quad \hat{\alpha}_1 = 0 + \epsilon 0 = 0.$$

Secondly, based on (4.27),

$$\hat{R}_0^1 = \begin{pmatrix} \hat{c}_1 & 0 & \hat{s}_1 \\ \hat{s}_1 & 0 & -\hat{c}_1 \\ 0 & 1 & 0 \end{pmatrix}, \quad \hat{R}_1^2 = \begin{pmatrix} \hat{c}_2 & 0 & \hat{s}_2 \\ \hat{s}_2 & 0 & -\hat{c}_2 \\ 0 & 1 & 0 \end{pmatrix},$$

$$\text{and} \quad \hat{R}_2^3 = \begin{pmatrix} \hat{c}_3 & -\hat{s}_3 & 0 \\ \hat{s}_3 & \hat{c}_3 & 0 \\ 0 & 0 & 1 \end{pmatrix},$$

where $\hat{c}_1 = \cos\hat{\theta}_1 = c_1 - \epsilon d_1 s_1$, $\hat{s}_1 = \sin\hat{\theta}_1 = s_1 + \epsilon d_1 c_1$, $\hat{c}_2 = c_2$, $\hat{s}_2 = s_2$, $\hat{c}_3 = c_3 = 1$ and $\hat{s}_3 = \epsilon d_3$. All the dual twist angles have also been taken care of in this special case with all $a_i = 0$ for $i = 1, 2, 3$.

Finally, after calculating $\hat{R}_0^3 = \hat{R}_0^1 \hat{R}_1^2 \hat{R}_2^3$, taking its dual part will yield the dual part of $\hat{R}_0^3$ as follows:

$$S_0^3 = \begin{pmatrix} -d_1 s_1 c_2 + d_3 s_1 & d_1 c_1 - d_3 c_1 c_2 & -d_1 s_1 s_2 \\ d_1 c_1 c_2 - d_3 c_1 & d_1 s_1 - d_3 s_1 c_2 & d_1 c_1 s_2 \\ 0 & d_3 s_2 & 0 \end{pmatrix}.$$

It can be directly verified that the skew-symmetric matrix of the position vector p_0^3 is

$$P_0^3 = S_0^3 R_3^0 = \begin{pmatrix} 0 & -d_1 + d_3 c_2 & d_3 s_1 s_2 \\ d_1 - d_3 c_2 & 0 & -d_3 c_1 s_2 \\ -d_3 s_1 s_2 & d_3 c_1 s_2 & 0 \end{pmatrix},$$

which is identical to (4.17).

Although the computational advantage is not quite clear in the above symbolical derivation for finding a dual rotation matrix, once MATLAB^{TM} could create a new internal subroutine that covers the dual-number calculus, the dual-number transformation would provide an exclusive advantage to numerical computations.

Let us look at an industrial robot model example, as shown in Figure 4.10, including all the link coordinate frames assignment based on the D-H convention. The D-H parameter table in a regular version can be constructed in Table 4.4. In contrast, by adopting the dual-number transformation, the D-H dual parameter table is given in Table 4.5.

Table 4.4 A regular D-H table for the industrial robot model

Joint Number	**Joint Angle** θ_i	**Joint Offset** d_i	**Twist Angle** α_i	**Link Length** a_i
1	θ_1	d_1	-90^0	a_1
2	θ_2	0	0	a_2
3	θ_3	0	-90^0	a_3
4	θ_4	d_4	90^0	0
5	θ_5	0	-90^0	0
6	θ_6	d_6	0	0

Thus, in terms of the dual angles, the dual D-H table has only two parameters for each joint/link versus four parameters in the regular D-H table. Based on the dual D-H table, all the six one-step special orthogonal dual matrices can be derived as follows:

$$\hat{R}_0^1 = \begin{pmatrix} \hat{c}_1 & -\epsilon a_1 s_1 & -\hat{s}_1 \\ \hat{s}_1 & \epsilon a_1 c_1 & \hat{c}_1 \\ 0 & -1 & \epsilon a_1 \end{pmatrix}, \quad \hat{R}_1^2 = \begin{pmatrix} c_2 & -s_2 & \epsilon a_2 s_2 \\ s_2 & c_2 & -\epsilon a_2 c_2 \\ 0 & \epsilon a_2 & 1 \end{pmatrix},$$

$$\hat{R}_2^3 = \begin{pmatrix} c_3 & -\epsilon a_3 s_3 & -s_3 \\ s_3 & \epsilon a_3 c_3 & c_3 \\ 0 & -1 & \epsilon a_3 \end{pmatrix}, \quad \hat{R}_3^4 = \begin{pmatrix} \hat{c}_4 & 0 & \hat{s}_4 \\ \hat{s}_4 & 0 & -\hat{c}_4 \\ 0 & 1 & 0 \end{pmatrix},$$

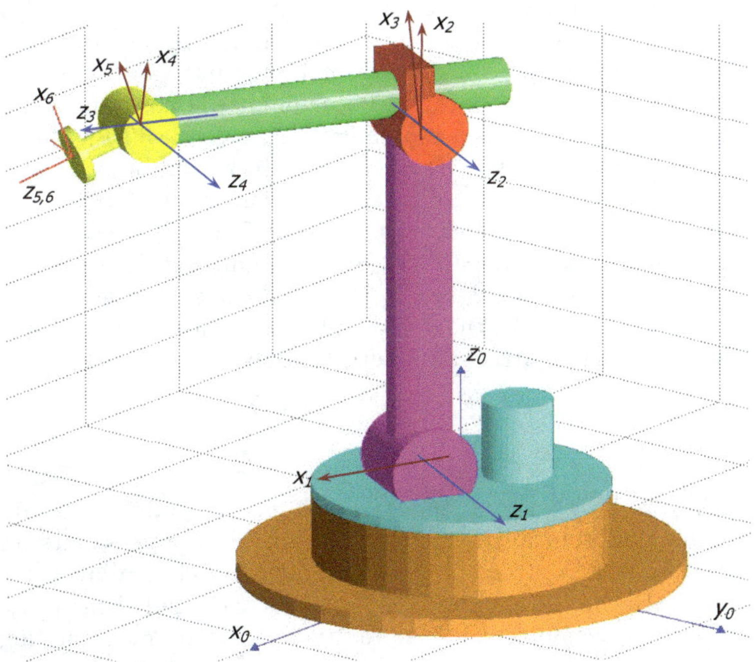

Fig. 4.10 An industrial robot model with coordinate frames assignment

Table 4.5 A dual D-H table for the industrial robot model

Joint Number	Dual Joint Angle $\hat{\theta}_i = \theta_i + \epsilon d_i$	Dual Twist Angle $\hat{\alpha}_i = \alpha_i + \epsilon a_i$
1	$\theta_1 + \epsilon d_1$	$-90^0 + \epsilon a_1$
2	θ_2	ϵa_2
3	θ_3	$-90^0 + \epsilon a_3$
4	$\theta_4 + \epsilon d_4$	90^0
5	θ_5	-90^0
6	$\theta_6 + \epsilon d_6$	0

and

$$\hat{R}_4^5 = \begin{pmatrix} c_5 & 0 & -s_5 \\ s_5 & 0 & c_5 \\ 0 & -1 & 0 \end{pmatrix}, \quad \hat{R}_5^6 = \begin{pmatrix} \hat{c}_6 & -\hat{s}_6 & 0 \\ \hat{s}_6 & \hat{c}_6 & 0 \\ 0 & 0 & 1 \end{pmatrix},$$

where $\hat{c}_i = c_i - \epsilon d_i s_i$ and $\hat{s}_i = s_i + \epsilon d_i c_i$ for each $i = 1, \cdots, 6$. With all the one-step 3 by 3 dual-number transformations prepared, any step dual-number transformation matrix $\hat{R}_i^j$ can be determined by their multiplications.

Therefore, when modeling a robot using the dual-number transformation, each transformation is a 3 by 3 special orthogonal dual matrix $\hat{R}_i^j \in SO(3, d)$, instead of 4 by 4 for a homogeneous transformation. However, you have to handle dual-number manipulations during the computation.

Furthermore, as discussed earlier, a Jacobian matrix has to be constructed for a robot via equation (4.14), especially for a numerical solution in many application cases. It involves a lot of cross-product operations. Now, we can take advantage of the dual part of every dual-number transformation matrix, which is formed as a moment of each axis of the frame. In other words, since for a dual rotation matrix $\hat{R} = R + \epsilon S$, each column of its dual part $S = PR$ is already the cross-product $s_i = p \times r_i$ for the corresponding column r_i in the real part R. Therefore, if the dual-number transformation is adopted, the formation of Jacobian matrix can be immediately reduced to

$$J_{(n)} = \begin{pmatrix} s_n^0 & \cdots & s_n^{n-1} \\ r_n^0 & \cdots & r_n^{n-1} \end{pmatrix}, \tag{4.28}$$

where each 3 by 1 vector s_n^i is the third column of the dual part S_n^i of the corresponding dual rotation matrix $\hat{R}_n^i$ if all the joints are revolute. In other words, $\hat{r}_n^i = r_n^i + \epsilon s_n^i$ is the third column of the dual rotation matrix $\hat{R}_n^i$, and we just take the third column out from each $\hat{R}_n^i$ and then split its real and dual parts to form the Jacobian matrix. Therefore, manipulating dual-number transformation matrices, instead of homogeneous transformations, will speed up the computation to determine the Jacobian matrix $J_{(n)}$ both symbolically and numerically for a given robotic system.

Finally, as discussed in Chapter 2 on the useful $k - \phi$ procedure, the conversion between a rotation matrix R and the corresponding unit vector k and angle ϕ is given by equation (2.8), i.e.,

$$R = I + \sin\phi K + (1 - \cos\phi)K^2.$$

If we now replace the rotating angle ϕ by a dual angle $\hat{\phi} = \phi + \epsilon d$, where the dual part d is a translation along the unit vector k so that $\hat{\phi}$ represents a screw motion about/along the unit vector k, then, the rotation matrix will immediately become a special orthogonal dual matrix $\hat{R} \in SO(3, d)$, i.e.,

$$\hat{R} = I + \sin(\phi + \epsilon d)K + (1 - \cos(\phi + \epsilon d))K^2$$

$$= I + \sin\phi K + (1 - \cos\phi)K^2 + \epsilon(d\cos\phi K + d\sin\phi K^2) = R + \epsilon S.$$

Recalling the properties: $K^3 = -K$ and $K^4 = -K^2$ from Chapter 2, we can readily prove that

$$P = SR^T = d(\cos\phi K + \sin\phi K^2)(I - \sin\phi K + (1-\cos\phi)K^2) = d \cdot K.$$

This means that the skew-symmetric matrix P for the translation is just equal to $d \cdot K$, i.e., shifting the frame a distance d along the unit vector k in addition to rotating the frame by ϕ about the same k.

4.6 Robotic Statics

After developing a general equation to construct a Jacobian matrix in (4.14) or (4.28) that is referred to the last frame for an n-joint open serial-chain robot, we now study its *duality*. Suppose that a Cartesian force/moment, commonly called a *wrench*, is applied on the robot end-effector. It is conceivable that to balance against the external force/moment, each joint of the robot has to produce an extra torque. Let the extra joint torque vector be $\tau = (\tau_1 \ \cdots \ \tau_n)^T$. Then, the total mechanical power at a time instant is $P = \tau^T \dot{q}$ that is seen in joint space. The same power can also be calculated in Cartesian space by $F_{(n)}^T V_{(n)}$, where the 6 by 1 Cartesian force/moment column vector (wrench) is defined by

$$F_{(n)} = (f_x \ f_y \ f_z \ m_x \ m_y \ m_z)^T \tag{4.29}$$

with three components for force and three components for Cartesian moment (or Cartesian torque), and $V_{(n)} = \begin{pmatrix} v_{(n)} \\ \omega_{(n)} \end{pmatrix}$ is the 6 by 1 Cartesian velocity, as defined in (4.13). Both $F_{(n)}$ and $V_{(n)}$ are referred to the last frame of the robot. Based on the principle of energy conservation, the above two power forms seen in different spaces must be equal to each other so that $\tau^T \dot{q} = F_{(n)}^T V_{(n)}$. By substituting the Jacobian equation (3.22) into it, and noticing that this power equation is always true for any $\dot{q}$, we finally arrive at

$$\tau = J_{(n)}^T F_{(n)}. \tag{4.30}$$

This is a duality equation of the kinematic Jacobian equation $V = J\dot{q}$, and is referred to as the **robotic statics** [2]–[5].

Note that the Cartesian force/moment vector $F_{(n)}$ must physically act on frame n, i.e., the last frame of the robot arm. However, it may be re-projected to any of the robotic link frames, say frame k, by the following mapping:

$$F_{(k)} = \begin{pmatrix} R_k^n & O_{3\times 3} \\ O_{3\times 3} & R_k^n \end{pmatrix} F_{(n)}.$$

Most typically $k = 0$, i.e., the Cartesian force/moment vector $F_{(0)}$ is projected onto the base. If the Cartesian force/moment vector is re-projected onto frame

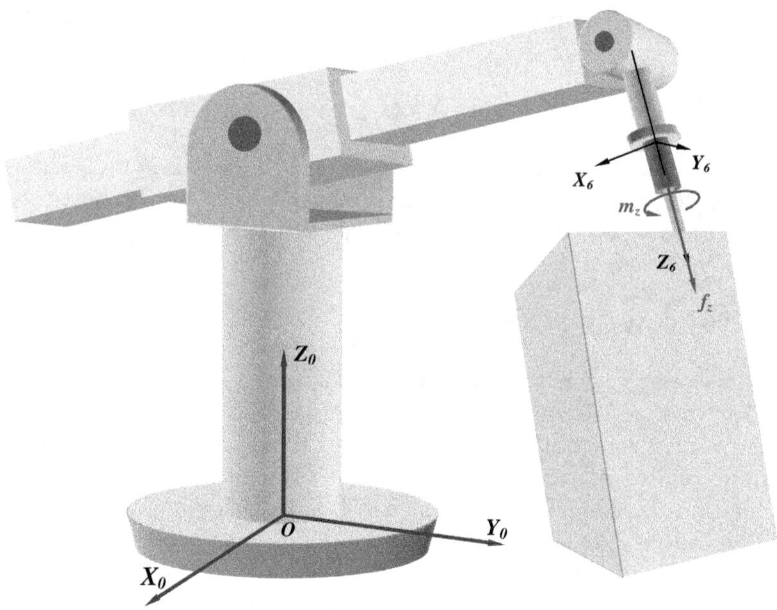

Fig. 4.11 The Stanford-type robot is driving a screw into a workpiece

k, the Jacobian matrix must be re-projected onto the same frame accordingly via equation (4.15). In other words, the statics equation (4.30) must have the same projection frame between J and F.

A real example is depicted in Figure 4.11, where the end-effector of the Stanford-Type robot arm carries a screw driver to work on a workpiece. The driving force f_z and moment m_z are all referred to frame 6 of the robot. Thus, the Cartesian force/moment vector should be

$$F_{(6)} = (0 \;\; 0 \;\; f_z \;\; 0 \;\; 0 \;\; m_z)^T.$$

With the Jacobian matrix $J_{(3)}$ of the Stanford-Type robot in equation (4.16), by re-projecting it to frame 6, i.e.,

$$J_{(6)} = \begin{pmatrix} R_6^3 & O_{3\times 3} \\ O_{3\times 3} & R_6^3 \end{pmatrix} J_{(3)},$$

we can immediately calculate the extra static joint torque vector $\tau = J_{(6)}^T F_{(6)}$.

Note that since f_z in this example has a distance away from the base origin, do not count the torque $p_0^6 \times (0 \;\; 0 \;\; f_z)^T$ into the Cartesian moment vector. It must be clarified and emphasized that the Cartesian force/moment is always acting on and referred to frame 6, the last frame, instead of the base!

The equation of robotic statics in (4.30) will be very useful. In particular, the statics equation reveals a distribution of all the joint torques whenever a Cartesian force/moment is imposed on the robotic hand, and this joint torque distribution is clearly a function of the instantaneous configuration (or posture) of the robot arm, which is uniquely represented by the Jacobian matrix.

Let us examine another example that illustrates how a Cartesian force/ moment vector $F_{(n)}$ is determined, and associate it with the robotic statics. This robot is a 3-joint RRR type arm with all three revolute joints, as shown in Figure 4.12. A simple pendulum hangs down from the robotic tip point w with its string length $l = 0.6$ m. and the point mass $m = 0.5$ Kg.

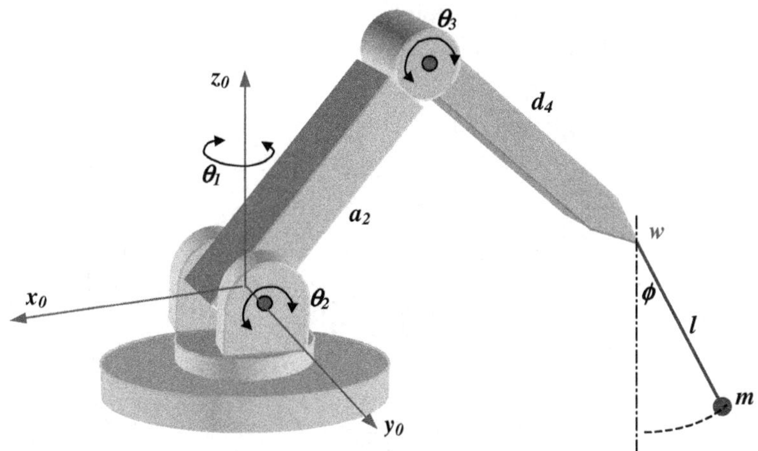

Fig. 4.12 A 3-joint RRR robot hanging a simple pendulum

We now ask the following two robotic statics questions:

1. If the robot arm is currently motionless at $\theta_1 = 0$, and the pendulum swings on the $z_0 - x_0$ plane of the base and reaches the maximum swing angle $\phi = 30^0$ at a time instant, find all three joint torques;
2. Also, determine all three joint torques when the point mass is swinging down to just pass through the $\phi = 0$ position.

First, the 3-joint robot can be modeled by the D-H convention to have a parameter D-H table in Table 4.6, where $a_2 = 1$ and $d_4 = 0.8$ in meters. In the D-H table, we purposely define four joint angles in order to extend the origin of the last frame reaching to the robotic tip point w, but only the first three joint angles are variables in the kinematic model with $\theta_4 = 0$, i.e., the joint position vector $q = (\theta_1\ \theta_2\ \theta_3)^T$.

Table 4.6 A D-H table for the 3-Joint RRR robot arm

Joint Angle	Joint Offset	Twist Angle	Link Length
θ_i	d_i	α_i	a_i
θ_1	0	-90^0	0
θ_2	0	0	a_2
θ_3	0	90^0	0
$\theta_4 = 0$	d_4	0	0

Based on the D-H table, all four one-step homogeneous transformation matrices can be determined below:

$$A_0^1 = \begin{pmatrix} c_1 & 0 & -s_1 & 0 \\ s_1 & 0 & c_1 & 0 \\ 0 & -1 & 0 & 0 \\ 0 & 0 & 0 & 1 \end{pmatrix}, \quad A_1^2 = \begin{pmatrix} c_2 & -s_2 & 0 & a_2 c_2 \\ s_2 & c_2 & 0 & a_2 s_2 \\ 0 & 0 & 1 & 0 \\ 0 & 0 & 0 & 1 \end{pmatrix},$$

$$A_2^3 = \begin{pmatrix} c_3 & 0 & s_3 & 0 \\ s_3 & 0 & -c_3 & 0 \\ 0 & 1 & 0 & 0 \\ 0 & 0 & 0 & 1 \end{pmatrix}, \quad A_3^4 = \begin{pmatrix} 1 & 0 & 0 & 0 \\ 0 & 1 & 0 & 0 \\ 0 & 0 & 1 & d_4 \\ 0 & 0 & 0 & 1 \end{pmatrix}.$$

Thus,

$$A_0^4 = A_0^1 A_1^2 A_2^3 A_3^4 = \begin{pmatrix} c_1 c_{23} & -s_1 & c_1 s_{23} & a_2 c_1 c_2 + d_4 c_1 s_{23} \\ s_1 c_{23} & c_1 & s_1 s_{23} & a_2 s_1 c_2 + d_4 s_1 s_{23} \\ -s_{23} & 0 & c_{23} & -a_2 s_2 + d_4 c_{23} \\ 0 & 0 & 0 & 1 \end{pmatrix},$$

where $c_i = \cos\theta_i$, $s_i = \sin\theta_i$, $c_{23} = \cos(\theta_2 + \theta_3)$ and $s_{23} = \sin(\theta_2 + \theta_3)$.

It is obvious that the last column of A_0^4 is the position vector p_0^w of the robotic tip point w with respect to the base. Then, the Jacobian matrix projected onto the base can be found by

$$J_{(0)} = \frac{\partial p_0^w}{\partial q} = \begin{pmatrix} -a_2 s_1 c_2 - d_4 s_1 s_{23} & -a_2 c_1 s_2 + d_4 c_1 c_{23} & d_4 c_1 c_{23} \\ a_2 c_1 c_2 + d_4 c_1 s_{23} & -a_2 s_1 s_2 + d_4 s_1 c_{23} & d_4 s_1 c_{23} \\ 0 & -a_2 c_2 - d_4 s_{23} & -d_4 s_{23} \end{pmatrix}. \tag{4.31}$$

If the joint angles are set to be $\theta_1 = 0$, $\theta_2 = -120^0$ and $\theta_3 = 0$, then, the Jacobian matrix becomes

$$J_{(0)} = \begin{pmatrix} 0 & 0.4660 & -0.4000 \\ -1.1928 & 0 & 0 \\ 0 & 1.1928 & 0.6928 \end{pmatrix}.$$

To answer the first question, at the time instant $\phi = 30^0$, the point mass is acted on by the gravity force mg vertically downward, where $g = 9.81$

m./sec^2. Since the string is soft, the robotic tip point w is exerted by a pulling strain force that is passed through the soft string only by the component $mg\cos\phi$, instead of the full gravity force mg. Thus, the Cartesian force that is projected onto the base frame should be

$$f_{(0)} = \begin{pmatrix} -mg\cos\phi\sin\phi \\ 0 \\ -mg\cos^2\phi \end{pmatrix}.$$

Because this particular robot does not include a wrist, it cannot offer a balance to any Cartesian moment. Therefore, the 6 by 1 Cartesian force/moment vector is downsized to the 3 by 1 $f_{(0)}$. After substituting the parameters and $\phi = 30^0$,

$$f_{(0)} = \begin{pmatrix} -2.1239 \\ 0 \\ -3.6788 \end{pmatrix} \quad \text{in Newtons,}$$

such that the robot joint torque distribution turns out to be

$$\tau = J_{(0)}^T f_{(0)} = \begin{pmatrix} 0 \\ -5.3779 \\ -1.6991 \end{pmatrix}.$$

To answer the second question, although at the time instant $\phi = 0$, the full gravity mg will pass to the robotic tip point w through the string, since the point mass has also a linear speed v, the tip point of the robot arm will be exerted by an additional centrifugal force $f_c = mv^2/l$. Based on the energy conservation, since the potential energy difference between $\phi = 30^0$ and $\phi = 0$ is determined by $\Delta P = mgl(1 - \cos\Delta\phi) = mgl(1 - \cos(30^0))$, the kinetic energy difference $\Delta K = \frac{1}{2}mv^2 - 0 = \Delta P = mgl(1 - \cos(30^0))$ so that the centrifugal force

$$f_c = \frac{mv^2}{l} = 2mg(1 - \cos(30^0)).$$

Hence, the Cartesian force at the $\phi = 0$ moment should be

$$f_{(0)} = \begin{pmatrix} 0 \\ 0 \\ -mg - 2mg(1 - \cos(30^0)) \end{pmatrix} = \begin{pmatrix} 0 \\ 0 \\ -6.2193 \end{pmatrix},$$

and the joint torque distribution of the robot becomes

$$\tau = J_{(0)}^T f_{(0)} = \begin{pmatrix} 0 \\ -7.4185 \\ -4.3089 \end{pmatrix}.$$

Suppose now the joint angles are changed to be $\theta_1 = 0$, $\theta_2 = -100^0$ and $\theta_3 = 30^0$, and the Cartesian force remains the same in both cases. The value of the Jacobian matrix will be changed to

$$J_{(0)} = \begin{pmatrix} 0 & 1.2584 & 0.2736 \\ -0.9254 & 0 & 0 \\ 0 & 0.9254 & 0.7518 \end{pmatrix},$$

and the new joint torque distribution in the case of $\phi = 30^0$ is

$$\tau = J_{(0)}^T f_{(0)} = \begin{pmatrix} 0 \\ -6.0771 \\ -3.3467 \end{pmatrix},$$

and in the case of $\phi = 0$ is

$$\tau = J_{(0)}^T f_{(0)} = \begin{pmatrix} 0 \\ -5.7553 \\ -4.6754 \end{pmatrix}.$$

Therefore, the extra joint torque distribution of a robot due to a Cartesian force/moment acting on the robotic hand fully depends on the robotic configuration (or posture) that is uniquely determined by the Jacobian matrix at a time instant, whenever the Cartesian force/moment is a constant.

Moreover, if the force acting point is specified anywhere along a robot arm other than at the robotic end-effector, such as point C in Figure 4.13, according to the formation of Jacobian matrix in (4.14), a Jacobian matrix of point C can always be constructed, as long as a new frame c is defined and originated at point C:

$$J_{(c)} = \begin{pmatrix} p_c^0 \times r_c^0 & \cdots & p_c^{j-1} \times r_c^{j-1} & \cdots & p_c^{c-1} \times r_c^{c-1} \\ r_c^0 & \cdots & r_c^{j-1} & \cdots & r_c^{c-1} \end{pmatrix}. \tag{4.32}$$

For our representation convenience, here we assume that every joint for the robot of interest is revolute as default. This Jacobian matrix of point C covers the robot arm from the base frame 0 up to the frame right before point C, and thus it may be called a *sub-Jacobian*. If point C is so picked that frame $c-1$ is frame k, $(0 < k < n)$, then the Jacobian matrix of point C is 6 by $k+1$. Such a sub-Jacobian matrix becomes very useful in finding a joint torque distribution if a force is applied at point C, instead of at the robot end-effector.

Suppose that a force $f \in \mathbb{R}^3$ and a Cartesian moment $m \in \mathbb{R}^3$ are both applied at point C. Then, the 6 by 1 Cartesian force vector F can be written as $F = \begin{pmatrix} f \\ m \end{pmatrix}$ and both f and m are projected on frame c. The joint torque distribution should obey the common robotic statics equation:

$$\tau^c = J_{(c)}^T F, \tag{4.33}$$

and the resulting joint torque vector τ^c is only $k+1$ by 1, not the full-size n by 1. In other words, this Cartesian force F acting at point C will not effect any joint above point C, or beyond frame k.

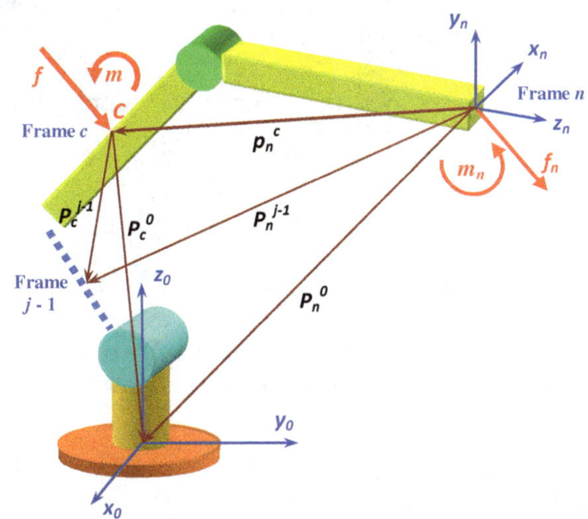

Fig. 4.13 A robot arm is exerted by a force f and a moment m at point C on the body

Furthermore, Figure 4.13 shows the relationship of all the position vectors tailed at frame c and the last end-effector frame n. At the origin of frame $j-1$ with $0 < j \leq k$, $p_n^{j-1} = p_{c(n)}^{j-1} + p_n^c$, where $p_{c(n)}^{j-1}$ is the vector p_c^{j-1} projected onto frame n. Also, at the origin of the base, $p_n^0 = p_{c(n)}^0 + p_n^c$. Substituting these position vector relations into the Jacobian matrix $J_{(n)}$ for frame n:

$$J_{(n)} = \begin{pmatrix} p_n^0 \times r_n^0 & \cdots & p_n^{j-1} \times r_n^{j-1} & \cdots & p_n^{n-1} \times r_n^{n-1} \\ r_n^0 & \cdots & r_n^{j-1} & \cdots & r_n^{n-1} \end{pmatrix},$$

the j-th column can be split into two columns:

$$\begin{pmatrix} p_n^{j-1} \times r_n^{j-1} \\ r_n^{j-1} \end{pmatrix} = \begin{pmatrix} p_{c(n)}^{j-1} \times r_n^{j-1} \\ r_n^{j-1} \end{pmatrix} + \begin{pmatrix} p_n^c \times r_n^{j-1} \\ 0_3 \end{pmatrix}.$$

We now seek the Cartesian force $F_n = \begin{pmatrix} f_n \\ m_n \end{pmatrix}$ to be applied on frame n that can be equivalent to the Cartesian force F applied at point C? Clearly, the j-th joint torque can be determined by

$$\tau_j = J_{(n)}(:, j)^T F_n = (p_{c(n)}^{j-1} \times r_n^{j-1})^T f_n + (r_n^{j-1})^T m_n + (p_n^c \times r_n^{j-1})^T f_n.$$

In the last (third) term, let the skew-symmetric matrix of p_n^c be $S(p_n^c) = p_n^c \times$ such that $p_n^c \times r_n^{j-1} = S(p_n^c) r_n^{j-1}$. The third term of the above equation becomes

$$(p_n^c \times r_n^{j-1})^T f_n = [S(p_n^c) r_n^{j-1}]^T f_n = -(r_n^{j-1})^T S(p_n^c) f_n = -(r_n^{j-1})^T (p_n^c \times f_n).$$

In comparison with the same j-th joint torque if using equation (4.33) with both $J_{(c)}$ and F re-projected onto frame n, i.e.,

$$\tau_j^c = J_{c(n)}(:, j)^T F = (p_{c(n)}^{j-1} \times r_n^{j-1})^T f + (r_n^{j-1})^T m,$$

it can be clearly seen that if

$$f_n = f \quad \text{and} \quad m_n = m + p_n^c \times f,$$

then, $\tau_j = J_{(n)}(:, j)^T F_n = \tau_j^c$.

Because the above justification is applicable to every $j \in (0,\, k]$, it shows that to determine a joint torque distribution if a Cartesian force F is applied at any point C along the robot arm other than the end-effector, then we always have two ways to go:

1. Formally define a frame originated at C and construct a Jacobian matrix of point C as formulated in (4.32), then using the statics equation in (4.33) to find the joint torque distribution τ^c;
2. Instead of creating a new Jacobian matrix, use the existing Jacobian matrix $J_{(n)}$ defined on the end-effector frame n, and let an equivalent Cartesian force F_n acting on frame n be created by

$$F_n = \begin{pmatrix} f \\ m + p_n^c \times f \end{pmatrix}, \tag{4.34}$$

where both f and m must be projected on frame n. Then, the joint torque distribution τ^c can be determined by

$$\tau^c = J_{(n)}(:, 1\!:\!k)^T F_n. \tag{4.35}$$

The second approach will be very useful in evaluating joint torque distributions for a digital human model in Chapter 9.

Let us continue to investigate the previous 3-joint RRR type robot, as given in Figure 4.12. Instead of a force acting at the robotic tip point w, a new force

$$f_{(0)}^e = \begin{pmatrix} -mg\cos\phi\sin\phi \\ 4 \\ -mg\cos^2\phi \end{pmatrix} = \begin{pmatrix} -2.1239 \\ 4 \\ -3.6788 \end{pmatrix}$$

is to act at the robotic elbow point e. This is the same force under $\phi = 30^0$ but adding 4 Newtons to the y component. Since the elbow position vector of the robot is

$$p_0^2 = \begin{pmatrix} a_2c_1c_2 \\ a_2s_1c_2 \\ -a_2s_2 \end{pmatrix},$$

the sub-Jacobian matrix of point e (elbow) becomes

$$J_{(0)}^e = \frac{\partial p_0^2}{\partial q} = \begin{pmatrix} -a_2s_1c_2 & -a_2c_1s_2 & 0 \\ a_2c_1c_2 & -a_2s_1s_2 & 0 \\ 0 & -a_2c_2 & 0 \end{pmatrix}.$$

Thus, at the second set of joint angle values $\theta_1 = 0$, $\theta_2 = -100^0$ and $\theta_3 = 30^0$, the joint torque distribution turns out to be

$$\tau_e = J_{(0)}^{eT} f_{(0)}^e = \begin{pmatrix} -0.6946 \\ -2.7305 \\ 0 \end{pmatrix}$$

with the zero 3rd joint torque.

Now, we keep using the original Jacobian matrix $J_{(0)}$ and applying equation (4.34) to find the equivalent 6 by 1 Cartesian force/moment vector. Since p_n^c in this example is the vector $p_w^e = -d_4 r_0^3$, where r_0^3 is the unit vector of the z_3-axis and is the 3rd column of the rotation matrix $R_0^3 = R_0^4$ in this special case, we have

$$p_w^e = -d_4 \begin{pmatrix} c_1s_{23} \\ s_1s_{23} \\ c_{23} \end{pmatrix}.$$

Therefore, according to (4.34), the equivalent Cartesian force/moment vector should be as follows:

$$F_{w(0)} = \begin{pmatrix} f_{(0)}^e \\ p_w^e \times f_{(0)}^e \end{pmatrix}.$$

However, the Jacobian matrix $J_{(0)}$ of the tip point w in (4.31) is the Jacobian of the linear velocity part, and it has to be augmented by the Jacobian of the angular velocity part:

$$J_\omega = (r_0^0 \; r_0^1 \; r_0^2) = \begin{pmatrix} 0 & -s_1 & -s_1 \\ 0 & c_1 & c_1 \\ 1 & 0 & 0 \end{pmatrix}$$

to form a 6 by 3 full Jacobian matrix:

$$J_0 = \begin{pmatrix} J_{(0)} \\ J_\omega \end{pmatrix} = \begin{pmatrix} -a_2s_1c_2 - d_4s_1s_{23} & -a_2c_1s_2 + d_4c_1c_{23} & d_4c_1c_{23} \\ a_2c_1c_2 + d_4c_1s_{23} & -a_2s_1s_2 + d_4s_1c_{23} & d_4s_1c_{23} \\ 0 & -a_2c_2 - d_4s_{23} & -d_4s_{23} \\ 0 & -s_1 & -s_1 \\ 0 & c_1 & c_1 \\ 1 & 0 & 0 \end{pmatrix}$$

before applying the equivalent statics equation (4.35). Under the second set of joint angle values, i.e., $\theta_1 = 0$, $\theta_2 = -100^0$ and $\theta_3 = 30^0$, we finally reach

$$\tau^e = J_0^T F_{w(0)} = \begin{pmatrix} -0.6946 \\ -2.7305 \\ 0 \end{pmatrix},$$

which exactly agrees with τ_e in the above sub-Jacobian matrix approach.

More applications of the robotic statics lie in force control. A robot installed with a force sensor on its wrist or other joint(s) can use a position/force hybrid control to adapt the complex environment. All the conversions of the force/torque between the Cartesian work space and the robotic joint space are done by the Jacobian matrix based on the statics equation (4.30). Typical examples include

1. Let a robot carry an eraser to wipe a chalkboard. Before the eraser reaches the board, the robot is on position control. Once it touches the board, the control system will immediately switch the robot to force control, and needs a force sensor to feedback the actual force acting on the board to the controller and compare it with the desired force value to further adjust and update for a new force output;
2. A robot arm can also assist a patient in a rehabilitation clinic to sense the force/torque applied on the patient's arm and feedback it to compare with the desired data for better serving the patient.
3. Robotic optimal configuration control, also called the optimal posture control, is another typical application that relies on the robotic statics, such as a humanoid robot walking on the floor while lifting a heavy load, which requires a better posture to perform the task. The robotic statics equation is the foundation to determine the best posture at each time instant.

Figure 4.14 depicts a block diagram of typical hybrid position/force control of a robot with a force sensor near the end-effector. The position or force control is switchable by the task decision based on its natural and artificial constraints, where S and $I - S$ are the opposite switch matrices. One can evidently see that the Jacobian matrix and its transpose play a pivotal role in both robotic kinematics and statics, and even more significant in robotic dynamics and control [2, 4].

In summary, a Jacobian matrix J for a robotic system or a digital human model, no matter whether it is a genuine Jacobian matrix or not, can perform the following unique functions:

1. Perform a linear transformation in tangent space between joint and Cartesian velocities;
2. Perform a dual transformation between joint torques and Cartesian forces;
3. The determinant $\det(J)$ or $\det(JJ^T)$ can directly be used to determine kinematic singularity;

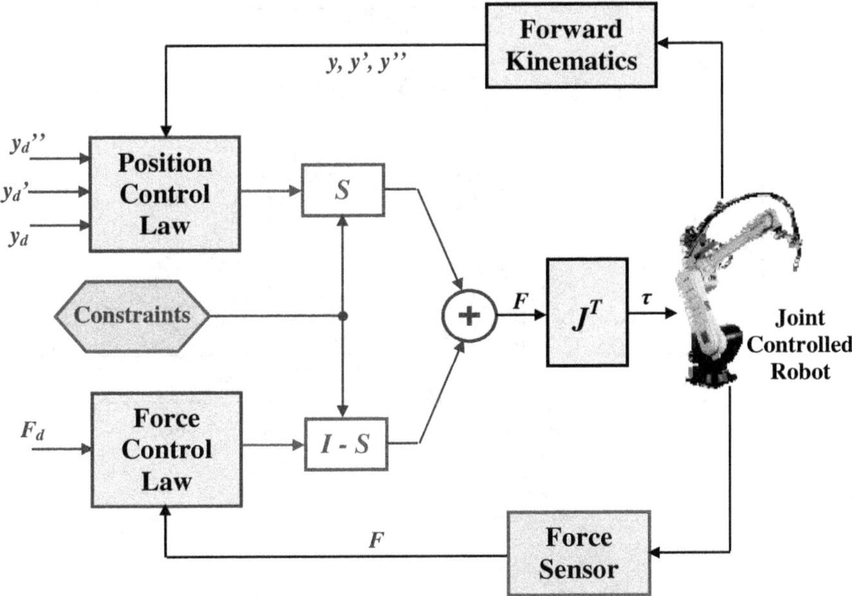

Fig. 4.14 A block diagram of robotic hybrid position/force control

4. As a unique representation of the instantaneous configuration for a robot or the instantaneous posture for a digital human;
5. The determinant $\det(JJ^T)$ can also be considered as a measure of the manipulability for a robotic system;
6. The trace $\mathrm{tr}(JJ^T)$ can be a measure of posture optimization criterion in terms of the uniform joint torque distribution for both a robot and a digital human model;
7. The Jacobian and sub-Jacobian matrices exclusively play a pivotal role in structuring robotic or digital human dynamics, such as in the determination of inertial matrices.

The trace of the Jacobian matrix and its applications will be formally introduced and further discussed in the next Chapter. More detailed concepts and formulations of the sub-Jacobian matrices and the inertial matrix determination will be addressed in Chapter 7.

4.7 Computer Projects and Exercises of the Chapter

4.7.1 Stanford Robot Motions

The first computer simulation project is assigned below to practice all the three robotic motion planning approaches: Cartesian motion, differential

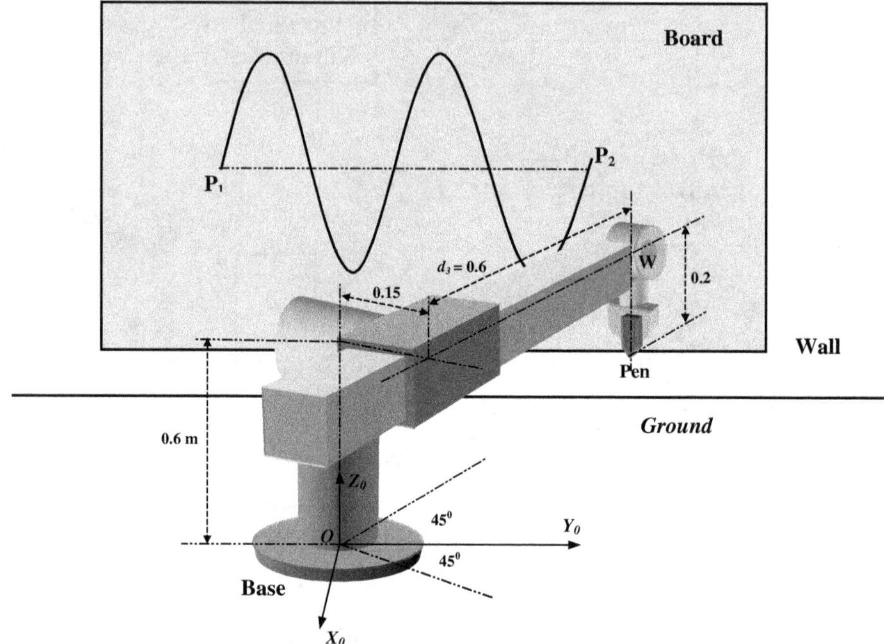

Fig. 4.15 A Stanford robot is sitting at the Home position and ready to move and draw on a board

motion and joint motion, on a 6-joint Stanford robot. The inverse kinematics is shown in Algorithm 1 in the previous section.

A Stanford robot is situated at its **Home** position, as shown in Figure 4.15. A board is fixed on the vertical wall behind the robot. We want the robot hand to grip a pen and draw a sine wave on the board.

1. All parameters of the Stanford robot as well as the initial data for the **Home** position are given in Figure 4.15 (all the distance units are in **meters**). The sine waveform is drawn from a starting point $P_1 = (-0.6,\ -0.2,\ 0.8)$ to an ending point $P_2 = (-0.6,\ 0.8,\ 0.8)$ with 2 periods and amplitude equal to 0.3 m.
2. The entire trajectory can be divided into the following three paths:
   a. From **Home** to P_1, using **Cartesian motion**, let the wrist center point W follow a straight line and match the hand orientation at P_1 to allow the pen to be perpendicular to the board with the number of sampling points $N_1 = 20$;
   b. Using **differential motion**, let the pen draw the specified sine wave on the board from P_1 to P_2, and always keep the hand orientation invariant until it reaches P_2 with $N_2 = 400$, as shown in Figure 4.16;

c. Then, using **joint motion**, from P_2 to come back to the **Home** with $N_3 = 20$.

3. In this project, you are not required to spline the path corners at P_1 and P_2, where the type of motion is switched.

The project can be further procedurized into three phases:

Phase 1. Complete all derivations. Find the initial values of all six joint positions for the Stanford robot at the **Home** with respect to the base. Then, applying the symbolical inverse kinematics, check and compare your frame assignment and D-H table to determine which combination of the I-K solutions will fit for your case.

Phase 2. Program and run your algorithm and print out all the resulting data in MATLAB^{TM}. The detailed requirement of the output is given as follows:

1. Modeling, analysis and derivations for the three motions;
2. Print all the $N_1 = 20$ sets of the joint vectors q along Path #1;
3. Plot the actual curve that the robot hand draws on the board in Path #2 from P_1 to P_2;
4. For the last path, just print what the initial and final positions and orientations of the robot hand situates.

Phase 3. Same as Phase 2, only to replace the symbolical I-K algorithm for the first Cartesian motion from **Home** to P_1 by a numerical solution with the fsolve function-call in MATLAB^{TM}. Verify the results with Phase 2.

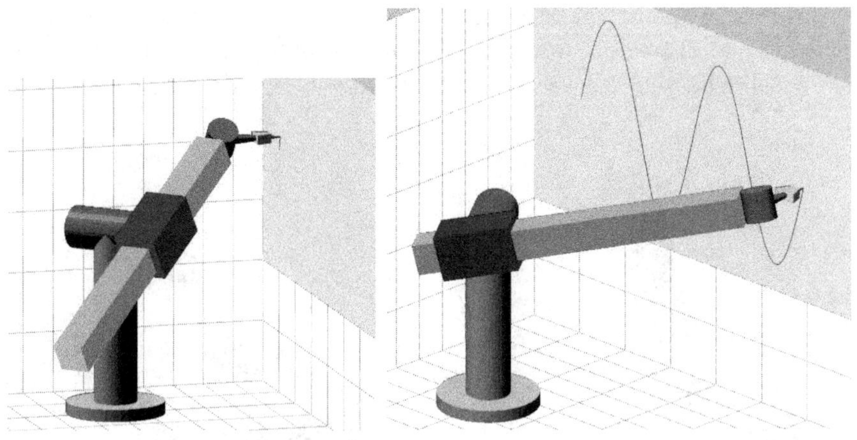

Fig. 4.16 The Stanford robot is drawing a sine wave on the board

4.7.2 The Industrial Robot Model and Its Motions

This project is to model and simulate kinematics and motion planning for an industrial robot model, as shown in part (b) of Figure 4.6. The symbolical inverse kinematics (I-K) is given by Algorithm 2 in the previous section. The project is divided into the following three phases:

Phase 1

1. Using D-H Convention, determine a D-H table and assign the joint/link coordinate frames for the industrial robot model, where $d_1 = 0.4$, $a_1 = 0.15$, $a_2 = 1.2$, $a_3 = 0.1$, $d_4 = 1.0$ and $d_6 = 0.25$, all in meter;
2. Given a Starting point and an Ending point (both p_0^6 and R_0^6) as follows:

$$p_0^6(Start) = \begin{pmatrix} -0.25 \\ 1.5 \\ 1.25 \end{pmatrix}, \quad R_0^6(Start) = \begin{pmatrix} 1 & 0 & 0 \\ 0 & -1 & 0 \\ 0 & 0 & -1 \end{pmatrix},$$

$$p_0^6(End) = \begin{pmatrix} 0.5 \\ 1.75 \\ 0.8 \end{pmatrix}, \quad R_0^6(End) = \begin{pmatrix} 0 & 0 & 1 \\ 0 & -1 & 0 \\ 1 & 0 & 0 \end{pmatrix}.$$

Program the symbolical I-K algorithm into MATLAB^{TM} and find the two sets of joint angles in degree at the Starting and Ending hand points. You may just pick + for $\tilde{c}_3$ and pick θ_4, instead of $\theta_4 = \theta_4 \pm 180^0$.

Phase 2

From the Starting point to the Ending point, as shown in Figure 4.17, let the position of the origin of Frame #6 moves along a straight line and the orientation of Frame #6 rotates using $k - \phi$ procedure. If the number of sampling points is $N = 10$, calculate and print the 10 sets of joint angles

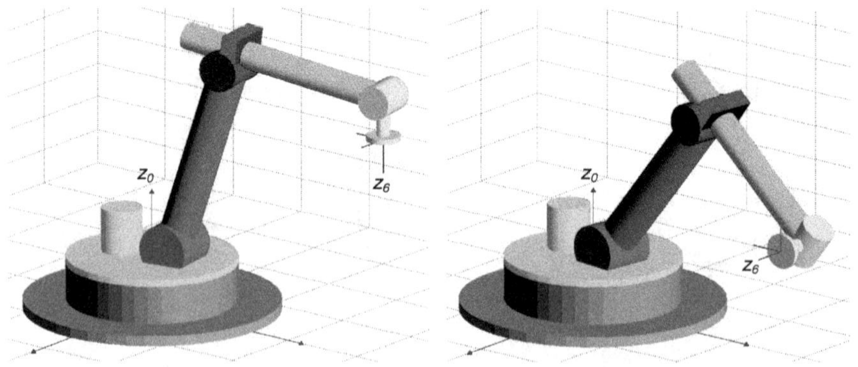

Fig. 4.17 The industrial robot model at the Starting and Ending positions

in degree for the industrial robot model. You may also plot each joint angle versus sampling points as time t.

Phase 3

Now, write a program using a numerical solution with the fsolve function-call to generate the required Cartesian motion. Also, plot the 10 sets of joint angles for the robot and compare them with the results in Phase 2.

4.7.3 Exercise Problems

1. For the three robot arms in Figures 4.18–4.20,
    a. Determine the D-H parameter table for each of them;
    b. Write all of their one-step homogeneous transformation matrices;
    c. Find the position vectors of the last frames for the 2nd and 3rd robots, i.e., the tip point positions with respect to their bases in symbolical form;
    d. Find the Jacobian matrices of the 2nd and 3rd robot arms;
    e. Determine the inverse kinematics solution symbolically in the 3rd robot case.

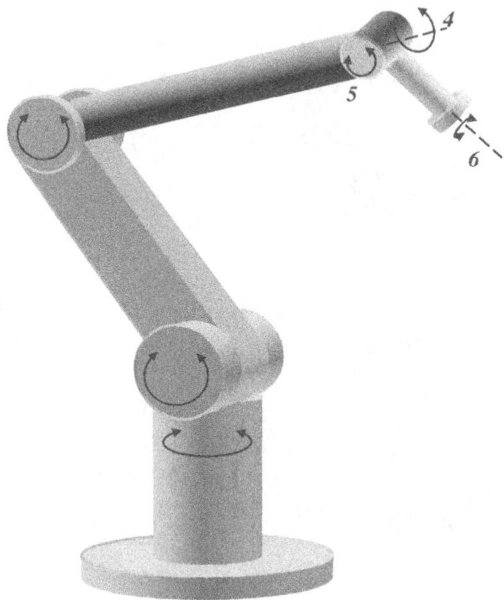

Fig. 4.18 Robot 1

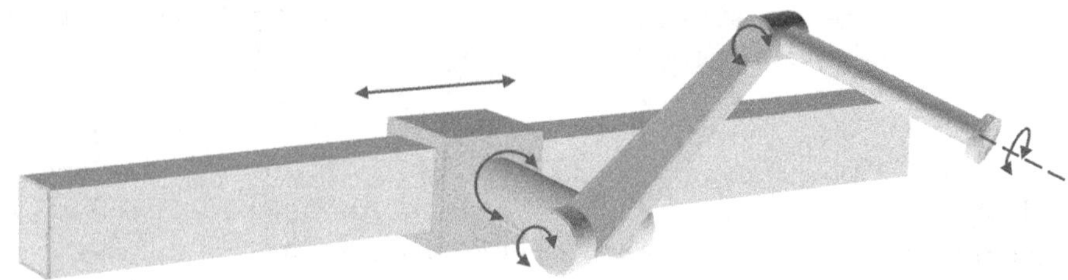

Fig. 4.19 Robot 2

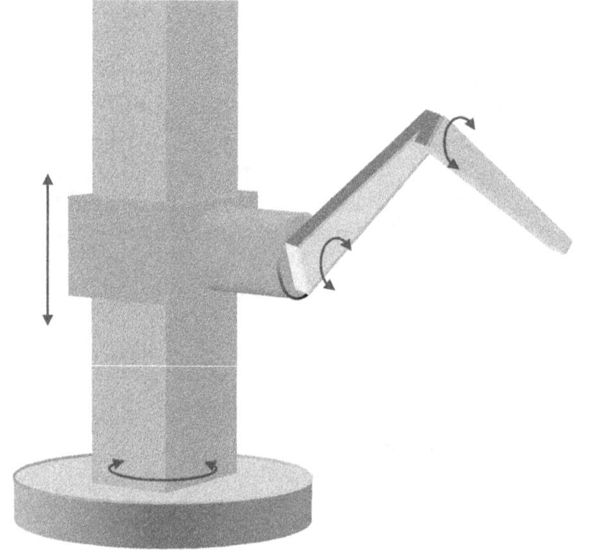

Fig. 4.20 Robot 3

2. Three frames and the base are given by the following homogeneous transformations:

$$H_0^1 = \begin{pmatrix} 1 & 0 & 0 & 0 \\ 0 & \frac{1}{2} & -\frac{\sqrt{3}}{2} & 2 \\ 0 & \frac{\sqrt{3}}{2} & \frac{1}{2} & 0 \\ 0 & 0 & 0 & 1 \end{pmatrix}, \quad H_1^2 = \begin{pmatrix} 0 & 1 & 0 & 3 \\ -1 & 0 & 0 & 0 \\ 0 & 0 & 1 & 0 \\ 0 & 0 & 0 & 1 \end{pmatrix},$$

and

$$H_2^3 = \begin{pmatrix} 0 & 0 & 1 & 0 \\ 0 & 1 & 0 & 0 \\ -1 & 0 & 0 & 1 \\ 0 & 0 & 0 & 1 \end{pmatrix}.$$

If we want to move a robot hand from frame 3 to frame 1 along a straight line for position and using the $k - \phi$ procedure for orientation, and to sample the path by $N = 10$ points evenly,

a. Determine the position vector of the hand w.r.t. the base at the **fourth** sampling point;
b. Determine the orientation of the hand frame w.r.t. the base at the **sixth** sampling point.

3. We wish a robot hand to draw a parabola on the y_0-z_0 plane of the base with the x-coordinate fixed as a constant x_0. The parabola is given by

$$z = c\,(y-2)^2 + d,$$

where c and d are constant parameters. It will start from a point $P_1 = (x_0 \;\; 1 \;\; 1)^T$ at $t = 0$ and end at a point $P_2 = (x_0 \;\; 2 \;\; 2)^T$ at $t = 4$ sec. The y-coordinate motion is linear of time t, i.e. $y = m\,t + b$.

a. Determine the parametric equation for all the three coordinates of the parabolic drawing;
b. Find the Cartesian linear velocity v with respect to the base.

4. A destination frame is created by first rotating the base frame about the z_0-axis by -30^0 and then rotating the new frame about the new x-axis by 40^0. If the first rotation is also adding a slide along the z_0 by 0.6 m. and the second rotation is adding a slide along the new x-axis by -0.4 m., using the dual-number transformation, find both the position and orientation of the destination frame w.r.t. the base.
5. A new frame is created by rotating an original reference frame about a vector $v = (2 \;\; 3 \;\; -1)^T$ by an angle $\phi = 70^0$, and also sliding it along v by $d = 1.2$ m. Find the position and orientation of the new frame w.r.t. the reference frame using (a) homogeneous transformation, and (b) dual-number transformation.
6. For the industrial robot model, as shown in part (b) of Figure 4.6,

a. Determine its D-H table and one-step homogeneous transformations;
b. Using Algorithm 2 to find the inverse kinematics solution if both p_0^6 and R_0^6 are given numerically in Project 2 at the starting position and orientation;
c. Using Matlab^{TM}, calculate all the related homogeneous transformations and determine the numerical Jacobian matrix $J_{(0)}$ at the position and orientation given in (b);
d. If a Cartesian force vector $F_{(0)}$ acting on the end-effector is given by

$$F_{(0)} = (0 \;\; 0 \;\; -100 \;\; 0 \;\; 30 \;\; 40)^T,$$

find the joint torque distribution.

7. For a two-joint planar robot arm, as shown in Figure 4.21, where $a_1 = 0.2$ and $a_2 = 0.8$ in meter,

   a. Determine its D-H parameter table;
   b. Find the symbolical form of the homogeneous transformation A_0^2 ;
   c. If the robot hand position $p_0^2 = \begin{pmatrix} p_y \\ p_z \end{pmatrix} = \begin{pmatrix} 0.8 \\ 1.0 \end{pmatrix}$ w.r.t. the base, derive the inverse kinematics solution of the robot;
   d. Determine the symbolical Jacobian matrix $J_{(0)}$ by taking derivative of p_0^2 w.r.t. the robotic joint positions;
   e. If the hand tip Cartesian velocity is given by $v = \begin{pmatrix} 0 \\ 1 \end{pmatrix}$, find the joint velocity vector $\dot{q}$;
   f. If a load of $m = 25$ Kg. is hanged down at the hand tip point of the arm, find the joint force/torque.

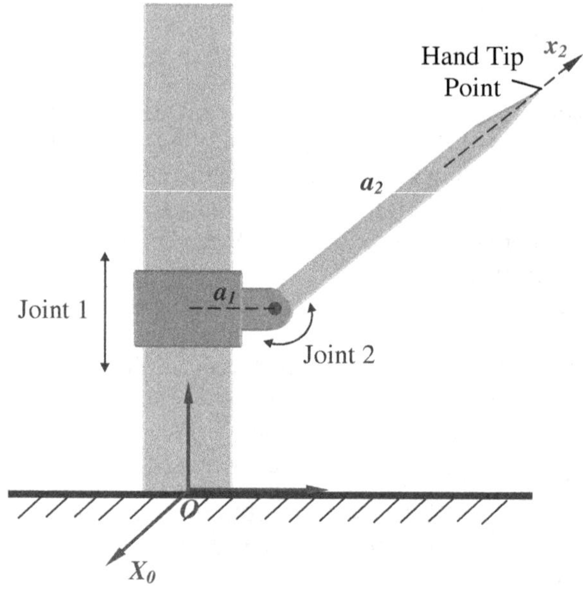

Fig. 4.21 A 2-joint prismatic-revolute planar arm

8. A three-link robot arm has three joint variables: θ_1, d_2 and θ_3, as shown in Figure 4.22.

   a. Determine a D-H table of the robot;
   b. Find the homogeneous transformation H_0^3 for the specified hand frame w.r.t. the base;
   c. Determine a 3 by 3 Jacobian matrix if only the Cartesian position of the hand is considered;

d. At what configuration does the robot arm encounter a singularity?
e. If $a_3 = 10$ in decimeter, and the hand has a position $p_0^h = (0\ \ 10\ \ 7)^T$, find the three joint values;
f. If at the above hand position, the hand instantaneous moving velocity is $v = (-2\ \ 0\ \ -3)^T$ w.r.t. the base, determine the three joint speeds;
g. If the hand is now pulled by a unit force towards the origin of the base as the robot arm is standstill at the position given in (e), what is each joint torque/force due to the Cartesian pulling force?

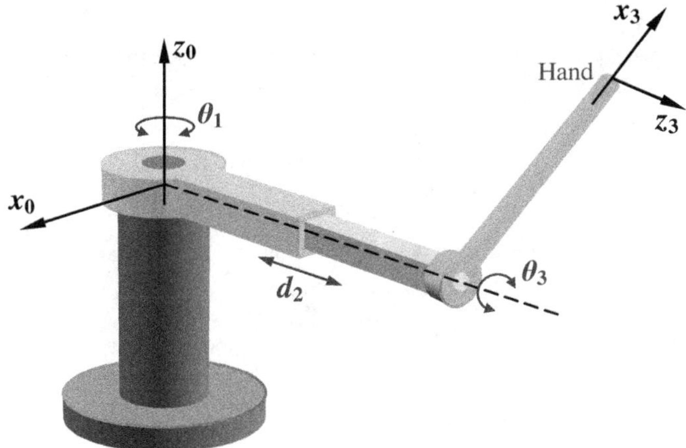

Fig. 4.22 A 3-joint RPR robot arm

9. For a given 3-joint robot arm with an overhead beam and two revolute joints, as shown in Figure 4.23, where $d_2 = 0.8$ m. and $a_3 = 1$ m., answer the following questions:

   a. Determine its D-H parameter table;
   b. Find the symbolical form of the homogeneous transformation A_0^3 ;
   c. For a given numerical position $p_0^3 = (-1.4\ \ 0\ \ 1)^T$ (m.) of the origin of frame 3 w.r.t. the base, find the three joint positions d_1, θ_2 and θ_3 of the robot;
   d. Determine the Jacobian matrix $J_{(0)}$ by taking derivative of the symbolical p_0^3 w.r.t. the robotic joint positions;
   e. Find all possible singular points;
   f. If the central axis of link 3 is parallel to the overhead beam during the motion of the robot, does this configuration fall into a singularity case? Explain why;
   g. If $\dot{d}_1 = 0.5$ m./sec., $\dot{\theta}_2 = 0.5$ rad/sec., and $\dot{\theta}_3 = -0.3$ rad/sec., when the robot passes at the position in (3), find the instantaneous velocity vector v w.r.t. the base for the origin of frame 3, i.e., the tip point of the robot;

h. we are planning to move frame 3 from $(-1.6 \;\; 0.6 \;\; 0.4)^T$ to $(-0.6 \;\; -0.4 \;\; 1.4)^T$ along a straight line by $N = 10$ sampling points, find the position vector at the 3rd sampling point w.r.t. the base;
i. We are now planning to move frame #3 from Orientation I (at $\theta_2 = 90^0$ and $\theta_3 = 0^0$ for the robot) to Orientation II (at $\theta_2 = 0^0$ and $\theta_3 = -90^0$) by using the k-ϕ algorithm, determine the unit vector k and the rotation angle ϕ ;

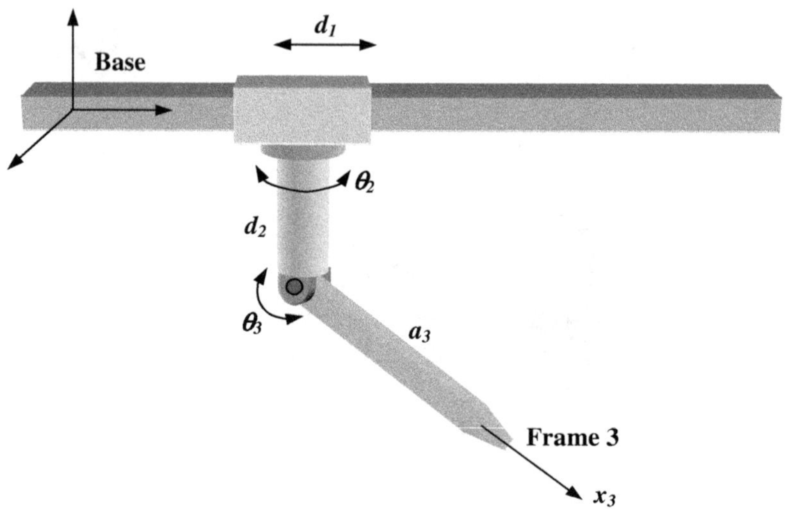

Fig. 4.23 A beam-sliding 3-joint robot

References

1. Fu, K., Gonzalez, R., Lee, C.: Robotics: Control, Sensing, Vision and Intelligence. McGraw-Hill, New York (1987)
2. Spong, M., Vidyasagar, M.: Robot Dynamics and Control. John Wiley & Sons, New York (1989)
3. Murray, R., Li, Z., Sastry, S.: A Mathematical Introduction to Robotic Manipulation. CRC Press, Boca Raton (1994)
4. Craig, J.: Introduction to Robotics: Mechanics and Control, 3rd edn. Pearson Prentice Hall, New Jersey (2005)
5. Duffy, J.: Statics and Kinematics with Applications to Robotics. Cambridge University Press, New-York (1996)
6. Gu, E., Luh, J.: Dual-Number Transformation and Its Applications to Robotics. IEEE Journal of Robotics and Automation RA-3(6), 615–623 (1987)

Chapter 5
Redundant Robots and Hybrid-Chain Robotic Systems

5.1 The Generalized Inverse of a Matrix

According to linear algebra [1, 2], a linear multi-variable simultaneous equation can always be written in a matrix form below:

$$Ax = b, \tag{5.1}$$

where the coefficient matrix A is m by n if there are n variables in x and m known values in b, or $x \in \mathbb{R}^n$ and $b \in \mathbb{R}^m$. If A is a square matrix, i.e., $m = n$, and also non-singular, then equation (5.1) has a unique solution $x = A^{-1}b$, where A^{-1} is known as the inverse of the square matrix A.

However, in many cases, either A is square but singular, or A is non-square, i.e., $m \neq n$, can we still solve equation (5.1) for x? Let a matrix be denoted by A^- and be defined such that $x = A^-b$ is a solution to (5.1). By substituting this nominal solution into the equation, we ask ourselves whether $AA^-b = b$? Since $AA^-b = AA^-Ax = b = Ax$, if the answer is yes, then $AA^-A = A$. Therefore, to solve equation (5.1) in a more general sense, we need to introduce a so-called **generalized inverse** A^- that is n by m for an m by n matrix A such that

$$AA^-A = A. \tag{5.2}$$

With a further test, we find that such a generalized inverse is not unique [3, 4]. For further improvement, we may test the solution in the reversed direction. Namely, if $x = A^-b$ is a solution, then $x = A^-b = A^-Ax = A^-AA^-b$. This implies that $A^-AA^- = A^-$, which can be imposed as the second condition to narrow down the non-unique solutions. To this end, let us re-define a so called **reflexive generalized inverse** $A^\#$ that is also n by m for an m by n matrix A such that

$$AA^\#A = A \quad \text{and} \quad A^\#AA^\# = A^\#. \tag{5.3}$$

E.Y.L. Gu, *A Journey from Robot to Digital Human*,
Modeling and Optimization in Science and Technologies 1,
DOI: 10.1007/978-3-642-39047-0_5,

This new definition of the generalized inverse under two conditions is, indeed, getting closer to uniqueness, but is still not quite unique yet.

Finally, two mathematicians Moore and Penrose proposed a so-called **pseudo-inverse** A^+ that is also n by m for any m by n matrix A such that all the following four conditions hold:

$$AA^+A = A,\ A^+AA^+ = A^+,\ (A^+A)^T = A^+A, \text{ and } (AA^+)^T = AA^+. \tag{5.4}$$

This pseudo-inverse A^+, or called Moore-Penrose inverse, is proven to be unique for any kind of matrix A, and it can always have a unique explicit form in each of the following cases:

1. If A is square, n by n and non-singular, then $A^+ = A^{-1}$;
2. If A is m by n with $m < n$, called a "short" matrix, then $A^+ = A^T(AA^T)^{-1}$;
3. If A is m by n with $m > n$, called a "tall" matrix, then $A^+ = (A^TA)^{-1}A^T$;
4. If A is square and n by n but singular with rank$(A) = k < n$, first, let its maximum-rank decomposition be $A = BC$, where B is n by k and C is k by n with both rank(B) =rank$(C) = k$. The pseudo-inverse of A becomes $A^+ = C^T(CC^T)^{-1}(B^TB)^{-1}B^T$.

Note that if either AA^T in case 2 or A^TA in case 3 is singular, then it has to apply the maximum-rank decomposition on it, like case 4, before finding its pseudo-inverse. The reader can verify without difficulty that the pseudo-inverse determined in each of the above cases for A satisfies all the four conditions in (5.4). In MATLAB^{TM}, there is an internal function pinv$(\cdot)$ to calculate the pseudo-inverse of $(\cdot)$ numerically, and this function will bring a lot of convenience to our future programming.

Though the formation of the pseudo-inverse for a matrix is unique, based on linear algebra, the solution of equation (5.1) itself is still not unique if the m by n matrix A is "short", i.e., $m < n$. Since n is the number of unknown variables x and m is the number of equations, the obvious question is how can one uniquely solve for more unknown variables by less equations? Nevertheless, the general solution can be written in terms of the pseudo-inverse A^+ as follows:

$$x = A^+b + (I - A^+A)z, \tag{5.5}$$

where I is the n by n identity, $z \in \mathbb{R}^n$ is an arbitrary vector. Because of the arbitrary choice of z, the number of distinct solutions can go to infinity. By substituting (5.5) into (5.1) and noticing all the conditions in (5.4), we will immediately see that it is a true solution, no matter what z is.

The geometrical meaning of the general solution (5.5) is quite significant. First, the two terms in (5.5) are always orthogonal to each other, i.e.,

$$z^T(I - A^+A)^TA^+b \equiv 0$$

for any $z \in \mathbb{R}^n$. In fact, according to the conditions in (5.4), we have

$$(I - A^+A)^T A^+ = (I - A^+A)A^+ = A^+ - A^+AA^+ = O,$$

the n by n zero matrix. This means that the n-dimensional solution space can be decomposed into two orthogonal subspaces: one is called a **rank space** $R(A)$, and the other one is called a **null space** $N(A)$.

Let y be an n-dimensional arbitrary vector. Then, $A^+Ay \in R(A)$ and $(I - A^+A)y \in N(A)$, which mean that both n by n matrices A^+A and $I - A^+A$ play a common role as a **projector**, and the former one projects y onto the rank subspace $R(A)$, while the latter one projects y onto the null subspace $N(A)$.

Moreover, any projector P must be *idempotent*, i.e., $P^2 = P$. In other words, after projecting an arbitrary vector y onto a targeting subspace, it becomes $z = Py$, and projecting z once again onto the same subspace will result in the same vector z, i.e., $Pz = P(Py) = P^2y = z = Py$, because z has already been inside the targeting subspace after the first projection. Using the conditions in (5.4) once again, one can readily show that both $A^+A = P_r$ and $I - A^+A = P_n$ are, indeed, the two projectors. Namely, $P_r^2 = P_r$ and $P_n^2 = P_n$.

Now, based on the general solution in (5.5), $x = A^+b + (I - A^+A)z = A^+Ax + (I - A^+A)z = x_r + x_n$, which implies that the first term $x_r = A^+b = A^+Ax$ is a projection of the general solution x onto the rank subspace by the projector $P_r = A^+A$, and the second term $x_n = (I - A^+A)z$ is the projection of an arbitrary vector z onto the null subspace by the projector $P_n = I - A^+A$. The two terms: the rank solution $x_r = A^+b$ and the null solution $x_n = (I - A^+A)z$ are always orthogonal to each other, $x_r \perp x_n$. Since $x_r + x_n = x$ with x_r to be an orthogonal projection of x onto the rank subspace, $\|x_r\| \leq \|x\|$. This shows that the rank solution $x_r = A^+b$ is always the **minimum-norm solution** over all general solutions x in (5.5) for equation (5.1).

Furthermore, since $AA^+ = AA^T(AA^T)^{-1} = I$ but $A^+A = A^T(AA^T)^{-1}A \neq I$, we can directly see that $Ax_r = AA^+b = b$ and $Ax_n = A(I - A^+A)z = (A - AA^+A)z \equiv 0$. Figure 5.1 depicts the geometric interpretation of such an orthogonal decomposition for the general solution, which will be a very useful mathematical foundation in the next modeling and analysis of redundant robotic systems.

5.2 Redundant Robotic Manipulators

A redundant robot has a number of joints that exceeds its output degrees of freedom (d.o.f), i.e., $n > m$. The excessive number $n - m = r$ is referred to as a **degree of redundancy**. For a robot with redundancy, its Jacobian matrix J is no longer square. Instead, it is an m by n "short" matrix. The solution to its Jacobian equation

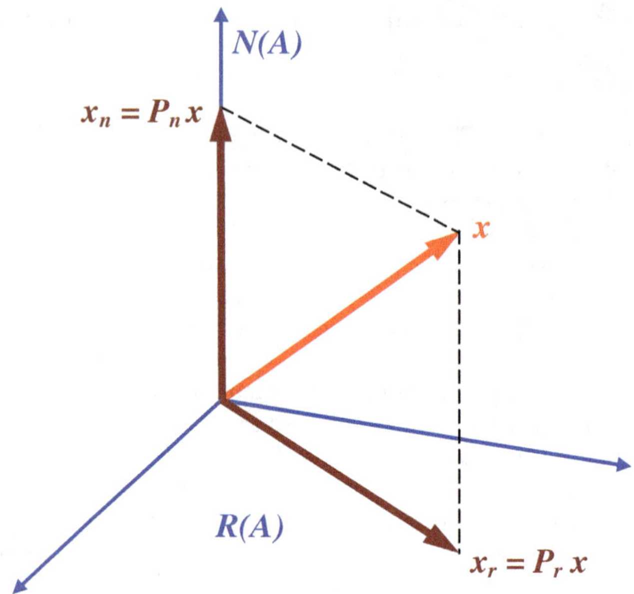

Fig. 5.1 Geometrical decomposition of the general solution

$$V = \begin{pmatrix} v \\ \omega \end{pmatrix} = J\dot{q} \tag{5.6}$$

is accordingly no longer unique. We now utilize the Moore-Penrose pseudo-inverse of the Jacobian matrix to represent the general inverse-kinematics (I-K) solution for a redundant robot:

$$\dot{q} = J^+V + (I - J^+J)z, \tag{5.7}$$

where $J^+ = J^T(JJ^T)^{-1}$ is the pseudo-inverse of J, I is the n by n identity and $z \in \mathbb{R}^n$ is an arbitrary vector.

In this general solution, the first term, $\dot{q}_r = J^+V \in R(J)$ is, again, called a rank solution that determines the robotic main task operation described by a Cartesian velocity V. In contrast, the second term in (5.7), $\dot{q}_n = (I - J^+J)z \in N(J)$ is called a null solution that may carry a subtask operation described by the vector z. Since the rank and null solutions are always orthogonal to each other, i.e. $\dot{q}_r \perp \dot{q}_n$ or $(\dot{q}_r)^T\dot{q}_n \equiv 0$, the subtask operation will never interfere with the main task execution [4, 5, 6].

Based on the theory of the pseudo-inverse discussed in the last section, $\dot{q} = \dot{q}_r = J^+V$ in the case $z = 0$ is the minimum-norm solution. The minimum-norm I-K solution is simple in computation, but its corresponding motion generated is not quite naturally looking, because the robot arm will maneuver in such a way that its lower joints will move much further than the upper joints in order to maintain $\|\dot{q}\| \to \min$. The reason is intuitively clear that a smaller angle change of a lower joint (closer to the robot base) will often contribute more linear motion of the robotic end-effector. Thus, in general applications, it is necessary to explore a null solution for possible improvement of both the redundant robot kinematic motion and task performance.

In fact, a major step for the inclusion of a null solution is to define an appropriate vector $z \in \mathbb{R}^n$ that can represent a desired subtask for optimization. Let a *scalar potential function* $p = p(q)$ be defined to describe a desired subtask to be either maximized or minimized for its time-rate $\dot{p}$. Since

$$\dot{p} = \frac{\partial p}{\partial q}\dot{q} = \eta^T \dot{q},$$

we can show that if z in (5.7) is set to be $z = k\eta$, where $\eta = \partial p/\partial q$ is the **gradient vector** (column vector) of $p = p(q)$, then p is monotonically increasing if $k > 0$ and p is monotonically decreasing if $k < 0$ [4, 5].

In fact, the definition of $p(q)$ is in effect only on the null solution, which is now $\dot{q}_n = k(I - J^+J)\eta$ so that the time-derivative of the potential function on the null solution side becomes

$$\dot{p}_n = \eta^T \dot{q}_n = k\eta^T(I - J^+J)\eta.$$

Note that the projector $P_n = I - J^+J$ is idempotent and also symmetric. Hence, if $k > 0$,

$$\dot{p}_n = k\eta^T(I - J^+J)\eta = k\eta^T P_n^2 \eta = k\eta^T P_n^T P_n \eta \geq 0,$$

because of the fact that the matrix $P_n^T P_n$ is always semi-positive-definite. Clearly, if $k < 0$, $\dot{p}_n \leq 0$. Therefore, in order to represent a subtask and its optimization when the redundant robot is operating a given main task, the key step is to define a potential function $p(q)$.

In summary, the general kinematic solution for a redundant robot in differential motion can be written as

$$\dot{q} = J^+V + (I - J^+J)k\eta, \tag{5.8}$$

where η is the gradient column vector of a scalar potential function $p(q)$, and k is a gain constant that is positive if one wants the value of $p(q)$ to be monotonically increasing, or is negative if one wants the value of $p(q)$ to be monotonically decreasing.

For example, in order to *avoid singularity*, one often defines the robotic "manipulability" as a potential function:

$$p = \sqrt{\det(JJ^T)},$$

because, in general, we wish the robot would always be distant from the zero determinant of the Jacobian matrix J, or the singular points [6]. However, this definition, though meaningful in concept, causes an unmanageable symbolical derivation in order to further find its gradient vector, and is unfeasible in robotic applications.

On the other hand, if one emphasizes only the 6th joint position in the wrist of a $(6+1)$-joint robot that is a 6-joint arm sitting on a linear track to escape from its rotational singularity as fast as possible, we may simply define $p = \sin^2\theta_6$, instead of $p = \det(JJ^T)$. Then, the symbolical derivation of its gradient vector for such a simple but effective potential function becomes much easier to handle. In this case,

$$\eta = \frac{\partial p}{\partial q} = \begin{pmatrix} 0_{5\times 1} \\ 2\sin\theta_6\cos\theta_6 \\ 0 \end{pmatrix} = \begin{pmatrix} 0_{5\times 1} \\ \sin 2\theta_6 \\ 0 \end{pmatrix},$$

and the coefficient $k > 0$, where $0_{5\times 1}$ is the 5 by 1 zero column vector.

Another typical example of the subtask is to *avoid being out-of-range* for each joint position during a motion. If we know the center of each robotic joint position range, and let all such center values form an n-dimensional constant vector q_c, the following potential function can be used to best represent this particular subtask for minimization:

$$p(q) = \frac{1}{2}\|q - q_c\|^2 = \frac{1}{2}(q - q_c)^T(q - q_c). \tag{5.9}$$

Then, its gradient vector can be immediately calculated as follows:

$$\eta = \frac{\partial p}{\partial q} = q - q_c, \tag{5.10}$$

and the coefficient here should be $k < 0$ so that with $z = k\eta$, the p value will be monotonically decreasing to make each joint position approach as close to its center as possible.

The third example is *collision avoidance*. If a redundant robot arm is situated in an environment with some obstacles nearby, we have to define a potential function to represent a subtask of avoiding possible collisions with the obstacles. Suppose that the robotic elbow is considered most likely to collide with an obstacle. If the most dangerous corner point of the obstacle for collision is determined and has a constant Cartesian position vector $p_0^{ob} \in \mathbb{R}^3$ with respect to the world base, and the position vector of the robotic elbow is p_0^{el} that is a function of q, then, the potential function can be defined by

$$p(q) = \|p_0^{el} - p_0^{ob}\|^2 = (p_0^{el} - p_0^{ob})^T(p_0^{el} - p_0^{ob}),$$

which may be a function of just the first two or three joint values, depending on which joint the elbow locates at. Once we have an explicit form of p_0^{el}

determined by the homogeneous transformations of the robot, it becomes relatively straightforward to find the gradient vector η, and, of course, $k > 0$ in this collision avoidance case.

The potential function $p(q)$ for collision avoidance can also be defined to approach a virtual point that is a short distance away from the obstacle as a safe position. If this virtual position is defined as p_0^v, then p_0^{ob} in the above potential function form is replaced by p_0^v, and set $k < 0$ to allow the elbow point to be as close to the *virtual safe point* as possible to avoid hitting the obstacle. Such an alternative approach to collision avoidance will be illustrated by a simulation study later in this section.

The fourth subtask optimization case is to *automatically approach the best posture* for either a robot arm or a digital human. We are human, but often overlook the question about the best posture. Although each of us well knows what is the best posture in performing a specific task, such as to pick up a heavy load or to lift and move a table, not everyone can tell why. To mathematically describe and model the best posture, we have to seek an explicit potential function $p = p(q)$ to represent a measure of the posture to be optimized. Since a Jacobian matrix for a robot or a digital human is the most complete and also unique quantity to determine each instantaneous posture, the desired potential function for posture optimization is closely related to the Jacobian matrix J.

A further study shows that a certain posture is comfortable for a human doing some task if every joint torque can be uniformly distributed over the entire body, instead of having some joints suffering from a higher torque while the others are exerting lower or no torque. An injury around some joint of a human body is often cumulatively caused by such an uneven and spiking torque distribution.

Before we further develop this notion, let us review a basic mathematical inequality. For n positive real numbers: $a_1, \cdots, a_n$ with each $a_i > 0$, it is well-known that their arithmetic mean value is always greater than or equal to their geometric mean value, i.e.,

$$\frac{a_1 + \cdots + a_n}{n} \geq \sqrt[n]{a_1 \cdots a_n},$$

and it becomes equal if and only if all the positive numbers a_i's are equal to each other, i.e., $a_1 = \cdots = a_n$.

Now, let a weighted joint torque norm square for a robot or a human body be defined as

$$\tau^T W \tau = w_1 \tau_1^2 + \cdots + w_n \tau_n^2,$$

where the joint torque vector is $\tau \in \mathbb{R}^n$ and the weight W is an n by n diagonal matrix with each diagonal element $w_i > 0$. Therefore, it must obey that

$$\frac{\tau^T W \tau}{n} \geq \sqrt[n]{w_1 \tau_1^2 \cdots w_n \tau_n^2},$$

and likewise, they are equal if and only if the weighted joint torques are uniformly distributed, i.e., $w_1\tau_1^2 = \cdots = w_n\tau_n^2$.

On the other hand, based on the robotic statics, the joint torque vector $\tau = J^T F$, where J is the Jacobian matrix of the robot or digital human, and $F \in \mathbb{R}^m$ is a Cartesian force vector (wrench) representing the load imposed on the robot or digital human at one or more end-effectors. Thus, we obtain a weighted quadratic form in terms of the Jacobian matrix:

$$\tau^T W \tau = F^T J W J^T F.$$

Therefore, the weighted joint torque norm square for a robot depends on the external Cartesian force (wrench), and this may bring a complication to the focus on configuration (posture) optimization.

However, based on the Rayleigh Quotient Theorem from linear algebra [1, 2],

$$\lambda_{min} \leq \frac{F^T J W J^T F}{F^T F} \leq \lambda_{max},$$

where λ_{min} and λ_{max} are the minimum and maximum eigenvalues of the positive-definite matrix JWJ^T for any vector F. If we just use the normalized load force vector $F^T F = 1$ to test the joint torque distribution, then, the above equation is reduced to

$$\lambda_{min} \leq F^T J W J^T F = \tau^T W \tau \leq \lambda_{max}.$$

This means that the weighted joint torque norm square is always upper-bounded by λ_{max} and lower-bounded by λ_{min} of the m by m positive-definite weighted Jacobian matrix JWJ^T, no matter what the unity external Cartesian force F is.

Since each eigenvalue λ_i of JWJ^T is a positive number as long as J is full-ranked, they also satisfy

$$\frac{\lambda_1 + \cdots + \lambda_m}{m} \geq \sqrt[m]{\lambda_1 \cdots \lambda_m}.$$

Because $\lambda_1 + \cdots + \lambda_m = \mathrm{tr}(JWJ^T)$ and $\lambda_1 \cdots \lambda_m = \det(JWJ^T)$, according to linear algebra, the inequality can be rewritten by

$$\frac{\mathrm{tr}(JWJ^T)}{m} \geq \sqrt[m]{\det(JWJ^T)}, \tag{5.11}$$

and they are equal if and only if $\lambda_{min} = \lambda_{max}$, or all the eigenvalues are squeezed to be uniformly distributed.

Since the right-hand side of equation (5.11) is related to the robotic manipulability, under a fixed manipulability, minimizing $\mathrm{tr}(JWJ^T)$ will make all the eigenvalues of JWJ^T equal. This is also the same method to make the weighted joint torques approach being evenly distributed. Therefore, the potential function for *the best posture in the sense of uniform joint torque distribution* can be defined as

$$p(q) = \mathrm{tr}(JWJ^T). \tag{5.12}$$

If the weight W is the identity matrix, this means that each joint has equal "opportunity", then,

$$p(q) = \mathrm{tr}(JJ^T).$$

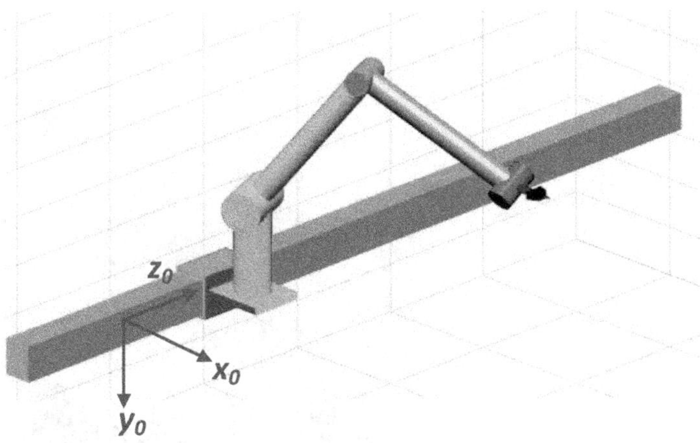

Fig. 5.2 A 7-joint redundant robot arm

Figures 5.2, 5.3 and 5.4 show a 7-joint redundant robot that is created by a regular 6-revolute-joint robot mounted on a linear track, and sometimes it is called a $(6+1)$-joint robot. We will perform a detailed kinematic analysis later, and also make a digital mock-up and animation for this typical redundant robot in next chapter. Here, we just introduce it to demonstrate the correlation between the trace $\mathrm{tr}(JWJ^T)$ and the weighted joint torque distribution.

Within 400 sampling points of simulation, this robot is forced to have both the position and orientation of its last frame #7 fixed without motion while changing and adjusting its configuration (posture) by a null solution $\dot{q} = (I - J^+J)k\eta$ to update its joint positions at each sampling point. The potential function is $p(q) = (d_1 - c)^2$, where d_1 is the first prismatic joint value that is the sliding displacement along the linear track, and c is a constant destination position on the linear track for the regular 6-joint robot arm sliding to. Thus, such a potential function has only one variable d_1 so that its gradient η can easily be determined, and the constant gain k should be negative in this case to make $p(q)$ as small as possible. We plan to let the robot make a round-trip while keeping its end-effector stationary but acting

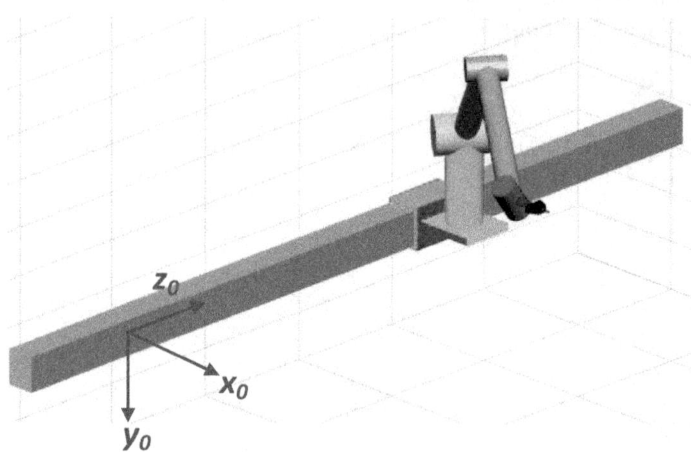

Fig. 5.3 A 7-joint redundant robot arm

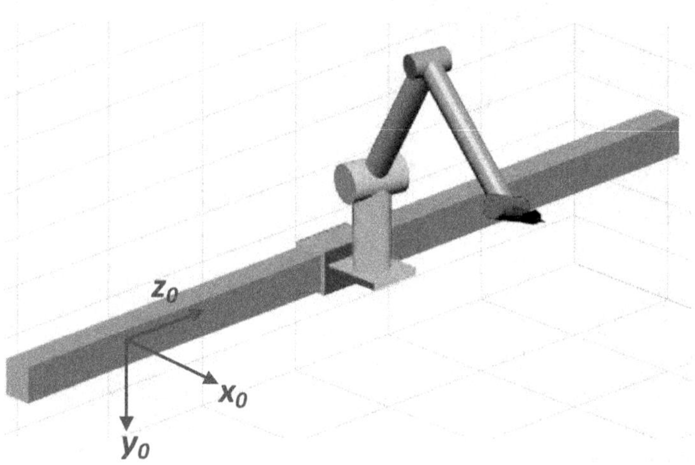

Fig. 5.4 A 7-joint redundant robot arm

on a Cartesian force given by $F = (-1 \;\; 2 \;\; 0 \;\; 0 \;\; 0 \;\; 0)^T$ with respect to the robot base frame #0. The 7 by 7 weighting matrix W is defined as

$$W = \mathrm{diag}(0.5 \;\; 0.05 \;\; 1 \;\; 1 \;\; 1 \;\; 1 \;\; 1)$$

along its diagonal.

As a result, Figure 5.2 shows the regular 6-joint robot on the linear track at the starting position, Figure 5.3 shows it near the destination c, and Figure 5.4 displays the robot near the middle on the linear track as it is coming back to

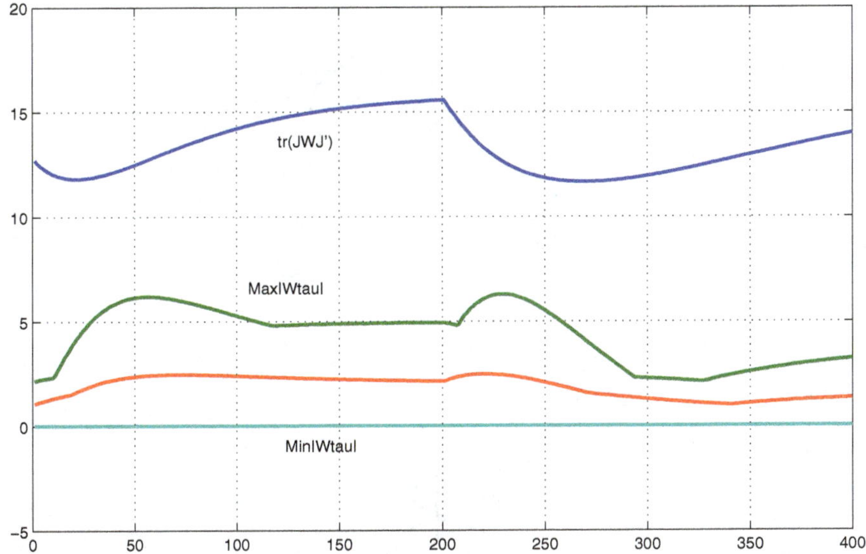

Fig. 5.5 A 7-joint redundant robot arm

the home position. The values of $\text{tr}(JWJ^T)$ and the maximum, the minimum, and the average absolute values over the seven joint torques are plotted in Figure 5.5. It can be clearly seen that if the maximum and minimum of the joint torques at each sampling point are getting closer to each other, $\text{tr}(JWJ^T)$ is approaching a smaller value. This evidently verifies that the potential function (5.12) is valid and also effective. We will apply it for much larger and more complex digital human models for their posture optimization in Chapter 9.

Let us now look at a three-revolute-joint planar arm, as shown in Figure 5.6. If the tip point of the robot is required only to draw a curve in 2D space without considering its orientation of the last frame, then the number of d.o.f. $m = 2$ and the number of joints $n = 3 > m$ so that this planar arm is a robot with one degree of redundancy. Under the link frames assignment given in Figure 5.6, its D-H table can immediately be determined as follows:

θ_i	d_i	α_i	a_i
θ_1	0	0	a_1
θ_2	0	0	a_2
θ_3	0	0	a_3

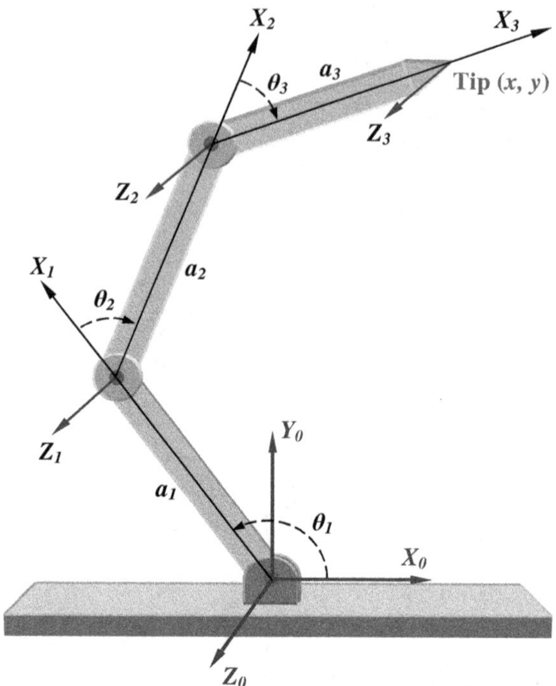

Fig. 5.6 A three-joint RRR planar redundant robot arm

From the D-H table, the one-step homogeneous transformations are determined by

$$A_0^1 = \begin{pmatrix} c_1 & -s_1 & 0 & a_1c_1 \\ s_1 & c_1 & 0 & a_1s_1 \\ 0 & 0 & 1 & 0 \\ 0 & 0 & 0 & 1 \end{pmatrix}, \quad A_1^2 = \begin{pmatrix} c_2 & -s_2 & 0 & a_2c_2 \\ s_2 & c_2 & 0 & a_2s_2 \\ 0 & 0 & 1 & 0 \\ 0 & 0 & 0 & 1 \end{pmatrix},$$

$$\text{and} \quad A_2^3 = \begin{pmatrix} c_3 & -s_3 & 0 & a_3c_3 \\ s_3 & c_3 & 0 & a_3s_3 \\ 0 & 0 & 1 & 0 \\ 0 & 0 & 0 & 1 \end{pmatrix}.$$

Then, we can find $A_0^3 = A_0^1 A_1^2 A_2^3$ and its last column should be the tip position vector p_0^3 for the robot, but we take only the first two nonzero elements due to $m = 2$, i.e.,

$$p_0^3 = \begin{pmatrix} a_1c_1 + a_2c_{12} + a_3c_{123} \\ a_1s_1 + a_2s_{12} + a_3s_{123} \end{pmatrix}.$$

Hence, the Jacobian matrix that is projected onto the base becomes

$$J = \frac{\partial p_0^3}{\partial q} = \begin{pmatrix} -a_1 s_1 - a_2 s_{12} - a_3 s_{123} & -a_2 s_{12} - a_3 s_{123} & -a_3 s_{123} \\ a_1 c_1 + a_2 c_{12} + a_3 c_{123} & a_2 c_{12} + a_3 c_{123} & a_3 c_{123} \end{pmatrix},$$

which is a 2 by 3 "short" matrix.

After the above preparation, we program them into Matlab^{TM} for a simulation study. Let $a_1 = a_2 = a_3 = 1$ m. Suppose that the initial joint values are $\theta_1 = 110^0$, $\theta_2 = -40^0$ and $\theta_3 = -30^0$ such that the initial tip position becomes $p_0^3(t = 0) = (0.7660 \;\; 2.5222)^T$. Now, starting from this initial Cartesian position, we want the robot tip point to follow a linear trajectory specified by

$$\begin{cases} x(t) = 0.2t + 0.7660 \\ y(t) = -0.4t + 2.5222. \end{cases}$$

Thus, the Cartesian velocity becomes $v = \begin{pmatrix} \dot{x} \\ \dot{y} \end{pmatrix} = \begin{pmatrix} 0.2 \\ -0.4 \end{pmatrix}$ in the unit of m./sec. and referred to the base.

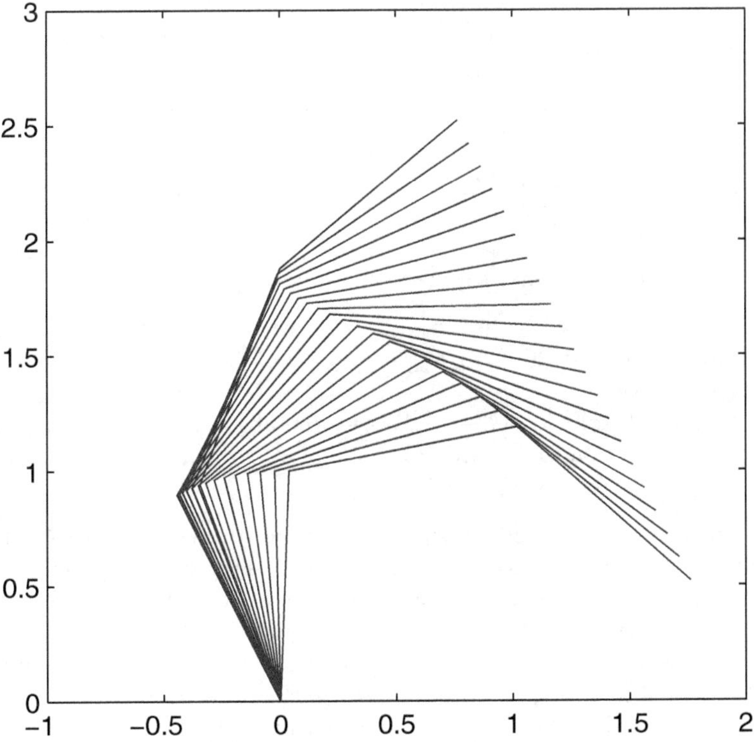

Fig. 5.7 Simulation results - only the rank (minimum-Norm) solution

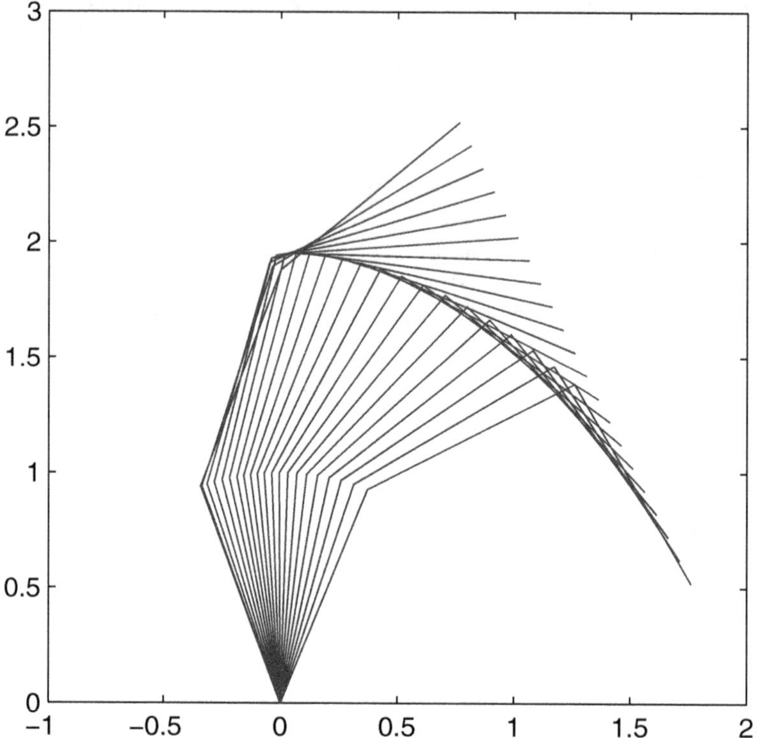

Fig. 5.8 Simulation results - both the rank and null solutions

In the simulation study, we made the following two cases:

1. Using only the rank solution $\dot{q} = J^+ v$ that is also the minimum-norm solution to update the joint angles $q_{new} = q_{current} + \dot{q}dt$ as the first-order approximation;
2. Adding a null solution $(I - J^+ J)k\eta$ to the above rank solution, where the potential function is defined by $p = \sin^2 \theta_3$ so that its gradient becomes $\eta = (0 \;\; 0 \;\; \sin 2\theta_3)^T$ with $k > 0$.

Note that the reason to define such a potential function in case 2 is to make the third joint angle be always as close to $\pm 90^0$ as possible. Figures 5.7 and 5.8 show its common motion output as the main task, but different instantaneous configurations between the two cases, as specified by the subtask. Clearly, the result in case 2 with $k = 1$ is, indeed, making link 3 be as perpendicular to link 2 as possible. In other words, $\theta_3 \to -90^0$ as the tip point is tracking the straight line for this three-revolute-joint redundant robot.

Another example with simulation study is to utilize the same 7-joint redundant robot as in Figure 5.2, but adding a post next to the robot for collision avoidance, as depicted in Figures 5.9 and 5.10.

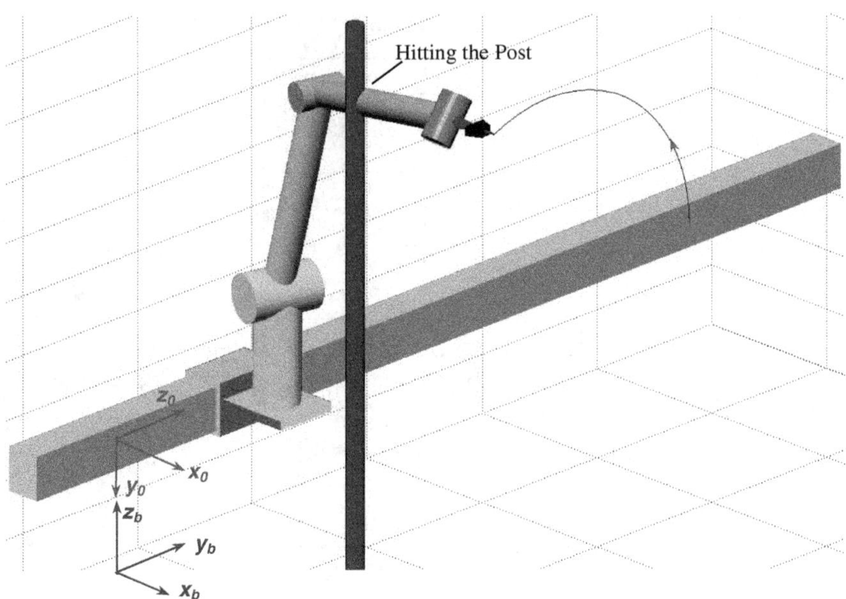

Fig. 5.9 The 7-joint robot arm is hitting a post when drawing a circle

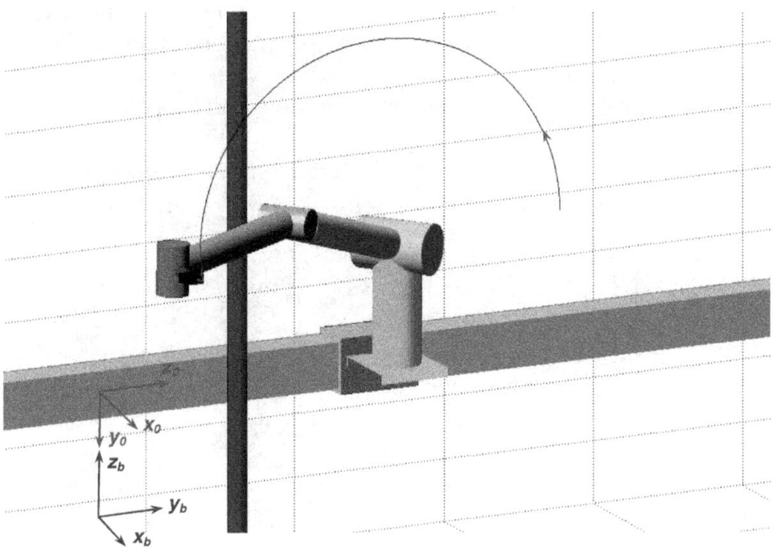

Fig. 5.10 The 7-joint robot is avoiding a collision by a potential function optimization

The D-H table for this 7-joint redundant robot arm is given as follows:

θ_i	d_i	α_i	a_i
$\theta_1 = 0$	d_1	90^0	a_1
θ_2	d_2	90^0	0
θ_3	0	0	a_3
θ_4	0	90^0	0
θ_5	d_5	-90^0	0
θ_6	0	90^0	0
θ_7	d_7	0	0

The above D-H table is started from the robot base frame 0. There is a world base frame b as a reference for ongoing path planning. The relation between frame 0 and frame b is given by

$$A_b^0 = \begin{pmatrix} 1 & 0 & 0 & 0 \\ 0 & 0 & 1 & 0 \\ 0 & -1 & 0 & 7 \\ 0 & 0 & 0 & 1 \end{pmatrix}, \tag{5.13}$$

where all the length units are in decimeters (dm.). Since for a robot with both revolute and prismatic joints, its joint position vector as well as the Jacobian matrix mix both angles in radians and displacements in length unit, they should have very close numerical values to avoid unnecessary numerical error when finding its I-K solution by inverting the Jacobian matrix. A length or displacement in dm. and an angle in radians are close to each other in value, and thus, we adopt dm. as the length unit here for every d_i and a_i in the above D-H table.

Based on the D-H table, the first three one-step homogeneous transformation matrices can be found as follows:

$$A_0^1 = \begin{pmatrix} 1 & 0 & 0 & a_1 \\ 0 & 0 & -1 & 0 \\ 0 & 1 & 0 & d_1 \\ 0 & 0 & 0 & 1 \end{pmatrix}, \quad A_1^2 = \begin{pmatrix} c_2 & 0 & s_2 & 0 \\ s_2 & 0 & -c_2 & 0 \\ 0 & 1 & 0 & d_2 \\ 0 & 0 & 0 & 1 \end{pmatrix},$$

and

$$A_2^3 = \begin{pmatrix} c_3 & -s_3 & 0 & a_3c_3 \\ s_3 & c_3 & 0 & a_3s_3 \\ 0 & 0 & 1 & 0 \\ 0 & 0 & 0 & 1 \end{pmatrix}. \tag{5.14}$$

Thus, the homogeneous transformation of frame 3 at the robot elbow with respect to the robot base can be calculated by

$$A_0^3 = A_0^1 A_1^2 A_2^3 = \begin{pmatrix} c_2c_3 & -c_2s_3 & s_2 & a_1 + a_3c_2c_3 \\ -s_3 & -c_3 & 0 & -d_2 - a_3s_3 \\ s_2c_3 & -s_2s_3 & -c_2 & d_1 + a_3s_2c_3 \\ 0 & 0 & 0 & 1 \end{pmatrix}.$$

Therefore, the position vector of the robot elbow p_0^3 is the 4th column of this A_0^3 referred to frame 0. Now, the vertical post with a radius = 0.3 dm. is standing at $x_b = 7$ and $y_b = 7.5$ in decimeters with respect to the world base, or $x_0 = 7$ and $z_0 = 7.5$ referred to the robot base frame 0. Let us define a virtual safe point at a position: $x_0 = 10$ and $z_0 = 20$ dm. and control the robot elbow to approach to it as closely as possible, as shown in Figure 5.11. This will have the same effect as avoiding the collision to the post.

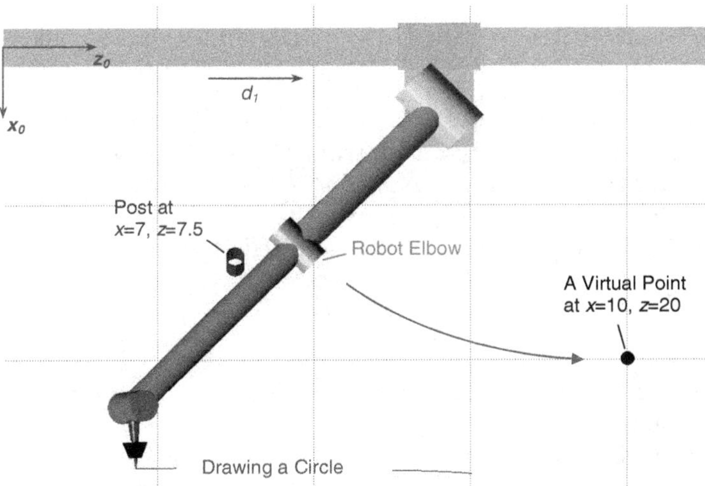

Fig. 5.11 A top view of the 7-joint redundant robot with a post and a virtual point

With the virtual safe point defined, the potential function to be minimized can be the distance between the robot elbow and the virtual point. Namely, let $\mu_x = p_{0x}^3 - 10 = a_1 + a_3c_2c_3 - 10$ and $\mu_z = p_{0z}^3 - 20 = d_1 + a_3s_2c_3 - 20$. Then,

$$p(q) = \frac{1}{2}(\mu_x^2 + \mu_z^2).$$

Its 7 by 1 gradient vector becomes

$$\eta = \frac{\partial p}{\partial q} = \begin{pmatrix} \mu_z \\ \mu_z a_3c_2c_3 - \mu_x a_3s_2c_3 \\ -\mu_z a_2s_2s_3 - \mu_x a_3c_2s_3 \\ 0 \\ 0 \\ 0 \\ 0 \end{pmatrix}.$$

If J_b is the Jacobian matrix of the robot and V_b is the end-effector Cartesian velocity vector for a circle drawing, both referred to the world base, and let $k = -1$ for minimization of the potential function $p(q)$, then, based on equation (5.7), the differential motion solution turns out to be

$$\dot{q} = J_b^+ V_b - (I - J_b^+ J_b)\eta.$$

More specifically, based on equation (4.14) in Chapter 4, the 6 by 7 Jacobian matrix $J_{(7)}$ that is projected to the last frame 7 for the 7-joint redundant robot is obtained by

$$J_{(7)} = \begin{pmatrix} r_7^0 & p_7^1 \times r_7^1 & \cdots & p_7^6 \times r_7^6 \\ 0 & r_7^1 & \cdots & r_7^6 \end{pmatrix},$$

where each p_7^{i-1} and each r_7^{i-1} for $i = 1, \cdots, 7$ are the 4th and 3rd columns of the homogeneous transformation matrix A_7^{i-1},respectively, which, as well as the cross-product between each pair of p_7^{i-1} and r_7^{i-1}, can all be numerically calculated in a MATLAB^{TM} program. After $J_{(7)}$ is prepared, according to equation (4.15), the Jacobian matrix projected onto the base can further be found by

$$J_b = \begin{pmatrix} R_b^7 & O_{3\times3} \\ O_{3\times3} & R_b^7 \end{pmatrix} J_{(7)},$$

where the rotation matrix R_b^7 is the upper-left 3 by 3 block of the homogeneous transformation matrix A_b^7.

To draw a circle of radius r on the $y_b - z_b$ base coordinates plane without any orientation change for frame 7, since the circle equation is given by

$$\begin{pmatrix} x \\ y \\ z \end{pmatrix} = \begin{pmatrix} 0 \\ r\cos(\omega t) + y_c \\ r\sin(\omega t) + z_c \end{pmatrix},$$

where y_c and z_c are the coordinates of the circle center with respect to the base and ω is the angular velocity of circle drawing, the Cartesian velocity should be the time-derivative of the above circle equation, i.e.,

$$V_b = \begin{pmatrix} v_b \\ \omega_b \end{pmatrix} = \begin{pmatrix} \dot{x} \\ \dot{y} \\ \dot{z} \\ \omega_x \\ \omega_y \\ \omega_z \end{pmatrix} = \begin{pmatrix} 0 \\ -r\omega\sin(\omega t) \\ r\omega\cos(\omega t) \\ 0 \\ 0 \\ 0 \end{pmatrix}.$$

As a result, Figure 5.9 shows that the elbow is hitting the post if $\dot{q}$ just takes the first term, i.e. the rank solution without any collision avoidance consideration. Once $\dot{q}$ also includes the second term, i.e. the null solution with the above gradient vector η, we can evidently see the effect of collision avoidance in Figure 5.10.

The above two simulation-based examples demonstrate the effectiveness of both main task execution and subtask optimization for a robot with only one degree of redundancy. In many cases, a robot may have higher degrees of redundancy, depending on how many joints it has and what the main task d.o.f. is. In such cases, one may add more potential functions to optimize them simultaneously.

Let an n-joint open serial-chain robot perform a main task that requires m d.o.f. with $m < n$. The degree of redundancy is $r = n - m$, and it should have an r-dimensional null space $N(J)$ to fill up to r independent subtasks for simultaneous optimization. Once each potential function $p_i(q)$ with $i = 1, \cdots, r$ is defined, their gradient vectors $\eta_1, \cdots, \eta_r$ will immediately be determined. Thus, the general solution turns out to be

$$\dot{q} = \dot{q}_r + \dot{q}_n = J^+V + (I - J^+J)(k_1\eta_1 + \cdots + k_r\eta_r).$$

In this multi-subtask case, we may interpret that each k_i is a weight on subtask i and all the k_i's can be dynamically controlled to reach the best overall subtask performance [7, 8]. Based on this notion, the above general solution can be re-organized as follows:

$$\dot{q} = J^+V + (I - J^+J)(\eta_1 \ \cdots \ \eta_r)\begin{pmatrix} k_1 \\ \vdots \\ k_r \end{pmatrix} = J^+V + (I - J^+J)Nk, \quad (5.15)$$

with an n by r matrix $N = (\eta_1 \ \cdots \ \eta_r)$ and an r by 1 column vector $k = (k_1 \ \cdots \ k_r)^T$. Since all the terms on the right-hand side of equation (5.15) are functions of q except the vector k, we may compare it with the standard form of nonlinear state-space equation:

$$\dot{x} = f(x) + \sum_{i=1}^{r} g_i(x)u_i = f(x) + G(x)u.$$

It is now quite clear that the robotic joint position vector q can be defined as a state vector $x \in \mathbb{R}^n$, the rank solution J^+V is considered as an n-dimensional tangent field $f(x)$, $(I - J^+J)N$ is the coefficient matrix $G(x)$ of the input, and vector $k \in \mathbb{R}^r$ is the control input u of the redundant robot kinematic system. Therefore, the r-dimensional control input k is no longer a constant. Instead, it should be determined and updated in a more dynamic fashion towards an overall optimization for all the r subtasks.

Starting with the above control system model, let an output vector $y \in \mathbb{R}^r$ be defined by

$$y = \begin{pmatrix} p_1(q) \\ \vdots \\ p_r(q) \end{pmatrix} = h(q)$$

that augments all the potential functions $p_i(q)$ for $i = 1, \cdots, r$, so that $h(q)$ is an r-dimensional output function of q. Since according to (5.15),

$$\dot{y} = \frac{\partial h}{\partial q}\dot{q} = \frac{\partial h}{\partial q}(J^+V + (I - J^+J)Nk),$$

and

$$\frac{\partial h}{\partial q} = \begin{pmatrix} \eta_1^T \\ \vdots \\ \eta_r^T \end{pmatrix} = N^T,$$

we have

$$\dot{y} = N^T J^+ V + N^T(I - J^+J)Nk = N^T J^+ V + Dk.$$

If the above r by r square matrix $D = N^T(I - J^+J)N$ is non-singular, then the control input can be resolved by

$$k = D^{-1}\dot{y} - D^{-1}N^T J^+ V.$$

Furthermore, if we wish each potential function $p_i(q)$ and its time-derivative $\dot{p}_i$ would approach a desired value p_i^d and $\dot{p}_i^d$, and by augmenting every p_i^d and $\dot{p}_i^d$ to form a desired output vector y^d and $\dot{y}^d$, respectively, then, we may define

$$\dot{y} = \dot{y}^d + K(y^d - y)$$

for a constant scalar or an r by r diagonal matrix $K > 0$ such that the control law can be determined by

$$k = D^{-1}[\dot{y}^d + K(y^d - y)] - D^{-1}N^T J^+ V.$$

This is conventionally called a *nonlinear state-feedback control*, and it can be readily justified that the above control law for the gain vector k can make each potential function $p_i(q)$ converge to its desired value p_i^d asymptotically. Clearly, it becomes feasible if the square matrix $D = N^T(I - J^+J)N$, called a *decoupling matrix*, is non-singular at each sampling point. This implies that if r is the dimension of the null space $N(J)$ of the Jacobian matrix J, then the number of independent potential functions to be controlled must be less than or equal to r. The concept and theory of such a nonlinear feedback control will be formally developed and further discussed in Chapter 8.

In fact, a regular robot arm with $m = n$ can be mounted on a wheeled or walking mobile cart/vehicle to extend its motion flexibility and working envelope. As shown in Figure 5.12, the degree of redundancy is usually equal to the number of axes that the cart or vehicle can offer. For the Stanford-Type robot arm sitting on a wheel-cart, since the wheel-cart can move in x and y directions and spin about the z-axis with respect to the world base frame b, the Stanford-Type arm is extended by three more joints. However, because the waist joint θ_1 of the robot arm shares the same axis with the

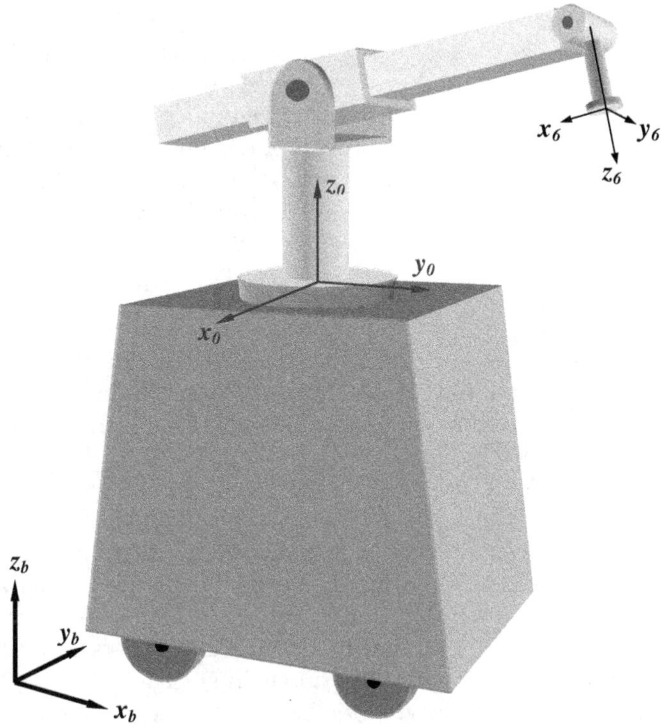

Fig. 5.12 The Stanford-type robot arm is sitting on a wheel mobile cart

cart spin, the net number of the independent axes added is reduced to 2 so that the degree of redundancy has to be $r = 2$.

If the robot arm could be mounted on the cart with an angle leaning away from the vertical axis, then the degree of redundancy would recover to $r = 3$. In either case, the entire **mobile robot** can be modeled as a highly redundant robotic system. Moreover, it is a reality that the three independent axes of the cart motion are just a theoretical model, and they are not directly controlled by three joint actuators. In other words, unlike a regular open serial-chain robot, the mobile cart/vehicle has no joint actuator at each axis of motion. Instead, the cart or the vehicle motion is driven indirectly by the four wheels with their steering system. Therefore, due to the indirect motion control fact, the dynamics, control and even kinematics of a mobile robot are often more complicated than a regular open serial-chain robot arm.

If the degree of redundancy is significantly high, it is often called a **hyper-redundant robotic system**. A snake-type or elephant (nose) trunk-like long serial-chain flexible robot can have up to 40 joints so that the degree of redundancy will be at least $r = 40 - 6 = 34$. In this special case, the subtask operation is even more significant than the main task execution, because the r-dimensional null space can provide a huge "room" to be filled with many desirable subtasks or sub-operations [9, 10].

The extensive research on redundant robotic systems kinematics, dynamics, control and design has a three-decade long history. A large volume of literature on this topic can be further referred to find the past, present and future trends in both theoretical developments and wide-spectrum applications [11, 12]. This section is just providing a summary of the mathematical principles, major concepts, algorithms and simulation studies in the kinematic modeling aspects of the robots with redundancy.

5.3 Hybrid-Chain Robotic Systems

An open-chain or a closed-chain multi-joint robot arm can be structured either in series or in parallel, or in the form of serial-parallel hybrid-chain mechanism. Figure 5.13 shows a serial-parallel hybrid-chain planar robot. The well-known Stewart platform, as shown in Figure 5.14, is the most typical 6-prismatic-joint parallel robot, which is serving in the U.S. Army Laboratories for tank vibration routine tests. The most typical hybrid-chain robot is our human body. If one wants to model a human body digitally, the four serial-chain limbs: two legs and two arms that are all connected in parallel to the human trunk can be integrated and grouped as a multi-joint hybrid-chain robotic system. Even for a human hand with five fingers connected to the common palm, it is also a typical hybrid-chain system.

In fact, a robotic system having two serial-chain arms that are connected in parallel with a common torso, like a human upper body without head, has appeared in the market as a heavy-duty dexterous industrial manipulator, called a dual-arm robot. In order to mimic a real human torso with two arms, each arm must be very flexible and dexterous. Figure 5.15 shows a single-arm 7-axis dexterous manipulator RRC K-1207 and a dual-arm 17-axis dexterous robot RRC K-2017 designed and developed by Robotics Research Corporation, Cincinnati, OH.

Before further exploring the kinematics for hybrid-chain robotic systems, let us first introduce and study how to determine the mobility, or the net degrees of freedom that a hybrid-chain mechanism can offer. In mechanics, the well-known **Grübler's formula** can directly predict the number m of net degrees of freedom for almost every open or closed hybrid-chain mechanical system [11, 13, 15].

Let l and n be the numbers of movable links and joints, respectively, for a system of interest, i.e., the fixed base must be excluded from the number of links l. Let f_i be the total number of independent axes that joint i can move, for $i = 1, \cdots, n$. For instance, f_i for a ball-joint without spin, or called a *spherical joint*, is two, while it is equal to three if including the link spin, called a *universal joint*. Then, the Grübler's formula is given by

$$m = D(l - n) + \sum_{i=1}^{n} f_i, \tag{5.16}$$

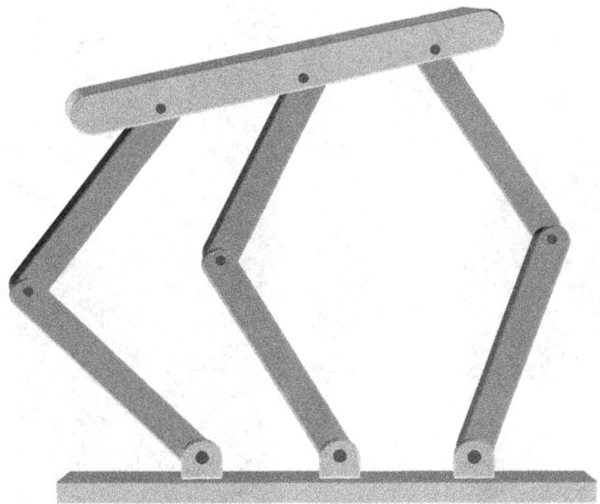

Fig. 5.13 A hybrid-chain planar robot

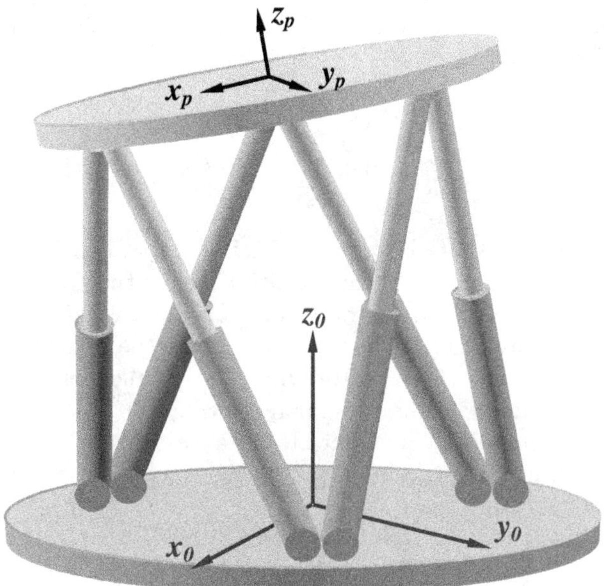

Fig. 5.14 Stewart platform - a typical 6-axis parallel-chain system

where D is the maximum d.o.f that the motion space of interest can offer. For instance, $D = 3$ for the motion on a 2D plane, while $D = 6$ in 3D space. By inspection, for the hybrid-chain planar robotic system in Figure 5.13, $D = 3$, $l = 7$ that exclude the fixed base, $n = 9$ and each joint offers $f_i = 1$ as a single axis for each joint. Then, according to (5.16),

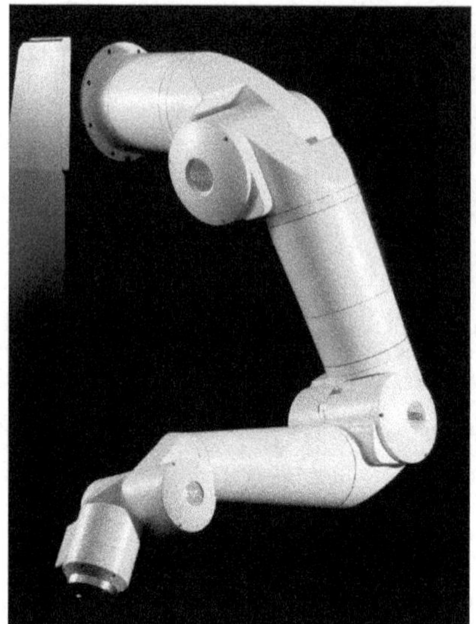

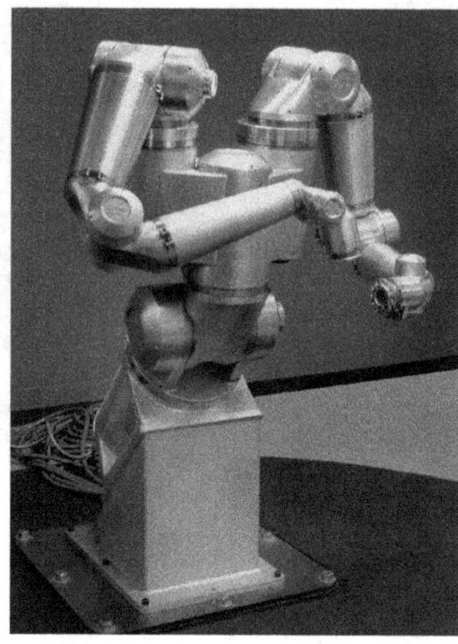

Fig. 5.15 A 7-axis dexterous manipulator RRC K-1207 and a dual-arm 17-axis dexterous manipulator RRC K-2017. Photo courtesy of Robotics Research Corporation, Cincinnati, OH.

$$m = 3(7-9) + \sum_{i=1}^{9} 1 = -3 \times 2 + 9 = 3.$$

This result shows that the top bar of the system can have 3 d.o.f. movement: translations of x and y, and rotation about the axis perpendicular to the plane. Therefore, based on the result, one can install three, and at most three, motors at any three of the 9 joints to drive the planar robot for 3 d.o.f. motion. Typically, the three motors may be installed on the three bottom joints to control uniquely the top bar motion in a 2D plane.

Note that a universal joint offers three axes of rotation, as a member of $SO(3)$ group. Since a spin belongs to $SO(2)$ group, based on topology, the quotient space

$$SO(3)/SO(2) \simeq S^2$$

is topologically equivalent or homeomorphic to a 2-sphere. This means that a universal (U-type) joint after eliminating its spin will be reduced to a spherical (S-type) joint. A conventional ball joint is a typical spherical joint.

For a Stewart platform that is also called a hexapod [13, 18], as shown in Figure 5.14, $D = 6$ in 3D motion space, and we can count its total number of links to be $l = 13$, including the top disc but excluding the fixed bottom base disc. The total number of joints is counted as $n = 18$. Among the 18 joints, suppose that each prismatic (P-type) joint offers one sliding axis, and each joint that connects each prismatic joint, or piston, to the top mobile disc is of spherical type that offers 2 axes each, while each joint connecting the piston to the base, i.e. the bottom disc, is of universal type offering 3 axes. This becomes a UPS-type structure for each of the six parallel legs. Thus, all the 6 prismatic joints provide 6 axes, and top 6 spherical joints provide $6 \times 2 = 12$ axes, while the bottom 6 universal joints provide $6 \times 3 = 18$ axes. The total 12 U/S-type joints provide $12 + 18 = 30$ axes. Finally, the net degrees of freedom for the Stewart platform turn out to be

$$m = 6(13 - 18) + 6 + 30 = -30 + 36 = 6.$$

Therefore, the top mobile disc can be driven by 6 actuators on the 6 prismatic joints to offer complete 6 d.o.f. motion.

It is also conceivable that the 6 d.o.f motion envelope for a serial-chain robot is, in general, much bigger than that of a parallel-chain robot. In contrast, the payload is just the opposite, and this was the primary reason why the U.S. Army laboratory uses the parallel-chain Stewart platform for their extremely heavy tank vibration test.

If we denote the second term of the Grübler's formula in (5.16) by $F = \sum_{i=1}^{n} f_i$ to represent the net amount of axes that all the joints of a system

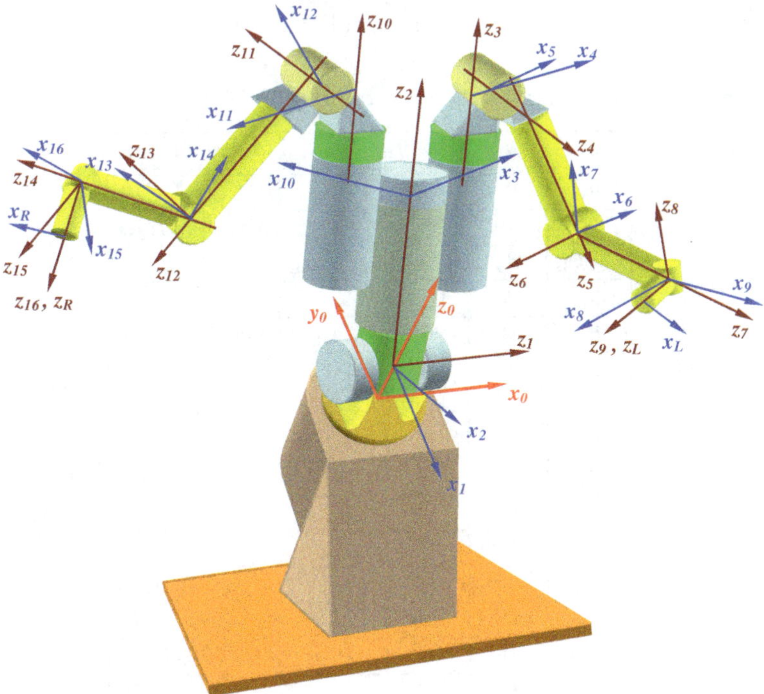

Fig. 5.16 Kinematic model of the two-arm 17-joint hybrid-chain robot

can offer, then, the Grübler's formula becomes $m = D(l-n)+F$. We classify most hybrid-chain mechanisms into two major categories:

1. If $n = l$, the number of joints is equal to the number of links, then $m = F$. This implies that the total number of axes offered by all the joints can be fully controlled to produce the same d.o.f. motion for such a system. Typical systems in this category are most of the open serial-chain robots, where each joint is actuated by a motor.
2. If $n > l$, the number of joints exceeds the number of links, then $m < F$. This means that there are $F - m$ excessive axes to be passive without control. Most full or partial closed parallel-chain mechanisms belong to this category.

It will be observed in the later development and analysis that the excessive axes for those systems in the second category often cause additional difficulty in solving their forward kinematics (F-K).

Let us now start investigating how to model a hybrid-chain robot kinematics by taking the dual-arm industrial robot, as shown in Figure 5.15, as a case study. Figure 5.16 depicts a two-arm robot theoretical model that was inspired by the industrial dual-arm dexterous robot designed and developed by Robotics Research Corporation along with all the link frames assignment by the D-H convention. Based on all the z_i and x_i-axes definitions for link i, we can readily find the D-H table for the two-arm hybrid-chain robot model in Table 5.1.

Table 5.1 The D-H table for the two-arm robot model

θ_i	d_i	α_i	a_i	**Joint Name**
θ_1	d_1	-90^0	0	
θ_2	0	90^0	0	Waist on Torso
θ_{3l}, θ_{3r}	d_3	0	$a_{3l} = a_{3r}$	
θ_4	d_4	90^0	a_4	
θ_5	0	90^0	a_5	Left Shoulder
θ_6	d_6	-90^0	0	
θ_7	0	90^0	0	Left Elbow
θ_8	d_8	90^0	0	
θ_9	0	90^0	0	Left Wrist
θ_{10}	d_{10}	0	0	
θ_{11}	d_{11}	90^0	a_{11}	
θ_{12}	0	90^0	a_{12}	Right Shoulder
θ_{13}	d_{13}	-90^0	0	
θ_{14}	0	90^0	0	Right Elbow
θ_{15}	d_{15}	90^0	0	
θ_{16}	0	90^0	0	Right Wrist
θ_{17}	d_{17}	0	0	

It can be seen that the common waist on the torso consists of three joint angles that are shared by both the left and right arms. The first two joint angles: θ_1 and θ_2 each has its individual joint value, while the value of the third one θ_3 has a constant difference between the left and right transitions from the torso to the two arms. Since the x_3 and x_{10} axes in Figure 5.16 are separated with a constant angle β, the relationship between θ_{3l} and θ_{3r} is clearly given by $\theta_{3r} = \theta_{3l} + \beta$. Due to the mechanical structure symmetry, it is true that the link lengths $a_{3l} = a_{3r}$ on the torso, and also $a_4 = a_{11}$ and $a_5 = a_{12}$ on the two shoulders. Similarly, the joint offsets $d_4 = d_{11}$ on the two shoulders, $d_6 = d_{13}$ on the two upper arms and $d_8 = d_{15}$ on the two forearms. Although the two end-effector offsets d_{10} and d_{17} look equal, too, we have to leave the two parameters determined by their specific applications. Since this is a typical *multiple end-effector* case, each end-effector may carry a different tool, each of d_{10} and d_{17} is finally determined by the total length, including the tool length, along the z_L axis and z_R axis, respectively.

Once the D-H parameter table is completed, it is easy now to find all the 17 one-step homogeneous transformations. The first three on the common torso are given as follows:

$$A_0^1 = \begin{pmatrix} c_1 & 0 & -s_1 & 0 \\ s_1 & 0 & c_1 & 0 \\ 0 & -1 & 0 & d_1 \\ 0 & 0 & 0 & 1 \end{pmatrix}, \quad A_1^2 = \begin{pmatrix} c_2 & 0 & s_2 & 0 \\ s_2 & 0 & -c_2 & 0 \\ 0 & 1 & 0 & 0 \\ 0 & 0 & 0 & 1 \end{pmatrix},$$

$$\text{and} \quad A_2^3 = \begin{pmatrix} c_3 & -s_3 & 0 & a_3c_3 \\ s_3 & c_3 & 0 & a_3s_3 \\ 0 & 0 & 1 & d_3 \\ 0 & 0 & 0 & 1 \end{pmatrix}.$$

Note that the angle θ_3 in A_2^3 is θ_{3l} as transiting to the left arm, and it is $\theta_{3r} = \theta_{3l} + \beta$ as transiting to the right arm, but they both share the same symbolical form of the homogeneous transformation A_2^3. Due to the constant difference β angle, it is quite significant that their joint velocities $\dot{\theta}_{3l} = \dot{\theta}_{3r} = \dot{\theta}_3$.

The remaining one-step homogeneous transformations for the joints/links on the two arms are straightforward, especially each twist angle α_i is either $\pm 90^0$ or 0 for this particular robot. Once we have all the 17 A_i^{i+1}'s ready, we need further to iteratively compute $A_8^{10} = A_8^9 A_9^{10}$, $A_7^{10} = A_7^8 A_8^{10}$, $A_6^{10} = A_6^7 A_7^{10}$, $\cdots$, until A_0^{10} for the left side of the robot. Likewise, continue to iteratively compute $A_{15}^{17} = A_{15}^{16} A_{16}^{17}$, $\cdots$, $A_{3r}^{17} = A_{3r}^{11} A_{11}^{17}$, $A_2^{17} = A_2^{3r} A_{3r}^{17}$, $\cdots$, until A_0^{17} for the right side of the robot. The index $3r$ just indicates the computation along the right arm, and as mentioned above, A_2^{3r} uses θ_{3r} but in the same symbolical form of A_2^3. After that, we have to invert each of the homogeneous transformation matrices to determine A_{10}^0, $\cdots$, A_{10}^9 for the torso plus left arm, as well as A_{17}^0, $\cdots$, A_{17}^{16} for the torso plus right arm. Taking the 3rd and 4th columns from each of the A_{10}^i's and A_{17}^j's, we are ready to find all the necessary Jacobian matrices for the two-arm hybrid-chain robot model.

Based on equation (4.14) in Chapter 4, the Jacobian matrices of the torso transiting to the left arm and to the right arm can be constructed respectively as follows:

$$J_{10}^{torso} = \begin{pmatrix} p_{10}^0 \times r_{10}^0 & p_{10}^1 \times r_{10}^1 & p_{10}^2 \times r_{10}^2 \\ r_{10}^0 & r_{10}^1 & r_{10}^2 \end{pmatrix} = \begin{pmatrix} s_{10}^0 & s_{10}^1 & s_{10}^2 \\ r_{10}^0 & r_{10}^1 & r_{10}^2 \end{pmatrix},$$

and

$$J_{17}^{torso} = \begin{pmatrix} p_{17}^0 \times r_{17}^0 & p_{17}^1 \times r_{17}^1 & p_{17}^2 \times r_{17}^2 \\ r_{17}^0 & r_{17}^1 & r_{17}^2 \end{pmatrix} = \begin{pmatrix} s_{17}^0 & s_{17}^1 & s_{17}^2 \\ r_{17}^0 & r_{17}^1 & r_{17}^2 \end{pmatrix},$$

and each of them is 6 by 3. The second matrix in each equation is a dual-number representation if you wish to derive the above Jacobian matrices using the dual-number transformation in lieu of the homogeneous transformation. Similarly, we can compute the two arms' Jacobian matrices:

$$J_{10}^{larm} = \begin{pmatrix} p_{10}^3 \times r_{10}^3 & p_{10}^4 \times r_{10}^4 & \cdots & p_{10}^9 \times r_{10}^9 \\ r_{10}^3 & r_{10}^4 & \cdots & r_{10}^9 \end{pmatrix} = \begin{pmatrix} s_{10}^3 & s_{10}^4 & \cdots & s_{10}^9 \\ r_{10}^3 & r_{10}^4 & \cdots & r_{10}^9 \end{pmatrix},$$

and

$$J_{17}^{rarm} = \begin{pmatrix} p_{17}^{3r} \times r_{17}^{3r} & p_{17}^{11} \times r_{17}^{11} & \cdots & p_{17}^{16} \times r_{17}^{16} \\ r_{17}^{3r} & r_{17}^{11} & \cdots & r_{17}^{16} \end{pmatrix} = \begin{pmatrix} s_{17}^{3r} & s_{17}^{11} & \cdots & s_{17}^{16} \\ r_{17}^{3r} & r_{17}^{11} & \cdots & r_{17}^{16} \end{pmatrix},$$

and each of them is 6 by 7. Since all the above computations are quite tedious in symbolical derivation, you may use computer programming, such as MATLAB^{TM}, to do them numerically, especially for the dual-number approach.

Let V_L and V_R be the 6 by 1 Cartesian velocities of the two end-effectors, and each augments both the linear velocity v and angular velocity ω, as defined in equation (4.13). Also, let $q = (\theta_1 \; \cdots \; \theta_{10} \; \theta_{11} \; \cdots \; \theta_{17})^T$ be the 17-dimensional joint position vector for the robot. Then, based on the Jacobian equation (3.22) and the matrix multiplication rule, we obtain

$$\begin{pmatrix} V_L \\ V_R \end{pmatrix} = J_{end}\,\dot{q} = \begin{pmatrix} J_{10}^{torso} & J_{10}^{larm} & O_{6\times 7} \\ J_{17}^{torso} & O_{6\times 7} & J_{17}^{rarm} \end{pmatrix} \dot{q}, \tag{5.17}$$

where $O_{6\times 7}$ is the 6 by 7 zero matrix. The 12 by 17 matrix J_{end} is called the **augmented Jacobian matrix** for such a hybrid-chain robot, which is now projected on the two respective end-effector frames. It can also be projected onto the common base by adopting equation (4.15), and both the two Cartesian velocities V_L and V_R for the two end-effectors can be planned with respect to the common base. Namely,

$$J_0 = \begin{pmatrix} R_0^{10} & O & O & O \\ O & R_0^{10} & O & O \\ O & O & R_0^{17} & O \\ O & O & O & R_0^{17} \end{pmatrix} J_{end},$$

where each O is the 3 by 3 zero matrix.

The 12 by 17 augmented Jacobian matrix J_0 is obviously "short" and possesses a $17 - 12 = 5$-dimensional null space. In other words, the two-arm hybrid-chain robot model is a redundant robot with the degree of redundancy $r = 5$. We may impose up to 5 subtasks for their simultaneous optimization while the two end-effectors are operating a specified main task. In fact, the two end-effectors (hands) can operate either two independent main tasks, or a single but coordinated main task. In the case of coordination, the two Cartesian velocities V_L and V_R will be related to each other, depending on the specification of the coordinated task for two hands.

This two-arm robot model will be digitally mocked up in MATLAB^{TM}. By further developing a differential motion based path/task planning procedure using the augmented Jacobian equation in (5.17), it will also be animated in the computer in Chapter 6.

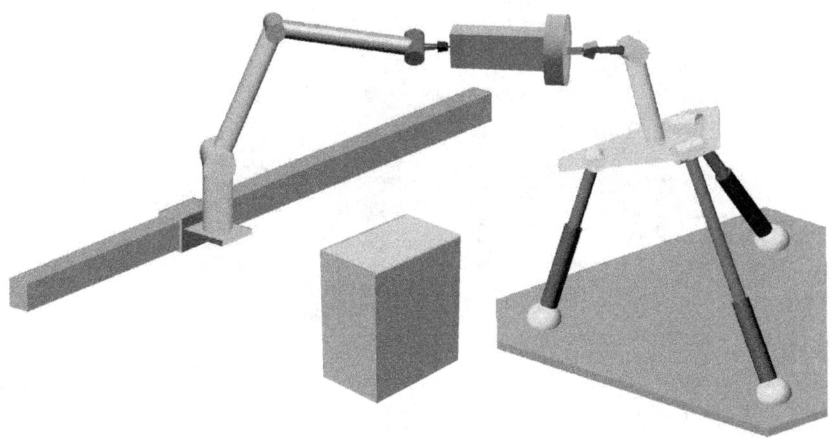

Fig. 5.17 A two-robot coordinated system

In addition to a hybrid-chain multiple end-effector robotic system, many research laboratories developed programs to allow multiple regular robot arms to work cooperatively [11]. Those multi-robot coordination applications often have no common torso or common waist. In this case, the augmented Jacobian matrix becomes decoupled, i.e.,

$$J = \begin{pmatrix} J_1 & O \\ O & J_2 \end{pmatrix},$$

if it is a two-robot coordinated system, where each J_i is the Jacobian matrix for robot i. However, the augmented Jacobian matrix J can be 12 by 12 and is clearly not a redundant case, unless a number of coordination constraints are imposed on the two Cartesian velocities to gain some redundancy. Often in such a decoupled coordinated system, two or more end-effectors operate a common task under certain coordination. Thus, the Cartesian velocities V_1 and V_2 are closely related by a requirement of the coordination. Figure 5.17 shows a typical two-robot coordinated system in working on a common task. More modeling, theories and applications in the areas of multi-robot coordination can be found in the literature [11].

Fig. 5.18 A Nao-H25 humanoid robotic system. Photo courtesy of Aldebaran Robotics, Paris, France.

Furthermore, a robotic hand with five fingers and a complete humanoid robot in Figure 5.18 are two most typical hybrid-chain multiple end-effector robotic systems. A humanoid robot can have four independent end-effectors: two hands and two feet. Both the above two cases are the redundant robotic systems after all of their Jacobian matrices are augmented. The reader is now able to try modeling each of them by using the procedures and kinematic algorithms that we have just discussed previously in this section.

5.4 Kinematic Modeling for Parallel-Chain Mechanisms

5.4.1 Stewart Platform

A general 6-axis parallel-chain hexapod is shown in Figure 5.19. Since both the bottom base plate and top mobile disc have six separated joints on each to connect the six pistons, it is often called a 6-6 parallel-chain hexapod, or Stewart platform [13]. If each pair of two adjacent pistons is geometrically merged to a single joint center underneath the top disc, then the top disc has only three joint points, but each of which is still a spherical type and offers two axes independently for each of the two adjacent pistons. If the base still has six independent universal joints without merging, then, it is called a 6-3 parallel Stewart platform. However, if the six universal joints on the base are also merged to three, then it is referred to as a 3-3 Stewart platform. Merging the spherical or universal joints is not an easy job for practical design, but it may ease the process of kinematic modeling and analysis. Therefore, let us first study a 3-3 Stewart platform model, and then extend it to 6-3 and 6-6 versions of parallel-chain mechanism.

Fig. 5.19 A 6-axis 6-6 parallel-chain hexapod system

To model a 3-3 Stewart platform, as a typical parallel-chain robotic system with six prismatic joints (pistons), let us define the base frame #0 on the base plate with the origin at the geometric center point of the three universal joints, as shown in Figure 5.20. Suppose that each pair of the adjacent pistons shares a single joint center that offers two-axis rotation for each end of the two pistons on the top and three-axis rotation on the base. The top three S-type joint points form three vertices A_6, B_6 and C_6 of an equilateral triangle underneath the top mobile disc. Likewise, the bottom three U-type joint points sit at the three vertices A_0, B_0 and C_0 of another equilateral triangle on the base plate.

Let frame #6 be originated at the center of the top disc with respect to A_6, B_6 and C_6. Then, the vector $p_0^6 \in \mathbb{R}^3$ that connects from the base origin to the origin of frame #6 becomes a position vector of the top mobile plate with respect to the base, while the rotation matrix $R_0^6 \in SO(3)$ determines the orientation of frame #6 with respect to the base. If each of the six P-type pistons is represented by a position vector $l_0^i \in \mathbb{R}^3$ for $i = 1, \cdots, 6$, then both the length and direction of each piston is completely determined by the corresponding position vector l_0^i. Furthermore, the vectors p_0^a, p_0^b and p_0^c that are all tailed at the base origin and arrow-pointing to the U-type joints A_0, B_0 and C_0, respectively, should be the three constant vectors referred to the base. Similarly, the vectors p_6^a, p_6^b and p_6^c that are all tailed at the origin of frame #6 and arrow-pointing to A_6, B_6 and C_6, respectively, are also the three constant vectors if they are referred to frame #6, see Figure 5.20 in detail. Then, we can immediately see that

$$l_0^1 = R_0^6 p_6^a + p_0^6 - p_0^a.$$

Likewise,

$$l_0^2 = R_0^6 p_6^a + p_0^6 - p_0^b, \quad l_0^3 = R_0^6 p_6^b + p_0^6 - p_0^b,$$
$$l_0^4 = R_0^6 p_6^b + p_0^6 - p_0^c, \quad l_0^5 = R_0^6 p_6^c + p_0^6 - p_0^c,$$

and

$$l_0^6 = R_0^6 p_6^c + p_0^6 - p_0^a. \tag{5.18}$$

Therefore, to determine the inverse kinematics (I-K) for this 3-3 Stewart platform, if p_0^6 and R_0^6 are given as a desired pair of position and orientation for the top disc with respect to the base, the above six equations in (5.18) can uniquely find each piston position vector l_0^i that includes both its length and direction. We will use the six inverse kinematics (I-K) equations to draw and animate a 3-3 Stewart platform in MATLAB^{TM} in the next chapter.

However, in terms of the complexity, such a straightforward I-K solution in (5.18) will make a huge contrast to its forward kinematics (F-K). An F-K problem for a Stewart platform is to find both p_0^6 and R_0^6 if the length l_i for each of its six pistons is given. In other words, only the six prismatic joint values in $q = (l_1 \ \cdots \ l_6)^T$ are given without their directions. Intuitively,

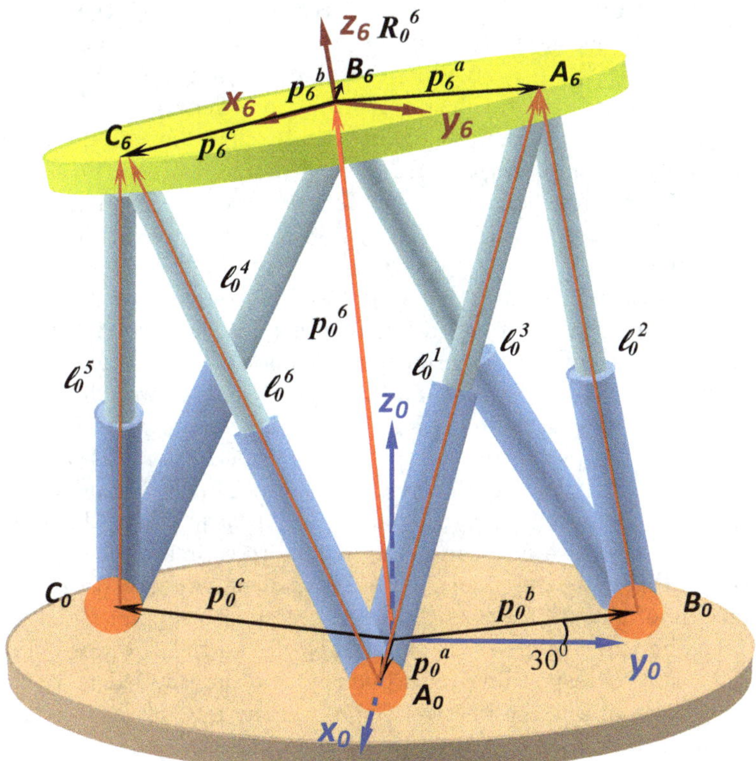

Fig. 5.20 Kinematic model of a 3-3 Stewart platform

giving six joint values to solve 6 d.o.f. Cartesian output is supposed to be sufficiently and uniquely solvable. However, as we classified earlier, most such parallel-chain or partial parallel-chain robotic systems belong to the second category that often contains many excessive axes.

Based on the Grübler's formula, the net d.o.f. of mobility for a mechanism $m = D(l - n) + F$ with $F = \sum_{i=1}^{n} f_i$ as the total number of axes that the system can offer. In most open serial-chain robots, the first term is often equal to zero due to $l = n$ so that the number of axes F can directly determine its net d.o.f. of mobility. In other words, there is no excessive (passive) joint in most open serial-chain robots in nature. This is the major reason why almost every open serial-chain robot can have a systematic kinematic model, such as the D-H convention to solve both the I-K and F-K problems.

In contrast, in most robotic systems with a parallel-chain mechanism, the number of joints is often greater than the number of links, $n > l$, so that the first term $D(l-n) < 0$. This causes the second term F to be greater than the net d.o.f. m of the system so that more excessive joints have to remain passive in such a parallel-chain or partial parallel-chain robotic system. For instance, as we predicted by using the Grübler formula for the Stewart platform earlier, the net d.o.f. was only $m = 6$, but the total number of axes was counted as $F = \sum_{i=1}^{n} f_i = 36$. Therefore, because of the $F - m = 36 - 6 = 30$ excessive joints, even if all the six prismatic joint lengths in $q = (l_1 \ \cdots \ l_6)^T$ are given, it will be extremely difficult to find a closed form to solve its $m = 6$ net d.o.f. output, i.e., both p_0^6 and R_0^6. As a matter of fact, the six I-K equations in (5.18) cannot be reversed to directly resolve p_0^6 and R_0^6 in terms of only the six known piston lengths $l_i = \|l_0^i\|$ for $i = 1, \cdots, 6$ without knowing their directions.

Nevertheless, we can at least seek a set of equations for a 3-3 Stewart platform to represent its forward kinematics (F-K). As we can see from Figure 5.21, because all the lengths l_i's are known, the shapes of the three triangles: $\triangle A_0B_0A_6$, $\triangle B_0C_0B_6$ and $\triangle C_0A_0C_6$ can be well determined, but the orientation of each triangle is still an unknown variable. Let the height h_1 from A_6 be perpendicular to the base line $\overline{A_0B_0}$ for $\triangle A_0B_0A_6$, let the height h_2 from B_6 be perpendicular to $\overline{B_0C_0}$ for $\triangle B_0C_0B_6$, and let the height h_3 from C_6 be perpendicular to $\overline{C_0A_0}$ for $\triangle C_0A_0C_6$. Then, the angle θ_1 between h_1 and the base disc can be a variable to represent the orientation of $\triangle A_0B_0A_6$. Likewise, the angles θ_2 and θ_3 can be variables to represent the orientations of $\triangle B_0C_0B_6$ and $\triangle C_0A_0C_6$, respectively, as depicted in Figure 5.21. The three intersection points Q_1, Q_2 and Q_3 between the heights h_i's and their corresponding base lines can be determined in terms of the segments b_1, b_2 and b_3, respectively, through the Law of Cosine on each triangle, and they will be illustrated in a later numerical example. Once b_1, b_2 and b_3 are found, each height h_i can exactly be determined accordingly.

Furthermore, the top vertices of the three triangles: A_6, B_6 and C_6 are varying and tracking along three circles, each of which is centered at the foot point Q_i of each height h_i's and has a radius equal to h_i for $i = 1, 2, 3$. We may find three equations to describe the three circles with respect to the base frame #0 in such a way that we can write a 3D parametric equation for Circle 2, because its base line $\overline{B_0C_0}$ is parallel to the y_0-axis of the base. Namely,

$$p_0^{b6} = \begin{pmatrix} x_2 \\ y_2 \\ z_2 \end{pmatrix} = \begin{pmatrix} x_{q2} + h_2 \cos\theta_2 \\ y_{q2} \\ h_2 \sin\theta_2 \end{pmatrix}, \tag{5.19}$$

where x_{q2} and y_{q2} are two constant x and y-coordinates of the center point Q_2 of Circle 2 with respect to the base.

Next, in order to find the equation for Circle 3, let us imagine that the entire $\triangle C_0A_0C_6$ was originally sitting at the same position as $\triangle B_0C_0B_6$ to find the coordinates x_{q3} and y_{q3} for the center point Q_3 of Circle 3 with respect to the base, similar to what we did for the determination of Circle 2. Then, rotate this imaginary $\triangle C_0A_0C_6$ to its actual position by $+120^0$ about the z_0-axis. Namely,

$$p_0^{c6} = \begin{pmatrix} x_3 \\ y_3 \\ z_3 \end{pmatrix} = \begin{pmatrix} \cos(120^0) & -\sin(120^0) & 0 \\ \sin(120^0) & \cos(120^0) & 0 \\ 0 & 0 & 1 \end{pmatrix} \begin{pmatrix} x_{q3} + h_3 \cos\theta_3 \\ y_{q3} \\ h_3 \sin\theta_3 \end{pmatrix}. \quad (5.20)$$

For Circle 1, by applying the same imagination on $\triangle A_0 B_0 A_6$, but rotating it about the z_0-axis by -120^0, instead of $+120^0$, we reach the following equation for Circle 1:

$$p_0^{a6} = \begin{pmatrix} x_1 \\ y_1 \\ z_1 \end{pmatrix} = \begin{pmatrix} \cos(120^0) & \sin(120^0) & 0 \\ -\sin(120^0) & \cos(120^0) & 0 \\ 0 & 0 & 1 \end{pmatrix} \begin{pmatrix} x_{q1} + h_1 \cos\theta_1 \\ y_{q1} \\ h_1 \sin\theta_1 \end{pmatrix}. \quad (5.21)$$

In fact, x_{q1}, x_{q2} and x_{q3} should be the same constant due to $\overline{B_0 C_0}$ always being parallel to the y_0-axis with a constant distance behind the y_0-axis, as depicted in Figure 5.21.

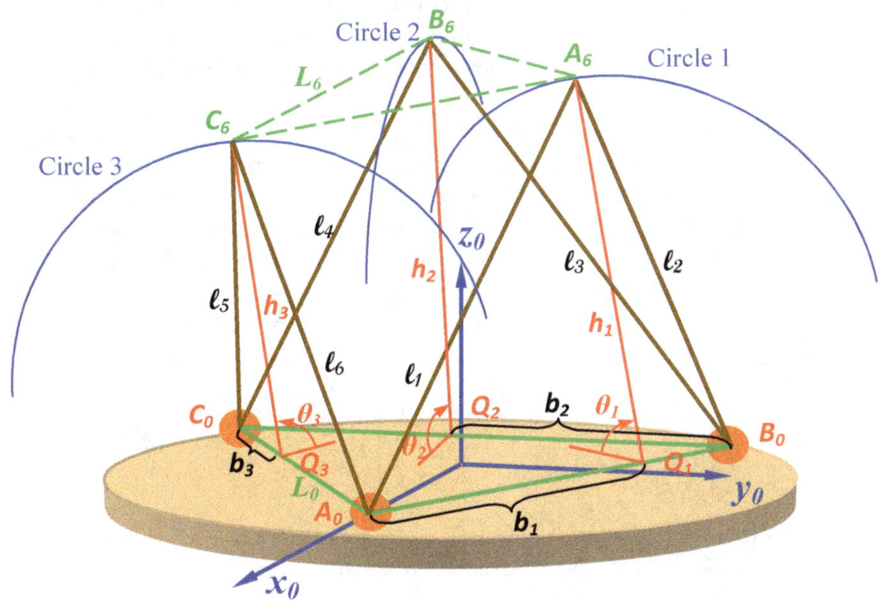

Fig. 5.21 Solution to the forward kinematics of the Stewart platform

Clearly, each of the three circular parametric equations contains only one single variable θ_i for $i = 1, 2, 3$ if all the six piston lengths l_i's are given. With such a detailed geometric interpretation, the F-K problem for the 3-3 Stewart platform can now be rephrased to determine three points: A_6 on Circle 1, B_6 on Circle 2 and C_6 on Circle 3 such that the distances between each pair of them are all equal to a common fixed length $\overline{A_6 B_6} = \overline{B_6 C_6} = \overline{C_6 A_6} = L_6$ that is the distance between two of the three spherical joints underneath the top mobile disc of the platform. In other words, if we make a hard paper equilateral triangle with each side equal to L_6 and drop it on the top of the three circles, where will the three vertices of the equilateral triangle touch the three circles with one on each? In mathematical language, under the above three parametric equations of the three circles with respect to the base as three conditions, we wish to solve the following three simultaneous equations:

$$\begin{cases} (x_1 - x_2)^2 + (y_1 - y_2)^2 + (z_1 - z_2)^2 = L_6^2 \\ (x_2 - x_3)^2 + (y_2 - y_3)^2 + (z_2 - z_3)^2 = L_6^2 \\ (x_3 - x_1)^2 + (y_3 - y_1)^2 + (z_3 - z_1)^2 = L_6^2. \end{cases} \quad (5.22)$$

Intuitively, there are only three variables θ_1, θ_2 and θ_3 to be solved from the three simultaneous equations in (5.22) under the three conditions of (5.19), (5.20) and (5.21). It seems to be solvable. Actually, it is not so easy, because the three equations in (5.22) are all quadratic, and every variable in the three conditions is involved in either sine or cosine functions. Therefore, there is no closed form for the solution, and one may have to prepare a recursive subroutine and call it in a computer to resolve the F-K problem at each sampling point for the 3-3 Stewart platform of such a typical fully parallel-chain system.

Once the three angles θ_1, θ_2 and θ_3 could be resolved at each sampling point by whatever algorithm or program in a computer, the final solution for p_0^6 and R_0^6 would not be far away. In fact, after substituting each θ_i into equations (5.19), (5.20) and (5.21), the radial vectors p_0^{b6}, p_0^{c6} and p_0^{a6} that are pointing to B_6, C_6 and A_6, respectively, and tailed at the common base origin are well determined. Then, by applying the well-known method of finding the center of gravity for a rigid body to the top equilateral triangle disc, the position vector should be

$$p_0^6 = \frac{1}{3}\left[\begin{pmatrix} x_1 \\ y_1 \\ z_1 \end{pmatrix} + \begin{pmatrix} x_2 \\ y_2 \\ z_2 \end{pmatrix} + \begin{pmatrix} x_3 \\ y_3 \\ z_3 \end{pmatrix}\right] = \frac{1}{3}(p_0^{a6} + p_0^{b6} + p_0^{c6}). \quad (5.23)$$

After the position vector p_0^6 is found, by inspecting Figure 5.20 more closely, the I-K equations in (5.18) can be reduced to

$$R_0^6 p_6^a + p_0^6 = p_0^{a6}, \quad R_0^6 p_6^b + p_0^6 = p_0^{b6},$$

and

$$R_0^6 p_6^c + p_0^6 = p_0^{c6}.$$

Augmenting the above three equations together, we obtain

$$R_0^6 (p_6^a \;\; p_6^b \;\; p_6^c) = (p_0^{a6} - p_0^6 \;\; p_0^{b6} - p_0^6 \;\; p_0^{c6} - p_0^6).$$

Unfortunately, since p_6^a, p_6^b and p_6^c are all referred to frame #6 and laying on the same top disc plane so that the last element of each vector must be zero, R_0^6 cannot be solved by taking the inverse of this singular matrix $(p_6^a\ p_6^b\ p_6^c)$. We have to replace any one of the three columns in the matrix by a cross-product between the other two. For instance, let p_6^b be replaced by the cross-product $p_6^a \times p_6^c$. Thus, the new equation becomes

$$R_0^6(p_6^a \quad p_6^c \quad p_6^a \times p_6^c) = (p_0^{a6} - p_0^6 \quad p_0^{c6} - p_0^6 \quad (p_0^{a6} - p_0^6) \times (p_0^{c6} - p_0^6)),$$

where a property of cross-product transformation given in (3.14) was applied. Since the new 3 by 3 matrix $P_6 = (p_6^a \quad p_6^c \quad p_6^a \times p_6^c)$ next to R_0^6 on the left-hand side of the above new equation is now nonsingular, we can finally resolve the orientation of the top mobile disc, R_0^6, for the 3-3 Stewart platform. Namely, if we denote

$$(p_0^{a6} - p_0^6 \quad p_0^{c6} - p_0^6 \quad (p_0^{a6} - p_0^6) \times (p_0^{c6} - p_0^6)) = P_0,$$

then,

$$R_0^6 = P_0 P_6^{-1}. \tag{5.24}$$

Let us give an example and try to call a recursive algorithm programmed in MATLAB^{TM} to numerically solve such a difficult Stewart platform F-K problem, even though it is just a 3-3 type hexapod. Suppose that for a 3-3 Stewart platform, as shown in Figure 5.21, the base disc equilateral triangle has each side $L_0 = 1.2$ and the top disc equilateral triangle has each side $L_6 = 1$ in meters. If the six prismatic joint lengths are given by $l_1 = 1$, $l_2 = 0.8$, $l_3 = 1.2$, $l_4 = 1.1$, $l_5 = 0.9$ and $l_6 = 1$, all in meters, then, we can apply the Law of Cosine on each triangle to find all the three angles $\angle A_6 A_0 B_0 = A_0$, $\angle B_6 B_0 C_0 = B_0$ and $\angle C_6 C_0 A_0 = C_0$ in the following cosine form:

$$\cos(A_0) = \frac{l_1^2 + L_0^2 - l_2^2}{2 l_1 L_0}, \quad \cos(B_0) = \frac{l_3^2 + L_0^2 - l_4^2}{2 l_3 L_0},$$

and

$$\cos(C_0) = \frac{l_5^2 + L_0^2 - l_6^2}{2 l_5 L_0}.$$

Their sine values can be determined directly by $\sin(A_0) = \sqrt{1 - \cos^2(A_0)}$ because the range of each angle is in $(0,\ 180^0)$ so that the sine value of each angle is always positive. Hence,

$$b_1 = l_1 \cos(A_0) \quad \text{and} \quad h_1 = l_1 \sin(A_0),$$

$$b_2 = l_3 \cos(B_0) \quad \text{and} \quad h_2 = l_3 \sin(B_0),$$

and

$$b_3 = l_5 \cos(C_0) \quad \text{and} \quad h_3 = l_5 \sin(C_0).$$

Since x_{q2}, as an x-coordinate of the point Q_2, is a distance behind the y_0-axis, it should be negative $\frac{1}{3}$ of the height of the base disc equilateral triangle, i.e., $x_{q2} = -\frac{\sqrt{3}}{6} L_0$. Based on the rotation idea, the other two: x_{q3} and x_{q1}

are all the same as x_{q2} in equations (5.19), (5.20) and (5.21). Whereas the y-coordinates of Q_i are different, and each can be determined by $y_{qi} = \frac{L_0}{2} - b_i$ for $i = 1, 2, 3$.

After the above preparation, we can now write a MATLAB^{TM} program to find solutions for p_0^6 and R_0^6. Because of no closed form of solutions for θ_1, θ_2 and θ_3 in equations (5.19), (5.20) and (5.21) along with (5.22), we have to use a three-while-loop based recursive algorithm to search and determine all the angles so that the radial vectors p_0^{a6}, p_0^{b6} and p_0^{c6} can all be resolved. Each angle θ_i in the algorithm is starting from 10^0 and making $N = 200$ sampling points with each step size $\Delta\theta = 0.8^0$ up to the maximum 170^0. When the search process finds a solution, it will automatically stop and print out both p_0^6 and R_0^6 through equations (5.23) and (5.24). The MATLAB^{TM} code is given as follows:

```
L6=1; L0=1.2;
l1=1; l2=0.8; l3=1.2; l4=1.1; l5=0.9; l6=1;    % Input Data

x6=sqrt(3)/3*L6;
p6c=[x6; 0; 0];   p6a=[-x6*sin(pi/6); x6*cos(pi/6); 0];

xq=-sqrt(3)*L0/6;

ca=(l1^2+L0^2-l2^2)/(2*l1*L0);
cb=(l3^2+L0^2-l4^2)/(2*l3*L0);
cc=(l5^2+L0^2-l6^2)/(2*l5*L0);    % The Law of Cosine

sa=sqrt(1-ca^2);   sb=sqrt(1-cb^2);   sc=sqrt(1-cc^2);

h1=l1*sa;   h2=l3*sb;   h3=l5*sc;
b1=l1*ca;   b2=l3*cb;   b3=l5*cc;

yq1=L0/2-b1;   yq2=L0/2-b2;   yq3=L0/2-b3;

th0=pi/36;   % Search Starting Angle
a=2*pi/3;    % +120 and -120 Degrees Rotations

i=0; j=0; k=0;

while i<=200
    th1=th0+i*0.8*pi/180;
    pa6=[cos(a) sin(a) 0; -sin(a) cos(a) 0; 0 0 1]* ...
                   [xq+h1*cos(th1); yq1; h1*sin(th1)];
    while j<=200
        th2=th0+j*0.8*pi/180;
        pb6=[xq+h2*cos(th2); yq2; h2*sin(th2)];
        if abs(norm(pb6-pa6)-L6) < 0.01
            while k<=200
                th3=th0+k*0.8*pi/180;
```

```
                    pc6=[cos(a) -sin(a) 0
                         sin(a) cos(a) 0
                          0 0 1]*...
                      [xq+h3*cos(th3); yq3; h3*sin(th3)];
                    if abs(norm(pc6-pa6)-L6) < 0.01 && ...
                          abs(norm(pc6-pb6)-L6) < 0.01
                        i=201;  k=201;  j=201;
                    end
                    k=k+1;
                end
            end
            j=j+1; k=0;
        end
        i=i+1; j=0; k=0;
end

theta=[th1 th2 th3]*180/pi
          % Print all the resulting thetas in degree

p06=(pa6+pb6+pc6)/3

R06=[pa6-p06 pc6-p06 cross(pa6-p06, pc6-p06)]/ ...
 [p6a p6c cross(p6a, p6c)]  % The F-K Solutions
```

The final results are printed out below:

```
theta =

   99.4000  109.8000  106.6000


p06 =

   -0.1236
   -0.0495
    0.7586


R06 =

    0.5769    0.7749    0.2724
   -0.8170    0.5644    0.0801
   -0.0956   -0.2673    0.9587
```

If the input of this F-K problem is changed to $l_1 = 0.8$, $l_2 = 0.6$, $l_3 = 1$, $l_4 = 1.2$, $l_5 = 0.7$ and $l_6 = 0.9$, then the output will immediately pop out in the MATLAB^{TM} working window as follows:

```
theta =

  121.0000  104.2000  101.0000


p06 =

   -0.0994
    0.0825
    0.5661


R06 =

    0.3584    0.7331    0.5657
   -0.9272    0.3336    0.1382
   -0.0934   -0.5765    0.7993
```

One can test and verify the above two results by substituting each pair of p_0^6 and R_0^6 into the I-K equation in (5.18), and the norm of each vector l_0^i will agree exactly with the input of the above F-K program.

We have discussed thus far both the I-K and F-K for a 3-3 Stewart platform. It can be easily extended to 6-3 and even 6-6 Stewart platforms for the I-K formulation given in (5.18). Since each of the top and bottom discs has now six geometric joint points for a 6-6 Stewart platform, we may split each p_0^a into p_0^{a1} and p_0^{a2} to respond the two different joint points at A_{01} and A_{02}. Applying the same splitting procedure to A_6 as well as to the rest of the joint points B_0, C_0, B_6 and C_6, we can have a more general I-K solution for a 6-6 Stewart platform:

$$l_0^1 = R_0^6 p_6^{a1} + p_0^6 - p_0^{a2}, \quad l_0^2 = R_0^6 p_6^{a2} + p_0^6 - p_0^{b1},$$

$$l_0^3 = R_0^6 p_6^{b1} + p_0^6 - p_0^{b2}, \quad l_0^4 = R_0^6 p_6^{b2} + p_0^6 - p_0^{c1},$$

and

$$l_0^5 = R_0^6 p_6^{c1} + p_0^6 - p_0^{c2}, \quad l_0^6 = R_0^6 p_6^{c2} + p_0^6 - p_0^{a1}. \tag{5.25}$$

In fact, all the six joint points on either the base or the top disc may not necessarily be forming a symmetric shape. Instead, they can be arbitrary and only some constant parameters, such as p_0^{ai} or p_6^{ai}, need to be re-measured, the I-F formulation will remain the same.

The F-K algorithm for the 3-3 Stewart platform can also be extended to a 6-3 type one, because splitting each of A_0, B_0 and C_0 on the base does not destroy each triangle formed by two adjacent pistons along with the base line L_0, provided that the top A_6, B_6 and C_6 are kept without splitting. There are only a few parameters, such as x_{qi} and y_{qi} for each $i = 1, 2, 3$, needed to be redefined, the search algorithm and program will remain the same. However, if the system is of 6-6 type, i.e., the top three joint points are also split to six different geometric points, then the above F-K algorithm will no longer be valid. In this general case, each triangle becomes a polygon with four vertices, and they may not always stay on a common plane.

5.4.2 Jacobian Equation and the Principle of Duality

Let us now turn our attention to investigating the kinematic behavior in tangent space for a general 6-6 Stewart platform. According to the I-K solution in (5.25), let the superscript i of each p_6^i or p_0^i be $a1 = 1$, $a2 = 2$, $b1 = 3$, $b2 = 4$, $c1 = 5$ and $c2 = 6$ for the sake of short notation. Taking time-derivatives for both sides of the i-th equation yields

$$\dot{l}_0^i = \dot{R}_0^6 p_6^i + \dot{p}_0^6,$$

because p_0^i and p_6^i are all the constant vectors. Recalling equation (3.10) from Chapter 3, $\Omega_0^6 = \omega_0^6 \times = \dot{R}_0^6 R_6^0$, the skew-symmetric matrix of the angular velocity ω_0^6 of the top disc, and noticing that $\dot{p}_0^6 = v_0^6$, the linear velocity of the top disc, we further have

$$\dot{l}_0^i = \Omega_0^6 R_0^6 p_6^i + v_0^6.$$

Since the vector l_0^i for the i-th prismatic leg can be rewritten as $l_0^i = q_i r_0^i$, where $r_0^i = l_0^i / \|l_0^i\|$ is the unit vector of l_0^i so that $q_i = l_i$ is the length of the i-th piston leg, its time-derivative becomes

$$\dot{l}_0^i = \dot{q}_i r_0^i + q_i \dot{r}_0^i. \tag{5.26}$$

Let $R_0^6 p_6^i = p_{6(0)}^i$ and its skew-symmetric matrix $P_{6(0)}^i = p_{6(0)}^i \times$. Then,

$$\Omega_0^6 R_0^6 p_6^i = \omega_0^6 \times p_{6(0)}^i = -P_{6(0)}^i \omega_0^6.$$

Moreover, since $r_0^{iT} r_0^i \equiv 1$,

$$\dot{r}_0^{iT} r_0^i + r_0^{iT} \dot{r}_0^i = 0.$$

On the other hand, the transpose of a scalar is equal to the scalar itself, i.e.,

$$\dot{r}_0^{iT} r_0^i = (\dot{r}_0^{iT} r_0^i)^T = r_0^{iT} \dot{r}_0^i.$$

Hence, $r_0^{iT} \dot{r}_0^i \equiv 0$. Namely, a unit vector is always perpendicular to its time-derivative.

Premultiplying r_0^{iT} to both sides of (5.26) and then substituting the above two identities $r_0^{iT} r_0^i \equiv 1$ and $r_0^{iT} \dot{r}_0^i \equiv 0$ into it, we obtain

$$\dot{q}_i = -r_0^{iT} P_{6(0)}^i \omega_0^6 + r_0^{iT} v_0^6 = \begin{pmatrix} r_0^{iT} & -r_0^{iT} P_{6(0)}^i \end{pmatrix} \begin{pmatrix} v_0^6 \\ \omega_0^6 \end{pmatrix}.$$

Now, by augmenting all the 6 prismatic joint velocities $\dot{q}_1, \cdots, \dot{q}_6$ together, we achieve a new transformation in tangent space:

$$\dot{q} = \begin{pmatrix} \dot{q}_1 \\ \vdots \\ \dot{q}_6 \end{pmatrix} = \begin{pmatrix} r_0^{1T} & -r_0^{1T} P_{6(0)}^1 \\ \vdots & \vdots \\ r_0^{6T} & -r_0^{6T} P_{6(0)}^6 \end{pmatrix} \begin{pmatrix} v_0^6 \\ \omega_0^6 \end{pmatrix}. \tag{5.27}$$

Let a Jacobian matrix of the 6-6 Stewart platform be defined by

$$J_0 = \begin{pmatrix} r_0^1 & \cdots & r_0^6 \\ p_{6(0)}^1 \times r_0^1 & \cdots & p_{6(0)}^6 \times r_0^6 \end{pmatrix}, \tag{5.28}$$

and let a 6 by 1 Cartesian velocity of the top disc of the Stewart platform be

$$V_0 = \begin{pmatrix} v_0^6 \\ \omega_0^6 \end{pmatrix}.$$

Then, equation (5.27) is actually a Jacobian equation for the closed parallel-chain Stewart platform, i.e.,

$$\dot{q} = J_0^T V_0. \tag{5.29}$$

Comparing the definition of the 6 by 6 Jacobian matrix J_0 in (5.28) with the following definition for a 6-revolute-joint serial-chain robot:

$$J_{(0)} = \begin{pmatrix} p_{6(0)}^0 \times r_0^0 & \cdots & p_{6(0)}^5 \times r_0^5 \\ r_0^0 & \cdots & r_0^5 \end{pmatrix},$$

according to equations (4.14) and (4.15) in Chapter 4, we can immediately see that they both have a common format, but just flip over between the top and bottom three rows, or just premultiply either one of the two by a linear transformation:

$$\begin{pmatrix} O & I \\ I & O \end{pmatrix},$$

where O and I are the 3 by 3 zero matrix and identity, respectively.

The geometrical meanings for the vectors inside the two Jacobian matrices are also similar. For instance, $p_{6(0)}^1$ in (5.28) is a radial vector tailed at the origin of frame 6 that is fixed on the top disc of the Stewart platform, and is arrow-pointing at the spherical joint center A_{61} underneath the top mobile disc. Whereas r_0^1 in (5.28) is the unit vector along the piston leg 1. In the serial-chain robotic Jacobian matrix $J_{(0)}$, $p_{6(0)}^j$ is also a radial vector tailed at the origin of frame 6 and arrow-pointing to the origin of frame j, while r_0^j is the unit vector of the z_j-axis of frame j for $j = 0, \cdots, 5$. Both the two Jacobian matrices are projected onto the base, i.e., frame 0.

However, the Jacobian equation (5.29) is different in form from the Jacobian equation $V_{(0)} = J_{(0)}\dot{q}$ for the serial-chain robot. Not only is J_0 transposed in (5.29), but also the joint velocity $\dot{q}$ and the Cartesian velocity V_0 are swapped.

Furthermore, let us borrow the 6 by 1 Cartesian force (wrench) vector definition from the serial-chain robotic statics, i.e.,

$$F_0 = \begin{pmatrix} f_0 \\ m_0 \end{pmatrix},$$

where f_0 is a 3 by 1 force vector and m_0 is a 3 by 1 moment (torque) vector, and both act on the top mobile disc of the Stewart platform and are projected on the base. Then, $F_0^T V_0 = P$, the mechanical power of the top mobile disc.

On the other hand, the power at the joint level should be $P = \tau^T \dot{q}$, where $\tau = (f_1 \ \cdots \ f_6)^T$ is a joint force/torque vector that lists all the joint forces along every piston leg. Based on the principle of energy conservation, $\tau^T \dot{q} = P = F_0^T V_0$. Substituting (5.29) into the power equation yields $\tau^T J_0^T V_0 = F_0^T V_0$, and this is valid for any V_0. Therefore, we reach a statics equation for the closed parallel-chain robotic systems:

$$F_0 = J_0 \tau. \tag{5.30}$$

In comparison with the statics of serial-chain robots $\tau = J_{(0)}^T F_{(0)}$, not only the Jacobian is non-transposed in (5.30), but also the joint torque and Cartesian force/moment (wrench) vectors are swapped, too. Actually, the parallel-chain robotic statics in (5.30) looks like the serial-chain robotic kinematic Jacobian equation $V_{(0)} = J_{(0)}\dot{q}$, while the parallel-chain robotic kinematic Jacobian equation in (5.29) looks like the serial-chain robotic statics $\tau = J_{(0)}^T F_{(0)}$. This phenomenon is known as a **Principle of Duality** between the open serial and closed parallel-chain mechanisms [13, 17, 19].

It can further be observed that the Jacobian equation in (5.27) or (5.29) for a 6-6 Stewart platform can be used to solve an I-K problem in tangent space without need to invert the Jacobian matrix J_0. Namely, it can directly find each prismatic joint speed $\dot{q}_i$ if a position p_0^6, an orientation R_0^6 and their velocities V_0 at the top disc are given. However, since the Jacobian matrix J_0 in (5.28) is a function of both the position and orientation of the top mobile disc, to solve the Cartesian velocity V_0 of the top disc in terms of each prismatic joint length q_i as an F-K problem in tangent space may still remain difficult, but offer a relief in numerical solution.

Let us look at a numerical example to illustrate how to determine a Jacobian matrix J_0 for a 6-6 Stewart platform and how to solve a differential motion-based F-K problem. If we specify the position and orientation of the top mobile disc at a time instant to be

$$p_0^6 = \begin{pmatrix} 0 \\ -0.2 \\ 1 \end{pmatrix}, \quad \text{and} \quad R_0^6 = \begin{pmatrix} 0.6428 & -0.6943 & 0.3237 \\ 0.7660 & 0.5826 & -0.2717 \\ 0 & 0.4226 & 0.9063 \end{pmatrix},$$

which is generated by two successive rotations of the base about z_0-axis by 50^0 and then about the new x-axis by 25^0.

Although the constant vectors p_0^i and p_6^i can be arbitrary for a general 6-6 Stewart platform, here we define each p_6^i and p_0^i around an equilateral triangle on the top and bottom discs, respectively, as shown in Figure 5.22. Once all the constant vectors as well as p_0^6 and R_0^6 are specified, each piston leg vector l_0^i can be determined by the I-K equations in (5.25) with each prismatic joint length $q_i = \|l_0^i\|$ and each unit vector $r_0^i = l_0^i/\|l_0^i\|$.

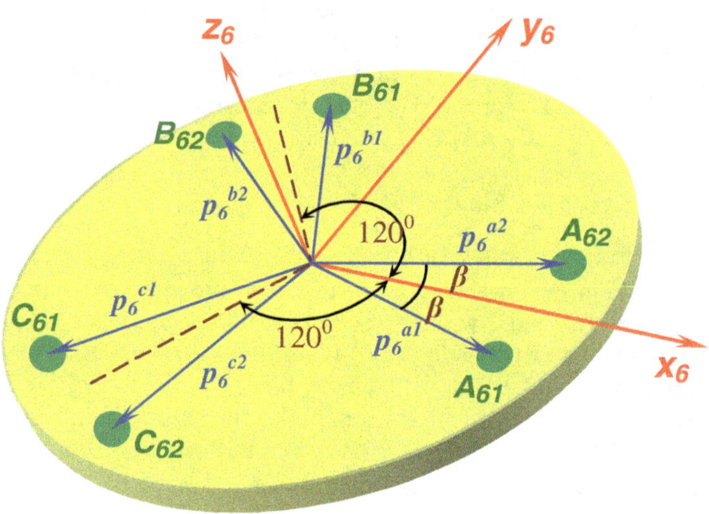

Fig. 5.22 The definitions of p_6^i's on the top mobile disc. They are also applicable to p_0^i's on the base disc of the 6-6 Stewart platform.

Due to each $p_{6(0)}^i = R_0^6 p_6^i$, we can readily find each $s_0^i = p_{6(0)}^i \times r_0^i$ so that the Jacobian matrix J_0 will be formed by (5.28). By further specifying a Cartesian velocity vector:

$$V_0 = \begin{pmatrix} v_0^6 \\ \omega_0^6 \end{pmatrix} = (0.1 \ \ 0.2 \ \ 0 \ \ -0.4 \ \ 0 \ \ -0.5)^T,$$

where v_0^6 is in meter/sec. and ω_0^6 is in rad./sec., the joint velocity $\dot{q}$ can be found by (5.29). A MATLAB^{TM} program is given as follows:

```
p06=[0; -0.2; 1];
al=50*pi/180;  be=25*pi/180;
R06=[cos(al) -sin(al) 0; sin(al) cos(al) 0; 0 0 1]* ...
    [1 0 0; 0 cos(be) -sin(be); 0 sin(be) cos(be)];
    % Cartesian Position/Orientation Inputs

bet=15*pi/180;  gam=120*pi/180;
p0=[1.2; 0; 0];  p6=[0.8; 0; 0];
         % To Define All the Constant Vectors
         % On Both Top and Bottom Discs
alp=-bet;
AR=[cos(alp) -sin(alp) 0; sin(alp) cos(alp) 0; 0 0 1];
p0a1=AR*p0;  p6a1=AR*p6;

alp=bet;
AR=[cos(alp) -sin(alp) 0; sin(alp) cos(alp) 0; 0 0 1];
p0a2=AR*p0;  p6a2=AR*p6;

alp=gam-bet;
AR=[cos(alp) -sin(alp) 0; sin(alp) cos(alp) 0; 0 0 1];
p0b1=AR*p0;  p6b1=AR*p6;

alp=gam+bet;
AR=[cos(alp) -sin(alp) 0; sin(alp) cos(alp) 0; 0 0 1];
p0b2=AR*p0;  p6b2=AR*p6;

alp=-gam-bet;
AR=[cos(alp) -sin(alp) 0; sin(alp) cos(alp) 0; 0 0 1];
p0c1=AR*p0;  p6c1=AR*p6;

alp=-gam+bet;
AR=[cos(alp) -sin(alp) 0; sin(alp) cos(alp) 0; 0 0 1];
p0c2=AR*p0;  p6c2=AR*p6;

l1=R06*p6a1+p06-p0a2;  l2=R06*p6a2+p06-p0b1;
l3=R06*p6b1+p06-p0b2;  l4=R06*p6b2+p06-p0c1;
l5=R06*p6c1+p06-p0c2;  l6=R06*p6c2+p06-p0a1;

q=[norm(l1);norm(l2);norm(l3);norm(l4);norm(l5);norm(l6)];
          % The Joint Position Vector

r1=l1/norm(l1);  r2=l2/norm(l2);
r3=l3/norm(l3);  r4=l4/norm(l4);
r5=l5/norm(l5);  r6=l6/norm(l6);

s1=cross(R06*p6a1, r1);  s2=cross(R06*p6a2, r2);
s3=cross(R06*p6b1, r3);  s4=cross(R06*p6b2, r4);
s5=cross(R06*p6c1, r5);  s6=cross(R06*p6c2, r6);

J0=[r1 r2 r3 r4 r5 r6; s1 s2 s3 s4 s5 s6];
        % The Jacobian Matrix of the Stewart Platform
```

```
    % The I-K Differential Motion Algorithm %

dt=0.01;     % The Sampling Interval 10 Milliseconds

V0=[0.1; 0.2; 0; -0.4; 0; -0.5];
             % Given a Cartesian Velocity
dq=J0'*V0;   % The Jacobian Equation

qnew = q+dq*dt;    % Update the Joint Values

    % The F-K Differential Motion Algorithm %

dq=[0.4; -0.5; -0.2; 0.6; -0.4; 0.5]; % Given a Joint Velocity

V0 = J0'\dq;       % Inverse Jacobian Equation to Find V0

p06new = p06+V0(1:3)*dt;    % Update the Position Vector

dphi=norm(V0(4:6));  k=V0(4:6)/dphi;
K=[0 -k(3) k(2); k(3) 0 -k(1); -k(2) k(1) 0];

R6d=eye(3)+sin(dphi*dt)*K+(1-cos(dphi*dt))*K^2;

R06new = R06*R6d;  % Update the Orientation of the Top Disc
```

To update the joint positions in the differential motion-based I-K algorithm, the first-order approximation is adopted,

$$q(j+1) = q(j) + \dot{q}dt = q(j) + J_0^T V_0 dt,$$

where the sampling interval is set to be $dt = 0.01$ in seconds, i.e., 10 milliseconds.

In contrast, for the differential motion-based F-K algorithm, by arbitrarily specifying a new joint velocity

$$\dot{q} = (0.4 \ \ -0.5 \ \ -0.2 \ \ 0.6 \ \ -0.4 \ \ 0.5)^T,$$

the Cartesian velocity V_0 can be solved by the inverse Jacobian equation of (5.29), i.e.,

$$V_0 = J_0^{-T}\dot{q}.$$

To update the position of the top mobile disc, the first-order approximation is also adopted with the same dt,

$$p_0^6(j+1) = p_0^6(j) + v_0^6 dt.$$

However, to update the orientation R_0^6 of the top disc, we cannot directly use $\dot{R}_0^6$, because taking the time-derivative of a rotation matrix will destroy its membership of the $SO(3)$ group. Instead, since based on equation (3.8) from Chapter 3,

$$\omega_0^6 = \dot{\phi} k,$$

with a unit vector k, we can immediately calculate

$$\dot{\phi} = \|\omega_0^6\| \quad \text{and also} \quad k = \frac{\omega_0^6}{\|\omega_0^6\|}.$$

Under the first-order approximation, $\Delta\phi \approx d\phi = \dot{\phi} dt$. Then, according to equation (2.8) in Chapter 2, the orientation "increment" can be determined by

$$R_6^\delta = I + \sin \Delta\phi K + (1 - \cos \Delta\phi) K^2,$$

where $K = S(k) = k\times$ is the skew-symmetric matrix of the unit vector k. Therefore, the new orientation of the top mobile disc turns out to be

$$R_0^6(j+1) = R_0^6(j) R_6^\delta.$$

The above MATLAB^{TM} program has also implemented this updating algorithm in Cartesian space as an F-K solution, and the final outputs are printed below:

```
J0 =              % The Jacobian Matrix

  -0.4938    0.4645    0.1164    0.0679    0.3968   -0.6698
  -0.0374   -0.4526   -0.4922    0.4015    0.2292   -0.4416
   0.8688    0.7612    0.8626    0.9133    0.8888    0.5969
   0.4062    0.5820    0.4123   -0.1908   -0.6233   -0.5076
  -0.5132   -0.2280    0.6156    0.7070   -0.1207   -0.0220
   0.2088   -0.4907    0.2957   -0.2966    0.3094   -0.5859

    % The I-K Differential Motion Algorithm Result

q =

    1.0503    1.4286    1.5378    1.3567    0.8561    1.1282

qnew =              % The Updated Joint Positions

    1.0471    1.4283    1.5338    1.3598    0.8579    1.1316

    % The F-K Differential Motion Algorithm Result

p06 =

         0   -0.2000    1.0000

p06new =            % The Updated Position Vector
```

```
   -0.0099   -0.1929    1.0010

R06 =

    0.6428   -0.6943    0.3237
    0.7660    0.5826   -0.2717
         0    0.4226    0.9063

R06new =           % The Updated Orientation

    0.6441   -0.6915    0.3271
    0.7650    0.5846   -0.2704
   -0.0043    0.4244    0.9055

J0new =            % The Updated Jacobian Matrix

  -0.5018    0.4613    0.1113    0.0609    0.3830   -0.6771
  -0.0317   -0.4505   -0.4872    0.4060    0.2372   -0.4341
   0.8644    0.7644    0.8662    0.9119    0.8928    0.5942
   0.4035    0.5824    0.4143   -0.1915   -0.6252   -0.5045
  -0.5082   -0.2320    0.6149    0.7037   -0.1150   -0.0168
   0.2156   -0.4882    0.2927   -0.3005    0.2988   -0.5872
```

Because the Jacobian matrix J_0 for a general 6-6 Stewart platform is a function of the position p_0^6 and orientation R_0^6 of the top mobile disc, once both p_0^6 and R_0^6 are updated at sampling point j, the Jacobian matrix value $J_0(j)$ will be updated to $J_0(j+1)$ accordingly, as illustrated in the above MATLAB^{TM} program and results. Then, a new round of updating begins. This is a typical differential motion-based F-K algorithm for a general 6-6 Stewart platform starting with a given initial $p_0^6(0)$, an initial $R_0^6(0)$, and a desired joint trajectory $q(t)$ with its time-derivative $\dot{q}(t)$.

It is also interesting that the statics equation in (5.30) can be utilized to find a joint force distribution over the six piston legs if a 6 by 1 Cartesian force is acting on the top disc statically [16, 17]. However, unlike the open serial-chain robots, based on (5.30), the Jacobian inverse is required, i.e.,

$$\tau = \begin{pmatrix} f_1 \\ \vdots \\ f_6 \end{pmatrix} = J_0^{-1} F_0 = J_0^{-1} \begin{pmatrix} f_0 \\ m_0 \end{pmatrix},$$

where both f_0 and m_0 are 3 by 1 and referred to the base.

For example, suppose that a small vehicle of the mass $M = 250$ Kilograms is loaded on the top disc of the Stewart platform and also a twisting moment of $m_z = 500$ in Newton-meter is applied about the z_6-axis of frame 6. Then, the force and moment vectors become

$$f_0 = \begin{pmatrix} 0 \\ 0 \\ -Mg \end{pmatrix} \quad \text{and} \quad m_0 = R_0^6 m_6 = R_0^6 \begin{pmatrix} 0 \\ 0 \\ m_z \end{pmatrix},$$

and both must be projected onto the common base before being augmented to form the 6 by 1 Cartesian force vector F_0. With the same J_0 as the previous numerical example, the joint force distributions under the load f_0 and with and without the twisting moment m_0 are calculated by the statics equation (5.30) in MATLAB^{TM} and printed as follows:

```
M = 250;    g = 9.81;    mz = 500;

f0 = [0; 0; -M*g];         % The Load Force
m0 = R06*[0; 0; mz];       % The Twisting Moment

tau1 = J0\[f0; m0]
                     % Joint Forces with Both the Load and Twist

-550.8253 -549.3252 -42.1803 -842.5501 -571.0821 -405.9182


tau2 = J0\[f0; zeros(3,1)]
               % Joint Forces with the Load But Without the Twist

-795.0905 -358.6778 -256.8914 -593.4673 -695.3522 -179.3081
```

If we add all the six joint forces together in the second case of only the vehicle loaded without the twisting moment applied, then, $\sum f_i = -2878.8$ Newtons. Comparing it with the loaded vehicle weight $-Mg = -250 \times 9.81 = -2452.5$ Newtons, the absolute value of the former is a little bigger than the latter, because the six legs are not all perpendicular to the base.

In summary, the Stewart platform, as a fully parallel-chain robotic system, has been widely used in many applications, especially in military sectors for warfighter compartment tests and military vehicle vibration tests. The advantage is that it offers an overwhelmingly high payload over any open serial-chain robot. However, in terms of the motion dynamic range and work envelope, it is very small in comparison with the open serial-chain robots. In order to acquire the advantages from both types of robots, we may make a compromise and create a hybrid robotic system, the first three links of which are parallel and the last two or three are serial. In industry, this kind of robots has already been developed and commercialized. One of the typical 3+2 hybrid industrial robot manipulators, called Exechon, was developed by Optikos, Inc.

5.4.3 Modeling and Analysis of 3+3 Hybrid Robot Arms

A 3 d.o.f. mechanical system, called Delta, was the first model for a parallel mechanism with three joints [11, 14]. As shown in Figure 5.23, there are two different types of such a three-axis parallel platforms. The left one is using three prismatic joints (pistons) to drive the top mobile disc and control its position. While the right one is the original Delta model, where the three axes on the legs are all revolute. Based on the Grübler's formula, since the left one has total number of links $l = 7$ that excludes the fixed base plate, and the number of joints $n = 9$, the first term of the Grübler's formula becomes $D(l - n) = 6(7 - 9) = -12$. Thus, among the $n = 9$ joints, there must be six joints of spherical type or S-type that can offer two axes each in order to have a net d.o.f. $m = 3$. Therefore, each of the base and top plates may install three S-type joints to connect the three pistons and form an SPS (Spherical-Prismatic-Spherical) type structure for each leg. Or, the three joints on the base may utilize the universal type or U-type that can offer three axes of rotation for each, so that each of the three joints on the top mobile disc may be just a one-axis revolute (R-type) joint, to form a UPR (Universal-Prismatic-Revolute) type structure. On the right-side picture of the Delta parallel platform in Figure 5.23, the most common structure of each leg is URR (Universal-Revolute-Revolute) with the central axis of the upper arm perpendicular to the axis of the revolute joint on the top disc.

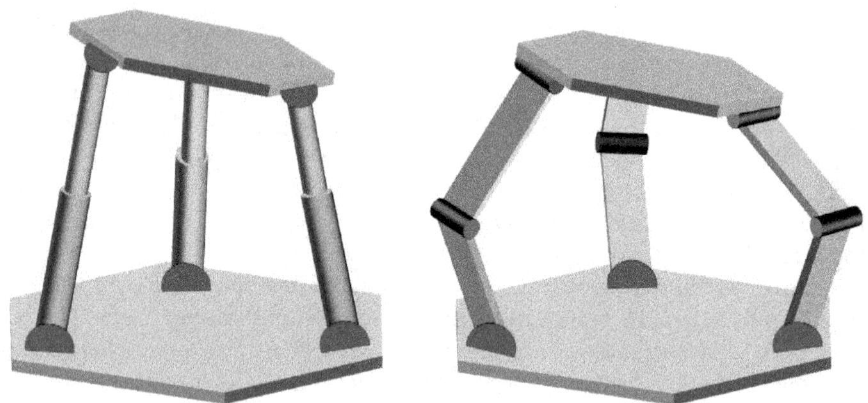

Fig. 5.23 Two types of the 3-parallel mechanism

To model and develop a kinematic motion algorithm for a 3+3 hybrid robot arm: the bottom three parallel links form a platform and the top three links constitute a 3-joint serial-chain robot sitting on the platform. Intuitively, the parallel-chain platform is primarily to produce a required position while the top serial-chain arm is to meet the specified orientation. Let us start our study on such a 3-parallel-joint platform with a UPR structure. Of course, the inverse kinematics (I-K) problem of the entire 3+3 hybrid robot is to find three prismatic joint lengths of the platform and three revolute (for an RRR-type in most cases) joint values of the top serial-chain arm such that the last frame #6 of the entire robot can meet a desired position vector p_0^6 and a desired orientation R_0^6. For a better analysis, let us decompose the entire I-K problem into two steps:

1. find the orientation R_0^3 of the platform if the entire robot can meet the specified position p_0^6, and
2. determine how much more rotation must be made up by the top serial-chain arm to meet the final orientation requirement R_0^6.

At the beginning, consider that $p_0^t \in \mathbb{R}^3$ is a required position vector for the Top point that is vertically located above the mobile disc with respect to the base frame #0, as shown in Figure 5.24. We try to answer what is the "passive" orientation R_0^3 of frame #3 on the mobile disc if the top point can meet a required p_0^t ? Since the platform is just a 3-active-joint parallel-chain robot, if the 3 d.o.f. position given by the required p_0^t is achieved by controlling the three active joints, the orientation R_0^3 of the platform has to be passive, i.e, uncontrollable. One will soon realize that it is still not so easy to find such a passive orientation for this 3 d.o.f. parallel-chain platform even if the target is reduced just to meet p_0^t, instead of p_0^6. At this point, solving the I-K problem for such a 3-leg platform may be even more difficult than that of a 6-6 Stewart platform.

Let η_0^α be a 3 by 1 vector tailed at the universal joint point A_0 on the base and with its arrow pointing to the Top point, as shown in Figure 5.24. Similarly, let η_0^β and η_0^γ be other two 3 by 1 vectors from points B_0 and C_0 to the Top point, respectively. Since we assumed that the revolute joint axis $R_0^3 p_3^\alpha$ with respect to the base frame #0 around the joint point A_3 is always perpendicular to the piston central axis l_0^a, we can prove that η_0^α is perpendicular to $R_0^3 p_3^\alpha$, too. In fact, p_3^α, as a unit vector of the revolute joint axis, is clearly perpendicular to both vectors p_3^a and $d_4 z_3$ with respect to frame #3, where $z_3 = (0\ 0\ 1)^T$ is the unit vector of the z_3-axis of frame #3 and d_4 is a height from the origin of frame #3 to the Top point. Thus, $p_3^{aT} p_3^\alpha = 0$, $z_3^T p_3^\alpha = 0$ and $l_0^{aT} R_0^3 p_3^\alpha = 0$. Because $\eta_0^\alpha = l_0^a + R_0^3 p_3^a + d_4 R_0^3 z_3$, see Figure 5.24, we reach to

$$\eta_0^{\alpha T} R_0^3 p_3^\alpha = 0,$$

and in the same token,

$$\eta_0^{\beta T} R_0^3 p_3^\beta = 0 \quad \text{and} \quad \eta_0^{\gamma T} R_0^3 p_3^\gamma = 0. \tag{5.31}$$

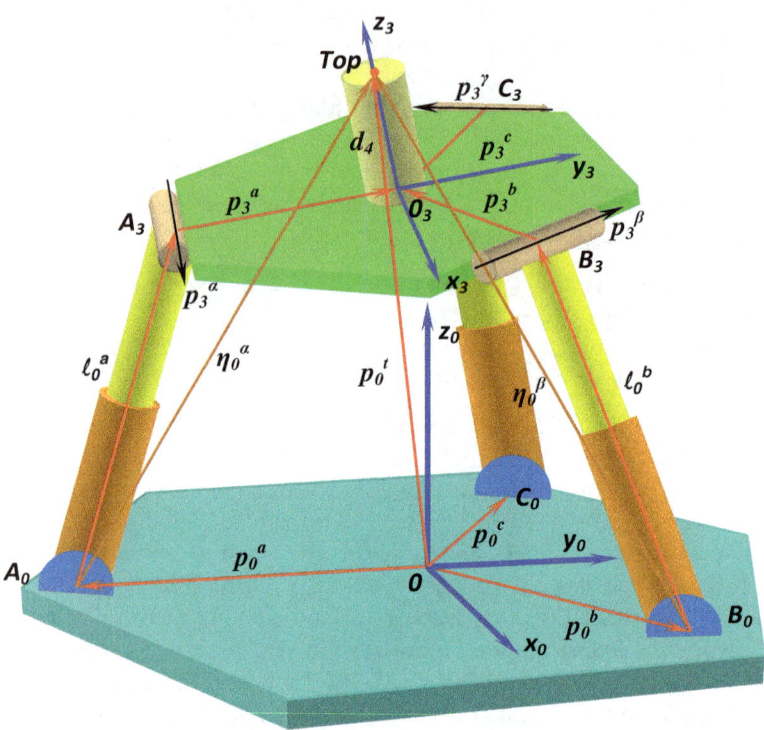

Fig. 5.24 Kinematic analysis of a 3-leg UPS platform

Furthermore, since all three revolute joint axis unit vectors p_3^α, p_3^β and p_3^γ lay on the $x_3 - y_3$ coordinate plane, they can be expressed in the following linear combination form with respect to frame #3:

$$p_3^\alpha = a_{11}x_3 + a_{12}y_3, \quad p_3^\beta = a_{21}x_3 + a_{22}y_3 \quad \text{and} \quad p_3^\gamma = a_{31}x_3 + a_{32}y_3,$$

where each a_{ij} is a constant coordinate of the corresponding revolute joint axis unit vector projected onto frame #3 axis $x_3 = (1\ 0\ 0)^T$ or $y_3 = (0\ 1\ 0)^T$ for $i = 1, 2, 3$ and $j = 1, 2$. For instance, if $p_3^\alpha \parallel x_3$ in the same direction, then $a_{11} = 1$ and $a_{12} = 0$.

Substituting the above linear combination form into the orthogonal equations in (5.31) and augmenting them together, we can write it in the following compact matrix form:

$$AR_0^3 x_3 + BR_0^3 y_3 = AR_0^3 \begin{pmatrix} 1 \\ 0 \\ 0 \end{pmatrix} + BR_0^3 \begin{pmatrix} 0 \\ 1 \\ 0 \end{pmatrix} = Ax_0^3 + By_0^3 = 0, \qquad (5.32)$$

where the 3 by 3 coefficient matrices

$$A = \begin{pmatrix} a_{11}\eta_0^{\alpha T} \\ a_{21}\eta_0^{\beta T} \\ a_{31}\eta_0^{\gamma T} \end{pmatrix} \quad \text{and} \quad B = \begin{pmatrix} a_{12}\eta_0^{\alpha T} \\ a_{22}\eta_0^{\beta T} \\ a_{32}\eta_0^{\gamma T} \end{pmatrix}, \tag{5.33}$$

and x_0^3 and y_0^3 are the projections of x_3 and y_3 onto the base, i.e.,

$$x_0^3 = R_0^3 x_3 = R_0^3 \begin{pmatrix} 1 \\ 0 \\ 0 \end{pmatrix} \quad \text{and} \quad y_0^3 = R_0^3 y_3 = R_0^3 \begin{pmatrix} 0 \\ 1 \\ 0 \end{pmatrix}.$$

It can further be seen from Figure 5.24 that

$$\eta_0^\alpha = p_0^t - p_0^a, \quad \eta_0^\beta = p_0^t - p_0^b, \quad \text{and} \quad \eta_0^\gamma = p_0^t - p_0^c. \tag{5.34}$$

Since p_0^a, p_0^b and p_0^c are all constant vectors laying on the base plate, if the position vector of the Top point referred to the base p_0^t is given, and the configuration of all the three revolute joints on the top disc is specified, then both the matrices A and B in (5.32) and (5.33) are known. Therefore, the first step of the decomposed I-K problems is to solve for the passive orientation R_0^3 that is now sandwiched inside each of the two terms of the above homogeneous equation (5.32).

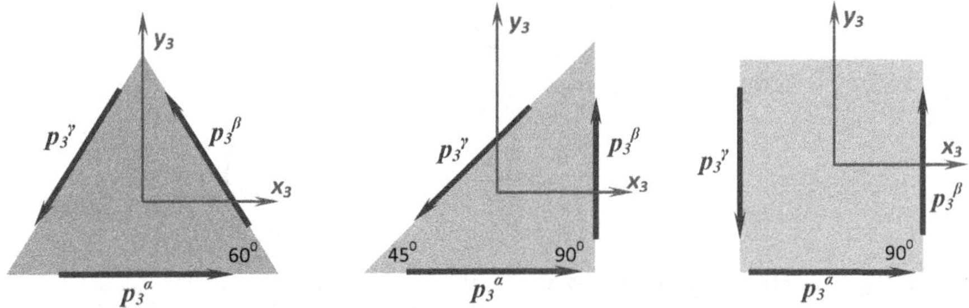

Fig. 5.25 Top revolute-joint configurations

Let us now discuss in more details the three typical cases of revolute joint configuration design, as shown in Figure 5.25. In the first (leftmost) one, the three unit vectors p_3^α, p_3^β and p_3^γ form an equilateral triangle so that each corner angle is 60^0. The second (middle) one is to form a right isosceles triangle so that the left bottom corner angles is 45^0 and the right bottom corner one is 90^0. The last one (rightmost) is to form a rectangle so that each corner angle is 90^0. Thus, all the three configurations have their linear combination coefficients given in the following table:

R-Joint Unit Vector	p_3^α		p_3^β		p_3^γ	
Configurations	a_{11}	a_{12}	a_{21}	a_{22}	a_{31}	a_{32}
Equilateral △	1	0	$-\frac{1}{2}$	$\frac{\sqrt{3}}{2}$	$-\frac{1}{2}$	$-\frac{\sqrt{3}}{2}$
Right Isosceles △	1	0	0	1	$-\frac{\sqrt{2}}{2}$	$-\frac{\sqrt{2}}{2}$
Rectangle □	1	0	0	1	0	-1

It can be seen that the rectangular configuration design will have the simplest kinematic model, and the right isosceles triangle is the second simplest. The equilateral triangle design is the most complex one for I-K solution in comparison with the other two. Accordingly, the 3 by 3 matrices A and B in (5.33) are given as follows:

1. For the equilateral triangle configuration:

$$A = \begin{pmatrix} \eta_0^{\alpha T} \\ -\frac{1}{2}\eta_0^{\beta T} \\ -\frac{1}{2}\eta_0^{\gamma T} \end{pmatrix} \quad \text{and} \quad B = \begin{pmatrix} 0_{1\times 3} \\ \frac{\sqrt{3}}{2}\eta_0^{\beta T} \\ -\frac{\sqrt{3}}{2}\eta_0^{\gamma T} \end{pmatrix};$$

2. For the right isosceles triangle design:

$$A = \begin{pmatrix} \eta_0^{\alpha T} \\ 0_{1\times 3} \\ -\frac{\sqrt{2}}{2}\eta_0^{\gamma T} \end{pmatrix} \quad \text{and} \quad B = \begin{pmatrix} 0_{1\times 3} \\ \eta_0^{\beta T} \\ -\frac{\sqrt{2}}{2}\eta_0^{\gamma T} \end{pmatrix};$$

3. For the rectangle configuration:

$$A = \begin{pmatrix} \eta_0^{\alpha T} \\ 0_{1\times 3} \\ 0_{1\times 3} \end{pmatrix} \quad \text{and} \quad B = \begin{pmatrix} 0_{1\times 3} \\ \eta_0^{\beta T} \\ -\eta_0^{\gamma T} \end{pmatrix},$$

where $0_{1\times 3}$ is the 1 by 3 zero row vector.

In order to solve the homogeneous equation (5.32) for the orientation R_0^3 of the top mobile disc, let R_0^3 be decomposed into two Euler angles of rotation: first rotate about the y_3-axis by ψ and then rotate about the new x_3-axis by ϕ. Namely,

$$R_0^3 = R(y,\psi)R(x,\phi) = \begin{pmatrix} \cos\psi & 0 & \sin\psi \\ 0 & 1 & 0 \\ -\sin\psi & 0 & \cos\psi \end{pmatrix} \begin{pmatrix} 1 & 0 & 0 \\ 0 & \cos\phi & -\sin\phi \\ 0 & \sin\phi & \cos\phi \end{pmatrix}. \tag{5.35}$$

Since R_0^3 is a passive orientation and each of the three revolute joints connecting to the top mobile disc is mechanically constrained by the orthogonality between the joint axis and the central axis of each leg for such a UPR type parallel mechanism, the top disc has no chance to twist itself about the z_0-axis so that it suffices to define two successive rotations about the y_0 and x_0-axis without a spin about the z_0-axis for R_0^3.

Substituting (5.35) into equation (5.32) yields

$$AR(y,\psi)R(x,\phi)\begin{pmatrix}1\\0\\0\end{pmatrix}+BR(y,\psi)R(x,\phi)\begin{pmatrix}0\\1\\0\end{pmatrix}=0.$$

It can be observed that for any one of the three different revolute joint design configurations, the first row of B is always zero so that

$$\eta_0^{\alpha T}R(y,\psi)R(x,\phi)\begin{pmatrix}1\\0\\0\end{pmatrix}=0.$$

Since

$$R(x,\phi)\begin{pmatrix}1\\0\\0\end{pmatrix}=\begin{pmatrix}1\\0\\0\end{pmatrix},$$

the above equation can be further reduced to

$$\eta_0^{\alpha T}\begin{pmatrix}\cos\psi\\0\\-\sin\psi\end{pmatrix}=b_1^\alpha\cos\psi-b_3^\alpha\sin\psi=0,$$

where $\eta_0^{\alpha T}=(b_1^\alpha\ \ b_2^\alpha\ \ b_3^\alpha)$, and the same definitions for $\eta_0^{\beta T}=(b_1^\beta\ \ b_2^\beta\ \ b_3^\beta)$ and $\eta_0^{\gamma T}=(b_1^\gamma\ b_2^\gamma\ b_3^\gamma)$.

If both the rotation angles ψ and ϕ about y-axis and x-axis, respectively, are limited within $(-90^0,90^0)$, then,

$$\psi=\arctan\left(\frac{b_1^\alpha}{b_3^\alpha}\right). \tag{5.36}$$

After ψ is found, for the second and third cases of configuration, they have the same second row:

$$\eta_0^{\beta T}\begin{pmatrix}\sin\psi\sin\phi\\\cos\phi\\\cos\psi\sin\phi\end{pmatrix}=b_1^\beta\sin\psi\sin\phi+b_2^\beta\cos\phi+b_3^\beta\cos\psi\sin\phi=0.$$

Thus, the angle ϕ can also be solved as well,

$$\phi=\arctan\left(\frac{-b_2^\beta}{b_1^\beta\sin\psi+b_3^\beta\cos\psi}\right). \tag{5.37}$$

Finally, substituting the two rotation angles into (5.35), we solve the passive orientation of the top mobile disc in terms of the given position vector p_0^t.

However, for the first equilateral triangle configuration case, two simultaneous equations from the last two rows can be found and are needed to solve the second rotation angle ϕ, and they are

$$\begin{cases} -\frac{1}{2}(b_1^\beta \cos\psi - b_3^\beta \sin\psi) + \frac{\sqrt{3}}{2}(b_1^\beta \sin\psi \sin\phi + b_2^\beta \cos\phi + b_3^\beta \cos\psi \sin\phi) = 0 \\ -\frac{1}{2}(b_1^\gamma \cos\psi - b_3^\gamma \sin\psi) - \frac{\sqrt{3}}{2}(b_1^\gamma \sin\psi \sin\phi + b_2^\gamma \cos\phi + b_3^\gamma \cos\psi \sin\phi) = 0. \end{cases}$$

It can be further reduced to

$$\begin{pmatrix} \beta_{11} & \beta_{12} \\ \gamma_{11} & \gamma_{12} \end{pmatrix} \begin{pmatrix} \sin\phi \\ \cos\phi \end{pmatrix} = \begin{pmatrix} \beta_{13} \\ \gamma_{13} \end{pmatrix}, \tag{5.38}$$

where $\beta_{11} = \frac{\sqrt{3}}{2}(b_1^\beta \sin\psi + b_3^\beta \cos\psi)$, $\beta_{12} = \frac{\sqrt{3}}{2}b_2^\beta$, $\gamma_{11} = -\frac{\sqrt{3}}{2}(b_1^\gamma \sin\psi + b_3^\gamma \cos\psi)$, $\gamma_{12} = -\frac{\sqrt{3}}{2}b_2^\gamma$, $\beta_{13} = \frac{1}{2}(b_1^\beta \cos\psi - b_3^\beta \sin\psi)$ and $\gamma_{13} = \frac{1}{2}(b_1^\gamma \cos\psi - b_3^\gamma \sin\psi)$. Then, ϕ can be determined as long as the first 2 by 2 matrix is nonsingular.

Once we finish the first step of the I-K problem for a 3+3 hybrid robot, we progress to the second step: given p_0^6 and R_0^6, find all the six joint values, including both the parallel-chain platform with three prismatic joints l_1, l_2 and l_3 and the top serial-chain arm with three revolute joints θ_4, θ_5 and θ_6. Suppose that the top arm has a regular RRR configuration with some joint offset along each revolute joint axis, as shown in Figure 5.26.

The D-H table for the top 3-joint arm can be easily deduced via the D-H convention and is given below:

Joint Angle	**Joint Offset**	**Twist Angle**	**Link Length**
θ_i	d_i	α_i	a_i
θ_4	d_4	-90^0	0
θ_5	d_5	90^0	0
θ_6	d_6	0	0

Then, the one-step homogeneous transformation matrices are found as follows:

$$A_3^4 = \begin{pmatrix} c_4 & 0 & -s_4 & 0 \\ s_4 & 0 & c_4 & 0 \\ 0 & -1 & 0 & d_4 \\ 0 & 0 & 0 & 1 \end{pmatrix}, \quad A_4^5 = \begin{pmatrix} c_5 & 0 & s_5 & 0 \\ s_5 & 0 & -c_5 & 0 \\ 0 & 1 & 0 & d_5 \\ 0 & 0 & 0 & 1 \end{pmatrix},$$

and

$$A_5^6 = \begin{pmatrix} c_6 & -s_6 & 0 & 0 \\ s_6 & c_6 & 0 & 0 \\ 0 & 0 & 1 & d_6 \\ 0 & 0 & 0 & 1 \end{pmatrix}.$$

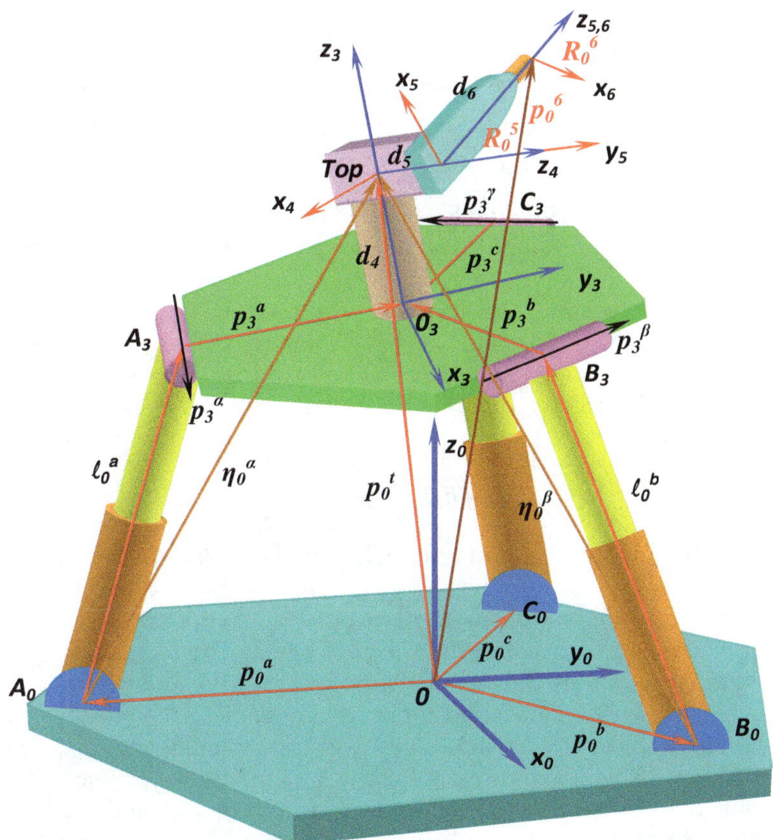

Fig. 5.26 Solve the I-K problem for a 3+3 hybrid robot

The total homogeneous transformation of the top arm can be calculated by $A_3^6 = A_3^4 A_4^5 A_5^6$, i.e.,

$$A_3^6 = \begin{pmatrix} c_4c_5c_6 - s_4s_6 & -c_4c_5s_6 - s_4c_6 & c_4s_5 & d_6c_4s_5 - d_5s_4 \\ s_4c_5c_6 + c_4s_6 & -s_4c_5s_6 + c_4c_6 & s_4s_5 & d_6s_4s_5 + d_5c_4 \\ -s_5c_6 & s_5s_6 & c_5 & d_6c_5 + d_4 \\ 0 & 0 & 0 & 1 \end{pmatrix}. \quad (5.39)$$

If both the position vector p_0^6 and the orientation R_0^6 of frame #6 with respect to the base are given, according to Figure 5.26, we can immediately see that

$$p_0^t = p_0^6 - d_6 z_0^6 - d_5 y_0^6 \mid_{\theta_6=0}, \quad (5.40)$$

where the last term is $d_5 y_0^6$ under the condition $\theta_6 = 0$, and y_0^6 and z_0^6 are the second and third columns of the given orientation R_0^6, respectively. Even if the conditional term $d_5 y_0^6 \mid_{\theta_6=0}$ is unknown, but we know the joint value of θ_6, we can determine

$$R_0^5 R(z_6, \theta_6) = R_0^6 \quad \text{so that} \quad R_0^5 = R_0^6 \begin{pmatrix} c_6 & s_6 & 0 \\ -s_6 & c_6 & 0 \\ 0 & 0 & 1 \end{pmatrix}.$$

Then,

$$p_0^t = p_0^6 - d_6 z_0^6 - d_5 y_0^5, \tag{5.41}$$

and the conditional term is replaced by $d_5 y_0^5$. In fact, for such an RRR type arm, the z_6-axis and z_5-axis are aligned so that $z_0^6 = z_0^5$, which is independent of θ_6.

Once p_0^t is determined, we can follow the above procedure to find the passive orientation R_0^3. Through the rotation matrix R_0^3, the three vectors p_3^a, p_3^b and p_3^c laying on the mobile disc can be found directly by the projections on frame #3. Therefore, the three prismatic joint vectors can be readily determined as

$$l_0^i = p_0^t - p_0^a - d_4 z_0^3 - R_0^3 p_3^i \quad \text{for } i = a, b, c, \tag{5.42}$$

where z_0^3 is the third column of R_0^3. Clearly, the norm of each vector is the joint value $l_i = \|l_0^i\|$.

Since p_3^6 can be found from $R_0^3 p_3^6 = p_0^6 - p_0^t$, comparing it with the last column of the above A_3^6 under $d_4 = 0$, plus comparing $R_3^6 = R_3^0 R_0^6 = R_0^{3T} R_0^6$ with the upper left 3 by 3 block of A_3^6, we can solve θ_4 and θ_5 under the pre-specified θ_6 at $t = 0$, as an initial tool spinning angle. It has to be recognized that without knowing the initial value of the last spinning angle θ_6, we cannot solve such an I-K problem directly for the 3+3 hybrid robot due to the issue of causality. In other words, if p_0^6 and R_0^5 are given, instead of R_0^6, the I-K problem can be completely resolved. We will apply all the above I-K computations to draw and further to animate such a 3+3 UPR+RRR type hybrid robotic system in the equilateral triangle configuration into MATLAB^{TM} in the next chapter.

In summary, the algorithm to solve for the I-K for such a 3+3 hybrid-chain robot can be procedurized as follows:

Given p_0^6 and R_0^6 with a known $\theta_6(0)$ at $t = 0$,

1. First, find p_0^t through equation (5.41) with the previous value of θ_6;
2. Then, calculate the matrices A and B via equation (5.33) along with (5.34) at each sampling point;
3. Solve equation (5.32) to determine R_0^3 by the solutions in (5.36) and (5.38) via (5.35);
4. The vectors l_0^i for $i = a, b, c$ of the three piston legs can be found by equation (5.42);
5. Calculate $R_3^6 = R_3^0 R_0^6 = (R_0^3)^T R_0^6$;
6. By comparing R_3^6 with the symbolical form of A_3^6 in equation (5.39), the last three joint angles can be determined by

$$\theta_4 = \text{atan2}(R_3^6(2,3),\ R_3^6(1,3)),$$
$$\theta_5 = \text{atan2}(R_3^6(1,3)/\cos\theta_4,\ R_3^6(3,3)),$$
$$\theta_6 = \text{atan2}(R_3^6(3,2),\ -R_3^6(3,1)).$$

Regarding the forward kinematics (F-K), the top RRR arm is straightforward because of the open serial-chain mechanism. However, the UPR type 3-leg parallel platform is as hard as that of the 6-leg Stewart platform. Since in this 3-leg parallel robot, the axis of each top revolute joint is perpendicular to the central axis of each leg, i.e., each $l_0^a \perp p_3^\alpha$, we may interpret that each l_0^i for $i = a, b, c$ is similar to the height h_i for corresponding $i = 1, 2, 3$ in Figure 5.21, and just make upside down. In other words, the top mobile disc of the 3-leg system is treated as a bottom one of the Stewart platform, while the base plate is treated as the top disc in Figure 5.21. Then, the angle between l_0^a and p_3^α becomes θ_1, and the other two become θ_2 and θ_3.

Under such an upside down comparison, the three universal joint points A_0, B_0 and C_0 for the 3-leg parallel robot in Figure 5.26 are imagined on the three circles with their centers at A_3, B_3 and C_3 and radii $h_1 = l_1 = \|l_0^a\|$, $h_2 = l_2 = \|l_0^b\|$ and $h_3 = l_3 = \|l_0^c\|$. Hence, the similar question is asked: where are the points A_0 on Circle 1, B_0 on Circle 2 and C_0 on Circle 3 such that the distance between each pair of the three points is equal to the real distance for the corresponding $\overline{A_0B_0}$, $\overline{B_0C_0}$ and $\overline{C_0A_0}$? This clearly shows that the F-K problem for the UPR type 3-leg parallel robot is the same as that for a 3-3 or 6-3 Stewart platform system, and we can also call the same algorithm to recursively search and find the solution, but just need to make an upside down imagination.

The original design by Delta was the URR-type on each of the three legs [14]. Since both the 3-leg systems of URR-type and UPR-type are structured with an orthogonality between the top revolute joint axis and the central axis of the upper leg (thigh), they are interchangeable. We can stay with the modeling and analysis of the UPR-type one, as we have just studied. Once each prismatic joint vector l_0^i for $i = a, b, c$ is determined by the I-K algorithm of the UPR-type, to find an equivalent URR-type joint value that is the angle around the knee point of each leg becomes straightforward.

If one wants to control a real URR-type platform in laboratory, it suffices to find the equivalent angle of the revolute joint of each knee. This can be directly converted from the resulting prismatic joint length $l_i = \|l_0^i\|$ by the Law of Cosine, because all the upper and lower leg lengths a_1 and a_2 are known, as shown in Figure 5.27. Namely,

$$\cos\angle A_a = \frac{a_1^2 + a_2^2 - l_1^2}{2a_1a_2}.$$

However, if one wishes to graphically draw and simulate such a URR-type parallel-chain system in MATLAB^{TM}, only solving and knowing the revolute joint angle of each knee is far not enough. Often, you have to tell MATLAB^{TM} the location and orientation of each link to be drawn, not just

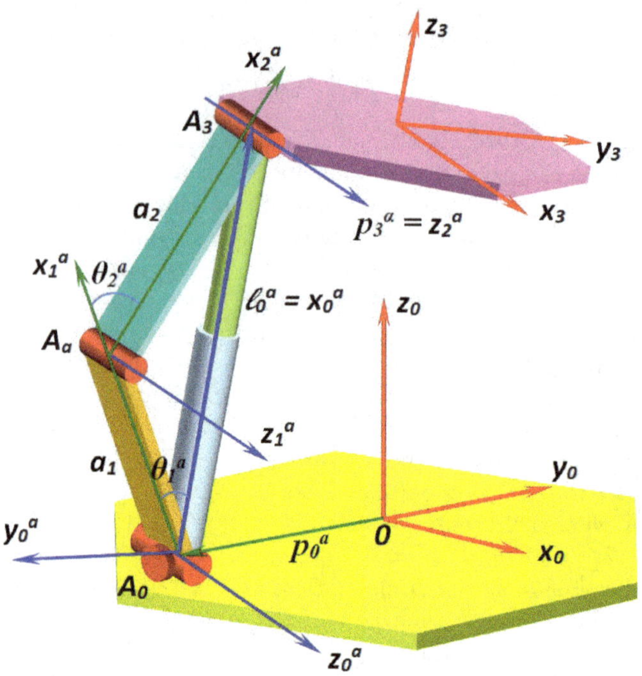

Fig. 5.27 Delta URR vs. UPR 3-leg parallel system

a size. Therefore, we need a more detailed study on the geometric relation between the UPR-type and URR-type systems.

It can be clearly seen from Figure 5.27 that the triangle $\triangle A_0 A_a A_3$ is formed by the prismatic joint length l_1 from the UPR-type one and the lower leg length a_1 and upper leg length a_2 from the URR-type system. Because of the orthogonality, the top revolute axis p_3^α must be a normal unit vector to the triangle. Thus, we may define a new coordinate system, called frame a, whose x_0^a-axis is just the normalized l_0^a: $l_0^a/\|l_0^a\| = l_0^a/l_1$, and whose z_0^a-axis is parallel to p_3^α. Furthermore, let $y_0^a = z_0^a \times x_0^a$ be the y_0^a-axis of the new frame. Therefore, the orientation of the new frame referred to the base can be fully determined by

$$R_0^a = \begin{pmatrix} \frac{l_0^a}{l_1} & R_0^3 p_3^\alpha \times \frac{l_0^a}{l_1} & R_0^3 p_3^\alpha \end{pmatrix}.$$

After finding the orientation, we now shift its origin to the universal joint point A_0 from the base origin by the constant vector p_0^a so that frame a can further be determined completely by the following homogeneous transformation with respect to the base:

$$H_0^a = \begin{pmatrix} R_0^a & p_0^a \\ 0_{1\times 3} & 1 \end{pmatrix}.$$

This homogeneous transformation matrix H_0^a will be useful in 3D drawing for every link involved in the leg. Using the same method, H_0^b and H_0^c for the other two legs at B_0 and C_0 can be found as well.

In fact, after the two different types of leg are put together in one modeling picture, as shown in Figure 5.27, we can see that the URR-type leg is a two-link planar arm sitting in the new frame H_0^a with respect to the base. Its tip point touches the point A_3 so that the position vector p_a^2 of this two-link arm is just the prismatic joint vector l_0^a but referred to the new frame, i.e., $p_a^2 = (l_1 \ \ 0 \ \ 0)^T$. Now, applying the D-H convention on the two-link arm, we have the following D-H table:

Joint Angle	**Joint Offset**	**Twist Angle**	**Link Length**
θ_i	d_i	α_i	a_i
θ_1^a	0	0	a_1
θ_2^a	0	0	a_2

Its one-step homogeneous transformations are computed as

$$A_a^1 = \begin{pmatrix} c_1 & -s_1 & 0 & a_1c_1 \\ s_1 & c_1 & 0 & a_1s_1 \\ 0 & 0 & 1 & 0 \\ 0 & 0 & 0 & 1 \end{pmatrix} \quad \text{and} \quad A_1^2 = \begin{pmatrix} c_2 & -s_2 & 0 & a_2c_2 \\ s_2 & c_2 & 0 & a_2s_2 \\ 0 & 0 & 1 & 0 \\ 0 & 0 & 0 & 1 \end{pmatrix},$$

where $c_i = \cos\theta_i^a$ and $s_i = \sin\theta_i^a$ for $i = 1, 2$. Multiplying them together yields

$$A_a^2 = A_a^1 A_1^2 = \begin{pmatrix} c_{12} & -s_{12} & 0 & a_1c_1 + a_2c_{12} \\ s_{12} & c_{12} & 0 & a_1s_1 + a_2s_{12} \\ 0 & 0 & 1 & 0 \\ 0 & 0 & 0 & 1 \end{pmatrix},$$

where $c_{12} = \cos(\theta_1^a + \theta_2^a)$ and $s_{12} = \sin(\theta_1^a + \theta_2^a)$ for short notation again.

By comparing the last column of A_a^2 with p_a^2, we have

$$\begin{pmatrix} a_1c_1 + a_2c_{12} \\ a_1s_1 + a_2s_{12} \\ 0 \end{pmatrix} = \begin{pmatrix} l_1 \\ 0 \\ 0 \end{pmatrix}$$

such that

$$a_1c_1 + a_2c_{12} = l_1 \quad \text{and} \quad a_1s_1 + a_2s_{12} = 0.$$

Squaring both the two equations and adding them together will reach to the same result from the Law of Cosine:

$$a_1^2 + a_2^2 + 2a_1a_2c_2 = l_1^2,$$

and the only difference is the definition between their angles: $\theta_2^a = \angle A_a - 180^0$. Since θ_2^a is defined from x_1^a to x_2^a according to the D-H convention, it is desired to have $\theta_2^a < 0$ in order to keep the knee A_a outward. Thus, θ_2^a should be in the range of $-180^0 < \theta_2^a < 0$ for such a URR-type system, and just use $-\arccos(\cdot)$ to solve the above equation for θ_2^a and set $s_2 = -\sqrt{1-c_2^2}$.

If we multiply c_1 to the first equation and multiply s_1 to the second one, and then add them together, we obtain

$$a_1 + a_2 c_2 = l_1 c_1.$$

If we now multiply s_1 to the first equation and multiply c_1 to the second one, and then subtract them together, we have

$$-a_2 s_2 = l_1 s_1.$$

Thus, the angle θ_1^a can be determined by calling the 4-quadrant arc tangent function atan2$(\cdot, \cdot)$:

$$\theta_1^a = \text{atan2}(-a_2 s_2,\ a_1 + a_2 c_2).$$

Once both two joint angles θ_1^a and θ_2^a are solved, the homogeneous transformations A_a^1 and A_a^2 are well determined, and the first link of this two-link arm is situated at the position and orientation given by $H_0^a A_a^1$ with respect to the base frame #0. Likewise, the second link has its position and orientation referred to the base by $H_0^a A_a^2$. Therefore, we have not only solved the equivalence between the UPR-type and URR-type 3-leg parallel-chain platforms for their conversion, but also made every necessary transformation ready for graphical drawing and simulation.

5.5 Computer Projects and Exercises of the Chapter

5.5.1 *Two Computer Simulation Projects*

1. A 3-joint RPR planar robot is sitting near a wall-floor corner, as shown in Figure 5.28. If we only consider the x and y coordinates of the tip-point w.r.t. the base as the output, this arm is a redundant robot with $n = 3 > m = 2$. Let the robotic tip-point draw a circle that is centered at (1.2, 1.0) with a radius $R = 0.6$ in meters, starting at the point (1.8, 1.0). The angular speed of the circular drawing is $\omega = 0.5$ rad/sec. counterclockwise. The total length of the sliding link is $L_2 = 1.8$ m., see Figure 5.28, and its back-end point B is desired to never collide with the vertical wall or the floor at any time during the circle drawing. Develop a complete algorithm for the redundant planar robot to draw the specified circle as a main task and to avoid the collision as a subtask, and then program it into Matlab^{TM} to make a 2D animation.
2. A 3+3 hybrid robot with a rectangle configuration on the top mobile plate has the following parameters:

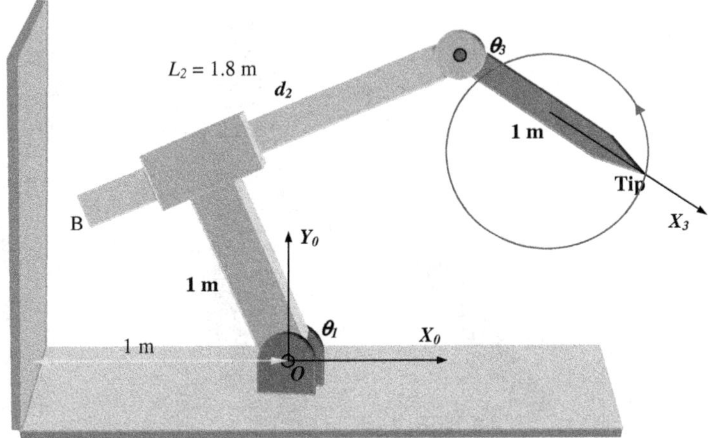

Fig. 5.28 A three-joint RPR planar robot arm

$$p_0^a = \begin{pmatrix} 1.2 \\ 0 \\ 0 \end{pmatrix}, \quad p_0^b = \begin{pmatrix} 0 \\ 0.8 \\ 0 \end{pmatrix}, \quad p_0^c = \begin{pmatrix} 0 \\ -0.8 \\ 0 \end{pmatrix},$$

on the base plate with respect to the base frame;

$$p_3^a = \begin{pmatrix} 0.6 \\ 0 \\ 0 \end{pmatrix}, \quad p_3^b = \begin{pmatrix} 0 \\ 0.4 \\ 0 \end{pmatrix}, \quad p_3^c = \begin{pmatrix} 0 \\ -0.4 \\ 0 \end{pmatrix},$$

on the top mobile plate referred to frame 3; and $d_4 = 0.6$, $d_5 = 0.2$ and $d_6 = 0.4$ all in meter for the last 3-revolute-joint serial-chain arm sitting on the top plate, as depicted in Figure 5.29.

At the home position, $\theta_6(h) = 0$, and

$$p_0^6(h) = \begin{pmatrix} 0.8 \\ 0.4 \\ 2.2 \end{pmatrix} \quad \text{and} \quad R_0^6(h) = \begin{pmatrix} 1 & 0 & 0 \\ 0 & 1 & 0 \\ 0 & 0 & 1 \end{pmatrix}.$$

The last tool frame 6 is required to travel linearly from the home to the following destination:

$$p_0^6(d) = \begin{pmatrix} -1 \\ -1 \\ 2 \end{pmatrix} \quad \text{and} \quad R_0^6 = \begin{pmatrix} 0 & 0 & -1 \\ 0 & 1 & 0 \\ 1 & 0 & 0 \end{pmatrix},$$

with a total number of sampling points $N = 100$ and the sampling interval $\Delta t = 0.01$ sec.

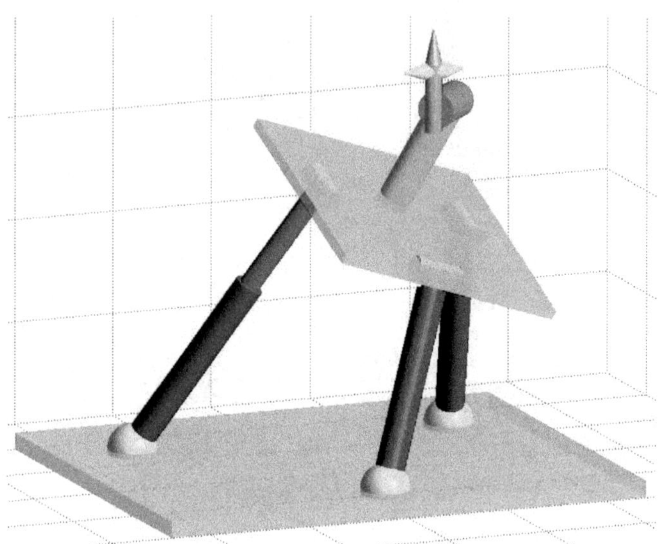

Fig. 5.29 A 3+3 hybrid robot in rectangle configuration

a. Following the procedure of the 3+3 hybrid-chain robot I-K solution, find the passive orientation R_0^3 of the top plate at the home position;
b. Determine the three prismatic joint vectors l_0^i for $i = a, b, c$ at the home;
c. Determine θ_4 and θ_5 at the home position;
d. Write a MATLAB^{TM} program to finally plot all the 6 joint profiles versus time from the joint lengths l_1, l_2, l_3 to the revolute joint angles θ_4, θ_5 and θ_6 over the $N = 100$ sampling points.

5.5.2 Exercise Problems

1. For a given 4-joint robot arm with an overhead beam and two revolute joints plus one prismatic joint, as shown in Figure 5.30, answer the following questions:

   a. Determine a D-H parameter table for the robot;
   b. Find a symbolical form of the homogeneous transformation A_0^4 ;
   c. Determine the Jacobian matrix $J_{(0)}$ by taking derivative of the symbolical position vector p_0^4 w.r.t. the robotic joint positions;
   d. Find the singular point(s) without the 4th joint, i.e., to find the zero points of the determinant of the first three columns of $J_{(0)}$;
   e. If only the 3 d.o.f. of the robot tip-point position is considered in motion-planning, find the minimum-norm solution of the joint velocities if the tip point is moving along the positive direction of the y_0-axis at a speed of 1 m./sec. when $d_1 = 1$ m., $\theta_2 = -120^0$, $\theta_3 = 60^0$ and $d_4 = 1.5$ m.

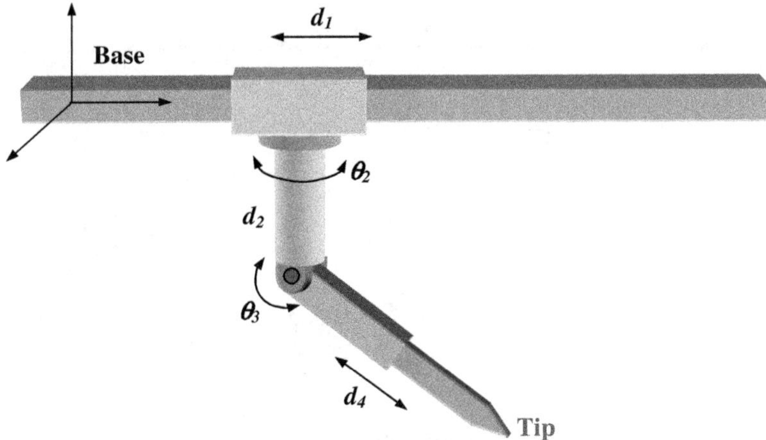

Fig. 5.30 A 4-joint beam-hanging PRRP robot

f. Find a general solution of the 4-joint velocities to track the same trajectory but adding a singularity avoidance subtask based on (d).

2. A 3-joint RRP planar robot is shown in Figure 5.31, where $a_1 = 1$ m.

   a. Find the 2D position vector p_0^w of the wrist point w with respect to the base, and determine the Jacobian matrix $J_{(0)}$;
   b. Find all the singular points;
   c. If the robot is motionless and the wrist point w is touching the inner wall of the bowl at (0.8, 0) with a pressing force $f = 12$ N along the z_3 direction, find all the three joint torques/force in terms of the joint positions θ_1, θ_2 and d_3;

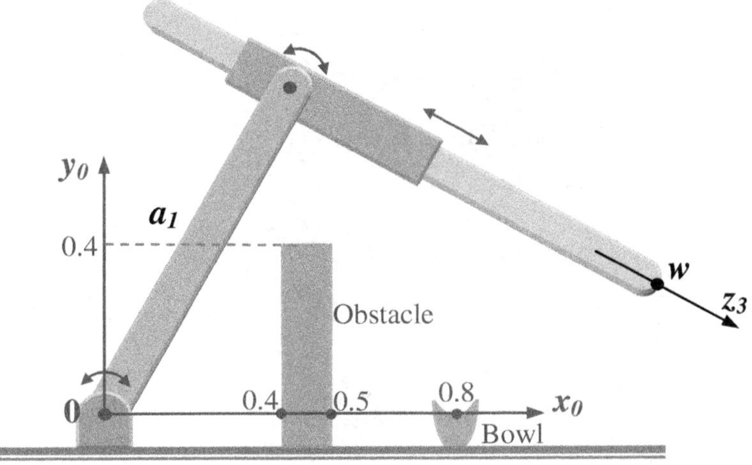

Fig. 5.31 An RRP 3-joint planar robot to touch a bowl

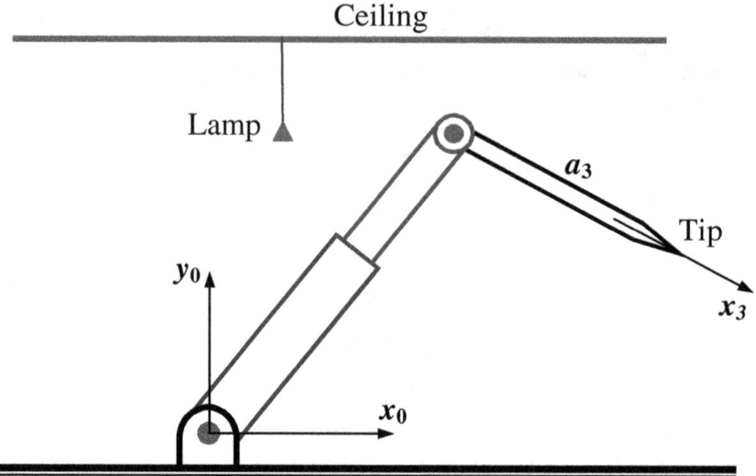

Fig. 5.32 An RPR 3-joint planar robot

d. Find a joint velocity solution $\dot{q}$ if point w is going to approach to the bowl **linearly** from the initial point (1, 1) and avoiding any collision with the obstacle.

3. A 3-joint RPR planar arm is shown in Figure 5.32, where $a_3 = 1$ in meter. A ceiling lamp is located at (0.4, 1.4) in meter that is referred to the base.

   a. If the robotic tip position vector is defined by $p_0^3 = \begin{pmatrix} x \\ y \end{pmatrix}$, where x and y are coordinates of the tip w.r.t. the base in 2D space, find the Jacobian matrix J_0 ;

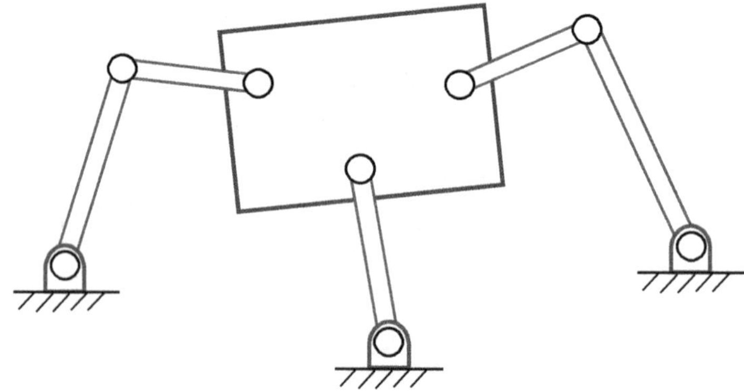

Fig. 5.33 A planar mechanism

b. If the tip point starts traveling from $p_0^3 = (0.8 \;\; 0.4)^T$ under $\theta_1 + \theta_3 = -90^0$ at $t = 0$, determine the three joint positions θ_1, d_2 and θ_3 ;

c. Does the elbow position touch the ceiling lamp at $t = 0$?

d. Is the joint velocity vector $\dot{q} = \begin{pmatrix} -\sqrt{3} \\ 1.6 \\ -1.5 \end{pmatrix}$ a null solution at $t = 0$? Why?

e. If the arm tip point hangs down a weight of 2 Kg at $t = 0$, find each joint torque/force;

f. If the arm's tip point is to travel linearly from the above starting point to the destination $p_0^3 = (1.4 \;\; 0.7)^T$ within $T = 3$ sec., and also to avoid the elbow point hitting the ceiling lamp, find a complete differential motion solution.

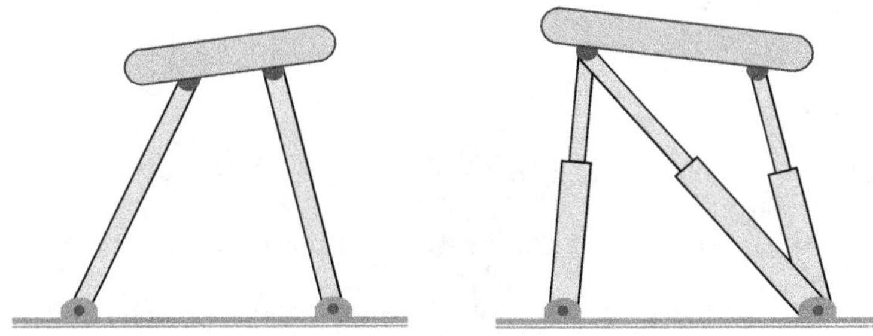

Two Planar Parallel-Chain Systems

A 3D Platform: All joints connected to both the top and base discs are ball-joint, and each prismatic joint can also be spinning.

Fig. 5.34 Three parallel-chain systems

4. A planar system has three legs, as shown in Figure 5.33, determine the net d.o.f. m.
5. Determine the net d.o.f. for each of the mechanisms shown in Figure 5.34.
6. Implement the given F-K recursive algorithm for a 3-3 type Stewart platform into MATLAB^{TM}, and then define your input set for the six prismatic leg lengths l_1 through l_6 to run the program and determine the top disc position p_0^6 and orientation R_0^6. You may also redefine a different set of parameters to extend the F-K algorithm to a new algorithm for a 6-3 Stewart platform system, and then run the new program again.
7. Using all the parameters that you defined in the last problem for either a 3-3 or a 6-3 type Stewart platform, and also based on the results after running the F-K algorithm in MATLAB^{TM}, find the Jacobian matrix J_0 for this closed parallel-chain system.

References

1. Bellman, R.: Introduction to Matrix Analysis. McGraw-Hill Book Inc., New York (1960)
2. Horn, R., Johnson, C.: Matrix Analysis. Cambridge University Press, New York (1985)
3. Boullion, T., Odell, P.: Generalized Inverse Matrices. John Wiley and Sons, New York (1971)
4. Nakamura, Y.: Advanced Robotics: Redundancy and Optimization. Addison Wesley, MA (1991)
5. Klein, C., Huang, C.: Review of Pseudoinverse Control for Use with Kinematically Redundant Manipulators. IEEE Transactions on Systems, Man, and Cybernetics 13(3), 245–250 (1983)
6. Yoshikawa, T.: Analysis and Control of Robot Manipulators with Redundancy. In: The 1st International Symposium of Robotics Research, Bretten Woods, New Hampshire, August 28-September 2 (1983)
7. Chan, J., Gu, E.: The Design and Kinematic Control of a 9-Joint Robotic Manipulator for Car-Interior Applications. In: Proc. 1992 IEEE Conference on Control Applications, Dayton, OH, pp. 300–305 (September 1992)
8. Chan, J., Gu, E.: Nonlinear Kinematic Control for a Robotic System with High Redundancy. In: Proc. 31st IEEE Conference on Decision and Control, Tucson, Arizona, pp. 614–619 (1992)
9. Hiroya, Y., Shigeo, H.: Development of practical 3-dimentinal active cord mechanism ACM-R4. Journal of Robotics and Mechatronics 18(3), 305–311 (2006)
10. Makoto, M., Shigeo, H.: Locomotion of 3D snake-like robots; shifting and rolling control of active cord mechanism ACM-R3. Journal of Robotics and Mechatronics 18(5), 521–528 (2006)
11. Siciliano, B., Khatib, O. (eds.): Springer Handbook of Robotics. Springer (2008)
12. Patel, R., Shadpey, F.: Control of Redundant Robot Manipulators, Theory and Experiments. LNCIS, vol. 316. Springer, Heidelberg (2005)
13. Merlet, J.: Parallel Robots, 2nd edn. Springer, The Netherlands (2006)
14. Vischer, P., Clavel, R.: Kinematic Calibration of Parallel Delta Robot. Robotica 16, 207–218 (1998)

15. Angeles, J.: Fundamentals of Robotic Mechanical Systems. Springer, New York (2002)
16. Kumar, V., Waldron, K.: Force Distribution in Closed Kinematic Chains. IEEE Trans. on Robotics and Automation 4(6), 657–664 (1988)
17. Ling, S., Huang, M.: Kinestatic Analysis of General Parallel Manipulators. Transactions: ASME Journal of Mechanical Design 117, 601–606 (1995)
18. Mohamed, M., Duffy, J.: A Direct Determination of the Instantaneous Kinematics of Fully Parallel Robot Manipulators. Transactions: ASME Journal of Mech. Transm. Automation Design 107, 226–229 (1985)
19. Waldron, K., Hunt, K.: Series-Parallel Dualities in Actively Coordinated Mechanisms. International Journal of Robotics Research 10(5), 473–480 (1991)

Chapter 6
Digital Mock-Up and 3D Animation for Robot Arms

6.1 Basic Surface Drawing and Data Structure in MATLAB^{TM}

In order to draw a multi-joint robot arm, the first thing that we must do is to know how to create surfaces of the 3D basic parts in MATLAB^{TM}. The 3D basic parts may include but are not limited to a cylinder and its variations, a sphere or an ellipsoid, a rectangular block, and a torus. Since in MATLAB^{TM}, each 3D part drawing is always initiated at the base coordinate frame, we need to be not only aware of how to create it, but also familiar of its data structure in order to modify it and/or send it to a desired destination (both position and orientation). The detailed instructions, procedures and examples can be found either in the MATLAB^{TM} online help manual, or refer to the literature [1]–[4]. Let us here start with a surface of cylinder drawing.

MATLAB^{TM} offers an internal function $[x, y, z]$ =cylinder(r, n), where the first input dummy element r can be specified either by a single radius for creating a regular cylinder, or by an array for drawing a cylinder with variable radii. The second dummy element n is to specify the number of planar faces to form the entire side surface of the cylinder. The user is not allowed to specify a height of the cylinder and it is always 1 as default. The output coordinates $[x, y, z]$ are an m by $n+1$ array for each of the x, y and z, where m is the dimension of the radius array r, which is at least 2 if r is a single constant for a regular cylinder. For example, $[x, y, z]$ =cylinder$(1, 10)$ creates a set of coordinate arrays x, y and z of 2 by 11 each for a cylinder side surface of radius equal to 1 and formed by 10 equal-size planar faces with a unity height, as shown in Figure 6.1. To draw this side surface, it must call a 3D surface drawing internal function surf$(\cdots)$ as follows:

```
>> [x,y,z] = cylinder(1,10);

>> surf(x,y,z,'FaceColor','c','Facealpha',0.6, ...
            'EdgeColor','r','AmbientStrength',0.4),
```

E.Y.L. Gu, *A Journey from Robot to Digital Human*,
Modeling and Optimization in Science and Technologies 1,
DOI: 10.1007/978-3-642-39047-0_6,

```
grid, axis([-1.5 1.5 -1.5 1.5 0 2]),
daspect([1 1 1]), grid, view([1 1 0.4]),
light('Position',[1 -0.2 1],'Style','infinite'),
```

The internal function surf($\cdots$), in addition to accepting all the output coordinate arrays x, y and z from the cylinder function, allows us to specify many drawing properties, including the face color, the degree of transparency in 'Facealpha' between 0 and 1 with smaller number more transparent, the edge color and the ambient strength. Under certain definitions of axis, view and lighting condition, a resulting semi-transparent cylindrical surface that is sitting on the base frame is given in Figure 6.1.

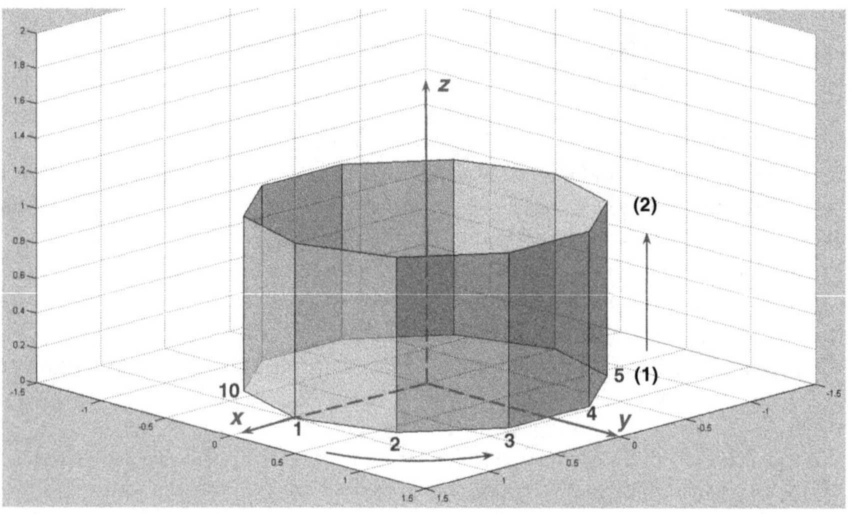

Fig. 6.1 Data structure of a cylinder drawing in MATLAB^{TM}

The data structure in each array of x, y and z is defined in MATLAB^{TM} by a certain counting rule: the number of rows is counted as a latitude of the cylindrical side surface with its bottom corresponding to the first row and its top corresponding to the last row in each of the x, y and z arrays. The columns of each array count the vertices counterclockwise starting at the point $(1, 0, 0)$ and ending to the same point after a round trip to form the first row of each coordinate array as a longitudinal measure of the cylinder. Then, starting at $(1, 0, 1)$ and ending at the same point to form the second row for each of the 2 by 11 arrays x, y and z.

In Figure 6.1, the number without parenthesis indicates the counting order of the vertices as a longitude, and the number with parenthesis counts the latitude from bottom up. For instance, if the radius $r = 1$ is replaced by an array $r = [0.5\ 1\ 0.5\ 0]$ and n remains 10 inside the cylinder function, then the resulting surface looks like a diamond shape, as depicted in the left picture of Figure 6.3. In this case, the number of latitude layers is 4 so that the dimension of each coordinate array x, y or z becomes 4 by 11.

However, the cylinder created by the above program in MATLAB^{TM} does not include the top and bottom planar covering surfaces. We have to call another internal function patch($\cdots$) if the covering planes are needed. Namely, with the above regular cylinder example, by adding the following two more lines:

```
patch('Vertices',[x(1,:)' y(1,:)' z(1,:)'], ...
          'Faces',[1:10],'FaceColor','c', ...
          'EdgeColor','r','AmbientStrength',0.4),

patch('Vertices',[x(2,:)' y(2,:)' z(2,:)'], ...
          'Faces',[1:10],'FaceColor','c', ...
          'EdgeColor','r','AmbientStrength',0.4),
```

both the top and bottom of the cylinder in Figure 6.1 will immediately be covered by the same color non-transparent planar surfaces (the 'Facealpha' is not specified here), where the first line covers the bottom and the second line covers the top.

The 'Vertices' inside the function patch($\cdots$) requires a number of vertices in order to be patched by a surface. The format is that this must be an m by 3 array in 3D drawing space if the number of vertices is m. Each row of the m by 3 array describes a vertex with its x, y and z-coordinates. The 'Faces' specifies which vertices are to be spanned to form a covering surface. In the above example, since $x(1,:)$ is the first row of the x-coordinate array of the cylinder$(1,10)$, which is 1 by 11, $[x(1,:)'\ y(1,:)'\ z(1,:)']$ should be an 11 by 3 array. Therefore, the number of rows corresponds to the number of vertices on the bottom of the cylinder. If the 'Faces' is specified by a row array $[1:10]$, this means to patch a surface that covers from the first to the tenth vertex, which are described by the first ten rows of the 11 by 3 vertices array $[x(1,:)'\ y(1,:)'\ z(1,:)']$. In fact, the 'Faces' in the function patch($\cdots$) can be specified by an n by k array to draw n surfaces in total, with each spanning k vertices defined by each row of that array.

Similarly, by calling the internal function $[x,y,z]$ =sphere(n), a set of unity spherical surface coordinate arrays with resolution n will be created. If $n = 10$, each of the x, y and z coordinate arrays is 11 by 11 in dimension. Figure 6.2 illustrates the number of rows with parenthesis and the number of columns without parenthesis for each 11 by 11 coordinate array. The sphere created is always sitting in the base frame with the center at the origin and the radius equal to 1 as default.

To deform the unity sphere to a non-unity sphere or an ellipsoid, one can just alter any one of the x, y and z inputs in the surf($\cdots$) function by multiplying a number. For instance, if only z is multiplied by 2, i.e., $z*2$, and the other two arrays x and y inside the function surf$(\cdots)$ are not changed, then, the new surface is an ellipsoid, as depicted on the right picture of Figure 6.3. However, the center of this ellipsoid is still at the origin of the base frame.

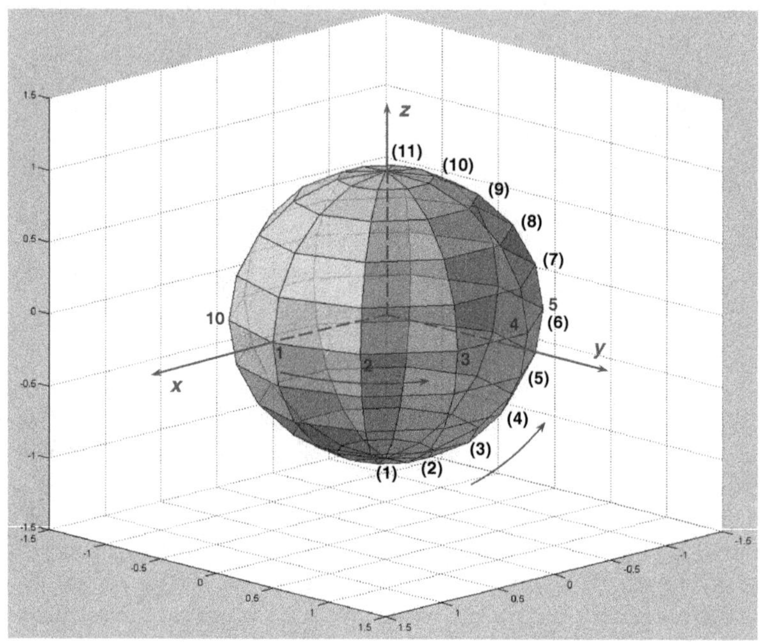

Fig. 6.2 Data structure of a sphere drawing in MATLAB^{TM}

In order to create a 3D rectangular block, the best way is using the internal function patch$(\cdots)$. Since a rectangular block has 8 vertices and 6 planar surfaces with each covering 4 vertices, we must prepare an 8 by 3 array for the 'Vertices' and a 6 by 4 array for the 'Faces' specification. A function subroutine that can generate two such arrays as well as a main program to draw the specified rectangular block are illustrated as follows:

```
[vert,face] = rectangular(5,3,2);

patch('Vertices',vert,'Faces',face,'FaceColor','y', ...
      'Facealpha',0.6,'EdgeColor','b','AmbientStrength',0.6),
grid, axis([-4 4 -3 3 0 3]), daspect([1 1 1]), grid,
view([1 1 0.6]),

% ----- The Main Program Above -------- %
```

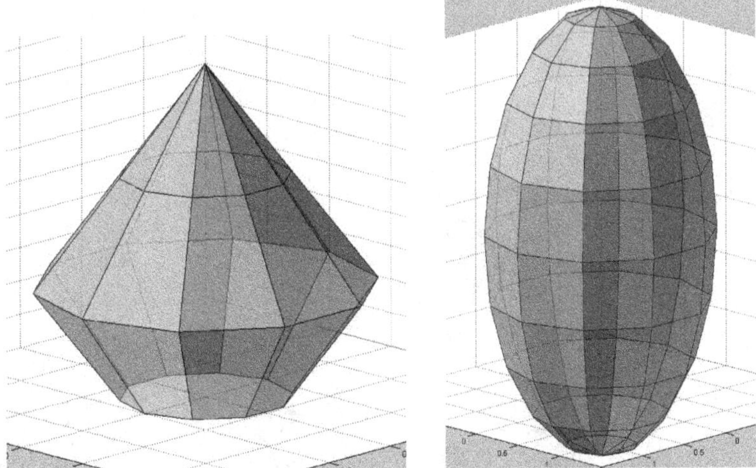

Fig. 6.3 A diamond and an ellipsoid drawing in MATLAB^{TM}

```
% ----- Function Subroutine Below ----- %

function [vertex,face]=rectangular(w,d,h)

   vertex = [w/2 -d/2 0; w/2 d/2 0; w/2 d/2 h;
             w/2 -d/2 h; -w/2 -d/2 0; -w/2 d/2 0;
            -w/2 d/2 h; -w/2 -d/2 h];

   face = [1 2 3 4; 5 6 7 8; 1 2 6 5;
           3 4 8 7; 2 6 7 3; 1 5 8 4];
end
```

In the above function subroutine, rectangular(w, d, h) requires input of width w, depth d, and height h for a block to be created. After calling the function, the two arrays $vert$ and $face$ are generated to meet the requirements of the patch$(\cdots)$ function in order to draw the specified rectangular block. With $w = 5$, $d = 3$ and $h = 2$, the rectangular block is drawn in Figure 6.4.

In MATLAB^{TM}, not every familiar surface can be drawn by the existing internal functions. For example, if we wish to create a torus, we may have to write our own function subroutine to generate its necessary data before calling the surf$(\cdots)$ function. Since a 2-torus is topologically equivalent (homeomorphic) to a direct product of two 1-circles, i.e., $T^2 \simeq S^1 \times S^1$, a regular T^2 surface that is embedded in $\mathbb{R}^3$ Euclidean space can have a following parametric equation of two angular parameters θ and ϕ:

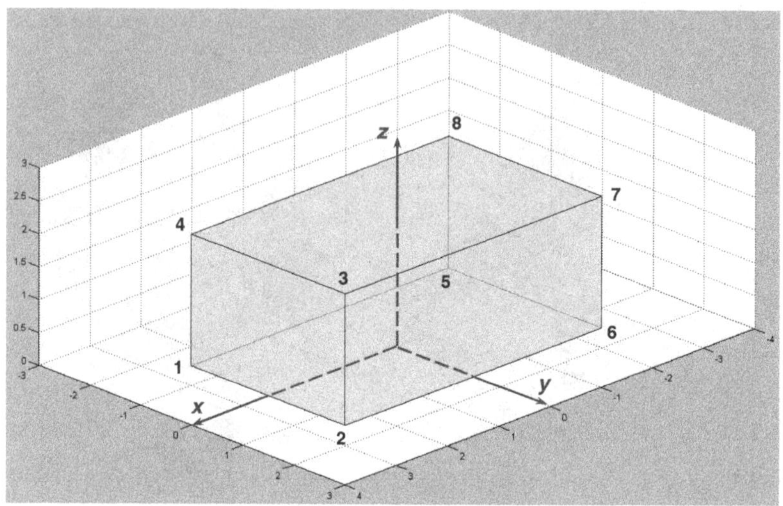

Fig. 6.4 Create a rectangular surface in MATLAB^{TM}

$$\begin{cases} x = (r\cos\theta + d)\cos\phi \\ y = (r\cos\theta + d)\sin\phi \\ z = r\sin\theta, \end{cases}$$

where r is the radius of the torus tube, called the minor radius, and d is the distance between the tube center and the center of the torus, called the major radius, under $r < d$. The main program and the function subroutine are given below:

```
[x,y,z] = torus(1,3,1);

surf(x,y,z,'FaceColor','c','Facealpha',0.6, ...
          'EdgeColor','r','AmbientStrength',0.4),
grid, daspect([1 1 1]), grid,

% ----- The Main Program Above -------- %

% ----- Function Subroutine Below ----- %

function [x,y,z] = torus(r,d,p)

   [th,phi] = meshgrid(-3.1:0.1:3.2);

   x = (r*cos(th/p)+d).*cos(phi);
   y = (r*cos(th/p)+d).*sin(phi);
   z = r*sin(th/p);

end
```

The resultant torus is created in Figure 6.5 if $p = 1$ for a full torus, and in Figure 6.6 if $p = 0.5$ for a half surface of torus. Often, we have to call an internal function meshgrid($\cdots$) in MATLAB^{TM} to standardize the data structure of the three coordinate arrays x, y and z through the standard array definitions of the two parameters θ and ϕ for a torus surface.

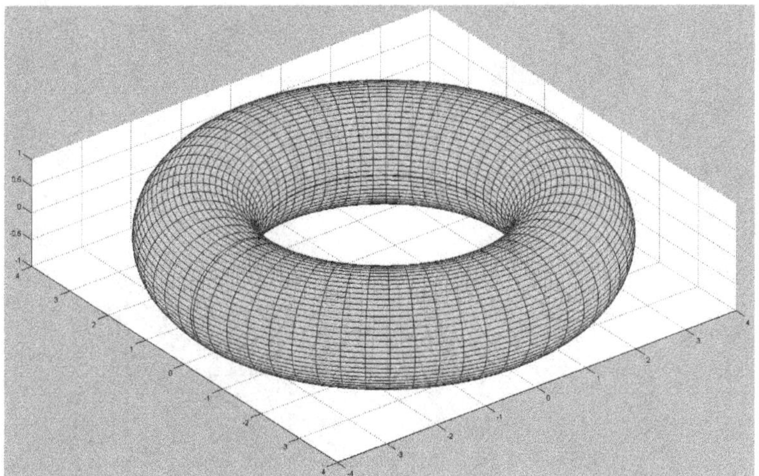

Fig. 6.5 Create a full torus surface in MATLAB^{TM}

After the coordinate data of a basic surface are generated either by the existing internal functions or by mathematically created functions in MATLAB^{TM}, we need to do two more things before digitally assembling every part to build a robot arm:

1. **Sculpturing**: to make a necessary local deformation on a basic surface;
2. **Transformation**: to send it to a desired destination.

To sculpture a basic surface, such as a cylinder, since we are already familiar with the data structure after calling the function cylinder(r, n), it is time to manipulate the three coordinate arrays x, y and z towards any number of desired local smooth deformations. For example, the following program can sculpture a cylindrical surface to make two local deformations:

```
[x,y,z] = cylinder(0.8:-0.005:0.6,40);

for i = 4:34
   for j = 1:19
      y(i,j) = y(i,j)+0.2*sin((i-4)*pi/30)^2* ...
                              sin((j-1)*pi/18)^2;
```

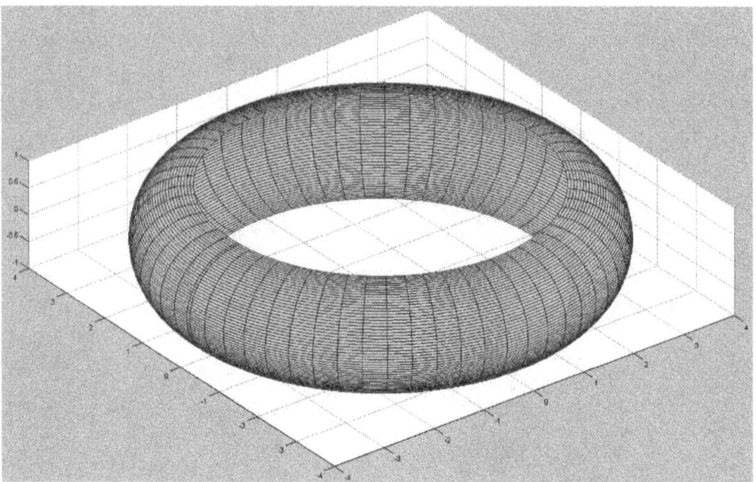

Fig. 6.6 Create a half torus surface in MATLAB^{TM}

```
    end
end            % The First Deformation on y

for i = 32:36
    for j = 35:46
        jj = mod(j,41)+1;
        x(i,jj) = x(i,jj)+0.2;
    end
end            % The Second Deformation on x

surf(x,y,z,'FaceColor','c','FaceLighting','Phong', ...
           'Facealpha',0.8,'EdgeColor','None', ...
           'AmbientStrength',0.5),
axis([-1 1 -1 1 0 1]), grid, daspect([1 1 1]),
view([1 0.6 0.2]), grid,
light('Position',[1 0.4 2],'Style','infinite');
```

The deformed surface is shown in Figure 6.7. The first deformation locally adds a product of two sine square function values into the y-coordinate array between the latitudes 4 and 34 and between the longitudes 1 and 19. In order to make a smooth deformation, the original cylindrical surface must have a higher resolution both in latitudinal and longitudinal directions. Here we create a cone-like cylinder with a varying radius $r = 0.8 : -0.005 : 0.6$ that generates $(0.8 - 0.6)/0.005 = 40$ latitudinal layers, and with $n = 40$ longitudinal grids. The second deformation is to add a constant to the

x-coordinate array between latitudes 32 and 36 and between the longitudes 36 and 41 and continuously between 1 and 6. Due to the break point for the number of columns of the x array at 41 and then back to 1 as a closed loop, here we simply define a new column index $jj = \text{mod}(j, 41) + 1$ to automatically bypass the break point.

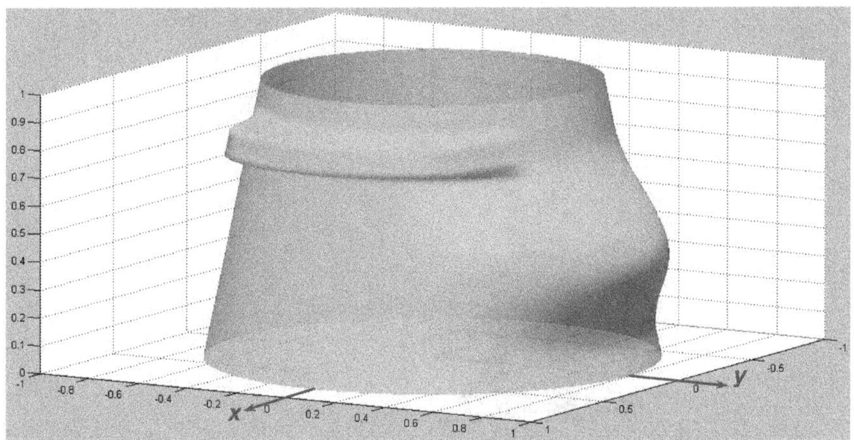

Fig. 6.7 Making a local deformation for a cylindrical surface in MATLAB^{TM}

A local deformation on either a cylindrical or spherical surface will be useful to make a better and more realistic look, especially in digital human modeling, where each segment of an arm or leg often has a muscle curve, and the details will be introduced in Chapter 10. If the amplitude of the sine square function in the above deformation program is set to be another sine or cosine function of time t and placed into a for-loop, you will see a dynamic deformation effect, and this may be useful to create a digital human local muscle movement.

Since each creation of surfaces is always initiated at the base frame, we have to send each of them to its desired destination, including desired position and orientation. A homogeneous transformation H_0^1 is the best way to send each surface from the base frame 0 to frame 1 for both position and orientation simultaneously. Figure 6.8 demonstrates such a result for an object created initially at the base frame. As an example, let

$$H_0^1 = \begin{pmatrix} \cos\theta & 0 & \sin\theta & -2 \\ 0 & 1 & 0 & 2 \\ -\sin\theta & 0 & \cos\theta & 4 \\ 0 & 0 & 0 & 1 \end{pmatrix},$$

where angle $\theta = -60^0$ to rotate the object about the y-axis of the base by θ, and also to translate it from the origin of the base to a new position $(-2\ 2\ 4)^T$ with respect to the base.

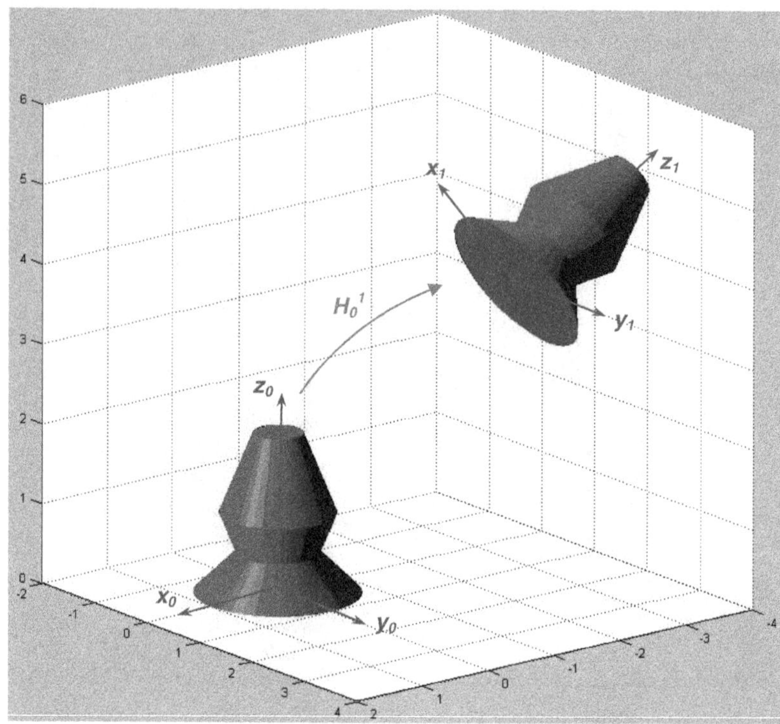

Fig. 6.8 Sending an object from the base to a desired destination

In order to multiply each set of x, y and z coordinates by the homogeneous transformation H_0^1, we have to rearrange the data and add a 1 at the bottom of each $(x\ y\ z)^T$ coordinate column, as required by equation (3.3) in Chapter 3. The program is given as follows:

```
[x0,y0,z0] = cylinder([1 0.5 0.7 0.5 0.3],40);
z0 = z0*2;

th = -pi/3;
H01 = [cos(th) 0 sin(th) -2
        0 1 0 2
       -sin(th) 0 cos(th) 4
        0 0 0 1];

x1 = []; y1 = []; z1 = []; % To Initialize New Arrays

for i = 1:5
    cylin = H01*[x0(i,:); y0(i,:); z0(i,:); ones(1,41)];
                                                    % To Add 41 Ones
    x1 = [x1; cylin(1,:)];
```

```
    y1 = [y1; cylin(2,:)];
    z1 = [z1; cylin(3,:)];
end
```

Once both $[x_0, y_0, z_0]$ and $[x_1, y_1, z_1]$ are prepared, by calling the surf($\cdots$) and patch($\cdots$) functions, the graphics will be created and show clearly sending the object to the destination by this particular homogeneous transformation H_0^1. Obviously, if the homogeneous transformation contains a variable angle and variable position coordinates that are all placed inside a for-loop or a while-loop of the main program, the object will dynamically move in 6 d.o.f. This is just the mechanism of digitally animating a robot arm in motion.

6.2 Digital Modeling and Assembling for Robot Arms

In this section, let us first digitally model and assemble an open serial-chain 7-joint redundant robot in MATLAB^{TM}, as introduced and discussed in Chapter 5. The assignment of all the coordinate frames based on the D-H convention is depicted in Figure 6.9.

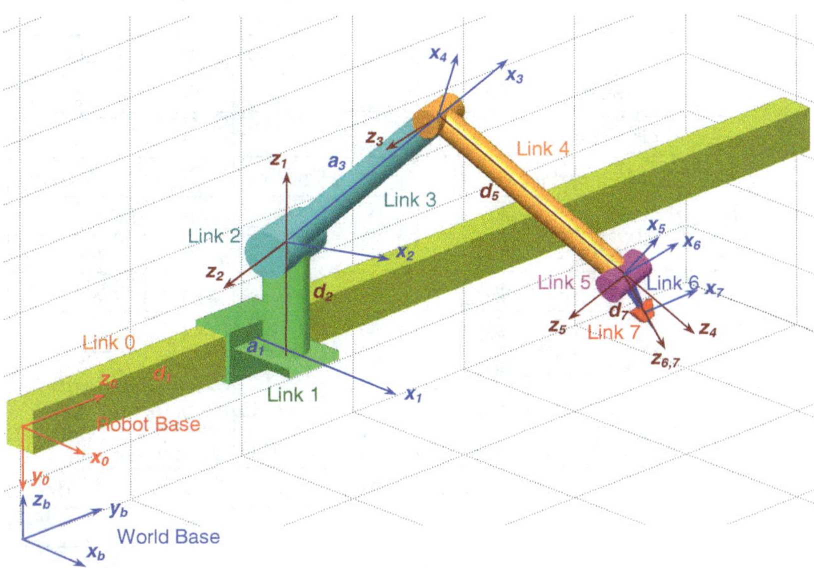

Fig. 6.9 D-H modeling of the 7-joint redundant robot

We have listed its complete D-H table in Chapter 5 and also derived the first three one-step homogeneous transformation matrices A_0^1, A_1^2 and A_2^3 in equation (5.14). Now, the remaining matrices are given below:

$$A_3^4 = \begin{pmatrix} c_4 & 0 & s_4 & 0 \\ s_4 & 0 & -c_4 & 0 \\ 0 & 1 & 0 & 0 \\ 0 & 0 & 0 & 1 \end{pmatrix}, \quad A_4^5 = \begin{pmatrix} c_5 & 0 & -s_5 & 0 \\ s_5 & 0 & c_5 & 0 \\ 0 & -1 & 0 & d_5 \\ 0 & 0 & 0 & 1 \end{pmatrix},$$

and

$$A_5^6 = \begin{pmatrix} c_6 & 0 & s_6 & 0 \\ s_6 & 0 & -c_6 & 0 \\ 0 & 1 & 0 & 0 \\ 0 & 0 & 0 & 1 \end{pmatrix}, \quad A_6^7 = \begin{pmatrix} c_7 & -s_7 & 0 & 0 \\ s_7 & c_7 & 0 & 0 \\ 0 & 0 & 1 & d_7 \\ 0 & 0 & 0 & 1 \end{pmatrix}. \tag{6.1}$$

Once all the one-step homogeneous transformation matrices are ready, write them into a MATLAB^{TM} program under the set of initial joint positions specified, and then allow MATLAB^{TM} to recursively compute $A_b^1 = A_b^0 A_0^1$, $A_b^2 = A_b^1 A_1^2$, $A_b^3 = A_b^2 A_2^3$, $A_b^4 = A_b^3 A_3^4$, $A_b^5 = A_b^4 A_4^5$, $A_b^6 = A_b^5 A_5^6$, and $A_b^7 = A_b^6 A_6^7$ numerically. With every homogeneous transformation prepared, we now begin to assemble all the necessary parts to construct such a 7-joint redundant robot arm in a digital environment. Table 6.1 details how to generate each part surface and how to send each part to the destination with some adjustments.

Table 6.1 Link drawing and sending information for the 7-joint redundant robot

Link Number	Part Surface (surf or patch)	Sent By A_b^j	Data Adjustments		
			x	y	z
0	rectangular(1.2, 1.5, 35)	A_b^0	x	y	z
1	rectangular(1.5, 1.8, 2.6)	A_b^1	$x - a_1$	y	$z - h/2$
	rectangular(2.5, 2.2, 0.3)	A_b^1	x	y	z
	cylinder(1, 40)	A_b^1	$x * 0.75$	$y * 0.75$	$z * d_2$
2	cylinder(1, 40)	A_b^2	$x * 0.8$	$y * 0.8$	$z * 2.4 - 1.2$
3	cylinder(1, 40)	A_b^3	$-z * a_3$	$y * 0.5$	$x * 0.5$
	cylinder(1, 40)	A_b^3	$x * 0.5$	$y * 0.5$	$z * 1.6 - 0.8$
4	cylinder(1, 40)	A_b^4	$x * 0.45$	$y * 0.45$	$z * d_5$
5	cylinder(1, 40)	A_b^5	$x * 0.5$	$y * 0.5$	$z * 1.6 - 0.8$
6	cylinder([0.24 0.02], 40)	A_b^6	x	y	$z * d_7$
7	endef(0.15, 0.4, 0.25, 0.6)	A_b^7	x	y	z

As a detailed explanation, note that in Table 6.1,

1. Link 1 consists of three parts. The first two are rectangular and the last one is cylindrical. Also, link 3 consists of two cylinders.
2. To create the required data for a rectangular surface, the previous function subroutine rectangular(w, d, h) has been called, where w is the width in the x-direction, d is the depth in the y-direction and h is the height in the

z-direction, all referred to the world base frame b. The output is an 8 by 3 vertex array and a 6 by 4 face array due to a 3D rectanglar block having 8 vertices and 6 faces.

3. In order to send the rectangular object to its destination, since the vertex array has to be premultiplied by a 4 by 4 homogeneous transformation A_b^j, the 8 by 3 vertex array must be concatenated with the 8 by 1 column ones(8,1) to form an 8 by 4 matrix $Vert$. Let $A_b^j * Vert' = V_j$. Then, the input into the 'Vertices' inside the patch function should be $V_j(1:3,:)'$, the transpose of the first three rows of the 4 by 8 matrix V_j.
4. In general, the side surface of each cylinder is created with the unity radius and 40-face resolution. The actual radius can be generated by multiplying the x and y coordinate arrays with a constant. Likewise, the actual height of a cylinder can be generated by multiplying the z array with a constant.
5. The first cylinder of link 3 has switched the order of the x, y and z array positions, because the destination of the cylinder is on frame 3 with the x_3-axis as the central axis of the cylinder, instead of the default z_3-axis.
6. Link 6 is created as a cone with the length equal to d_7.
7. The new function endef(a, b, h, e) generates the data for vertices and faces for a trapezoid surface drawn as link 7 or the end-effector, where a is in the x-direction, b is along the y-axis and h and e are in the z-direction. The center of this trapezoid is expected to match the origin of frame 7 by adjusting both h and e along the z-axis.
8. Since the surf function only draws the side surface for a cylinder, the cylinders of link 2 and link 5 as well as the second cylinder of link 3 need to call the patch function to cover the two open terminals for each.

After calling the surf and patch functions by entering all the prepared data, and also by setting every necessary graphics parameter, such as the axis, grid, view, and light conditions, the 7-joint redundant robot drawing will show up on the screen at the initially specified configuration (posture).

The second example of digitally modeling and assembling robot arms is a 3-3 Stewart platform, which has also been introduced and discussed in Chapter 5. The Stewart platform is one of the most typical 6-actuated-joint closed parallel-chain robotic systems, as shown in Figure 6.10.

The inverse kinematics (I-K) of the 3-3 Stewart platform has been developed in Chapter 5. Given both the position p_0^t and orientation R_0^t of frame t fixed on the top mobile platform with respect to the base frame 0, the directional vector l_0^i for each of the six prismatic joint/link can be readily determined by equations 5.18. Let us re-write them here once again:

$$l_0^1 = R_0^t p_t^a + p_0^t - p_0^a, \quad l_0^2 = R_0^t p_t^a + p_0^t - p_0^b,$$

$$l_0^3 = R_0^t p_t^b + p_0^t - p_0^b, \quad l_0^4 = R_0^t p_t^b + p_0^t - p_0^c,$$

and

$$l_0^5 = R_0^t p_t^c + p_0^t - p_0^c, \quad l_0^6 = R_0^t p_t^c + p_0^t - p_0^a. \tag{6.2}$$

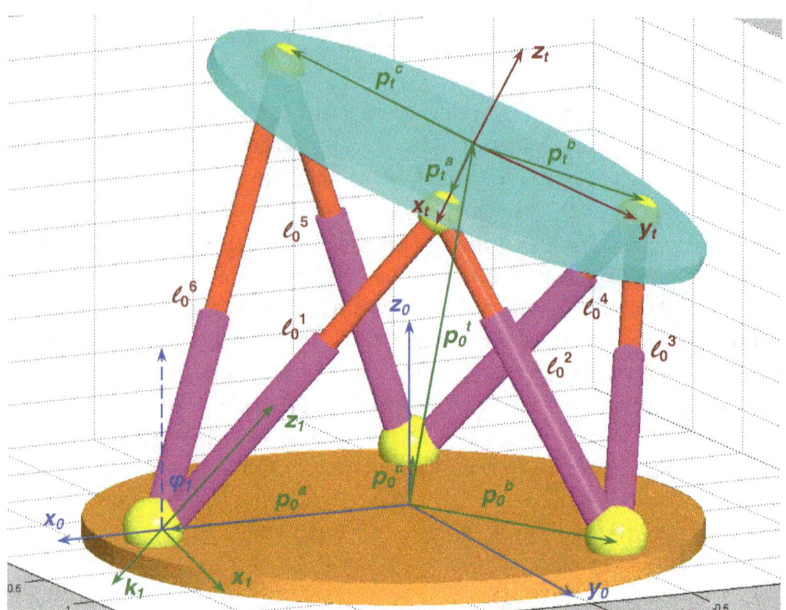

Fig. 6.10 A Stewart platform and coordinate frames assignment

Note that the above set of I-K solutions is only applied for the 3-3 type of Stewart platform to be simulated here, i.e., each adjacent pair of the prismatic legs is universally and spherically jointed to a common point on the base and the bottom of the top disc, respectively, as displayed in Figure 6.10. If it is a 6-3 or 6-6 type, each of p_0^a, p_0^b and p_0^c should be split to two distinct but close vectors to reflect two separated universal or spherical joints between each prismatic leg and the base or the top mobile platform.

To draw such a closed parallel-chain mechanism, in addition to the set of I-K equations, we have to also find the frame of each piston leg with respect to the base. In other words, the unit vector of l_0^i can be defined as a z_i-axis of frame i for $i = 1, \cdots, 6$. However, only knowing the z_i-axis is not sufficient enough for drawing. Thus, to determine the orientation of frame i with respect to the base, R_0^i, one may adopt either one of the following two approaches:

1. First, let the unit vector of l_0^i be $\hat{l_0^i} = l_0^i/||l_0^i|| = z_i$, where $||l_0^i||$ is the norm (or length) of the vector l_0^i. Then, let a unit vector be defined by

$$k_i = \frac{\hat{l_0^i} \times z_0}{||\hat{l_0^i} \times z_0||}.$$

In fact, since $z_0 = (0 \ \ 0 \ \ 1)^T$,

$$\hat{l_0^i} \times z_0 = -\begin{pmatrix} 0 & -1 & 0 \\ 1 & 0 & 0 \\ 0 & 0 & 0 \end{pmatrix} \begin{pmatrix} \hat{l_x^i} \\ \hat{l_y^i} \\ \hat{l_z^i} \end{pmatrix} = \begin{pmatrix} \hat{l_y^i} \\ -\hat{l_x^i} \\ 0 \end{pmatrix}.$$

The norm of the above cross-product becomes

$$\|\hat{l_0^i} \times z_0\| = \sqrt{\hat{l_y^i}^2 + \hat{l_x^i}^2} = \sin\phi_i,$$

where ϕ_i is the angle between z_i and z_0. Similarly, the inner product is

$$\hat{l_0^i}^T z_0 = \hat{l_z^i} = \cos\phi_i.$$

Therefore, according to the $k-\phi$ procedure in equation (2.8) from Chapter 2, the orientation matrix can directly be found by:

$$R_0^i = I + \sin\phi_i K_i + (1 - \cos\phi_i)K_i^2 = I + \sqrt{\hat{l_y^i}^2 + \hat{l_x^i}^2} K_i + (1 - \hat{l_z^i})K_i^2,$$

where $K_i = S(k_i)$, the skew-symmetric matrix of the unit vector k_i. Since the distance between frame i and frame 0 is the vector p_0^j, where j is the corresponding index of a, b or c, as depicted in Figure 6.10, the homogeneous transformation from frame 0 to frame i turns out to be

$$A_0^i = \begin{pmatrix} R_0^i & p_0^j \\ 0_{1\times 3} & 1 \end{pmatrix}.$$

2. First, let the x_i-axis be the unit vector of the cross-product

$$z_i \times x_0 = \hat{l_0^i} \times x_0 = \begin{pmatrix} 0 \\ \hat{l_z^i} \\ -\hat{l_y^i} \end{pmatrix}.$$

Then, the y_i-axis can be found by $z_i \times x_i = y_i$ so that the orientation matrix becomes $R_0^i = (x_i \ y_i \ z_i)$. Using the same method as the first approach, the homogeneous transformation A_0^i for each $i = 1, \cdots, 6$ will be finally determined and ready for drawing.

The above two approaches to defining a coordinate frame on each of the six piston legs may result in different orientations, although their z_i-axis has to be identical along the central axis of the prismatic joint i. Since the appearance of a cylindrical surface is independent of how the x and y axes orient and only depends on the direction of the z-axis, choosing either one of the two approaches will not alter the cylinder drawing. However, if each piston leg is required to be rectangular, or under an orientation constraint, a certain definition of the coordinate frame must follow.

Table 6.2 details all the parts creating and drawing data. Since each prismatic leg consists of two cylindrical links that are connected back to back, and all six legs adopt the same dimension of cylindrical surfaces, the table just gives the i-th leg homogeneous transformation A_0^i to send it to the corresponding destination along with their x, y and z coordinate arrays adjustment.

Note that A_0^j for each base ball in the table means A_0^a, A_0^b or A_0^c, and each of them, for instance A_0^a, is a constant homogeneous transformation matrix given by

$$A_0^a = \begin{pmatrix} I_{3\times 3} & p_0^a \\ 0_{1\times 3} & 1 \end{pmatrix}.$$

Likewise, A_t^j for each top ball in the table also means one of the three constant homogeneous transformation matrices A_t^a, A_t^b or A_t^c that can send each ball from frame t to the corresponding spherical joint center underneath the top platform. As a final transformation, each of the three top balls must be sent from the base to the top mobile disc by the product $A_0^t A_t^j$.

Table 6.2 Link/part drawing and sending information for the Stewart platform

Link Name	Part Surface (surf and patch)	Sent By A_0^j	Data Adjustments x	y	z
Base Disc	cylinder$(1, 40)$	I	$x * 1.4$	$y * 1.4$	$z * 0.1 - 0.1$
Top Disc	cylinder$(1, 40)$	A_0^t	$x * 1.2$	$y * 1.2$	$z * 0.08$
Piston	cylinder$(1, 40)$	A_0^i	$x * 0.07$	$y * 0.07$	$z * 1.2$
Leg i	cylinder$(1, 40)$	A_0^i	$x * 0.05$	$y * 0.05$	$\|\|l_0^i\|\| - z * 1.4$
Base Ball j	sphere(40)	A_0^j	$x * 0.14$	$y * 0.14$	$z * 0.14$
Top Ball j	sphere(40)	$A_0^t A_t^j$	$x * 0.1$	$y * 0.1$	$z * 0.1$

Similarly, after calling the surf and patch functions by entering all the data that have been prepared in MATLAB^{TM}, and specifying the initial values of both the position p_0^t and orientation R_0^t, such a 3-3 type Stewart platform will immediately show up on the computer screen at its initial configuration (posture).

6.3 Motion Planning and 3D Animation

Actually, after preparing and drawing the 7-joint redundant robot arm and the Stewart platform by tracking Tables 6.1 and 6.2, respectively, each of them is ready for animation. Because each homogeneous transformation A_b^j for the 7-joint robot is a function of all the joint positions in vector $q \in \mathbb{R}^7$, and each A_0^i for the Stewart platform is also a function of the Cartesian position p_0^t and orientation R_0^t of the top mobile platform via each piston leg vector $l_0^i \in \mathbb{R}^3$, they can easily be driven in a dynamic motion by varying their joint or Cartesian values. This can be done by placing the procedures of sending every part to their destinations inside a for-loop or a while-loop in a MATLAB^{TM} program.

To update each joint or Cartesian position/orientation in every sampling interval, write a path planning routine and apply an I-K algorithm. As developed and discussed in Chapters 4 and 5, a robotic system, no matter if it belongs to a regular or redundant open serial-chain, or a hybrid serial/parallel-chain kinematic structure, can always be driven for motion in three major different ways: joint motion, Cartesian motion and differential motion.

In the 7-joint redundant robot case, the motion of drawing a circle uses a differential motion path planning. Although it can also be planned by a Cartesian motion to solve its I-K equations either symbolically or by a numerical solver, such as fsolve($\cdots$) or fmincon($\cdots$) with constraints or fgoalattain($\cdots$) carrying some objective functions in MATLAB^{TM}, a differential motion plan is more convenient to handle not only the circle drawing as a main task execution, but also the collision avoidance as a subtask optimization, as demonstrated in Figures 5.9 and 5.10.

In contrast, regarding the motion planning for the Stewart platform, such a closed parallel-chain structure, since its I-K is so easy and straightforward as given in equations (6.2), a Cartesian motion planning is directly employed. To sample both the Cartesian position p_0^t and orientation R_0^t of the top mobile disc, we utilize equations (4.5) and (4.6) along with the $k-\phi$ procedure from Chapter 4. For instance, Figure 6.10 shows a starting configuration of the Stewart platform, where

$$p_0^t = \begin{pmatrix} -0.5000 \\ -0.4000 \\ 1.5000 \end{pmatrix} \quad \text{and} \quad R_0^t = \begin{pmatrix} 0.5000 & -0.7977 & -0.3372 \\ 0.8660 & 0.4605 & 0.1947 \\ 0 & -0.3894 & 0.9211 \end{pmatrix},$$

while Figure 6.11 exhibits a new configuration, where

$$p_0^t = \begin{pmatrix} 0.4800 \\ 0.3000 \\ 1.9200 \end{pmatrix} \quad \text{and} \quad R_0^t = \begin{pmatrix} 0.5000 & -0.8273 & 0.2559 \\ 0.8660 & 0.4777 & -0.1478 \\ 0 & 0.2955 & 0.9553 \end{pmatrix}.$$

Thus, the differences between the new position/orientation and the old ones are given by

$$\Delta p_0^t = p_0^{newt} - p_0^{oldt} = \begin{pmatrix} 0.9800 \\ 0.7000 \\ 0.4200 \end{pmatrix},$$

and

$$R_{oldt}^{newt} = R_{oldt}^0 R_0^{newt} = \begin{pmatrix} 1.0000 & -0.0000 & 0.0000 \\ -0.0000 & 0.7649 & -0.6442 \\ 0.0000 & 0.6442 & 0.7649 \end{pmatrix}.$$

Actually, the above difference of orientation for the top mobile disc is formed by rotating the old frame t about the x_0-axis of the base by a certain angle to reach the new one, as a special but simple case.

By applying equation (4.5) to the position difference Δp_0^t with $N = 36$ sampling points, and applying equation (4.6) to the orientation difference R_{oldt}^{newt} through the $k - \phi$ procedure, the Stewart platform will showcase a realistic animation of moving itself point by point from the initial configuration to the destination, as given in Figure 6.11.

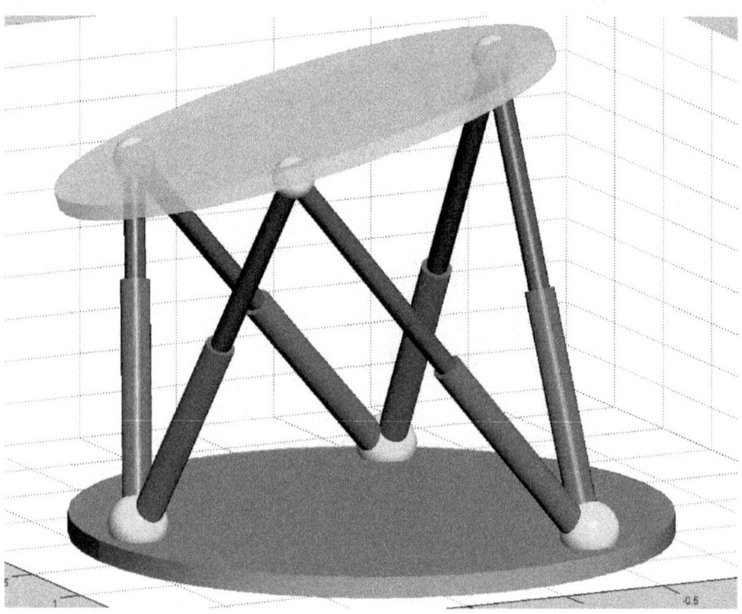

Fig. 6.11 The Stewart platform in motion

Two more realistic animation examples are illustrated here, and they are the two-arm robot to pick up a disc for hanging it on the wall and the $3 + 3$ hybrid robot to draw a sine wave. The two-arm robot has been modeled in Figure 5.16 along with its D-H table in Table 5.1, and also extensively discussed with its Jacobian equation (5.17) in Chapter 5. It has 17 revolute joints and two end-effectors, and thus its augmented Jacobian matrix J_{end} is 12 by 17, see the right-hand side of (5.17). The path planning was done by calling a differential motion algorithm based on the Jacobian equation (5.17). At the beginning, this two-arm robot is sitting at a home position given in Figure 6.12. It then bends the entire torso down to allow its two arms and end-effectors to reach the floor and to grip and pick up a disc, as shown in Figure 6.13. After picking up the disc, by using the same differential motion planning again, its torso turns up and moves aside to allow the two

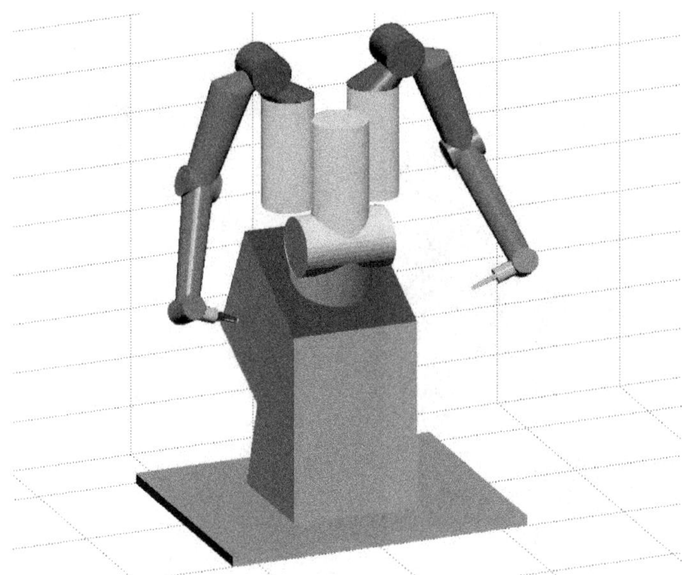

Fig. 6.12 A two-arm robot at its Home position

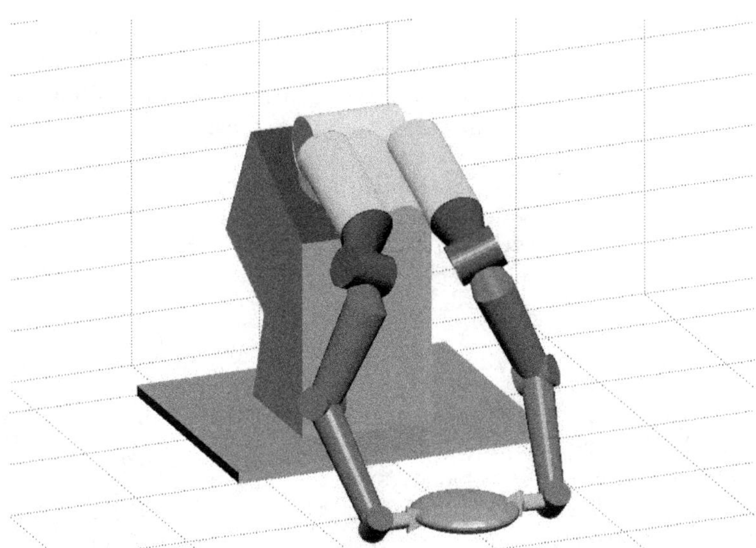

Fig. 6.13 A two-arm robot is picking up a disc from the floor

end-effectors with the disc gripped to reach a wall and hang the disc on the wall, as realistically demonstrated in Figure 6.14.

Since the entire path planning series for the two-arm robot to bow down and pick up a disc and then to stand up and hang the disc on the wall

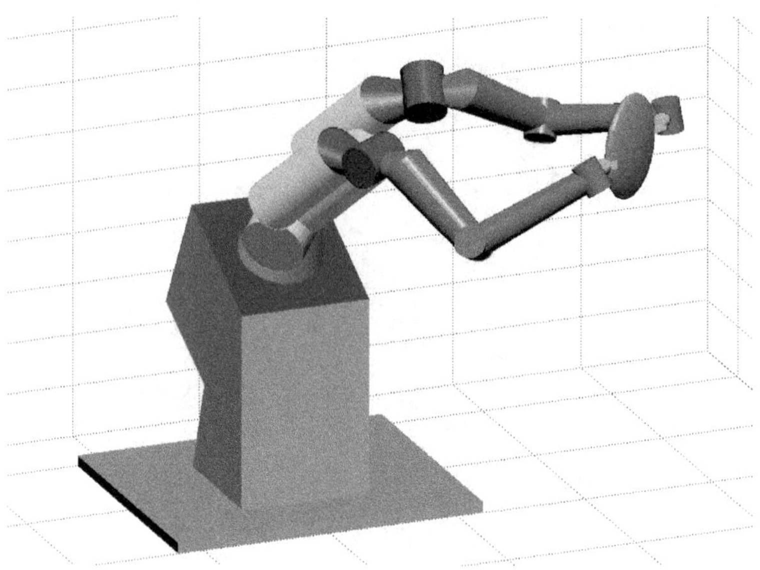

Fig. 6.14 A two-arm robot is hanging the disc on the wall

utilizes the differential motion, each path segment must be specified with a 6 by 1 Cartesian velocity V_0 for each end-effector. The linear velocity portion is relatively simple if each position path is a straight line. First, find the difference between the starting and ending positions:

$$\Delta p_0^{e_i} = p_0^{e_i}(end) - p_0^{e_i}(start),$$

where $p_0^{e_i}$ is the position vector of end-effector i for $i = 1, 2$. Then, the linear velocity in numerical computation becomes

$$v_0 = \frac{\Delta p_0^{e_i}}{N \Delta t},$$

where N is the total number of sampling points within the particular path segment and Δt is the sampling interval.

In contrast, the angular velocity is determined as follows: first find the difference between the ending orientation R_0^e and starting orientation R_0^s for each of the end-effector frames, i.e., frame L and frame R in Figure 5.16, by $R_s^e = R_s^0 R_0^e = (R_0^s)^T R_0^e$. Then, apply equations (2.9) and (2.11) from Chapter 2 to the orientation difference R_s^e to find both ϕ and K, and the angular velocity $\omega_s = \dot{\phi} k$ that is projected on each starting end-effector frame. Finally, the angular velocity ω_0 that is projected onto the base turns out to be

$$\omega_0 = R_0^s \omega_s = \dot{\phi} R_0^s k.$$

Note that $K = S(k)$ is the skew-symmetric matrix of the unit vector k, and $\dot{\phi} = \phi/\Delta t$ in the numerical computation.

Another example of robotic animation is the 3 + 3 hybrid robotic system, the modeling and inverse kinematics of which have been extensively studied in Chapter 5 with Figure 5.26. The animation of such a serial-parallel hybrid chain robotic system is planned to start at a home position given in Figure 6.15, and then to draw a sine wave, as displayed in Figures 6.16 and 6.17.

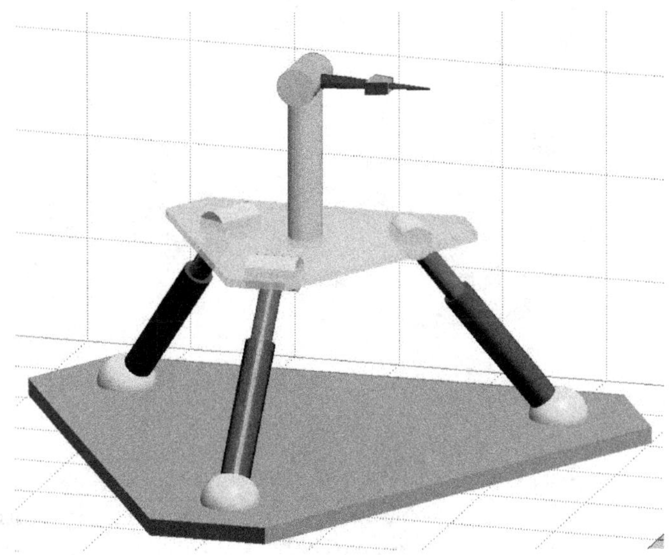

Fig. 6.15 A 3+3 hybrid robot with equilateral triangle configuration at its Home position

In this particular case, the top mobile disc that connects to each of the three piston legs with a revolute joint belongs to the category of equilateral triangle configuration, as depicted on the left most of Figure 5.25 in Chapter 5. For its motion planning, the specified sine wave to be drawn can be described by the following parametric equation:

$$\begin{cases} x(t) = x_0 \\ y(t) = mt + y_0 \\ z(t) = A_m \sin(\omega t) + z_0, \end{cases}$$

where m is the slope of the linear movement in y-direction, A_m and ω are the amplitude and angular frequency of the sine wave, respectively, and x_0, y_0 and z_0 are constants that determine the plane location for the sine wave drawing. In this particular simulation study, the orientation of the end-effector does not change so that the desired R_0^6 is a constant rotation matrix. Therefore,

Fig. 6.16 The 3+3 hybrid robot with equilateral triangle configuration starts drawing a sine wave

both the position vector $p_0^6 = (x(t)\ y(t)\ z(t))^T$ and the orientation $R_0^6(t)$ are specified at each sampling point along the entire sine wave drawing trajectory.

The challenging step now remains: to solve the three prismatic joint values of the bottom parallel-chain platform as well as the three revolute joint angles θ_4, θ_5 and θ_6 of the top serial-chain arm. Based on the study of this special inverse kinematics (I-K) in Chapter 5, the last joint angle θ_6 value at the beginning $t = 0$, i.e., $\theta_6(0)$ must be known. Then, the I-K solving procedure at each sampling point can be summarized again as follows:

Given p_0^6 and R_0^6 with a known $\theta_6(0)$ at $t = 0$,

1. Find p_0^t through equation (5.41) with the previous value of θ_6;
2. Calculate A and B via equation (5.33) and (5.34);
3. Solve equation (5.32) to determine R_0^3 by (5.36) and (5.38) via (5.35);
4. The vectors l_0^i for $i = a, b, c$ can be found by equation (5.42);
5. Calculate $R_3^6 = R_3^0 R_0^6 = (R_0^3)^T R_0^6$;
6. The last three joint angles can be determined by

$$\theta_4 = \text{atan2}(R_3^6(2,3),\ R_3^6(1,3)),$$
$$\theta_5 = \text{atan2}(R_3^6(1,3)/\cos(\theta_4),\ R_3^6(3,3)),$$
$$\theta_6 = \text{atan2}(R_3^6(3,2),\ -R_3^6(3,1)).$$

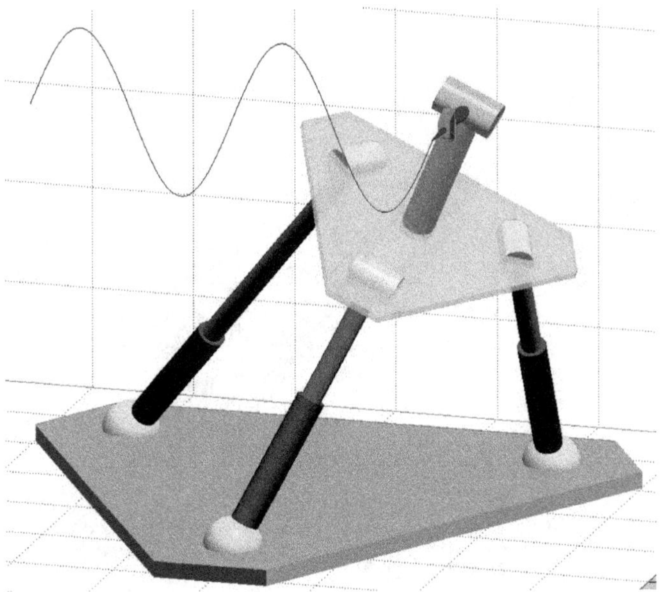

Fig. 6.17 The 3+3 hybrid robot with equilateral triangle configuration ends the drawing

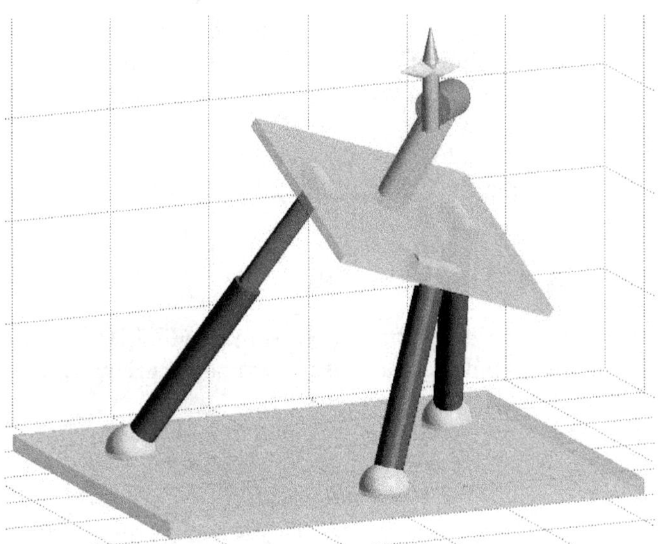

Fig. 6.18 A 3+3 hybrid robot with rectangle configuration at its Home position

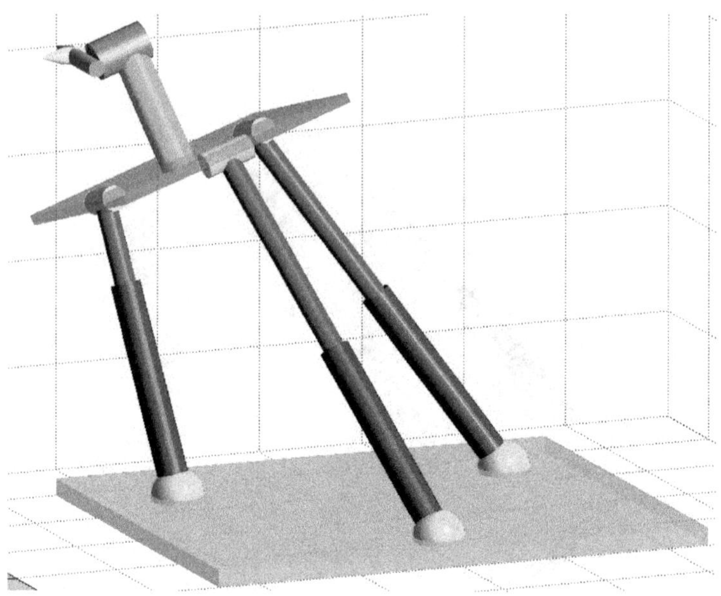

Fig. 6.19 The 3+3 hybrid robot in rectangle configuration is reaching a wall

Finally, as a counterpart, Figures 6.18 and 6.19 show a 3+3 hybrid-chain robot in rectangular configuration at its home position and at a moment of reaching a wall. Both the drawing, assembling, and motion generation are all the same as the one in the above equilateral triangle configuration case in terms of the procedures. The only difference between them is the I-K solution, see the right most of Figure 5.25 and the I-K derivations in Chapter 5.

6.4 Exercises of the Chapter

1. Practice to draw a variety of polyhedrons in MATLAB^{TM}, such as a tetrahedron, octahedron, pentagonal prism, hexagonal prism, right or non-right pyramid, etc.
2. Create one quarter of a torus, one half of a cylinder, including the top, bottom and side covers, and one quarter of a spherical ball in MATLAB^{TM}.
3. Draw a hex bolt and also a hex nut in MATLAB^{TM}. Then, send the hex nut to the end of the hex bolt, and ready to twist the nut onto the bolt.
4. Referring to Figure 6.9 and Table 6.1, draw and assemble the 7-joint redundant robot arm in MATLAB^{TM}. Then, try to program either a joint motion, or a Cartesian motion, or a differential motion algorithm to maneuver and control the created robot to do a specified task.

5. Using the I-K algorithm and the modeling procedures given in this chapter for a 3-3 type Stewart platform, as shown in Figure 6.11, practice to prepare all the parts drawing and necessary transformations, and then assemble them to mock up this system in MATLAB^{TM}. By applying the I-K algorithm and following the Cartesian motion planning, as discussed in this chapter, animate this 3-3 type Stewart platform in motion.

References

1. Siciliano, A.: MATLAB: Data Analysis and Visualization. World Scientific Publishing Co. Pte. Ltd., Hackensack (2008)
2. Hanselman, D., Littlefield, B.: Mastering MATLAB. Prentice Hall, New Jersey (2011)
3. Gilat, A.: MATLAB: An Introduction with Applications, 4th edn. John Wiley and Sons, New York (2010)
4. Pratap, R.: Getting Started with MATLAB: A Quick Introduction for Scientists and Engineers. Oxford University Press, New York (2009)

Chapter 7
Robotic Dynamics: Modeling and Formulations

7.1 Geometrical Interpretation of Robotic Dynamics

Dynamics is the study of motion response to force and torque. At this point, kinematics is a structure of dynamics for every dynamic system. To further interpret the dynamics of a robotic system, we need a deeper insight into the relationship between the dynamics and kinematics before developing dynamic models, formulations, and computational algorithms.

Dynamic modeling and control of nonlinear and highly coupled multi-body systems have been extensively investigated in the past decades. All the studies and developments follow closely with the advance of mathematics [1, 2, 3]. One of the most typical areas of this research is robotics. Since the early 80's, the research and development of robotic dynamics and control theories have been one of the hottest fields drawing a great deal of attention.

A robot manipulator is recognized to be one of the most typical multi-body mechanical systems. Over the past half a century, the Lagrangian and Newton-Euler formulations were the two main streams in the robotic dynamics research [4, 5, 6, 7]. Since the later 80's, a variety of alternative formulations and control algorithms have been developed to improve the modeling quality, computational speed, control accuracy and robustness, theoretical generalizability, and the breadth and depth of physical insight into robotic kinematics and dynamics. Many of them were based on more advanced mathematical theories than the early Lagrangian formulation of four decades ago. Brockett first utilized the Lie group, Lie algebra and their exponential maps to deal with robot kinematic models and made some extension to new dynamic models [8, 9, 10]. Many other researchers followed to develop their new ideas more broadly and specifically. The representative work includes seeking an isometry to linearize the dynamic equations [11], finding a special output function to factorize the inertial matrix and "flatten" the metric surfaces [12, 13], applying classical groups to develop recursive dynamic algorithms [14], using a spatial operator algebra to simplify dynamic formulations [15, 16], analysis and design of redundant robots under isotropy conditions

E.Y.L. Gu, *A Journey from Robot to Digital Human*,
Modeling and Optimization in Science and Technologies 1,
DOI: 10.1007/978-3-642-39047-0_7,

[17, 18], exploring a canonical transformation for curvature-vanishing [19], and using a Lie group and product-of-exponential formulation to reduce dynamic models via an extended inertial matrix factorization [20, 21]. More Lie algebraic representation methods with geometrical formulations can be found in [22, 7]. Such a large volume of the extensive studies have brought much deeper interpretations with advanced mathematical theories into robotics research.

In this chapter, we intend to present and show the following two mathematical interpretations on robotic kinematics and dynamics:

1. Kinematics determines topology of dynamic motion constraints so that it is a structure of dynamics;
2. The architecture of dynamics is a linear combination of kinematics by dynamic parameters so that it determines the geometry of motion constraints.

Note that all the dynamic motion constraints mentioned here are holonomic, i.e., they are integrable in general, unless they will be specified otherwise in future discussions.

More specifically, the major concepts and theories that underlie the above interpretations include

1. A **configuration manifold (C-manifold) embedding model** as a new promising approach to robot dynamic modeling and control;
2. An explicit representation of **isometric embedding** that can send a C-manifold into Euclidean space with one-to-one and Riemannian metric preservation;
3. A smooth **topological equivalence** (or a diffeomorphism) between the minimum embeddable C-manifold and combined C-manifold for an open serial-chain robotic system;
4. A process of **isometrization** between two diffeomorphic C-manifolds;
5. A **lower-bound of dynamic model reduction** that is a system with the minimum embeddable C-manifold in the sense of topology; and
6. A **compact dynamic equation** that is equivalent to the Lagrange equation but offers a great advantage to the robotic adaptive control formulation.

To further understand and answer what is the essence behind an n-th order (n-body) mechanical dynamic system, such as an open serial-chain robot arm with n joints, let us introduce the concept of *configuration manifold*, or *C-Manifold* that is an n-dimensional differentiable (smooth) generalized surface as a unique dynamic motion constraint for the system. More precisely, every dynamic motion trajectory of the robotic system is constrained on this invisible C-manifold. In other words, such a C-manifold, denoted by M^n, becomes the identity of the system. Similar to the personality of a human, each type of robotic systems has own individual C-manifold that distinguishes itself from the others [23, 24].

Furthermore, the C-manifold for a particular robotic system has a unique topological structure that can be directly determined by the kinematics. The C-manifold can also be endowed at each point with an arc-length that is related to the kinetic energy of the motion as a Riemannian metric to "personalize" its geometry. In mathematics, a space is called a manifold if it is globally non-Euclidean, but it can be locally Euclidean. In more abstract mathematical language, a small open neighborhood around each point on a manifold can be homeomorphic to an open region in a Euclidean space of the same dimension. Namely, every small open non-Euclidean neighborhood on the manifold can be topologically equivalent to the same dimensional open region of the Euclidean space. A coordinate system can then be defined in the topologically equivalent Euclidean space, and it is often called the **local coordinates** at each point of the manifold.

Under such a mathematical perspective, it is now reasonable to interpret all the joint positions in $q = (q_1 \cdots q_n)^T$ for an n-joint robotic system as a set of the local coordinates on its C-manifold. Although all the joint variables in q may not be orthogonal to each other, they are all linearly independent to form a basis of the locally homeomorphic Euclidean space. In fact, there are many ordinary examples of local coordinates to describe a non-Euclidean manifold in mathematics, such as the well known spherical coordinates and cylindrical coordinates in calculus, etc.

As discussed in the previous chapters, a robotic system can have a very different kinematics as well as a different Jacobian matrix from the others if their mechanical structures are different. For instance, the Stanford robot arm has a different kinematics and Jacobian matrix from the Fanuc S-900 early model robot, because the former has a prismatic joint while the latter does not. However, the Fanuc S-900 clearly possesses a similar kinematic model with the other 6-revolute-joint robots, such as the Fanuc R-2000iB/165F and Fanuc M-900iA/600 industrial manipulators, as shown in Figure 7.1.

A further study on the theory of C-manifolds and their embeddings shows that for any two n-joint robot arms with a common structural configuration, their C-manifolds are topologically equivalent (homeomorphic). However, they may have different geometrical details due to the different sizes and dimensions in their mechanical structures. Therefore, it is clear that the topology of the C-manifold for a dynamic system is determined by its kinematics, while the geometry of the C-manifold is determined by its dynamics.

The most simple but typical examples are the two-link robots: one has two revolute joints (RR-type) and the other one is of revolute-prismatic (RP) type, as shown in Figure 7.2. The C-manifold of the RR-type arm is homeomorphic to the surface of 2-torus $T^2 \simeq S^1 \times S^1$, while the second one has a C-manifold topologically equivalent to a 2D cylindrical surface $S^1 \times I^1$, as shown in Figure 7.3, where $I^1 = [0, 1]$ is the 1-dimensional unity interval.

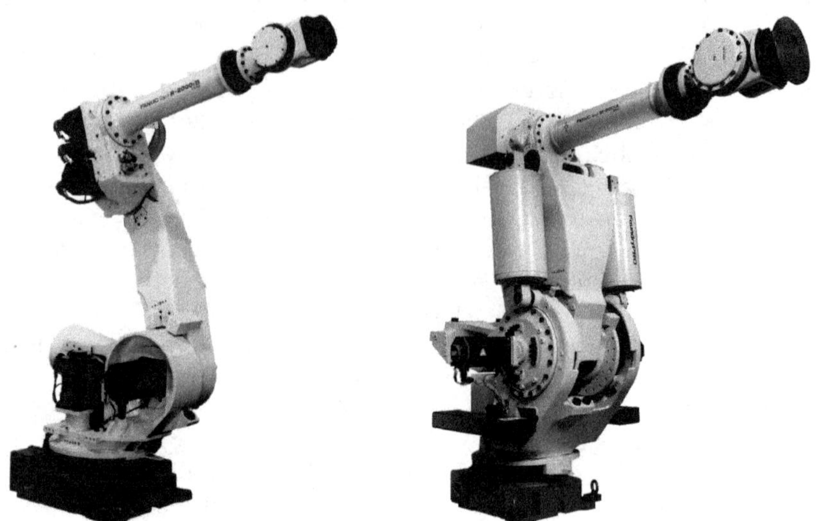

Fig. 7.1 Two 6-revolute-joint industrial robots: Fanuc R-2000iB (left) and Fanuc M-900iA (right). Photo courtesy of Fanuc Robotics, Inc.

In addition, for each type of the two-link robot arms, a variation of the kinematic parameters, such as the link length and joint off-set, and a variation of the dynamic parameters, such as the mass, mass center, and the moment of inertia, will cause each individual C-manifold to have a different geometry, while the topology remains invariant. In the later sections of this chapter, we will further discuss how to represent both the topology and geometry for a C-manifold.

In order to more explicitly reveal the geometrical details for a C-manifold owned by an n-joint robot dynamic system, as aforementioned, let us first define a **local coordinate system** at each point of the C-manifold M^n by

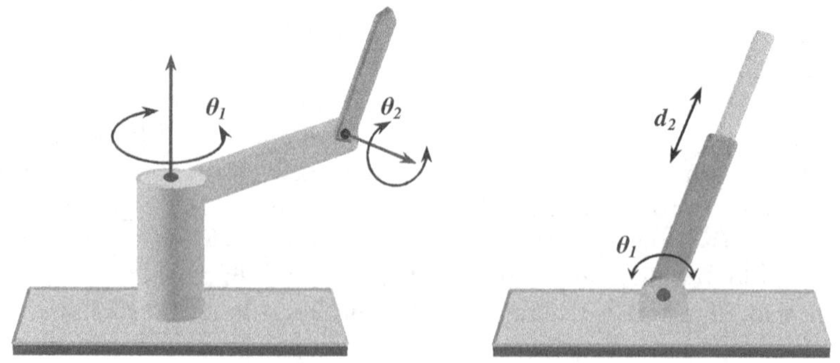

Fig. 7.2 RR-type and RP-type 2-link robots

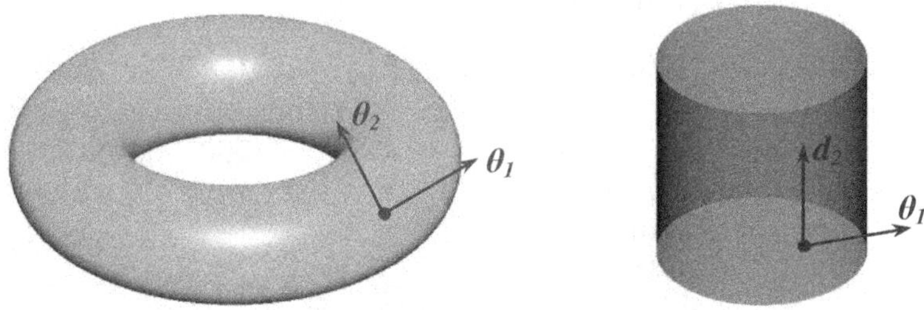

Fig. 7.3 C-manifolds for RR-type and RP-type 2-link robots

adopting all the n joint positions $\{q_1, \cdots, q_n\}$. It will be shown in a later section that the total kinetic energy K for the robotic system can be written in the following quadratic form of the joint velocity $\dot{q}$:

$$K = \frac{1}{2}\dot{q}^T W \dot{q}, \tag{7.1}$$

where all the coefficients in the quadratic form are augmented to an n by n positive-definite symmetric matrix W, called the **inertial matrix** of the robot.

At a mathematical point of view, the collection of all the elements w_i^j in the inertial matrix W constitutes a set of Riemannian metrics that are endowed on the C-manifold M^n. Based on the Riemannian metric geometry, equation (7.1) can be rewritten in a contravariant tensor form:

$$K = \frac{1}{2} w_{ij} \dot{q}^i \dot{q}^j,$$

and by the Einstein convention in both mathematics and physics, the summation symbol for every pair of the corresponding indices is omitted, but it must be automatically operated in the computation. This tensor form exhibits that each w_{ij} plays a role like a weight on the corresponding product between $\dot{q}_i$ and $\dot{q}_j$, and thus we use W to denote the inertial matrix of a robotic system. Note that the inertial matrix W for an open serial-chain robotic system is, in general, a nonlinear function of every joint variable in q, but it is a linear function of all the dynamic parameters.

It can be further interpreted that each change to a dynamic parameter will deform the C-manifold, i.e., will alter its geometry so that the kinetic energy K, as an arc-length at each point of the C-manifold, varies accordingly. However, such a deformation will not alter the topology of the C-manifold so that the robot kinematic model remains invariant. This shows that kinematics is virtually a structure of every robotic dynamic motion, while the detailed motion trajectory has to be determined by the dynamics.

7.2 The Newton-Euler Algorithm

Before we introduce in this section the well-known Newton-Euler algorithm in robotics research and application history, a number of key concepts from classical mechanics must be reviewed first. For a rigid body with its mass center c, the 3D linear velocity v at a point of the rigid body with respect to the fixed base is determined by

$$v = v_c + \omega \times r, \tag{7.2}$$

where v_c is the linear velocity of the mass center referred to the base, ω is the angular velocity of the rigid body with respect to the base and r is the radial vector from the mass center to the point of interest and projected on the base. This equation also shows that *any motion of a rigid body can be fully decoupled into a translation of its mass center and a rotation about the mass center.*

More precisely, let $p_0^i \in \mathbb{R}^3$ be a position vector for a point i other than the mass center of the rigid body with respect to the base. If $p_0^c \in \mathbb{R}^3$ is the mass center position vector that is also referred to the base, then

$$p_0^i = p_0^c + R_0^c r_c^i,$$

where R_0^c is the orientation of frame c with respect to the base to project the radial vector r_c^i onto the base, i.e., $r = R_0^c r_c^i$. After everyone is projected onto the base, taking time-derivatives for both sides yields

$$\dot{p}_0^i = \dot{p}_0^c + R_0^c \dot{r}_c^i + \dot{R}_0^c r_c^i.$$

Since the distance between any two distinct points in a rigid body is always a constant, the radial vector r_c^i that is always fixed on the rigid body frame c should be time-invariant, i.e., $\dot{r}_c^i = 0$. In addition, according to equation (3.10), $\dot{R}_0^c R_c^0 = \Omega = \omega \times$. Due to $\dot{p}_0^i = v$ and $\dot{p}_0^c = v_c$, we reach an agreement with equation (7.2):

$$v = \dot{p}_0^i = \dot{p}_0^c + \dot{R}_0^c R_c^0 R_0^c r_c^i = v_c + \omega \times r.$$

If a rigid body is pivoted at a fixed point, the torque τ due to a force f acting at a point of the body is given by

$$\tau = r \times f,$$

where r is the radial vector tailed at the pivoting point with its arrow at the force acting point.

The inertia tensor for a rigid body with respect to an arbitrary frame a is defined in classical mechanics by

$$\Gamma_a = \begin{pmatrix} I_{xx} & -I_{xy} & -I_{xz} \\ -I_{yx} & I_{yy} & -I_{yz} \\ -I_{zx} & -I_{zy} & I_{zz} \end{pmatrix}, \tag{7.3}$$

where

$$I_{xx} = \int_M (y^2 + z^2)dm, \quad I_{yy} = \int_M (z^2 + x^2)dm, \quad I_{zz} = \int_M (x^2 + y^2)dm,$$

and

$$I_{xy} = I_{yx} = \int_M xydm, \quad I_{yz} = I_{zy} = \int_M yzdm, \quad I_{zx} = I_{xz} = \int_M zxdm.$$

All the values of x, y and z in the above integrals are the coordinates at each locating point of the differential mass dm to be integrated over the rigid body with respect to frame a. If the origin of frame a is defined at the mass center c, then we denote this special inertia tensor by Γ_c.

Furthermore, let $r = (x \ y \ z)^T$ be a radial vector with its arrow at the locating point of each dm and its tail at the origin of frame a. Then, the above 3 by 3 inertia tensor (7.3) can be written in a more compact form:

$$\Gamma_a = \int_M (r^T r I_3 - rr^T)dm, \tag{7.4}$$

where I_3 is the 3 by 3 identity.

Once we define the inertia tensor Γ_a that is referred to frame a for a rigid body, the scalar moment of inertia about any unit vector ξ_a that is projected on frame a can be immediately determined by

$$I_{\xi_a} = \xi_a^T \Gamma_a \xi_a \tag{7.5}$$

as a quadratic form. In a special case, if $\xi_a = (1 \ 0 \ 0)^T$, i.e. the unit vector of the x-axis of frame a, then, by substituting the inertia tensor definition (7.3) into (7.5), we obtain $I_{\xi_a} = I_{xx}$. Similarly, $I_{\xi_a} = I_{yy}$ and I_{zz} if $\xi_a = (0 \ 1 \ 0)^T$ and $\xi_a = (0 \ 0 \ 1)^T$, respectively.

In classical mechanics, it is well known that the above moment of inertia I_{ξ_a} can also be determined by a "parallel shift" for the moment of inertia I_{ξ_c} about the axis ξ_c that is parallel to ξ_a but passes through the mass center of the rigid body, i.e.,

$$I_{\xi_a} = I_{\xi_c} + md^2,$$

where m is the mass of the rigid body and d is the shortest distance between the two parallel axes ξ_a and ξ_c. This property is commonly referred to as the **Parallel Axis Theorem**.

In fact, we may extend the above Parallel Axis Theorem from a scalar moment of inertia to a 3 by 3 inertia tensor defined by (7.3). Let $c_a = (c_x \ c_y \ c_z)^T$ be a vector that is projected on frame a with its arrow pointing to the mass center of the rigid body and tail at the origin of frame a. Let $C_a = S(c_a) = c_a \times$ be the skew-symmetric matrix of the vector c_a. It can be shown without major difficulty that the following **Tensor Parallel Axis Theorem** is true:

$$\mathit{\Gamma}_a = \mathit{\Gamma}_c + mC_aC_a^T = \mathit{\Gamma}_c + mC_a^TC_a = \mathit{\Gamma}_c - mC_a^2. \tag{7.6}$$

This extension offers a useful procedure of inertia tensor determination when the reference frame is parallel-shifted. For instance, if frame a is parallel-shifted to frame b by a position vector p_a^b that is projected on frame a, then the vector tailed at the origin of frame b and pointing to the mass center becomes $c_b = c_a - p_a^b$, and its skew-symmetric matrix is $C_b = S(c_b) = c_b \times$, as shown in Figure 7.4. Based on the Tensor Parallel Axis Theorem in equation (7.6),

$$\mathit{\Gamma}_b = \mathit{\Gamma}_c + mC_bC_b^T = \mathit{\Gamma}_a - mC_aC_a^T + mC_bC_b^T = \mathit{\Gamma}_a + m(C_a^2 - C_b^2). \tag{7.7}$$

In other words, $\mathit{\Gamma}_a + mC_a^2 = \mathit{\Gamma}_b + mC_b^2 = \mathit{\Gamma}_c$. This means that the sum $\mathit{\Gamma}_a + mC_a^2$ is invariant under any parallel-shift of frame a.

If the reference frame is now rotated from frame a to frame b by a rotation matrix R_a^b without any shift of the origin, then,

$$\mathit{\Gamma}_b = R_b^a \mathit{\Gamma}_a R_a^b = (R_a^b)^T \mathit{\Gamma}_a R_a^b. \tag{7.8}$$

In general, if the reference frame is both translated and rotated simultaneously, which can be described by a homogeneous transformation, or a dual rotation matrix from frame a to frame b:

$$H_a^b = \begin{pmatrix} R_a^b & p_a^b \\ 0_3 & 1 \end{pmatrix}, \quad \text{or} \quad \hat{R}_a^b = R_a^b + \epsilon S_a^b,$$

as shown in Figure 7.4, then, by combining equations (7.7) and (7.8) together, we obtain

$$\mathit{\Gamma}_b = R_b^a[\mathit{\Gamma}_a + m(C_a^2 - C_b^2)]R_a^b, \tag{7.9}$$

where $C_a - C_b = P_a^b = S(p_a^b) = p_a^b \times$.

As a special case, if frame a is originated at the mass center of the rigid body, then $c_a = 0$ and $\mathit{\Gamma}_a = \mathit{\Gamma}_c$ so that equation (7.9) is back to the Tensor Parallel Axis Theorem with a rotation R_c^b, i.e.,

$$\mathit{\Gamma}_b = R_b^c(\mathit{\Gamma}_c - mC_b^2)R_c^b = R_b^c(\mathit{\Gamma}_c + mC_b^TC_b)R_c^b, \tag{7.10}$$

This equation shows a clear fact that any shift of the reference frame away from the mass center will create an additional linear velocity.

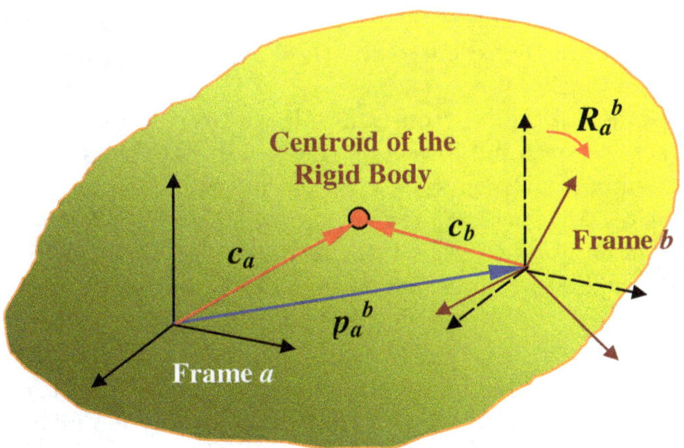

Fig. 7.4 A rigid body and its reference frame changes

Once we have the concept of inertia tensor, it becomes more convenient to represent the kinetic energy for a single rigid body in 6 d.o.f. motion, i.e., three for the mass center translation and three for its rotation about the mass center. Let $v^c_{0(a)} \in \mathbb{R}^3$ and $\omega^c_{0(a)} \in \mathbb{R}^3$ be the mass center linear velocity and angular velocity of the rigid body, respectively, both are referred to the base but projected on frame a. The kinetic energy of a single rigid body can be written as

$$K = \frac{1}{2}m(v^c_{0(a)})^T v^c_{0(a)} + \frac{1}{2}(\omega^c_{0(a)})^T \mathit{\Gamma}_{c(a)}\omega^c_{0(a)}. \tag{7.11}$$

After all of the above preparations, we now introduce, but just briefly, the well-known **Newton-Euler Recursive Algorithm** to determine the inverse dynamics for a robotic system. Namely, given a robotic kinematics model, find each joint torque τ_i if it is revolute, or joint force f_i if it is prismatic for $i = 1, \cdots, n$.

In history, people often preferred using the Lagrange equation to develop a robot dynamic model, because it does offer a global and systematic approach to dynamic modeling and analysis over the other formulations. However, the computations of a Lagrange equation are time-consuming with the computational complexity up to the order of $O(n^4)$, where n is the total number of robotic joints.

In the early 1980's, due to limited computer power, the robotics research community had to make extensive studies in developing more efficient inverse dynamics algorithms. The Newton-Euler algorithm was one of the most representative achievements at that time for a significant computational reduction [30, 4, 6]. By taking advantage of the computer recursion, the algorithm can offer only $O(n)$ in its computational complexity. This remarkable advantage has made itself very popular in robotic industrial applications. However, it is just a computer recursive numerical algorithm, and though fast, cannot be used for globally modeling a robotic dynamics. With the exponential growth of computer power in the recent years, the Lagrangian formulation has been regaining its popularity.

The central idea of the Newton-Euler algorithm is to update each force or torque that acts on each link based on the Newton's Second and Third Laws plus the D'Alembert's Principle in classical mechanics. It is observable that by directly inspecting a robot arm, as shown in Figure 7.5, the direction of getting kinematically "busier" is from frame 0 (the base) forward to frame n (the hand), while the direction of getting dynamically "busier" is just opposite. Therefore, we should first calculate both the linear and angular velocities as well as the accelerations for each link from link 1 recursively to link n as an *outward procedure*, and then compute each joint torque or force from joint n recursively back to joint 1 as an *inward procedure*.

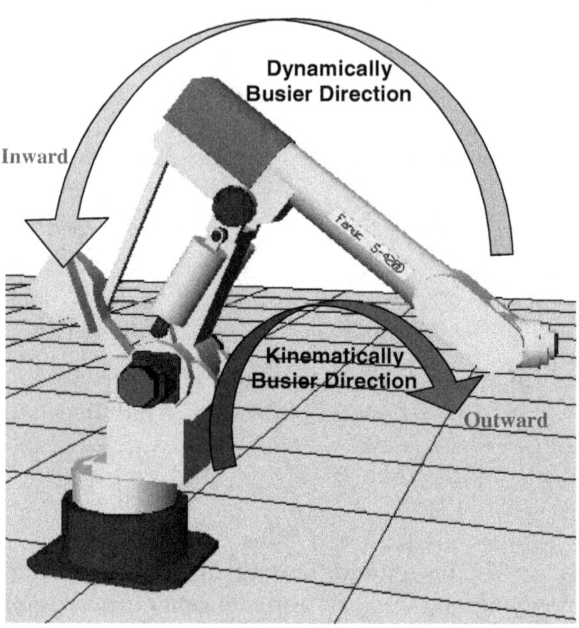

Fig. 7.5 Getting-busier directions for kinematics and dynamics

Figure 7.6 illustrates a schematic force and torque analysis of link i that is interacted by both two neighboring links $i-1$ and $i+1$. Based on the D'Alembert's Principle, we can derive in detail the following simultaneous equations of force balance and torque balance:

$$\begin{cases} f^i_{i-1} - R^i_{i-1} f^{i+1}_i + m_i R^0_{i-1}\mathbf{g} - m_i \dot{v}^{c_i}_{0(i-1)} = 0 \\ N^i_{i-1} - R^i_{i-1} N^{i+1}_i - c^i_{i-1} \times f^i_{i-1} + R^i_{i-1}(c^i_i \times f^{i+1}_i) - \\ \qquad -\Gamma_{c_i} \dot{\omega}^i_{0(i-1)} - \omega^i_{0(i-1)} \times \Gamma_{c_i} \omega^i_{0(i-1)} = 0, \end{cases} \quad (7.12)$$

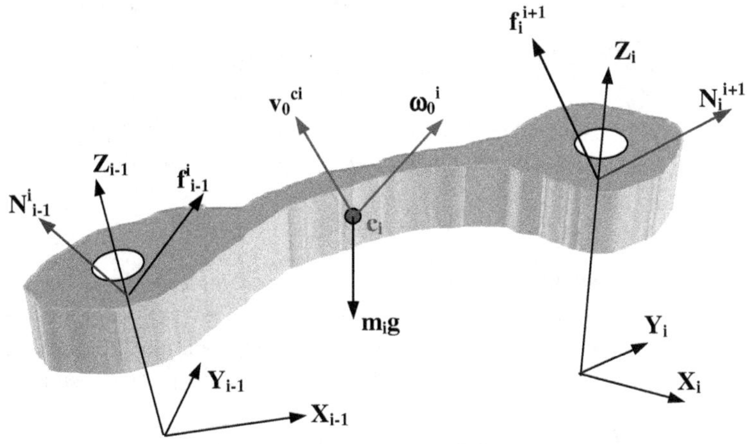

Fig. 7.6 Force/torque analysis of link i

where by applying the Newton's third law,

$$R_i^{i+1} f_{i+1}^i = -f_i^{i+1} \quad \text{and} \quad R_i^{i+1} N_{i+1}^i = -N_i^{i+1},$$

and the gravitational acceleration vector $\mathbf{g} = (0 \;\; 0 \;\; -g)^T$ if the z_0-axis of the base is defined vertically upward. In the above equation and hereafter, we always use a subscript without a parenthesis for a variable to represent which frame it is physically referred to, and that with a parenthesis $(\cdot)$ to represent which frame it is projected onto.

Equation (7.12) underlies the foundation to develop the outward procedure of recursion. However, it requires knowing both the linear and angular accelerations of the link. Through a further analysis of the link motion kinematics for the robot, we first obtain the following velocity recursion:

$$\begin{cases} \omega_{0(i)}^i = R_i^{i-1}(\omega_{0(i-1)}^{i-1} + \dot{\theta}_i z) \\ v_{0(i)}^i = R_i^{i-1}(v_{0(i-1)}^{i-1} + \dot{d}_i z) + \omega_{0(i)}^i \times p_{i-1(i)}^i, \end{cases} \tag{7.13}$$

where $z = (0 \;\; 0 \;\; 1)^T$.

To determine the acceleration vectors, it should be emphasized once again that every velocity vector must be projected onto the base before taking the time-derivative. For instance, to find the angular acceleration $\alpha_{0(i)}^i$, we must find $\omega_{0(0)}^i = R_0^i \omega_{0(i)}^i$ first, then take the time-derivative and further project it back to frame i, i.e.,

$$\alpha_{0(i)}^i = R_i^0 \dot{\omega}_{0(0)}^i = \dot{\omega}_{0(i)}^i + R_i^0 \dot{R}_0^i \omega_{0(i)}^i.$$

Likewise, we can have the linear acceleration vector as follows:

$$a^i_{0(i)} = R^0_i \dot{v}^i_{0(0)} = \dot{v}^i_{0(i)} + R^0_i \dot{R}^i_0 v^i_{0(i)}.$$

Note that since $\dot{R}^i_0 R^0_i = \Omega^i_0$ so that $\dot{R}^i_0 \omega^i_{0(i)} = \Omega^i_0 R^i_0 \omega^i_{0(i)} = \Omega^i_0 \omega^i_0 = \omega^i_0 \times \omega^i_0 = 0$, the above angular acceleration can be reduced to

$$\alpha^i_{0(i)} = \dot{\omega}^i_{0(i)},$$

which becomes a special case only for the angular velocity. In other words, it reinforces the fact that for a rigid body in motion, every point on the body has the same angular velocity, but it is not true for the linear velocity. Instead, according to the above derivation of the angular and linear accelerations,

$$a^i_{0(i)} = R^0_i \dot{v}^i_{0(0)} = \dot{v}^i_{0(i)} + R^0_i \dot{R}^i_0 v^i_{0(i)} = \dot{v}^i_{0(i)} + \omega^i_{0(i)} \times v^i_{0(i)}.$$

Once all the velocities and accelerations are prepared for the outward recursion, i.e., from $i = 1$ to $i = n$, we turn to perform the inward recursion from $i = n$ backward to $i = 1$ using equation (7.12) to calculate every f^i_{i-1} and N^i_{i-1}. Note that the final solution for each joint torque τ_i is given by the last (the z-) component of N^i_{i-1} if the joint is revolute, or for a joint force f_i by the last (the z-) component of f^i_{i-1} if it is prismatic.

Summing up all the above derivations, we conclude the following complete Newton-Euler Recursive Algorithm:

Algorithm of the Newton-Euler Inverse Dynamics

With $v^0_{0(0)} = \omega^0_{0(0)} = \alpha^0_{0(0)} = 0$, and $a^0_{0(0)} = \mathbf{g} = (0 \;\; 0 \;\; -g)^T$,

1. The Outward Procedure: $i = 1$ to n:

$$\begin{aligned}
\omega^i_{0(i)} &= R^{i-1}_i(\omega^{i-1}_{0(i-1)} + \dot{\theta}_i z), \\
v^i_{0(i)} &= R^{i-1}_i(v^{i-1}_{0(i-1)} + \dot{d}_i z) + \omega^i_{0(i)} \times p^i_{i-1(i)}, \\
\alpha^i_{0(i)} &= R^{i-1}_i(\alpha^{i-1}_{0(i-1)} + \ddot{\theta}_i z + \omega^{i-1}_{0(i-1)} \times \dot{\theta}_i z), \\
a^i_{0(i)} &= R^{i-1}_i(a^{i-1}_{0(i-1)} + \ddot{d}_i z + 2\omega^{i-1}_{0(i-1)} \times \dot{d}_i z) + \alpha^i_{0(i)} \times p^i_{i-1(i)} + \\
&\quad + \omega^i_{0(i)} \times (\omega^i_{0(i)} \times p^i_{i-1(i)}), \\
a^{c_i}_{0(i)} &= a^i_{0(i)} + C^T_i \alpha^i_{0(i)} + \omega^i_{0(i)} \times C^T_i \omega^i_{0(i)}, \\
F_i &= m_i a^{c_i}_{0(i)}, \\
H_i &= \Gamma_{c_i} \alpha^i_{0(i)} + \omega^i_{0(i)} \times \Gamma_{c_i} \omega^i_{0(i)}.
\end{aligned}$$

where $z = (0 \;\; 0 \;\; 1)^T$.

With $f^{n+1}_n = N^{n+1}_n = 0$,

2. The Inward Procedure: $i = n$ to 1:

$$f_{i-1}^i = R_{i-1}^i (f_i^{i+1} + F_i),$$
$$N_{i-1}^i = R_{i-1}^i (N_i^{i+1} + C_i F_i + H_i) + p_{i-1}^i \times f_{i-1}^i.$$

3. The Result: $i = 1, \cdots, n$,

$$\tau_i = (N_{i-1}^i)^T z \quad \text{for revolute joint,}$$
$$f_i = (f_{i-1}^i)^T z \quad \text{for prismatic joint.}$$

7.3 The Lagrangian Formulation

Instead of formulating the robotic dynamics via a direct force/torque analysis, we can also model it for such an open serial-chain high-dimensional and high-coupling nonlinear system in a more global and systematic fashion. As we can see from many real examples, every force-free mechanical system, no matter how simple or how complex, will always follow the direction of minimum effort spontaneously. The Lagrange equation was developed based on the *Principle of Minimum Effort.* It was originally derived by starting from the integral Hamilton's Principle for a conservative system, which states that *the motion of the system from time t_1 to time t_2 is such that the line integral $I = \int_{t_1}^{t_2} L dt$, where $L = K - P$, the difference between kinetic energy K and potential energy P, is a minimum for the path of motion.*

This line integral I can reasonably be interpreted as the "effort" of the system in motion. Since the kinetic energy of a moving system is a scalar function of both position and velocity $K = K(q, \dot{q})$, as can be seen in (7.1), while the conservative potential energy is a scalar function of position only, i.e., $P = P(q)$, the scalar function L inside the line integral I becomes $L = L(q, \dot{q}) = K(q, \dot{q}) - P(q)$, called a **Lagrangian** for a given system. In general, q and $\dot{q}$ are collectively referred to as a set of **generalized coordinates**, which actually span a local coordinate system plus its tangent space for the configuration manifold (C-manifold) of the system.

Through the well-known Calculus of Variation in mathematics, the Lagrange equation of motion that minimizes the effort I for a conservative force-free system is given by

$$\frac{d}{dt}\left(\frac{\partial L}{\partial \dot{q}}\right) - \frac{\partial L}{\partial q} = 0. \tag{7.14}$$

This is a homogeneous equation written in a vector form, i.e., each term of (7.14) is a vector field of the same dimension as the local coordinate system $q \in \mathbb{R}^n$. If now an external force/torque vector $\tau \in \mathbb{R}^n$ is imposed on the system, then the Lagrange equation becomes nonhomogeneous:

$$\frac{d}{dt}\left(\frac{\partial L}{\partial \dot{q}}\right) - \frac{\partial L}{\partial q} = \tau. \tag{7.15}$$

For an n-joint robot as a typical high-order dynamic system, substituting both the total kinetic energy $K = K(q, \dot{q})$ given in (7.1) and potential energy $P = P(q)$ into the Lagrange equation (7.15) yields

$$W\ddot{q} + \dot{W}\dot{q} - \frac{1}{2}W_d\dot{q} + \frac{\partial P}{\partial q} = \tau, \tag{7.16}$$

where the n by n *derivative matrix* W_d for the n by n inertial matrix W is defined by

$$W_d = \begin{pmatrix} \dot{q}^T \frac{\partial W}{\partial q_1} \\ \vdots \\ \dot{q}^T \frac{\partial W}{\partial q_n} \end{pmatrix}. \tag{7.17}$$

Since $\dot{W}$ is symmetric while W_d is not, $\dot{W} \neq W_d^T$. However, we can show the following useful identity:

$$\dot{W}\dot{q} \equiv W_d^T \dot{q} \tag{7.18}$$

that holds for any q and $\dot{q}$. In fact, because

$$\dot{W} = \frac{\partial W}{\partial q_1}\dot{q}_1 + \cdots + \frac{\partial W}{\partial q_n}\dot{q}_n,$$

$$\dot{W}\dot{q} = \frac{\partial W}{\partial q_1}\dot{q}\dot{q}_1 + \cdots + \frac{\partial W}{\partial q_n}\dot{q}\dot{q}_n = \begin{pmatrix} \frac{\partial W}{\partial q_1}\dot{q} & \cdots & \frac{\partial W}{\partial q_n}\dot{q} \end{pmatrix} \dot{q}.$$

By comparing the last matrix with the definition of W_d in (7.17) and noticing that W is symmetric, we reach the identity (7.18).

If we consider only the gravitational potential energy involved in the robotic system, by defining an additional joint torque $\tau_g = -\frac{\partial P}{\partial q}$ due to gravity, we obtain the following compact dynamic equation for the robotic system:

$$W\ddot{q} + (W_d^T - \frac{1}{2}W_d)\dot{q} - \tau_g = W\ddot{q} + \frac{1}{2}(\dot{W} + W_d^T - W_d)\dot{q} - \tau_g = \tau. \tag{7.19}$$

In mathematical history, on the other hand, many mathematicians in differential geometry had independently studied the problem of finding the shortest trajectory on a non-Euclidean n-surface, such as a C-manifold M^n for an n-joint robot, and achieved the following well-known **Geodesic Equation** written in tensor form:

$$\ddot{q}^j + \Gamma^j_{ki}\dot{q}^k\dot{q}^i = 0, \quad j = 1, \cdots, n, \tag{7.20}$$

where q^j is the j-th element of the local coordinates $q \in \mathbb{R}^n$ on M^n, and Γ^j_{ki} is called the Christoffel's symbol and given as follows: if the Riemannian metric $W = \{w^j_i\}$ is nonsingular,

$$\Gamma^j_{ki} = \frac{1}{2} w^{jl} \left(\frac{\partial w_{li}}{\partial q^k} + \frac{\partial w_{kl}}{\partial q^i} - \frac{\partial w_{ki}}{\partial q^l} \right).$$

The geodesic equation for a given manifold M^n can predict a curve on M^n with the minimum arc length by solving the equation under a given initial point $q(0)$ and initial velocity $\dot{q}(0)$, from which the curve is taking off on M^n.

Since the Riemannian metric $W = \{w^j_i\}$ is mathematically endowed on the C-manifold M^n and is also physically the inertial matrix of the robot, it is always nonsingular. By premultiplying the metric w_{lj} on both sides of the geodesic equation (7.20), we have

$$w_{lj}\ddot{q}^j + \frac{1}{2} \left(\frac{\partial w_{li}}{\partial q^k} + \frac{\partial w_{kl}}{\partial q^i} - \frac{\partial w_{ki}}{\partial q^l} \right) \dot{q}^k \dot{q}^i = 0, \quad l = 1, \cdots, n. \tag{7.21}$$

Comparing this tensor form with the robot dynamic equation (7.19) in the matrix form, we can immediately see that everything is the same except the torque terms τ and τ_g, because equation (7.21) did not consider any external force/torque and gravity. This coincidence shows a beautiful consistency between physics and mathematics. In fact, the minimum effort and the shortest arc length are, indeed, synonymous in meaning, though different in language.

In summary, an n-joint robotic system consists of up to n scalar second-order differential equations in its dynamic model, as seen in (7.21). All the terms containing $\ddot{q}^j$ are called the inertial forces, while all the terms containing $\dot{q}^k \dot{q}^i$ are called the centrifugal forces if $k = i$ and called the Coriolis forces if $k \neq i$. Therefore, it is quite reasonable to denote

$$C(q, \dot{q})\dot{q} = c(q, \dot{q}) = \left(W_d^T - \frac{1}{2} W_d \right) \dot{q}$$

as a collection of all the centrifugal and Coriolis forces/torques to shorten the robot dynamic equation (7.19).

Furthermore, it can be directly proven that for a skew-symmetric matrix S, the scalar quadratic form $x^T S x \equiv 0$ for any vector x with the dimension compatible to S. Since $W_d^T - W_d$ is clearly a skew-symmetric matrix so that $\dot{q}^T (W_d^T - W_d)\dot{q} \equiv 0$, the *kinetic power* turns out to be

$$\dot{q}^T \tau = \dot{q}^T W \ddot{q} + \frac{1}{2} \dot{q}^T \dot{W} \dot{q} - \dot{q}^T \tau_g. \tag{7.22}$$

This property of kinetic power will be very useful in justifying the stability of robotic control systems in the next chapter.

7.4 Determination of Inertial Matrix

Having introduced the Lagrangian formulation in modeling robotic dynamics, it remains to find a symbolical form of the inertial matrix W, or the Riemannian metric $\{w_i^j\}$ on the C-manifold for a given robotic system. Basically, we have the following two approaches:

1. Determine a symbolical form of kinetic energy for each link, from link 1 through link n, by a sequential velocity analysis of each link to find the total kinetic energy $K = K_1 + \cdots + K_n$. Then, extract all the coefficients of $\dot{q}_i\dot{q}_j$ terms to form the W matrix according to the quadratic form of $\dot{q}_i$'s in equation (7.1);
2. Directly compute either a symbolical or numerical form of W by following a global procedure that will be developed later in this section.

The first approach, though straightforward, works only in relatively simple cases, such as a planar robot or a 3D robot with up to 3 joints, because a robot with more links will exponentially increase the complexity of its velocity analysis and difficulty of geometric inspection. In contrast, the second approach will offer a global solution that allows us to determine a larger inertial matrix W both symbolically and numerically for most types of robotic systems and even for a large-scale digital human model [4, 12, 13].

Let us now look at a three-joint planar robot arm, as shown in Figure 7.7, to illustrate the first approach to finding its 3 by 3 inertial matrix W via a total kinetic energy derivation. This planar robot consists of two revolute joints and one prismatic joint as an RPR type robot arm. Let $q = (\theta_1 \;\; d_2 \;\; \theta_3)^T$ be the joint position vector. Each of the three links has its mass m_i and the moment of inertia I_i about the axis passing through the mass center of each link and perpendicular to the plane for $i = 1, 2, 3$. In Figure 7.7, each l_{ci} is the length between the mass center of link i and the origin of frame i.

Based on equation (7.11), the kinetic energy K_1 of link 1 can be written as

$$K_1 = \frac{1}{2}m_1 l_{c1}^2 \dot{\theta}_1^2 + \frac{1}{2}I_1\dot{\theta}_1^2,$$

because the mass center velocity is given by $v = \omega r = l_{c1}\dot{\theta}_1$.

Since the motion of link 2 is imposed by the rotation of link 1 in addition to its own sliding with a linear velocity $\dot{d}_2$, the kinetic energy K_2 of link 2 becomes

$$K_2 = \frac{1}{2}m_2[(d_2 - l_{c2})^2\dot{\theta}_1^2 + \dot{d}_2^2] + \frac{1}{2}I_2\dot{\theta}_1^2.$$

Note that the two velocities $(d_2 - l_{c2})\dot{\theta}_1$ and $\dot{d}_2$ for link 2 happen to be perpendicular to each other so that the resultant velocity square is equal to the sum of the two velocity squares. If they are not perpendicular, we have to project each of them onto a local orthogonal frame originated at the common acting point and then sum up their squares.

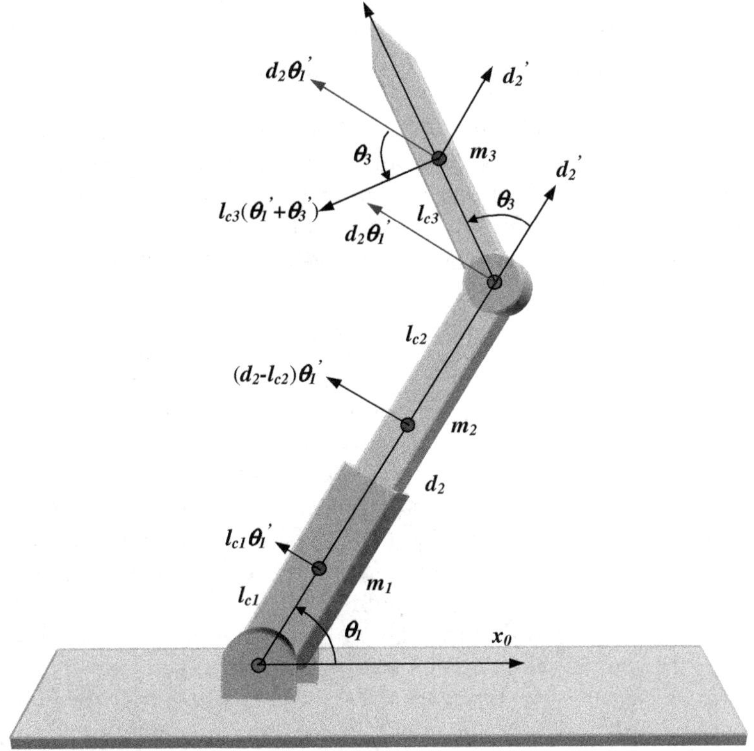

Fig. 7.7 Velocity analysis of a three-joint planar robot arm

For link 3, there are three linear velocities: the self-rotation linear velocity $l_{c3}(\dot{\theta}_1 + \dot{\theta}_3)$ because the total angular velocity of link 3 is now $\dot{\theta}_1 + \dot{\theta}_3$, plus two imposed velocities $d_2\dot{\theta}_1$ and $\dot{d}_2$ by the motion of link 2, which are all delivered to the mass center of link 3, see Figure 7.7.

Since the last two imposed velocity vectors are perpendicular to each other, we may utilize them as the local orthogonal frame just for the purpose of projection convenience. Thus, we have

$$K_3 = \frac{1}{2}m_3[(d_2\dot{\theta}_1 + l_{c3}(\dot{\theta}_1 + \dot{\theta}_3)c_3)^2 + (\dot{d}_2 - l_{c_3}(\dot{\theta}_1 + \dot{\theta}_3)s_3)^2] + \frac{1}{2}I_3(\dot{\theta}_1 + \dot{\theta}_3)^2$$
$$= \frac{1}{2}m_3[d_2^2\dot{\theta}_1^2 + \dot{d}_2^2 + 2d_2l_{c3}\dot{\theta}_1(\dot{\theta}_1 + \dot{\theta}_3)c_3 - 2l_{c3}\dot{d}_2(\dot{\theta}_1 + \dot{\theta}_3)s_3 +$$
$$+l_{c3}^2(\dot{\theta}_1 + \dot{\theta}_3)^2] + \frac{1}{2}I_3(\dot{\theta}_1^2 + \dot{\theta}_3^2 + 2\dot{\theta}_1\dot{\theta}_3),$$

where $c_3 = \cos\theta_3$ and $s_3 = \sin\theta_3$.

Once a symbolical form of the total kinetic energy $K = K_1 + K_2 + K_3$ is achieved, the 3 by 3 inertial matrix W can directly be found by extracting and augmenting all the coefficients from K. Since based on (7.1),

$$K = \frac{1}{2}\dot{q}^T W \dot{q} = \frac{1}{2}\begin{pmatrix}\dot{\theta}_1 & \dot{d}_2 & \dot{\theta}_3\end{pmatrix}\begin{pmatrix} w_{11} & w_{12} & w_{13} \\ w_{21} & w_{22} & w_{23} \\ w_{31} & w_{32} & w_{33}\end{pmatrix}\begin{pmatrix}\dot{\theta}_1 \\ \dot{d}_2 \\ \dot{\theta}_3\end{pmatrix},$$

the sum of all the coefficients of $\dot{\theta}_1^2$ will be w_{11}, the sum of all the coefficients of $\dot{d}_2^2$ will be w_{22} and the sum of all the coefficients of $\dot{\theta}_3^2$ will be w_{33}.

However, only one half of the sum of the coefficients of $\dot{\theta}_1\dot{\theta}_3$ will be w_{13} and the other half will fill into w_{31} due to the matrix multiplication rule and the symmetry of W, i.e., $w_{13} = w_{31}$ and so are the remaining off-diagonal elements of W. Finally,

$$W = \begin{pmatrix} w_{11} & -m_3 l_{c3} s_3 & m_3(d_2 l_{c3} c_3 + l_{c3}^2) + I_3 \\ -m_3 l_{c3} s_3 & m_2 + m_3 & -m_3 l_{c3} s_3 \\ m_3(d_2 l_{c3} c_3 + l_{c3}^2) + I_3 & -m_3 l_{c3} s_3 & m_3 l_{c3}^2 + I_3 \end{pmatrix}, \tag{7.23}$$

where $w_{11} = m_1 l_{c1}^2 + m_2(d_2 - l_{c2})^2 + m_3(d_2^2 + 2d_2 l_{c3} c_3 + l_{c3}^2) + I_1 + I_2 + I_3$.

After the inertial matrix W is determined symbolically, finding its derivative matrix W_d based on equation (7.17) just needs to take partial derivatives of W with respect to each joint position q_i. Then, let a computer program finish the rest of the calculations numerically, because after W_d there will be no more derivatives needed to take on W for completing the robot dynamic modeling as well as the control algorithm development.

It can be clearly seen throughout the above example that the first approach is limited to the robotic systems with relatively simple and geometrically visualizable structures. To break this limitation, we now introduce the second approach to finding a symbolical or numerical form of the inertial matrix W for almost every kind of open serial-chain robotic systems.

Let us start with equation (7.11). We apply it on link j of an n-joint robot and project both its linear velocity v_0^c and angular velocity ω_0^c onto frame j that is attached on link j for $j = 1, \cdots, n$ so that

$$K_j = \frac{1}{2}m_j(v_{0(j)}^c)^T v_{0(j)}^c + \frac{1}{2}(\omega_{0(j)}^j)^T \Gamma_{c(j)} \omega_{0(j)}^j. \tag{7.24}$$

Based on equation (7.2), we have

$$v_{0(j)}^j = v_{0(j)}^c - \omega_{0(j)}^j \times c_j,$$

where c_j is the vector tailed at the origin of frame j with the arrow pointing at the mass center of link j so that it is opposite to the radial vector r in (7.2). By defining a skew-symmetric matrix $C_j = S(c_j) = c_j\times$, we continue to have

$$v_{0(j)}^c = v_{0(j)}^j + \omega_{0(j)}^j \times c_j = v_{0(j)}^j + C_j^T \omega_{0(j)}^j.$$

Substituting this into the first term of (7.24) yields

$$m_j(v_{0(j)}^c)^T v_{0(j)}^c = m_j(v_{0(j)}^j + C_j^T \omega_{0(j)}^j)^T(v_{0(j)}^j + C_j^T \omega_{0(j)}^j)$$

$$= m_j \left[v_{0(j)}^{jT} v_{0(j)}^{j} + 2 v_{0(j)}^{jT} C_j^T \omega_{0(j)}^{j} + \omega_{0(j)}^{jT} C_j C_j^T \omega_{0(j)}^{j} \right].$$

On the other hand, based on the Tensor Parallel Axis Theorem in (7.6), the second term of (7.24) can be converted to

$$(\omega_{0(j)}^{j})^T \Gamma_{c(j)} \omega_{0(j)}^{j} = \omega_{0(j)}^{jT} (\Gamma_j - m_j C_j C_j^T) \omega_{0(j)}^{j}.$$

Finally, the kinetic energy of link j becomes

$$K_j = \frac{1}{2} m_j v_{0(j)}^{jT} v_{0(j)}^{j} + m_j v_{0(j)}^{jT} C_j^T \omega_{0(j)}^{j} + \frac{1}{2} \omega_{0(j)}^{jT} \Gamma_j \omega_{0(j)}^{j}. \tag{7.25}$$

Let us further define a following 6 by 6 symmetric *dynamic parameter matrix* for link j:

$$U_j = \begin{pmatrix} m_j I_3 & m_j C_j^T \\ m_j C_j & \Gamma_j \end{pmatrix}, \tag{7.26}$$

where I_3 is the 3 by 3 identity. Also, by defining a 6-dimensional Cartesian velocity of frame j as $V_j = \begin{pmatrix} v_{0(j)}^{j} \\ \omega_{0(j)}^{j} \end{pmatrix}$, equation (7.25) can be reduced to

$$K_j = \frac{1}{2} V_j^T U_j V_j$$

so that the total kinetic energy of the n-joint robot turns out to be

$$K = K_1 + \cdots + K_n = \frac{1}{2} \sum_{j=1}^{n} V_j^T U_j V_j. \tag{7.27}$$

Now, it is time to introduce a concept of **subrobot** and **sub-Jacobian** for the n-joint robot. The j-th subrobot, $j = 1, \cdots, n$, is a robot arm with only the first j links of the original robot as if the rest links beyond link j disappear. Therefore, frame j that is fixed on link j is considered to be the hand (the last) frame of the j-th subrobot. The Jacobian matrix of this subrobot, which is projected onto frame j, is thus 6 by j dimensional. Furthermore, by augmenting it with the 6 by $(n-j)$ zero matrix, we obtain the following 6 by n matrix that is called a sub-Jacobian of the j-th subrobot if all the first j joints are revolute:

$$J_j = \begin{pmatrix} p_j^0 \times r_j^0 & \cdots & p_j^{j-1} \times r_j^{j-1} & O_{3\times(n-j)} \\ r_j^0 & \cdots & r_j^{j-1} & O_{3\times(n-j)} \end{pmatrix}. \tag{7.28}$$

Based on the Jacobian matrix formation procedure and equation (4.14) in the previous chapter, if the i-th joint between joints 1 and j is prismatic, then similarly, the i-th column of the sub-Jacobian in (7.28) must be replaced by $\begin{pmatrix} r_j^{i-1} \\ 0_3 \end{pmatrix}$ with the bottom 3 by 1 zero column 0_3.

Once all the n sub-Jacobians $J_1, \cdots, J_n$ for the n-joint robot are found, the sub-Jacobian equation for each subrobot can be written by

$$V_j = J_j \dot{q}$$

that is in terms of the common n-dimensional joint velocity vector $\dot{q}$ over all the n subrobots.

Substituting each sub-Jacobian equation into the total kinetic energy equation (7.27), we reach

$$K = \frac{1}{2} \sum_{j=1}^{n} \dot{q}^T J_j^T U_j J_j \dot{q} = \frac{1}{2} \dot{q}^T \left(\sum_{j=1}^{n} J_j^T U_j J_j \right) \dot{q}. \tag{7.29}$$

Comparing (7.29) with the general kinetic energy equation (7.1), we conclude that

$$W = \sum_{j=1}^{n} J_j^T U_j J_j. \tag{7.30}$$

This equation provides us with a global and systematic algorithm to determine either a symbolical form or a numerical solution of the inertial matrix W for a given n-joint robotic system. Although the symbolical derivations will soon be lengthy as n increases, the algorithm can at least break the limitation in comparison with the first approach. We will apply such a compact form of computing W in (7.30) to programming and solving a 47 by 47 very large numerical inertial matrix for a digital human dynamic model at each sampling time instant in Chapter 11.

If each term in equation (7.30) is denoted by $W_j = J_j^T U_j J_j$ for $j = 1, \cdots, n$, then,

$$W = W_1 + \cdots + W_n.$$

It can be interpreted that each W_j is the contribution of link j as a part of the entire inertial matrix W. In other words, the total inertial matrix W of an n-joint robotic system is a sum of the contributions from all the n links.

In fact, you do not have to fill $n - j$ zero columns into each sub-Jacobian matrix J_j. Then, each component $W_j = J_j^T U_j J_j$ is just a j by j square matrix. The final inertial matrix W can be formed by stacking all the n j by j square matrices W_j's together, starting from and lining up at the upper-left corner of each matrix, as depicted in Figure 7.8. This can save a lot of memory in MATLAB^{TM} programming.

Figure 7.8 also realistically reveals a fact that the last link of an open serial-chain robot makes the most contribution to the inertial matrix W formation, while the first link makes the least contribution to W. The reason is quite clear that the last link for the robot is always the busiest body in motion, in contrast to the first link that is movable only by joint 1. This fact once again reinforces the concept that *kinematics determines the structure of dynamics.*

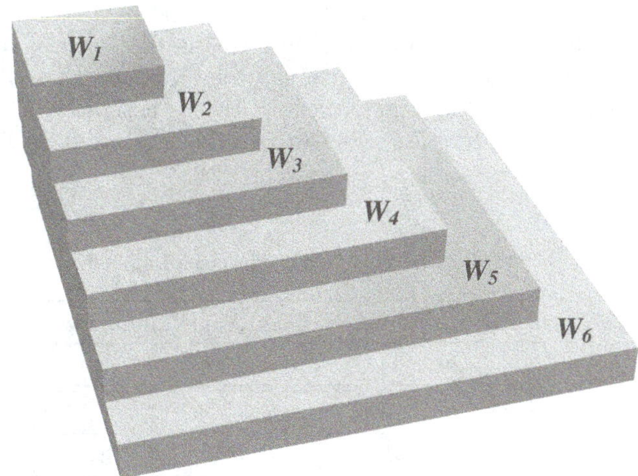

Fig. 7.8 An inertial matrix W is formed by stacking every W_j together

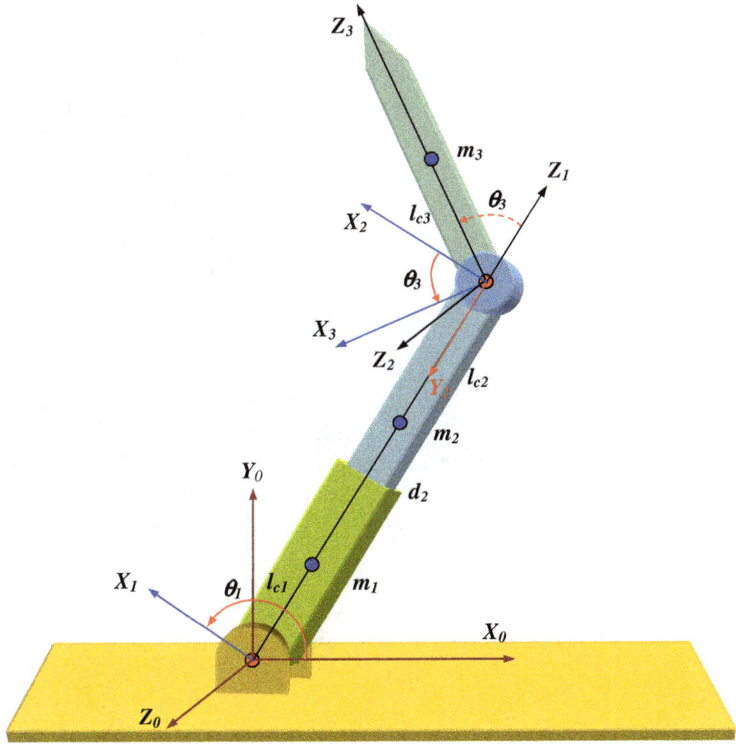

Fig. 7.9 Axes assignment of the three-joint planar robot

To verify the second approach, let us look back at the same example as illustrated for the first approach to finding the inertial matrix W. This time we apply equation (7.30), which obviously requires the D-H convention to explicitly model the robotic kinematics. With the assignment of coordinate frames, as shown in Figure 7.9, we can readily determine its D-H table in Table 7.1.

Table 7.1 The D-H table of the three-joint planar robot in Figure 7.9

Joint Angle	Joint Offset	Twist Angle	Link Length
θ_i	d_i	α_i	a_i
θ_1	0	$+90^0$	0
$\theta_2 = 0$	d_2	-90^0	0
θ_3	0	$+90^0$	0

Then, all the one-step homogeneous transformation matrices can be derived as follows:

$$A_0^1 = \begin{pmatrix} c_1 & 0 & s_1 & 0 \\ s_1 & 0 & -c_1 & 0 \\ 0 & 1 & 0 & 0 \\ 0 & 0 & 0 & 1 \end{pmatrix}, \quad A_1^2 = \begin{pmatrix} 1 & 0 & 0 & 0 \\ 0 & 0 & 1 & 0 \\ 0 & -1 & 0 & d_2 \\ 0 & 0 & 0 & 1 \end{pmatrix},$$

$$A_2^3 = \begin{pmatrix} c_3 & 0 & s_3 & 0 \\ s_3 & 0 & -c_3 & 0 \\ 0 & 1 & 0 & 0 \\ 0 & 0 & 0 & 1 \end{pmatrix}.$$

To prepare the necessary homogeneous transformations towards the three sub-Jacobian determinations, first, it should be clarified that the first subrobot is only link 1, the second subrobot consists of both link 1 and link 2, while the third subrobot is the entire three-joint planar robot. Thus, with some calculations, and noticing that $A_j^i = (A_i^j)^{-1}$ and the inverse equation (3.5) in Chapter 3, we have

$$A_1^0 = \begin{pmatrix} c_1 & s_1 & 0 & 0 \\ 0 & 0 & 1 & 0 \\ s_1 & -c_1 & 0 & 0 \\ 0 & 0 & 0 & 1 \end{pmatrix}, \quad A_2^0 = \begin{pmatrix} c_1 & s_1 & 0 & 0 \\ -s_1 & c_1 & 0 & d_2 \\ 0 & 0 & 1 & 0 \\ 0 & 0 & 0 & 1 \end{pmatrix},$$

$$A_2^1 = \begin{pmatrix} 1 & 0 & 0 & 0 \\ 0 & 0 & -1 & d_2 \\ 0 & 1 & 0 & 0 \\ 0 & 0 & 0 & 1 \end{pmatrix},$$

$$A_3^0 = \begin{pmatrix} c_{13} & s_{13} & 0 & d_2 s_3 \\ 0 & 0 & 1 & 0 \\ s_{13} & -c_{13} & 0 & -d_2 c_3 \\ 0 & 0 & 0 & 1 \end{pmatrix}, \quad A_3^1 = \begin{pmatrix} c_3 & 0 & -s_3 & d_2 s_3 \\ 0 & 1 & 0 & 0 \\ s_3 & 0 & c_3 & -d_2 c_3 \\ 0 & 0 & 0 & 1 \end{pmatrix},$$

$$A_3^2 = \begin{pmatrix} c_3 & s_3 & 0 & 0 \\ 0 & 0 & 1 & 0 \\ s_3 & -c_3 & 0 & 0 \\ 0 & 0 & 0 & 1 \end{pmatrix},$$

where $s_{13} = \sin(\theta_1 + \theta_3)$ and $c_{13} = \cos(\theta_1 + \theta_3)$, once again as the short notations.

Based on equation (7.28), and noticing that the second joint of the robot is prismatic, the three sub-Jacobian matrices turn out to be

$$J_1 = \begin{pmatrix} 0 & 0 & 0 \\ 0 & 0 & 0 \\ 0 & 0 & 0 \\ 0 & 0 & 0 \\ 1 & 0 & 0 \\ 0 & 0 & 0 \end{pmatrix}, \quad J_2 = \begin{pmatrix} d_2 & 0 & 0 \\ 0 & -1 & 0 \\ 0 & 0 & 0 \\ 0 & 0 & 0 \\ 0 & 0 & 0 \\ 1 & 0 & 0 \end{pmatrix}, \quad J_3 = \begin{pmatrix} d_2 c_3 & -s_3 & 0 \\ 0 & 0 & 0 \\ d_2 s_3 & c_3 & 0 \\ 0 & 0 & 0 \\ 1 & 0 & 1 \\ 0 & 0 & 0 \end{pmatrix}.$$

To determine the three dynamic parameter matrices U_j's for the three robot links, by referring to each individual link frame assigned, the mass center coordinates vector of each link is given by $c_1 = (0 \;\; 0 \;\; l_{c1})^T$, $c_2 = (0 \;\; l_{c2} \;\; 0)^T$ and $c_3 = (0 \;\; 0 \;\; l_{c3})^T$. After converting them into their skew-symmetric matrices $C_j = S(c_j)$ for $j = 1, 2, 3$, based on (7.26), we obtain

$$U_1 = \begin{pmatrix} m_1 & 0 & 0 & 0 & m_1 l_{c1} & 0 \\ 0 & m_1 & 0 & -m_1 l_{c1} & 0 & 0 \\ 0 & 0 & m_1 & 0 & 0 & 0 \\ 0 & -m_1 l_{c1} & 0 & I_{x1} & 0 & 0 \\ m_1 l_{c1} & 0 & 0 & 0 & I_{y1} & 0 \\ 0 & 0 & 0 & 0 & 0 & I_{z1} \end{pmatrix},$$

$$U_2 = \begin{pmatrix} m_2 & 0 & 0 & 0 & 0 & -m_2 l_{c2} \\ 0 & m_2 & 0 & 0 & 0 & 0 \\ 0 & 0 & m_2 & m_2 l_{c2} & 0 & 0 \\ 0 & 0 & m_2 l_{c2} & I_{x2} & 0 & 0 \\ 0 & 0 & 0 & 0 & I_{y2} & 0 \\ -m_2 l_{c2} & 0 & 0 & 0 & 0 & I_{z2} \end{pmatrix},$$

$$U_3 = \begin{pmatrix} m_3 & 0 & 0 & 0 & m_3 l_{c3} & 0 \\ 0 & m_3 & 0 & -m_3 l_{c3} & 0 & 0 \\ 0 & 0 & m_3 & 0 & 0 & 0 \\ 0 & -m_3 l_{c3} & 0 & I_{x3} & 0 & 0 \\ m_3 l_{c3} & 0 & 0 & 0 & I_{y3} & 0 \\ 0 & 0 & 0 & 0 & 0 & I_{z3} \end{pmatrix},$$

where only the diagonal elements in each inertia tensor Γ_j are nonzero if we assume that each link is geometrically symmetric and mass-density uniformly distributed.

Now, multiplying each J_j with U_j together yields

$$W_1 = J_1^T U_1 J_1 = \begin{pmatrix} I_{y1} & 0 & 0 \\ 0 & 0 & 0 \\ 0 & 0 & 0 \end{pmatrix},$$

$$W_2 = J_2^T U_2 J_2 = \begin{pmatrix} m_2 d_2^2 - 2m_2 l_{c2} d_2 + I_{z2} & 0 & 0 \\ 0 & m_2 & 0 \\ 0 & 0 & 0 \end{pmatrix},$$

and $W_3 = J_3^T U_3 J_3 =$

$$\begin{pmatrix} m_3 d_2^2 + 2m_3 l_{c3} d_2 c_3 + I_{y3} & -m_3 l_{c3} s_3 & m_3 l_{c3} d_2 c_3 + I_{y3} \\ -m_3 l_{c3} s_3 & m_3 & -m_3 l_{c3} s_3 \\ m_3 l_{c3} d_2 c_3 + I_{y3} & -m_3 l_{c3} s_3 & I_{y3} \end{pmatrix}.$$

We now add them together, i.e., $W = W_1 + W_2 + W_3$, and notice that I_{y1}, I_{z2} and I_{y3} here are about the y_1-axis, z_2-axis and y_3-axis, respectively. According to the Parallel Axis Theorem, $I_{y1} = I_1 + m_1 l_{c1}^2$, $I_{z2} = I_2 + m_2 l_{c2}^2$ and $I_{y3} = I_3 + m_3 l_{c3}^2$, each of which can be converted to referring to the mass center. Therefore, by comparing the current $W = W_1 + W_2 + W_3$ with equation (7.23), we conclude that the two symbolical results of the inertial matrix W for the three-joint planar robot are exactly same, although the approaches are quite different.

Equation (7.30) provides us with a more systematic way than the first approach to determine the inertial matrix W for an open serial-chain robotic system. Not only can it avoid unnecessary human error due to the direct geometrical inspection, but is also programmable for higher-dimensional dynamic systems, such as a digital human model. However, a dynamic model also requires finding the centrifugal and Coriolis terms in the form of $(W_d^T - \frac{1}{2}W_d)\dot{q}$. In other words, the n by n derivative matrix W_d has to be computed and programmed as well. According to equation (7.17), W_d requires taking partial derivatives of W with respect to each q_i, which will be too late to do after the inertial matrix W is numerically calculated in the computer. Nevertheless, as we have introduced and discussed in Chapters 2 and 4, a dual-number transformation would be capable of offering an accurate numerical solution to the derivatives if all the dual number operations could be built in MATLAB^{TM}.

Moreover, it can also be observed that based on equation (7.30), every dynamic parameter of each robotic link is part of the 6 by 6 matrix U_j. By counting the total number of independent dynamic parameters in each $W_j = J_j^T U_j J_j$ for link j, we discover that since each sub-Jacobian J_j only contains the kinematic parameters without any dynamic parameter involved, there are up to 10 dynamic parameters in each U_j: one for the mass m_j, three for the mass center coordinates in the 3 by 3 skew-symmetric matrix C_j, and six in the 3 by 3 inertia tensor block. Furthermore, the 10 dynamic parameters inside each U_j are all linearly related to the inertial matrix W. Such a linear parameterization is a key property owned by almost every open serial-chain robotic system and will be useful for the robotic adaptive control development in the next chapter.

Furthermore, if we keep using the kinetic energy equation in (7.24), instead of (7.25), then we have to redefine a new sub-Jacobian matrix that is referred to the mass center point of link j, instead of the origin of frame j. By defining a new frame cj that is created by parallel-shifting frame j to the mass center of link j, each 3 by 1 position vector p_j^{i-1} in (7.28), which is tailed at the origin of frame j and pointing to the origin of frame $i-1$ with $0 < i \le j$ is now replaced by p_{cj}^{i-1} that is tailed at the mass center of link j or at the origin of frame cj. A new sub-Jacobian matrix is then formed by

$$J_{cj} = \begin{pmatrix} p_{cj}^0 \times r_j^0 & \cdots & p_{cj}^{j-1} \times r_j^{j-1} & O_{3\times(n-j)} \\ r_j^0 & \cdots & r_j^{j-1} & O_{3\times(n-j)} \end{pmatrix}, \tag{7.31}$$

where

$$p_{cj}^{i-1} = p_j^{i-1} - c_j \quad \text{for} \quad 0 < i \le j,$$

and the mass center coordinates vector c_j that was defined earlier is a vector tailed at the origin of frame j and pointing to the mass center of link j. Clearly, c_j is a variable vector, but it is independent of each i in the sub-Jacobian matrix equation (7.31).

In this new sub-Jacobian, due to the parallel shift, every r_j^{i-1} does not need any change. Based on the kinetic equation in (7.24), we can also have the following inertial matrix equation similar to (7.30):

$$W = \sum_{j=1}^{n} J_{cj}^T U_{cj} J_{cj}, \tag{7.32}$$

where

$$U_{cj} = \begin{pmatrix} m_j I_3 & O_3 \\ O_3 & \Gamma_{c(j)} \end{pmatrix}, \tag{7.33}$$

with the 3 by 3 zero matrix O_3. This concludes that the inertial matrix W for a robot can also be determined by referring the tail point of each position vector to the mass center of the corresponding link, instead of the origin of that link frame. Such a parallel-shift alternative will be found very useful if we wish to calculate numerically both the inertial matrix W and the joint torque τ_g due to gravity at the same time, especially in the digital human dynamic modeling in Chapters 9 and 11.

To verify the new inertial matrix equation in (7.32), let us revisit the same 3-joint RPR planar arm as illustrated before. Since $c_1 = (0 \;\; 0 \;\; l_{c1})^T$, $c_2 = (0 \;\; l_{c2} \;\; 0)^T$ and $c_3 = (0 \;\; 0 \;\; l_{c3})^T$,

$$p_{c1}^0 = p_1^0 - c_1 = \begin{pmatrix} 0 \\ 0 \\ 0 \end{pmatrix} - \begin{pmatrix} 0 \\ 0 \\ l_{c1} \end{pmatrix} = \begin{pmatrix} 0 \\ 0 \\ -l_{c1} \end{pmatrix},$$

$$p_{c2}^0 = p_2^0 - c_2 = \begin{pmatrix} 0 \\ d_2 \\ 0 \end{pmatrix} - \begin{pmatrix} 0 \\ l_{c2} \\ 0 \end{pmatrix} = \begin{pmatrix} 0 \\ d_2 - l_{c2} \\ 0 \end{pmatrix},$$

$$p_{c3}^0 = p_3^0 - c_3 = \begin{pmatrix} d_2 s_3 \\ 0 \\ -d_2 c_3 \end{pmatrix} - \begin{pmatrix} 0 \\ 0 \\ l_{c3} \end{pmatrix} = \begin{pmatrix} d_2 s_3 \\ 0 \\ -d_2 c_3 - l_{c3} \end{pmatrix},$$

and

$$p_{c3}^2 = p_3^2 - c_3 = \begin{pmatrix} 0 \\ 0 \\ 0 \end{pmatrix} - \begin{pmatrix} 0 \\ 0 \\ l_{c3} \end{pmatrix} = \begin{pmatrix} 0 \\ 0 \\ -l_{c3} \end{pmatrix}.$$

Here you do not need to find p_{c2}^1 and p_{c3}^1, because the second joint of this particular robot is prismatic.

After substituting all the above p_{cj}^{i-1} into equation (7.31), we obtain

$$J_{c1} = \begin{pmatrix} l_{c1} & 0 & 0 \\ 0 & 0 & 0 \\ 0 & 0 & 0 \\ 0 & 0 & 0 \\ 1 & 0 & 0 \\ 0 & 0 & 0 \end{pmatrix}, \quad J_{c2} = \begin{pmatrix} d_2 - l_{c2} & 0 & 0 \\ 0 & -1 & 0 \\ 0 & 0 & 0 \\ 0 & 0 & 0 \\ 0 & 0 & 0 \\ 1 & 0 & 0 \end{pmatrix},$$

$$J_{c3} = \begin{pmatrix} d_2 c_3 + l_{c3} & -s_3 & l_{c3} \\ 0 & 0 & 0 \\ d_2 s_3 & c_3 & 0 \\ 0 & 0 & 0 \\ 1 & 0 & 1 \\ 0 & 0 & 0 \end{pmatrix}.$$

Accordingly, the three dynamic parameter matrices U_{cj}'s that are now referred to the mass center of each link, based on (7.33), will be the same as the previous U_j's except that all the off-diagonal 3 by 3 blocks are zeros and each moment of inertia is referred to the mass center of the corresponding link. Through equation (7.32), the final result of W will exactly agree with equation (7.23).

Therefore, in terms of the inertial matrix W result, both (7.30) and (7.32) will make no difference. However, if we go through equation (7.32) to find W, every sub-Jacobian matrix prepared is referred to the mass center of each link, instead of the origin of each link frame. This will have an exclusive advantage in determining the joint torque distribution due to gravity over either a robot body or an entire digital human body by applying the robotic statics equation (4.30) in Chapter 4.

Finally, it should be pointed out that to determine the inertial matrix W for a robot dynamic system or a digital human dynamic model, either equation (7.30) or (7.32) adds every contribution W_j made by link j together. However, if some links or segments have zero mass in their dynamic model, then those links have no contributions to the inertial matrix W and the summation can simply skip over those mass-less links to save a lot of computations. This phenomenon occurs, in particular, quite often in a digital human dynamic model, where a number of adjacent axes form a universal-type or spherical-type joint to drive a single segment, such as a shoulder joint to drive the upper arm in 3-axis movement, see Chapters 9 and 11 for details.

7.5 Configuration Manifolds and Isometric Embeddings

The concept of a configuration manifold, or C-manifold, though invisible and abstract, is quite profound, which reveals a fundamental principle for every dynamic system in motion. To better understand and visually represent such a high-dimensional non-Euclidean hyper-surface, we have to embed it into a certain larger Euclidean space before further exploring and unveiling its geometrical meaning and physical insight. To this end, let us first introduce a mathematical theory regarding the immersion and embedding between two manifolds, and then discuss how to embed a non-Euclidean manifold into an ambient Euclidean space.

7.5.1 *Metric Factorization and Manifold Embedding*

An n-dimensional smooth surface situated in a Euclidean m-space $\mathbb{R}^m$ with $m \geq n$ can be viewed as a differentiable manifold M^n of dimension n. Let $q = (q^1 \ \cdots \ q^n)^T$ be a set of local coordinates defined on M^n along with a homeomorphism ϕ (often called a chart of the manifold) that maps q in an open neighborhood U of each point on M^n to an open region U' of a Euclidean n-space $\mathbb{R}^n$ [26, 29].

Let $z = (z^1 \ \cdots \ z^m)^T$ be a coordinate system of the Euclidean m-space $\mathbb{R}^m$. Then, the surface M^n can be represented by

$$z = \zeta(q). \tag{7.34}$$

The Jacobian matrix of (7.34) is given by

$$J = \frac{\partial \zeta}{\partial q} = (g_1 \ \cdots \ g_n), \tag{7.35}$$

which is m by n and $g_i = \partial\zeta/\partial q^i$ is the i-th column of J for each $i = 1, \cdots, n$. The manifold M^n is said to be **non-singular** if its Jacobian matrix has full rank, i.e., rank$(J) = n$ at every point on the manifold. Based on differential geometry [26, 29], all the n columns $g_1, \cdots, g_n$ of the Jacobian matrix J for a non-singular manifold M^n span a **tangent space** $T_q(M^n)$ at each point $q \in U \subset M^n$.

In order to determine the arc length l of a curve given by $z(t) = \zeta(q(t))$ for $t \in [a, b]$ on M^n situated in $\mathbb{R}^m$, we have to integrate the following inner product:

$$\dot{z}^T \dot{z} = \langle \dot{z}, \dot{z} \rangle_\delta = \sum_{i,j=1}^{m} \delta_{ij} \dot{z}^i \dot{z}^j = \sum_{i,j=1}^{n} w_{ij} \dot{q}^i \dot{q}^j = \langle \dot{q}, \dot{q} \rangle_w, \tag{7.36}$$

where the Kronecker delta in the first summation is defined by $\delta_{ij} = 1$ if $i = j$ and $\delta_{ij} = 0$ if $i \neq j$. The second summation in (7.36) is represented in terms of all the tangent vectors of the local coordinates on M^n, and thus it produces a set of new weights w_{ij} in the last inner product. Equation (7.36) is obviously positive-definite, and is known as a **Riemannian metric** for a differentiable manifold [26, 28].

With such a coordinate transformation between q and z, we obtain

$$w_{ij} = \sum_{k,l=1}^{m} \delta_{kl} \frac{\partial \zeta^k}{\partial q^i} \frac{\partial \zeta^l}{\partial q^j}, \tag{7.37}$$

for each $i, j = 1, \cdots, n$. Equation (7.37) is written in a tensor form and can be rewritten in the following matrix form:

$$W = J^T J. \tag{7.38}$$

This clearly exhibits that a Riemannian metric W endowed on a non-singular smooth manifold M^n situated in an ambient Euclidean space $\mathbb{R}^m$ with $m \geq n$ can always be factorized by the Jacobian matrix J of M^n.

In fact, the concepts of local coordinates and Riemannian metrics on a globally non-Euclidean but locally Euclidean manifold have already been frequently used in calculus. The well-known cylindrical and spherical coordinate systems are the basic and most typical local coordinate systems to deal with the integration over a cylindrical or spherical region. The relationship between the Cartesian and cylindrical coordinates in 3D space is known as

$$\begin{cases} x = r\cos\phi = rc\phi \\ y = r\sin\phi = rs\phi \\ z = z, \end{cases} \tag{7.39}$$

where r and ϕ are the polar coordinates on the x-y plane under the constraints $r \geq 0$ and $0 \leq \phi < 2\pi$. The relationship between the Cartesian and spherical coordinates is given by

$$\begin{cases} x = r\sin\theta\cos\phi = rs\theta c\phi \\ y = r\sin\theta\sin\phi = rs\theta s\phi \\ z = r\cos\theta = rc\theta, \end{cases} \tag{7.40}$$

where r is a radial variable, θ is an elevation angle and ϕ is an azimuth angle under $r \geq 0$, $0 \leq \theta \leq \pi$ and $0 \leq \phi < 2\pi$, and all the three variables form a local coordinate system at an open neighboring region of each point, see Figure 7.10.

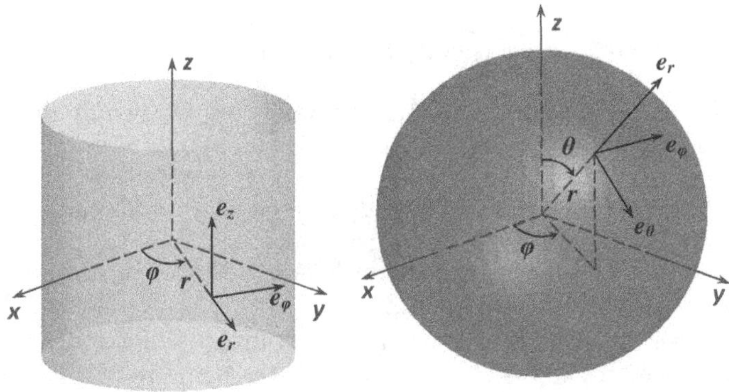

Fig. 7.10 The cylindrical and spherical local coordinate systems

According to the cylindrical coordinates definition, if e_r is the first unity axis, e_ϕ is the second one and e_z is the third one of the 3D basis, the Jacobian matrix of (7.39) can be calculated as follows:

$$J = \frac{\partial(x,y,z)}{\partial(r,\phi,z)} = \begin{pmatrix} c\phi & -rs\phi & 0 \\ s\phi & rc\phi & 0 \\ 0 & 0 & 1 \end{pmatrix}.$$

Then, based on (7.38), the Riemannian metrics are given by

$$W = J^T J = \begin{pmatrix} 1 & 0 & 0 \\ 0 & r^2 & 0 \\ 0 & 0 & 1 \end{pmatrix},$$

which reflects every coefficient of the arc lengths. Its determinant $\det W = r^2$ such that $\det J = \sqrt{\det W} = r$.

Likewise, let e_θ be the first unity axis, e_ϕ be the second one and e_r be the third one of the spherical coordinate basis. The Jacobian matrix of the spherical coordinates (7.40) can be derived as

$$J = \frac{\partial(x,y,z)}{\partial(\theta,\phi,r)} = \begin{pmatrix} rc\theta c\phi & -rs\theta s\phi & s\theta c\phi \\ rc\theta s\phi & rs\theta c\phi & s\theta s\phi \\ -rs\theta & 0 & c\theta \end{pmatrix}.$$

Thus, the Riemannian metrics of the spherical coordinates turn out to be

$$W = J^T J = \begin{pmatrix} r^2 & 0 & 0 \\ 0 & r^2 s^2\theta & 0 \\ 0 & 0 & 1 \end{pmatrix}.$$

Clearly, the determinant $\det J = \sqrt{\det W} = r^2 \sin\theta$.

Recalling the exterior calculus introduced in Chapter 2, based on equation (2.25), we have

$$dx \wedge dy \wedge dz = (\det J)\, dr \wedge d\phi \wedge dz = r\; dr \wedge d\phi \wedge dz$$

for the differential volume conversion between the Cartesian and cylindrical coordinates. Similarly,

$$dx \wedge dy \wedge dz = (\det J)\, d\theta \wedge d\phi \wedge dr = r^2 \sin\theta\; d\theta \wedge d\phi \wedge dr$$

for the differential element conversion between the Cartesian and spherical coordinates inside a volume integral.

For instance, if we wish to calculate the volume V of a sphere with radius $r = a$, which satisfies $x^2 + y^2 + z^2 \le a^2$, denoted as a region Ω, then,

$$V = \iiint_{\Omega} dxdydz = \iiint_{\Omega} r^2 \sin\theta\, drd\theta d\phi.$$

Since the Cartesian coordinates x, y and z are coupled to each other in this spherical region, it is not easy to evaluate the triple integration. In contrast, the spherical coordinates θ, ϕ and r are totally decoupled in this particular region Ω so that

$$V = \int_0^a r^2\, dr \cdot \int_0^{\pi} \sin\theta\, d\theta \cdot \int_0^{2\pi} d\phi = \frac{4}{3}\pi a^3.$$

Furthermore, suppose we wish to find the volume V for an ellipsoid:

$$\frac{x^2}{a^2} + \frac{y^2}{b^2} + \frac{z^2}{c^2} \le 1,$$

where a, b and c are the semi-axes of the ellipsoid along the respective x, y and z-axes. Then, by defining a new coordinate system for a scale change:

$$\xi = \frac{x}{a}, \quad \eta = \frac{y}{b}, \quad \text{and} \quad \zeta = \frac{z}{c},$$

the ellipsoid can be seen as a unity sphere in the new coordinate system, i.e., $\xi^2 + \eta^2 + \zeta^2 \le 1$. However, the differential volume inside the integral has to be changed accordingly by a factor that should be the determinant of the

Jacobian matrix between $\{x, y, z\}$ and $\{\xi, \eta, \zeta\}$, and this Jacobian matrix can be easily found as

$$J = \frac{\partial(x, y, z)}{\partial(\xi, \eta, \zeta)} = \begin{pmatrix} a & 0 & 0 \\ 0 & b & 0 \\ 0 & 0 & c \end{pmatrix}.$$

Its determinant is obviously equal to $\det J = abc$. Based on (2.25), and noticing that the volume of the unity sphere is $\frac{4}{3}\pi$, we finally obtain the volume of the ellipsoid as follows:

$$V = \iiint_{\Omega} dxdydz = \iiint_{\Omega} abc\ d\xi d\eta d\zeta = \frac{4}{3}\pi abc.$$

If a 2D spherical surface S^2 is studied for surface integration, only θ and ϕ are the local coordinates in the neighboring region of each point on the surface.

Moreover, any n-dimensional smooth and compact manifold M^n can be endowed with a Riemannian metric [26], and such a manifold M^n is often called a **Riemannian manifold**. Now, a fundamental question is: what is, in general, the least dimension m of a Euclidean space $\mathbb{R}^m$ such that M^n can be immersed or embedded into $\mathbb{R}^m$? To answer the question, let us first go over the definitions of **the immersion and embedding** between manifolds [27, 28].

Definition 6. A manifold M^n is said to be immersed into a manifold Y^m with $m \geq n$, if there is a smooth mapping $f : M \to Y$ such that the induced mapping $\bar{f} : T_x(M) \to T_{f(x)}(Y)$ at each point $x \in M^n$ is a one-to-one mapping of the tangent plane. The mapping f is called an immersion of M^n into Y^m. The immersion f is further called an embedding if f itself is one-to-one.

Hassler Whitney, an American mathematician, first showed the following well-known immersion/embedding theorem [31, 32, 27, 28]:

Theorem 2. *(Whitney) A smooth and compact manifold M of dimension n can be smoothly immersed in $\mathbb{R}^{2n}$, and can also be smoothly embedded into $\mathbb{R}^{2n+1}$.*

Based on this theorem, any smooth and compact manifold M^n can be embedded into Euclidean $(2n+1)$-space $\mathbb{R}^{2n+1}$ to preserve its topology. In many cases, however, the necessary dimension m of an ambient Euclidean space can be less than $2n$ (or $2n+1$) for an immersion (or embedding), depending on the topological structure for a particular manifold M^n to be immersed (or embedded).

For example, let us take picture for a soft (deformable) one-dimensional closed loop. If this soft loop can be deformed to any closed shape with no intersection between any two distinct points on the loop, then this loop can always be homeomorphic to a 1-circle S^1 that is compact. The image of this loop onto a Euclidean 1-space $\mathbb{R}^1$ is obviously a closed interval that is homeomorphic to $I^1 = [0, 1]$. Since the tangent to this image at each of the two ends is degenerated, the two ends become singular points for the image so that the mapping $f_1 : S^1 \to \mathbb{R}^1$ is neither an immersion, nor an embedding, as shown in Figure 7.11.

If the closed loop is sent into a Euclidean 2-space $\mathbb{R}^2$, we may have three different cases for this mapping [24]. In Figure 7.11, since f_{21} maps S^1 to a loop in $\mathbb{R}^2$ with a singular point at its rightmost end, this is also not an immersion due to a "bad angle" of the picture-taking. However, the image of f_{22} has no singularity but has an intersection point. Clearly, f_{22} is just an immersion but is not an embedding, because the intersection point makes f_{22} violate the one-to-one condition so that such a so-called figure-eight image cannot be homeomorphic to the original S^1. In contrast, $f_{23} : S^1 \to \mathbb{R}^2$ is clearly an embedding due to a "perfect angle" of the picture-taking. Finally, the map $f_3 : S^1 \to \mathbb{R}^3$, as shown in Figure 7.11, is always an embedding, because in this case, $n = 1$ and $m = 3 = 2n+1$ so that f_3 meets the condition of the Whitney Theorem.

The second example is a regular 2-torus T^2. The surface equation represented in $\mathbb{R}^3$ is given by

$$\begin{cases} z^1 = (a + r\cos\theta_2)\cos\theta_1 \\ z^2 = (a + r\cos\theta_2)\sin\theta_1 \\ z^3 = r\sin\theta_2, \end{cases} \tag{7.41}$$

where $a > 0$ and $r > 0$ are two constant parameters, called the major and minor radii, respectively, and the two local coordinates $q^1 = \theta_1$ and $q^2 = \theta_2$, as shown on the left of Figure 7.12. It can be easily observed that if $a = 0$, equation (7.41) represents a 2-dimensional spherical surface S^2 of radius r, while if $a > r$, this is a smooth and compact surface of 2-torus T^2. Topologically, this torus is viewed as a product of two 1-circles that are corresponding to the two rotation angles θ_1 and θ_2, i.e., $T^2 \simeq S^1 \times S^1$ [27, 28]. The Jacobian matrix J of (7.41) becomes

$$J = \frac{\partial \zeta}{\partial q} = \begin{pmatrix} -(a + r\cos\theta_2)\sin\theta_1 & -r\sin\theta_2\cos\theta_1 \\ (a + r\cos\theta_2)\cos\theta_1 & -r\sin\theta_2\sin\theta_1 \\ 0 & r\cos\theta_2 \end{pmatrix}. \tag{7.42}$$

Thus, the Riemannian metric on the 2-torus is a diagonal matrix,

$$W = J^T J = \begin{pmatrix} (a + r\cos\theta_2)^2 & 0 \\ 0 & r^2 \end{pmatrix} \tag{7.43}$$

which is always non-singular as $a > r > 0$.

Therefore, the mapping $\zeta : T^2 \to \mathbb{R}^3$ given by (7.41) is an embedding. In fact, it can be shown [27] that all the known compact and orientable 2-dimensional manifolds can be embedded into $\mathbb{R}^3$. However, if T^2 is mapped into a Euclidean 2-space $\mathbb{R}^2$, which is, for instance, spanned only by z^1 and z^2 axes, as shown on the right of Figure 7.12, the explicit form of this mapping

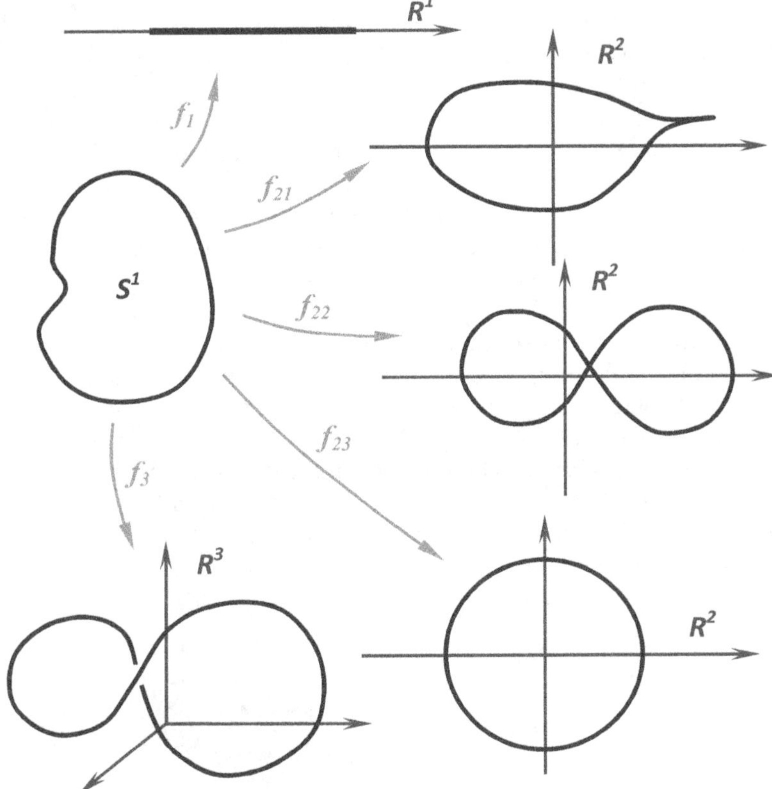

Fig. 7.11 Different mapping cases from S^1 to Euclidean spaces

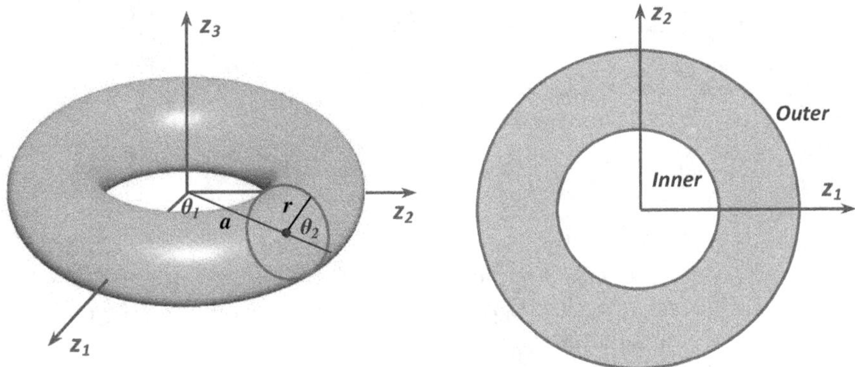

Fig. 7.12 A 2D torus T^2 situated in Euclidean spaces $\mathbb{R}^3$ and $\mathbb{R}^2$

ζ_{2d} is just the first two equations of (7.41), and its Jacobian matrix takes the first two rows of (7.42), as if the third equation of z^3 in (7.41) and the third row for the Jacobian matrix in (7.42) disappear. Then, the determinant of this 2 by 2 Jacobian matrix is $r(a + r\cos\theta_2)\sin\theta_2$, which vanishes at $\theta_2 = 0$ or π. This evidently shows that on the 2D image of the torus T^2, each point on either the inner or the outer circular contour is a singular point. Therefore, the mapping $\zeta_{2d} : T^2 \to \mathbb{R}^2$ is neither an immersion, nor an embedding.

We can further observe from the 2D squeezed image of T^2 that each point (z_1, z_2) inside the region of the annulus between the inner and outer contours corresponds to two distinct sets of the local coordinates, i.e., $(\theta_1,\ \theta_2)$ and $(\theta_1,\ -\theta_2)$ for every $0 \le \theta_1 < 2\pi$ and $0 < \theta_2 < \pi$. This phenomenon is referred to as a **multi-configuration** issue that makes the mapping ζ_{2d} directly violate the one-to-one condition. Therefore, taking a 2D picture for the 2-torus T^2 will damage its topology, while taking a 3D stereo picture for T^2 can preserve all of its topological properties, such as its two "holes" at a homotopical point of view. The multi-configuration issue will be very important for our next discussion on robotic dynamics and adaptive control, and the above example is just an explicit interpretation at both the geometrical and topological standpoints.

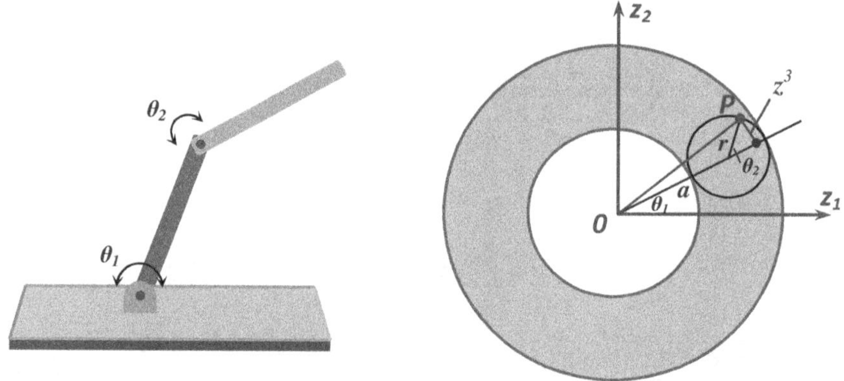

Fig. 7.13 A planar RR-type arm and its C-manifold as a flatted torus

Unfortunately, for an RR-type 2-link planar arm, or a robot having two revolute joints, whose axes are parallel to each other, its C-manifold becomes a flatted torus, as shown in Figure 7.13. This is also similar to the above case of squeezing T^2 into the 2D plane, resulting in an annulus in Figure 7.12. The z^3 component of the original T^2 is now pushed down into the 2D plane to impose an additional projection on each of the first two components z^1 and z^2. According to Figure 7.13 on the right at point P, those additional projections are $-z^3\sin\theta_1 = -r\sin\theta_2\sin\theta_1$ for z^1, and $z^3\cos\theta_1 = r\sin\theta_2\cos\theta_1$ for z^2 such that the new equation for the flatted torus turns out to be

$$\begin{cases} z^1 = (a + r\cos\theta_2)\cos\theta_1 - r\sin\theta_2\sin\theta_1 = a\cos\theta_1 + r\cos(\theta_1+\theta_2) \\ z^2 = (a + r\cos\theta_2)\sin\theta_1 + r\sin\theta_2\cos\theta_1 = a\sin\theta_1 + r\sin(\theta_1+\theta_2). \end{cases} \tag{7.44}$$

This new equation coincidentally looks like the position vector p_0^2 of the RR-type 2-link planar robot in symbolical form. The Jacobian matrix J can be immediately calculated as

$$J = \frac{\partial \zeta}{\partial q} = \begin{pmatrix} -as_1 - rs_{12} & -rs_{12} \\ ac_1 + rc_{12} & rc_{12} \end{pmatrix},$$

where $s_i = \sin\theta_i$ and $c_i = \cos\theta_i$ for $i = 1, 2$, and $s_{12} = \sin(\theta_1 + \theta_2)$ and $c_{12} = \cos(\theta_1 + \theta_2)$. Its determinant is $\det(J) = ars_2$, which vanishes at $\sin\theta_2 = 0$, or $\theta_2 = 0$ or $\pm\pi$. This becomes the similar singularity and multi-configuration issues as squeezing the regular torus T^2 into a 2D plane. In other words, for each given tip point position p_0^2 of the planar RR-type 2-link arm, there are always two distinct configurations $(\theta_1^{(1)}, \theta_2)$ and $(\theta_1^{(2)}, -\theta_2)$ that are symmetric around its singular point $\theta_2 = 0$ or $\pm\pi$ with the different values of θ_1. In this sense, both the singularity and multi-configuration issues exposed by the C-manifold are exactly the same as those issues occurred in its real kinematics.

Moreover, the Riemannian metric becomes

$$W = J^T J = \begin{pmatrix} a^2 + r^2 + 2arc_2 & r^2 + arc_2 \\ r^2 + arc_2 & r^2 \end{pmatrix}, \tag{7.45}$$

which is no longer a 2 by 2 diagonal matrix in comparison with the Riemannian metric W for the regular non-flatted 2-torus T^2 in (7.43).

To further compare the above Riemannian metric of the flatted T^2 with the real inertial matrix W for the RR-type 2-link planar robot, by following the same procedure as discussed in the last section, we can derive it and find its symbolical form below:

$$W = \begin{pmatrix} w_{11} & m_2 l_2^2 + I_{z2} + m_2 a_1 l_2 c_2 \\ m_2 l_2^2 + I_{z2} + m_2 a_1 l_2 c_2 & m_2 l_2^2 + I_{z2} \end{pmatrix}, \tag{7.46}$$

where $w_{11} = m_1 l_1^2 + I_{z1} + m_2 l_2^2 + I_{z2} + m_2 a_1 + 2m_2 a_1 l_2 c_2$, and m_1 and m_2 are the masses of link 1 and link 2, while l_1 and l_2 are the mass center coordinates of link 1 and link 2, respectively. In addition, a_1 is the length of link 1, and I_{z1} and I_{z2} are the moments of inertia for the two links. It is observable that the above Riemannian metric $W = J^T J$ in (7.45) is very close to the inertial matrix in (7.46), but still has a small difference between the two W matrices, no matter how you adjust the two torus parameters a and r in (7.45). In other words, no matter how you deform the torus, such a C-manifold for the RR-type planar robot within the range of equivalent topology, the Riemannian metric cannot completely match with the inertial matrix of the system. Therefore, the ultimate goal is not only to seek an embedding, but is also desirable to find an isometric embedding that can preserve both the topology and geometry. Namely, if $z = \zeta(q)$ in equation (7.34) can be found as an *isometric embedding* for the C-manifold of some robotic system, then $J^T J$ is already the inertial matrix W of the system, and you don't have to go through the lengthy procedure of either (7.30) or (7.32) to determine it again.

In general, for a robot having six revolute joints, its C-manifold is supposed to possess a topology of 6-torus $T^6 = S^1 \times S^1 \times S^1 \times S^1 \times S^1 \times S^1$. However, it is just an ideal inference in mathematics, but is not true in reality. If any two of the six revolute joints for the robot are permanently rotating about two parallel axes, or any two of their z-axes are parallel in structure, then at least one product $S^1 \times S^1$ will be degenerated to a flatted 2-torus, or an annulus that is homeomorphic to a closed cylindrical surface $S^1 \times I^1$, where I^1 is the unity interval $[0, 1]$.

In mathematics, the topology of the 6-torus T^6 can be equivalent to the following quotient space:

$$T^6 \simeq \mathbb{R}^6/\mathbb{Z}^6,$$

where $\mathbb{Z}^6$ is a 6-dimensional topological space formed by a direct product of the integer additive groups. This means that there are six "holes" to make the 6-torus T^6 non-contractible. Intuitively, it is not true for the 6-dimensional C-manifold of a real robot.

In fact, the first two revolute joints of a typical 6-revolute-joint open serial-chain robots produce a 2-spherical surface S^2 to the C-manifold topology. If the third joint axis is parallel to the second one, then the first three joints will contribute a solid sphere $S^2 \times I^1 = B^3$ that is a compact subspace $B^3 \subset \mathbb{R}^3$. Among the last three joints in the robotic wrist, similarly, the first two produce a topology of S^2. The last joint rotation is virtually a spin $SO(2)$ so that the last three joints in the wrist will contribute a topology of $S^2 \times SO(2) \simeq SO(3)$. Therefore, the topology of the C-manifold for such a 6-revolute-joint robot becomes $B^3 \times SO(3)$, which is not homeomorphic to T^6 at all.

7.5.2 Isometric Embedding of C-Manifolds

A mechanical dynamic system with its kinetic energy K and potential energy P obeys the Lagrange equation (7.15). The kinetic energy K is always a positive-definite quadratic form of velocities. Namely,

$$K = \frac{1}{2}\langle \dot{q},\, \dot{q} \rangle_w = \frac{1}{2}\dot{q}^T W \dot{q} \tag{7.47}$$

with a weighting coefficient matrix W, called the **inertial matrix** of the system [4, 5, 6]. Based on the classical mechanics, any rigid motion can be represented in terms of the special Euclidean group $SE(3) = \mathbb{R}^3 \times SO(3)$ [1, 7]. If the motion is finite, its translation part will be limited in a bounded and closed (or can make it closed) subspace $T(3) \subset \mathbb{R}^3$ so that the topological space $T(3) \times SO(3)$ is compact [28]. With the above interpretation, let us formally define a **configuration manifold**, or C-manifold:

Definition 7. A Riemannian manifold M^n is call a configuration manifold, or C-manifold of an n-th order mechanical dynamic system if

1. A certain n-tuple local coordinate system $q = \{q^1, \cdots, q^n\}$ is defined on M^n;
2. The basis of the topology of M^n is a collection of open subsets of $T(3) \times SO(3)$; and
3. The kinetic energy K of the system given by (7.47) is endowed on M^n as its Riemannian metric.

For the first condition of the above definition, all the joint positions of an n-joint robot arm may be the most convenient definition of the local coordinates on its C-manifold. The second and third conditions of Definition 7 ensure that the C-manifold of a given mechanical dynamic system is the unique motion constraint surface that is hidden behind the system. Namely, every motion trajectory of the system is constrained on the C-manifold without exception.

In order to explicitly represent the C-manifold for a given system, we need to seek a **global isometric embedding** that can not only embed the C-manifold into a Euclidean space $\mathbb{R}^m$ of dimension $m \geq n$, but also preserve its Riemannian metric, i.e. the kinetic energy of the system.

Consider a free-motion 3D rigid body with $p_0^c \in \mathbb{R}^3$ as its mass center (the center of gravity) position vector with respect to a fixed base coordinate frame. The kinetic energy of this rigid body can be written by

$$K = \frac{1}{2} m \dot{p}_0^{cT} \dot{p}_0^c + \frac{1}{2} \omega^T \Gamma_c \omega, \tag{7.48}$$

where m is the mass of the body and ω is a 3-dimensional angular velocity [25, 4, 6]. The 3 by 3 symmetric matrix Γ_c is known as the inertia tensor of the body with respect to the mass center, and based on (7.3) and (7.4), it is given by

$$\Gamma_c = \int_M (r_c^T r_c I_3 - r_c r_c^T) dm = \begin{pmatrix} I_{xx} & I_{xy} & I_{xz} \\ I_{yx} & I_{yy} & I_{yz} \\ I_{zx} & I_{zy} & I_{zz} \end{pmatrix}, \tag{7.49}$$

where the integral with the differential mass dm is calculated over the entire body M, I_3 is the 3 by 3 identity, and $r_c \in \mathbb{R}^3$ is a radial vector tailed at the mass center with its arrow pointing to each point of the body with respect to some coordinate frame F_c, the origin of which is fixed at the mass center [1, 25].

Among the total six degrees of freedom, three are translational with a linear velocity $v_0^c = \dot{p}_0^c$ that is the time-derivative of p_0^c, while the other three are rotational and the angular velocity ω is, in general, not a time-derivative of some vector. In other words, the differential 1-form $\sigma = \omega dt \in \Lambda^1$ is not *exact* in general, as addressed in Chapter 3. Therefore, we have to find an alternative way to represent the rotational velocity for a rigid body to be an **exact differential 1-form** that is necessary for finding a desired isometric embedding.

Intuitively, the most promising way is to try two non-trivial radial vectors r_c^1 and r_c^2, both of which are referred to the frame F_c and tailed at the mass center to uniquely represent the body rotation. It can be shown that two distinct arrow points that are unaligned with the mass center can uniquely represent the rotation with respect to the mass center. By including the mass center itself, it is called a **three-point model** [1], where p_0^c determines the position of the rigid body and r_c^1 and r_c^2 determine its orientation. Therefore, by setting the following equation:

$$\sum_{i=1}^{2} (r_c^{iT} r_c^i I_3 - r_c^i r_c^{iT}) = \Gamma_c, \tag{7.50}$$

one can determine the two arrow points of r_c^i's with $2 \times 3 = 6$ elements in total in terms of up to six independent components I_{xx}, I_{yy}, I_{zz}, $I_{xy} = I_{yx}$, $I_{xz} = I_{zx}$ and $I_{yz} = I_{zy}$ of the symmetric inertia tensor Γ_c given by (7.49) with respect to the frame F_c.

If a rigid body has both a symmetric shape and uniform mass-distribution about each axis of the frame F_c, then Γ_c can be reduced to a diagonal matrix with only three non-zero components. In this case, a *two-point model* may be sufficient to determine the body rotation and translation.

Once the 6 d.o.f. motion of a rigid body is modeled by three non-trivial vectors p_0^c, r_c^1 and r_c^2, we achieve an exact differential 1-form to represent both the rotational and translational velocities. This progress is the key step towards a qualified isometric embedding. We now show the following theorem:

Theorem 3. *The C-manifold M^6 of a rigid body under six degrees of freedom finite motion can be smoothly and isometrically embedded (globally) into Euclidean 9-space $\mathbb{R}^9$ by the following mapping $\zeta : M^6 \to \mathbb{R}^9$:*

$$z = \zeta(q) = \begin{pmatrix} \sqrt{m} p_0^c(q) \\ r_c^1(q) \\ r_c^2(q) \end{pmatrix}, \tag{7.51}$$

where $q = \{q^1, \cdots, q^6\}$ is a 6-tuple local coordinate system defined at each point on M^6 such that the rank of the Jacobian matrix J of (7.51) is full, i.e., rank$\left(\frac{\partial \zeta}{\partial q}\right) = 6$ at each point of M^6, and $r_c^1(q)$ and $r_c^2(q)$ are determined by (7.50).

Proof. Clearly, M^6 for the rigid body under finite motion has a topology of $T(3) \times SO(3)$ that is compact. Since $T(3)$ has already been sitting in $\mathbb{R}^3$, this compact Euclidean subspace can be directly embedded into $\mathbb{R}^3$ by the first three components of (7.51).

Now, the 3-dimensional compact topological space $SO(3)$, at first glance, can be smoothly embedded into Euclidean $2\times3+1=7$-space $\mathbb{R}^7$ according to Theorem 2 by Whitney. However, the dimension of embedding for this special rotation group can be reduced, because the quotient space $SO(3)/SO(2)\simeq S^2$ is a 2-dimensional spherical surface, which can be smoothly embedded into $\mathbb{R}^3$. Because the Lie group $SO(2)$ just represents a 1-dimensional spin, it can be embedded into a Euclidean 2-space $\mathbb{R}^2$. Thus, the 3-dimensional compact topological space $SO(3)\simeq S^2\times SO(2)$ can be smoothly embedded into up to $\mathbb{R}^5$.

Now, equation (7.51) suggests that a manifold in $SO(3)$ be embedded into Euclidean 6-space $\mathbb{R}^6$ by using the model of two radial vectors, i.e., the last six components of (7.51) in order to meet the isometric condition beyond the embedding. Therefore, it suffices to show that (7.51) is also an isometry for the C-manifold M^6 of the rigid body. In other words, the Riemannian metric $\frac{1}{2}\langle\dot{z},\,\dot{z}\rangle_\delta=\frac{1}{2}\dot{z}^T\dot{z}$ after the embedding will be equal to the kinetic energy K of the body given by (7.48). According to (7.51),

$$\dot{z}^T\dot{z}=m\dot{p}_0^{cT}\dot{p}_0^c+\sum_{i=1}^{2}\dot{r}_c^{iT}\dot{r}_c^i. \tag{7.52}$$

Since for a rigid body, the distance between every pair of distinct points in the body must be time-invariant, each r_c^i should satisfy

$$\dot{r}_c^i=\omega\times r_c^i$$

so that

$$\dot{r}_c^{iT}\dot{r}_c^i=(\omega\times r_c^i)^T(\omega\times r_c^i)=\omega^T(r_c^{iT}r_c^iI_3-r_c^ir_c^{iT})\omega.$$

Substituting this into (7.52) yields

$$\dot{z}^T\dot{z}=m\dot{p}_0^{cT}\dot{p}_0^c+\omega^T\sum_{i=1}^{2}(r_c^{iT}r_c^iI_3-r_c^ir_c^{iT})\omega.$$

Therefore, by invoking equation (7.50), we finally have

$$\frac{1}{2}\dot{z}^T\dot{z}=\frac{1}{2}m\dot{p}_0^{cT}\dot{p}_0^c+\frac{1}{2}\omega^T\Gamma_c\omega=K.$$

□

It can be seen that equation (7.51) proposes an explicit form of the isometric embedding for a rigid body C-manifold, which also depends upon the definition of local coordinates on the C-manifold. Namely, the isometric embedding of a manifold under a certain local coordinate system could be damaged by redefining a different local coordinate system.

7.5.3 Combined Isometric Embedding and Structure Matrix

Consider now an open serial-chain robot arm with n joints/links. Let all the joint positions in $q = (q_1 \ \cdots \ q_n)^T$ of the robot form a local coordinate system $\{q^1, \cdots, q^n\}$ on its n-dimensional C-manifold M_c^n. If we temporarily treat each link of the robot as if it has up to 6 d.o.f. rigid motion and also satisfies Theorem 3, then, an isometric embedding of link i can be expressed as $z_i = \zeta_i(q)$ by using (7.51), which sends the C-manifold M_i of link i to $\mathbb{R}^9$.

Suppose that the robot has been assigned a coordinate system for each link by the D-H convention, see Chapter 4. It can be seen that for link i,

$$p_0^{ci} = p_0^i + R_0^i b_i, \quad \text{and} \quad r_{ci}^j = R_0^i a_i^j, \quad j = 1, 2. \tag{7.53}$$

In the first equation, p_0^{ci} is the radial vector from the base origin to the mass center of link i, while p_0^i is the radial vector pointing to the origin of frame i that is fixed on link i, and b_i is a 3 by 1 vector of dynamic parameters such that $R_0^i b_i$ represents a vector from the origin of frame i to the mass center of link i as a linear combination of the three columns of R_0^i. Similarly in the second equation of (7.53), each r_{ci}^j can also be viewed as a linear combination of the three columns of R_0^i by a 3 by 1 dynamic parameters column a_i^j for $j = 1, 2$. Substituting (7.53) into (7.51) for link i, we obtain

$$z_i = \zeta_i(q) = \begin{pmatrix} \sqrt{m_i} p_0^{ci} \\ r_{ci}^1 \\ r_{ci}^2 \end{pmatrix} = \begin{pmatrix} R_0^i & p_0^i & & O \\ & & R_0^i & \\ O & & & R_0^i \end{pmatrix} \begin{pmatrix} \sqrt{m_i} b_i \\ \sqrt{m_i} \\ a_i^1 \\ a_i^2 \end{pmatrix}. \tag{7.54}$$

Let the 9 by 10 block-diagonal matrix on the right-hand side of (7.54) be denoted by Z_0^i, i.e.,

$$Z_0^i = \begin{pmatrix} R_0^i & p_0^i & & O \\ & & R_0^i & \\ O & & & R_0^i \end{pmatrix}.$$

The first 3 by 4 block $(R_0^i \ \ p_0^i)$ is exactly the first three rows of the 4 by 4 homogeneous transformation matrix A_0^i between frame i and the base according to the robotic kinematics. The last two blocks of Z_0^i are two copies of the rotation matrix $R_0^i \in SO(3)$. Therefore, the matrix Z_0^i is a collection of the kinematic transformations that are members of the Lie group $SE(3) = \mathbb{R}^3 \times SO(3)$.

Let the last 10 by 1 column vector in (7.54) be denoted by ξ_i that contains up to 10 dynamic parameters and matches with the 10 independent parameters in the dynamic parameter matrix U_i of link i, see equations (7.29) and (7.30). Then, (7.54) can be rewritten as

$$z_i = Z_0^i \xi_i.$$

By augmenting all such mappings from link 1 through link n, we obtain a global **combined isometric embedding** that sends the C-manifold M_c^n of the robot into Euclidean $9n$-space $\mathbb{R}^{9n}$:

$$z = \zeta(q) = \begin{pmatrix} z_1 \\ \vdots \\ z_n \end{pmatrix} = \begin{pmatrix} Z_0^1 & & O \\ & \ddots & \\ O & & Z_0^n \end{pmatrix} \begin{pmatrix} \xi_1 \\ \vdots \\ \xi_n \end{pmatrix} = Z\xi. \tag{7.55}$$

Therefore, the **global structure matrix** Z in (7.55) becomes a collection of all the kinematic transformations for the n-joint robot, while ξ is a vector that lists all the dynamic parameters of the n links. This demonstrates that *the combined isometric embedding (7.55) that is augmented by every (7.54) is a linear combination of all the columns of the kinematic transformation collection Z by all the dynamic parameters ξ, provided that the local coordinates q on the C-manifold M^n are defined by using the robotic joint positions.*

Moreover, due to the above augmentation, the total Riemannian metric can be determined by

$$W = J^T J = J_1^T J_1 + \cdots + J_n^T J_n, \tag{7.56}$$

where each J_i is the Jacobian matrix of the i-th isometric embedding $z_i = \zeta_i(q)$ for $i = 1, \cdots, n$.

It must be clarified that in the global structure matrix Z, each p_0^i is a position vector that may also contain some parameters, such as a link length and/or a joint offset of link i. However, they are all the *kinematic parameters* that are measurable, and thus they do not belong to the category of *dynamic parameters*. Therefore, the matrix Z can also be interpreted as a structural frame of the robotic system, which will be "coated" linearly by all the dynamic parameters in the vector ξ.

We have so far made substantial progress in achieving an explicit form of the combined isometric embedding $\zeta : M^n \to \mathbb{R}^{9n}$ for an n-joint serial-chain robot. However, the dimension of such an isometric embedding is $9n = 54$ if $n = 6$, which is obviously too high to be acceptable.

In history, the study of isometric embedding has become an active mathematical research topic since Nash initiated it in the 1950's [33]. Greene improved the Nash's Theorem towards a much better result in 1970 [34]. Their research was then followed by many other mathematicians [35]. The best known result is the Nash-Greene Theorem which showed that a compact Riemannian manifold of dimension n can be isometrically embedded (globally) into

$$m = \frac{n(n+1)}{2} + n$$

dimensional Euclidean space. Based on this, with $n = 6$, M_c^6 is supposed to be isometrically embeddable into $\mathbb{R}^{27}$ that is much smaller than $\mathbb{R}^{54}$ in the above augmentation (7.55).

However, all the existing mathematical theories have never provided us with any explicit approach to finding such a minimum isometric embedding. In particular, we have also a *restriction of local coordinates definition* to study a robotic system, i.e., choosing the joint positions is most preferred, which may even reduce the chance to find the minimum isometric embedding. Nevertheless, Nash-Greene Theorem does make a challenging prediction: there is an opportunity of model reduction to be further explored.

In fact, the motion of link i for an open serial-chain robot with $i = 1, \cdots, n$ only depends on the first i joint positions $q_1, \cdots, q_i$, and is independent of the rest. In other words, the links proximate to the base do not have all the 6 d.o.f. of rigid motion. For instance, link 1 has only 1 d.o.f motion. This phenomenon may help us for a further attempt of model reduction.

7.5.4 The Minimum Isometric Embedding and Isometrization

To further reduce the dimension of a combined isometric embedding given in (7.55), let us now pay more attention to the last link, i.e., the kinematically "busiest" link of an n-joint open serial-chain robot. Obviously, the isometric embedding of link n, $z_n = \zeta_n(q)$ can embed the C-manifold M_n of link n into $\mathbb{R}^9$ by (7.54) with $i = n$. Intuitively, because any joint movement will contribute a non-trivial part to the last link motion, the dimensions of M_n for the last link and M_c for the entire robot should both be equal to n. Now the question is can the mapping $z_n = \zeta_n(q)$ of the last link also embed the C-manifold M_c of the entire robot into $\mathbb{R}^9$? If so, we would achieve a significant model reduction.

Let $z_n|_{M_n}$ and $z|_{M_c}$ be, respectively, a point on the last link C-manifold M_n sitting in $\mathbb{R}^9$ and a point on the entire robot C-manifold M_c in $\mathbb{R}^{9n}$. If we define a common local coordinate system by using the n robotic joint positions $q = (q_1 \; \cdots \; q_n)^T$ on both M_n and M_c, then we seek a continuous one-to-one in both directions between M_n and M_c via the common local coordinates q, i.e.,

$$z_n|_{M_n} \longleftrightarrow q|_{\mathbb{Z}} \longleftrightarrow z|_{M_c}.$$

We can directly observe that through the common q under the quotient operation by the integer group $\mathbb{Z}$ for every joint angle, each point on M_n determined by $z_n = \zeta_n(q) = Z_0^n(q)\xi_n$ can respond to a unique point on M_c determined by $z = \zeta(q) = Z(q)\xi$, and vice verso, as long as the mapping $\zeta_n : M_n \to \mathbb{R}^9$ of the last link is a smooth embedding under the robotic joint positions in q as the local coordinate system. This reaches the following lemma:

Lemma 3. *For an open serial-chain robotic system with n joints, let the n joint positions in $q = (q_1 \ \cdots \ q_n)^T$ be defined as a common n-tuple local coordinate system on both the C-manifold M_n of the last link and C-manifold M_c of the entire robot. If (7.54) with $i = n$ can be a smooth embedding that sends M_n into $\mathbb{R}^9$, then M_n is diffeomorphic to M_c, i.e., $M_n \simeq M_c$.*

Note that if the number of joints of the robot arm is $n \leq 6$ in the above lemma, it may have more chances for the mapping ζ_n to be a smooth embedding. If $n > 6$, such an embedding condition on M_n would not be possible. In fact, $n > 6$ is a redundant robot case, as discussed in Chapter 5, and every redundant robotic system has a continuum of multi-configurations. In the redundant robot case, the combined C-manifold M_c of the entire robot may still be 6-dimensional and so is the C-manifold M_n of the last link, unless additional subtasks are augmented to the 6 d.o.f. rigid motion as a redundant kinematic output. In addition, Lemma 3 reveals a differentiable (smooth) **topological equivalence** between the two C-manifolds $M_n \simeq M_c$ under the condition that (7.54) with $i = n$ is a smooth embedding if all the joint positions of the robot constitute a local coordinate system.

A number of non-redundant robot arms can directly satisfy the above condition due to their special mechanical structures. A robot having more independent prismatic joints may have more chances to satisfy the smooth embedding condition. If, in general, the smooth embedding condition cannot hold by the last link mapping alone, we have to augment the Euclidean z-space by adding one or more links until the augmented mapping is qualified to be a smooth embedding.

Obviously, if the final augmentation covers all n links of the robot, it must be smoothly embeddable, because the total Riemannian metric $W = J^T J$ that is also the inertial matrix of the robot should always be non-singular. However, in most cases, only a small number of links need to be selected, say $1 < k < n$, which, of course, include the last link to form an augmented mapping into $\mathbb{R}^{9k}$ to meet the smooth **embeddability** condition. Such an augmented smooth embedding ζ_{nk} with the smallest k is referred to as **the minimum embedding** of the C-manifold M_c for the robotic system. Therefore, based on the principle of Lemma 3, under the smallest number k, it is always true that $M_{nk} \simeq M_c$ for every open serial-chain robotic system, even for a redundant robot.

Now, the new question is how to find such a minimum embedding? Recalling Definition 6, a mapping between two manifolds can be an embedding if it is a one-to-one immersion. Thus, by intuition, this one-to-one condition holds if no more **multi-configuration** case can be found among the remaining $n - k$ links ($k < n$) whenever the selected k links are stationary.

It is well known in robotic kinematics that a 6-joint PUMA robot with a shoulder offset (i.e., $d_2 \neq 0$) has up to eight different configurations (i.e., eight different sets of I-K solutions) if the last link (link 6) is fixed. Whereas for the 6-joint PUMA-like arm without the shoulder offset ($d_2 = 0$), the number of different configurations is down to four. In contrast, a 6-joint Stanford robot with a shoulder offset ($d_2 \neq 0$) has in total four different configurations whenever link 6 is motionless. These four different sets of the I-K solutions for the Stanford arm have been animated in MATLAB^{TM}, as realistically depicted in Figures 7.14 and 7.15.

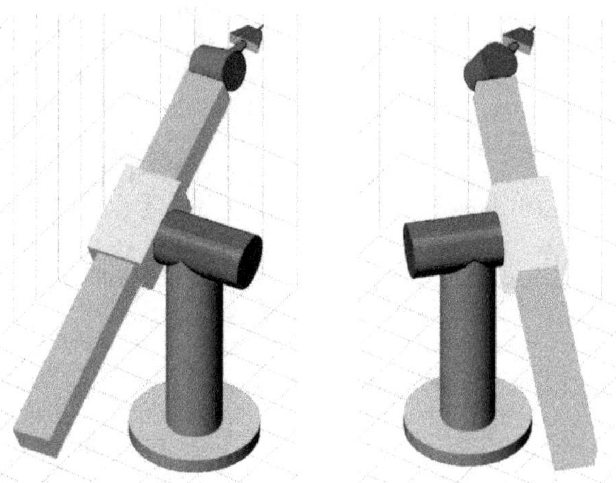

Fig. 7.14 The first and second of four I-K solutions for a Stanford arm

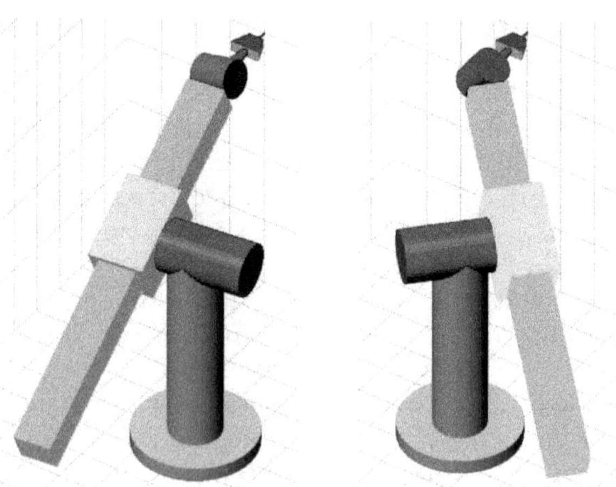

Fig. 7.15 The third and forth of four I-K solutions for a Stanford arm

Therefore, to only pick the last link of the 6-joint Stanford arm (or PUMA arm) is insufficient to constitute a smooth embedding for the C-manifold M_c if all the 6 joint positions are defined as the local coordinates on the last link C-manifold M_6. However, it is also observable that if we further select link 1 and link 4, in addition to link 6 for the Stanford arm to augment a relatively larger mapping for M_c, then it becomes embeddable because all the possible multi-configurations disappear whenever the selected three links are stationary.

In this case, the minimum embedding ζ_{nk} is determined by augmenting the mappings (7.54) of link 6, link 4 and link 1, and thus $k = 3$. The total dimension of this minimum embedding is now $9k = 9 \times 3 = 27$ that is much less than $9n = 9 \times 6 = 54$ in the regular augmentation (7.55), and just touches the dimension predicted by the Nash-Greene Theorem for isometrically embedding the C-manifold M_c of the Stanford arm. Furthermore, if each link can adopt the two-point model, i.e., only one radial vector r_c^1 for each link with a diagonal inertia tensor, then the dimension of the minimum embedding will be further reduced to $6k = 18$.

Actually, to test whether a mapping $\zeta : M^n \to \mathbb{R}^m$ for an n-dimensional C-manifold M^n is an embedding, we may just check its Jacobian matrix $J = \partial\zeta/\partial q$ to see if $\text{rank}(J) = n$, the full rank at every point on M^n. If the Jacobian matrix of a mapping has full rank, then every multi-configuration case will disappear. Therefore, to find the minimum embedding of the C-manifold M_c for a robot by the above k-augmentation procedure, testing its Jacobian matrix J to see if it is full-ranked is the easiest way to forecast the result.

Moreover, two smoothly topologically equivalent (diffeomorphic) Riemannian manifolds in a common ambient Euclidean space can always be smoothly deformed to make them isometric to each other. Even though their ambient Euclidean spaces are in different dimensions, it may still be possible, but not always. We refer to this smooth deformation process as an **isometrization** [23].

Since the Riemannian metric of any smooth C-manifold is given by $W = J^T J$, where J is the Jacobian matrix of the C-manifold, based on (7.55) and (7.56), each component of W can be written as

$$w_j^i = g_i^T g_j = \xi^T \frac{\partial Z}{\partial q^i}^T \frac{\partial Z}{\partial q^j} \xi = \text{tr}\left(\frac{\partial Z^T}{\partial q^i} \frac{\partial Z}{\partial q^j} \Psi \right), \quad \text{for} \quad i, j = 1, \cdots, n, \tag{7.57}$$

where $\Psi = \xi\xi^T$ is called a parameter matrix and $\text{tr}(\cdot)$ is the trace of a square matrix $(\cdot)$. We now give a formal definition regarding the isometrization below:

Definition 8. Let two n-dimensional C-manifolds M_a^n and M_b^n be smoothly embedded into $\mathbb{R}^{m_a}$ and $\mathbb{R}^{m_b}$, respectively, on both of which a common local coordinate system $\{q^1, \cdots, q^n\}$ is defined. Then, M_a^n is said to be **isometrizable** to M_b^n if for each element $(w_a)_j^i$ of the metric W_a on M_a^n, i.e.,

$$(w_a)^i_j = \xi^T_a \frac{\partial Z^T_a}{\partial q^i} \frac{\partial Z_a}{\partial q^j} \xi_a = \mathrm{tr}\left(\frac{\partial Z^T_a}{\partial q^i} \frac{\partial Z_a}{\partial q^j} \Psi_a \right),$$

there is a real parameter function $D(\cdot)$ such that the new parameter matrix $\Psi'_a = D(\Psi_b)$ can make the following equation hold globally:

$$(w'_a)^i_j = \mathrm{tr}\left(\frac{\partial Z^T_a}{\partial q^i} \frac{\partial Z_a}{\partial q^j} \Psi'_a \right) = \mathrm{tr}\left(\frac{\partial Z^T_b}{\partial q^i} \frac{\partial Z_b}{\partial q^j} \Psi_b \right) = (w_b)^i_j$$

for each $i, j = 1, \cdots, n$.

We call this real parameter function $D(\cdot)$ a **deformer**. Therefore, to respond to the model reduction challenge, the most promising approach is to expect the minimum embedding ζ_{nk} of the C-manifold M_{nk} to be isometrizable to the combined C-manifold M_c through the above deformation with a certain deformer. However, M_c has been isometrically embedded into $\mathbb{R}^{9n}$, and may possess more geometrical details than M_{nk} that is embedded into a smaller Euclidean space $\mathbb{R}^{9k}$ with $k < n$. In other words, $\mathbb{R}^{9k}$ may not be spatial enough to allow M_{nk} to be isometrizable to M_c even though topologically $M_{nk} \simeq M_c$.

Nevertheless, we are dealing with a special open serial-chain robot case. As commonly experienced, when trying to derive a symbolical form of kinetic energy K_i for link i of an n-joint serial-chain robotic system with $(1 < i \leq n)$, we must take into account every velocity of both translation and rotation of link $i - 1$, in addition to all the terms contributed by the motion of link i itself, because the latter is often imposed by the former. Therefore, the sum K_{nk} of all the kinetic energies K_i's for the k links involved in the C-manifold M_{nk} that is associated with the minimum embedding ζ_{nk} should be able to cover every factor contained in the total kinematic energy $K = \sum_{i=1}^n K_i$ of the robot. In other words, we can always adjust each dynamic parameter in K_{nk} to make K_{nk} equal to K. Because the coefficient of each $\dot{q}^i \dot{q}^j$ in K given by (7.47) is just the metric element w_{ij}, this parameter adjustment is virtually a smooth deformation mentioned in Definition 8. Hence, we reach the following theorem:

Theorem 4. *For an n-joint open serial-chain robot dynamic system, its minimum embeddable C-manifold M_{nk} is diffeomorphic to the combined C-manifold M_c of the entire robotic system. If all the robotic joints are revolute, then, M_{nk} is also isometrizable to M_c.*

This theorem underlies **a fundamental principle of robot dynamic model reduction**. Namely, *the lower-bound of dynamic model reduction in the sense of topology is a subsystem with the minimum embeddable C-manifold for every open serial-chain robotic system.* In other words, the robot dynamic model cannot be further reduced below the lower-bound, otherwise both the embeddability and isometrizability will no longer be guaranteed, and may even cause a catastrophe of topological structure destruction [38, 39].

Theorem 4 indicates that if every joint is revolute for a robotic system, then the minimum embeddable C-manifold M_{nk} is not only diffeomorphic to the combined C-manifold M_c, but is also isometrizable to M_c. If one of the joints is prismatic, M_{nk} is still diffeomorphic to M_c, but M_{nk} may have to further include the prismatic link in order to meet the isometrizability. The reason is that since the variable for a prismatic joint is a length d_i, instead of an angle, as one of the local coordinates, it will often be mixed with the dynamic parameters to be deformed together in the Riemannian metrics $\{w_{ij}\}$, causing unavailability of an effective deformer to distinguish the joint variable d_i from the dynamic parameters.

As a first example, let us look at a well-known inverted pendulum system that consists of only two links: a linearly moving cart and a rotating pole mounted on the cart, as shown in Figure 7.16. With the joint positions x_1 and θ_2 defined as two local coordinates $q = (x_1 \ \theta_2)^T$ for its 2-dimensional combined C-manifold, we can readily find its inertial matrix W by extracting all the coefficients of $\dot{q}^i\dot{q}^j$ terms for $i, j = 1, 2$ from its kinetic energy K, i.e.,

$$W = \begin{pmatrix} m_1 + m_2 & -m_2 l_{c2} s_2 \\ -m_2 l_{c2} s_2 & m_2 l_{c2}^2 + I_2 \end{pmatrix}, \tag{7.58}$$

where m_1 and m_2 are the masses of the cart and the top pole as link 1 and link 2, respectively, l_{c2} is the length between the revolute joint axis and the mass center of the pole, as link 2, and I_2 is the moment of inertia of link 2.

It can be directly seen that if link 2 is fixed, the cart (link 1) will have no chance to move. Thus, link 2 determines the minimum embeddable C-manifold M_{nk} of the system with $n = 2$ and $k = 1$. For this planar system, the position vector of the mass center of link 2 with respect to the base is

$$p_0^{c2} = \begin{pmatrix} a x_1 + b c_2 \\ b s_2 \end{pmatrix},$$

where a and b are two dynamic parameters of link 2. Since this pole rotates only about one single axis, we can simply use $h\theta_2$ with a new parameter h to represent the rotation of link 2 under $0 \le \theta_2 < 2\pi$. Therefore, the smooth embedding for M_{nk} can be written as

$$\begin{cases} z_2^1 = a x_1 + b c_2 \\ z_2^2 = b s_2 \\ z_2^3 = h\theta_2. \end{cases} \tag{7.59}$$

The Jacobian matrix of this minimum embedding $z_2 = \zeta_2(q)$ becomes

$$J_2 = \frac{\partial \zeta}{\partial q} = \begin{pmatrix} a & -b s_2 \\ 0 & b c_2 \\ 0 & h \end{pmatrix},$$

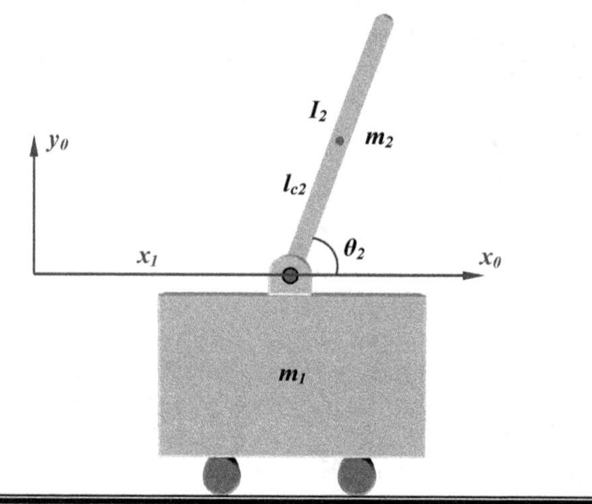

Fig. 7.16 An inverted pendulum system

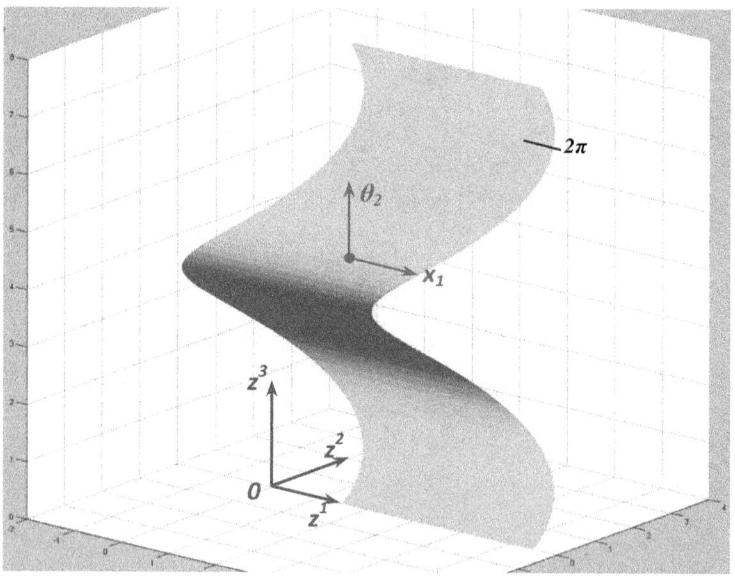

Fig. 7.17 The minimum embeddable C-manifold of the inverted pendulum system

the rank of which is obviously $\text{rank}(J) = 2$, as long as both $a \neq 0$ and $h \neq 0$. Its Riemannian metric turns out to be

$$W_2 = J_2^T J_2 = \begin{pmatrix} a^2 & -abs_2 \\ -abs_2 & b^2 + h^2 \end{pmatrix}.$$

Comparing this with the inertial matrix W in (7.58), we can see that if

$$a = \sqrt{m_1 + m_2}, \quad b = \frac{m_2 l_{c2}}{\sqrt{m_1 + m_2}} \quad \text{and} \quad h = \sqrt{\frac{m_1 m_2}{m_1 + m_2} l_{c2}^2 + I_2}, \tag{7.60}$$

then $W_2 = W$. This demonstrates that the minimum embeddable C-manifold given by (7.59) is isometrizable to the combined C-manifold M_c of the inverted pendulum system, and equation (7.60) is just its deformer $D(\Psi)$ during the isometrization process.

Figure 7.17 visualizes the minimum embeddable C-manifold M_{21} in $\mathbb{R}^3$ based on (7.59) when $a = 4$, $b = 1.5$ and $h = 1$ in MATLAB^{TM}. This 2-surface is diffeomorphic to the compact cylindrical surface $S^1 \times I^1$ [28, 29] after θ_2 in z_2^3 is confined within $[0, 2\pi)$ by a quotient operation $\mathbb{R}^1/\mathbb{Z}$, where $\mathbb{Z}$ is the integer additive group. In other words, once the z_2^3 component of the 2-surface reaches 2π, it should be imagined to return immediately back to the zero and start over again.

The second example is a three-revolute-joint (RRR-type) planar arm, as shown in Figure 7.18. According to equation (7.54), the last link $i = n = 3$ has a 3-dimensional C-manifold M_3 that can be sent to $\mathbb{R}^3$ by

$$\begin{cases} z_3^1 = ac_1 + bc_{12} + dc_{123} \\ z_3^2 = as_1 + bs_{12} + ds_{123} \\ z_3^3 = h(\theta_1 + \theta_2 + \theta_3) \end{cases} \tag{7.61}$$

with four non-zero parameters a, b, d and h, where $s_{ij} = \sin(\theta_i + \theta_j)$, $c_{ij} = \cos(\theta_i + \theta_j)$, $s_{ijk} = \sin(\theta_i + \theta_j + \theta_k)$ and $c_{ijk} = \cos(\theta_i + \theta_j + \theta_k)$ for $i, j, k = 1, 2, 3$. Its Jacobian matrix becomes

$$J_3 = \frac{\partial \zeta_3}{\partial q} = \begin{pmatrix} -as_1 - bs_{12} - ds_{123} & -bs_{12} - ds_{123} & -ds_{123} \\ ac_1 + bc_{12} + dc_{123} & bc_{12} + dc_{123} & dc_{123} \\ h & h & h \end{pmatrix}. \tag{7.62}$$

The determinant of this 3 by 3 Jacobian matrix is calculated as $\det J_3 = abhs_2 = abh \sin \theta_2$. This shows that M_3 is singular at $\theta_2 = 0$ or $\pm\pi$. In fact, we can see that for each fixed $z_3 = (z_3^1\ z_3^2\ z_3^3)^T \in \mathbb{R}^3$, θ_1 through θ_3 can have two different sets of I-K solution, as long as they keep the sum $\theta_1 + \theta_2 + \theta_3$ fixed while one uses θ_2 and the other one takes $-\theta_2$, as shown in Figure 7.18. This is a typical multi-configuration phenomenon.

We may thus construct a minimum embeddable C-manifold by adding link 1, which will clearly make the above multi-configuration disappear. Since

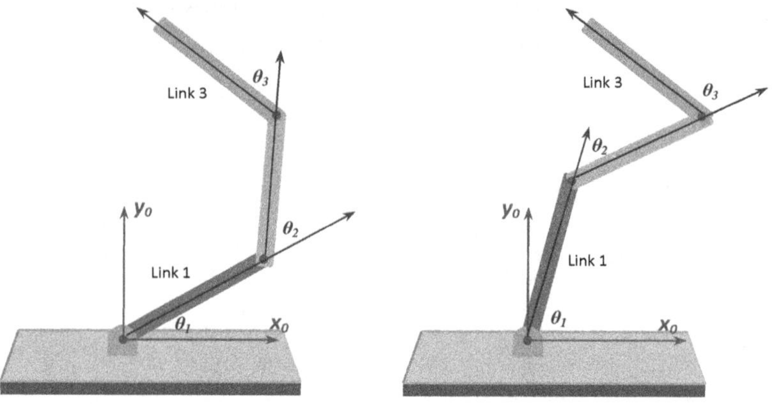

Fig. 7.18 An RRR-type planar robot and its multi-configuration

link 1 has only a single-axis rotation, we can augment (7.61) by $z_4 = h_1\theta_1$ so that a possible minimum embedding would be

$$\begin{cases} z^1 = ac_1 + bc_{12} + dc_{123} \\ z^2 = as_1 + bs_{12} + ds_{123} \\ z^3 = h(\theta_1 + \theta_2 + \theta_3) \\ z^4 = h_1\theta_1 \end{cases} \tag{7.63}$$

with a new non-zero parameter h_1.

It can be easily verified that the rank of the following new 4 by 3 Jacobian matrix J is always full:

$$J = \frac{\partial \zeta_{32}}{\partial q} = \begin{pmatrix} -as_1 - bs_{12} - ds_{123} & -bs_{12} - ds_{123} & -ds_{123} \\ ac_1 + bc_{12} + dc_{123} & bc_{12} + dc_{123} & dc_{123} \\ h & h & h \\ h_1 & 0 & 0 \end{pmatrix},$$

and the set of all four equations in (7.63) is also one-to-one. Thus, it shows that equation (7.63) is a minimum embedding with $n = 3$ and $k = 2$.

The Riemannian metric W_{32} on the minimum embeddable C-manifold can be found by

$$W_{32} = J^T J = \begin{pmatrix} w_{11} & w_{12} & w_{13} \\ w_{12} & b^2 + 2bdc_3 + d^2 + h^2 & bdc_3 + d^2 + h^2 \\ w_{13} & bdc_3 + d^2 + h^2 & d^2 + h^2 \end{pmatrix},$$

where

$$w_{11} = a^2 + b^2 + 2abc_2 + 2adc_{23} + 2bdc_3 + d^2 + h^2 + h_1^2,$$

and

$$w_{12} = b^2 + abc_2 + adc_{23} + 2bdc_3 + d^2 + h^2, \quad w_{13} = adc_{23} + bdc_3 + d^2 + h^2.$$

One can also verify without difficulty that by adjusting all the parameters a, b, d, h and h_1, the metric $W_{32} = J^T J$ can be equal to the inertial matrix W of this RRR-type planar robot. Such a parameter adjustment just plays the role of deformer $D(\Psi)$ in the C-manifold isometrization.

In contrast to the above RRR-type planar robot arm, let us revisit the RPR-type planar robot, as given in Figure 7.7 or 7.9. At first glance, it seems sufficient enough to pick up only the last link to represent the minimum embeddable C-manifold for this particular robot, because the configuration becomes unique whenever the last link is motionless. If so in this case, $n = 3$ and $k = 1$. Based on the previous kinematics analysis of this planar robot, the tip position vector and the position vector of frame 2 can be found as

$$p_0^t = \begin{pmatrix} d_2 s_1 + d_4 s_{13} \\ -d_2 c_1 - d_4 c_{13} \\ 0 \end{pmatrix}, \quad \text{and} \quad p_0^2 = \begin{pmatrix} d_2 s_1 \\ -d_2 c_1 \\ 0 \end{pmatrix}$$

where d_4 is the length of the last link. Since the z-component of p_0^t is zero due to the planar arm, while the number of joints is three, we must replace this zero by the orientation of the last link, which can be uniquely determined by $\theta_1 + \theta_3$. Thus, the proposed embedding is given as follows:

$$\begin{cases} z_3^1 = ad_2 s_1 + bs_{13} \\ z_3^2 = -ad_2 c_1 - bc_{13} \\ z_3^3 = h(\theta_1 + \theta_3) \end{cases} \tag{7.64}$$

with parameters a, b and h to be adjusted. Then, the Jacobian matrix becomes

$$J_3 = \frac{\partial \zeta_3}{\partial q} = \begin{pmatrix} ad_2 c_1 + bc_{13} & as_1 & bc_{13} \\ ad_2 s_1 + bs_{13} & -ac_1 & bs_{13} \\ h & 0 & h \end{pmatrix}.$$

Its determinant is $\det(J) = -a^2 h d_2$ and is never zero as long as the prismatic joint value $d_2 \neq 0$.

Now, the Riemannian metric for this minimum embeddable C-manifold M_{31} turns out to be

$$W_{31} = J_3^T J_3 = \begin{pmatrix} a^2 d_2 + b^2 + h^2 + 2abd_2 c_3 & -abs_3 & abd_2 c_3 + b^2 + h^2 \\ -abs_3 & a^2 & -abs_3 \\ abd_2 c_3 + b^2 + h^2 & -abs_3 & b^2 + h^2 \end{pmatrix}.$$

Comparing this with the inertial matrix W of this RPR arm derived in equation (7.23), we find that if

$$a^2 = m_2 + m_3, \quad ab = m_3 l_{c3}, \quad \text{and} \quad b^2 + h^2 = I_3 + m_3 l_{c3}^2,$$

every element w_{ij} in W can be matched by W_{31} except the first one w_{11} that contains a mixed term $m_2(d_2 - l_{c2})^2$ between the prismatic joint variable d_2 and the dynamic parameters. Therefore, while equation (7.64) is an embedding, it has not been isometrizable yet to exactly match the inertial matrix W.

This phenomenon reveals a special complication due to the prismatic joint in isometrization. By further augmenting the second prismatic link, the new embedding becomes:

$$\begin{cases} z_3^1 = a_1 d_2 s_1 + b_1 s_{13} \\ z_3^2 = -a_1 d_2 c_1 - b_1 c_{13} \\ z_3^3 = h_1(\theta_1 + \theta_3) \\ z_3^4 = a_2(d_2 - b_2)s_1 \\ z_3^5 = -a_2(d_2 - b_2)c_1 \\ z_3^6 = h_2\theta_1 \end{cases} \tag{7.65}$$

along with six parameters a_1, b_1, h_1, a_2, b_2 and h_2 to be adjusted, where $d_2 - b_2$ indicates a distance to converge to the mass center of link 2.

The Jacobian matrix of the new proposed embedding (7.65) can be derived as

$$J_3 = \frac{\partial \zeta_3}{\partial q} = \begin{pmatrix} a_1 d_2 c_1 + b_1 c_{13} & a_1 s_1 & b_1 c_{13} \\ a_1 d_2 s_1 + b_1 s_{13} & -a_1 c_1 & b_1 s_{13} \\ h_1 & 0 & h_1 \\ a_2(d_2 - b_2)c_1 & a_2 s_1 & 0 \\ a_2(d_2 - b_2)s_1 & -a_2 c_1 & 0 \\ h_2 & 0 & 0 \end{pmatrix}.$$

Thus, the Riemannian metric becomes

$$W_{32} = J_3^T J_3 = \begin{pmatrix} w_{11} & -a_1 b_1 s_3 & a_1 b_1 d_2 c_3 + b_1^2 + h_1^2 \\ -a_1 b_1 s_3 & a_1^2 + a_2^2 & -a_1 b_1 s_3 \\ a_1 b_1 d_2 c_3 + b_1^2 + h_1^2 & -a_1 b_1 s_3 & b_1^2 + h_1^2 \end{pmatrix},$$

where $w_{11} = a_1^2 d_2^2 + b_1^2 + h_1^2 + 2a_1 b_1 d_2 c_3 + a_2^2(d_2 - b_2)^2 + h_2^2$.

In comparison with equation (7.23) again, the new Riemannian metric W_{32} has now sufficient factors to match the inertial matrix W. Namely, let

$$a_1 = \sqrt{m_3}, \quad a_2 = \sqrt{m_2}, \quad b_1 = \sqrt{m_3} l_{c3}, \quad h_1 = \sqrt{I_3},$$

and

$$b_2 = l_{c2}, \quad h_2 = \sqrt{I_1 + I_2 + m_1 l_{c1}^2}.$$

Then every element w_{ij} inside the new metric W_{32} can exactly match with the inertial matrix W in (7.23).

The last example is to dynamically model a general fully parallel-chain mechanism, such as a 6-6 Stewart platform, as studied in [36, 37]. If we focus on the top mobile disc, according to equation (7.25), its kinetic energy can be written as follows:

$$K = \frac{1}{2} m v_0^{6T} v_0^6 + m v_0^{6T} C_6^T \omega_0^6 + \frac{1}{2} \omega_0^{6T} \Gamma_6 \omega_0^6,$$

where m and Γ_6 are the mass and inertia tensor with respect to frame 6 on the top disc, respectively, and C_6 is the skew-symmetric matrix of the mass center coordinates with respect to frame 6. If the origin of frame 6 is defined at the mass center of the top mobile disc, the second term of the kinetic energy K vanishes and $\Gamma_6 = \Gamma_c$.

Because the top disc is a single rigid body, Theorem 3 can be directly applied to it. Then, an isometric embedding that can send the 6-dimensional C-manifold M_6^6 of the top disc motion to Euclidean 9-space $\mathbb{R}^9$ is given by

$$z = \zeta(q) = Z\xi = \begin{pmatrix} R_0^6 & p_0^6 & & O \\ & & R_0^6 & \\ O & & & R_0^6 \end{pmatrix} \xi \tag{7.66}$$

with a 10 by 1 dynamic parameter vector ξ based on equation (7.54).

In order to find the Riemannian metric $W = J^T J$ endowed on the C-manifold M_6 for the top mobile disc of the 6-6 Stewart platform, it is required to know a local coordinate system q to be defined on M_6. If we pick the 6 prismatic joint lengths $q_i = l_i$'s to form the local coordinate system, then intuitively, the top disc will have no multi-configuration chance so that the mapping (7.66) is a minimum embedding for the C-manifold M_c of the entire Stewart platform. The Jacobian matrix of the minimum embedding should be $J = \partial\zeta/\partial q$. However, this turns back to the forward kinematics (F-K) problem. Namely, it is difficult or even impossible to find an explicit closed form for either R_0^6 or p_0^6 as a function of $q = (q_1 \; \cdots \; q_6)^T = (l_1 \; \cdots \; l_6)^T$ before taking derivatives to determine J and then W.

If we define a local coordinate system alternatively other than the 6 piston lengths, it may be possible to reach a Riemannian metric result. For instance, let the 6 local coordinates be defined by the variables in Cartesian space:

$$q^c = \begin{pmatrix} x \\ y \\ z \\ \phi \\ \theta \\ \psi \end{pmatrix} = \begin{pmatrix} p_0^6 \\ \rho_0^6 \end{pmatrix}, \tag{7.67}$$

where x, y and z are the three coordinate components of p_0^6, and ϕ, θ and ψ are the roll, pitch and yaw Euler angles to represent R_0^6, see equation (3.7) in Chapter 3. Namely,

$$p_0^6 = \begin{pmatrix} x \\ y \\ z \end{pmatrix}, \quad R_0^6 = \begin{pmatrix} c\phi c\theta & -s\phi c\psi + c\phi s\theta s\psi & s\phi s\psi + c\phi s\theta c\psi \\ s\phi c\theta & c\phi c\psi + s\phi s\theta s\psi & -c\phi s\psi + s\phi s\theta c\psi \\ -s\theta & c\theta s\psi & c\theta c\psi \end{pmatrix}.$$

Then, the 9 by 1 column of the embedding $z = \zeta(q^c) = Z\xi$ in (7.66) after the structure matrix Z is linearly combined by the dynamic parameters in ξ can be expressed explicitly in terms of the above 6 new local coordinates. The Jacobian matrix $J = \partial\zeta/\partial q^c$ as well as the Riemannian metric $W = J^T J$ can also be determined successfully.

However, under the new local coordinate system (7.67), due to the non-uniqueness between the roll, pitch and yaw Euler angles and the rotation matrix, see the example in Chapter 3, there will be at least one multi-configuration case. In other words, the mapping $z = \zeta(q^c)$ will no longer be a minimum embedding. Even if it could be remedied by imposing a constraint on the three Euler angles, once the Riemannian metric W on the C-manifold is determined, by substituting it into the Lagrange equation (7.15), the control input τ_c will no longer be the desired six piston forces. Instead, it should be a some Cartesian force vector in correspondence to the new local coordinates defined in Cartesian space and given by (7.67), which requires a conversion.

Naturally, we may take advantage of the statics equation (5.30) from Chapter 5 for the Stewart platform to convert the control input τ^c that is resolved by the Lagrange equation in Cartesian space to a piston joint force vector by $\tau = J_0^{-1} F_0$. To do so, according to equation (3.12) in Chapter 3, the angular velocity of the top disc is

$$\omega_0^6 = \begin{pmatrix} 0 & -s\phi & c\phi c\theta \\ 0 & c\phi & s\phi c\theta \\ 1 & 0 & -s\theta \end{pmatrix} \begin{pmatrix} \dot{\phi} \\ \dot{\theta} \\ \dot{\psi} \end{pmatrix} = D\dot{\rho}_0^6.$$

Hence, the 6 by 1 Cartesian velocity vector becomes

$$V_0 = \begin{pmatrix} v_0^6 \\ \omega_0^6 \end{pmatrix} = \begin{pmatrix} I & O \\ O & D \end{pmatrix} \begin{pmatrix} \dot{p}_0^6 \\ \dot{\rho}_0^6 \end{pmatrix} = B\dot{q}^c,$$

where the 6 by 6 coefficient matrix is denoted by B, and I and O are the 3 by 3 identity and zero matrix, respectively.

Based on the Jacobian equation (5.29) in Chapter 5 for the Stewart platform,

$$\dot{q} = J_0^T V_0 = J_0^T B\dot{q}^c. \tag{7.68}$$

This arrives at a kinematic conversion in tangent space between the prismatic joint position vector $q = (l_1 \ \cdots \ l_6)^T$ and the local coordinate system q^c defined by (7.67).

Furthermore, the mechanical power seen in joint space is $P = \dot{q}^T \tau$, where $\tau = (f_1 \ \cdots \ f_6)^T$ is the piston actuating force vector, while the same power seen in Cartesian space is $P = \dot{q}^{cT} \tau^c$ for a Cartesian force τ^c that is corresponding to the local coordinates in (7.67). Based on the principle of energy conservation, $\dot{q}^T \tau = \dot{q}^{cT} \tau^c$. Substituting (7.68) into here, we have $\dot{q}^{cT} B^T J_0 \tau = \dot{q}^{cT} \tau^c$ so that

$$\tau^c = B^T J_0 \tau. \tag{7.69}$$

This new statics equation is similar to (5.30) in Chapter 5. Thus, if a control law τ^c can be resolved through the Lagrange equation (7.15), then the piston actuating forces can be determined by the new statics equation (7.69), i.e.,

$$\tau = J_0^{-1} B^{-T} \tau^c,$$

provided that both the Jacobian matrix J_0 and the matrix B are non-singular.

Therefore, the conversion issue between τ^c and τ has been solved by the statics in (7.69), but the one-to-one, or multi-configuration issue still remains unfixed as a barrier of achieving the minimum isometrizable embedding.

In conclusion, if we insist in using the 6 piston lengths to define a local coordinate system on the C-manifold M_c of the Stewart platform, we will not be able to continue its Riemannian metric determination towards a successful adaptive control design until the F-K problem for the closed parallel-chain systems is resolved.

Finally, it should be pointed out that all the parameter adjustments in the above examples are just to illustrate the possibility of isometrization, and are not required to do so by hand at all. Instead, the computer program will automatically adjust the model parameters towards the isometrization if an effective adaptive control algorithm is implemented. The detailed introduction and discussion on adaptive control as well as a 3D robotic system example will be given in the next chapter.

7.6 A Compact Dynamic Equation

In this section, we continue to study and answer how the theory of embedding and isometrization for a C-manifold can reduce the dynamic formulation and benefit developing and realizing an adaptive control strategy for robotic systems.

Let the n-dimensional C-manifold of a dynamic system be smoothly and isometrically embedded into a Euclidean m-space ($m > n$) by $z = \zeta(q)$. The kinetic energy K of the system is given by (7.47), which obeys the following Lagrange equation [1, 2, 25]:

$$\frac{d}{dt}\left(\frac{\partial K}{\partial \dot{q}}\right) - \frac{\partial K}{\partial q} = \tau, \tag{7.70}$$

where $\tau = (\tau_1 \ \cdots \ \tau_n)^T$ is a torque vector over all the n local coordinates and covers all the possible external torques, including the conservative torque $\tau_g = -\partial P/\partial q$ due to gravity and the other external driving torques/forces τ_{ex}. Basically, we can write

$$\tau = \tau_g + \tau_{ex}$$

for the right-hand side of (7.70).

Theorem 5. *Let M^n be the n-dimensional smooth C-manifold of a dynamic system that obeys the Lagrange equation (7.70), and be smoothly and isometrically embedded into a Euclidean m-space ($m > n$) by $z = \zeta(q)$, where $z \in \mathbb{R}^m$ and q is a local coordinate system on M^n. If $J = \partial\zeta/\partial q$ is the Jacobian matrix of the embedding, then*

$$J^T \ddot{z} = \tau \tag{7.71}$$

is equivalent to (7.70).

This theorem can be directly shown by substituting $K = \frac{1}{2}\dot{q}^T W \dot{q} = \frac{1}{2}\dot{q}^T J^T J \dot{q}$ into (7.70) and noticing that J is a genuine mathematical Jacobian matrix so that

$$\dot{J} = \frac{d}{dt}\frac{\partial \zeta}{\partial q} = \frac{\partial \dot{\zeta}}{\partial q}.$$

In fact, since the Jacobian matrix of the embedding is independent of $\dot{q}$,

$$\frac{\partial K}{\partial \dot{q}} = J^T J \dot{q}.$$

Thus,

$$\frac{d}{dt}\left(\frac{\partial K}{\partial \dot{q}}\right) = J^T J \ddot{q} + J^T \dot{J} \dot{q} + \dot{J}^T J \dot{q}.$$

On the other hand, since $K = \frac{1}{2}\dot{z}^T \dot{z} = \frac{1}{2}\dot{\zeta}^T \dot{\zeta}$ after the C-manifold M^n is embedded into the Euclidean z-space $\mathbb{R}^m$,

$$\frac{\partial K}{\partial q} = \left(\frac{\partial \dot{\zeta}}{\partial q}\right)^T \dot{\zeta} = \dot{J}^T \dot{\zeta} = \dot{J}^T J \dot{q}.$$

Subtracting the above two equations yields the left-hand side of the Lagrange equation (7.70), i.e.,

$$\frac{d}{dt}\left(\frac{\partial K}{\partial \dot{q}}\right) - \frac{\partial K}{\partial q} = J^T J \ddot{q} + J^T \dot{J} \dot{q} = J^T \ddot{z},$$

which should be equal to the total external torque vector τ.

To further interpret this compact dynamic equation, let us look at the kinetic energy K and its time-derivative $\dot{K}$ that should be the kinetic power of the dynamic system. Since $K = \frac{1}{2}\dot{z}^T \dot{z}$ after M^n is embedded into the Euclidean z-space, by noticing $\dot{z} = J\dot{q}$, the kinetic power becomes $\dot{K} = \dot{z}^T \ddot{z} = \dot{q}^T J^T \ddot{z}$. On the other hand, at the local coordinate system standpoint, the power should be $\dot{K} = \dot{q}^T \tau$, where τ is the right-hand side of the Lagrange equation (7.70). Comparing these two forms of $\dot{K}$ seen in two different coordinate systems but representing the same energy rate of the system, we conclude that (7.71) holds and is equivalent to (7.70).

Equation (7.71) looks like a robotic statics equation (4.30) from Chapter 4 if $\ddot{z}$ is considered as a Cartesian force F. In fact, it should not be surprising that any complex dynamics originally represented in non-Euclidean space will be seen in appearance as simple as a Newton second law if it can be embedded into a Euclidean space.

Moreover, since $\dot{z} = J\dot{q}$ and $\ddot{z} = J\ddot{q} + \dot{J}\dot{q}$, substituting them into equation (7.71), we obtain

$$J^T J\ddot{q} + J^T \dot{J}\dot{q} = W\ddot{q} + J^T \dot{J}\dot{q} = \tau.$$

Comparing this with the robotic dynamic equation in (7.19), we can immediately see that

$$J^T \dot{J}\dot{q} = \left(W_d^T - \frac{1}{2}W_d\right)\dot{q},$$

which is just the centrifugal and Coriolis term but expressed in terms of the Jacobian matrix of the embedding. Therefore, if one can find an embedding for a robotic system C-manifold, its dynamic formulation will immediately become more compact and short [23].

In summary, the three approaches to modeling the dynamics for robotic systems have been introduced and discussed in a fair amount of details. Each of them has own advantage and weakness in terms of their applications, as highlighted in Table 7.2. More specifically, according to the comparison in Table 7.2,

1. The first approach can be applied, in addition to most open serial or hybrid-chain robotic systems, also for a number of relatively simple closed parallel-chain robots, though the forward kinematics (F-K) of the latter is a big challenge that may bring certain difficulty to the symbolical development of their kinetic energy formulation. Unfortunately, the second approach using equation (7.30) or (7.32) is almost not possible to handle dynamic modeling for the closed parallel-chain robots, mainly because it requires to determine every sub-Jacobian from the corresponding subrobot in sequence, which is unavailable in the closed parallel-chain robot cases. In contrast, for the third approach, the top platform (disc) of either a hexapod or a Delta closed parallel-chain system can be considered as the "busiest" link so that its C-manifold can represent the system dynamics, and can also be embedded into Euclidean 9-space $\mathbb{R}^9$ and isometrized to meet its inertial matrix W based on Theorem 3.
2. While the third approach offers a compact dynamic formulation, all the dynamic parameters involved have to wait for isometrization. Nevertheless, it is a recognizable fact that the dynamic modeling for almost every multi-body mechanical system, such as a robot manipulator, has a common issue of parameter uncertainty. For instance, one may possibly measure a mass for each link of a robot, but it is difficult to exactly find where the mass center is and what each moment of inertia involved in the inertia tensor will be. Thus, it actually becomes a reality that every dynamic parameter needs

to be adapted, not exactly measured. Therefore, the third approach of C-manifold and its isometric embedding will be able to provide the adaptive control of robotic systems with the best modeling and formulation tools. We will address this issue and develop a robotic adaptive control strategy based on the theory of C-manifold and embedding in more details in the next chapter.

3. The second approach with equation (7.30) or (7.32) is thus far the best approach for numerical solutions, especially for a large-scale robotic system, such as to model the dynamics for a digital human. In addition to the numerical computation of the inertial matrix W, with a possible dual-number operational algorithm built-in MATLAB^{TM}, the derivative matrix W_d would also be calculated so that all the centrifugal and Coriolis terms could be accurately determined numerically in the computer, no matter how large the robotic system scale is. This advantage will be demonstrated and further confirmed in Chapter 11 of digital human dynamic modeling.

Table 7.2 A comparative study of the dynamic modeling approaches

Approaches	Eqn's	Weakness	Applications
Directly find K via geometric analysis of link velocities	(7.23)	Limited in planar robots or with smaller number of joints	Planar or $n \leq 3$ 3D robots dynamic modeling
The algorithm of sub-Jacobians with subrobots	(7.30) (7.32)	Hard to extend to closed parallel-chain robotic systems	Best for numerical solutions to larger-scale robot dyn. modeling
The C-manifold isometric embedding	(7.51) (7.55) (7.66) (7.71)	Dynamic parameters are undetermined, but are ready for adaptation	Mid-scale robot dynamic modeling for adaptive control and extendable to parallel-chain mechanisms

7.7 Exercises of the Chapter

1. Determine an inertia tensor Γ_c with respect to the mass center for each of the two rigid bodies: a solid cylinder with the height h and radius r, and an ellipsoid with the semi-axes a, b and c, if both objects have a uniform mass density and the same mass m.
2. A two-joint robotic system has the following kinetic energies: $K_1 = \frac{1}{2}m_1 a^2\dot{\theta}_1^2 + \frac{1}{2}J_1\dot{\theta}_1^2$ and $K_2 = \frac{1}{2}m_2(2a\dot{\theta}_1 + \dot{d}_2)^2 + \frac{1}{2}J_2\dot{\theta}_1^2$. The total potential energy is $P = (m_1 + m_2)ga\cos\theta_1$, where θ_1 and d_2 are two joint positions of the robot, and m_1, m_2, a, J_1 and J_2 are all given constant dynamic parameters, and g is the gravitational acceleration. If a torque τ_1 and a force f_2, as the system inputs, are acting on joints 1 and 2, respectively, derive a dynamic model of the robot through the Lagrange equation, i.e., find the inertial matrix W, the centrifugal and Coriolis forces/torques $c(q, \dot{q})$, and the input vector u.

3. A solid ball with a mass m and radius r is grasped by a 3-joint RRP type robot gripper. The position and orientation of the robot gripper frame are given by equations (4.17) and (4.18) in Chapter 4, which are the functions of the robotic joint variables θ_1, θ_2 and d_3. Find the kinetic energy K_{ball} and gravitational potential energy P_{ball} for the ball as well as its inertial matrix W_{ball} in terms of the robotic joint variables.
4. Using the kinematic modeling procedures given in Chapters 5 and 6 for the 7-joint redundant robot, as shown in Figure 6.9, write a MATLAB^{TM} program to calculate the inertial matrix W of the robot by following equation (7.30) or (7.32). Appropriately define dynamic parameters for each link of the robot.
5. A 2-joint prismatic-revolute planar robot is given in Figure 4.21 from Chapter 4. All the parameters are defined by $a_1 = 0.2$, $a_2 = 0.8$ in meter, and the mass of link 1 is $m_1 = 2.5$ Kg., and the mass and moment of inertia of link 2 are $m_2 = 2$ Kg. and $I_2 = 0.1$ Kg-m^2. w.r.t. the mass center, respectively. Suppose that the mass center is same as the centroid of each link.

   a. Derive all the sub-Jacobian matrices J_j and the dynamic parameter matrices U_j for $j = 1, 2$;
   b. Using equation (7.30) to find the inertial matrix W of the robot;
   c. Also, calculate the derivative matrix W_d;
   d. Find a minimum embedding $z = \zeta(q)$ for the C-manifold of the robot;
   e. Determine its Riemannian metric $W = J^T J$ and compare it with the above inertial matrix W.

6. Follow the same questions as in the last problem to study the 3-joint RPR robot given in Figure 4.22 from Chapter 4. Appropriately define your own kinematic and dynamic parameters. If the symbolical derivation is lengthy, you may use MATLAB^{TM} for a numerical solution.
7. For a general 6-6 Stewart platform, let equation (7.67) be the local coordinate system on the C-manifold M_6 of the top mobile disc. Then, the mapping (7.66) will be the possible minimum embedding that sends the combined C-manifold M_c of the entire system to $\mathbb{R}^9$.

   a. By defining 10 unknown parameters $\xi = (\xi_1 \ \cdots \ \xi_{10})^T$, derive the Jacobian matrix J of the minimum embedding $z = Z\xi$ in symbolical form;
   b. Calculate the Riemannian metric $W = J^T J$ for the 6-6 Stewart platform.

References

1. Arnold, V.: Mathematical Methods of Classical Mechanics. Springer, New York (1978)
2. Abraham, R., Marsden, J.: Foundations of Mechanics. The Benjamin/ Cummings Publishing Company (1978)

3. Marsden, J., Ratiu, T.: Introduction to Mechanics and Symmetry. Springer, New York (1994)
4. Asada, H., Slotine, J.: Robot Analysis and Control. John Wiley & Sons, New York (1986)
5. Fu, K., Gonzalez, R., Lee, C.: Robotics: Control, Sensing, Vision, and Intelligence. McGraw-Hill (1987)
6. Spong, M., Vidyasagar, M.: Robot Dynamics and Control. John Wiley & Sons, New York (1989)
7. Murray, R., Li, Z., Sastry, S.: A Mathematical Introduction to Robotic Manipulation. CRC Press, Boca Raton (1994)
8. Brockett, R.: Robotic Manipulators and the Product of Exponentials Formula. In: Fuhrman, P. (ed.) Mathematical Theory of Networks and Systems. LNCIS, vol. 58, pp. 120–129. Springer, Heidelberg (1984)
9. Brockett, R.: Some Mathematical Aspects of Robotics. In: Robotics. AMS Short Course Lecture Notes, vol. 41. American Mathematical Society, Providence (1990)
10. Brockett, R., Stokes, A., Park, F.: A Geometrical Formulation of the Dynamic Equations Describing Kinematic Chains. In: Proc. IEEE Int. Conf. Robotics and Automation, Atlanta, GA, pp. 637–641 (1993)
11. Koditschek, D.: Robot Kinematics and Coordinate Transformations. In: Proc. 1985 IEEE Conf. on Decision and Control, Florida, pp. 1–4 (December 1985)
12. Gu, E., Loh, N.: Dynamic Modeling and Control by Utilizing an Imaginary Robot Model. IEEE Journal of Robotics and Automation 4(5), 532–540 (1988)
13. Gu, E.: Modeling and Simplification for Dynamic Systems with Testing Procedures and Metric Decomposition. In: Proc. 1991 IEEE International Conference on Systems, Man, and Cybernetics, Charlottesville, VA, pp. 487–492 (October 1991)
14. Featherstone, R.: Robot Dynamics Algorithms. Kluwer Publisher, Boston (1987)
15. Rodriguez, G., Jain, A., Kreutz-Delgado, K.: A Spatial Operator Algebra for Manipulator Modeling and Control. International Journal on Robotics Research 10, 371–381 (1991)
16. Rodriguez, G., Kreutz-Delgado, K.: Spatial Operator Factorization and Inversion of the Manipulator Mass Matrix. IEEE Transactions on Robotics and Automation 8, 65–76 (1992)
17. Angeles, J.: The Design of Isotropic Manipulator Architectures in the Presence of Redundancies. International Journal of Robotics Research 11(3), 196–201 (1992)
18. Angeles, J., Lopez-Cajun, C.: Kinematic Isotropy and Conditioning Index of Serial Robotic Manipulators. International Journal of Robotics Research 11(6), 560–571 (1992)
19. Spong, M.: Remarks on Robot Dynamics: Canonical Transformations and Riemannian Geometry. In: Proc. IEEE Int. Conf. Robotics and Automation, Nice, France (1992)
20. Park, F.: Computational Aspects of the Product-of-Exponentials Formula for Robot Kinematics. IEEE Transactions on Automatic Control 39(3), 643–647 (1994)
21. Park, F., Bobrow, J., Ploen, S.: A Lie Group Formulation of Robot Dynamics. The International Journal of Robotics Research 14(6), 609–618 (1995)
22. McCarthy, J.: Introduction to Theoretical Kinematics. MIT Press (1990)

23. Gu, E.: Configuration Manifolds and Their Applications to Robot Dynamic Modeling and Control. IEEE Transactions on Robotics and Automation 16(5), 517–527 (2000)
24. Gu, E.: A Configuration Manifold Embedding Model for Dynamic Control of Redundant Robots. International Journal of Robotics Research 19(3) (March 2000)
25. Landau, L., Lifshitz, E.: Mechanics. Addison-Wesley (1960)
26. Dubrovin, B., Fomenko, A., Novikov, S.: Modern Geometry – Methods and Applications, Part I - The Geometry of Surfaces, Transformation Groups, and Fields. Springer, New York (1992)
27. Dubrovin, B., Fomenko, A., Novikov, S.: Modern Geometry – Methods and Applications, Part II - The Geometry and Topology of Manifolds. Springer, New York (1992)
28. Bredon, G.: Topology and Geometry. Springer, New York (1993)
29. Berger, M., Gostiaux, B.: Differential Geometry: Manifolds, Curves, and Surfaces. Springer, New York (1988)
30. Luh, J., Walker, M., Paul, R.: On-Line Computational Scheme for Mechanical Manipulators. ASME Journal of Dynamic Systems, Measurement and Control 102, 69–76 (1980)
31. Whitney, H.: Differentiable Manifolds. Ann. of Math. 37(2), 645–680 (1936)
32. Warner, F.: Foundations of Differentiable Manifolds and Lie Groups. Springer, New York (1983)
33. Nash, J.: The Embedding Problem for Riemannian Manifolds. Ann. of Math. 63(2), 20–63 (1956)
34. Greene, R.: Isometric Embeddings of Riemannian and Pseudo-Riemannian Manifolds. Memories of the American Mathematical Society (97), iii+63 (1970)
35. Gromov, M.: Partial Differential Relations. Springer, New York (1986)
36. Khalil, W., Ibrahim, O.: General Solution for the Dynamics Modeling of Parallel Robots. Journal of Intelligent and Robotic Systems 49, 19–37 (2007)
37. Merlet, J.: Parallel Robots, 2nd edn. Springer, Dordrecht (2006)
38. Lu, Y.: Singularity Theory and an Introduction to Catastrophe Theory. Springer, New York (1976)
39. Gilmore, R.: Catastrophe Theory for Scientists and Engineers. John Wiley & Sons, New York (1981)

Chapter 8
Control of Robotic Systems

8.1 Path Planning and Trajectory Tracking

Having studied and discussed the modeling, formulations, and procedures of robotic kinematics and dynamics in a great detail, we now turn our attention to the main theme in robotics research and applications: the design considerations and strategies in robotic control systems.

It is quite clear that any motion of a robot arm is driven by part or all of its joint actuators simultaneously. In most industrial robotic systems, each joint is actuated by either a DC or a AC servo-motor. Only in some heavy application cases would a giant robot manipulator be driven by hydraulic mechanism. However, their overall control design considerations and control strategies are basically the same. In whatever circumstances of robotic control, one requirement is mandatory, i.e., each robotic joint velocity must be at least continuous during all the time of motion. In other words, no joint acceleration will go to infinity at any time. Otherwise, the entire robot arm will suffer from a sudden stop if any one of the joint velocities is discontinued, or has an abrupt change. This trouble dead point should always be avoided before the robot starts a movement.

For example, suppose that a robot hand is planned to travel from a starting point to a destination and one of its joints has a position profile $\theta_i(t)$ and a velocity profile $\dot{\theta}_i(t)$, both versus time t, as shown in Figure 8.1. In this case, although $\theta_i(t)$ is always continuous over a certain time period, it has two turning corners at both the starting and ending points so that the joint velocity $\dot{\theta}_i(t)$ will respond two abrupt jumps, either one of which will result in a sudden stop of the motion. In order to remedy it, one has to apply a suitable **spline function** to make the position profile $\theta_i(t)$ be smooth, or differentiable, not just continuous over the time axis [1, 2, 3, 4].

Let the i-th joint position be $\theta_i(t_0) = \theta_0$ at the starting point $t = t_0$ and $\theta_i(t_f) = \theta_f$ at the ending point $t = t_f$. Let the starting and ending joint velocities be $\dot{\theta}_i(t_0) = \dot{\theta}_0$ and $\dot{\theta}_i(t_f) = \dot{\theta}_f$, respectively. If we adopt a **cubic spline function** in the form of

E.Y.L. Gu, *A Journey from Robot to Digital Human*,
Modeling and Optimization in Science and Technologies 1,
DOI: 10.1007/978-3-642-39047-0_8,

$$\theta_i(t) = a_3 t^3 + a_2 t^2 + a_1 t + a_0, \tag{8.1}$$

then, by substituting all the initial and final conditions into it, we have

$$\begin{cases} \theta_0 = a_3 t_0^3 + a_2 t_0^2 + a_1 t_0 + a_0 \\ \theta_f = a_3 t_f^3 + a_2 t_f^2 + a_1 t_f + a_0 \\ \dot{\theta}_0 = 3a_3 t_0^2 + 2a_2 t_0 + a_1 \\ \dot{\theta}_f = 3a_3 t_f^2 + 2a_2 t_f + a_1. \end{cases} \tag{8.2}$$

This simultaneous equation set can be rewritten in the following matrix form:

$$\begin{pmatrix} t_0^3 & t_0^2 & t_0 & 1 \\ t_f^3 & t_f^2 & t_f & 1 \\ 3t_0^2 & 2t_0 & 1 & 0 \\ 3t_f^2 & 2t_f & 1 & 0 \end{pmatrix} \begin{pmatrix} a_3 \\ a_2 \\ a_1 \\ a_0 \end{pmatrix} = \begin{pmatrix} \theta_0 \\ \theta_f \\ \dot{\theta}_0 \\ \dot{\theta}_f \end{pmatrix}. \tag{8.3}$$

By inverting the 4 by 4 *timing matrix* on the left-hand side of (8.3) and pre-multiplying the inverse to the right-hand side of (8.3), we can solve uniquely for all the coefficients a_3, a_2, a_1 and a_0 of the cubic spline function in (8.1), as long as the 4 by 4 timing matrix is invertible (it is true in most cases).

As an example, let $t_0 = 0$, $t_f = 2$ in seconds, $\theta_0 = 0$, $\theta_f = 1$ in radians, and let both $\dot{\theta}_0 = 0$ and $\dot{\theta}_f = 0$. Substituting the values into (8.3) yields

$$\begin{pmatrix} 0 & 0 & 0 & 1 \\ 8 & 4 & 2 & 1 \\ 0 & 0 & 1 & 0 \\ 12 & 4 & 1 & 0 \end{pmatrix} \begin{pmatrix} a_3 \\ a_2 \\ a_1 \\ a_0 \end{pmatrix} = \begin{pmatrix} 0 \\ 1 \\ 0 \\ 0 \end{pmatrix}.$$

Solving the equation, we get $a_3 = -0.25$, $a_2 = 0.75$, $a_1 = 0$ and $a_0 = 0$. Thus, the cubic spline equation becomes $\theta(t) = -0.25t^3 + 0.75t^2$ to meet all the initial and final conditions.

The cubic spline approach clearly offers a very convenient and effective way to make each joint movement smooth without any dead corner. However, since it has to modify the entire trajectory in each segment of the robotic motion due to the cubic-spline alteration on each joint path, the trajectory-tracking accuracy may cause a concern in many robotic application cases. To minimize the path alteration, the second typical spline approach, called the **linear function with parabolic blends**, provides a partial spline effect in each joint path segment.

Let $\theta(0) = \theta_0$ for a joint position at $t = 0$ and $\theta(t_f) = \theta_f$ as the final value of the joint. If both $\dot{\theta}(0) = 0$ and $\dot{\theta}(t_f) = 0$, then we may define a small time length $t_b < \frac{t_f}{2}$ such that the entire joint travel from $t = 0$ to t_f is divided into three segments: the first one is a parabolic curve between $t = 0$ and t_b, the second one is a straight line between $t = t_b$ and $t_f - t_b$ and the last one is a reversed parabolic curve, as shown in Figure 8.2(a). The derivative of such a parabolic-blended linear function for the joint position is a joint velocity

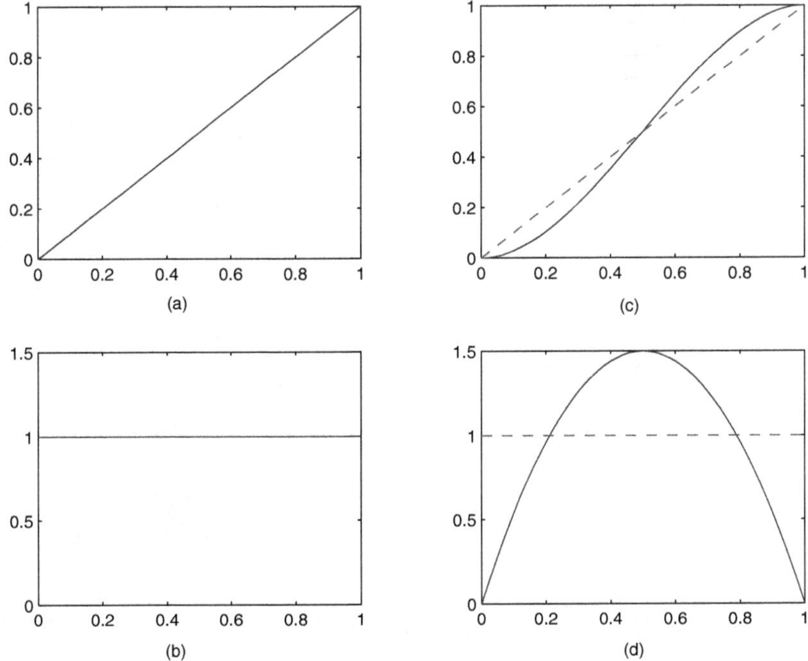

Fig. 8.1 A joint path example without and with cubic spline function

with a trapezoid-shaped profile, as depicted in Figure 8.2(b). For the sake of solution uniqueness, the acceleration a within the two corners of the joint motion is a common constant to be determined. Let the mid-point of the joint angle be θ_h at $t = t_h = \frac{t_f}{2}$. It can be seen from Figure 8.2 that

$$at_b = \dot{\theta}_h = \frac{\theta_h - \theta_b}{t_h - t_b}. \tag{8.4}$$

On the other hand, for this joint motion under the constant acceleration a between $t = 0$ and t_b,

$$\theta_b = \theta_0 + \frac{1}{2}at_b^2. \tag{8.5}$$

Since $\theta_h = \frac{\theta_f + \theta_0}{2}$ and $t_f = 2t_h$, substituting them and also (8.5) into (8.4) yields

$$at_b = \frac{\theta_f + \theta_0 - 2\theta_b}{2t_h - 2t_b} = \frac{\theta_f - \theta_0 - at_b^2}{t_f - 2t_b}.$$

Hence, we reach the following quadratic equation for t_b:

$$at_b^2 - at_f t_b + \Delta\theta = 0, \tag{8.6}$$

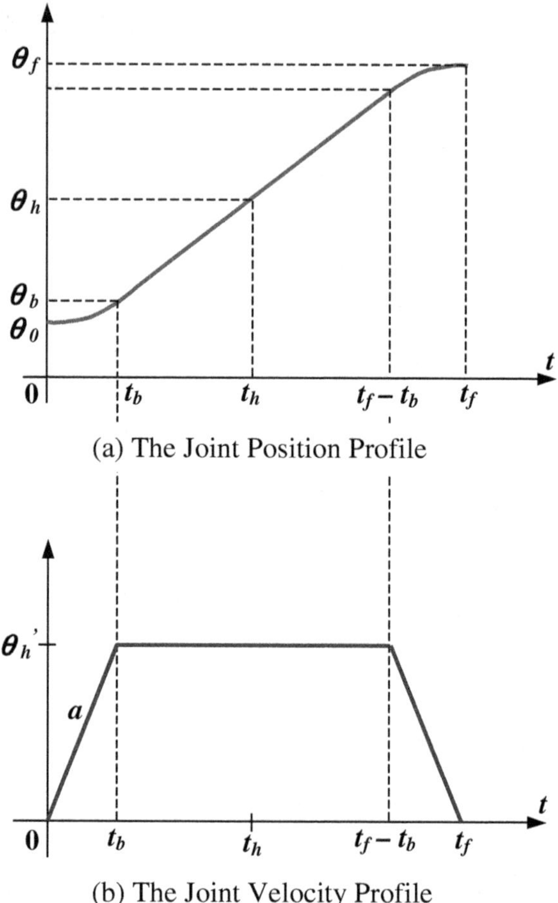

Fig. 8.2 Joint position and velocity profiles for the second spline function

where $\Delta\theta = \theta_f - \theta_0$ is the total joint angle change during the entire travel. The solution of (8.6) clearly becomes

$$t_b = \frac{at_f - \sqrt{a^2 t_f^2 - 4a\Delta\theta}}{2a}, \tag{8.7}$$

where another possible root has been dropped out because it must obey $t_b < \frac{t_f}{2}$.

Since everything inside the square root of (8.7) must be non-negative, we obtain

$$a \geq \frac{4\Delta\theta}{t_f^2} = \frac{4(\theta_f - \theta_0)}{t_f^2}. \tag{8.8}$$

This inequality sets a *lower bound* of picking a common constant acceleration a for both the two parabolic spline curves, though it can be arbitrary. After a is chosen, using equation (8.7), we can determine the corresponding blending time length t_b. As we can further observe, a higher value of the acceleration a will shorten the parabolic spline interval t_b so that the linear portion of the entire joint travel will be longer and the path alteration will be less.

Let us look at a numerical example. Suppose that $\theta_0 = 0$ at $t = 0$ and $\theta_f = 1$ rad. at $t_f = 2$ sec. Then, based on the lower-bound equation for the acceleration in (8.8), $a \geq 1$. Let us pick $a = 2$ rad./sec^2. Using (8.7), we can determine how long each parabolic blend will be over the entire $\Delta t = t_f - t_0 = 2$ sec., which is $t_b = (4 - \sqrt{8})/4 = 0.29$ sec. In addition, we can also determine at what speed the joint angle will travel along the linear segment by equation (8.4), i.e., $\dot{\theta}_h = at_b = 2 \times 0.29 = 0.58$ rad./sec. In fact, the ratio of the two parabolic blends to the total time length is $2t_b/\Delta t = 0.58/2 = 29\%$. In other words, 71% is on the linear segment without alteration.

The above linear function with parabolic blends can also be extended to the cases of non-zero joint initial velocity $\dot{\theta}(0) = \dot{\theta}_0 \neq 0$ or non-zero final velocity $\dot{\theta}(t_f) = \dot{\theta}_f \neq 0$, or both. In either case, equation (8.5) has to be added with a term $\dot{\theta}_0 t_b$, and the starting and ending parabolic curves may have either different accelerations a to pick, or different blending time lengths t_b to set.

8.2 Independent Joint-Servo Control

After path planning for each robotic joint has been done, the desired new joint position θ^d will become an input to each joint-servo controller that excites the individual motor to drive the actual joint position $\theta(t)$ towards θ^d. Since each desired joint position θ^d can be determined through the robotic inverse kinematics under the path planning requirement, it suffices to control each robotic joint independently. In other words, each joint has its own servo control loop with its individual input and output channels. We will show in later sections that such an **independent joint control** pattern for any multi-joint robotic system is ideal to perform a **set-point** task, but is not accurate enough for a **trajectory-tracking** application [3, 4].

Before we start developing a complete model for the servo control of each joint of a robot arm, we have to first review what a DC motor model looks like. A DC or AC servo motor is a typical electromechanical system with electrical input and mechanical output, as seen in Figure 8.3. Based on the principle of electric circuit analysis and Newton's second law, we have

$$\begin{cases} L_a \frac{di_a}{dt} + R_a i_a + e_b = v_a \\ J_m \frac{d\omega_m}{dt} + B_m \omega_m + \tau_{ex} = \tau_m, \end{cases} \tag{8.9}$$

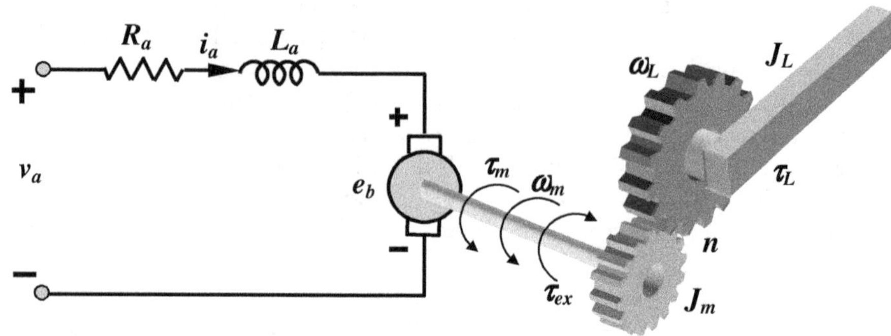

Fig. 8.3 A DC-motor electrical and mechanical model

where e_b is called a *back emf* (electromotive force) across the motor armature circuit, which, based on the Faraday's law, is proportional to the motor shaft speed, i.e., $e_b = K_a\omega_m$ with an armature constant K_a if a separately-excited DC motor or a PM (Permanent Magnet) DC motor is considered.

Due to the law of energy conservation, the power between the electrical and mechanical sides must be equal to each other, i.e., $e_b i_a = \tau_m\omega_m$ so that the motor produced internal torque becomes $\tau_m = K_a i_a$ with the same armature constant K_a if the SI metric system (the International System of Units) is adopted. However, under the English unit system, the two internal relations $e_b = K_b\omega_m$ and $\tau_m = K_a i_a$ have different proportional constants $K_a \neq K_b$, but we will use the SI unit in the later discussion as a default. The parameters J_m and B_m in the second equation of (8.9) are the motor shaft moment of inertia and rotational viscous friction (damping) coefficient, respectively. τ_{ex} is a total external torque that is referred to the motor shaft side, including an equivalent torque due to a load through the gearbox and a motor shaft rotating friction torque τ_{loss}.

Let the gear ratio of the gearbox be $n = \frac{\omega_m}{\omega_L}$ which is greater than 1 if the gearbox offers a speed reduction. Due to the law of energy conservation, the mechanical power between two sides of the gearbox must also be equal to each other, i.e., $\tau_m\omega_m = \tau_L\omega_L$. Therefore,

$$n = \frac{\omega_m}{\omega_L} = \frac{\tau_L}{\tau_m}. \tag{8.10}$$

This exhibits a common fact that *to gain a torque amplification through a gearbox, the cost one has to pay is a speed reduction.* Thus, by referring to the motor shaft side, the external torque in (8.9) turns out to be

$$\tau_{ex} = \frac{\tau_L}{n} + \tau_{loss}. \tag{8.11}$$

On the other hand, based on the Newton's second law, we have

$$J_L \frac{d\omega_L}{dt} + B_L \omega_L = \tau_L$$

that is referred to the load side, where we have merged all the rotation frictions into a single term τ_{loss} in (8.11) that is referred to the motor shaft side, and J_L and B_L are the load moment of inertia and load rotational viscous coefficient, respectively. Substituting the above equation into (8.11) and then into the second equation of (8.9) yields

$$J_m \frac{d\omega_m}{dt} + B_m \omega_m + \frac{1}{n}\left(J_L \frac{d\omega_L}{dt} + B_L \omega_L\right) + \tau_{loss} = \tau_m.$$

By further noticing the gear ratio equation (8.10), we obtain

$$J_{ef} \frac{d\omega_m}{dt} + B_{ef} \omega_m + \tau_{loss} = \tau_m, \tag{8.12}$$

where

$$J_{ef} = J_m + \frac{1}{n^2} J_L \quad \text{and} \quad B_{ef} = B_m + \frac{1}{n^2} B_L \tag{8.13}$$

are called the *effective moment of inertia* and *effective viscous coefficient* on the motor shaft side, respectively. Equation (8.12) demonstrates that if J_m and B_m in the second equation of (8.9) are replaced by J_{ef} and B_{ef}, respectively, then the load torque τ_L will disappear, as shown in (8.12).

Because all the differential equations in the DC motor model are linear, we can take Laplace transformations for both sides of the first equation of (8.9) and equation (8.12) to convert them to the following two simultaneous algebraic equations in s-domain:

$$\begin{cases} sL_a I_a + R_a I_a + K_a \Omega_m = V_a \\ sJ_{ef} \Omega_m + B_{ef} \Omega_m + T_{loss} = K_a I_a, \end{cases} \tag{8.14}$$

where I_a, V_a, Ω_m and T_{loss} are the Laplace transforms of i_a, v_a, ω_m and τ_{loss}, respectively.

Based on equation (8.14), we can readily draw an equivalent block diagram for the DC-motor model by defining the armature terminal voltage V_a as an input and the motor shaft speed Ω_m as an output, as shown in Figure 8.4. Another input to the motor model is the torque loss T_{loss}. Integrating the speed can directly find the motor shaft angular position, i.e., $\Theta_m(s) = \frac{\Omega_m(s)}{s}$. Figure 8.4 reveals a negative feedback from Ω_m to the input V_a via the armature constant K_a. This internal feedback loop represents an inherent physical property of the back emf $E_b = K_a \Omega_m$ that is proportional to the motor shaft speed. According to this block diagram, we can immediately achieve the following transfer function if $\tau_{loss} = 0$:

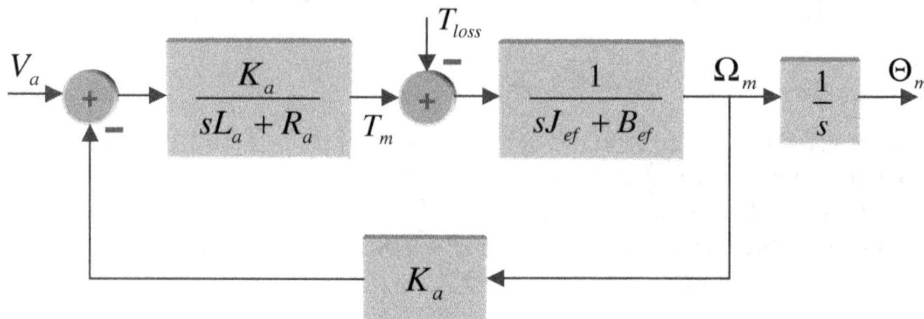

Fig. 8.4 A block diagram of the DC-motor model

$$G_m(s) = \frac{\Omega_m}{V_a} = \frac{K_a}{J_{ef}L_a s^2 + (J_{ef}R_a + B_{ef}L_a)s + B_{ef}R_a + K_a^2}. \tag{8.15}$$

Therefore, a DC motor is basically a second-order linear system between its input voltage and shaft speed output [5, 6, 7].

To analyze a second-order linear system, it is conventional to compare its normalized characteristic equation in s-domain with the following standard form:

$$s^2 + 2\zeta\omega_n s + \omega_n^2 = 0, \tag{8.16}$$

where ζ is called a damping ratio and ω_n is called a natural frequency. Now, comparing this standard template with the normalized denominator of the motor model transfer function (8.15), we obtain

$$\omega_n = \sqrt{\frac{B_{ef}R_a + K_a^2}{J_{ef}L_a}} \quad \text{and} \quad \zeta = \frac{J_{ef}R_a + B_{ef}L_a}{2\sqrt{J_{ef}L_a(B_{ef}R_a + K_a^2)}}.$$

As a numerical example, let $J_{ef} = 0.1$ Kg-m^2, $B_{ef} = 0.01$ N-m-sec/rad., $L_a = 0.01$ Henry, $R_a = 0.1\ \Omega$, and $K_a = 0.5$ N-m/A. for a given DC motor. Then, with the above equations, we find $\omega_n = 15.843$ rad/sec., and $\zeta = 0.32$.

Since every coefficient in the second-order characteristic polynomial for a DC motor model given by the denominator of (8.15) is positive, the system is stable. However, it is not good enough to be just stable for control of robotic systems via every joint motor. According the classical control theory [6, 7], if the damping ratio $0 < \zeta < 1$, such as in the above numerical example, the second-order linear system is *underdamping* so that its step response has an overshooting and a damping oscillation within a certain sustain time t_s. The overshooting and oscillation for a motor to drive a link of the robot will cause the robot hand to shake, which is unacceptable.

In order to improve the motor transient performance, we must insert a controller to compensate the characteristics for an overshooting reduction and response delay time t_d shortening. The most typical compensator is known as the PID (Proportion-Integral-Differential) controller with the following transfer function:

$$G_c(s) = K_p + \frac{K_i}{s} + K_d s,$$

where K_p, K_i and K_d are the proportional, integral and differential gain constants, respectively, and are usually all positive. With such a controller inserted, the block diagram of a typical DC-motor position-feedback control loop is depicted in Figure 8.5.

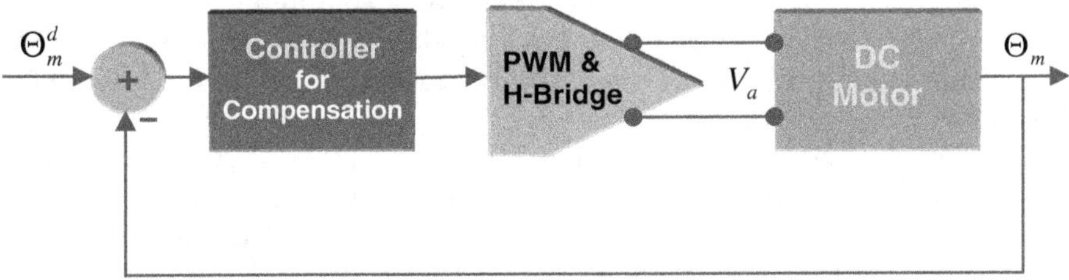

Fig. 8.5 A block diagram of DC-motor position-feedback control

This unity feedback loop allows the motor to be always excited by a signal proportional to the error e_m between the motor desired angular position θ_m^d and motor actual position θ_m. The advantage of such a feedback control scheme is to provide each joint of the robot with a dynamically high stiffness. Namely, the motor actual position θ_m will not be easily changed by any external mechanical force until the input desired position command θ_m^d is modified. At this point of view, the primary purpose of inserting a controller is to improve the performance of convergence $e_m(t) = \theta_m^d - \theta_m(t) \to 0$ as $t \to \infty$. Therefore, the total closed-loop transfer function of such a complete motor feedback control system becomes

$$T_m(s) = \frac{\Theta_m}{\Theta_m^d} = \frac{G_c(s)AG_m(s)\frac{1}{s}}{1 + G_c(s)AG_m(s)\frac{1}{s}} = \frac{G_c(s)AG_m(s)}{s + G_c(s)AG_m(s)}, \tag{8.17}$$

where A is a constant voltage gain for the PWM (Pulse-Width-Modulation) and H-Bridge block that will be further explained later [8]. As we can see, both the numerator and denominator of (8.17) contain a controller transfer function $G_c(s)$ that allows us to reassign both zeros and poles of the motor system by adjusting the gain constants K_p, K_i and/or K_d in $G_c(s)$ towards a better and more desirable overall control performance [7].

Since a motor actual position θ_m can be directly measured by an optical encoder that is mounted co-axially on the motor shaft, it becomes more conventional to adopt the format of θ_m-feedback, instead of the robotic link position θ_L-feedback. For this reason, the desired motor position command θ_m^d has to be acquired from the desired joint position θ_L^d by

$$\theta_m^d = n\theta_L^d$$

with a gear ratio n, while θ_L^d is the signal directly from the robotic inverse kinematics (I-K) and path planner to update the value at each sampling point.

In the motor feedback control loop, as depicted in Figure 8.5, a power amplifier must be included between the controller output and motor armature input port. Because the entire robotic independent joint control system is often implemented digitally, one of the best formats to convert a digital signal from the controller to an analog driving signal for the motor armature excitation is the PWM-based power amplifier. The PWM (Pulse-Width-Modulation) means that the armature voltage v_a value is modulated on a constant peak-to-peak pulse train by a constant period but a variable pulse width that is proportional to the armature voltage v_a. If v_a is an output of the controller, which is currently a digital signal, then using a digital counter and a digital comparator, or a special chip of PWM generator, one can easily achieve such a desirable format of signal ready to switch on/off a special bi-directional power amplifier, called the *H-Bridge*. With the H-bridge circuit, only the average voltage of the PWM signal is filtered and directly drives the DC-motor in either direction [8]. However, the entire circuit of both the PWM generator and H-bridge power amplifier can, in general, be modeled as a block with a flat (constant) voltage gain A, as given in equation (8.17).

Finally, each of the n joints for a robot arm is driven through a gear by an individual DC-motor with its controller to move itself independently to a desired joint position instructed by the robotic inverse kinematics (I-K) and path planner that are pre-programmed inside the robot control box. In other words, after receiving a command of the desired joint value at each sampling point, each joint is controlled and moved individually without any coupling with the other joints. This **independent joint control scheme** has been proven to be feasible and quite effective for a **set-point task** operation.

However, the motor model requires to know the mechanical parameters J_{ef} and B_{ef} in order to design a PID controller and power amplifier, and they obviously depend on the joint load according to equation (8.13). In fact, the load of each joint is varying during the entire robot body moving with the configuration change, and this has already been revealed by the concept of robotic inertial matrix W that is a function of almost every joint position in the last chapter of robot dynamic modeling. Therefore, the independent joint control scheme without any dynamics compensation will lead to a noticeable deviation in operating a **trajectory-tracking** task, especially at relatively faster tracking speeds. The trajectory-tracking accuracy issue will be further addressed in the next sections. The solution to improve the trajectory-tracking accuracy lies in a more global control modeling strategy by taking robotic dynamics into consideration, instead of just the above joint-by-joint independent control scheme.

8.3 Input-Output Mapping and Systems Invertibility

Starting with this section, we will model and study a *complete nonlinear control system* that contains both input (control) and output vectors in a more global fashion. The general model for a complete nonlinear autonomous control system that is affine of the inputs can be written as

$$\begin{cases} \dot{x} = f(x) + \sum_{i=1}^{r} g_i(x)u_i = f(x) + G(x)u \\ y = h(x), \end{cases} \tag{8.18}$$

where $G(x) = (g_1(x) \ \cdots \ g_r(x))$ is an n by r coefficient matrix of the input $u = (u_1 \ \cdots \ u_r)^T$. For instance, a complete linear time-invariant control system is a special case of (8.18), where $f(x) = Ax$, $G(x) = B$ and $h(x) = Cx$, where A, B and C are n by n, n by r and m by n constant matrices, respectively. Thus, the complete linear time-invariant control system equation becomes

$$\begin{cases} \dot{x} = Ax + Bu \\ y = Cx. \end{cases} \tag{8.19}$$

8.3.1 The Concepts of Input-Output Mapping and Relative Degree

A complete control system model either in (8.18) or in (8.19) represents a certain but indirect relationship between the input (control) u and the output y through the state vector x, unless $y = h(x, u)$ is defined to contain a feedforward portion. However, we do not consider any feedforward case in our discussion here and only treat the input and output relationship to be purely indirect through the state. Such an indirect relationship between u and y is often referred to as an *input-output mapping* [10]. For the linear time-invariant system (8.19), we can take Laplace transformations for both sides of the equations, and obtain the following simultaneous algebraic equations in s-domain:

$$\begin{cases} sX(s) = AX(s) + BU(s) \\ Y(s) = CX(s), \end{cases} \tag{8.20}$$

where we assume that the initial state $x(0) = 0$, and $X(s)$, $U(s)$ and $Y(s)$ are the Laplace transforms of $x(t)$, $u(t)$ and $y(t)$, respectively. Solving the two s-domain algebraic equations in (8.20) yields

$$Y(s) = C(sI - A)^{-1}BU(s). \tag{8.21}$$

This solution shows that the input-output mapping of the complete linear time-invariant system is given by an m by r *Transfer Matrix* $T(s) = C(sI - A)^{-1}B$ in s-domain. In general, each element of the transfer matrix $T(s)$, called a transfer function, is an s-domain polynomial fraction of the form:

$$\frac{b_0 s^q + b_1 s^{q-1} + \cdots + b_q}{s^p + a_1 s^{p-1} + \cdots + a_p}, \tag{8.22}$$

where the highest power of the polynomial in the denominator, p, can be up to n, the dimension of the state space, while the highest order of the numerator is $q < p$ in general as a proper transfer function case, and all the coefficients a_i's and b_j's are real. According to the linear systems theory, the highest-order difference between the denominator and numerator of a transfer function is equal to the difference between the number of poles and number of zeros. We call the difference integer $r_d = p - q$ the *relative degree* (or the *relative order*) of the system.

In general control design cases, we often demand to find a control law, i.e., an input function such that the system output can satisfy a desired performance objective. Intuitively, the control problem can also be interpreted as *inverting* the input-output system. In other words, given a desired output function, the control problem is to determine an input function in terms of the desired output under the system equation as a constraint. Therefore, whether an input-output system is invertible and what is the bottom line of the invertibility become two fundamental questions requiring for a further investigation.

In qualitatively speaking, for a single-input/single-output (SISO) linear system, its transfer function is often defined as a ratio of the output Laplace transform $Y(s)$ to the input Laplace transform $U(s)$ and is usually in the form given by (8.22). The number of poles p is greater than the number of zeros q, $p > q$, in a *proper system* case. Now, if we want to invert the system, i.e., to find $U(s)$ in terms of $Y(s)$, the new transfer function after the inversion becomes the reciprocal of (8.22). In this case, since the zeros and poles are swapped over, the number of zeros in the inverted system is now greater than the number of poles so that the inverted system becomes improper.

It is also well known from the linear control theory that each pole (if it is a real number) contributes a slope of -20 db/decade to the frequency response of the transfer function magnitude, while each zero contributes $+20$ db/decade to it in a steady-state analysis of the linear system, i.e., $s = j\omega$. An improper transfer function with more zeros than poles will cause energy divergence as frequency goes to infinity in the system frequency-spectrum. Therefore, we need to add more zeros to the original transfer function before inverting the system. What would be the least number of zeros to add on? The answer is now quite obvious that it at least needs $r_d = p - q$ more zeros.

In order to add r_d more zeros into the original system transfer function, the simplest way is to multiply the output function $Y(s)$ by s^{r_d}, which will result in the same numbers for both zeros and poles in the inverted system transfer function $U(s)/[s^{r_d}Y(s)]$. Because the inverse Laplace transform of $s^{r_d}Y(s)$ is equivalent to differentiating $y(t)$ by r_d times if zero initial conditions of the output are assumed, we conclude that to invert an input-output system with a relative degree r_d, the output function must be differentiated up to the r_d-th order.

For a more rigorous explanation, let us visit the SISO linear system given by (8.19) with scalar u and y. First, the output equation $y = Cx$ does not explicitly contain the input u so that the current output equation cannot immediately be used for the system inversion. We now take time-derivatives for both sides of the output equation, and then substitute the state equation to obtain

$$\dot{y} = C\dot{x} = CAx + CBu.$$

If $CB \neq 0$, then the input u appears in the above equation so that we can invert the system by solving u in terms of y and x. However, if $CB = 0$, we have to take the second-order time-derivative for y, i.e.,

$$\ddot{y} = CA\dot{x} = CA^2x + CABu.$$

If $CAB \neq 0$, we say the system is invertible with $r_d = 2$. Otherwise, we have to continuously take the higher order of time-derivatives until the input u shows up in the equation.

Now, we turn our attention to the SISO complete nonlinear control system given by (8.18) with $r = m = 1$, i.e.,

$$\begin{cases} \dot{x} = f(x) + g(x)u \\ y = h(x). \end{cases} \tag{8.23}$$

Since there are no constant coefficient matrices A, B and C in (8.23), the Laplace transformation is no longer effective. Nevertheless, we can follow the same procedure as presented for the linear system case, and adopt the Lie derivative concept that is defined by

$$L_f h(x) = \frac{\partial h(x)}{\partial x} f(x)$$

as a generalized directional derivative to reach a higher-order differentiation of the output function. For the first-order time-derivative,

$$\dot{y} = \frac{\partial h(x)}{\partial x}\dot{x} = L_f h(x) + (L_g h(x))u.$$

Likewise, if $L_g h(x) = 0$, the second-order time-derivative is needed for y and becomes

$$\ddot{y} = \frac{\partial \dot{y}}{\partial x}\dot{x} = L_f^2 h(x) + (L_g L_f h(x))u.$$

The differentiation keeps going on until the input u shows up in the equation. Therefore, the relative degree r_d for an SISO nonlinear system can be defined formally as follows:

Definition 9. The system (8.23) is said to have a relative degree r_d in some $\mathcal{M} \subset \mathbb{R}^n$ if for any $x \in \mathcal{M}$,

$$L_g L_f^k h(x) = 0, \quad 0 \le k < r_d - 1, \quad \text{and}$$

$$L_g L_f^{r_d - 1} h(x) \neq 0.$$

The definition of relative degrees can be extended to an MIMO complete nonlinear system given in (8.18) [10, 12]:

Definition 10. The system (8.18) is said to have relative degrees (vector) $\{r_1, \cdots, r_m\}$ at a point $x = x_0 \in \mathcal{M}$ if

1. $L_{g_j} L_f^k h_i(x) = 0$, for $1 \le j \le r$, $1 \le i \le m$, $0 \le k < r_i - 1$ and $x \in U(x_0) \subset \mathcal{M}$, and
2. the matrix

$$D(x) = \begin{pmatrix} L_{g_1} L_f^{r_1 - 1} h_1(x) & \cdots & L_{g_r} L_f^{r_1 - 1} h_1(x) \\ \vdots & \cdots & \vdots \\ L_{g_1} L_f^{r_m - 1} h_m(x) & \cdots & L_{g_r} L_f^{r_m - 1} h_m(x) \end{pmatrix}$$

has m independent rows at $x = x_0$.

The matrix $D(x)$ is called a *Decoupling Matrix* [10, 14]. If $r = m$, i.e., the input and output vectors have the same dimension, we call such a system a *square input-output system*, or just a *square system*, and the second condition of the above definition can be re-stated that the matrix $D(x)$ is nonsingular at $x = x_0$. Moreover, in the MIMO case, where the system has m output channels y_1 through y_m, and each channel has its individual relative degree r_i for $i = 1, \cdots, m$, the *total relative degree* r_d of the MIMO system is defined by the sum of all the individual r_i's, i.e.,

$$r_d = r_1 + \cdots + r_m.$$

Let us take the following control system to illustrate the determination of relative degree for an SISO system:

$$\begin{cases} \dot{x}_1 = x_2 \\ \dot{x}_2 = -\sin x_2 + u. \end{cases} \tag{8.24}$$

In the first case, let the output function be $y = h(x) = x_1$. Based on Definition 9, we calculate

$$L_g h(x) = \frac{\partial h}{\partial x} g(x) = (1 \;\; 0) \begin{pmatrix} 0 \\ 1 \end{pmatrix} = 0.$$

Then, we need to calculate the second-order Lie derivative $L_g L_f h(x)$. Since

$$L_f h(x) = (1 \;\; 0) \begin{pmatrix} x_2 \\ -\sin x_2 \end{pmatrix} = x_2,$$

and

$$L_g L_f h(x) = L_g(x_2) = (0 \;\; 1) \begin{pmatrix} 0 \\ 1 \end{pmatrix} = 1 \neq 0,$$

the relative degree $r_d = 2$ if the output is defined by $y = x_1$. In fact, according to (8.24), $\ddot{x}_1 = \dot{x}_2 = -\sin x_2 + u$. This shows that by taking up to the second time-derivative for the output $y = x_1$, the input u shows up.

In the second case, let the output be re-defined by $y = h(x) = x_2$. Using the same procedure,

$$L_g h(x) = (0 \;\; 1) \begin{pmatrix} 0 \\ 1 \end{pmatrix} = 1 \neq 0,$$

thus, $r_d = 1$. It can also be observed from the second equation of (8.24) that only taking the first time-derivative for $y = x_2$ is sufficient enough to make the input u show up, and it confirms that the relative degree $r_d = 1$. This example tells us an interesting fact that the relative degree for a system also depends on the definition of the output function.

We now look at a following fifth-order MIMO system:

$$\dot{x} = \begin{pmatrix} -x_1^3 + x_2 \\ x_1 x_3 \\ -x_1 + x_3 \\ x_2 \\ x_5 + x_3^2 \end{pmatrix} + \begin{pmatrix} 0 \\ 1 \\ 1 \\ 0 \\ x_2 \end{pmatrix} u_1 + \begin{pmatrix} 0 \\ 1 \\ 0 \\ 0 \\ 0 \end{pmatrix} u_2, \quad \text{and} \quad y = \begin{pmatrix} x_3 \\ x_4 \end{pmatrix}. \qquad (8.25)$$

According to Definition 10 for an MIMO system, we need to compute the decoupling matrix $D(x)$ in order to determine r_d for (8.25). Since

$$L_{g_1} h_1(x) = \frac{\partial x_3}{\partial x} g_1 = (0 \;\; 0 \;\; 1 \;\; 0 \;\; 0) \begin{pmatrix} 0 \\ 1 \\ 1 \\ 0 \\ x_2 \end{pmatrix} = 1,$$

and

$$L_{g_2}h_1(x) = (0\ \ 0\ \ 1\ \ 0\ \ 0)\begin{pmatrix} 0 \\ 1 \\ 0 \\ 0 \\ 0 \end{pmatrix} = 0,$$

we complete the first row of $D(x)$, because one of the above two components has been non-zero.

Next, let us calculate

$$L_{g_1}h_2(x) = (0\ \ 0\ \ 0\ \ 1\ \ 0)\begin{pmatrix} 0 \\ 1 \\ 1 \\ 0 \\ x_2 \end{pmatrix} = 0,$$

and

$$L_{g_2}h_2(x) = (0\ \ 0\ \ 0\ \ 1\ \ 0)\begin{pmatrix} 0 \\ 1 \\ 0 \\ 0 \\ 0 \end{pmatrix} = 0.$$

Since both are zeros, we have to continue calculating

$$L_f h_2(x) = (0\ \ 0\ \ 0\ \ 1\ \ 0)f = x_2,$$

$$L_{g_1}L_f h_2(x) = (0\ \ 1\ \ 0\ \ 0\ \ 0)\begin{pmatrix} 0 \\ 1 \\ 1 \\ 0 \\ x_2 \end{pmatrix} = 1,$$

and

$$L_{g_2}L_f h_2(x) = (0\ \ 1\ \ 0\ \ 0\ \ 0)\begin{pmatrix} 0 \\ 1 \\ 0 \\ 0 \\ 0 \end{pmatrix} = 1.$$

Thus, the decoupling matrix becomes

$$D(x) = \begin{pmatrix} 1 & 0 \\ 1 & 1 \end{pmatrix}$$

which is always non-singular. Therefore, we conclude that $r_1 = 1$ for the first channel of the output $y = h(x)$, and $r_2 = 2$ for the second channel of the output. The total relative degree of the fifth-order MIMO system (8.25) becomes $r_d = r_1 + r_2 = 3$, which is less than the system dimension $n = 5$.

8.3.2 Systems Invertibility and Applications

With the definitions of relative degree introduced, we now formally state that an input-output system is *invertible* if its relative degree $r_d \le n$, the dimension of the state space. Note that in this *invertibility* definition, if the system is MIMO, then r_d should be the total relative degree, i.e., $r_d = r_1 + \cdots + r_m \le n$.

In contrast, if $r_d > n$, the input-output system is *non-invertible* [9, 10]. Therefore, to determine the invertibility for a complete system, one needs only to calculate the (total) relative degree r_d and compare it with n, the dimension of the state space.

The physical meaning of the system invertibility can be viewed as a *sufficiency of output channels.* In other words, if a system has sufficient output channels, through which all the input variables can be seen via the internal state variables, then this system must be invertible, or say, the inputs can all be determined in terms of the outputs and states.

The previous two examples in the last subsection are both invertible. More specifically, the first example includes two cases of the output definitions: one has $r_d = 2 = n$, and the other one has $r_d = 1 < n = 2$. The second example is an MIMO system with $n = 5$. Since the total relative degree $r_d = r_1 + r_2 = 1 + 2 = 3 < n = 5$, it is also invertible.

In order to gain more physical insight into the relative degree and invertibility, let us take a 6-joint fully-actuated open serial-chain industrial robot as a realistic nonlinear MIMO system to test its invertibility. Note that the adjective "fully-actuated" means that each joint of the robot is driven by a motor. In other words, the joint torque vector τ of such a fully-actuated robot has the same dimension as the joint position vector q. Let us now derive the decoupling matrix $D(x)$ and determine its total relative degree r_d.

For the 6-joint robot dynamic system, $x = \begin{pmatrix} q \\ \dot{q} \end{pmatrix} \in \mathbb{R}^{12}$ is the state vector that augments the joint position vector q and joint velocity vector $\dot{q}$ together. According to the robot dynamic equation given by (7.19) from Chapter 7, a general state-space equation (8.18) can be expressed as

$$f(x) = \begin{pmatrix} \dot{q} \\ W^{-1}((\frac{1}{2}W_d - W_d^T)\dot{q} + \tau_g) \end{pmatrix} \quad \text{and} \quad G(x) = \begin{pmatrix} O \\ W^{-1} \end{pmatrix}, \tag{8.26}$$

where O is the 6 by 6 zero matrix.

Let the six degrees of freedom position vector (three for position and three for orientation) in a 3D Cartesian space be defined as an output function $y = h(x) \in \mathbb{R}^6$. Such an output definition $y = h(x)$, in most robotic application cases, is a function of the joint position q only, and is independent of $\dot{q}$ so that $y = h(x) = h(q)$ for most robotic systems.

It is also presumable that the six output channels, $h_1(q)$ through $h_6(q)$, represent all the six degrees of freedom motion, and each channel should have an "equal opportunity" in motion that is controlled by all the six joint torques. Therefore, it is reasonable to assume that each output channel for the 6-joint fully-actuated robotic system has a common relative degree, i.e., $r_1 = \cdots = r_6 = r_{cd}$. This is referred to as an equal-RD (relative-degree) hypothesis.

Under the equal-RD hypothesis, every row in the targeting decoupling matrix $D(x)$ will arrive at the non-zero status concurrently. Therefore, we can compute the common relative degree r_{cd} through a compact but more global derivation, instead of repeating six times to compute each individual relative degree r_i once for each output channel.

To do so, let us first compute $L_{g_1}h(q)$ through $L_{g_6}h(q)$, which can actually be augmented to form a 6 by 6 matrix and denoted by $L_G h(q)$ for a notation convenience, where $G(x) = (g_1(x) \ \cdots \ g_6(x))$ is the 12 by 6 coefficient matrix of the input u in equations (8.18) and (8.26). In fact,

$$L_G h(q) = \frac{\partial h(q)}{\partial x} G(x) = \begin{pmatrix} \frac{\partial h(q)}{\partial q} & \frac{\partial h(q)}{\partial \dot{q}} \end{pmatrix} G(x) = (J \ \ O) \begin{pmatrix} O \\ W^{-1} \end{pmatrix} = O,$$

where we have defined $\partial h(q)/\partial q = J$, the 6 by 6 kinematic Jacobian matrix of the robot, and $\partial h(q)/\partial \dot{q} = O$, the 6 by 6 zero matrix. This zero result suggests that the common relative degree r_{cd} for each output channel be greater than one according to Definition 10 for MIMO systems in the last subsection.

We now proceed to the second-order differentiation, and further calculate that $L_f h(q) = (J \ \ O) f(x) = J\dot{q}$. Thus,

$$L_G L_f h(q) = \begin{pmatrix} \frac{\partial J\dot{q}}{\partial q} & \frac{\partial J\dot{q}}{\partial \dot{q}} \end{pmatrix} \begin{pmatrix} O \\ W^{-1} \end{pmatrix} = JW^{-1},$$

because $\partial J\dot{q}/\partial \dot{q} = J$. Since the inertial matrix W of the robot is always non-singular, the above result clearly exhibits that if the Jacobian matrix J is also non-singular, then the decoupling matrix

$$D(x) = L_G L_f h(q) = JW^{-1}$$

can be non-singular so that the common relative degree $r_{cd} = 2$. In this case, the total relative degree $r_d = 6\,r_{cd} = 12$ that is exactly equal to the dimension of the state space.

Therefore, we conclude that a 6-joint fully-actuated industrial robot with the six-dimensional Cartesian position plus orientation to form an output vector $h(q)$ is invertible if the 6 by 6 Jacobian matrix $J = \partial h(q)/\partial q$ does not fall in singularity. In other words, only at a singular point of J, the fully-actuated robotic system will damage its invertibility.

It should be pointed out that the above justification is under the assumption of forming all 6 d.o.f. with a single 6 by 1 Cartesian position vector to be ready to take time-derivatives. As we have already investigated in Chapters 3 and 4, to find an unified 6 by 1 vector and its time-derivative to uniquely represent the 6 d.o.f. motion: three for position/translation and three for orientation/rotation is almost unfeasible. Therefore, defining a 6-dimensional output function $y = h(q)$ in the above 6-joint industrial robot example is just a conceptual model.

The above example also reveals a fact that to invert a robot dynamic system, or equivalently, to design a control law for the robotic system to meet a desired output objective, it is required to compute up to the second-order time-derivative for an output function. In other words, we need *acceleration* information on the output in order to resolve the global control design problem for a robot dynamic system.

8.4 The Theory of Exact Linearization and Linearizability

After the concept of systems invertibility is introduced, we now continue to discuss how to test a nonlinear system to see if it is exactly linearizable and how to linearize it exactly without error. To address the *linearizability* issue for a nonlinear system, two mathematical concepts must be introduced at the beginning: *involutivity* and *complete integrability.* There are two major procedures: *input-state and input-output linearizations* in the theory of exact (or state-feedback) linearization that have been developed in the past three decades. Since not every nonlinear system is exactly linearizable, we have to decompose an unlinearizable nonlinear system into two parts: a linearizable subsystem and an unlinearizable subsystem, which, however, may not be decoupled from each other. Furthermore, after a system is decomposed in terms of the linearizability, the overall system stability must be re-tested.

8.4.1 Involutivity and Complete Integrability

Let us first make a mathematical preparation before commencing the exact linearization theory and procedure development. Suppose that there are k smooth n-dimensional tangent vector fields $f_1, \cdots, f_k$, where $k \leq n$. A *distribution* Δ is defined as a vector space that is spanned by all these tangent vector fields, i.e.,

$$\Delta = span\{f_1, \cdots, f_k\}.$$

After taking the Lie bracket for each pair of the vector fields in the distribution Δ, if every resultant vector field is still contained in Δ, we say the distribution is *involutive.* Formally, a distribution Δ of dimension k is said to be involutive if for each pair $f_i \in \Delta$ and $f_j \in \Delta$, $(1 \leq i, j \leq k)$, $[f_i, f_j] \in \Delta$, where the Lie bracket is defined by

$$[f_i, f_j] = ad_{f_i} f_j = \frac{\partial f_j}{\partial x} f_i - \frac{\partial f_i}{\partial x} f_j$$

that was also defined in equation (2.4) from Chapter 2, and can be rewritten as an adjoint operator ad_{f_i} to apply on f_j.

There are two trivial involutive distribution cases. One is a simple distribution spanned by only one tangent vector field, say f. Since $[f, f] = 0$ and $[f, 0] = 0$, and the zero vector should always be a member of any distribution, $\Delta = span\{\ f\ \}$ is obviously involutive. Another trivial involutive distribution is the entire vector space $\mathbb{R}^n$.

The second concept is called *complete integrability*. Let the *complementary distribution* of a distribution Δ, denoted by $\Delta^\perp$, be defined by $\Delta^\perp =$

$$\{span\{f_{k+1}, \cdots, f_n\} \mid \langle f_j, f_i \rangle = 0,\ k+1 \le j \le n,\ 1 \le i \le k \text{ and } f_i \in \Delta\},$$

where $\langle\ \cdot\ ,\ \cdot\ \rangle$ is the inner product of two vector fields. Clearly, the complementary distribution $\Delta^\perp$ collects all the remaining vector fields from the entire space $\mathbb{R}^n$, excluding the vector fields that have already been defined in Δ. Furthermore, every member in $\Delta^\perp$ must be orthogonal to each member of Δ.

The distribution Δ is said to be *completely integrable* if each member $f_j(x)$ of $\Delta^\perp$ is a gradient vector of some scalar field $\mu_j(x)$, i.e., $f_j(x) = \partial \mu_j(x)/\partial x$ for $j = k+1, \cdots, n$. This concept implies that a distribution Δ is completely integrable if there exist $n-k$ scalar fields $\mu_{k+1}(x)$ through $\mu_n(x)$ such that

$$\frac{\partial \mu_j(x)}{\partial x} f_i(x) = L_{f_i}\mu_j(x) = 0, \quad \text{for } 1 \le i \le k, \text{ and } k+1 \le j \le n.$$

For example, if we have two vector fields in $\mathbb{R}^3$ to span a distribution Δ:

$$f_1(x) = \begin{pmatrix} x_1 \\ -x_2 \\ 0 \end{pmatrix} \quad \text{and} \quad f_2(x) = \begin{pmatrix} 0 \\ 0 \\ x_3 \end{pmatrix},$$

then with a simple calculation, we obtain

$$[f_1, f_2] = \frac{\partial f_2}{\partial x} f_1 - \frac{\partial f_1}{\partial x} f_2 = 0.$$

Since the above $[f_1, f_2] = 0 \in \Delta$, the distribution Δ is involutive. Furthermore, let a scalar function be $\mu(x) = x_1 x_2$. Its gradient vector becomes

$$\frac{\partial \mu}{\partial x} = (x_2\ \ x_1\ \ 0).$$

This row vector is obviously orthogonal to either f_1 or f_2, and is a member of the complementary distribution $\Delta^\perp$. Based on the above definition, the distribution $\Delta = span\{f_1,\ f_2\}$ is also completely integrable.

It can be further observed that to determine the involutivity for a distribution Δ, we need only to calculate the Lie bracket of each pair of the vector fields. Whereas to determine the complete integrability, we have to seek all such scalar fields. Thus, between the two properties for a distribution Δ, the involutivity is much easier to be tested than the complete integrability. Nevertheless, it is quite exciting that a direct relationship between the two properties has been found so that the testing difficulty for the complete integrability is alleviated. This direct relationship is known as the Frobenius Theorem:

Theorem 6. Frobenius Theorem
A distribution Δ is completely integrable if and only if Δ is involutive.

Using the Frobenius Theorem, one can readily test whether Δ is completely integrable by repeatedly calculating the Lie brackets to see if Δ is involutive. This will be the key step forward to the success of our linearization theory development.

8.4.2 The Input-State Linearization Procedure

Intuitively, to possibly linearize a nonlinear system, one may find and operate a certain nonlinear coordinate transformation to cancel out the nonlinearity. Here the linearization means to *exactly* linearize a nonlinear system without any residue term to be truncated. To this end, there are two fundamental questions that must be answered:

1. Is a given nonlinear system linearizable? and
2. If linearizable, what is the suitable nonlinear transformation?

In this subsection, we intend to develop a general theory to answer those two fundamental questions.

Consider a *single-input* nonlinear control system modeled by

$$\dot{x} = f(x) + g(x)u, \quad x \in \mathcal{M} \subset \mathbb{R}^n, \text{ and } u \in \mathbb{R}^1, \tag{8.27}$$

and both $f(x)$ and $g(x)$ are n-dimensional smooth vector fields. Our objective is to find a possible coordinate transformation that is a differentiable and one-to-one mapping: $T: \mathcal{M} \to \mathcal{Z} \subset \mathbb{R}^n$, whose inverse is also differentiable and one-to-one. Mathematically, it is called a *diffeomorphism*.

Let us first study a special distribution that is spanned by $g(x)$ and up to the $(n-2)$-th order Lie brackets between $f(x)$ and $g(x)$:

$$\Delta_l = span\{g(x), ad_f g(x), \cdots, ad_f^{n-2} g(x)\}. \tag{8.28}$$

Note that in this distribution Δ_l, there are at most $n-1$ independent vectors. Suppose that Δ_l is completely integrable, i.e., there exists a scalar field $\mu(x)$ such that its gradient $\partial\mu(x)/\partial x$ is orthogonal to each of the $n-1$ independent members in Δ_l. Since

$$\frac{\partial\mu(x)}{\partial x} ad_f^i g(x) = L_{ad_f^i g}\mu(x)$$

for each $i = 0, \cdots, n-2$, based on the orthogonality, the following equations should hold:

$$L_g\mu(x) = L_{ad_f g}\mu(x) = \cdots = L_{ad_f^{n-2}g}\mu(x) = 0. \tag{8.29}$$

Furthermore, according to the Jacobi identity in the preceding Lie algebra definition in Chapter 2, it can be shown [10, 14] that for any scalar field $\mu(x)$ and vector fields $f, g \in \mathbb{R}^n$,

$$L_{[f,g]}\mu(x) \equiv L_{ad_f g}\mu(x) \equiv L_f L_g \mu(x) - L_g L_f \mu(x). \tag{8.30}$$

Using this identity, we can further prove that equation (8.29) implies that

$$L_g\mu(x) = L_g L_f \mu(x) = \cdots = L_g L_f^{n-2}\mu(x) = 0. \tag{8.31}$$

If the nonlinear single-input system in (8.27) has property (8.31), to find a set of appropriate new coordinates $\{z_1, \cdots z_n\}$, as the n components of a promising diffeomorphism $z = T(x)$, one can readily start with the definition $z_1 = \mu(x)$ so that

$$\dot{z}_1 = \frac{\partial \mu}{\partial x}\dot{x} = L_f\mu + (L_g\mu)u = L_f z_1.$$

Let the second component of $z = T(x)$ be $z_2 = L_f z_1$. Then, by noticing (8.31) again, we obtain

$$\dot{z}_2 = \frac{\partial L_f \mu}{\partial x}\dot{x} = \frac{\partial L_f \mu}{\partial x}(f + gu) = L_f^2\mu + (L_g L_f \mu)u = L_f^2 z_1 = L_f z_2.$$

If repeating to define $z_3 = L_f z_2 = L_f^2 z_1, \cdots$, until reaching the last component z_n of $z = T(x)$, which yields $z_n = L_f^{n-1} z_1$, then,

$$\dot{z}_n = \frac{\partial L_f^{n-1}\mu}{\partial x}\dot{x} = L_f^n\mu + (L_g L_f^{n-1}\mu)u.$$

Therefore, to find a nonlinear transformation candidate $z = T(x)$ for possible exact linearization, one can start with $z_1 = \mu(x)$, and then repeatedly take time-derivatives to achieve a new state-space equation in the z coordinate system:

$$\begin{aligned}
\dot{z}_1 &= z_2 \\
&\vdots \\
\dot{z}_{n-1} &= z_n \\
\dot{z}_n &= L_f^n\mu + (L_g L_f^{n-1}\mu)u.
\end{aligned} \tag{8.32}$$

Let a new input v be defined by $v = L_f^n \mu(x) + (L_g L_f^{n-1} \mu(x))u$. We can thus rewrite the above equation into the following matrix form:

$$\dot{z} = Az + Bv, \tag{8.33}$$

where the new state $z = (z_1 \ \cdots \ z_n)^T$ and

$$A = \begin{pmatrix} 0 & 1 & 0 & \cdots & 0 \\ \vdots & \vdots & \vdots & \cdots & \vdots \\ 0 & 0 & 0 & \cdots & 1 \\ 0 & 0 & 0 & \cdots & 0 \end{pmatrix}, \quad B = \begin{pmatrix} 0 \\ \vdots \\ 0 \\ 1 \end{pmatrix}.$$

This is known as a *controllable canonical form* seen in the new state space $\mathcal{Z} \subset \mathbb{R}^n$ with a new input v that is directly related to the old input u and old state x. It is now clear that if one stands in the new state space $\mathcal{Z}$ along with the new input v, then the entire control system can be viewed as a strictly linear time-invariant controllable system given by (8.33). We refer to this procedure as an *exact linearization.* Since the original input u can be resolved algebraically in terms of the original state x and the new input v by

$$u = -\frac{L_f^n \mu(x)}{L_g L_f^{n-1} \mu(x)} + \frac{1}{L_g L_f^{n-1} \mu(x)} v = \alpha(x) + \beta(x)v, \tag{8.34}$$

we also call this procedure an *input-state linearization* [9, 10, 14, 13].

Based on the above development, we can now formally define the *linearizability.*

Definition 11. A single-input system (8.27) is said to be *input-state linearizable* if there exists a differentiable scalar function $\mu(x)$ along with its Lie derivatives:

$$z = T(x) = \begin{pmatrix} \mu(x) \\ L_f \mu(x) \\ \vdots \\ L_f^{n-1} \mu(x) \end{pmatrix}$$

such that it can transform (8.27) into the linear controllable canonical system in (8.33) with a new input v, as viewed in z-space.

According to this definition as well as the above discussions, including the Frobenius Theorem, we now formally state the following theorem to summarize the testing criterion of the input-state linearizability for a *single-input* control system:

Theorem 7. *An n-dimensional single-input control system (8.27) is input-state linearizable if*

1. *rank*$\{g, ad_f g, \cdots, ad_f^{n-1} g\} = n$, *and*
2. *the distribution* $\Delta_l = span\{g, ad_f g, \cdots, ad_f^{n-2} g\}$ *is involutive.*

In this theorem, the first condition is related to the controllability criterion, while the second one, based on the Frobenius Theorem, guarantees that the distribution Δ_l is completely integrable so that a desired scalar function $\mu(x)$ exists and the linearization procedure can always be fulfilled. A block diagram that illustrates the input-state linearization for a nonlinear system is shown in Figure 8.6.

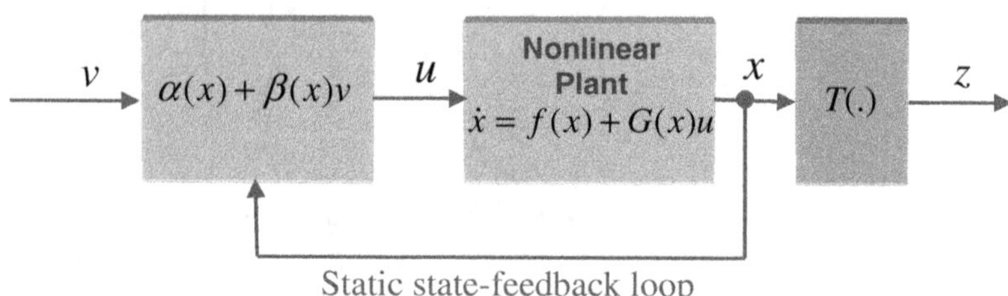

Fig. 8.6 A block diagram for an input-state linearized system

As an example, consider a nonlinear control system with a single input u as follows:

$$\dot{x} = \begin{pmatrix} x_2 \\ (1 - \ln x_3)x_2 \\ -2x_1x_3 \end{pmatrix} + \begin{pmatrix} 0 \\ 0 \\ 1 \end{pmatrix} u, \tag{8.35}$$

where $x_1 > 0$, $x_2 > 0$ and $x_3 > 0$. In order to linearize it, we have to first test its linearizability by applying the above Theorem 7. Since $g(x) = (0\ \ 0\ \ 1)^T$ is a constant vector, the first-order Lie bracket between $f(x)$ and $g(x)$ becomes

$$ad_f g = \frac{\partial g}{\partial x} f - \frac{\partial f}{\partial x} g = -\begin{pmatrix} 0 & 1 & 0 \\ 0 & 1 - \ln x_3 & -\frac{x_2}{x_3} \\ -2x_3 & 0 & -2x_1 \end{pmatrix} \begin{pmatrix} 0 \\ 0 \\ 1 \end{pmatrix} = \begin{pmatrix} 0 \\ \frac{x_2}{x_3} \\ 2x_1 \end{pmatrix}.$$

To test the first condition of the theorem, we need to further calculate

$$ad_f^2 g = \frac{\partial (ad_f g)}{\partial x} f - \frac{\partial f}{\partial x} ad_f g = \begin{pmatrix} -\frac{x_2}{x_3} \\ \frac{4x_1x_2}{x_3} \\ 2x_2 + 4x_1^2 \end{pmatrix}.$$

Thus, the 3 by 3 matrix $(g\ \ ad_f g\ \ ad_f^2 g)$ has its determinant $-x_2^2/x_3^2 \neq 0$. Therefore, the first condition of Theorem 7 holds.

To check the second condition, we have to see if the distribution Δ_l spanned only by g and $ad_f g$ due to $n = 3$ is involutive. To do so, let us compute the Lie bracket $[g, ad_f g]$ to find out if either it is linearly dependent on g and/or $ad_f g$, or the 3 by 3 matrix $H = (g\ \ ad_f g\ \ [g, ad_f g])$ has rank$(H) = 2$, i.e., H is singular. Because

$$[g, ad_f g] = \frac{\partial (ad_f g)}{\partial x} g - \frac{\partial g}{\partial x} ad_f g = \begin{pmatrix} 0 & 0 & 0 \\ 0 & \frac{1}{x_3} & -\frac{x_2}{x_3^2} \\ 2 & 0 & 0 \end{pmatrix} \begin{pmatrix} 0 \\ 0 \\ 1 \end{pmatrix} = \begin{pmatrix} 0 \\ -\frac{x_2}{x_3^2} \\ 0 \end{pmatrix},$$

the matrix H becomes

$$H = \begin{pmatrix} 0 & 0 & 0 \\ 0 & \frac{x_2}{x_3} & -\frac{x_2}{x_3^2} \\ 1 & 2x_1 & 0 \end{pmatrix},$$

which has a zero first row and is obviously singular. Therefore, the second condition of Theorem 7 also holds and the system is input-state linearizable.

In order to find the first new state component z_1 as a starting point, let us examine the two vector fields $g = (0 \ \ 0 \ \ 1)^T$ and $ad_f g = (0 \ \ x_2/x_3 \ \ 2x_1)^T$. Since they both have a zero as the first element, one may define a scalar function $z_1 = T_1(x)$ such that its gradient has only the first element being non-zero while the last two are all zeros, i.e., $\partial T_1/\partial x = (\times \ \ 0 \ \ 0)$ with some non-zero "×". Such a gradient vector can always be orthogonal to both g and $ad_f g$, and can be realized by defining the first new state component z_1 as a function of x_1 only. For the sake of simplicity, let $z_1 = x_1$. Then, according to (8.32) and the system equation (8.35), $z_2 = \dot{z}_1 = \dot{x}_1 = x_2$, and $z_3 = \dot{z}_2 = \dot{x}_2 = (1 - \ln x_3)x_2$. Finally,

$$\dot{z}_3 = \dot{x}_2(1 - \ln x_3) - \frac{x_2}{x_3}\dot{x}_3 = (1 - \ln x_3)^2 x_2 + 2x_1 x_2 - \frac{x_2}{x_3}u.$$

Let the new input v be

$$v = (1 - \ln x_3)^2 x_2 + 2x_1 x_2 - \frac{x_2}{x_3}u. \tag{8.36}$$

Then, under the new state $z = (z_1 \ z_2 \ z_3)^T$ and the new input v, the system is linearized to a controllable canonical system. Therefore, after the original nonlinear system is linearized, the old input u can be resolved in terms of the new input v as well as the old state variables x_1 through x_3, i.e.,

$$u = \alpha(x) + \beta(x)v, \tag{8.37}$$

where, for this particular example,

$$\alpha(x) = (1 - \ln x_3)^2 x_3 + 2x_1 x_3 \quad \text{and} \quad \beta(x) = -\frac{x_3}{x_2}.$$

Because equation (8.37) provides an algebraic relation between u and v and it does not contain any term with time-derivatives of u or v, equation (8.37) is further called a *static state-feedback control.*

Based on equations (8.32) and (8.34), a more general form for the new input, i.e., the static state-feedback control is given by

$$\alpha(x) = -\frac{L_f^n \mu(x)}{L_g L_f^{n-1} \mu(x)}, \quad \text{and} \quad \beta(x) = \frac{1}{L_g L_f^{n-1} \mu(x)}. \tag{8.38}$$

Therefore, to determine a static state-feedback control for a single-input nonlinear system, it is required to find a certain scalar function $\mu(x)$. Moreover, it can also be observed from the above example that $z_1 = \mu(x)$ is not unique in general. However, one can always choose a desired state such that its gradient vector is orthogonal to all the vector fields in the distribution Δ_l, as illustrated in the above example.

In order to extend the input-state linearization from a single-input case to a multi-input case, where the input has more than one coefficient vectors, i.e., $g_1(x), \cdots, g_r(x)$, one has to examine the involutivity for every sub-distribution $\Delta_{l0} = span\{g_1, \cdots, g_r\}$, $\Delta_{l1} = span\{g_1, \cdots, g_r, ad_f g_1, \cdots, ad_f g_r\}$, $\cdots$, $\Delta_{li} = span\{ad_f^k g_j : 0 \le k \le i, 1 \le j \le r\}$ [10]. This becomes a very lengthy and tedious task, and could even be impractical. Therefore, beyond the single-input case, the input-state linearization approach is often replaced by a more practical input-output linearization procedure.

8.4.3 The Input-Output Linearization Procedure

The main motivation for developing this *input-output linearization* procedure is to take advantage of an existing output function that is often specified by the physical requirements of a given nonlinear system to possibly generate a nonlinear transformation for the linearization. Clearly, this provides us with the following two major advantages: (1) to avoid the most difficult step of seeking some scalar functions as required by the input-state linearization procedure, especially for a multi-input nonlinear system; and (2) the new state of the linearized system can immediately be employed to represent any physical task required by the original system.

As was discussed in the last section, an MIMO complete nonlinear autonomous system can be generally expressed by

$$\begin{cases} \dot{x} = f(x) + \sum_{i=1}^{r} g_i(x) u_i \\ y = h(x). \end{cases} \tag{8.39}$$

If the dimension m of the output function $h(x)$ is equal to the input dimension r, the system in (8.39) is called a *square nonlinear system*. For such a square system, the decoupling matrix $D(x)$, as formulated in Definition 10, is an m by m square matrix ($m = r$). If $D(x)$ at a point of interest is non-singular, then the total relative degree r_d of the square system should be either less than or equal to n, the dimension of the state vector x.

In general,

Definition 12. A complete nonlinear control system (8.39) is said to be input-output linearizable if its input-output mapping is invertible and the total relative degree $r_d = n$.

Under this definition, let us derive an input-output relation that is required for the linearization by taking the time-derivative for each component of the output $y = h(x)$. Since for the j-th component $y_j = h_j(x)$, $\dot{y}_j = (\partial h_j(x)/\partial x)\dot{x}$, substituting the first equation of (8.39) yields

$$\dot{y}_j = \frac{\partial h_j(x)}{\partial x}[f(x) + \sum_{i=1}^{m} g_i(x)u_i] = L_f h_j(x) + \sum_{i=1}^{m} (L_{g_i} h_j(x))u_i.$$

Continuing to take higher time-derivatives until reaching the r_j-th order, where r_j is the relative degree of the j-th output channel $h_j(x)$, we obtain

$$y_j^{(r_j)} = L_f^{r_j} h_j(x) + \sum_{i=1}^{m} (L_{g_i} L_f^{r_j-1} h_j(x))u_i.$$

Augmenting all the output components together, one can finally express

$$\begin{aligned}\begin{pmatrix} y_1^{(r_1)} \\ \vdots \\ y_m^{(r_m)} \end{pmatrix} &= \begin{pmatrix} L_f^{r_1} h_1(x) \\ \vdots \\ L_f^{r_m} h_m(x) \end{pmatrix} + \\ &+ \begin{pmatrix} L_{g_1} L_f^{r_1-1} h_1(x) & \cdots & L_{g_r} L_f^{r_1-1} h_1(x) \\ \vdots & \cdots & \vdots \\ L_{g_1} L_f^{r_m-1} h_m(x) & \cdots & L_{g_r} L_f^{r_m-1} h_m(x) \end{pmatrix} \begin{pmatrix} u_1 \\ \vdots \\ u_m \end{pmatrix} \\ &= b(x) + D(x)u,\end{aligned}$$

where $b(x) = (L_f^{r_1} h_1(x) \ \cdots \ L_f^{r_m} h_m(x))^T$ is an m by 1 vector, $D(x)$ is just the m by m decoupling matrix ($r = m$), and $u = (u_1 \ \cdots \ u_m)^T$ is the input vector of the square system. By defining a new input $v = (y_1^{(r_1)} \ \cdots \ y_m^{(r_m)})^T$, we arrive at the desired destination:

$$v = b(x) + D(x)u, \tag{8.40}$$

where the old input u can be resolved in terms of the new one v through the internal states by

$$u = \alpha(x) + \beta(x)v = -D^{-1}(x)b(x) + D^{-1}(x)v. \tag{8.41}$$

This form is identical to the static state-feedback control law given in (8.37) except that both u and v are now two vectors. Because the above control law requires inverting $D(x)$, the nonlinear system should be at least invertible before it is possibly input-output linearizable.

With the new input v defined, we now try to find a new state vector. Since the input or output dimension m is, in general, less than (at most equal to) the state dimension n, and the input-output linearization also requires the total relative degree $r_d = r_1 + \cdots + r_m = n$, we may divide the n promising new state variables z_1 through z_n into m groups, with the i-th group having r_i members ($i = 1, \cdots, m$). Each member of the i-th group is defined by the i-th output component $y_i = h_i(x)$ and its time-derivatives.

In order to more explicitly illustrate the definition of z, without loss of generality, let us assume that $m = 2$, $n = 5$, $r_1 = 2$ and $r_2 = 3$. Then, defining $z_1 = y_1$, $z_2 = \dot{y}_1$, $z_3 = y_2$, $z_4 = \dot{y}_2$ and $z_5 = \ddot{y}_2$, and noticing that the proper definition of the new input v in this case is $v = (y_1^{(r_1)} \; y_2^{(r_2)})^T = (y_1^{(2)} \; y_2^{(3)})^T$, the linearized system becomes

$$\dot{z} = Az + Bv, \tag{8.42}$$

where $z = (z_1 \; \cdots \; z_5)^T$ and

$$A = \begin{pmatrix} 0 & 1 & 0 & 0 & 0 \\ 0 & 0 & 0 & 0 & 0 \\ 0 & 0 & 0 & 1 & 0 \\ 0 & 0 & 0 & 0 & 1 \\ 0 & 0 & 0 & 0 & 0 \end{pmatrix}, \quad \text{and} \quad B = \begin{pmatrix} 0 & 0 \\ 1 & 0 \\ 0 & 0 \\ 0 & 0 \\ 0 & 1 \end{pmatrix}.$$

It can be observed from the above definition of z that the total number of the new state variables is equal to r_d, which is equal to the old state space dimension. This is the main reason why the input-output linearization should also require $r_d = n$.

Because an output function and its time-derivatives can be chosen to form a set of new state variables in an input-output linearizable system, any physical task-planning for a real system can directly be represented in terms of the new state variables. A typical application is to carry out a *trajectory-tracking* task for a dynamic system, such as for an industrial robot manipulator to track a desired path. For instance, the above input-output linearizable example was assumed to have $m = 2$, $n = 5$, $r_1 = 2$ and $r_2 = 3$. If a desired output trajectory $y_d(t) = (y_{1d}(t) \; y_{2d}(t))^T$ is specified for the actual system output to follow, then the following PD control law for the new input v can guarantee the trajectory-tracking convergence:

$$v = \begin{pmatrix} \ddot{y}_{1d} + k_{1v}\Delta\dot{y}_1 + k_{1p}\Delta y_1 \\ y_{2d}^{(3)} + k_{2a}\Delta\ddot{y}_2 + k_{2v}\Delta\dot{y}_2 + k_{2p}\Delta y_2 \end{pmatrix}, \tag{8.43}$$

where, and hereafter $y^{(3)}$ is the third time-derivative of $y(t)$, $\Delta y_1 = y_{1d} - y_1$ and $\Delta y_2 = y_{2d} - y_2$ are the two output error components between the desired and actual output trajectories, and k_{1v}, k_{1p}, k_{2a}, k_{2v} and k_{2p} are positive gain constants. In fact, since $v = (\ddot{y}_1 \;\; y_2^{(3)})^T$, substituting it into equation (8.43), with some manipulation, we obtain

$$\begin{pmatrix} \Delta\ddot{y}_1 + k_{1v}\Delta\dot{y}_1 + k_{1p}\Delta y_1 \\ \Delta y_2^{(3)} + k_{2a}\Delta\ddot{y}_2 + k_{2v}\Delta\dot{y}_2 + k_{2p}\Delta y_2 \end{pmatrix} = 0. \tag{8.44}$$

This clearly demonstrates that the output error vector Δy can asymptotically converge to zero if the gain constants for each component of the vector in (8.44) can make each equation be Hurwitz, i.e., all the roots of their characteristic equations have strictly negative real parts.

To illustrate the input-output linearization procedure, we look at the following 3-dimensional nonlinear control system:

$$\begin{cases} \dot{x}_1 = x_2 \\ \dot{x}_2 = -x_1 + x_3^2 + u_1 + 3u_2 \\ \dot{x}_3 = -x_1^3 + 2u_1 + 4u_2. \end{cases} \tag{8.45}$$

If an output vector is defined by $y = h(x) = \begin{pmatrix} x_1 \\ x_2 \end{pmatrix}$, then we follow the procedure to find its relative degree. In this case, the input coefficients are $g_1 = (0 \;\; 1 \;\; 2)^T$ and $g_2 = (0 \;\; 3 \;\; 4)^T$, and the first-order Lie derivatives become

$$\begin{aligned} L_{g_1}h_1(x) &= \frac{\partial h_1}{\partial x}g_1(x) = (1 \;\; 0 \;\; 0)g_1(x) = 0, \\ L_{g_2}h_1(x) &= \frac{\partial h_1}{\partial x}g_2(x) = (1 \;\; 0 \;\; 0)g_2(x) = 0, \\ L_{g_1}h_2(x) &= \frac{\partial h_2}{\partial x}g_1(x) = (0 \;\; 1 \;\; 0)g_1(x) = 1, \\ L_{g_2}h_2(x) &= \frac{\partial h_2}{\partial x}g_2(x) = (0 \;\; 1 \;\; 0)g_2(x) = 3. \end{aligned}$$

Because the first two Lie derivatives for $h_1(x)$ are all zeros and the last two for $h_2(x)$ are non-zero, we should stop differentiating $h_2(x)$ while continuing to take higher-order Lie derivatives for $h_1(x)$. Since

$$\begin{aligned} L_f h_1(x) &= \frac{\partial h_1}{\partial x}f(x) = (1 \;\; 0 \;\; 0)f(x) = x_2, \\ L_{g_1}L_f h_1(x) &= \frac{\partial L_f h_1}{\partial x}g_1(x) = (0 \;\; 1 \;\; 0)g_1(x) = 1, \\ L_{g_2}L_f h_1(x) &= \frac{\partial L_f h_1}{\partial x}g_2(x) = (0 \;\; 1 \;\; 0)g_2(x) = 3, \end{aligned}$$

both of which are non-zero, but the decoupling matrix

$$D(x) = \begin{pmatrix} 1 & 3 \\ 1 & 3 \end{pmatrix}$$

is singular. Therefore, the system with $y = \begin{pmatrix} x_1 \\ x_2 \end{pmatrix}$ is not invertible, nor is it input-output linearizable.

If the output vector is now redefined to be $y = h(x) = \begin{pmatrix} x_1 \\ x_3 \end{pmatrix}$, then we can repeat the above derivations again and show that the relative degrees $r_1 = 2$ and $r_2 = 1$, and the decoupling matrix

$$D(x) = \begin{pmatrix} 1 & 3 \\ 2 & 4 \end{pmatrix}$$

is obviously non-singular. Therefore, with the output redefined, due to $r_d = r_1 + r_2 = 3 = n$, the system is input-output linearizable.

It is time now to define the new state variables $z_1 = h_1(x) = x_1$, $z_2 = \dot{x}_1$ and $z_3 = h_2(x) = x_3$, and define a new input $v = (\ddot{y}_1\ \dot{y}_2)^T$. Based on equation (8.41), the old input u can be resolved by calculating the vector $b(x)$ and the decoupling matrix inverse $D^{-1}(x)$. Since $b(x) = (L_f^2 h_1(x)\ L_f h_2(x))^T = (-x_1 + x_3^2\ \ -x_1^3)^T$, we obtain

$$\alpha(x) = -D^{-1}(x)b(x) = \begin{pmatrix} -2x_1 + 2x_3^2 + \frac{3}{2}x_1^3 \\ x_1 - x_3^2 - \frac{1}{2}x_1^3 \end{pmatrix}$$

and

$$\beta(x) = D^{-1}(x) = \begin{pmatrix} -2 & \frac{3}{2} \\ 1 & -\frac{1}{2} \end{pmatrix}.$$

Finally, the static state-feedback control for the system given in (8.45) with $y = \begin{pmatrix} x_1 \\ x_3 \end{pmatrix}$ is determined by $u = \alpha(x) + \beta(x)v$.

As a second realistic example, let us revisit the fully-actuated 6-joint robot arm. It has been shown in the previous section that the robotic system along with the end-effector Cartesian position and orientation as a 6-dimensional output function is invertible with the total relative degree $r_d = 12 = n$ if the Jacobian matrix J is non-singular, and thus, it is input-output linearizable. Although we have also derived its decoupling matrix $D(x) = JW^{-1}$ under the equal-RD assumption, the vector $b(x) = L_f^2 h(q)$ is quite difficult to calculate symbolically. Because a fully-actuated serial-chain robot arm is always a typical mechanical system with the relative degree $r_{cd} = 2$ for each output channel, we may alternatively develop its static state-feedback control law from its dynamic equation in a more compact manner.

Since $y = h(q)$ and $\dot{y} = J\dot{q}$, one can find $\ddot{y} = J\ddot{q} + \dot{J}\dot{q}$ so that $\ddot{q} = J^{-1}\ddot{y} - J^{-1}\dot{J}\dot{q}$. On the other hand, the dynamic equation of the robotic system can be expressed as

$$W\ddot{q} + c(q, \dot{q}) - \tau_g = \tau.$$

Substituting the above $\ddot{q}$ into the robot dynamic equation, we obtain

$$WJ^{-1}\ddot{y} - WJ^{-1}\dot{J}\dot{q} + c(q, \dot{q}) - \tau_g = \tau.$$

If we define the old input $u = \tau$ and the new input $v = \ddot{y}$, then the above equation becomes

$$u = WJ^{-1}v - WJ^{-1}\dot{J}\dot{q} + c(q, \dot{q}) - \tau_g.$$

Comparing this with equation (8.41), we conclude that

$$\alpha(x) = -WJ^{-1}\dot{J}\dot{q} + c(q, \dot{q}) - \tau_g, \quad \text{and} \quad \beta(x) = WJ^{-1}. \tag{8.46}$$

This straightforward and compact derivation showcases that instead of the required computation for both $D(x)$ and $b(x)$, we can determine a static state-feedback control law directly from the robot dynamic equation.

It can also be seen that the new state vector, after the input-output linearization is applied to the robotic system, is $z = \begin{pmatrix} y \\ \dot{y} \end{pmatrix}$. If a trajectory-tracking task is planned for the robot, then we can define the following PD control law for the linearized system:

$$v = \ddot{y}_d + K_v \Delta\dot{y} + K_p \Delta y, \tag{8.47}$$

where $y_d = y_d(t)$ represents a desired output trajectory, $\Delta y = y_d - y$ is the output error vector between the desired and actual output vectors, and K_v and K_p are two positive-definite diagonal constant gain matrices. It is quite interesting that the entire robotic control system, as shown in Figure 8.7, contains an *inner nonlinear feedback loop* representing the static state-feedback control law and an *outer linear feedback loop* that represents the PD control law (8.47).

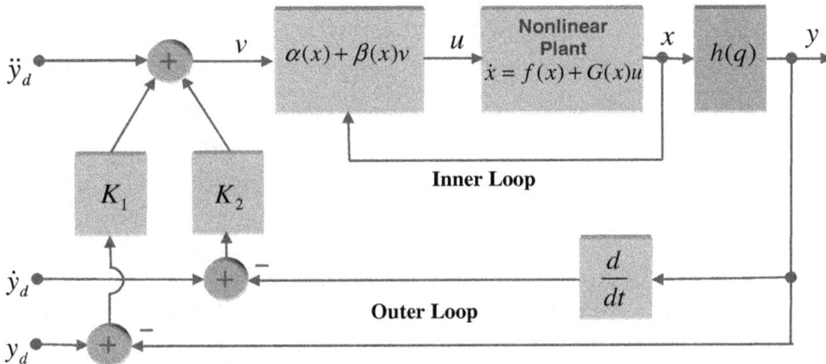

Fig. 8.7 A block diagram for an input-output linearized trajectory-tracking system

8.4.4 Dynamic Extension for I/O Channels

When the determination of relative degree is extended from an SISO system to an MIMO system, the decoupling matrix $D(x)$ is required to be not only non-zero, but also non-singular if it is a square system. To resolve the input, based on the input-output linearization theory, the decoupling matrix $D(x)$ has to be inverted. What should we conclude about the relative degree, invertibility and input-output linearizability for a given MIMO complete system if $D(x)$ is non-zero but is singular? In fact, if we alternatively define the output $y = \begin{pmatrix} x_1 \\ x_3 \end{pmatrix}$ for the previous fifth-order system in (8.25), using the same procedure, we will readily reach the result that $r_1 = 2$ and $r_2 = 1$ and the total relative degree $r_d = r_1 + r_2 = 3 < n = 5$. However, the decoupling matrix becomes

$$D(x) = \begin{pmatrix} 1 & 1 \\ 1 & 1 \end{pmatrix},$$

which is obviously singular. Based on the definition, it has failed to determine both the relative degree and invertibility for this particular MIMO system.

With further study, we discover that a singular decoupling matrix is, in most cases, caused by either insufficient input channels or insufficient output channels, or both. In many such cases, however, it can be remedied by adding or re-defining more input or output functions. The remedy procedure is often referred to as a *dynamic extension* [14]. Let us look at the following example: a 3rd-order nonlinear system is given by

$$f(x) = \begin{pmatrix} -x_1^2 - x_3 \\ x_1 x_2 \\ -x_2 + x_1 x_3 \end{pmatrix}, \quad g_1(x) = \begin{pmatrix} 1 \\ 1 \\ 0 \end{pmatrix} \quad \text{and} \quad g_2(x) = \begin{pmatrix} 1 \\ 0 \\ 1 \end{pmatrix} \tag{8.48}$$

for two inputs u_1 and u_2. If the output is $y = \begin{pmatrix} x_1 \\ x_2 + x_3 \end{pmatrix}$, then, we can immediately calculate that $L_{g_1} h_1(x) = 1$ and $L_{g_2} h_1(x) = 1$. Hence, $r_1 = 1$ for the first output channel. Now, by calculating the second output channel, we have $L_{g_1} h_2(x) = 1$ and $L_{g_2} h_2(x) = 1$, and $r_2 = 1$, too. Therefore, the new input $v = \begin{pmatrix} \dot{y}_1 \\ \dot{y}_2 \end{pmatrix}$ and the decoupling matrix turns out to be

$$D(x) = \begin{pmatrix} 1 & 1 \\ 1 & 1 \end{pmatrix}$$

that is, however, singular with $\text{rank}(D) = 1$.

This dilemma is often due to a delay of either one of the inputs or outputs. To remedy it, there are two different approaches available: one is to add an input integrator and the other one is to redefine an output.

Let us first add an input integrator. Before doing so, let the input vector $\begin{pmatrix} u_1 \\ u_2 \end{pmatrix} = T \begin{pmatrix} w_1 \\ w_2 \end{pmatrix}$ that is congruently transformed to a new input w by a non-singular constant matrix T. According to equation (8.40), the last term $D(x)u = D(x)Tw$. For this particular system, let

$$T = \begin{pmatrix} 1 & -1 \\ 0 & 1 \end{pmatrix} \quad \text{such that} \quad D(x)T = \begin{pmatrix} 1 & 0 \\ 1 & 0 \end{pmatrix}.$$

The new decoupling matrix $D(x)T$ with respect to the input $w = \begin{pmatrix} w_1 \\ w_2 \end{pmatrix}$ becomes only one non-zero column to explicitly reflect $\mathrm{rank}(D) = 1$. This also reveals a fact that the second input channel is delayed to show up in comparison with the first one (the second column of DT is zero while the first one is not). By taking one more time-derivative on both sides of equation (8.40), and invoking the state equation in (8.39), we obtain

$$\dot{v} = \begin{pmatrix} \ddot{y}_1 \\ \ddot{y}_2 \end{pmatrix} = \dot{b}(x) + DT\dot{w} = L_f b(x) + L_{\bar{g}_1} b(x) w_1 + L_{\bar{g}_2} b(x) w_2 + \begin{pmatrix} 1 \\ 1 \end{pmatrix} \dot{w}_1, \tag{8.49}$$

where $\bar{g}_1$ and $\bar{g}_2$ in the Lie derivatives are the coefficient vectors of the input w_1 and w_2 after they are congruently transformed. Namely,

$$(\bar{g}_1 \;\; \bar{g}_2) = (g_1 \;\; g_2)T = \begin{pmatrix} 1 & 1 \\ 1 & 0 \\ 0 & 1 \end{pmatrix} \begin{pmatrix} 1 & -1 \\ 0 & 1 \end{pmatrix} = \begin{pmatrix} 1 & 0 \\ 1 & -1 \\ 0 & 1 \end{pmatrix}.$$

With a further computation, we have

$$L_f b(x) = \begin{pmatrix} 2x_1^3 + x_1 x_3 + x_2 \\ -2x_1 x_2 - x_2 x_3 - x_3^2 \end{pmatrix}, \quad L_{\bar{g}_1} b(x) = \begin{pmatrix} -2x_1 \\ x_1 + x_2 + x_3 - 1 \end{pmatrix}$$

and

$$L_{\bar{g}_2} b(x) = \begin{pmatrix} -1 \\ 1 \end{pmatrix}.$$

Therefore, the new state-feedback (8.49) becomes

$$\dot{v} = \eta(x,\, w_1) + E(x) \begin{pmatrix} \dot{w}_1 \\ w_2 \end{pmatrix},$$

where $\eta(x,\, w_1) = L_f b(x) + L_{\bar{g}_1} b(x) w_1$ and $E(x) = \begin{pmatrix} 1 & -1 \\ 1 & 1 \end{pmatrix}$. Since this *extended decoupling matrix* $E(x)$ is absolutely non-singular, we can resolve the inputs w_1 and w_2 by the following differential equation:

$$\begin{pmatrix} \dot{w}_1 \\ w_2 \end{pmatrix} = E^{-1}(\dot{v} - \eta(x,\, w_1)). \tag{8.50}$$

After solving the differential equation in (8.50), the original inputs of the system is determined by $u = Tw$. Because the input w has to be resolved from a differential equation, instead of an algebraic equation in the static state-feedback linearization case, we call equation (8.50) the *dynamic state-feedback control.* Moreover, since the new input $\dot{v}$ has raised one more order of time-derivative after the dynamic extension, it looks like the new relative degree $r_d = 2 + 2 = 4$, and the linearized system dimension will also be extended to be higher than $n = 3$.

The second approach is to redefine a new output. With the same system as above, let $z = \dot{y}_1 - \dot{y}_2$, and $\dot{z} = \ddot{y}_1 - \ddot{y}_2$. This is virtually to subtract the second element of $\dot{v}$ in (8.49) from the first one so that the $\dot{w}_1$ term will be canceled out due to its coefficient matrix $\begin{pmatrix} 1 \\ 1 \end{pmatrix}$. Thus,

$$\dot{z} = \dot{v}_1 - \dot{v}_2 = \phi_1(x) + \phi_2(x)w_1 - 2w_2,$$

where $\phi_1(x) = 2x_1^3 + x_1x_3 + x_2 + 2x_1x_2 + x_2x_3 + x_3^2$ and $\phi_2(x) = 1 - 3x_1 - x_2 - x_3$. If we now augment $v_1 = \dot{y}_1$ and $\dot{z}$ together, we can define a 2-dimensional extended new input $\bar{v}$ as follows:

$$\bar{v} = \begin{pmatrix} \dot{y}_1 \\ \dot{z} \end{pmatrix} = \begin{pmatrix} b_1(x) + w_1 \\ \phi_1(x) + \phi_2(x)w_1 - 2w_2 \end{pmatrix}, \tag{8.51}$$

where $b_1(x) = L_f h_1(x) = -x_1^2 - x_3$ is the first element of the vector $b(x)$ in (8.40). Clearly, the new extended decoupling matrix

$$\bar{E}(x) = \begin{pmatrix} 1 & 0 \\ \phi_2(x) & -2 \end{pmatrix},$$

which is the coefficient of the input vector w and is non-singular for this particular example. Therefore, we finally resolve the input w in terms of the new extended input $\bar{v}$ by

$$w = \begin{pmatrix} w_1 \\ w_2 \end{pmatrix} = \bar{E}^{-1} \left(\begin{pmatrix} \dot{y}_1 \\ \dot{z} \end{pmatrix} - \begin{pmatrix} b_1(x) \\ \phi_1(x) \end{pmatrix} \right) = \bar{E}^{-1}(\bar{v} - \bar{b}(x)). \tag{8.52}$$

By $u = Tw$, we can also resolve the original input u for the given system (8.48).

In summary, it can clearly be seen that the input w in the second approach of dynamic extension is not involved in a differential equation in contrast to the first input-integrator approach. In other words, equation (8.52) is an algebraic equation and thus it belongs to the static state-feedback category. This is considered as a major advantage over the first approach.

Moreover, the output functions in $\bar{v}$, though requiring higher-order time-derivatives, do not change the dimension of $\bar{v}$ in the second output-redefining approach. The only drawback is that since z is redefined as a linear combination between the original outputs y_1 and y_2, its physical meaning will no longer be straightforward in comparison with the individual y_1 and y_2. This disadvantage may cause some difficulty to determine a PD controller for $\dot{z}$ in a trajectory-tracking task planning case.

Finally, the above two approaches of dynamic extension to remedy a singular but nonzero decoupling matrix are both effective for the systems with constant decoupling matrices. However, in many dynamic systems, a decoupling matrix $D(x)$ is often a function of the state x. It could be singular only at a few isolated points so that it is unable to apply either one of the two approaches to resolve the feedback control input at a singular point of $D(x)$. For instance, a 6-joint robotic system has its nonlinear static state-feedback in (8.46). Its 6 by 6 decoupling matrix is $D(x) = JW^{-1}$. Since the inertial matrix W and its inverse W^{-1} are never singular, the decoupling matrix $D(x)$ becomes singular only at the singular points of the Jacobian matrix J. If J reaches its singularity, not only the dynamic control, but also the entire kinematic motion will all be degenerated. As discussed in Chapter 4, one cannot eliminate the singularity, but can only avoid it before occurring.

8.4.5 Linearizable Subsystems and Internal Dynamics

As we have just discussed, for a complete nonlinear control system, the invertibility and the total relative degree $r_d = n$ imply that the system is input-output linearizable. However, in many practical cases, a system may be just invertible, but $r_d < n$. This situation, though excluded by the definition of input-output linearizability, can still be linearized *partially*. In other words, an invertible system with $r_d < n$ can be decomposed into two portions: one is an input-output linearizable subsystem of dimension r_d, and the other one is an unlinearizable subsystem of dimension $n - r_d$, which is conventionally called an *internal dynamics* [10, 12, 14].

Recalling the fifth-order nonlinear control system given in (8.25), let us repeat the equation here:

$$\dot{x} = \begin{pmatrix} -x_1^3 + x_2 \\ x_1 x_3 \\ -x_1 + x_3 \\ x_2 \\ x_5 + x_3^2 \end{pmatrix} + \begin{pmatrix} 0 \\ 1 \\ 1 \\ 0 \\ x_2 \end{pmatrix} u_1 + \begin{pmatrix} 0 \\ 1 \\ 0 \\ 0 \\ 0 \end{pmatrix} u_2, \quad \text{and} \quad y = \begin{pmatrix} x_3 \\ x_4 \end{pmatrix}. \tag{8.53}$$

As we have calculated earlier, the relative degrees $r_1 = 1, r_2 = 2$ and the total relative degree $r_d = 3 < n$. Therefore, the entire system can be decomposed into a 3-dimensional linearizable subsystem and a 2-dimensional internal dynamics. For the linearizable one, the new state variables are $z_1 = y_1 = x_3$, $z_2 = y_2 = x_4$, and $z_3 = \dot{y}_2 = \dot{x}_4 = x_2$ that is based on the fourth equation of (8.53). Therefore, with $z = (z_1 \;\; z_2 \;\; z_3)^T = (x_3 \;\; x_4 \;\; x_2)^T$, the 3-dimensional linearized subsystem can be described by $\dot{z}_1 = v_1$, $\dot{z}_2 = z_3$ and $\dot{z}_3 = v_2$, or

$$\dot{z} = \begin{pmatrix} 0 & 0 & 0 \\ 0 & 0 & 1 \\ 0 & 0 & 0 \end{pmatrix} z + \begin{pmatrix} 1 & 0 \\ 0 & 0 \\ 0 & 1 \end{pmatrix} \begin{pmatrix} v_1 \\ v_2 \end{pmatrix}, \tag{8.54}$$

where the new input

$$\begin{aligned} v = \begin{pmatrix} v_1 \\ v_2 \end{pmatrix} &= b(x) + D(x)u = \begin{pmatrix} L_f h_1(x) \\ L_f^2 h_2(x) \end{pmatrix} + D(x)u \\ &= \begin{pmatrix} -x_1 + x_3 \\ x_1 x_3 \end{pmatrix} + \begin{pmatrix} 1 & 0 \\ 1 & 1 \end{pmatrix} \begin{pmatrix} u_1 \\ u_2 \end{pmatrix}. \end{aligned} \tag{8.55}$$

By taking aside the second, third and fourth rows of equation (8.53), which correspond to the new states of the linearized subsystem, the remaining two rows are supposed to represent the internal dynamics. If we denote $\zeta_1 = x_1$ and $\zeta_2 = x_5$, then the first and the fifth rows in (8.53) can be rewritten by

$$\begin{cases} \dot{\zeta}_1 = -\zeta_1^3 + z_3 \\ \dot{\zeta}_2 = \zeta_2 + z_1^2 + z_3 u_1, \end{cases} \tag{8.56}$$

where u_1 can be solved using the first row of (8.55), i.e., $u_1 = \zeta_1 - z_1 + v_1$.

In general, for an n-dimensional complete nonlinear control system with m-dimensional input and output, if it is invertible and the total relative degree $r_d < n$, then the new state vector z for the input-output linearizable subsystem is r_d-dimensional, while the new state vector ζ for the internal dynamics is $(n - r_d)$-dimensional. The entire control system sitting in the new coordinate system can be expressed as

$$\begin{cases} \dot{z} = Az + Bv \\ \dot{\zeta} = \phi(\zeta,\ z) + \psi(\zeta,\ z)v, \end{cases} \tag{8.57}$$

where A and B are r_d by r_d and r_d by m constant matrices, respectively, in a controllable canonical form, and $\phi(\,\cdot\,,\,\cdot\,)$ and $\psi(\,\cdot\,,\,\cdot\,)$ are $n - r_d$ by 1 and $n - r_d$ by m vector fields representing the unlinearizable subsystem, i.e., the internal dynamics. Equation (8.57) is called a **normal form** of the control system [10, 12, 14]. The m-dimensional original input u and the new input v are related by either $v = b(x) + D(x)u$, or $u = \alpha(x) + \beta(x)v$ that is still called a *static state-feedback control.* The block diagram that represents this decomposition and the state-feedback control is shown in Figure 8.8.

It should be noted that the above discussion presumes *square* nonlinear systems, i.e., the input u and output y have the same dimension so that the new input v is also of the same dimension. The main reason for the equal dimensions between u and v is that the decoupling matrix is a square matrix and can be uniquely inverted to resolve the original input u. If a practical system does not meet the same dimensional condition between its input and output, one may increase or decrease the number of output variables to match the number of inputs. This is a reasonable method in most applications,

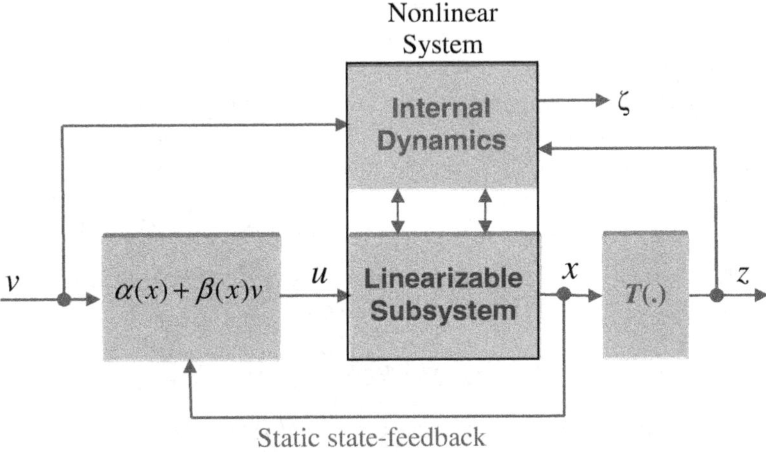

Fig. 8.8 A block diagram for a partially input-output linearized system

because a certain number of independent inputs, intuitively, are supposed to produce the same number of output responses.

In fact, the category of under-actuated robotic systems, such as a space robot or a floating base robot, inherently has a non-trivial internal dynamics. For example, if a 3-joint robot arm is mounted on a floating base that can free move in 3 d.o.f.: two translational axes and one rotational axis. Then, such a floating base robotic system can be modeled as a 6-joint robot with respect to the fixed base on the earth. The total dimension of the state space is $2 \times 6 = 12$ due to $x = \begin{pmatrix} q \\ \dot{q} \end{pmatrix}$. In this case, only three joints of the robot arm sitting on the 3-axis floating base can be driven by their motors so that the dimension of the input is three, i.e., $u = (\tau_4\ \tau_5\ \tau_6)^T$.

If an output is defined to represent the robot hand 3 d.o.f. motion, the dimension of the output $y = h(q)$ should also be three. Its Jacobian matrix can be determined by

$$J = \frac{\partial h(q)}{\partial q},$$

which is a 3 by 6 short matrix because the joint position vector q of this floating base robot model is 6-dimensional. Now, according to the equal-RD condition, each of the three output channels has a relative degree $r_i = 2$ so that the total relative degree is $r_d = 3r_i = 6 < n = 12$. Therefore, we can predict that the linearizable subsystem for this floating base robotic system referred to the fixed base on the earth is $r_d = 6$-dimensional and the dimension of the internal dynamics becomes $n - r_d = 6$. For this non-fully linearizable system, we cannot employ the same formula as derived in equation (8.46) to determine its state-feedback control law due to the non-square Jacobian matrix.

Although we can directly apply the general theory and procedures of the exact linearization to find a nonlinear state-feedback control, as expressed in equation (8.41) through a determination of the decoupling matrix $D(x)$ and $b(x)$, the derivation for such an under-actuated robotic system with a non-trivial internal dynamics will be very tedious. In order to seek a better and more compact symbolical derivation, let us start from the robot dynamic equation:

$$W\ddot{q} + c(q, \dot{q}) - \tau_g = \tau.$$

Since the inertial matrix W is always non-singular, pre-multiplying the W inverse on both sides of the equation yields

$$\ddot{q} + W^{-1}[c(q, \dot{q}) - \tau_g] = W^{-1}\tau.$$

For a certain robotic output function $y = h(q)$, $\dot{y} = J\dot{q}$ so that $\ddot{y} = J\ddot{q} + \dot{J}\dot{q}$. By further pre-multiplying the Jacobian matrix J, even though it is not a square matrix, to the above dynamic equation, we obtain

$$J\ddot{q} + JW^{-1}[c(q, \dot{q}) - \tau_g] = \ddot{y} - \dot{J}\dot{q} + JW^{-1}[c(q, \dot{q}) - \tau_g] = JW^{-1}\tau. \quad (8.58)$$

Since τ is a nominal n-dimensional external torque vector, in an under-actuated robot case, such as in the above floating base robotic system, only the last three components of τ are non-zero to form a non-zero input $u = (\tau_4\ \tau_5\ \tau_6)^T$, and the first three components of τ that belong to the floating base model are all zero. Therefore, without loss of generality, staying in the floating base robot case as a typical under-actuated robotic system, the 6 by 1 torque vector τ in the above dynamic equation (8.58) can be replaced by

$$\tau = \begin{pmatrix} 0_{3\times 1} \\ u \end{pmatrix}.$$

Let the last three columns of W^{-1} form a new 6 by 3 matrix B, i.e.,

$$B = W^{-1}(:, 4:6).$$

Then, equation (8.58) becomes

$$\ddot{y} - \dot{J}\dot{q} + JW^{-1}[c(q, \dot{q}) - \tau_g] = JBu.$$

Clearly, the 3 by 3 square coefficient matrix $JB = D$ of the input u is just the decoupling matrix. If $D = JB$ is non-singular, then, the static state-feedback control of the floating base robotic system can be resolved by

$$u = D^{-1}\ddot{y} - D^{-1}\dot{J}\dot{q} + D^{-1}JW^{-1}[c(q, \dot{q}) - \tau_g] = \alpha(x) + \beta(x)v, \quad (8.59)$$

where the new input $v = \ddot{y}$ and

$$\alpha(x) = -D^{-1}\dot{J}\dot{q} + D^{-1}JW^{-1}[c(q, \dot{q}) - \tau_g], \quad \text{and} \quad \beta(x) = D^{-1} = (JB)^{-1}.$$

While it is a success to achieve the above compact nonlinear feedback control law in (8.59) for an under-actuated robotic system, to find an explicit form of its internal dynamics may still be a tough job. In particular, the two subsystems: the linearizable one and the internal dynamics are often coupled to each other. Nevertheless, as long as the stability of the internal dynamics can be justified, finding its explicit form becomes unimportant.

8.4.6 Zero Dynamics and Minimum-Phase Systems

Having introduced the special decomposition for a partially input-output linearizable nonlinear control system, we now turn our attention to discussing the stability issue. It is now very clear that a linearizable subsystem can always be stabilized by designing a certain linear control law, for instance, using a PD controller to asymptotically stabilize the trajectory-tracking problem. However, what is the stability of the internal dynamics? Does it interfere the stabilized linearizable subsystem if it is unstable?

To answer those fundamental questions, let us first consider the previous example in (8.53). Its 3-dimensional linearized subsystem was given by (8.54) and the 2-dimensional internal dynamics was described in (8.56). If the system is to be applied for executing a trajectory-tracking task, under a given desired trajectory $y_d(t) = (y_{1d}(t) \;\; y_{2d}(t))^T$, we can always define a PD control law as follows:

$$v = \begin{pmatrix} \dot{y}_{1d} + k_{1p}\Delta y_1 \\ \ddot{y}_{2d} + k_{2v}\Delta\dot{y}_2 + k_{2p}\Delta y_2 \end{pmatrix}. \tag{8.60}$$

This ensures that the actual output vector $y(t)$ of the linearized subsystem will converge to the desired trajectory $y_d(t)$, as long as the constant gains k_{1p}, k_{2v}, and k_{2p} can make sure each component of (8.60) be Hurwitz.

However, the non-trivial internal dynamics (8.56) may interfere and destroy the entire system's stability even if it has a small unstable region. In general, according to equation (8.56), an internal dynamics may still be coupled with the linearized subsystem through the new state z. Although the linearizable subsystem and the internal dynamics can be *decomposed* in concept, they may not be thoroughly *decoupled* in reality for many nonlinear systems. This causes a difficulty in testing whether an internal dynamics is stable [15, 16].

To answer the fundamental questions, we need more insight into the concept of decomposition and the related stability. Let us go backwards to revisit the typical linear time-invariant SISO system, as given in equation (8.19). After taking the Laplace transformation under the zero initial condition, the entire system is converted to (8.20) so that the transfer function can be deduced by

$$G(s) = \frac{Y(s)}{U(s)} = C(sI - A)^{-1}B = \frac{b_0 s^k + b_1 s^{k-1} + \cdots + b_k}{s^n + a_1 s^{n-1} + \cdots + a_n} = \frac{N(s)}{D(s)},$$

where we assume that the dimension of the system is n. Clearly, the denominator $D(s)$ is an n-th order s-domain polynomial and the numerator $N(s)$ is defined as a k-th order polynomial with $k < n$ for a proper transfer function. In order words, the transfer function $G(s)$ has n poles and k zeros and the difference $n - k = r_d$ is just the relative degree of this linear system.

Let $D(s)$ be divided by $N(s)$ through a polynomial long division procedure. We can always expect to have the following general result:

$$D(s) = Q(s)N(s) + R(s),$$

where $Q(s)$ is the r_d-th order *quotient* polynomial and $R(s)$ is up to $k - 1$st order *residue* polynomial. For example, suppose that

$$G(s) = \frac{2s^2 + 2s + 1}{s^4 + 3s^3 + 4s^2 + s + 1}.$$

After a long division of the 4th order $D(s)$ by the 2nd order $N(s)$, i.e., $n = 4$ and $k = 2$, we obtain a quotient $Q(s) = 0.5s^2 + s + 0.75$ that is the 2nd order due to $r_d = n - k = 2$ and a residue $R(s) = -1.5s + 0.25$. With such a long division, in general, we can rewrite

$$G(s) = \frac{N(s)}{D(s)} = \frac{N(s)}{Q(s)N(s) + R(s)} = \frac{\frac{1}{Q(s)}}{1 + \frac{1}{Q(s)} \cdot \frac{R(s)}{N(s)}}. \tag{8.61}$$

In comparison to the well-known negative feedback closed-loop gain equation:

$$G(s) = \frac{H(s)}{1 + H(s)F(s)}$$

with an open-loop gain $H(s)$ and a feedback gain $F(s)$, we conclude that the transfer function $G(s)$ in (8.61) is virtually a negative feedback closed-loop gain with its open-loop gain $H(s) = \frac{1}{Q(s)}$ and the feedback gain $F(s) = \frac{R(s)}{N(s)}$, see Figure 8.9.

Since the open-loop gain $\frac{1}{Q(s)}$ has r_d poles and no zeros, this block is analog to the linearizable subsystem. Whereas the feedback gain $\frac{R(s)}{N(s)}$ has $k = n - r_d$ poles, the same number as the zeros of the closed-loop transfer function $G(s)$, because the numerator of $G(s)$ appears now in the denominator of the feedback $\frac{R(s)}{N(s)}$. Thus, this feedback block is analog to the internal dynamics. Conventionally, we say a linear system is *minimum phase* if both its zeros and poles are located on the left half of s-plane. Therefore, by borrowing the concept from the linear systems theory, we can state that *if the internal dynamics is asymptotically stable, then the entire nonlinear system should be asymptotically minimum-phase.*

A further study shows that we can determine the stability of an internal dynamics by "turning off" the linearized subsystem to ease the testing

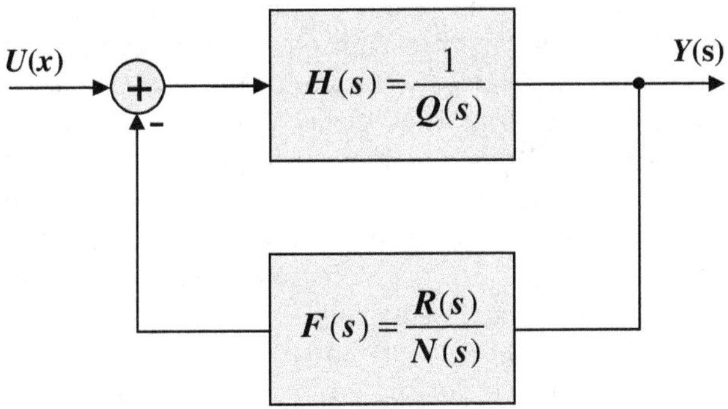

Fig. 8.9 The block diagram of a single feedback loop

difficulty. Namely, first set all the output variables and their time-derivatives to be zero and then test the internal dynamics stability under such a zero-output condition. We refer to the internal dynamics under the zero-output condition as the *zero dynamics* [10, 12, 14].

According to the normal form (8.57) in the sense of input-output linearization, a general internal dynamics can be written by

$$\dot{\zeta} = \phi(\zeta, z) + \psi(\zeta, z)v.$$

Since the zero-output condition implies that $z = 0$ and also $v = 0$, the corresponding zero dynamics becomes

$$\dot{\zeta} = \phi(\zeta,\ 0). \tag{8.62}$$

Therefore, with the zero dynamics as a reduced form of the original internal dynamics, we may have more successful chances to test its stability.

It can be further shown that a complete nonlinear control system (8.39) is locally (globally) asymptotically *minimum phase* if its zero dynamics is locally (globally) asymptotically stable [10, 12, 13, 14]. In addition, if (8.39) is locally (globally) asymptotically minimum phase, and the new input v of the linearized subsystem is defined by a PD trajectory-tracking control law, like equation (8.60) with each component being Hurwitz, then, the entire trajectory-tracking control system is at least locally asymptotically stable. This conclusion will be very useful in most practical cases for nonlinear trajectory-tracking control systems.

Let us now revisit the last example given in (8.53). We have found its internal dynamics described in (8.56) as 2-dimensional. Under the zero-output condition, $y_1 = 0$ and $\dot{y}_1 = 0$ due to $r_1 = 1$, and $y_2 = 0$, $\dot{y}_2 = 0$ and $\ddot{y}_2 = 0$ due to $r_2 = 2$, so that $z_1 = z_2 = z_3 = 0$ and $v = (\dot{y}_1\ \ \ddot{y}_2)^T = 0$ as well. Thus, the zero dynamics becomes

$$\begin{cases} \dot{\zeta}_1 = -\zeta_1^3 \\ \dot{\zeta}_2 = \zeta_2. \end{cases} \tag{8.63}$$

Obviously, the two equations in (8.63) are decoupled in this particular case. We can immediately show that the first equation is globally asymptotically stable at its equilibrium point $\zeta_1 = 0$, while the second one is unstable due to its solution $\zeta_2(t) = \zeta_2(0)e^t$ that grows exponentially to ∞ as $t \to \infty$. Therefore, the system (8.53) is non-minimum phase, nor is it stable.

However, if we replace x_5 in the last row of (8.53) by $-x_5$, then every thing will remain the same except that the second equation of the zero dynamics (8.63) will be changed to $\dot{\zeta}_2 = -\zeta_2$, which is obviously globally asymptotically stable. Therefore, under such a relatively simple modification in the model, the entire control system becomes globally asymptotically minimum phase, and with the PD trajectory-tracking control law (8.60) being Hurwitz, the remodeled system can be globally asymptotically stable.

To sum up all the above analyses and discussions on the concepts of the zero dynamics and its stability as well as the decomposition between the linearizable subsystem and internal dynamics in the sense of input-output linearization, let us look at a following interesting example that is called a ball-board control system. We want to control the board that is pivoted at the center point to allow itself only having two d.o.f. of rotation: pitch and yaw without roll (spin) rotation as well as no translation is allowed so that the ball on the board can track a desired sine wave trajectory.

In order to model both the kinematics and dynamics before developing a control strategy, we utilize the D-H convention in robotics to model such a non-robotic system. It will be immediately confirmed that the D-H convention borrowed from robotics is so convenient and effective for dealing with both the kinematic and dynamic modelings for this particular system. First, the entire ball-board control system can be modeled as a 4-joint under-actuated robot: two are revolute as the board's two d.o.f. rotation and the other two are prismatic to represent the ball rolling on the 2D board plane. The assignment of coordinate frames is given in Figure 8.10.

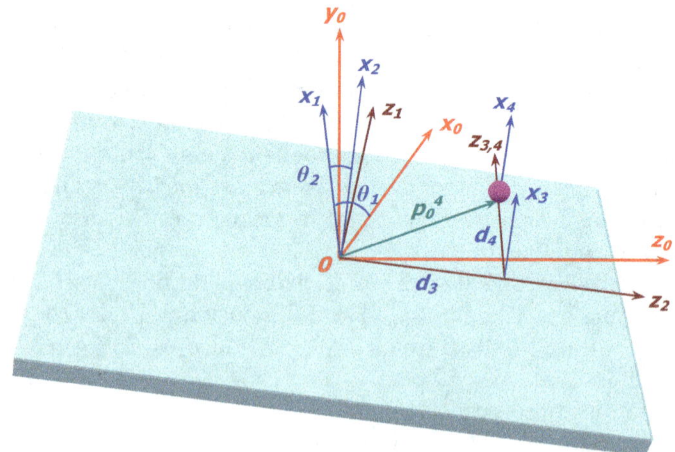

Fig. 8.10 Model a ball-board control system using the robotic D-H convention

Based on the D-H convention, the D-H parameter table for the ball-board system can be derived in Table 8.1. It has four joints: the first two are rotating the board about the z_0-axis of the base by θ_1 and rotating the board about the z_1-axis by θ_2 to reflect the board's two d.o.f restricted rotation. The last two are the orthogonal prismatic joints d_3 and d_4 to model the ball rolling in 2D space on the board. Therefore, with such an effective robotic modeling approach, the ball's instantaneous location can be uniquely represented by the position vector p_0^4 that is pointing to the origin of frame 4 and tailed at the origin of the base, see Figure 8.10 in detail.

Table 8.1 The D-H table of the ball-board system in Figure 8.10

Joint Angle	Joint Offset	Twist Angle	Link Length
θ_i	d_i	α_i	a_i
θ_1	0	$+90^0$	0
θ_2	0	-90^0	0
$\theta_3 = 0$	d_3	$+90^0$	0
$\theta_4 = 0$	d_4	0	0

With the D-H table, we can have four one-step homogeneous transformation matrices:

$$A_0^1 = \begin{pmatrix} c_1 & 0 & s_1 & 0 \\ s_1 & 0 & -c_1 & 0 \\ 0 & 1 & 0 & 0 \\ 0 & 0 & 0 & 1 \end{pmatrix}, \quad A_1^2 = \begin{pmatrix} c_2 & 0 & -s_2 & 0 \\ s_2 & 0 & c_2 & 0 \\ 0 & -1 & 0 & 0 \\ 0 & 0 & 0 & 1 \end{pmatrix},$$

and

$$A_2^3 = \begin{pmatrix} 1 & 0 & 0 & 0 \\ 0 & 0 & -1 & 0 \\ 0 & 1 & 0 & d_3 \\ 0 & 0 & 0 & 1 \end{pmatrix}, \quad A_3^4 = \begin{pmatrix} 1 & 0 & 0 & 0 \\ 0 & 1 & 0 & 0 \\ 0 & 0 & 1 & d_4 \\ 0 & 0 & 0 & 1 \end{pmatrix}.$$

Thus, the homogeneous transformation of frame 4 with respect to the base becomes

$$A_0^4 = A_0^1 A_1^2 A_2^3 A_3^4 = \begin{pmatrix} c_1c_2 & -c_1s_2 & s_1 & -d_3c_1s_2 + d_4s_1 \\ s_1c_2 & -s_1s_2 & -c_1 & -d_3s_1s_2 - d_4c_1 \\ s_2 & c_2 & 0 & d_3c_2 \\ 0 & 0 & 0 & 1 \end{pmatrix}.$$

Then, the position vector to trace the ball location on the 2D board with respect to the base is the last column of the above matrix:

$$p_0^4 = \begin{pmatrix} -d_3c_1s_2 + d_4s_1 \\ -d_3s_1s_2 - d_4c_1 \\ d_3c_2 \end{pmatrix}.$$

After modeling the system kinematics, we next find its dynamic model. Since the kinetic energy for the board can be expressed by

$$K_1 = \frac{1}{2} I_z \dot{\theta}_1^2 + \frac{1}{2} I_y \dot{\theta}_2^2,$$

because the board is only allowed to have two d.o.f. rotation without translation, where I_z and I_y are the moments of inertia for the board about its z_2-axis and y_2-axis of frame 2, respectively. Whereas the kinetic energy for the ball should be determined by

$$K_2 = \frac{1}{2} m_2 \dot{p}_0^{4T} \dot{p}_0^4 + \frac{1}{2} I_2 \left(\frac{\dot{d}_3^2 + \dot{d}_4^2}{r^2} \right),$$

where m_2 is the mass of the ball, and I_2 is the moment of inertia for the ball in any direction because it is a symmetric sphere. The first term of K_2 represents the translational kinetic energy. Since the relationship between the linear velocity v of the center and angular velocity ω for a spherical ball or a wheel with radius r is always equal to $v = \omega r$ for any rolling movement without slip, the second term of K_2 is the rotational kinetic energy due to the ball rolling on the board.

Now, the time-derivative of p_0^4 has a lengthy symbolical form:

$$\dot{p}_0^4 = \begin{pmatrix} -\dot{d}_3 c_1 s_2 + d_3 s_1 s_2 \dot{\theta}_1 - d_3 c_1 c_2 \dot{\theta}_2 + \dot{d}_4 s_1 + d_4 c_1 \dot{\theta}_1 \\ -\dot{d}_3 s_1 s_2 - d_3 c_1 s_2 \dot{\theta}_1 - d_3 s_1 c_2 \dot{\theta}_2 - \dot{d}_4 c_1 + d_4 s_1 \dot{\theta}_1 \\ \dot{d}_3 c_2 - d_3 s_2 \dot{\theta}_2 \end{pmatrix}.$$

However, once it is pre-multiplied by its transpose, $\dot{p}_0^{4T} \dot{p}_0^4$ will be much shorter due to a lot of cancellations. Finally, the 4 by 4 inertial matrix W for this ball-board system can be determined by extracting every coefficient of the quadratic form from the total kinetic energy $K = K_1 + K_2$, and is given as follows:

$$W = \begin{pmatrix} m_2 d_4^2 + m_2 d_3^2 s_2^2 + I_z & -m_2 d_3 d_4 c_2 & -m_2 d_4 s_2 & m_2 d_3 s_2 \\ -m_2 d_3 d_4 c_2 & m_2 d_3^2 + I_y & 0 & 0 \\ -m_2 d_4 s_2 & 0 & m_2 + \frac{I_2}{r^2} & 0 \\ m_2 d_3 s_2 & 0 & 0 & m_2 + \frac{I_2}{r^2} \end{pmatrix}.$$

Once the inertial matrix W is determined, to find the derivative matrix

$$W_d = \begin{pmatrix} \dot{q}^T \frac{\partial W}{\partial \theta_1} \\ \dot{q}^T \frac{\partial W}{\partial \theta_2} \\ \dot{q}^T \frac{\partial W}{\partial d_3} \\ \dot{q}^T \frac{\partial W}{\partial d_4} \end{pmatrix}$$

is just an extra job of taking four partial derivatives.

In the dynamics modeling process, it is important to find the joint torque due to gravity. Since the y_0-axis of the base is in upward direction, the second element of p_0^4 should be the relative height h of the ball with respect to the base, i.e.,

$$h = p_0^4(2) = -d_3 s_1 s_2 - d_4 c_1.$$

Because the board mass center is just at the pivoting point without any height change, the total gravitational potential energy becomes $P = m_2 g h$ and the joint torque due to gravity can be found by

$$\tau_g = -\frac{\partial P}{\partial q} = m_2 g \begin{pmatrix} d_3 c_1 s_2 - d_4 s_1 \\ d_3 s_1 c_2 \\ s_1 s_2 \\ c_1 \end{pmatrix}.$$

Now, every preparation required by the dynamic formulation:

$$W\ddot{q} + \left(W_d^T - \frac{1}{2} W_d \right) \dot{q} - \tau_g = \tau \tag{8.64}$$

are nearly completed. We turn our attention to developing a control strategy for this ball-board system and making everything ready for MATLAB^{TM} programming and realistic computer animation.

For this 4-joint robotic model, the state vector should be defined by

$$x = \begin{pmatrix} q \\ \dot{q} \end{pmatrix} \in \mathbb{R}^8.$$

Thus, the state space equation becomes

$$\dot{x} = f(x) + G(x)u = \begin{pmatrix} \dot{q} \\ W^{-1}[\tau_g - (W_d^T - \frac{1}{2}W_d)\dot{q}] \end{pmatrix} + \begin{pmatrix} O \\ W^{-1} \end{pmatrix} u, \tag{8.65}$$

where O is the 4 by 4 zero matrix and u is an input to the system. In this particular case, only the first two joints θ_1 and θ_2 of the board rotation can be controlled by two joint torques τ_1 about the z_0-axis and τ_2 about the z_1-axis. Whereas the last two prismatic joints d_3 and d_4 are just a virtual model and cannot be intervened by any external force except the possible ball rolling friction reacted from the board. Therefore, the input $u = \begin{pmatrix} \tau_1 \\ \tau_2 \end{pmatrix}$ is only 2-dimensional.

In order to meet the square system condition, the output $y = h(q)$ of the system model should be 2-dimensional as well. Since the control objective for this ball-board system is to make the ball track a desired trajectory on the board by the ball gravity, the joint displacements d_3 and d_4 are the best variables to form the output y. However, we also wish to make the output function depend on all the four joint variables, instead of just the last two. Therefore, let the output function be defined by a following mixed form:

$$y = h(q) = \begin{pmatrix} x_3 + x_2 \\ x_4 - x_1 \end{pmatrix} = \begin{pmatrix} d_3 + \theta_2 \\ d_4 - \theta_1 \end{pmatrix}.$$

The reason to have such a mixed definition is that the board counterclockwise rotation by θ_2 about the z_1-axis can enhance the ball sliding displacement d_3 in the z_2-direction due to gravity, while the board clockwise rotation by θ_1 about the z_0-axis can also enhance the ball sliding displacement d_4 in the z_3-direction, as can be clearly observed in Figure 8.10.

With the above 2-dimensional output definition, the Jacobian matrix will be a 2 by 4 constant matrix, i.e.,

$$J = \frac{\partial h}{\partial q} = \begin{pmatrix} 0 & 1 & 1 & 0 \\ -1 & 0 & 0 & 1 \end{pmatrix}.$$

Thus, at the state-space modeling standpoint, the state dimension of this ball-board system is $n = 8$, while the common dimension of both the input and output is $r = m = 2$. If each of the two output channels has relative degree $r_i = 2$, then the total relative degree becomes $r_d = 4 < n = 8$. Therefore, according to the nonlinear feedback control theory, only a 4-dimensional subsystem is exactly linearizable and another 4-dimensional subsystem becomes the internal dynamics of the system.

Intuitively, since the board can be directly controlled, while the ball cannot, the board is a physical entity of the linearizable subsystem, while the ball plays the role of the internal dynamics. However, in terms of the dynamic equation, they are highly coupled to each other, and are nondetachable. This intuition can be directly confirmed by the inertial matrix W, where many off-diagonal elements are nonzero and contain sine or cosine functions of the joint angle θ_2. The coupling factor may cause difficulty testing the stability of the internal dynamics to see if it would interfere the linearizable subsystem convergence.

Nevertheless, we can still develop an overall control law for the entire system. Let us start from the dynamics formulation in (8.64). Because the inertial matrix W is always non-singular, substituting $u = (\tau_1\ \tau_2)^T \in \mathbb{R}^2$ into (8.64) yields

$$\ddot{q} + W^{-1}(W_d^T - \frac{1}{2}W_d)\dot{q} - W^{-1}\tau_g = W^{-1}\tau = W^{-1}\begin{pmatrix} u \\ 0_{2\times 1} \end{pmatrix}.$$

Furthermore, similar to the derivation of equation (8.59), let the first two columns of the 4 by 4 W^{-1} be denoted by a 4 by 2 matrix $B = W^{-1}(:, 1:2)$. Note that $\dot{y} = J\dot{q}$ so that $\ddot{y} = J\ddot{q} + \dot{J}\dot{q} = J\ddot{q}$ if J is a constant matrix. By pre-multiplying the Jacobian matrix J to both sides of the above equation, we obtain

$$J\ddot{q} + JW^{-1}(W_d^T - \frac{1}{2}W_d)\dot{q} - JW^{-1}\tau_g = JBu.$$

If the 2 by 2 matrix $JB = D$ is non-singular, and let $\ddot{y} = J\ddot{q} = v$ be a new 2 by 1 input, then, we can readily resolve the input u by

$$u = D^{-1}v + D^{-1}JW^{-1}[(W_d^T - \frac{1}{2}W_d)\dot{q} - \tau_g] = \alpha(x) + \beta(x)v. \tag{8.66}$$

This reaches the expected nonlinear state-feedback control law for the ball-board system.

To specify the new input $v = \ddot{y}$ in order to indirectly control the ball to track a desired sine wave trajectory on the controlled board, based on (8.47), the desired output vector $y_d(t)$ can be defined in the following parametric form:

$$y_d(t) = \begin{pmatrix} at + b + \theta_2(0) \\ A\sin(\omega t) - \theta_1(0) \end{pmatrix},$$

where $\theta_1(0)$ and $\theta_2(0)$ are the initial values of the first two joints, and a and b are the slope and intercept of the straight line in time domain, and A and ω are the amplitude and angular frequency of the sine wave trajectory, respectively. Then, the desired velocity and acceleration becomes

$$\dot{y}_d = \begin{pmatrix} a \\ A\omega\cos(\omega t) \end{pmatrix}, \quad \text{and} \quad \ddot{y}_d = \begin{pmatrix} 0 \\ -A\omega^2\sin(\omega t) \end{pmatrix}.$$

By setting two appropriate positive-definite gain constant matrices for both K_p and K_v, the board can be driven by the control law (8.66) to indirectly maneuver the ball through the gravity to track the desired sine wave trajectory.

A simulation study in MATLAB^{TM} has shown that if the ball on the board without rolling friction, under the control law of the board, the ball will struggle to catch up the desired path for a few seconds, and then it will fly away. The main cause of the divergence is due to the unstable internal dynamics. Now, a rolling viscous friction is added in each of the two directions along the z_2-axis and z_3-axis on the board into the real-plant of the simulation study. Namely, the real system input τ is augmented to be 4 by 1 given by

$$\tau = \begin{pmatrix} u \\ -\dot{d}_3 \\ -\dot{d}_4 \end{pmatrix},$$

where u is the 2 by 1 control law from equation (8.66), and each viscous friction is proportional to the corresponding joint velocity. With the two PD gain constant matrices K_p and K_v specified, the ball will be able to stay on the track by taking advantage of the ball gravity through the two d.o.f. orientation control of the board. Therefore, with the friction added, the internal dynamics turns stable and the entire system becomes minimum-phase. Figure 8.11 shows the ball is at an initial position that is slightly away from the starting point of the desired sine wave trajectory. Figures 8.12, 8.13 and 8.14 sequentially demonstrate the ball in the trajectory-tracking phase in the computer animation, and Figure 8.15 is a snapshot of the ball at the ending point.

In fact, we can roughly test the stability of the zero dynamics by neglecting the centrifugal and Coriolis term in considering the ball in a very slow motion case near the initial position. Namely, let $v = 0$ due to the zero dynamics condition and $(W_d^T - \frac{1}{2}W_d)\dot{q} \approx 0$ so that

$$u = \alpha(x) \approx -D^{-1}JW^{-1}\tau_g.$$

Then, the last two rows of the state equation (8.65) should suffice to represent the zero dynamics, i.e.,

$$\begin{pmatrix} \ddot{d}_3 \\ \ddot{d}_4 \end{pmatrix} \approx W^{-1}(3:4,:)\tau_g + B(3:4,:)u$$

$$= W^{-1}(3:4,:)\tau_g - B(3:4,:)D^{-1}JW^{-1}\tau_g,$$

where $W^{-1}(3:4,:)$ and $B(3:4,:)$ are the last two rows of the 4 by 4 W^{-1} and 4 by 2 B matrices, respectively, as a common index notation for an array in MATLAB^{TM}. This equation of representing the ball rolling acceleration is, though approximated, reflecting a fact that even if the right-hand side is equal to zero, i.e., $\tau_g = 0$ without gravity, the zero dynamics is unstable in nature, let alone in the real case of $\tau_g \neq 0$ due to gravity. It can be clearly seen that $\ddot{d}_3 = 0$ has a solution $d_3 = C_1 t + C_2$ for two constants C_1 and C_2, which linearly goes to infinity as $t \to \infty$, and so is the equation $\ddot{d}_4 = 0$. Now, by adding a damping term, i.e., $\ddot{d}_3 = -\sigma\dot{d}_3$ with a positive constant coefficient $\sigma > 0$, the solution becomes

$$d_3 = C_1(1 - e^{-\sigma t}) + C_2,$$

so that d_3 is at least stable, though it may not be asymptotically stable. Therefore, the zero dynamics of the ball-board control system is unstable until it adds a viscous friction for each of the ball rolling displacements to substantially make the entire system be at least minimum-phase.

This realistic example from its kinematic and dynamic modelings to control strategy development with a MATLAB^{TM} animation demonstrates the power of robotic system modeling technology. Even for such a non-robotic system, by borrowing robotic modeling tools, including the D-H convention, homogeneous transformation, and dynamic formulation, many difficult modeling issues and control design challenges can be effectively tackled and resolved. As aforementioned, the ball-board system robotic model also belongs to the category of under-actuated robotic systems, where the number of

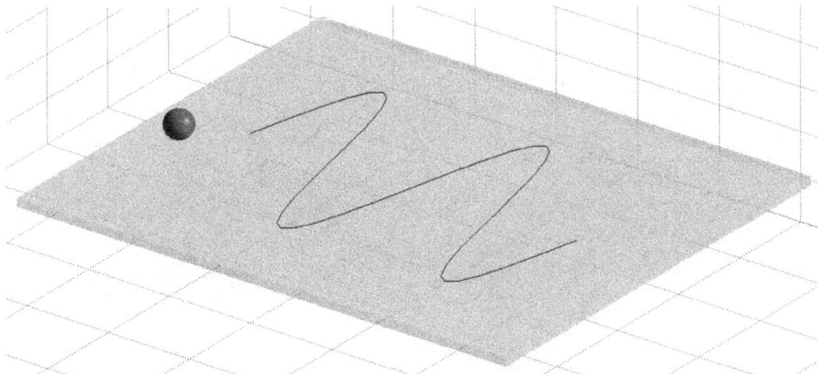

Fig. 8.11 The ball is at an initial position to start tracking a sine wave on the board

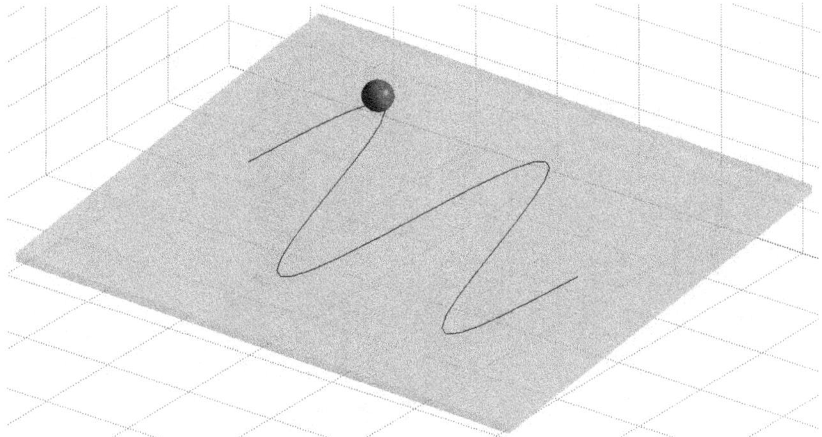

Fig. 8.12 The ball is catching up the track at early time

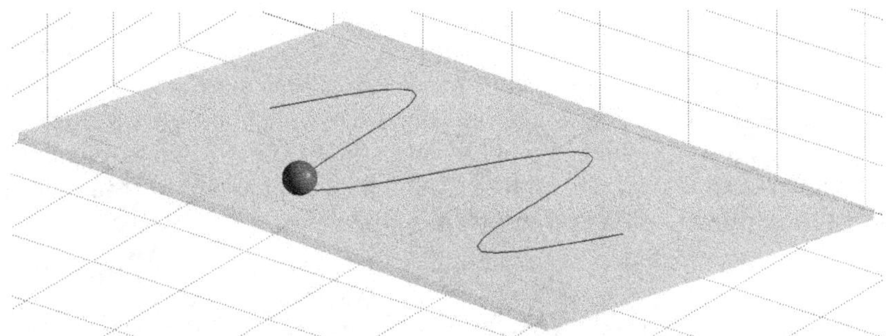

Fig. 8.13 The ball is now on the track by controlling the board orientation

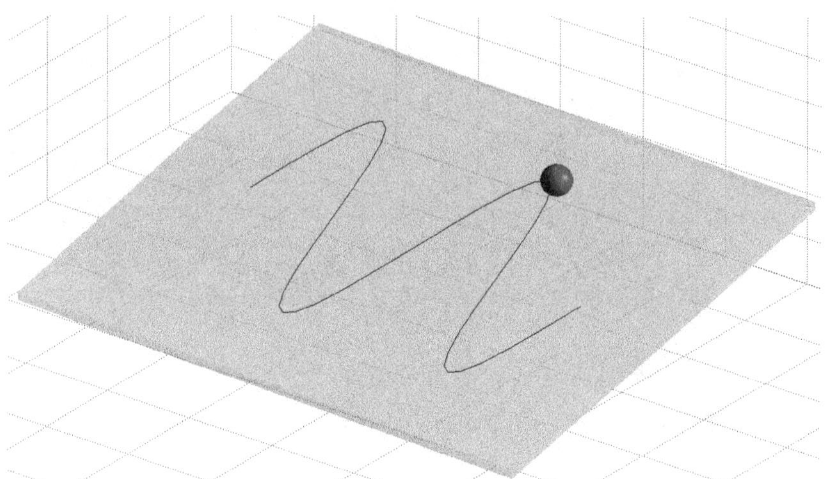

Fig. 8.14 The ball is well controlled to continue tracking the sine wave on the board

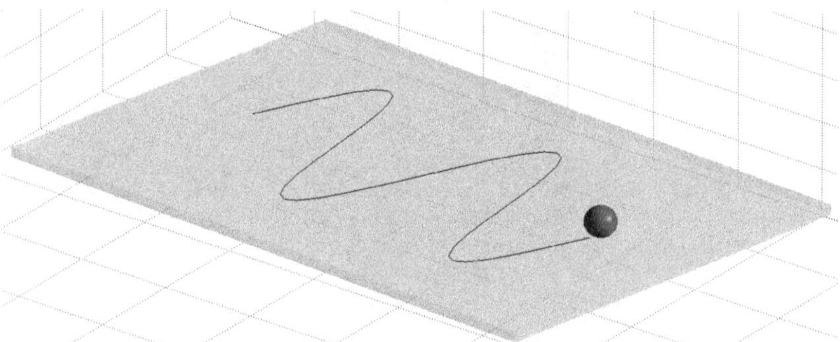

Fig. 8.15 The ball is successfully reaching the end of the sine wave on the board

controllable joints is less than the total number of joints. Almost every dynamic system in this category, such as a space robot or a robot with a floating base, possesses a non-trivial internal dynamics that is unlinearizable.

Finally, in order to help for better understanding how to implement and perform this animation, a complete MATLAB^{TM} program for the ball-board control system example is given below as a reference:

```
q=[pi/2; 0; -0.25; -0.05];  % The Initial Joint Values
dq=zeros(4,1);              % The Initial Joint Velocities
x=[q; dq];                  % Augment to Form a State Vector

A=0.15;      % The Amplitude of the Sine Wave Trajectory
r=0.016;  m=0.4;  % The Radius and Mass of the Ball
```

```
Iz=0.4;  Iy=0.4;  % The Moments of Inertia for the Board
g=9.8;

N=460;  dt=0.01;
Kp=diag([160 80]);  Kv=diag([4 2]);

[vb,fb]=rectangular(0.60,0.50,0.01);
      % Call a Subroutine for Board Drawing, See in Chapter 6
[xs,ys,zs]=sphere(20);

A2b=[0 0 1 0; 0 -1 0 0; 1 0 0 0; 0 0 0 1];
Ab0=[0 0 1 0;1 0 0 0;0 1 0 0;0 0 0 1];

for i=0:N
    s1=sin(x(1));  c1=cos(x(1));
    s2=sin(x(2));  c2=cos(x(2));
    d3=x(3);  d4=x(4);

    A02=[c1*c2 -s1 -c1*s2 0; s1*c2 c1 -s1*s2 0
         s2 0 c2 0; 0 0 0 1];

    t=i*dt;
    y=[d3+x(2); d4-x(1)];         % The Output Vector
    J=[[0 1; -1 0] eye(2)];       % The Jacobian Matrix
    p04=[-d3*c1*s2+d4*s1; -d3*s1*s2-d4*c1; d3*c2];
                                  % The Ball Position

    W=m*[d4^2+d3^2*s2^2+1 -d3*d4*c2 -d4*s2 d3*s2
         -d3*d4*c2 d3^2+1 0 0
         -d4*s2 0 1.4 0
         -d3*s2 0 0 1.4];

    Wd=m*[zeros(1,4)
          x(5:8)'*[2*d3^2*c2*s2 d3*d4*s2 -d4*c2 d3*c2
                   d3*d4*s2 0 0 0; -d4*c2 0 0 0; -d3*c2 0 0 0]
          x(5:8)'*[2*d3*s2^2 -d4*c2 0 s2; -d4*c2 2*d3 0 0
                   0 0 0 0; -s2 0 0 0]
          x(5:8)'*[2*d4 -d3*c2 s2 0; -d3*c2 0 0 0
                   -s2 0 0 0; 0 0 0 0]];

    WI=inv(W);

    yd=[0.1*t-0.22; A*sin(pi*t)-pi/2];
    dyd=[0.1; A*pi*cos(pi*t)];        % The Desired Trajectory
    ddyd=[0; -A*pi^2*sin(pi*t)];

    tg=m*g*[d3*c1*s2-d4*s1; d3*s1*c2; s1*s2; c1];
                              % The Joint Torque Due to Gravity
    v=ddyd+Kv*(dyd-J*x(5:8))+Kp*(yd-y);
                           % PD-Control for Trajectory-Tracking
    D=J*WI(:,1:2);  DI=inv(D);
```

```
    u=DI*v+DI*J*WI*((Wd'-Wd/2)*x(5:8)-tg);  % The Control Law

    ver=Ab0*A02*A2b*vb';  ver0=ver(1:3,:)';   % Start Drawing

    clf,
    patch('Vertices',ver0,'Faces',fb,'FaceColor','c',
     'FaceAlpha',0.4,'EdgeColor','None','AmbientStrength',0.6),
    grid, axis([-0.6 0.6 -0.5 0.5 -0.2 0.2]), daspect([1 1 1]),
    zoom(2), view([0.3 -0.4 0.25]),
    light('Position',[1 0.4 0.6],'Style','infinite'), hold on,

    x1=[];  y1=[];  z1=[];
    for ii=1:21
        cylin=Ab0*[eye(3) p04;0 0 0 1]*[xs(ii,:)*r
                   ys(ii,:)*r+0.01+r; zs(ii,:)*r; ones(1,21)];
        x1=[x1;cylin(1,:)];
        y1=[y1;cylin(2,:)];
        z1=[z1;cylin(3,:)];
    end
    surf(x1,y1,z1,'FaceColor','m','FaceLighting','Phong',
                  'EdgeColor','None','AmbientStrength',0.6),

    YYd=[];
    for tt=0:0.1:4
        yyd=Ab0*A02*A2b*[0.1*tt-0.2; A*sin(pi*tt); 0.01; 1];
        YYd=[YYd yyd];
    end
    if i > 0
        plot3(YYd(1,:),YYd(2,:),YYd(3,:),'r'),
    end

          % The Runge-Kutta Algorithm with f_bbn.m
    r1=f_bbn(x,u)*dt;
    r2=f_bbn(x+r1/2,u)*dt;
    r3=f_bbn(x+r2/2,u)*dt;
    r4=f_bbn(x+r3,u)*dt;
    x = x+(r1+2*r2+2*r3+r4)/6;
                          % Update the State
    if i == 0
        pause;
        plot3(YYd(1,:),YYd(2,:),YYd(3,:),'r'),
        pause;
    end
    drawnow,
end


   % -------The Function Subroutine ----------- %

function fs = f_bbn(x,u)      % Function for Real-Plant
```

```
    m=0.4;  g=9.8;   Iz=0.4;  Iy=0.4;

    s1=sin(x(1));   c1=cos(x(1));
    s2=sin(x(2));   c2=cos(x(2));
    d3=x(3);   d4=x(4);

    W=m*[d4^2+d3^2*s2^2+1 -d3*d4*c2 -d4*s2 d3*s2
          -d3*d4*c2 d3^2+1 0 0
          -d4*s2 0 1.4 0
          -d3*s2 0 0 1.4];

    Wd=m*[zeros(1,4)
          x(5:8)'*[2*d3^2*c2*s2 d3*d4*s2 -d4*c2 d3*c2
                   d3*d4*s2 0 0 0; -d4*c2 0 0 0; -d3*c2 0 0 0]
          x(5:8)'*[2*d3*s2^2 -d4*c2 0 s2; -d4*c2 2*d3 0 0
                   0 0 0 0; -s2 0 0 0]
          x(5:8)'*[2*d4 -d3*c2 s2 0; -d3*c2 0 0 0
                   -s2 0 0 0; 0 0 0 0]];

    c=(Wd'-Wd/2)*x(5:8);
    tg=m*g*[d3*c1*s2-d4*s1; d3*s1*c2; s1*s2; c1];

    fs=[x(5:8); W\(-c+tg+[u; -x(7); -x(8)])];
end
```

8.5 Dynamic Control of Robotic Systems

The independent joint control has been demonstrated to be effective for a robot arm in operating a set-point task in almost every industrial application. Now, we are going to show mathematically that an independent PD control law on each joint of a robot can asymptotically stabilize the robot to reach a desired destination. The major difference between a set-point task and a trajectory-tracking task lies in the fact that if $q^d(t)$ is a given desired joint position vector, then $\dot{q}^d = 0$ for the set-point task while $\dot{q}^d \neq 0$ for the trajectory-tracking. In other words, $q^d(t)$ represents a fixed destination point in the former case, while it describes a desired trajectory in the latter case. This also implies that a set-point task operation can only control the robot to reach a given desired destination point, but cannot guarantee what kind of actual trajectory it follows during the travel.

Let $\Delta q = q^d - q$ be a joint error vector for an n-joint robot in set-point task operations. Then, the time-derivative of the error becomes $\Delta\dot{q} = \dot{q}^d - \dot{q} = -\dot{q}$. Let an independent joint PD control law be defined as

$$u = \tau = K_p\Delta q + K_d\Delta\dot{q} - \tau_g = K_p\Delta q - K_d\dot{q} - \tau_g, \tag{8.67}$$

where both K_p and K_d are n by n diagonal positive-definite matrices representing the proportional and derivative gain constants, respectively. The input $u \in \mathbb{R}^n$ is a control signal to excite each joint of the robot as an n-dimensional joint torque vector $\tau = u$.

Let $x = \begin{pmatrix} q \\ \dot{q} \end{pmatrix} \in \mathbb{R}^{2n}$ be the state vector of the robotic system. According to the robot dynamic model developed in equation (7.19), we can rewrite it in the following state-space equation form:

$$\dot{x} = f(x) + G(x)u, \tag{8.68}$$

where

$$f(x) = \begin{pmatrix} \dot{q} \\ W^{-1}(\tau_g - C(q, \dot{q})\dot{q}) \end{pmatrix}, \quad \text{and} \quad G(x) = \begin{pmatrix} O \\ W^{-1} \end{pmatrix}$$

with the n by n zero matrix O in $G(x)$. The control objective can now be stated to make the norm of the state-error

$$\Delta x = \begin{pmatrix} \Delta q \\ \Delta \dot{q} \end{pmatrix} = \begin{pmatrix} \Delta q \\ -\dot{q} \end{pmatrix}$$

go to 0 as $t \to \infty$.

Before the discussion continues, let us introduce the well-known theory of stability in the Lyapunov sense as a preparation for the stability justification.

8.5.1 *The Theory of Stability in the Lyapunov Sense*

At the beginning, let an *equilibrium point* of a system be defined as a special state, at which the state velocity $\dot{x}$ is equal to zero. In other words, each equilibrium point of a system

$$\dot{x} = f(t, x) \tag{8.69}$$

is a root of the algebraic equation $f(t, x) = 0$. If the system has an input: $\dot{x} = f(t, x, u)$, then each equilibrium point is the solution of $f(t, x, 0) = 0$ by shutting off the input, i.e., $u = 0$.

It can be seen that a nonlinear system often has multiple *isolated equilibrium points*, because $f(t, x) = 0$ can have more than one solution if $f(\cdot)$ is nonlinear. In contrast, if it is a linear function, i.e., $f(t, x) = A(t)x$, the solution of $A(t)x = 0$ is only one at $x = 0$ if $A(t)$ is non-singular. However, if $A(t)$ is singular, the equation $A(t)x = 0$ has an infinite number of solutions inside the null space of $A(t)$, but they are continuous, not isolated from each other.

Now, the formal definition of stability around an equilibrium point $x(t) = 0$ for a general system in the form of (8.69) can be stated as follows:

Definition 13. The system (8.69) at its equilibrium point $x(t) = 0$ is said to be stable at $t = t_0$ if for each $\epsilon > 0$, there exists $\delta(t_0, \epsilon) > 0$, such that

$$\|x(t_0)\| < \delta(t_0, \epsilon) \Longrightarrow \|x(t)\| < \epsilon, \quad \text{for all } t \geq t_0.$$

The above definition of systems stability around an equilibrium point can be interpreted as a capacity from an initial state such that the system can transit to any destination close to the equilibrium point. This definition is conventionally called *stability in the Lyapunov sense*, which can be thought of as a bottom line for a system to be stable [13, 14, 11].

However, for many dynamic systems, only to be stable in the Lyapunov sense at an equilibrium point is not good enough. Instead, it is also desired that the system state can converge to the equilibrium point. This is called the *asymptotic stability.*

Definition 14. The system (8.69) about the equilibrium point $x(t) = 0$ is said to be asymptotically stable at $t = t_0$ if

1. it is stable at $t = t_0$, and
2. there exists a $\delta_1(t_0) > 0$ such that

$$\|x(t_0)\| < \delta_1(t_0) \Longrightarrow \|x(t)\| \to 0 \text{ as } t \to \infty.$$

Theoretically, to determine whether a system is stable can be done by directly solving the system equation to see if the solution is close to or converges to the specified equilibrium point. However, to find an explicit form of the solution is usually a very difficult task, and even impossible for many nonlinear systems. We are now faced with a challenging question: can we determine the stability for a given system without solving the equation? The answer is affirmative, and there is an effective approach to the stability test and determination, called the *Lyapunov direct method.*

A typical example of the stability test is a ball rolling around the inner wall of a basin in gravitational field. It can be seen that the ball starting from an initial point will always keep its total energy (kinetic energy plus potential energy) unchanged under a friction-free condition, or the total energy is reduced gradually until the ball stops at the bottom point of the basin if there is a rolling friction. As a counter situation, if the ball is placed on the top point of a smooth hill surface under gravity, then it will no longer be back to the original top point once the ball is tapped by a little initial force and slid down the hill. In this case, the ball will keep rolling down the hill and the kinetic energy can increase unboundedly. This simple example implies that every dynamic system commonly tends to the minimum energy state in order to stabilize itself. In other words, the lower the energy level is, the more stable the system is supposed to be. Therefore, we are motivated to explore an *energy-like function* associated with the system equation as a dynamic constraint for the stability determination.

To effectively determine a system stability, such an energy-like scalar function must be formalized with a number of conditions. For instance, consider a two-dimensional system with an equilibrium point at $x = 0$. Without loss of generality, we assume that the two state variables x_1 and x_2 of the system span a horizontal $x_1 - x_2$ phase plane, while the vertical $V-$axis that is perpendicular to the phase plane represents the value of the energy-like scalar function $V(x)$. Then, the scalar function $V(x)$ can geometrically be conceived like a bowl shape, and only its bottom point touches at the origin of the $x_1 - x_2$ phase plane, while all the other points on the surface are above the phase plane, as shown in Figure 8.16. Every phase trajectory on the horizontal phase plane can be thought of as a projection of some curve on the bowl surface. For the sake of clear discussion, we refer to such a curve on the bowl surface as a *V-lifted curve* of the corresponding phase trajectory.

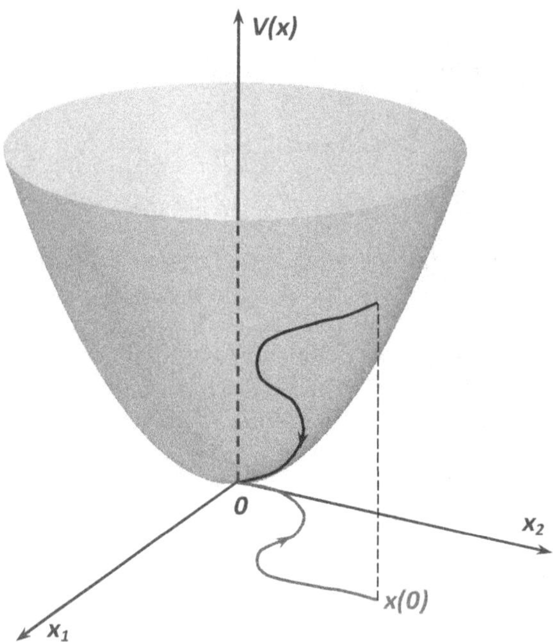

Fig. 8.16 An energy-like function $V(x)$ and a V-lifted trajectory

Suppose that a phase trajectory that obviously satisfies the system equation is specified by starting from a non-zero initial point and ending at the origin of the phase plane, i.e., the equilibrium point of the system. Then, the V-lifted curve can be viewed, at least in an open neighborhood of the origin, to slide down along the bowl surface until reaching the bottom point, as shown in Figure 8.16. In this case, the equilibrium point, i.e., the origin of the $x_1 - x_2$ plane, is asymptotically stable for the system.

Motivated by the above geometrical imagination and physical interpretation, we now formally introduce such an energy-like scalar function associated with a given system equation to determine the stability at an equilibrium point. First, we realize that the energy-like function is not just a positive function, but it should also satisfy some additional conditions to form a bowl-like geometrical surface. Such a special positive scalar function is called a *positive-definite function*, or a *p.d.f.* in abbreviation. Here we will focus our discussions only on the time-invariant cases, i.e., the functions do not contain time explicitly, or do not contain any time-varying parameter. The discussions on time-explicit positive-definite functions can be found in the literature [13, 11, 14].

Definition 15. A scalar continuous function $V(x)$ of $x \in \mathbb{R}^n$ is said to be a local positive-definite function (l.p.d.f.) if in some open ball $B(\delta) = \{x \mid \|x\| < \delta\}$, $V(0) = 0$ and $V(x) > 0$ for $x \neq 0$. Furthermore, $V(x)$ is said to be a positive-definite function (p.d.f.) if the ball limit $B(\delta)$ can be extended to the entire state space $\mathbb{R}^n$ and $V(x) \to \infty$ as $\|x\| \to \infty$.

It can be seen from the above definition that a positive scalar function is different from an l.p.d.f. or a p.d.f., and the latter requires stronger conditions than the former. For instance, if $x \in \mathbb{R}^1$, the scalar function $\exp(-x^2)$ (Gaussian) is always positive. However, it is neither an l.p.d.f. nor a p.d.f. The reason is quite obvious that this Gaussian function does not vanish at $x = 0$. In contrast to this, the function $V(x) = 1 - \exp(-x^2)$ is an l.p.d.f., due to $V(0) = 0$ and $V(x) > 0$ for any $x \neq 0$, but it is not a p.d.f., because $V(x) \to 1 < \infty$ as $x \to \infty$.

For a quadratic form of any state vector $x \in \mathbb{R}^n$, i.e., $V(x) = x^T M x$ with an n by n symmetric matrix, it can be proven that $V(x)$ is a p.d.f. if and only if the matrix M is positive-definite. Therefore, to determine if a quadratic form of x is a p.d.f., one can just test the coefficient matrix M of the quadratic form to see if $M > 0$. For instance, if $V(x) = 2x_1^2 - 2x_1x_2 + x_2^2$ as $x \in \mathbb{R}^2$, at first glance, it may hardly be judged whether it is a p.d.f. or not. Now, let it be rewritten by

$$V(x) = (x_1 \; x_2) \begin{pmatrix} 2 & -1 \\ -1 & 1 \end{pmatrix} \begin{pmatrix} x_1 \\ x_2 \end{pmatrix}.$$

This exhibits that M is the 2 by 2 middle matrix that is clearly positive-definite, because all of its diagonal principal minors have positive determinants. Namely, the first minor is 2 that is obviously a positive number, and the second one is the entire 2 by 2 matrix whose determinant $\det M = 2 \cdot 1 - (-1) \cdot (-1) = 1 > 0$. Therefore, the scalar function $V(x)$ is a p.d.f. Note that based on linear algebra, the matrix positive-definiteness can also be tested by its eigenvalues to see if they are all strictly positive.

Another example is $V(x) = x_1^2 + \sin^2 x_2$, where $x \in \mathbb{R}^2$. It can be justified that $V(x)$ is an l.p.d.f., but not a p.d.f. The reason is that first, $V(0) = 0$, and second, one can define an open ball $B(\delta)$ around $x = 0$ with a radius $\delta = \pi$. Then, for any $0 \neq x \in B(\delta)$, $V(x) > 0$. However, when one extends x to the entire space, i.e., $\delta \to \infty$, this $V(x)$ does not go to infinity if $\|x\| \to \infty$ along the x_2-direction. Therefore, the p.d.f. condition in Definition 15 does not hold for such a function $V(x)$.

The third example is $V(x) = x_1^2$, where $x = (x_1 \; x_2)^T \in \mathbb{R}^2$. It looks like a p.d.f., because $V(0) = 0$ and $V(x) = x_1^2 > 0$ when $x_1 \neq 0$. Actually, since this scalar function only contains x_1 and misses x_2, at every point on the straight line $x_1 = 0$ in the 2D state space, any non-zero value of x_2 will make the function value vanish. For instance, if $x = (0 \;\; 4)^T \neq 0$, then $V(x) = 0$. This violates the conditions for an l.p.d.f. or p.d.f., which require $V(x) = 0$ only at the single point $x = 0$ in the entire state space. Therefore, $V(x) = x_1^2$ in 2D state space is neither an l.p.d.f. nor a p.d.f.

Having introduced the formal definitions of both the l.p.d.f. and p.d.f., we now state the following stability theorem in the Lyapunov sense:

Theorem 8. *The equilibrium point $x = 0$ of a system $\dot{x} = f(x)$ is stable if there exists an l.p.d.f. $V(x)$ such that its time-derivative along any trajectory of the system is non-positive, i.e., $\dot{V}(x) \leq 0$ for each $x \in B(\delta) \subset \mathbb{R}^n$.*

This is commonly called the *Lyapunov Direct Method* of stability. In the above theorem, "along any trajectory" means that $f(x)$ should be substituted into every term containing $\dot{x}$ after taking the time-derivative on $V(x)$.

A more important question, which sometimes may not be so straightforward when this theorem is being applied, is how do we find such an l.p.d.f. $V(x)$ for a given system? The theorem does not directly address this issue. However, the system would be guaranteed to be stable in the Lyapunov sense if $V(x)$ could be found and satisfy all the conditions in the theorem. On the other hand, if one couldn't find a qualified $V(x)$, the system may still possibly be stable, because being unable to find a qualified function $V(x)$ does not imply that it does not exist! Therefore, using the Lyapunov direct method to determine a system stability requires effort and skill to find a right function $V(x)$, often called a *Lyapunov candidate*, such that all the conditions in Theorem 8 hold.

Note that Theorem 8 only provides a criterion for a system to be *stable*. In order to promote it to a local or global asymptotic stability, some of the conditions need to be made stronger. Based on Theorem 8, if

1. the condition "$\dot{V}(x) \leq 0$" is replaced by "$-\dot{V}(x)$ is an l.p.d.f. in an open ball $B(\delta)$", then the system at $x = 0$ is also *locally asymptotically stable*; and if
2. the condition "an l.p.d.f. $V(x)$" is replaced by "a p.d.f. $V(x)$" and the condition "$\dot{V}(x) \leq 0$" is replaced by "$-\dot{V}(x)$ is a p.d.f.", then the system at $x = 0$ is *globally asymptotically stable.*

For example, consider a 1-dimensional system $\dot{x} = -x$ with its equilibrium point at $x = 0$. Let $V(x) = x^2$ be a Lyapunov candidate. Obviously, $V(x)$ is a p.d.f in 1D state space. Then, $\dot{V}(x) = 2x\dot{x} = -2x^2$, which is not only non-positive, but also $-\dot{V}(x) = 2x^2$ is a p.d.f. Therefore, the system is globally asymptotically stable with respect to the equilibrium point $x = 0$. In fact, the solution of $\dot{x} = -x$ is $x(t) = x(0)e^{-t}$. Thus, for any finite $x(0)$ as a starting state, $x(t) \to 0$ as $t \to \infty$.

As a second example, let a 2-dimensional system be given by

$$\begin{cases} \dot{x}_1 = x_1(x_1^2 + x_2^2 - 1) - x_2 \\ \dot{x}_2 = x_1 + x_2(x_1^2 + x_2^2 - 1). \end{cases} \tag{8.70}$$

This system has an equilibrium point at $x_1 = 0$ and $x_2 = 0$. Let $V(x) = x_1^2 + x_2^2$ that is obviously a p.d.f. in the 2D phase plane. Since

$$\dot{V}(x) = 2x_1\dot{x}_1 + 2x_2\dot{x}_2 = 2(x_1^2 + x_2^2)(x_1^2 + x_2^2 - 1),$$

$-\dot{V}(x)$ is clearly an l.p.d.f. in the open unity ball $B(1)$, i.e., $x_1^2 + x_2^2 < 1$. Thus, the system is locally asymptotically stable at $x = (x_1\ x_2)^T = 0$.

A well-known simple-pendulum mechanical system can be modeled by

$$\begin{cases} \dot{x}_1 = x_2 \\ \dot{x}_2 = -bx_2 - k\sin x_1, \end{cases} \tag{8.71}$$

where $x_1 = \theta$ is the angular displacement of the pendulum and $x_2 = \dot{\theta}$ is its angular velocity. In such a second-order mechanical system, only picking the kinetic energy $K = \frac{1}{2}x_2^2$ may not be qualified as an l.p.d.f. or a p.d.f., because K contains only one component of $x = (x_1\ x_2)^T$. If we add a gravitational potential energy term $k(1 - \cos x_1)$ to form a Lyapunov candidate:

$$V(x) = k(1 - \cos x_1) + \frac{1}{2}x_2^2, \tag{8.72}$$

then $V(x)$ can be an l.p.d.f. as the angle $x_1 = \theta$ is confined within $(-\pi,\ \pi)$.

Since

$$\dot{V}(x) = k\sin x_1\dot{x}_1 + x_2\dot{x}_2 = k\sin x_1 x_2 - bx_2^2 - kx_2\sin x_1 = -bx_2^2 \le 0, \tag{8.73}$$

the system is stable at $x = 0$ when $b > 0$. In this case, $-\dot{V}(x) = bx_2^2 > 0$ is just positive, but is not an l.p.d.f.

However, in terms of the physical meaning, this system is supposed to be also locally asymptotically stable at $x = 0$ if $b > 0$. The reason is clear that with a positive viscous friction coefficient $b > 0$, the pendulum should converge to the equilibrium point $x = 0$ from any starting point within $-\pi < x_1 < \pi$.

La Salle had first discovered this insufficiency [13, 11], and showed the following important theorem to relax the conditions for an asymptotic stability:

Theorem 9. La Salle's Theorem
For an autonomous system $\dot{x} = f(x)$, suppose $V : \mathbb{R}^n \to \mathbb{R}$ is a continuously differentiable l.p.d.f. (p.d.f.), $\dot{V}(x) \le 0$ for $x \in B(\delta)$ ($x \in \mathbb{R}^n$), and the set $\mathcal{S} = \{x \in \mathbb{R}^n \mid \dot{V}(x) \equiv 0\}$ contains no non-trivial trajectories of the system. Then, $x = 0$ is a locally (globally) asymptotically stable equilibrium point.

This Theorem shows that only the condition $\dot{V}(x) \le 0$ may already imply an asymptotic stability, provided that the set $\mathcal{S}$ has only a trivial trajectory, i.e., the equilibrium point $x = 0$. Applying the La Salle's Theorem to the simple-pendulum system in (8.71), one may, at first glance, observe that the set $\mathcal{S}$ contains only $x_2 \equiv 0$ due to (8.73). However, by substituting $x_2 \equiv 0$ and also $\dot{x}_2 = 0$ into the two state equations in (8.71), we obtain $\dot{x}_1 = 0$ and $\sin x_1 = 0$. Since x_1 is confined to $(-\pi, \pi)$, this suffices to show that $x_1 \equiv 0$, too. Thus, the set $\mathcal{S}$ contains only the equilibrium point $x = 0$ that is trivial. Therefore, at $x = 0$, the simple-pendulum system is not only stable, but is also locally asymptotically stable if the damping coefficient $b > 0$.

8.5.2 Set-Point Stability and Trajectory-Tracking Control Strategy

After the stability theory is well prepared, we are now able to justify the stability for the robotic independent joint control systems. Based on the Lyapunov Direct Method, we may define the following *Lyapunov candidate* that is a positive-definite scalar function of the state error Δx:

$$V = \frac{1}{2}\dot{q}^T W \dot{q} + \frac{1}{2}\Delta q^T K_p \Delta q. \tag{8.74}$$

To prove that the independent PD control law (8.67) can asymptotically stabilize the robotic error system, i.e., $\|\Delta x\| \to 0$ as $t \to \infty$, it suffices to show that the time-derivative $\dot{V}$ of the Lyapunov candidate (8.74) along any system trajectory given by (8.68) is at least non-positive.

Taking the time-derivative for (8.74) yields

$$\dot{V} = \dot{q}^T W \ddot{q} + \frac{1}{2}\dot{q}^T \dot{W} \dot{q} + \Delta\dot{q}^T K_p \Delta q.$$

Substituting the robotic power equation (7.22) that was derived in Chapter 7 into here, and invoking the PD control law (8.67), we obtain

$$\begin{aligned} \dot{V} &= \dot{q}^T(\tau + \tau_g) - \dot{q}^T K_p \Delta q \\ &= \dot{q}^T(K_p \Delta q - K_d \dot{q}) - \dot{q}^T K_p \Delta q = -\dot{q}^T K_d \dot{q}. \end{aligned} \tag{8.75}$$

Since $K_d > 0$, $\dot{V}$ is shown to be at least non-positive. We can further justify that if $\Delta\dot{q} = -\dot{q} = 0$ such that $\dot{V} = 0$, then $\Delta q = 0$ according to the PD control law (8.67) and the robotic state equation (8.68). Therefore, based on the La Salle's Theorem, $\|\Delta x\| \to 0$ asymptotically as $t \to \infty$. In other words, $q \to q^d$ and $\dot{q} \to 0$ as $t \to \infty$. This rigorously shows that *an independent-joint PD control law given by (8.67) with knowledge of the joint torque τ_g due to gravity can perfectly drive the robotic system to perform any set-point task.*

However, as pointed out in the previous independent joint control section, due to the variations of J_{ef} and B_{ef} in each joint controller design, the robot will not accurately track a desired trajectory if dynamics is not considered. For this reason, we intend to develop a global dynamic control strategy for a robot to improve the accuracy and effectiveness in trajectory-tracking task operations.

Based on the nonlinear control theory introduced in the preceding sections, for a nonlinear dynamic system that can be described by (8.68), the key to linearize the system is to seek a nonlinear state mapping (a diffeomorphism) $z = \zeta(x)$ that can send $x \in \mathbb{R}^{2n}$ to $z \in \mathbb{R}^{2n}$ such that the system is exactly linearized in z space without error. For a robotic system, the forward kinematics $y = h(q)$ and its time-derivatives are well suitable to form such a diffeomorphism.

Furthermore, as we have illustrated with an example for the exact linearization procedure, if an n-joint ($n \leq 6$) fully-actuated robotic system represented by (8.68) along with an n-dimensional output $y = h(q)$ that is the forward kinematics and its n by n square Jacobian matrix $J = \partial h/\partial q$, then, the robotic system is exactly linearizable with a *relative degree* $r_i = 2$ in each output channel if J is non-singular.

Under the above conditions and due to $r_i = 2$ for $i = 1, \cdots, n$, we take the second time-derivative on $y = h(q)$, i.e.,

$$\ddot{y} = J\ddot{q} + \dot{J}\dot{q}.$$

Then, $\ddot{q} = J^{-1}\ddot{y} - J^{-1}\dot{J}\dot{q}$. Substituting this into the robot dynamic equation (7.19) yields

$$WJ^{-1}\ddot{y} - WJ^{-1}\dot{J}\dot{q} + C(q, \dot{q})\dot{q} - \tau_g = \tau = u.$$

If we define a new input $v = \ddot{y}$, then the following nonlinear state-feedback control law can be achieved:

$$u = \alpha(x) + \beta(x)v, \tag{8.76}$$

where

$$\alpha(x) = -WJ^{-1}\dot{J}\dot{q} + C(q, \dot{q})\dot{q} - \tau_g \quad \text{and} \quad \beta(x) = WJ^{-1}$$

as developed earlier in equation (8.46). The newly created input $\ddot{y} = v$ is clearly equivalent to

$$\dot{z} = \begin{pmatrix} \dot{y} \\ \ddot{y} \end{pmatrix} = \begin{pmatrix} O & I \\ O & O \end{pmatrix} z + \begin{pmatrix} O \\ I \end{pmatrix} v \tag{8.77}$$

as long as a new state is defined by $z = \begin{pmatrix} y \\ \dot{y} \end{pmatrix} \in \mathbb{R}^{2n}$, where O and I are the n by n zero matrix and identity, respectively. Equation (8.77) represents a very simple linear time-invariant system in the controllable canonical form. Therefore, under the nonlinear state-feedback control (8.76), the original nonlinear robotic system (8.68) is now exactly linearized into (8.77).

Since equation (8.76) does not contain any time-derivative of either u or v, it is also called a *static* state-feedback control law. Moreover, the centrifugal and Coriolis term $C(q, \dot{q})\dot{q}$ in $\alpha(x)$ of (8.76) can be directly calculated in terms of the derivative matrix W_d defined in equation (7.17) by

$$C(q, \dot{q})\dot{q} = \left(W_d^T - \frac{1}{2} W_d \right) \dot{q}. \tag{8.78}$$

It is observable that the new input v is involved in the linearized system (8.77) and can thus be designed by adopting any existing linear control scheme. The most typical one is the following PD control law for the input v:

$$v = \ddot{y}_d + K_2 \Delta\dot{y} + K_1 \Delta y, \tag{8.79}$$

where $y_d(t)$ is a desired Cartesian trajectory to be tracked, $\Delta y = y_d - y$ is a tracking error vector in Cartesian space, and K_1 and K_2 are two n by n positive-definite constant gain matrices. Because $v = \ddot{y}$, the PD control law (8.79) is equivalent to

$$\Delta\ddot{y} + K_2 \Delta\dot{y} + K_1 \Delta y = 0,$$

which guarantees the Cartesian tracking error $\Delta y \to 0$ as $t \to \infty$. In other words, the actual output $y(t)$ will converge to the desired trajectory $y_d(t)$ asymptotically even if there is a large non-zero initial error $\Delta y(0) \neq 0$. A block diagram for the entire robotic system to track a given desired trajectory by using the PD control law in (8.79) has been depicted in Figure 8.7.

This block diagram summarizes all the above discussions on the input-output linearization procedure for a robotic system to be globally controlled to track a desired trajectory. As aforementioned, basically there are two feedback loops in the block diagram. The *inner loop* represents an updating process for the nonlinear feedback control law (8.76) by substituting a fresh value of the state x at each sampling point. The *outer loop* describes the PD control law (8.79) that is applied directly on the exactly linearized system given in (8.77).

Such an exact linearization-based global robotic control strategy is, indeed, very effective for a robot with known dynamic parameters, but would cause a non-zero steady-state error or even divergence if the dynamic parameters are uncertain. One of the most promising remedies is a special nonlinear adaptive control strategy that will be introduced and developed in a later section of this chapter.

It can also be observed that to fulfill the input-output linearization procedure for a robot in performing a trajectory-tracking operation, it is desirable to have a unique representation $y = h(q)$ for up to 6 d.o.f. forward kinematics. If the required output y is only to track a positional trajectory, then a 3 by 1 position vector referred to the fixed base is perfect to be the output vector y. However, if a 3-dimensional orientation-trajectory is also required to be tracked, then it would be difficult to find a 3-dimensional vector y and its time-derivative $\dot{y}$ to uniquely represent the orientation and rotation velocity, respectively, as required in the PD control law (8.79). This thorny issue caused by the fact that a general angular velocity ω is not an exact differential 1-form, which has been addressed and discussed in Chapter 3.

Finally, if a robotic system is kinematically redundant, the Jacobian matrix J of $y = h(q)$ is not a square matrix so that one cannot solve $\ddot{y} = J\ddot{q} + \dot{J}\dot{q}$ uniquely for $\ddot{q}$. However, the simplest and most reasonable remedy is the minimum-norm solution for $\ddot{q}$, i.e.,

$$\ddot{q} = J^{+}\ddot{y} - J^{+}\dot{J}\dot{q}.$$

Thus, both $\alpha(x)$ and $\beta(x)$ in the state-feedback control law (8.76) will contain the pseudo-inverse $J^{+} = J^{T}(JJ^{T})^{-1}$ of the non-square Jacobian matrix J. Such a redundant case will be met again in the later modeling and analysis of digital human dynamics in Chapter 11.

8.6 Backstepping Control Design for Multi-Cascaded Systems

8.6.1 Control Design with the Lyapunov Direct Method

We have introduced and discussed thus far the exact linearization procedures and state-feedback control schemes. The procedures, though effective in many applications, they rely on a *cancellation of nonlinearity* between the system itself and the state-feedback control. In other words, the state-feedback control law given in (8.76) is virtually an inverse of the system equation. Sometimes we do not wish to cancel all the nonlinearities. Motivated by this notion, let us develop a direct approach to the control design without going through any linearization.

In a large number of control applications, we can directly design a control law during the course of testing the stability for a given control system through the Lyapunov direct method. For a better illustration, let us look at a so-called disease control system [13]. Suppose that there is a fatal disease spreading across a village. If any one is infected, then this person will either die, or be cured with permanent immunity from the disease.

To model this system, we may define the number of infected people in the village as the first state component x_1, while the number of non-infected healthy people is defined as the second state component x_2. Note that the number of people is often treated as a continuous and differentiable variable if this number is relatively large. This has been a common practice in modeling social and other human-related systems. Thus, the time-derivative $\dot{x}_1$ represents the infection rate, while $\dot{x}_2$ represents an increasing rate of the numbers for both the originally non-infected people and the cured/permanently immune individuals.

Since more infected people may cause either more deaths that reduce x_1, or more cures that increase x_2, the decreasing rate $-\dot{x}_1$ of the infected people should have a relation: $-\dot{x}_1 = ax_1 + \dot{x}_2$, where $a > 0$ is a proportional constant. Moreover, since both the larger number of the originally non-infected people x_2 and the larger number of the infected people x_1 will lead to more people being infected, the decreasing rate $-\dot{x}_2$ should be proportional to both x_1 and x_2 so that $\dot{x}_2 = -bx_1x_2$, where $b > 0$ is another proportional constant. Therefore, the entire system reaches a final model in the following state-equation form:

$$\begin{cases} \dot{x}_1 = -ax_1 + bx_1x_2 \\ \dot{x}_2 = -bx_1x_2. \end{cases} \tag{8.80}$$

Once again, the state vector x of this disease-spreading system is

$$x = \begin{pmatrix} x_1 \\ x_2 \end{pmatrix}, \quad \text{and} \quad f(x) = \begin{pmatrix} -ax_1 + bx_1x_2 \\ -bx_1x_2 \end{pmatrix}.$$

Clearly, the above system model is nonlinear. A more interesting issue is that since the above model does not have any input, we may add an input u on the first equation of (8.80) to achieve a so-called *disease control system model*:

$$\begin{cases} \dot{x}_1 = -ax_1 + bx_1x_2 + u \\ \dot{x}_2 = -bx_1x_2. \end{cases} \tag{8.81}$$

It can be easily found that the equilibrium point of the disease-control system under $u = 0$ is $x_1 = 0$. In other words, x_2 can be an arbitrary positive number, and thus the system has infinite number of equilibrium points that are continuously distributed on the positive half of the x_2-axis in the 2D phase plane.

Since $x_1(0) + x_2(0)$ represents the total number of initial lives at $t = 0$ in the village, during the disease spreading and developing period, the amount of people alive, including both infected and non-infected people, could be decreasing due to deaths. Therefore, we may set $x_2^d = x_1(0) + x_2(0)$ as a desired target to control the system towards the healthy people $x_2 \to x_2^d$ as $t \to \infty$. Under such a control objective, the Lyapunov candidate can be defined as

$$V(x) = \frac{1}{2}(x_1^2 + (x_2 - x_2^d)^2),$$

which is obviously a p.d.f. at the equilibrium point $x_e = \begin{pmatrix} 0 \\ x_2^d \end{pmatrix}$. Hence,

$$\dot{V} = x_1\dot{x}_1 + (x_2 - x_2^d)\dot{x}_2 = -ax_1^2 + bx_1^2x_2 + x_1u - b(x_2 - x_2^d)x_1x_2.$$

If we define a control law

$$u = -bx_1x_2 + b(x_2 - x_2^d)x_2 - kx_1 \tag{8.82}$$

with a gain constant $k > 0$, then $\dot{V} = -(a+k)x_1^2 \leq 0$. Furthermore, since the first equation of (8.81) after invoking the above control law (8.82) becomes $\dot{x}_1 = -(a+k)x_1 + b(x_2 - x_2^d)x_2$, by setting $x_1 \equiv 0$, we can see that only $x_2 = x_2^d$ or $x_2 = 0$ is in the set $\mathcal{S} = \{x \in \mathbb{R}^2 \mid \dot{V}(x) \equiv 0\}$. Therefore, based on Theorem 9 (La Salle), the disease control system (8.81) under the control law (8.82) has an asymptotically stable equilibrium point x_e.

After the above disease-control system is resolved by a Lyapunov direct control design procedure, a more general question arises, i.e., does the above control design procedure based on the Lyapunov direct method always work for every nonlinear control system? To answer this fundamental question, we need a further study of the theories behind the procedure.

First, for a general nonlinear system

$$\dot{x} = f(x, t), \quad x \in \mathbb{R}^n, \tag{8.83}$$

let us introduce the following two theorems regarding the stability:

Theorem 10. La Salle-Yoshizawa

Suppose that $x = 0$ is an equilibrium point of (8.83) and f is locally Lipschitz in x uniformly over t. Let $V(x)$ be a p.d.f. such that

$$\dot{V} = \frac{\partial V}{\partial x} f(x, t) \leq -\epsilon(x) \leq 0, \quad \forall t \geq 0, \ \forall x \in \mathbb{R}^n,$$

where $\epsilon(\cdot)$ is a continuous function. Then, all solutions of (8.83) are globally uniformly bounded and satisfy

$$\lim_{t\to\infty} \epsilon(x(t)) = 0.$$

In addition, if $\epsilon(x)$ is a p.d.f., then the equilibrium point $x = 0$ is globally uniformly asymptotically stable.

Before we state the second theorem, let us give a definition of *invariant set* for an autonomous system:

$$\dot{x} = f(x), \quad x \in \mathbb{R}^n. \tag{8.84}$$

Definition 16. A set M is called an invariant set of (8.84) if any solution $x(t)$ that belongs to M at some time instant t_1 must also belong to M for all the future and past time, i.e.,

$$x(t_1) \in M \Rightarrow x(t) \in M, \quad \forall t \in \mathbb{R}.$$

It is called a positively invariant set if only for the future time:

$$x(t_1) \in M \Rightarrow x(t) \in M, \quad \forall t \geq t_1.$$

Theorem 11. La Salle
Let Ω be a positively invariant set of (8.84). Let $V: \Omega \to \mathbb{R}_+$ be a differentiable function such that $\dot{V}(x) \leq 0$, $\forall x \in \Omega$. Let $\mathcal{S} = \{x \in \Omega \mid \dot{V}(x) = 0\}$, and M be the largest invariant set in $\mathcal{S}$. Then, every bounded solution $x(t)$ starting in Ω converges to M as $t \to \infty$.

Actually, this was the original theorem by La Salle and Theorem 9 was its corollary. After introducing the above two fundamental theorems to reinforce the Lyapunov stability theory, we now consider a nonlinear system with a control input to see how to take advantage of the control input to stabilize or improve the stability of the system. Given a single-input nonlinear control system

$$\dot{x} = f(x, u), \quad x \in \mathbb{R}^n, \quad u \in \mathbb{R}, \quad f(0, 0) = 0. \tag{8.85}$$

Definition 17. A smooth p.d.f. $V(x)$ is called a control Lyapunov function (c.l.f.) for (8.85) if

$$\inf_{u \in \mathbb{R}} \left\{ \frac{\partial V}{\partial x} f(x, u) \right\} < 0, \quad \forall x \neq 0.$$

This mathematical definition promotes the general p.d.f. requirement to a c.l.f. by adding a "hope" condition to find an input u such that $\dot{V}$ is at least non-positive [17], even if the "hope" is little.

If a general single-input system is affine of the input, then equation (8.85) is narrowed down to

$$\dot{x} = f(x) + g(x)u, \quad f(0) = 0. \tag{8.86}$$

We now start with (8.86) to interpret the above two fundamental stability theorems along with the c.l.f. definition. Consider a state-feedback control law $u = \alpha(x)$ available for the system (8.86). Then, a scalar function $V(x)$ to be c.l.f. implies that

$$\frac{\partial V}{\partial x} f(x) + \frac{\partial V}{\partial x} g(x)\alpha(x) \leq -\epsilon(x) \tag{8.87}$$

for a non-negative function or a p.d.f $\epsilon(x)$. The equilibrium point $x = 0$ is asymptotically stable under the control law $u = \alpha(x)$ if $\epsilon(x)$ can be a p.d.f. Therefore, the c.l.f. definition imposes a new requirement on the p.d.f. condition for a Lyapunov candidate $V(x)$. In other words, $V(x)$ is not only a p.d.f., but it is also able to offer a "hope" of finding a state-feedback control $u = \alpha(x)$ such that the scalar function $\epsilon(x)$ in (8.87) will at lease be non-negative.

For example, let us test a scalar system given by

$$\dot{x} = x^3 \cos x + x^2 u.$$

Since $x = 0$ is the only equilibrium point at the origin, we define $V(x) = \frac{1}{2}x^2$ that is a p.d.f. Then, $\dot{V} = x\dot{x} = x^4 \cos x + x^3 u \le -\epsilon(x)$. Suppose that we choose $\epsilon(x) = cx^4$ with a constant $c > 0$, because the highest power of $\dot{V}$ is the 4th order. This function $\epsilon(x) = cx^4$ is obviously a p.d.f., and $\dot{V} = x^4 \cos x + x^3 u \le -\epsilon(x) = -cx^4$. Let us solve this inequality for the input u by treating it as an equation to find a control law:

$$u = -cx - x\cos x = \alpha(x).$$

Thus, the Lyapunov function $V(x) = \frac{1}{2}x^2$ for this particular system is also a c.l.f.

Another example is a 2D system with a single input:

$$\begin{cases} \dot{x}_1 = x_1 x_2 + x_2^2 \\ \dot{x}_2 = -x_1^2 + x_2^2 + u. \end{cases} \tag{8.88}$$

Because the equilibrium point is at $x = \begin{pmatrix} x_1 \\ x_2 \end{pmatrix} = 0$, let $V(x) = \frac{1}{2}(x_1^2 + x_2^2)$ that is a p.d.f. Then,

$$\dot{V} = x_1\dot{x}_1 + x_2\dot{x}_2 = x_1 x_2^2 + x_2^3 + x_2 u \le -\epsilon(x).$$

Let $\epsilon(x) = cx_2^2$ with a constant $c > 0$. By equating $\dot{V} = x_1 x_2^2 + x_2^3 + x_2 u = -\epsilon(x) = -cx_2^2$, a control law can be immediately resolved by

$$u = \alpha(x) = -x_1 x_2 - x_2^2 - cx_2.$$

However, in this case, $\epsilon(x) = cx_2^2$ is just a positive function, instead of a p.d.f. Therefore, the control law can stabilize the 2D system, but whether it is asymptotic has to be answered by the La Salle's Theorem.

If the number of inputs in a system is more than one, say $u \in \mathbb{R}^r$ for $r > 1$, then we have the following general affine equation:

$$\dot{x} = f(x) + \sum_{i=1}^{r} g_i(x)u_i, \quad f(0) = 0. \tag{8.89}$$

The implication of the c.l.f. condition now becomes

$$\frac{\partial V}{\partial x} f(x) + \sum_{i=1}^{r} \frac{\partial V}{\partial x} g_i(x)\alpha_i(x) \le -\epsilon(x),$$

for each component of a control law $u_i = \alpha_i(x)$, $i = 1, \cdots, r$. Once a scalar function $\epsilon(x) \ge 0$ is chosen, the solutions of the inputs u_i's may not be unique in many cases, and one may select them to avoid any state variable x_i appearing in the denominator, because the equilibrium point is at $x_i = 0$ for each $i = 1 \cdots, n$. If finding a state-feedback control $u = \alpha(x)$ is hopeless, either the Lyapunov candidate $V(x)$ has not been a c.l.f. yet, or the system is uncontrollable. In the former case, one may either try a different non-negative function $\epsilon(x)$ to reduce its positiveness, or redefine a new Lyapunov candidate to try it again, or both. However, in the latter case, we have to remodel the system to be controllable. Regarding the controllability test procedure for nonlinear systems, one may refer to the literature [10, 11, 12].

For example, a 2-input system is given as follows:

$$\begin{cases} \dot{x}_1 = -x_3 + \exp(x_2)u_1 \\ \dot{x}_2 = x_1 + x_2^2 + \exp(x_2)u_2 \\ \dot{x}_3 = x_1 - x_2 - x_3. \end{cases} \tag{8.90}$$

Clearly, the equilibrium point of this system is $x = 0$. Let

$$V(x) = \frac{1}{2}(x_1^2 + x_2^2 + x_3^2),$$

which is obviously a p.d.f. Then,

$$\dot{V} = x_1x_2 - x_2x_3 + x_2^3 - x_3^2 + x_1 \exp(x_2)u_1 + x_2 \exp(x_2)u_2 \le -\epsilon(x)$$

for some function $\epsilon(x) \ge 0$. Let us more ambitiously choose a p.d.f. $\epsilon(x) = x_1^2 + x_2^2 + x_3^2$. We can thus solve the two inputs by

$$u_1 = \alpha_1(x) = -\exp(-x_2)(x_1+x_2) \text{ and } u_2 = \alpha_2(x) = \exp(-x_2)(x_3-x_2^2-x_2),$$

and none of them has any x_i appearing in a denominator. Since this $\epsilon(x)$ is a p.d.f., the control law $u = (\alpha_1(x) \;\; \alpha_2(x))^T$ can asymptotically stabilize the system.

8.6.2 Backstepping Recursions in Control Design

A backstepping approach for control design is to recursively "step back" until a control law is determined. This control design procedure is often used to deal with a cascaded control system [17]. It will be found very useful in modeling and control of human-machine dynamic interactive systems in Chapter 11.

To introduce the concepts and formulations of this interesting control design approach in a progressive fashion, let us start with a simple backstepping case.

First, let a single-input system (8.86) be augmented by an integrator:

$$\begin{cases} \dot{x} = f(x) + g(x)\xi \\ \dot{\xi} = u, \end{cases} \tag{8.91}$$

and let the first equation of (8.91) have a differentiable state-feedback virtual control law $\xi = \sigma(x)$. Namely, there are two p.d.f. $V(x)$ and $\epsilon(x)$ such that

$$\frac{\partial V}{\partial x} f(x) + \frac{\partial V}{\partial x} g(x)\sigma(x) \leq -\epsilon(x).$$

Then, by defining an error function $e(x, \xi) = \xi - \sigma(x)$, a new Lyapunov candidate is proposed by

$$V_a(x, \xi) = V(x) + \frac{1}{2}e^2 = V(x) + \frac{1}{2}(\xi - \alpha(x))^2,$$

which should be a c.l.f. for the augmented system (8.91). In other words, there exists a feedback control law $u = \alpha(x, \xi)$ such that $x = 0$ with $e = \xi - \sigma(x) = 0$ is a globally asymptotically stable equilibrium point of (8.91).

In fact, $\xi = e(x, \xi) + \sigma(x)$, and

$$\dot{V}_a(x, \xi) = \dot{V}(x) + e(\dot{\xi} - \dot{\sigma}(x)) = \dot{V}(x) + e(u - \dot{\sigma}(x)). \tag{8.92}$$

Notice that

$$\dot{V}(x) = \frac{\partial V}{\partial x} f(x) + \frac{\partial V}{\partial x} g(x)(e + \sigma) \leq -\epsilon(x) + \frac{\partial V}{\partial x} g(x)e,$$

and

$$\dot{\sigma}(x) = \frac{\partial \sigma}{\partial x} f(x) + \frac{\partial \sigma}{\partial x} g(x)\xi.$$

Substituting them into (8.92), it can be clearly seen that if

$$u = \alpha(x, \xi) = -c(\xi - \sigma(x)) + \frac{\partial \sigma(x)}{\partial x}(f(x) + g(x)\xi) - \frac{\partial V(x)}{\partial x} g(x), \tag{8.93}$$

with a constant $c > 0$, then the time-derivative of the new Lyapunov candidate becomes

$$\dot{V}_a(x, \xi) \leq -\epsilon(x) - ce^2,$$

and obviously, $\epsilon(x) + ce^2$ is also a p.d.f.

Let us look at the following example to illustrate the procedure of finding the control law $u = \alpha(x, \xi)$:

$$\begin{cases} \dot{x} = x^2 + x\xi \\ \dot{\xi} = u. \end{cases} \tag{8.94}$$

In the first equation, we have a p.d.f. $V(x) = \frac{1}{2}x^2$ defined and $\dot{V} = x\dot{x} = x^3 + x^2\xi \leq -\epsilon(x)$. Let $\epsilon(x) = x^2$ such that the *virtual control law* is $\xi = \sigma(x) = -x - 1$.

Now, defining an error state $e = \xi - \sigma(x) = \xi + x + 1$, we have

$$\begin{cases} \dot{x} = x^2 + x\xi = x^2 + x(e - x - 1) \\ \dot{e} = u + x^2 + x(e - x - 1). \end{cases} \tag{8.95}$$

Since the new state equation for both x and e has $x = 0$ and $e = 0$ as its equilibrium point, we choose $V_a = \frac{1}{2}(x^2 + e^2)$ that is a p.d.f. Thus,

$$\dot{V}_a = x\dot{x} + e\dot{e} = -x^2 + eu + x^2e + xe^2 - xe \leq -\epsilon_a(x, e).$$

Let $\epsilon_a(x, e) = x^2 + e^2$. We finally obtain a control law for (8.94):

$$u = \alpha(x, \xi) = -x^2 - xe + x - e = -2x^2 - x\xi - \xi - x - 1.$$

If we use the general form of feedback control law in (8.93) and pick $c = 1$, since $\frac{\partial \sigma}{\partial x} = -1$ and $\frac{\partial V(x)}{\partial x} = x$, we reach

$$u = -(\xi + x + 1) - (x^2 + x\xi) - x^2 = -2x^2 - x\xi - \xi - x - 1,$$

which is the same as the above $\alpha(x, \xi)$.

Second, we extend the backstepping design procedure for more complex cascaded control systems. Let the single-input system (8.86) be augmented by a linear system through its output:

$$\begin{cases} \dot{x} = f(x) + g(x)y, \quad f(0) = 0, \quad x \in \mathbb{R}^n, \quad y \in \mathbb{R} \\ \dot{\xi} = A\xi + Bu, \quad y = C\xi, \quad \xi \in \mathbb{R}^k, \quad u \in \mathbb{R}. \end{cases} \tag{8.96}$$

If the (second) linear system is minimum-phase with relative degree $r_d = 1$, i.e., $CB \neq 0$, and the first system satisfies the condition (8.87) for both p.d.f. $V(x)$ and $\epsilon(x)$ with a smooth virtual control law $y = \sigma(x)$, then there exists a feedback control $u = \alpha(x, \xi)$ to globally asymptotically stabilize the equilibrium point at $x = 0$ and $\xi = 0$ for the full system (8.96).

Similarly, by defining an error state

$$e(x, \xi) = y - \sigma(x) = C\xi - \sigma(x),$$

a new Lyapunov candidate can be proposed by

$$V_a(x, \xi) = V(x) + \frac{1}{2}e^2 = V(x) + \frac{1}{2}(C\xi - \sigma(x))^2.$$

Since the time-derivative of the output is $\dot{y} = C\dot{\xi} = CA\xi + CBu$, and the decoupling matrix $D = CB$ in this SISO case is a non-zero scalar due to $r_d = 1$ so that

$$u = \dot{y} - CA\xi = v - CA\xi, \tag{8.97}$$

where a new input $v = \dot{y}$ is defined. After taking the time-derivative on the new Lyapunov candidate, we obtain almost the same result as (8.92), except u in the last term of (8.92) is replaced by $\dot{y} = v$. Therefore, the new input v should have exactly the same formula as (8.93). Finally, substituting it into (8.97) yields

$$\begin{aligned} u = \alpha(x, \xi) = \frac{1}{CB}[-c(C\xi - \sigma(x)) - CA\xi + \\ + \frac{\partial \sigma(x)}{\partial x}(f(x) + g(x)C\xi) - \frac{\partial V(x)}{\partial x} g(x)], \quad c > 0. \end{aligned} \tag{8.98}$$

As a more general case, the system (8.86) is now augmented by a nonlinear system, instead of a linear system, through its output, i.e.,

$$\begin{cases} \dot{x} = f(x) + g(x)y, \quad f(0) = 0, \;\; x \in \mathbb{R}^n, \;\; y \in \mathbb{R} \\ \dot{\xi} = \beta(x, \xi) + \gamma(x, \xi)u, \quad y = h(\xi), \;\; h(0) = 0, \;\; \xi \in \mathbb{R}^k, \;\; u \in \mathbb{R}. \end{cases} \tag{8.99}$$

Under the similar conditions, the (second) nonlinear system has a constant relative degree $r_d = 1$ and its zero dynamics is asymptotically stable with respect to ξ and y as its output, while the first system satisfies the condition (8.87) for both p.d.f. $V(x)$ and $\epsilon(x)$ with a smooth virtual control law $y = \sigma(x)$. Then, there exists a feedback control $u = \alpha(x, \xi)$ to globally asymptotically stabilize the equilibrium point at $x = 0$ and $\xi = 0$ for the cascaded system (8.99).

By following the same procedure as the above linear augmentation case in equation (8.96),

$$\dot{y} = L_\beta h(\xi) + (L_\gamma h(\xi))u = \frac{\partial h(\xi)}{\partial \xi}\beta(x, \xi) + \frac{\partial h(\xi)}{\partial \xi}\gamma(x, \xi)u.$$

Since $L_\gamma h(\xi) \neq 0$ due to the condition $r_d = 1$, we can resolve the input u by

$$u = \left(\frac{\partial h(\xi)}{\partial \xi}\gamma(x, \xi)\right)^{-1}\left(v - \frac{\partial h(\xi)}{\partial \xi}\beta(x, \xi)\right),$$

where a new input $v = \dot{y}$ is defined here, too.

Since the remaining steps, including to define a new Lyapunov candidate and take its time-derivative, are exactly the same as the second case, we finally achieve a control law for the nonlinear cascaded system (8.99):

$$u = \left(\frac{\partial h(\xi)}{\partial \xi}\gamma(x, \xi)\right)^{-1}\left[-c(h(\xi) - \sigma(x)) - \frac{\partial h(\xi)}{\partial \xi}\beta(x, \xi) + \right.$$

$$+\frac{\partial\sigma(x)}{\partial x}(f(x)+g(x)h(\xi)) - \frac{\partial V(x)}{\partial x}g(x)\Big], \quad c>0. \tag{8.100}$$

As an example, let the 2D system in (8.88) be the first nonlinear equation of (8.99). Then, it is cascaded by another 2D nonlinear mechanical system through its output, i.e.,

$$\begin{cases} \dot{x}_1 = x_1x_2 + x_2^2 \\ \dot{x}_2 = -x_1^2 + x_2^2 + y, \end{cases} \quad y = a_1\xi_1 + a_2\xi_2, \quad \begin{cases} \dot{\xi}_1 = \xi_2 \\ \dot{\xi}_2 = -b\xi_2 - k\sin\xi_1 + u, \end{cases} \tag{8.101}$$

where a_1, a_2, b and k are all positive real constants. First, the x-system has been analyzed in the previous example to have a c.l.f.

$$V(x) = \frac{1}{2}(x_1^2 + x_2^2)$$

and a virtual control law

$$y = \sigma(x) = -x_1x_2 - x_2^2 - cx_2$$

for a constant $c > 0$. However, $\epsilon(x) = cx_2^2$ is just non-negative but is not a p.d.f. We can apply the La Salle's Theorem to find that it is also asymptotically stabilized by the virtual control law. Now, we turn to investigate the ξ-system with its output $y = h(\xi)$. Under this output, the relative degree $r_d = 1$, because

$$L_\gamma h(\xi) = (a_1 \;\; a_2)\begin{pmatrix}0\\1\end{pmatrix} = a_2 \neq 0.$$

Thus, the exactly linearized system is only 1-dimensional, i.e., $\dot{y} = v$ for a new input v based on the input-output linearization procedure in the previous section.

As a counterpart, the 1-dimensional internal dynamics that is unlinearizable can be defined by

$$\dot{\xi}_1 = \xi_2 = \frac{1}{a_2}(y - a_1\xi_1).$$

By turning off the output, i.e., $y = 0$, we obtain a zero dynamics

$$\dot{\xi}_1 = -\frac{a_1}{a_2}\xi_1$$

that is obviously asymptotically stable due to $a_1/a_2 > 0$.

After both the two cascaded systems are tested and all the conditions for (8.99) hold, we can utilize the suggested control law (8.100) to determine the input $u = \alpha(x, \xi)$ for the entire cascaded system given in (8.101), where

$$\beta(x,\xi) = \begin{pmatrix}\xi_2\\ -b\xi_2 - k\sin\xi_1\end{pmatrix}, \quad \gamma(x,\xi) = \begin{pmatrix}0\\1\end{pmatrix}, \quad \frac{\partial h(\xi)}{\partial\xi} = (a_1 \;\; a_2),$$

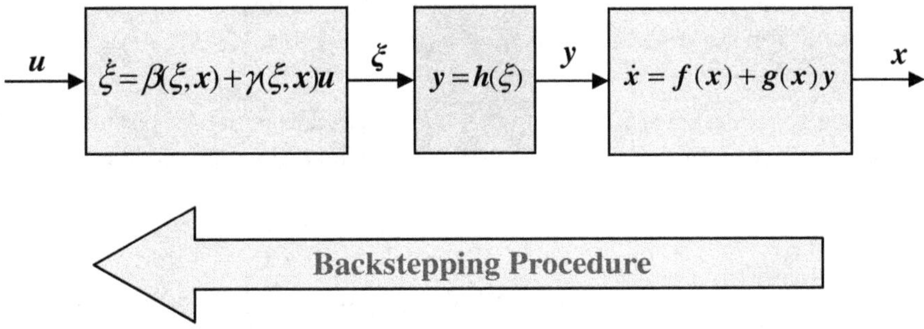

Fig. 8.17 A flowchart of the backstepping control design approach

and

$$\frac{\partial \sigma(x)}{\partial x} = (-x_2 \quad -x_1 - 2x_2 - c), \quad \text{and} \quad \frac{\partial V(x)}{\partial x} = (x_1 \;\; x_2).$$

As a summary, the flowchart of the backstepping control design approach is depicted in Figure 8.17.

Furthermore, the nonlinear augmented systems control design through the backstepping procedure can be extended to a nonlinear k-cascaded MIMO (Multi-Input Multi-Output) dynamic system that can be formulated in the following general form:

$$\begin{aligned}
\dot{x} &= f(x) + G(x)y_1 \\
y_1 &= h_1(x, \xi_1) \\
\dot{\xi}_1 &= \eta_1(x, \xi_1) + \Gamma_1(x, \xi_1)y_2 \\
y_2 &= h_2(\xi_1, \xi_2) \\
&\vdots \\
\dot{\xi}_{k-1} &= \eta_{k-1}(\xi_{k-2}, \xi_{k-1}) + \Gamma_{k-1}(\xi_{k-2}, \xi_{k-1})y_k \\
y_k &= h_k(\xi_{k-1}, \xi_k) \\
\dot{\xi}_k &= \eta_k(\xi_{k-1}, \xi_k) + \Gamma_k(\xi_{k-1}, \xi_k)u,
\end{aligned} \tag{8.102}$$

where $x \in \mathbb{R}^N$, $y_i \in \mathbb{R}^{l_i}$, $\xi_i \in \mathbb{R}^{m_i}$ for $i = 1, \cdots, k$, and $u \in \mathbb{R}^r$ is the overall control input. Since for the entire cascaded system, each stage is, in general, an MIMO subsystem, the coefficient $G(x)$ for the input y_1 in (8.102) is an N by l_1 matrix, i.e., $G(x) = (g_1(x) \; \cdots \; g_{l_1}(x))$, where each column is an N-dimensional vector field of x. Also, each $\Gamma_i(\xi_{i-1}, \xi_i)$ is an m_i by l_{i+1} matrix. We may intentionally define each y_i and the overall input u of (8.102) to have the same dimension, i.e., $y_i \in \mathbb{R}^l$ and $u \in \mathbb{R}^l$ to make each stage be a square MIMO subsystem [18, 19].

It can be interpreted that for the i-th ξ-subsystem, $y_i = h_i(\xi_{i-1}, \xi_i)$ is an output function for the equation $\dot{\xi}_i = \eta_i(\xi_{i-1}, \xi_i) + \Gamma_i(\xi_{i-1}, \xi_i)y_{i+1}$ with an input y_{i+1}. We assume that

1. For the top x-subsystem $\dot{x} = f(x) + G(x)y_1$, there exists a virtual control law $y_1 = \sigma_1(x)$ to asymptotically stabilize the x-subsystem at its equilibrium point $x = 0$; and
2. For the i-th input-output ξ-subsystem, the relative degree of each output channel is $r^i_{dj} = 1$ for $j = 1, \cdots, l_i$ so that the total relative degree is $r^i_d = l_i$, the dimension of the output y_i, and it is also an asymptotically minimum-phase subsystem for $i = 1, \cdots, k$.

The first assumption implies that there is a Lyapunov candidate $V_1(x)$ that is a p.d.f. such that

$$\dot{V}_1(x) = \frac{\partial V_1}{\partial x} f(x) + \frac{\partial V_1}{\partial x} G(x)\sigma_1(x) \le -\epsilon_1(x) \tag{8.103}$$

for a p.d.f. $\epsilon_1(x)$, or just a semi-positive-definite function $\epsilon_1(x) \ge 0$ if every condition of the La Salle's Theorem holds. The second assumption imposed on each input-output ξ-subsystem ensures that its feedback linearization requires no higher than the first-order time-derivative of the output plus its internal dynamics is asymptotically stable.

Afterwards, based on the theory of exact linearization, the first time-derivative of the output of the first ξ-subsystem ($i = 1$) is given by

$$\dot{y}_1 = \frac{\partial h_1}{\partial x} f(x) + \frac{\partial h_1}{\partial x} G(x)y_1 + \frac{\partial h_1}{\partial \xi_1}\eta_1(x, \xi_1) + \left(\frac{\partial h_1}{\partial \xi_1}\Gamma_1(x, \xi_1)\right) y_2, \tag{8.104}$$

and the coefficient of the input y_2 here must be an l by l non-singular matrix due to the second assumption, which is known as a decoupling matrix $D(x, \xi_1)$ in a certain neighborhood of the equilibrium point at $x = 0$ and $\xi_1 = 0$ for the first ξ-subsystem in (8.102). Therefore, the input of the first ξ-subsystem can be resolved by

$$y_2 = \left(\frac{\partial h_1}{\partial \xi_1}\Gamma_1(x, \xi_1)\right)^{-1} \left[v_1 - \frac{\partial h_1}{\partial x} f(x) - \frac{\partial h_1}{\partial x} G(x)y_1 - \frac{\partial h_1}{\partial \xi_1}\eta_1(x, \xi_1)\right] \tag{8.105}$$

with a new input $v_1 = \dot{y}_1$ to be further determined.

Since the output y_1 of the first ξ-subsystem is also the input of the top x-subsystem, we wish this output $y_1 = h_1(x, \xi_1)$ could be as close to the virtual control $\sigma_1(x)$ as possible. Namely, by defining an error function

$$e_1(x, \xi_1) = y_1 - \sigma_1(x) = h_1(x, \xi_1) - \sigma_1(x),$$

we wish e_1 would go to zero asymptotically. Thus, an *extended Lyapunov candidate* for the first set ($i = 1$) of cascaded subsystems can be defined as follows:

$$V_2(x, \xi_1) = V_1(x) + \frac{1}{2} e_1^T e_1. \tag{8.106}$$

Its time-derivative becomes

$$\begin{aligned}\dot{V}_2 = \dot{V}_1 + e_1^T \dot{e}_1 &= \frac{\partial V_1}{\partial x} f(x) + \frac{\partial V_1}{\partial x} G(x) y_1 + e_1^T (\dot{y}_1 - \dot{\sigma}_1) \\ &= \frac{\partial V_1}{\partial x} f(x) + \frac{\partial V_1}{\partial x} G(x)(\sigma_1 + e_1) + e_1^T (\dot{y}_1 - \dot{\sigma}_1) \\ &\leq -\epsilon_1(x) + \frac{\partial V_1}{\partial x} G(x) e_1 + e_1^T (\dot{y}_1 - \dot{\sigma}_1).\end{aligned} \tag{8.107}$$

Now, let

$$v_1 = \dot{y}_1 = -a_1 e_1 + \dot{\sigma}_1 - G^T \left(\frac{\partial V_1}{\partial x} \right)^T. \tag{8.108}$$

Substituting it into (8.107), the time-derivative of the extended Lyapunov function becomes

$$\dot{V}_2 \leq -\epsilon_1(x) - a_1 e_1^T e_1 \leq -\epsilon_2(x, \xi_1)$$

with a p.d.f. $\epsilon_2(x, \xi_1)$ and a positive gain constant $a_1 > 0$. Substituting equation (8.108) into (8.105), we obtain a new virtual control $y_2 = \sigma_2(x, \xi_1)$ for the first set of cascaded subsystem.

By repeating the same procedure for the second set ($i = 2$) of ξ-subsystem again, first, let the error

$$e_2(x, \xi_1, \xi_2) = y_2 - \sigma_2(x, \xi_1) = h_2(\xi_1, \xi_2) - \sigma(x, \xi_1).$$

Then, similar to (8.106), we can define a new extended Lyapunov candidate as follows:

$$V_3(x, \xi_1, \xi_2) = V_2(x, \xi_1) + \frac{1}{2} e_2^T e_2 = V_1(x) + \frac{1}{2} e_1^T e_1 + \frac{1}{2} e_2^T e_2. \tag{8.109}$$

Likewise, without need of further explanation, the new input y_3 to the second set of ξ-subsystem turns out to be

$$\begin{aligned} y_3 = &\left(\frac{\partial h_2}{\partial \xi_2} \Gamma_2(\xi_1, \xi_2) \right)^{-1} \times \\ &\times \left[v_2 - \frac{\partial h_2}{\partial \xi_1} \eta_1(x, \xi_1) - \frac{\partial h_2}{\partial \xi_1} \Gamma_1(x, \xi_1) y_2 - \frac{\partial h_2}{\partial \xi_2} \eta_2(\xi_1, \xi_2) \right]\end{aligned} \tag{8.110}$$

with a new input $v_2 = \dot{y}_2$, once again, to be further determined. To do so, similar to (8.107),

$$\dot{V}_3 \leq -\epsilon_2(x, \xi_1) + \frac{\partial V_2}{\partial \xi_1} \Gamma_1(x, \xi_1) e_2 + e_2^T (\dot{y}_2 - \dot{\sigma}_2). \tag{8.111}$$

Let the new input

$$v_2 = \dot{y}_2 = -a_2 e_2 + \dot{\sigma}_2 - \Gamma_1^T \left(\frac{\partial V_2}{\partial \xi_1} \right)^T . \tag{8.112}$$

Substituting it into (8.111) yields

$$\dot{V}_3 \le -\epsilon_2(x, \xi_1) - a_2 e_2^T e_2 \le -\epsilon_3(x, \xi_1, \xi_2)$$

with another p.d.f. $\epsilon_3(x, \xi_1, \xi_2)$ and another positive gain constant $a_2 > 0$. Substituting (8.112) into (8.110), we achieve a new virtual control $y_3 = \sigma_3(x, \xi_1, \xi_2)$ for the second set ($i = 2$) of cascaded subsystems.

Continuing the above recursion up to k steps, we will finally be able to reach and resolve the overall input u for the k-cascaded system control design. The final Lyapunov function is constructed by

$$V_{k+1}(x, \xi_1, \cdots, \xi_k) = V_1(x) + \frac{1}{2} \sum_{i=1}^{k} e_i^T e_i \tag{8.113}$$

with each error function $e_i = y_i - \sigma_i(x, \xi_1, \cdots, \xi_{i-1})$. It can also be observed from (8.107) and (8.111) that at each recursive step, the time-derivative of each extended Lyapunov function is always in the following form:

$$\dot{V}_{i+1} \le -\epsilon_i(x, \xi_1, \cdots, \xi_{i-1}) + \frac{\partial V_i}{\partial \xi_{i-1}} \Gamma_{i-1}(\xi_{i-2}, \xi_{i-1}) e_i + e_i^T (\dot{y}_i - \dot{\sigma}_i)$$

so that each new input is defined by

$$v_i = \dot{y}_i = -a_i e_i + \dot{\sigma}_i - \Gamma_{i-1}^T \left(\frac{\partial V_i}{\partial \xi_{i-1}} \right)^T .$$

This is due to the common fact that the "distance" between each output y_i and the corresponding virtual control σ_i demands converging to zero. In summary, Figure 8.18 shows the overall backstepping control design procedure for a k-cascaded dynamic system.

The k-cascaded systems modeling and backstepping control design procedure will be very useful in analysis and control of human-machine dynamic interaction systems [18, 19]. In Chapter 11, we will discuss in detail how to model a human-machine dynamic interaction and how to design a global control to meet certain desired objectives or criteria. Typical examples include the analysis of vehicle collision dynamics, active restraint systems with control design considerations, dynamic analysis and control of active suspension systems and military vehicle dynamics after a hit by an IED (Improvised Explosive Device), etc. All these typical application cases commonly have human drivers, passengers, and/or warfighters involved. Therefore, the k-cascaded

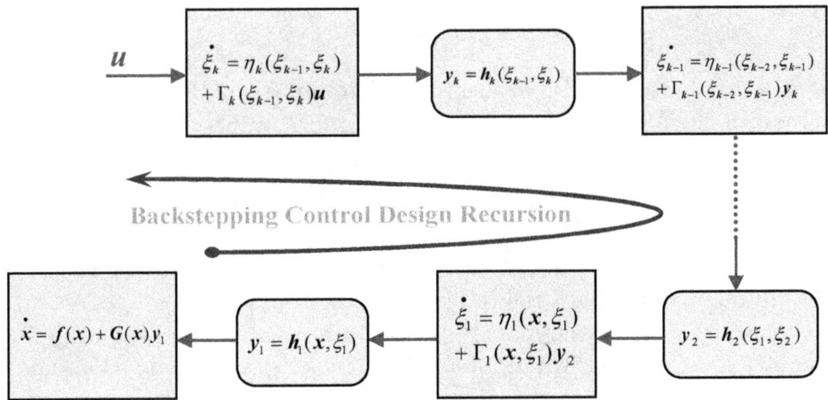

Fig. 8.18 A flowchart of backstepping control design for a k-cascaded dynamic system

systems modeling along with the backstepping control design procedure will well fit those applications to significantly improve the performance, safety, and ride quality, as well as to minimize any injuries.

8.7 Adaptive Control of Robotic Systems

Adaptive control plays a pivotal role in stabilizing a system with parameter uncertainty [22]. Although we can design a very robust controller to overwhelm the error due to the parameter deviation, such as using a high control gain to accelerate the attraction rate, etc., we cannot achieve a satisfactory convergence in many cases, especially in a nonlinear system. Therefore, in addition to designing a control law, it often requires a parameter *adaptation law* acting simultaneously on the system in a real-time fashion. A general parameter-emphasized system model is given as follows:

$$\dot{x} = f(x) + F(x)\phi + \sum_{i=1}^{r} g_i(x)u_i, \tag{8.114}$$

where $\phi \in \mathbb{R}^k$ is a parameter column vector to be adapted. We assume that all the coefficient $g_i(x)$'s of the input $u \in \mathbb{R}^r$ do not contain any uncertain parameter.

In addition, it is important that all the available nonlinear adaptive control methods thus far are under a *linear parametric condition*, i.e., the system equation is linear in parameters, as seen in (8.114). A typical example is the disease control system given by

$$\begin{cases} \dot{x}_1 = -ax_1 + bx_1x_2 + u \\ \dot{x}_2 = -bx_1x_2. \end{cases} \tag{8.115}$$

If both parameters a and b in the model are uncertain, then (8.115) can be rewritten in the following form:

$$\dot{x} = \begin{pmatrix} -x_1 & x_1x_2 \\ 0 & -x_1x_2 \end{pmatrix} \begin{pmatrix} a \\ b \end{pmatrix} + \begin{pmatrix} 1 \\ 0 \end{pmatrix} u.$$

Thus, $f(x) = 0$, $F(x)$ is the 2 by 2 matrix and the parameter vector $\phi = (a \ \ b)^T$ in comparison with the parameter-emphasized system model (8.114).

Another example is a nonlinear-spring mechanical system, called the Duffing's system, with a control input u added into its second equation, i.e.,

$$\begin{cases} \dot{x}_1 = x_2 \\ \dot{x}_2 = -\mu x_2 + \alpha x_1 - \beta x_1^3 + u. \end{cases} \tag{8.116}$$

In comparison with (8.114),

$$f(x) = \begin{pmatrix} x_2 \\ 0 \end{pmatrix}, \quad F(x) = \begin{pmatrix} 0 & 0 & 0 \\ -x_2 & x_1 & -x_1^3 \end{pmatrix}, \quad \phi = \begin{pmatrix} \mu \\ \alpha \\ \beta \end{pmatrix}, \quad g(x) = \begin{pmatrix} 0 \\ 1 \end{pmatrix}.$$

In this case, the parameter vector ϕ is 3 by 1. Since its equilibrium point is $x = \begin{pmatrix} x_1 \\ x_2 \end{pmatrix} = 0$ when $u = 0$, the simplest Lyapunov candidate can be defined as $V(x) = \frac{1}{2}x^T x$. However, we have to add a positive-definite term to represent the parameter ϕ for a possible convergence between the modeled parameter vector ϕ_m and the actual but unknown parameter ϕ. Let a parameter error vector be $\Delta\phi = \phi - \phi_m$. Therefore, a completed Lyapunov candidate for adaptive control should be

$$V(x) = \frac{1}{2}x^T x + \frac{1}{2}\Delta\phi^T \Delta\phi$$

that is clearly positive-definite if the state vector x is augmented by the parameter error vector $\Delta\phi$. Taking a time-derivative on it yields

$$\dot{V} = x^T \dot{x} + \Delta\phi^T \Delta\dot{\phi} = x^T (f(x) + F(x)\phi + g(x)u) + \Delta\phi^T \Delta\dot{\phi}.$$

By noticing that the actual parameter $\phi = \Delta\phi + \phi_m$, we obtain

$$\dot{V} = x^T f(x) + x^T F(x)\Delta\phi + x^T F(x)\phi_m + x^T g(x)u + \Delta\phi^T \Delta\dot{\phi}.$$

If we define an adaptation law in the form

$$\Delta\dot{\phi} = -F^T(x)x,$$

then,

$$\dot{V} = x^T f(x) + x^T F(x)\phi_m + x^T g(x)u \le -\epsilon(x, \Delta\phi),$$

where $\epsilon(x, \Delta\phi)$ is a scalar positive-definite function of the state x and the parameter-error $\Delta\phi$. Since $x^T f(x) = x_1 x_2$, $x^T F(x) = x_2(-x_2\ x_1\ \ -x_1^3)$ and $x^T g(x) = x_2$, with $\epsilon(x, \Delta\phi) = cx_2^2$ for $c > 0$, we obtain a control law

$$u = \alpha(x, \phi_m) = -x_1 - cx_2 - (-x_2\ x_1\ \ -x_1^3)\phi_m,$$

which requires to know the modeled parameter vector ϕ_m and is quite reasonable.

Even if $\epsilon(x, \Delta\phi)$ is not positive-definite, it can be shown that the above control law plus the adaptation law $\Delta\dot{\phi} = -F^T(x)x$ can still asymptotically stabilize the system based on the La Salle's Theorem. However, the choice of $\epsilon(x, \Delta\phi)$ does not take care of the parameter adaptation, and it is virtually impossible to include any $\Delta\phi^T \Delta\phi$ term into the definition of $\epsilon(x, \Delta\phi)$. Therefore, the modeled parameter ϕ_m is not guaranteed to converge to the actual real parameter ϕ, but at least it does not diverge according to the La Salle's Theorem and its Corollary. Nevertheless, in most application cases, we only care about the convergence of the state x and do not really need each parameter convergence. In fact, since the parameters are uncertain and also immeasurable, the desired target for each parameter is unavailable at all.

Moreover, when one implements both the control and adaptation laws into a system, it should be understood that in $\Delta\phi = \phi - \phi_m$, the actual real parameter ϕ is time-invariant, i.e., $\dot{\phi} = 0$ for every autonomous system, while the modeled parameter ϕ_m is being adjusted during the process of adaptation so that $\Delta\dot{\phi} = -\dot{\phi}_m$. In other words, the adaptation law can be rewritten as

$$\dot{\phi}_m = F^T(x)x.$$

The above adaptive control strategy on a single system can be extended to a cascaded control system by using an *adaptive backstepping procedure*, and the reader may refer to [17] for a detailed discussion. Figure 8.19 shows a block diagram to summarize the above adaptive control design procedure.

We now turn our attention to investigating how to adapt a state-feedback control system after exact linearization. Recalling the state-feedback control law in (8.76), each of $\alpha(x)$ and $\beta(x)$ contains a number of parameters, which may need adaptation if they are uncertain.

Let $\alpha_m(x)$ and $\beta_m(x)$ be the two functions of x but using all the modeled parameters ϕ_m, and $\alpha(x)$ and $\beta(x)$ without the subscript m denote the functions with the actual parameters ϕ. Thus, a control law must be determined and updated by $u = \alpha_m(x) + \beta_m(x)v$ with a new input v. In a trajectory-tracking control task, if each output channel of the system has a relative degree $r_i = 2$, then v can be defined by equation (8.47), i.e.,

$$v = \ddot{y}_d + K_v \Delta\dot{y} + K_p \Delta y. \tag{8.117}$$

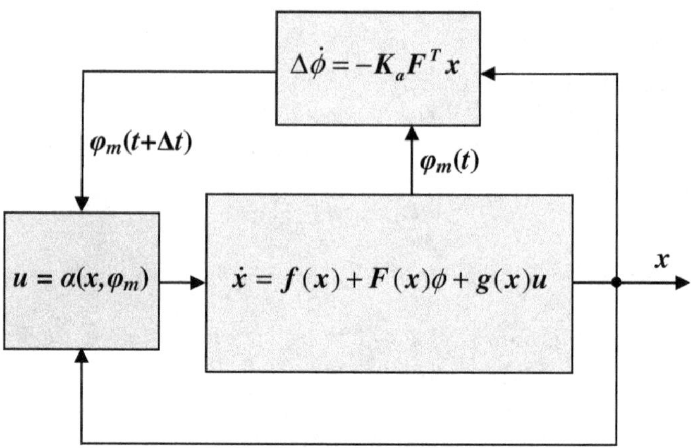

Fig. 8.19 A block diagram of adaptive control design

On the other hand, the modeled control law will be implemented to excite the real system that is described by $u = \alpha(x) + \beta(x)\ddot{y}$ with the actual parameters ϕ and the output acceleration vector $\ddot{y}$. Hence,

$$\alpha_m(x) + \beta_m(x)v = u = \alpha(x) + \beta(x)\ddot{y}.$$

Rearranging this equation and noticing that $v - \ddot{y} = \ddot{e} + K_v\dot{e} + K_p e$ due to (8.117), where $e = \Delta y = y_d - y$ is the output error vector, we have

$$\ddot{e} + K_v\dot{e} + K_p e = \beta_m^{-1}(x)[\alpha(x) - \alpha_m(x) + (\beta(x) - \beta_m(x))\ddot{y}].$$

This clearly shows that if the parameters are all precisely known, then the right-hand side of the above equation will vanish, and the left-hand side guarantees that the output tracking error e asymptotically converges to zero if both $K_p > 0$ and $K_v > 0$. However, due to the parameter uncertainty, the right-hand side cannot be zero. Instead, it will be a non-zero function of x, ϕ and even $\ddot{y}$, which jeopardizes the tracking convergence.

If the system is nonlinear in x, but is linear in the parameter ϕ, i.e., the linear parametric condition holds, the right-hand side of the above error equation can be factorized to

$$\ddot{e} + K_v\dot{e} + K_p e = F(x, \phi_m, \ddot{y})\Delta\phi, \tag{8.118}$$

for an m by k matrix F that is a function of the state x, modeled parameter ϕ_m and the output acceleration $\ddot{y}$. Since we wish both $e \to 0$ and $\dot{e} \to 0$ as $t \to \infty$, and also wish $\Delta\phi \to 0$, it is reasonable to construct a p.d.f. Lyapunov candidate as follows:

$$V(x) = \frac{1}{2}(\dot{e} + K_v e)^T(\dot{e} + K_v e) + \frac{1}{2}e^T K_p e + \frac{1}{2}\Delta\phi^T K_a^{-1}\Delta\phi,$$

where $K_a > 0$ is a k by k adaptation gain constant symmetric matrix. Thus,

$$\dot{V} = (\dot{e} + K_v e)^T(\ddot{e} + K_v\dot{e}) + e^T K_p\dot{e} + \Delta\phi^T K_a^{-1}\Delta\dot{\phi}.$$

According to (8.118), $\ddot{e} + K_v\dot{e} = F(x, \phi_m, \ddot{y})\Delta\phi - K_p e$. If we define an adaptation law:

$$\Delta\dot{\phi} = -\dot{\phi}_m = -K_a F^T(x, \phi_m, \ddot{y})(\dot{e} + K_v e), \tag{8.119}$$

then,

$$\dot{V} = -e^T K_v K_p e \le 0,$$

provided that $K_v K_p$ is also positive-definite. Equation (8.119) offers a general form of adaptation law that requires factorizing all the parameters from the state-feedback equation to find the matrix function F, and then to adapt (adjust and update) the modeled parameter ϕ_m by (8.119) at each sampling point. Namely, under the first-order approximation,

$$\phi_m(t + \Delta t) \approx \phi_m(t) + \Delta\phi_m = \phi_m(t) + K_a F(t)^T(\dot{e}(t) + K_v e(t))\Delta t.$$

The above discussion sounds successful in achieving an adaptation law as a value-added result to the state-feedback control for an exactly linearizable system. However, such an adaptation law given by (8.119) is actually unfeasible, because it requires us to know the output acceleration $\ddot{y}$ that is unavailable. In fact, the above Lyapunov function is constructed without invoking any dynamics of the system if it is a dynamic system, such as a mechanical system. Therefore, we have to seek a more feasible adaptation law for a class of high-dimensional and high-nonlinear complex dynamic control systems, such as a robotic system.

For an n-joint robotic system, if $q \in \mathbb{R}^n$ is a joint position vector, its general dynamic equation is given by

$$W(q)\ddot{q} + C(q, \dot{q})\dot{q} = u, \tag{8.120}$$

where $W(q) > 0$ is the n by n inertial matrix, $C(q, \dot{q})$ is an n by n matrix that contains all the centrifugal and Coriolis forces/torques as a coefficient of $\dot{q}$, and $u \in \mathbb{R}^r$ is the joint torque vector, including all the external torques and the torques due to gravity, and $r = n$ if the robot is fully-actuated. It has been developed in Chapter 7 that

$$C(q, \dot{q}) = \frac{1}{2}(\dot{W} + W_d^T - W_d),$$

where

$$W_d = \begin{pmatrix} \dot{q}^T \frac{\partial W}{\partial q_1} \\ \vdots \\ \dot{q}^T \frac{\partial W}{\partial q_n} \end{pmatrix}$$

is called an inertial derivative matrix in (7.17). Since $W_d^T - W_d$ is always skew-symmetric, for any vector $\xi \in \mathbb{R}^n$,

$$\xi^T C \xi = \frac{1}{2} \xi^T \dot{W} \xi. \tag{8.121}$$

Furthermore, both W and C satisfy the linear parametric condition.

After the above preparation on robotic systems modeling, consider that all the joint positions in q are defined to form an output, i.e., $y = q \in \mathbb{R}^n$, just for a discussion convenience. Let $s = \dot{e} + K_v e$ and $\eta = \dot{q}_d + K_v e$, where the error $e = y_d - y = q_d - q$. Then, $s = \eta - \dot{q}$ and $\dot{s} = \dot{\eta} - \ddot{q}$. Now, let a Lyapunov candidate be redefined by involving the robotic dynamics:

$$V(x) = \frac{1}{2} s^T W s + \frac{1}{2} \Delta\phi^T K_a^{-1} \Delta\phi.$$

Thus,

$$\dot{V} = s^T W \dot{s} + \frac{1}{2} s^T \dot{W} s + \Delta\phi^T K_a^{-1} \Delta\dot{\phi}.$$

Using (8.121), we further have

$$\begin{aligned} \dot{V} &= s^T W(\dot{\eta} - \ddot{q}) + s^T C s + \Delta\phi^T K_a^{-1} \Delta\dot{\phi} \\ &= s^T W \dot{\eta} - s^T W \ddot{q} + s^T C \eta - s^T C \dot{q} + \Delta\phi^T K_a^{-1} \Delta\dot{\phi}. \end{aligned}$$

Substituting the robotic dynamics equation (8.120) into here yields

$$\dot{V} = s^T W \dot{\eta} + s^T C \eta - s^T u + \Delta\phi^T K_a^{-1} \Delta\dot{\phi}.$$

Let a control law be defined by

$$u = W_m \dot{\eta} + C_m \eta + K s \tag{8.122}$$

with a constant gain matrix $K > 0$, and let an adaptation law be defined by

$$\Delta\dot{\phi} = -\dot{\phi}_m = -K_a Y^T s, \tag{8.123}$$

where Y is an n by k matrix that is determined by a factorization procedure between the parameters and variables as follows:

$$(W - W_m)\dot{\eta} + (C - C_m)\eta = Y \Delta\phi.$$

Obviously, the matrix Y is now a function of q, $\dot{q}$, q_d, $\dot{q}_d$ and $\ddot{q}_d$, and is independent of the actual acceleration $\ddot{q}$ and all the parameters. Finally, we achieve

$$\dot{V} = -s^T K s \leq 0,$$

and $s^T K s$ is a p.d.f. in the $x = \begin{pmatrix} q \\ \dot{q} \end{pmatrix}$ state space, but is not included in the parameter space.

By further testing both the control law in (8.122) and adaptation law in (8.123), it can be concluded that they are feasible. The only difficulty is to perform the above factorization symbolically by hand in order to extract the matrix Y from the equation, which will be very tedious and lengthy as n is large. A more compact way for such a factorization procedure as well as the development of both the control and adaptation laws can now be achieved using the configuration manifold (C-manifold) embedding theory.

We expect that through a control, both e and $\dot{e}$ can asymptotically converge to zero. This convergence is equivalent to the extended tracking error $s = \dot{e} + ke \to 0$ asymptotically [14, 20, 21]. If the C-manifold of a robotic system is embedded into an ambient Euclidean space by an embedding $z = \zeta(q)$, then $\dot{z} = J\dot{q}$, where J is the Jacobian matrix of the embedding $z = \zeta(q)$. We can similarly define $\gamma = J\eta$ such that $\dot{\gamma} = J\dot{\eta} + \dot{J}\eta$, where $\eta = \dot{q}_d + K_v e$ and $s = \eta - \dot{q}$.

Based on equation (7.55) in Chapter 7, a C-manifold embedding can naturally be factorized between its structure matrix Z and parameters by $z = Z\xi$. Therefore,

$$\gamma = J\eta = \left(\frac{\partial Z}{\partial q^1}\xi \ \cdots \ \frac{\partial Z}{\partial q^n}\xi \right) \eta = \left(\sum_{i=1}^{n} \eta^i \frac{\partial Z}{\partial q^i} \right) \xi = \Gamma \xi,$$

where a new structure matrix Γ for γ is defined and can be calculated by using $\dot{Z}$ but replacing each component of $\dot{q}$ in $\dot{Z}$ by the corresponding component of η, i.e.,

$$\Gamma = \dot{Z}|_{\dot{q}=\eta}.$$

On the other hand, according to Theorem 5 and equation (7.71) from Chapter 7, each component of τ can be determined by

$$\tau^i = g_i^T \ddot{z} = \xi^T \frac{\partial Z^T}{\partial q^i} \ddot{Z} \xi = \xi^T Q^i(q, \dot{q}, \ddot{q}) \xi, \tag{8.124}$$

where

$$Q^i(q, \dot{q}, \ddot{q}) = \frac{\partial Z^T}{\partial q^i} \ddot{Z}.$$

Equation (8.124) is a quadratic form of all k components of ξ if ξ is k-dimensional.

In order to avoid the actual acceleration $\ddot{q}$ involved, we replace $\ddot{z}$ in (8.124) by $\dot{\gamma}$ so that

$$\tau^i = g_i^T \dot{\gamma} = \xi^T \frac{\partial Z^T}{\partial q^i} \dot{\Gamma} \xi = \xi^T Q_\eta^i (q, \dot{q}, q_d, \dot{q}_d, \ddot{q}_d) \xi = \mathrm{tr}(Q_\eta^i \xi \xi^T), \tag{8.125}$$

where the new **torque structure matrix** Q_η^i is given by

$$Q_\eta^i = \frac{\partial Z^T}{\partial q^i} \dot{\Gamma} = \frac{\partial Z^T}{\partial q^i} \sum_{j=1}^{n} \left(\frac{\partial \dot{Z}}{\partial q^j} \eta^j + \frac{\partial Z}{\partial q^j} \dot{\eta}^j \right), \quad i = 1, \cdots, n. \tag{8.126}$$

Furthermore, let $\delta\Psi = \Psi_m - \Psi = \xi_m \xi_m^T - \xi\xi^T$ be the dynamic parameter deviation matrix between the model and real plant. Clearly, the entire adaptive control problem is to find a control law and a dynamic parameter adaptation law such that the extended tracking error $s = \dot{e} + ke$ can go to zero asymptotically. In other words, the equilibrium point of the entire tracking-error system at $e = 0$ and $\dot{e} = 0$ (i.e., $s = 0$) along with $\delta\Psi = 0$ is desired to be globally asymptotically stable [23, 24].

Lemma 4. *Let M^n be the combined C-manifold for an n-joint robotic system, which is smoothly and isometrically embedded into $\mathbb{R}^m$ $(m > n)$ by $z = \zeta(q)$. If $\delta\Psi$ is a dynamic parameter deviation matrix between the model and real plant, then the control law*

$$\tau^i = tr(Q_\eta^i \Psi_m) + k_m^i s, \quad for\, i = 1, \cdots, n, \tag{8.127}$$

along with the adaptation law

$$\delta\dot{\Psi} = \dot{\Psi}_m = \frac{k_a}{2} \sum_{i=1}^{n} (Q_\eta^i + Q_\eta^{iT}) s^i \tag{8.128}$$

can asymptotically stabilize the entire robotic tracking-error system in the Lyapunov sense, where k_m^i is the i-th row of an n by n positive-definite symmetric constant gain matrix K_m and $k_a > 0$ is a constant adaptation gain number.

Proof. Let a Lyapunov candidate be defined as follows:

$$V = \frac{1}{2} s^T W s + \frac{1}{2k_a} \mathrm{tr}(\delta\Psi^T \delta\Psi). \tag{8.129}$$

Taking a time-derivative for the above positive-definite function V yields

$$\begin{aligned}
\dot{V} &= s^T W \dot{s} + s^T J^T \dot{J} s + \frac{1}{k_a} \mathrm{tr}(\delta\dot{\Psi}^T \delta\Psi) \\
&= s^T W(\dot{\eta} - \ddot{q}) + s^T J^T \dot{J}(\eta - \dot{q}) + \frac{1}{k_a} \mathrm{tr}(\dot{\Psi}_m^T \delta\Psi)
\end{aligned}$$

$$= s^T(W\dot{\eta} + J^T\dot{J}\eta - \tau) + \frac{1}{k_a}\mathrm{tr}(\dot{\Psi}_m^T\delta\Psi). \tag{8.130}$$

Since $\gamma = J\eta$ and $\dot{\gamma} = J\dot{\eta} + \dot{J}\eta$, $J^T\dot{\gamma} = W\dot{\eta} + J^T\dot{J}\eta$ due to $W = J^TJ$. By noticing (8.125), each component of $J^T\dot{\gamma}$ is in the form of $g_i^T\dot{\gamma} = \xi^TQ_\eta^i\xi = \mathrm{tr}(Q_\eta^i\Psi)$. Therefore, substituting (8.127) and (8.128) into (8.130), we have

$$\begin{aligned}\dot{V} &= -\sum_{i=1}^{n} s^i\mathrm{tr}(Q_\eta^i\delta\Psi) - s^TK_ms + \frac{1}{2}\sum_{i=1}^{n} s^i\mathrm{tr}[(Q_\eta^i + Q_\eta^{iT})\delta\Psi] \\ &= -s^TK_ms. \end{aligned} \tag{8.131}$$

Therefore, $-\dot{V}$ is a positive-definite function in the phase subspace spanned by $s = \dot{e} + ke$. It can be further seen that the only point making $\dot{V} = 0$ is $s = 0$, and however, this results in $\delta\dot{\Psi} = 0$, instead of $\delta\Psi = 0$. Hence, we conclude that the proposed control law and adaptation law in the theorem guarantee the extended tracking error $s = \dot{e} + ke \to 0$ asymptotically as $t \to \infty$, although each element of $\delta\Psi$ may possibly stick around a non-zero but finite steady-state error. □

Now, utilizing the concept of the minimum embeddable C-manifold M_{nk} as the smallest dynamic model for any n-joint open serial-chain robotic system, the above regular adaptive control scheme can significantly be reduced. Based on Lemma 4 along with Theorem 4 in Chapter 7, we finally reach the following theorem:

Theorem 12. *The control law (8.127) and adaptation law (8.128) in Lemma 4 can be reduced by using only the minimum embedding ζ_{nk} of the C-manifold M_{nk} for an open serial-chain robotic system to determine its torque structure matrices Q_η^i for $i = 1, \cdots, n$. Then, the reduced control law and adaptation law can asymptotically stabilize, in the Lyapunov sense, the entire robotic tracking-error system in a global sense if there is no nonholonomic constraint involved.*

Note that in the above theorem, a nonholonomic constraint in the form $\omega(q, \dot{q}) = h(q)\dot{q} = 0$ for a dynamic system means non-integrable, i.e., the differential 1-form $\sigma = \omega dt = h(q)dq$ is not exact according to the exterior calculus introduced in Chapter 2. A control system that contains a non-trivial internal dynamics is a typical nonholonomic example, such as an under-actuated robotic system. Therefore, Theorem 12 can be applied only in the fully-actuated open serial-chain robotic systems.

It can also be seen from Theorem 12 that to implement the above adaptive control scheme, the key step is to determine the structure matrix Z through the kinematic homogeneous transformations based on (7.54) for each of the k links involved in the minimum embedding ζ_{nk} of the embeddable C-manifold M_{nk} for a given robotic system. Then, compute each torque structure matrix

Q^i_η for $i = 1, \cdots, n$ by (8.126). The parameter matrix in both the reduced control law and adaptation law should be $\Psi_{nk} = \xi_{nk}\xi^T_{nk}$ that consists of only the dynamic parameters of the k links involved in $z_{nk} = \zeta_{nk}(q)$. If one does this way, the parameter factorization will no longer be required, because a general embedding $z = \zeta(q)$ has already been factorized in nature, i.e., $z = Z\xi$.

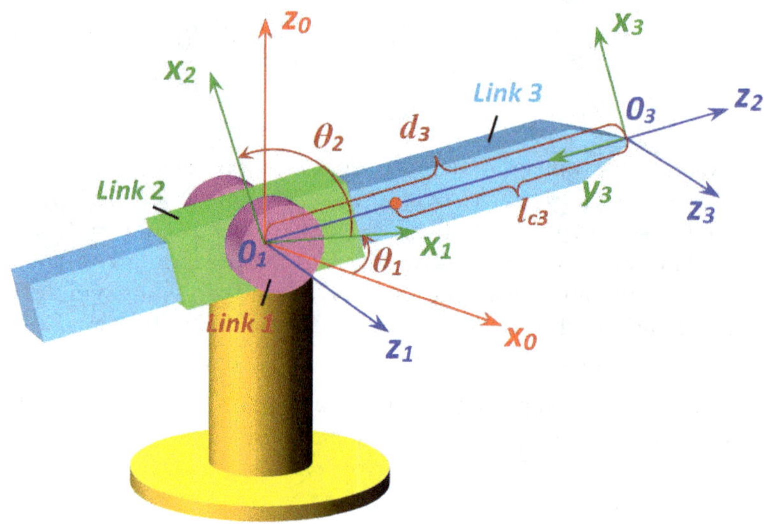

Fig. 8.20 An RRP type three-joint robot arm

As an illustrative example, let us look at a 3D RRP-type three-joint robot, as shown in Figure 8.20. The Denavit-Hartenberg (D-H) kinematic parameter table of this robot is given in Table 8.2.

Table 8.2 The D-H table for the RRP-type 3-joint robot arm

Joint Angle	**Joint Offset**	**Twist Angle**	**Link Length**
θ_i	d_i	α_i	a_i
θ_1	0	90^0	0
θ_2	0	90^0	0
$\theta_3 = 0$	d_3	-90^0	0

Based on the D-H table, we can readily find the homogeneous transformation A^3_0 between frame 3 of link 3 and the base (frame 0),

$$A^3_0 = \begin{pmatrix} R^3_0 & p^3_0 \\ 0 & 1 \end{pmatrix} = \begin{pmatrix} c_1c_2 & -c_1s_2 & s_1 & d_3c_1s_2 \\ s_1c_2 & -s_1s_2 & -c_1 & d_3s_1s_2 \\ s_2 & c_2 & 0 & -d_3c_2 \\ 0 & 0 & 0 & 1 \end{pmatrix}. \tag{8.132}$$

This matrix contains all elements needed to construct a structure matrix Z_0^3 for the C-manifold M_3 of the last link of the robot. It is observable that M_3 is also the minimum embeddable C-manifold M_{nk} for this particular robot with $n = 3$ and $k = 1$, because both link 1 and link 2 have no alternative positions and orientations once link 3 is fixed. We can adopt a two-point model if the inertia tensor would be diagonal. Here, in order to demonstrate the case more generally, let us stay in the complete three-point model for its embedding determination.

Let b_3, a_3^1 and a_3^2 be three 3 by 1 dynamic parameter vectors. Let m_3 be the mass of link 3. Then, the embedding $z_3 = \zeta_3(q)$ can be defined as follows:

$$z_3 = \zeta_3(q) = \begin{pmatrix} \sqrt{m_3}(p_0^3 + R_0^3 b_3) \\ R_0^3 a_3^1 \\ R_0^3 a_3^2 \end{pmatrix}. \tag{8.133}$$

Clearly, this embedding sends the 3-dimensional C-manifold M_3 into Euclidean 9-space $\mathbb{R}^9$. Its 9 by 3 Jacobian matrix can thus be found by

$$J_3 = \frac{\partial \zeta_3}{\partial q} = \begin{pmatrix} \sqrt{m_3}\left[J_0^3 + \left(\frac{\partial R_0^3}{\partial \theta_1} b_3 \;\; \frac{\partial R_0^3}{\partial \theta_2} b_3 \;\; \frac{\partial R_0^3}{\partial d_3} b_3\right)\right] \\ \begin{matrix} \frac{\partial R_0^3}{\partial \theta_1} a_3^1 & \frac{\partial R_0^3}{\partial \theta_2} a_3^1 & \frac{\partial R_0^3}{\partial d_3} a_3^1 \\ \frac{\partial R_0^3}{\partial \theta_1} a_3^2 & \frac{\partial R_0^3}{\partial \theta_2} a_3^2 & \frac{\partial R_0^3}{\partial d_3} a_3^2 \end{matrix} \end{pmatrix}, \tag{8.134}$$

where $J_0^3 = \partial p_0^3 / \partial q$ is the conventional kinematic Jacobian matrix for this three-joint robot, and the position vector p_0^3 is just the last column of A_0^3 given in (8.132).

The inertial matrix W of this robot can be derived directly from the equation of kinetic energy for the entire system under the assumption that the inertia tensor of each link is reduced to be diagonal. Such a derivation is only for preparing a *real-plant simulator* for the robotic system. Namely,

$$W = \begin{pmatrix} w_{11} & 0 & 0 \\ 0 & m_2 l_{c2}^2 + m_3(d_3 - l_{c3})^2 + I_{y2} + I_{z3} & 0 \\ 0 & 0 & m_3 \end{pmatrix}, \tag{8.135}$$

where $w_{11} = m_2 l_{c2}^2 s_2^2 + m_3(d_3 - l_{c3})^2 s_2^2 + I_{y1} + (I_{z2} + I_{y3})c_2^2 + (I_{x2} + I_{x3})s_2^2$. Here we have assumed that the mass center of link 3 is at a point on the y_3-axis of frame 3, i.e.,

$$p_0^{c3} = \begin{pmatrix} (d_3 - l_{c3})c_1 s_2 \\ (d_3 - l_{c3})s_1 s_2 \\ -(d_3 - l_{c3})c_2 \end{pmatrix}.$$

In this RRP type three-joint robot case, under the above assumption of the mass center of link 3, the parameter vector of link 3 to be adapted is

$$\xi_3 = \begin{pmatrix} \bar{b}_3^1 \\ \bar{b}_3^2 \\ a_3^1 \\ a_3^2 \end{pmatrix}$$

that is 8-dimensional. The 9 by 8 structure matrix Z_0^3 for the C-manifold M_3 of link 3, which is also the minimum embeddable C-manifold of the entire system, is given by

$$Z_0^3 = \begin{pmatrix} y_0^3 & p_0^3 & O & O \\ 0 & 0 & R_0^3 & O \\ 0 & 0 & O & R_0^3 \end{pmatrix}, \tag{8.136}$$

where $y_0^3 = (-c_1 s_2 \quad -s_1 s_2 \quad c_2)^T$ is the second column of A_0^3 in (8.132) that represents the unit vector of the y_3-axis of frame 3. After $\partial Z_0^3/\partial q^i$ for $i = 1, 2, 3$ and their time-derivatives are calculated, we finally obtain the torque structure matrices

$$Q_\eta^i = \frac{\partial Z_0^{3T}}{\partial q^i} \dot{\Gamma}_0^3$$

for $i = 1, 2, 3$, where

$$\Gamma_0^3 = \dot{Z}_0^3|_{\dot{q}=\eta}.$$

A simulation study has been done in MATLAB^{TM} to verify the proposed adaptive control scheme in Lemma 4 and Theorem 12 in performing a trajectory-tracking task by this three-joint RRP robot. In simulation programming, we use the following complete robotic dynamic equation to be the real-plant simulator:

$$W\ddot{q} + (W_d^T - \frac{1}{2}W_d)\dot{q} = \tau, \tag{8.137}$$

where W is the inertial matrix given by (8.135) and the 3 by 3 derivative matrix W_d is determined by

$$W_d = \begin{pmatrix} \dot{q}^T \frac{\partial W}{\partial \theta_1} \\ \dot{q}^T \frac{\partial W}{\partial \theta_2} \\ \dot{q}^T \frac{\partial W}{\partial d_3} \end{pmatrix}.$$

Note that both the control law and the input of the real-plant have combined the joint torque τ_g due to gravity and the joint torque τ for dynamic motion together, i.e., $u = \tau + \tau_g$. It is easy to split them by subtracting τ_g from u, and the equation for $\tau_g = -\partial P/\partial q$ in this particular robot case can easily be determined. Namely, if both link 1 and link 2 are pivoted at their mass centers, their mass center heights are all zero with respect to the base frame 0. Since the last z-component of the mass center position p_0^{c3} is just the height h_3 of the link 3 with respect to the bass, the total gravitational potential energy of the robot is

$$P = m_3 g h_3 = -m_3 g (d_3 - l_{c3}) c_2$$

so that

$$\tau_g = -\frac{\partial P}{\partial q} = m_3 g \begin{pmatrix} 0 \\ -(d_3 - l_{c3}) s_2 \\ c_2 \end{pmatrix}.$$

Therefore, in terms of the symbolical derivation, the modeled control law in the reduced form of (8.127) is developed based on (7.71) with only link 3 being structured, while the real-plant (8.137) is based on the complete Lagrange equation (7.70). Furthermore, in the real-plant simulator, all the dynamic parameters are defined appropriately (see the function subroutine in the MATLAB^{TM} program), while $\Psi_{m(nk)} = \xi_{m(nk)}\xi_{m(nk)}^T$ in both the modeled control law and adaptation law is initiated with 1 (it can be arbitrary, but better be non-zero) for each element of $\xi_{m(nk)}$.

The simulated RRP robot arm is to track a desired trajectory in 3-D Cartesian space, but specified by a desired joint vector q_d as a function of time t:

$$q_d = \begin{pmatrix} 0.2 + 0.2t \\ 1.8 + \sin t \\ 1.2 + 0.7 \sin t \end{pmatrix}.$$

At each sampling point, substituting the above desired joint values into the position vector p_0^3 yields a desired Cartesian output y_d at time t. It can also be seen that since the initial joint values in the simulation are $\theta_1(0) = 0$, $\theta_2(0) = 2$ in radians and $d_3 = 1$ in meters, there is an initial deviation that is purposely set between the desired q_d and the real q at the starting point $t = 0$.

With the control gain matrix defined by $K_m = 100I_3$, where I_3 is the 3 by 3 identity, the adaptation gain $k_a = 1$ and $k = 8$ for the extended error vector $s = \dot{e} + ke$, the simulation study has achieved a satisfactory convergence, as shown in Figure 8.21.

All the detailed settings and derivations can be seen in the following MATLAB^{TM} code, including a main program and a function subroutine to be called through the Runge-Kutta algorithm as a real-plant simulator:

```
q=[0; 2; 1];               % The Initial Joint Values
dq=[0; 0; 0];              % The Initial Joint Velocity
N=299;   dt=0.01;          % Total Number of Sampling
                           % Points and Time Interval
k=8;     ks=100;           % Gain Constants

xi=[1;1;1;1;1;1;1;1];      psi=xi*xi';
                         % The Initial Set of Parameters

Y=[];    Yd=[];    T=[];
```

```
for i=0:N
    t=i*dt;  T=[T t];
    s1=sin(q(1));  c1=cos(q(1));
    s2=sin(q(2));  c2=cos(q(2));
    d3=q(3);

    qd=[0.2+0.2*t; 1.8+sin(t);  1.2+0.7*sin(t)];
         % Joint Values Specified for Desired Trajectory
    dqd=[0.2; cos(t); 0.7*cos(t)];
    ddqd=[0; -sin(t); -0.7*sin(t)];

    Yd=[Yd [qd(3)*cos(qd(1))*sin(qd(2))
            qd(3)*sin(qd(1))*sin(qd(2))
                 -qd(3)*cos(qd(2))]];
                        % Calculate and Accumulate the
                        % Desired Cartesian Trajectory

    R03=[c1*c2 -c1*s2 s1; s1*c2 -s1*s2 -c1; s2 c2 0];
    p03=[d3*c1*s2; d3*s1*s2; -d3*c2];
    Y=[Y p03];          % The Actual Robot Position Output

    dR1=[-s1*c2 s1*s2 c1; c1*c2 -c1*s2 s1; 0 0 0];
    dR2=[-c1*s2 -c1*c2 0; -s1*s2 -s1*c2 0; c2 -s2 0];

    J03=[-d3*s1*s2 d3*c1*c2 c1*s2
          d3*c1*s2 d3*s1*c2 s1*s2
              0       d3*s2   -c2 ];
                        % The Kinematic Jacobian Matrix

    dJ03=[-d3*c1*s2*dq(1)-d3*s1*c2*dq(2)-dq(3)*s1*s2
          -d3*s1*s2*dq(1)+d3*c1*c2*dq(2)+dq(3)*c1*s2
                                    0];
    dJ03=[dJ03 [-d3*s1*c2*dq(1)-d3*c1*s2*dq(2)+...
                                         dq(3)*c1*c2
                 d3*c1*c2*dq(1)-d3*s1*s2*dq(2)+...
                                         dq(3)*s1*c2
                       d3*c2*dq(2)+dq(3)*s2]];
    dJ03=[dJ03 [-s1*s2*dq(1)+c1*c2*dq(2)
                 c1*s2*dq(1)+s1*c2*dq(2)
                        s2*dq(2)]];  % Time-Derivatives

    ddR1=[-c1*c2*dq(1)+s1*s2*dq(2) c1*s2*dq(1)+...
                                s1*c2*dq(2) -s1*dq(1)
          -s1*c2*dq(1)-c1*s2*dq(2) s1*s2*dq(1)-...
                                c1*c2*dq(2)  c1*dq(1)
                               0  0  0];
    ddR2=[ s1*s2*dq(1)-c1*c2*dq(2) s1*c2*dq(1)+...
                                      c1*s2*dq(2) 0
          -c1*s2*dq(1)-s1*c2*dq(2) -c1*c2*dq(1)+...
```

```
                                        s1*s2*dq(2) 0
          -s2*dq(2)   -c2*dq(2)   0];

O=zeros(3);
e=qd-q;  de=dqd-dq;  eta=dqd+k*e;
deta=ddqd+k*de;  s=eta-dq;

Zq1= [dR1(:,2) J03(:,1) O  O
      zeros(3,2) dR1  O
      zeros(3,2)  O  dR1];
Zq2= [dR2(:,2) J03(:,2) O  O
      zeros(3,2) dR2  O
      zeros(3,2)  O  dR2];
Zq3= [zeros(3,1) J03(:,3) O  O
      zeros(3,2)  O  O
      zeros(3,2)  O  O];
dZq1=[ddR1(:,2) dJ03(:,1) O  O
      zeros(3,2) ddR1  O
      zeros(3,2)  O  ddR1];
dZq2=[ddR2(:,2) dJ03(:,2) O  O
      zeros(3,2) ddR2  O
      zeros(3,2)  O  ddR2];
dZq3=[zeros(3,1) dJ03(:,3) O  O
      zeros(3,2)  O  O
      zeros(3,2)  O  O];    % The Structure Matrices

dGamma=Zq1*deta(1)+Zq2*deta(2)+Zq3*deta(3);
dGamma=dGamma+dZq1*eta(1)+dZq2*eta(2)+dZq3*eta(3);

Q1=Zq1'*dGamma;
Q2=Zq2'*dGamma;
Q3=Zq3'*dGamma;      % Torque Structure Matrices

psi=psi+((Q1+Q1')*s(1)+(Q2+Q2')*s(2)+(Q3+Q3')*s(3))*dt;
                       % The Parameter Adaptation Law

u=[trace(Q1*psi);trace(Q2*psi);trace(Q3*psi)]+ks*s;
                       % The control law

x=[q; dq];

    % Runge-Kutta Algorithm with f_tp.m

r1=f_tp(x,u)*dt;
r2=f_tp(x+r1/2,u)*dt;
r3=f_tp(x+r2/2,u)*dt;
r4=f_tp(x+r3,u)*dt;

x=x+(r1+2*r2+2*r3+r4)/6;     % Update the Old State
```

```
    q=x(1:3);    dq=x(4:6);

    plot3(Yd(1,:),Yd(2,:),Yd(3,:),Y(1,:),Y(2,:),Y(3,:)),
    grid,xlabel('x'),ylabel('y'),zlabel('z'),
    view([2 -1 0.4]),axis([0 1.5 0 0.8 0 2]);

    drawnow;
end

pause
axis;
plot(T',Yd(1,:)'-Y(1,:)',T',Yd(2,:)'-Y(2,:)',...
                            T',Yd(3,:)'-Y(3,:)'),grid

% ------------------------------------- %

function fs = f_tp(x,u)        % Function Subroutine

    q=x(1:3);  dq=x(4:6);

    m2=2;  m3=2.5;  lc2=.2;  lc3=0.6;
    Jy1=2.5;  Jx2=1.5;  Jy2=1.4;
    Jz2=0.8;  Jx3=4;  Jy3=0.6;  Jz3=4;

    s1=sin(q(1));  c1=cos(q(1));
    s2=sin(q(2));  c2=cos(q(2));
    d3=q(3);

    w11=m2*lc2^2*s2^2+m3*(d3-lc3)^2*s2^2+Jy1+...
                   (Jz2+Jy3)*c2^2+(Jx2+Jx3)*s2^2;
    w22=m2*lc2^2+m3*(d3-lc3)^2+Jy2+Jz3;
    W=[w11  0  0
        0  w22 0
        0   0  m3];

    Wd=[0    0    0
        dq'*[(m2*lc2^2+m3*(d3-lc3)^2+(Jx2+Jx3)-...
                            (Jz2+Jy3))*2*s2*c2  0  0
                0     0     0
                0     0     0]
        dq'*[2*m3*(d3-lc3)*s2^2  0  0
               0     2*m3*(d3-lc3)    0
               0     0          0]];

    fs=[dq; inv(W)*((Wd/2-Wd')*dq+u)];
end
```

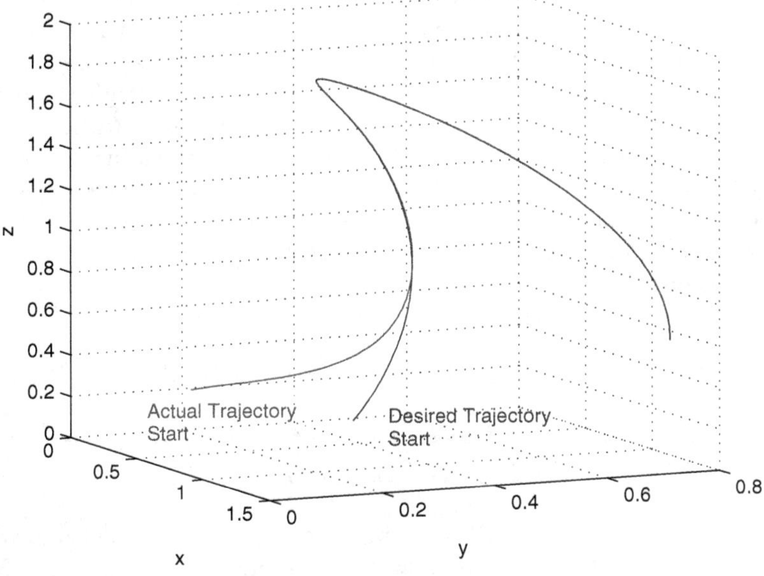

(a) The desired and actual trajectories in 3D Cartesian space

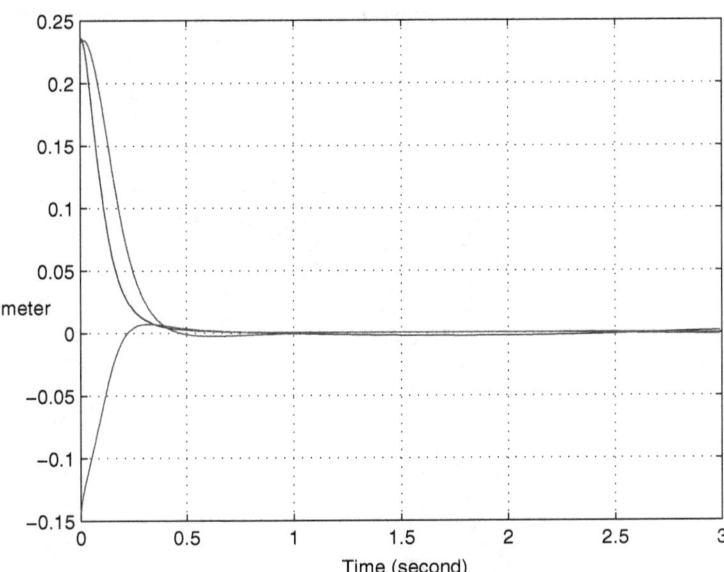

(b) The Cartesian tracking errors vs. time

Fig. 8.21 The simulation results with M_3 as the minimum embeddable C-manifold

The final result of the simulation study is shown in Figure 8.21. It can be evidently seen that the trajectory-tracking task converges within 0.5 second, and the adaptive control scheme, indeed, does not require any knowledge of dynamic parameters for a given robotic system. Although the algorithm may still require a certain symbolic derivation in the preparation phase, the adaptive control model based on the C-manifold embedding theory has already been factorized between the system structure and dynamic parameters in nature. In other words, it does not need extra effort to extract all the dynamic parameters from the dynamic equation, in contrast to the conventional adaptive control scheme for robotic systems.

8.8 Computer Projects and Exercises of the Chapter

8.8.1 Dynamic Modeling and Control of a 3-Joint Stanford-Like Robot Arm

The main objective of this computer project is to model and simulate a three-joint Stanford-like robot arm in computer and verify its nonlinear feedback control algorithm. The robot arm consists of two revolute and one prismatic joints, as shown in Figure 8.22, where the trunk height $d_1 = 0.6$ m. The project is divided into three phases:

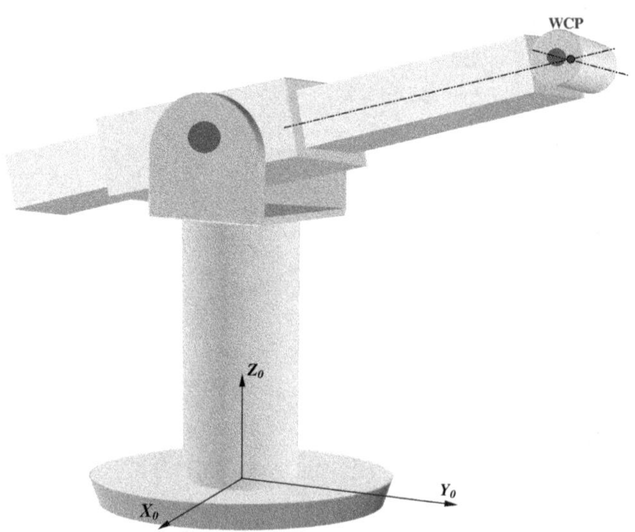

Fig. 8.22 A 3-joint Stanford-like robot arm

Phase 1. Derive all necessary matrices and equations to model both kinematics and dynamics of the robot arm, such as D-H table, homogeneous transformation matrices, the inertial matrix W and its derivative matrix W_d.

Phase 2. Develop a nonlinear state-feedback control law for the robot arm wrist center point (WCP) ready to track a following desired trajectory:

$$y_d(t) = \begin{pmatrix} 0.1t \\ 1.0 - 0.12t^2 \\ 0.8/(1+t^2) \end{pmatrix} \text{ m.}$$

The sampling interval is recommended to be $\Delta t = 0.01$ sec., and total number of sampling points is at least $N = 400$. The dynamic parameters of the robot between the real-plant and the model are given as follows:

System	I_{y1}	I_{x2}	I_{y2}	I_{z2}	m_3	I_{x3}^c	I_{y3}^c	I_{z3}^c	l_{c3}
Real	0.2	0.075	0.075	0.1	4	0.78	0.78	0.065	0.7
Model	0.18	0.08	0.07	0.12	4.1	0.76	0.8	0.07	0.67

Phase 3. Implement the nonlinear feedback control algorithm into your MATLAB^{TM} program. Define the robot WCP position as a 3-dimensional output vector $y = h(q)$. The state of the entire control system is defined by $x = \begin{pmatrix} q \\ \dot{q} \end{pmatrix}$, and the system input $u(t)$ is augmented by all the joint force/torque: τ_1, τ_2 and f_3.
In the program, the real plant is simulated using either the ode45$(\cdots)$ internal function or the Runge-Kutta algorithm with the Real set of parameters, while the control law is calculated using the Model set of parameters in order to simulate the control effect under the parameter uncertainty. Furthermore, to observe the control performance against any external disturbance, you may set an initial output position error at $t = 0$ as $\Delta\theta_1 = 0.25$ rad., $\Delta\theta_2 = -0.5$ rad. and $\Delta d_3 = 0.15$ m.

Your report should include an introduction, complete derivations, necessary block diagram and/or flowchart, computer program and all output plots, discussions and conclusion.

The following **Runge-Kutta Algorithm** can be applied to solve an equation $\dot{x} = f(x)$ numerically, where $x \in \mathbb{R}^n$ with $x(0) = x_0$:
First,

$$z_1 = f(x_k)\Delta t, \quad z_2 = f(x_k + \frac{z_1}{2})\Delta t,$$

$$z_3 = f(x_k + \frac{z_2}{2})\Delta t, \quad z_4 = f(x_k + z_3)\Delta t,$$

and then,

$$x_{k+1} = x_k + \frac{1}{6}(z_1 + 2z_2 + 2z_3 + z_4), \quad k = 0, 1, \cdots.$$

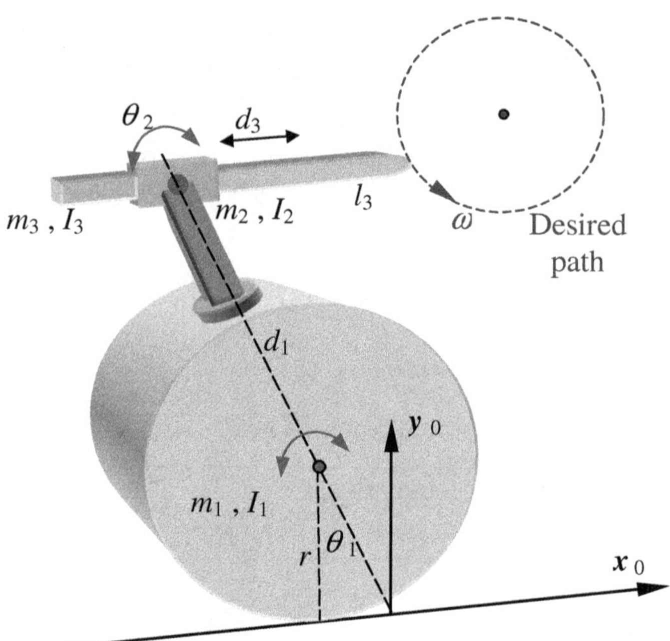

Fig. 8.23 A 2-joint robot arm sitting on a rolling log

8.8.2 Modeling and Control of an Under-Actuated Robotic System

This project is to model, simulate and verify a three-joint under-actuated mechanical system that consists of a 2-joint RP robot arm mounted on a rolling cylindrical log, as shown in Figure 8.23. The procedure of the project is specified as follows:

1. Derive a dynamic equation of the system, and determine $f(x)$, $g_1(x)$ and $g_2(x)$, where x is the state vector to be predefined and two-joint torque and force are considered as two inputs, while the cylindrical log is free-rolled on the ground without rolling friction;
2. If the tip-point position of the arm w.r.t. the fixed base is defined as a 2 by 1 output vector, find an explicit form of the output $y = h(x)$;
3. Find the relative degree $\{r_1,\ r_2\}$ for the two output channels $h_1(x)$ and $h_2(x)$;
4. Determine the input-output linearization for the linearizable subsystem, and also find its static state-feedback $u = \alpha(x) + \beta(x)v$;
5. If $r = r_1 + r_2 < n$, determine an internal dynamics and further discuss the system stability issue;
6. Design a trajectory-tracking controller for the entire system and draw an overall block diagram;

7. Develop a MATLAB^{TM} program to complete the simulation study if
   a. The desired trajectory for the arm tip-point is a circle, the center of which is at (1.2, 3, 0) and the radius is 0.5 m., starting from a point (0.7, 3, 0) at a rotating speed $\omega = 0.5$ rad./sec. in counterclockwise;
   b. The physical parameters: $m_1 = 4$, and $m_2 = m_3 = 1$ Kg., $r = 1$ m., $d_1 = 1.5$ m. and $l_3 = 0.8$ m. $I_2 = 0.3$ and $I_3 = 0.3$ Kg.-m^2. Also, assume that the mass center of the log is just at its centroid, and the first link of the robot is pivoted at its mass center;
   c. The real-plant should be simulated using the Runge-Kutta Algorithm given above.

8.8.3 Dynamic Modeling and Control of a Parallel-Chain Planar Robot

A planar parallel-chain system to be controlled is shown in Figure 8.24, where $a = 0.6$ and $b = 1$ in meter.

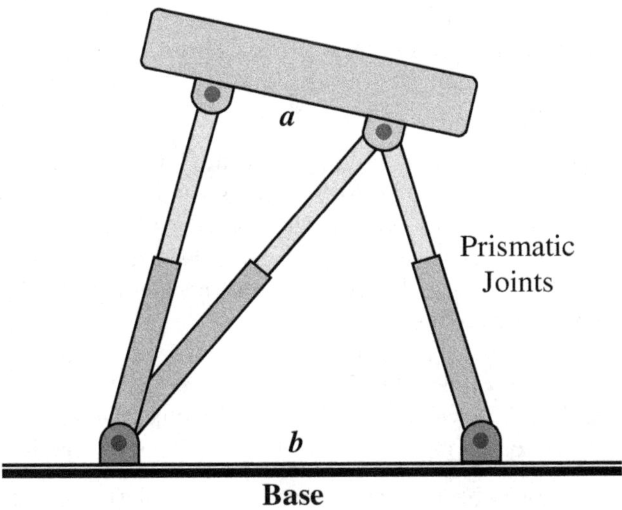

Fig. 8.24 A 3-piston parallel-chain planar robot

1. Using the Grübler's formula to determine the net d.o.f. m of this mechanism;
2. Under the net d.o.f. result, find a complete kinematic model of the top block in terms of the lengths of the three prismatic joints (pistons), i.e., its forward kinematics (F-K);
3. Determine a minimum isometrizable embedding $z = \zeta(q) = Z\xi$ and calculate its Jacobian matrix J towards finding the Riemannian metric $W = J^T J$ of the system (since this is a planar case, the dimension of $z = \zeta(q)$ will not exceed 3 or 4, see the examples in this chapter);

4. If the three prismatic joints are actuated as a control input vector u, develop an adaptive control strategy, including a control law and an adaptation law based on equations (8.127) and (8.128);
5. If the centroid of the top block is desired to travel along a straight line and also to rotate itself counterclockwise by a constant angular velocity ω, program the adaptive control algorithm into MATLAB^{TM} to simulate or animate the system in the desired motion, where all the dynamic parameters in ξ are initialized to be 1 for each at $t = 0$;
6. Plot the resultant trajectory versus time, or show a 2D graphical real-time animation of the controlled parallel-chain planar robot.

8.8.4 Exercise Problems

1. A robot revolute joint is planned to move its angle θ from its originally idled $\theta_0 = 20^0$ at $t \leq 0$ to $\theta_f = 80^0$ at $t = 2$ sec., and then keep its speed as a constant $\dot{\theta} = 10^0$/sec. for $t \geq 2$. Find a cubic spline function to smooth the entire joint trajectory without acceleration impulse, and also plot both the joint position and velocity profiles.
2. A robot arm is to move its TCP (Tool Center Point) along the following given trajectories from an initial point $P_0 = (-1,\ \ 5)$ to a final point $P_f = (5,\ \ 7)$ on a 2D $x - y$ work plane.

   a. The first trajectory is a straight line from P_0 to P_f with both zero initial and ending velocities. The total time is required to be $T = 6$ sec. The velocity profile of the TCP motion is an equilateral trapezoid with 2 seconds for each of its acceleration and deceleration periods. Find the maximum Cartesian velocity v_{max} and acceleration a_{max} of the TCP with respect to the base;
   b. The second trajectory is a minimum path from P_0 to P_f to avoid any collision with an obstacle disc given by $(x - 3)^2 + (y - 6)^2 < 4$. The initial and ending velocities of the TCP must be all zero. Determine your optimal path planning and plot its velocity profiles.

3. A PM DC-motor is used to drive and control the speed of a robotic link, as shown in Figure 8.25. It is coupled with a 4:1 reduction gear to the link that has a moment of inertia $J_L = 2$ Kg-m^2. The speed of the link is monitored by a tachometer with a generated voltage of 0.25 v. per radian/sec of speed. The difference between the tachometer voltage output v_t and a command voltage v_d is to excite a voltage amplifier of gain A. The motor has the following parameters: $R_a = 0.3\ \Omega$, $L_a = 2.3$ mH, $J_m = 0.07$ Kg-m^2, $K_a = 0.9$ N-m/A., and the viscous friction is negligible.

   a. Develop a transfer function related to the link speed $\Omega_L(s)$ and the command voltage $V_d(s)$;

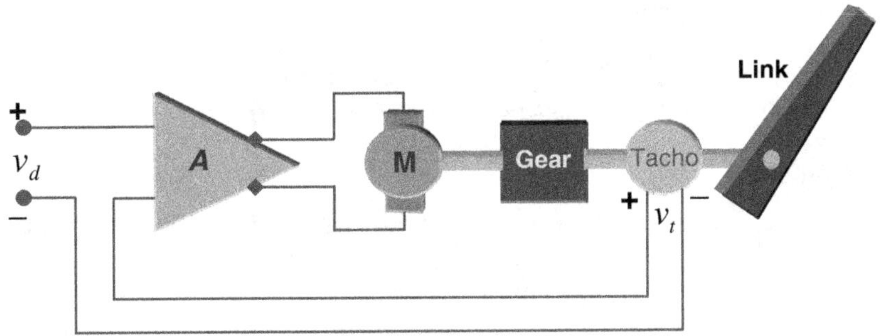

Fig. 8.25 A block diagram of the DC-motor in driving a robotic link

b. Determine the gain constant A such that the damping ratio ζ of the system is desired to be 0.707;
c. Estimate the sustain time t_s of the system step response by cutting off at the time of oscillation within 5% of the step. (Note: $t_s = -\frac{1}{\zeta\omega_n}\ln(0.05\sqrt{1-\zeta^2})$ or $t_s \approx \frac{3}{\zeta\omega_n}$ if ζ is small)

4. A robotic link is driven by a PM servo-motor through a reduction gear with ratio $n = 8$. The motor is energized by an amplifier with a voltage gain $A = 100$. The link has the following parameters: $J_L = 4$ Kg-m^2, $B_L = 2$ N-m-sec./rad. The parameters of the DC motor are $L_a = 0$, $R_a = 0.4\ \Omega$, $K_a = 0.8$ N-m./A., $J_m = 0.06$ Kg-m^2, and $B_m = 0.1$ N-m-sec./rad. For the link rotating position control purpose, the input is a target angle θ_L^d and the output is an actual angle θ_L of the link.

a. Draw a block diagram to represent this motor-link control system;
b. Find the transfer function $T(s) = \Theta_L/\Theta_L^d$;
c. Determine the natural frequency ω_n and the damping ratio ζ of the entire control system;
d. What is the steady-state value of the output if the input is a unit-step?

5. A *gradient system* satisfies the equation

$$\dot{x} = f(x) = -\frac{\partial\phi}{\partial x},$$

where $x = (x_1\ \cdots\ x_n)^T \in \mathbb{R}^n$, and $\partial\phi/\partial x$ is the gradient column vector for a scalar field $\phi(x)$. Suppose that $\phi(x)$ has a unique minimum at its equilibrium point x_e in $\mathbb{R}^n$.

a. Show that $V(x) = \phi(x) - \phi(x_e)$ is a Lyapunov candidate;
b. Using the above result, prove the stability for the following system:

$$\begin{cases} \dot{x}_1 = -2x_1(1 + x_2^2) \\ \dot{x}_2 = -2x_2(1 + x_1^2) \end{cases}$$

6. A nonlinear system is given by

$$\begin{cases} \dot{x}_1 = x_2 - x_1 f(x_1, x_2) \\ \dot{x}_2 = -x_1 - x_2 f(x_1, x_2) \end{cases}$$

with some scalar function $f(x_1, x_2)$.

   a. Using the Lyapunov direct method, discuss how the system stability depends on the scalar function $f(x_1, x_2)$;
   b. If (i) $f = e^{-x_1} + e^{-x_2}$, and (ii) $f = x_1^2 + x_2^2 - 1$, is the system stable?

7. A nonlinear system given by

$$\begin{cases} \dot{x}_1 = -x_2 + x_1 x_3 \\ \dot{x}_2 = x_1 + x_2 x_3 \\ \dot{x}_3 = -x_3 - (x_1^2 + x_2^2) + x_3^2 \end{cases}$$

has three equilibrium points $(0\ 0\ 0)^T$ and $(0\ 0\ \pm 1)^T$. Using the Lyapunov direct method, discuss the stability issue at **each** of the three equilibrium points.

8. Discuss the stability issue for the following nonlinear system:

$$\begin{cases} \dot{x}_1 = x_2 \\ \dot{x}_2 = x_1^2(1 - \exp(x_2)) - x_1 \exp(-x_1^2). \end{cases}$$

9. A nonlinear control system is given as

$$w(x)\ddot{x} + \frac{1}{2}\dot{w}\dot{x} + w(x)x = u$$

where $w(x) > 0$ is a scalar function of the 1-dimensional variable x and $\dot{w}$ is the time-derivative of $w(x)$. Determine a control u such that the entire system can be asymptotically stable.

10. A single-joint robot arm has the following dynamic equation:

$$(\theta + 2\sqrt{\theta} + 1)\ddot{\theta} + \frac{\sqrt{\theta} + 1}{2\sqrt{\theta}}\dot{\theta}^2 = u,$$

where $\theta > 0$ is the joint angle and u is the total joint torque as an input.

   a. By defining a state $x = \begin{pmatrix} \theta \\ \dot{\theta} \end{pmatrix}$, convert the above dynamic equation into the standard state equation form $\dot{x} = f(x) + G(x)u$;
   b. Find the total kinetic energy K of the robot;
   c. If an output is defined by $y = \left(\frac{2}{3}\sqrt{\theta} + 1\right)\theta$, determine its nonlinear state-feedback control law, i.e. find $\alpha(x)$ and $\beta(x)$ and the linearized new input v;
   d. If a PD control law is applied on the linearized system, what are the ranges of the gain constants K_p and K_d such that the robot arm can converge to a desired $y_d(t)$ in trajectory-tracking operation?

11. A nonlinear control system is given by

$$\begin{cases} \dot{x} = p(x,z) + u_1 \\ z^{(3)} = q(x,z) + u_2, \end{cases}$$

with an output equation $y = \begin{pmatrix} z \\ x \end{pmatrix}$, where p and q are two arbitrary scalar differentiable functions of x and z. Determine its relative degrees, and find the decoupling matrix D. Is this system invertible?

12. Given a nonlinear system:

$$\begin{cases} \dot{x}_1 = x_2^2 + x_2 u \\ \dot{x}_2 = 1 - u \\ \dot{x}_3 = x_1 + u \\ \dot{x}_4 = x_2 \end{cases}$$

   a. Is this single-input system input-state linearizable?
   b. If (a) is "Yes", find a nonlinear transformation to linearize it, and also determine the corresponding new input.

13. Given a nonlinear system:

$$\begin{cases} \dot{x}_1 = x_2^2 - x_1 + \dot{x}_2 + 2u_1 \\ \ddot{x}_2 + x_2\dot{x}_2 + x_1 x_2 = u_2 \end{cases}$$

   a. If the output $y = \begin{pmatrix} x_1 \\ x_2 \end{pmatrix}$ and input $u = \begin{pmatrix} u_1 \\ u_2 \end{pmatrix}$, determine the relative degrees;
   b. Is this system input-output linearizable? If "Yes", linearize it and find the static state-feedback control law.

14. Consider a nonlinear system $\dot{x} = f(x)$, where

$$f(x) = \begin{pmatrix} -x_2 - (2x_1 + x_3)^3 \\ x_1 \\ x_2 \end{pmatrix}.$$

   a. Determine its equilibrium point;
   b. Is the function $V(x) = x_1^2 + \frac{1}{2}x_2^2 + \frac{1}{2}x_3^2 + x_1 x_3$ an l.p.d.f. or a p.d.f.? Why?
   c. If (b) is yes, determine stability of the system.

15. A two-input nonlinear system is given as follows:

$$\begin{cases} \dot{x}_1 = x_2 + \exp(x_1 + 2x_2 + x_3) - 1 + u_1 \\ \dot{x}_2 = x_3 \\ \dot{x}_3 = u_2 \end{cases}$$

a. Find the equilibrium point;
b. Determine the stability when $u_1 = 0$ and $u_2 = 0$;
c. Can we design a state-feedback control for u_1 and u_2 to locally asymptotically stabilize this nonlinear system?

16. Consider a nonlinear system

$$\begin{cases} \dot{x}_1 = -x_1 + \frac{2+x_3^2}{1+x_3^2} u \\ \dot{x}_2 = x_3 \\ \dot{x}_3 = x_1 x_3 + u, \end{cases}$$

with an output $y = x_1$.

a. Determine its relative degree r_d;
b. Find the input-output linearized (sub)system and state-feedback control;
c. Determine its internal dynamics if $r_d < n$;
d. Is the system minimum-phase?

17. A cascaded system is given by

$$\begin{cases} \dot{x}_1 = x_2 \\ \dot{x}_2 = -x_1 + x_1 x_3 + x_1 y_1 \\ \dot{x}_3 = -x_3 - x_1 x_2 - x_2 y_2, \end{cases} \quad \begin{cases} y_1 = \xi_1 + \xi_2 \\ y_2 = \xi_3, \end{cases} \quad \begin{cases} \dot{\xi}_1 = \xi_2 \\ \dot{\xi}_2 = -\xi_3 + 3u_1 \\ \dot{\xi}_3 = \xi_1 - \xi_3^2 - 2u_2. \end{cases}$$

a. Find a virtual control law $y = \sigma(x)$ to asymptotically stabilize the first x-equation at its equilibrium point;
b. As the second y-equation is the output of the third ξ-equation, determine the relative degree r_i for each of the output channels y_i for $i = 1, 2$, and also find the decoupling matrix $D(\xi)$;
c. Is the ξ-equation along with its output y minimum-phase?
d. Using the backstepping procedure, design a global control law $u = (u_1\ u_2)^T$ to asymptotically stabilize the entire cascaded system.

18. A nonlinear system is given by

$$\begin{cases} \dot{x}_1 = \theta_1 x_1 + \theta_2 x_1 x_2 + u \\ \dot{x}_2 = -x_2 + \theta_3 x_3 \\ \dot{x}_3 = \theta_1 x_1^2 - \theta_4 x_3 + u, \end{cases}$$

where θ_1 through θ_4 are all uncertain parameters.

a. Determine its adaptive control standard form;
b. Design a control law and an adaptation law for the system. Please give a detailed design procedure.

References

1. Asada, H., Slotine, J.: Robot Analysis and Control. John Wiley & Sons, New York (1986)
2. Fu, K., Gonzalez, R., Lee, C.: Robotics: Control, Sensing, Vision, and Intelligence. McGraw-Hill (1987)
3. Spong, M., Vidyasagar, M.: Robot Dynamics and Control. John Wiley & Sons, New York (1989)
4. Craig, J.: Introduction to Robotics: Mechanics and Control, 3rd edn. Pearson Prentice Hall, New Jersey (2005)
5. Sarma, M.: Electric Machines, 2nd edn. West Publishing Company, Minneapolis (1994)
6. Ogata, K.: Modern Control Engineering, 5th edn. Prentice Hall, New York (2010)
7. Dorf, R., Bishop, R.: Modern Control Systems, 12th edn. Prentice Hall, Upper Saddle River (2010)
8. Hughes, A.: Electric Motors and Drives, Fundamentals, Types and Applications, 3rd edn. Elsevier Ltd., Oxford (2006)
9. Banks, S.: Mathematical Theories of Nonlinear Systems. Prentice Hall, New York (1988)
10. Isidori, A.: Nonlinear Control Systems: An Introduction, 3rd edn. Springer, New York (1995)
11. Khalil, H.: Nonlinear Systems, 2nd edn. Prentice Hall, New Jersey (1996)
12. Nijmeijer, H., Van der Schaft, A.: Nonlinear Dynamical Control Systems. Springer, New York (1990)
13. Vidyasagar, M.: Nonlinear Systems Analysis, 2nd edn. Prentice-Hall, Englewood Cliffs (1993)
14. Slotine, J., Li, W.: Applied Nonlinear Control. Prentice Hall, New Jersey (1991)
15. Gu, E.: A Direct Adaptive Control Scheme for Under-Actuated Dynamic Systems. In: Proc. 32nd IEEE Conference on Decision and Control, San Antonio, TX, pp. 1625–1627 (December 1993)
16. Gu, E., Xu, Y.: A Normal Form Augmentation Approach to Adaptive Control of Space Robot Systems. Journal of Dynamics and Control (June 1994)
17. Kristic, M., Kanellakopoulos, I., Kokotovic, P.: Nonlinear and Adaptive Control Design. John Wikey & Sons, New York (1995)
18. Gu, E.: Modeling of Human-Vehicle Dynamic Interactions and Control of Vehicle Active Systems. International Journal on Vehicle Autonomous Systems 10(4), 297–314 (2012)
19. Gu, E., Das, M.: Backstepping Control Design for Vehicle Active Restraint Systems. Transactions: ASME Journal of Dynamic Systems, Measurement and Control 135(1), 011012 (2013)
20. Slotine, J., Li, W.: Adaptive Manipulator Control: A Case Study. IEEE Transactions on Automatic Control 33(11) (1988)
21. Slotine, J., Li, W.: Composite Adaptive Control of Robot Manipulators. Automatica 25(4) (1989)

22. Sastry, S., Bodson, M.: Adaptive Control: Stability, Convergence, and Robustness. Prentice Hall (1989)
23. Gu, E.: Configuration Manifolds and Their Applications to Robot Dynamic Modeling and Control. IEEE Transactions on Robotics and Automation 16(5), 517–527 (2000)
24. Gu, E.: A Configuration Manifold Embedding Model for Dynamic Control of Redundant Robots. International Journal of Robotics Research 19(3), 289–304 (2000)

Chapter 9
Digital Human Modeling: Kinematics and Statics

9.1 Local versus Global Kinematic Models and Motion Categorization

Digital human modeling has become an active research and development field since 1980's [1]–[9]. The primary applications include engineering design and validation, digital ergonomic assessment and evaluation, digital simulation and verification of product design and assembling, etc. In history, to simulate a human and motion in a digital environment, people often hired a real person and installed a number of sensors, called a flog of birds, over the entire human body for motion captures. In other words, each sensor would be capable of sensing and transmitting a 3D-coordinate signal at each time instant, and generate a sequence of points as a trajectory by a recorder or a camera. If all the sensors are installed to cover every major marker point over the human body, then a complete motion is captured and ready to playback in a computer. Therefore, the motion capture can reproduce a realistic motion and most Hollywood computer-created movies are made by this technology. The drawback is obvious that each capture can record and generate only one single motion, and has to capture another live motion again for any update or change.

In summary, there were/are three major approaches to reproducing a human motion in the computer:

1. Motion capture by a real human with a flog of birds;
2. Data insertion between the live motion capture data using a mathematical interpolation and/or auto-regression;
3. Digital human modeling and motion creation by mathematical algorithms.

In order to develop a complete mathematical model for a human body and further to produce a realistic motion, the first step is to understand the major movable joints that a real human possesses. Figure 9.1 depicts in total about 19 major movable joints with their types in a biomedical perspective: a neck joint to move the head, two (left and right) clavicle joints, two shoulder

E.Y.L. Gu, *A Journey from Robot to Digital Human*,
Modeling and Optimization in Science and Technologies 1,
DOI: 10.1007/978-3-642-39047-0_9, © Springer-Verlag Berlin Heidelberg 2013

joints, two elbow joints and two wrist joints for the two arms, a thoracic joint and a lumbar joint to cover the vertebral column or spine, and two hip joints, two knee joints, two ankle joints and two toe joints along the two legs [10]. The above count excludes the joints involved in each hand, and modeling a hand will be specifically introduced and discussed in a later section of this chapter.

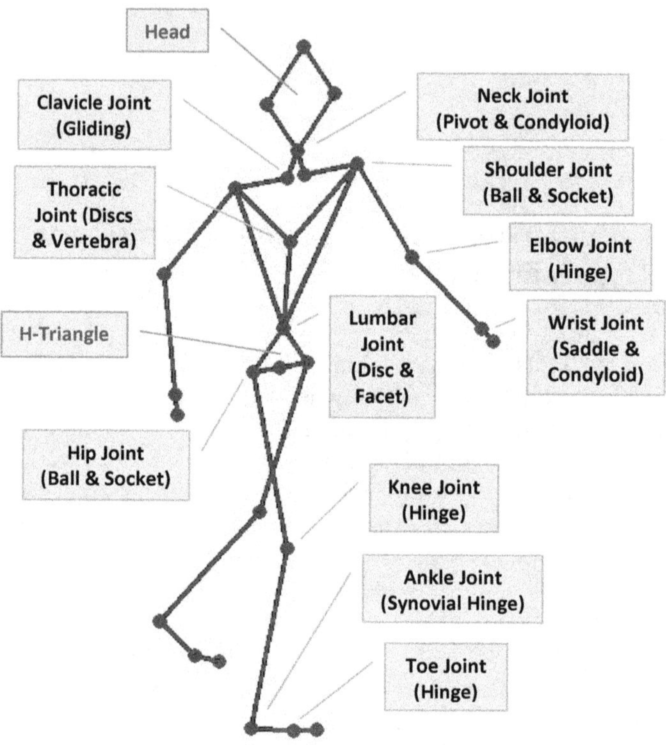

Fig. 9.1 Major joints and types over an entire human body

In fact, the vertebral column, or spine of a real human consists of multiple discs: the group of the top seven is called the cervical area, the group of the next 12 discs is the thoracic or chest region, and it is followed by the group of 5 lumbar vertebrae [11]. The last tail part is called the sacrum and coccyx, as shown in Figure 9.2. The cervical movement will be modeled as a 3-axis neck/head motion, the 12-disc thoracic movement will be combined as a 3-axis thorax motion, and the 5-disc lumbar movement will be modeled as a 3-axis lumbar joint motion. Defining further more joints in modeling both the thoracic and lumbar regions may not be helpful for better digital human motion generation, and may even cause more complications for kinematic solutions.

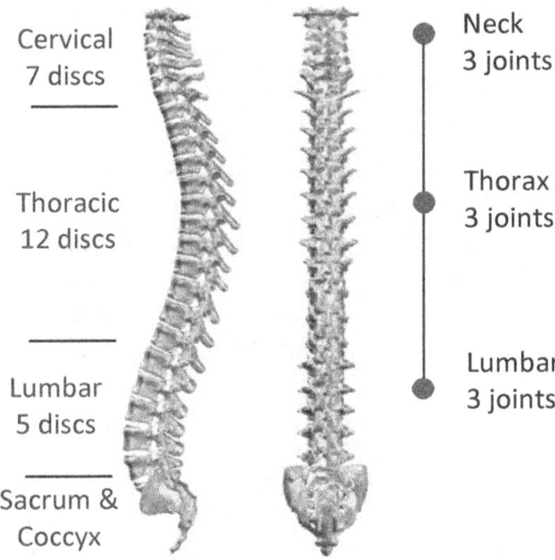

Fig. 9.2 The real human vertebral column and its modeling

At the robotic modeling standpoint, however, the neck, each shoulder, each wrist, the thorax and lumbar, and each hip are considered to be a universal-type of joint, i.e., each offers three independent joints to generate a complete roll-pitch-yaw rotation under the $SO(3)$ group. Each clavicle and each ankle are treated as a spherical-type of joint that offers two independent revolute axes at each, while each elbow and each knee have only one single revolute joint for each. Under such a joint distribution, the total amount of movable joints over the entire human body reaches 47 joints, which again exclude about 20 more joints on each hand. The 47 joints will be sufficient enough to represent a complete human posture at any time instant.

Figure 9.3 exhibits an assignment of the 47 joint positions. The first three prismatic joints d_1, d_2 and d_3 are the human body global translations along x_b, y_b and z_b-axis of the world base, respectively. The next three revolute joints θ_4, θ_5 and θ_6 are the global rotations: spin, fall over and fall aside, respectively. The joint angles θ_7, θ_8 and θ_9 represent the lumbar flexion/extension, lumbar lateral left/right and lumbar rotation. Similarly, θ_{10} – θ_{12} represent the thoracic flexion/extension, lateral left/right and rotation.

After the thorax, the sequence of joints splits into two parallel branches: the left clavicle and arm, and the right clavicle and arm. θ_{13} and θ_{14} are the left clavicle flexion/extension and elevation/depression. θ_{15}, θ_{16} and θ_{17} are the left shoulder flexion/extension, abduction/adduction and medial/lateral rotation, respectively. θ_{18} and θ_{19} are the joint angles for the left elbow flexion/extension and pronation/supination. θ_{20} and θ_{21} are the joint angles for the left wrist flexion/extension and radial/ulnar deviation, respectively. Note

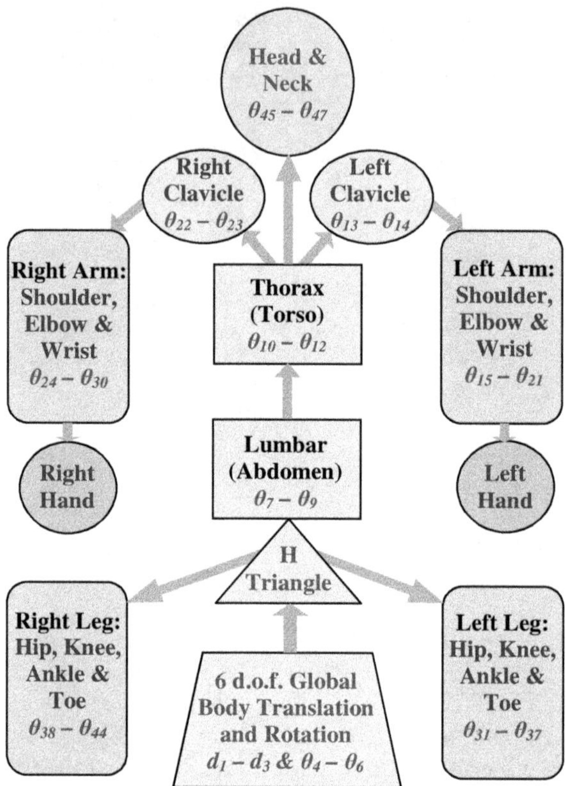

Fig. 9.3 A block diagram of digital human joint distribution

that θ_{19} may be counted as one of the three joints on the wrist, although it is referred to as the left elbow pronation/supination.

Another branch after the thorax is going to the right clavicle and arm. Similarly, θ_{22} and θ_{23} are assigned for the right clavicle, $\theta_{24} - \theta_{26}$ are the joint angles defined for the right shoulder, and the right elbow has a joint angle θ_{27} and it is followed by the right wrist with θ_{28}, θ_{29} and θ_{30}.

After the first six joints assigned for the human body global translations and rotations, in addition to forwarding to the lumbar joints, it also departs into the two legs immediately. On the left leg, θ_{31}, θ_{32} and θ_{33} are the joint angles for the left hip flexion/extension, abduction/adduction and medial/lateral rotation, respectively. θ_{34} and θ_{35} are the joint angles of the left knee flexion/extension and medial/lateral rotation. θ_{36} and θ_{37} are the joint angles of the left ankle dorsiflexion/plantar flexion and the left toe flexion/extension, respectively.

On the right leg, similar to the left one, $\theta_{38} - \theta_{40}$ are the joint angles for the right hip. The right knee has θ_{41} and θ_{42} joint angles, and the right ankle joint angle is θ_{43}, which is followed by the right toe joint angle θ_{44}.

The last three joint angles are assigned for the neck/head motion: θ_{45} is the head turning angle, θ_{46} is the head nodding angle, while θ_{47} is the head tilting angle, as shown in Figure 9.3.

All the 47 joints covering the entire human body should be further defined in a formal and precise fashion based on the Denavit-Hartenberg (D-H) convention in robotic kinematics. A complete assignment of segment/link coordinate frames is shown in Figure 9.4. To keep consistent to the D-H convention, the digital human base frame 0 has to be defined not same as the world base coordinate system frame b. The transformation between their orientations can be easily seen as

$$R_b^0 = \begin{pmatrix} 0 & 0 & 1 \\ 0 & -1 & 0 \\ 1 & 0 & 0 \end{pmatrix}, \tag{9.1}$$

which happens to be symmetric so that $R_b^0 = R_0^b$.

Moreover, Figure 9.4 does not give any kinematic parameters, such as the joint offsets d_i and link lengths a_i, but they are all precisely defined in the following D-H tables according to the D-H convention:

θ_i	d_i	α_i	a_i	$\theta_i(home)$	**Joint Names**
0	d_1	90^0	0	0	Translation x_b
90^0	d_2	90^0	0	90^0	Translation y_b
0	d_3	0	0	0	Translation z_b
θ_4	0	-90^0	0	0	Body Spin
θ_5	0	-90^0	0	-90^0	Fall Over
θ_6	0	90^0	a_6	0	Fall Aside

θ_i	d_i	α_i	a_i	$\theta_i(home)$	**Joint Names**
θ_7	0	90^0	0	0	Lumbar Flex/Ext
θ_8	0	90^0	0	-90^0	Lumbar Lateral
θ_9	$-d_9$	90^0	0	-90^0	Lumbar Rotation
θ_{10}	0	90^0	0	-90^0	Thoracic Flex/Ext
θ_{11}	0	90^0	0	-90^0	Thoracic Lateral

θ_i	d_i	α_i	a_i	$\theta_i(home)$	**Joint Names**
θ_{12}	$-d_{12}$	0	$-a_{12l}$	0	Thorax Rotation A_{11}^{12}
θ_{13}	0	90^0	0	0	L.Clavicle Flex/Ext
θ_{14}	0	-90^0	0	-90^0	L.Clavicle Elev/Depr
θ_{15}	$-d_{15}$	90^0	0	0	L.Shoulder Flex/Ext
θ_{16}	0	90^0	0	-90^0	L.Shoulder Abd/Add
θ_{17}	d_{17}	-90^0	0	90^0	L.Shoulder Med/Lat

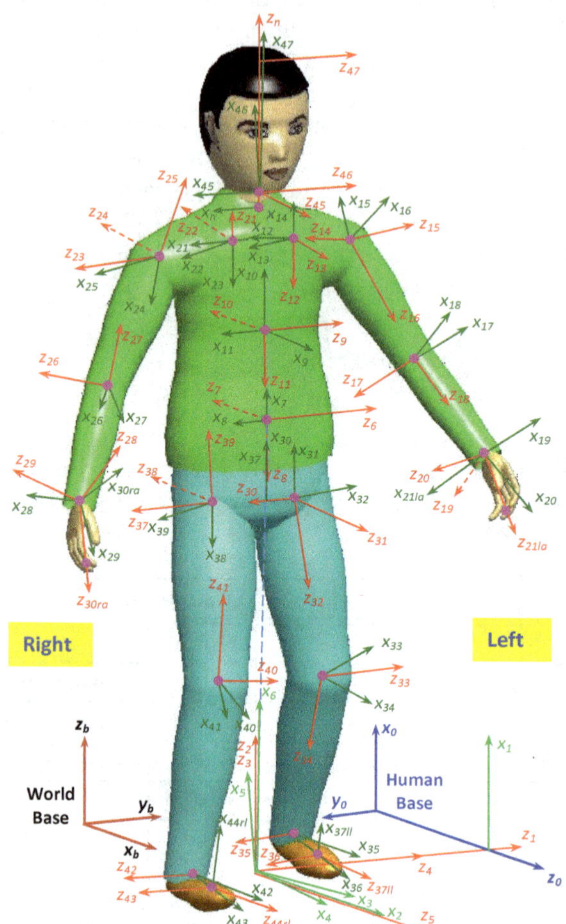

Fig. 9.4 Coordinate frame assignment on a digital mannequin

θ_i	d_i	α_i	a_i	$\theta_i(home)$	**Joint Names**
θ_{18}	0	90^0	0	0	L.Elbow Flex/Ext
θ_{19}	d_{19}	90^0	0	-90^0	L.Elbow Pron/Supin
θ_{20}	0	-90^0	0	90^0	L.Wrist Flex/Ext
θ_{21}	0	-90^0	0	-90^0	L.Wrist Rad/Uln

θ_i	d_i	α_i	a_i	$\theta_i(home)$	**Joint Names**
θ_{12}	$-d_{12}$	180^0	a_{12r}	0	Thorax Rotation A_{11}^{21}
θ_{22}	0	90^0	0	0	R.Clavicle Flex/Ext
θ_{23}	0	-90^0	0	-90^0	R.Clavicle Elev/Depr
θ_{24}	$-d_{24}$	90^0	0	0	R.Shoulder Flex/Ext
θ_{25}	0	-90^0	0	90^0	R.Shoulder Abd/Add
θ_{26}	$-d_{26}$	90^0	0	90^0	R.Shoulder Med/Lat

θ_i	d_i	α_i	a_i	$\theta_i(home)$	**Joint Names**
θ_{27}	0	-90^0	0	0	R.Elbow Flex/Ext
θ_{28}	$-d_{28}$	-90^0	0	-90^0	R.Elbow Pron/Supin
θ_{29}	0	90^0	0	90^0	R.Wrist Flex/Ext
θ_{30}	0	90^0	0	90^0	R.Wrist Rad/Uln

θ_i	d_i	α_i	a_i	$\theta_i(home)$	**Joint Names**
θ_h	0	180^0	$-a_h$	0	H-Triangle A_6^h
θ_{31}	$-d_{31}$	90^0	0	0	L.Hip Flex/Ext
θ_{32}	0	90^0	0	-90^0	L.Hip Abd/Add
θ_{33}	d_{33}	90^0	0	90^0	L.Hip Med/Lat
θ_{34}	0	-90^0	0	0	L.Knee Flex/Ext
θ_{35}	d_{35}	-90^0	0	0	L.Knee Med/Lat
θ_{36}	0	0	a_{36}	0	L.Ankle Dors/Plan
θ_{37}	0	90^0	0	90^0	L.Toe Flex/Ext

θ_i	d_i	α_i	a_i	$\theta_i(home)$	**Joint Names**
θ_h	0	180^0	$-a_h$	0	H-Triangle A_6^h
θ_{38}	d_{38}	90^0	0	180^0	R.Hip Flex/Ext
θ_{39}	0	-90^0	0	90^0	R.Hip Abd/Add
θ_{40}	$-d_{40}$	-90^0	0	90^0	R.Hip Med/Lat
θ_{41}	0	90^0	0	0	R.Knee Flex/Ext
θ_{42}	$-d_{42}$	90^0	0	0	R.Knee Med/Lat
θ_{43}	0	0	a_{43}	0	R.Ankle Dors/Plan
θ_{44}	0	90^0	0	90^0	R.Toe Flex/Ext

θ_i	d_i	α_i	a_i	$\theta_i(home)$	**Joint Names**
θ_{12}	$-d_n$	180^0	0	0	Thorax Rotation A_{11}^n
θ_{45}	d_{45}	-90^0	0	0	Head Turn
θ_{46}	0	90^0	0	-90^0	Head Nod
θ_{47}	0	0	a_{47}	0	Head Tilt

Note that the joint/link frame assignment and the corresponding D-H tables are not unique, and they can be defined in slightly different way. No matter in which way, the frame assignment and D-H tables must be self-consistent. Because a digital human is a typical open serial/parallel hybrid-chain structure, a common trunk will split to four limbs: two arms and two legs. After frame 11, there are two ways apart: one goes to frame 12 towards the left arm, and the other one goes to frame 21 towards the right arm. At the H-triangle, after frame 6, there are also two ways apart via a common constant transformation A_6^h: one is going to frame 30 forward to the left leg, and the other one is going to frame 37 forward to the right leg.

According to the D-H convention that was introduced and discussed in Chapter 4, we can readily write all the 47 one-step homogeneous transformation matrices based on equation (4.1). For example, the first six matrices to represent the human global movement are given as follows:

$$A_0^1 = \begin{pmatrix} 1 & 0 & 0 & 0 \\ 0 & 0 & -1 & 0 \\ 0 & 1 & 0 & d_1 \\ 0 & 0 & 0 & 1 \end{pmatrix}, \quad A_1^2 = \begin{pmatrix} 0 & 0 & 1 & 0 \\ 1 & 0 & 0 & 0 \\ 0 & 1 & 0 & d_2 \\ 0 & 0 & 0 & 1 \end{pmatrix},$$

$$A_2^3 = \begin{pmatrix} 1 & 0 & 0 & 0 \\ 0 & 1 & 0 & 0 \\ 0 & 0 & 1 & d_3 \\ 0 & 0 & 0 & 1 \end{pmatrix}, \quad A_3^4 = \begin{pmatrix} c_4 & 0 & -s_4 & 0 \\ s_4 & 0 & c_4 & 0 \\ 0 & -1 & 0 & 0 \\ 0 & 0 & 0 & 1 \end{pmatrix},$$

$$A_4^5 = \begin{pmatrix} c_5 & 0 & -s_5 & 0 \\ s_5 & 0 & c_5 & 0 \\ 0 & -1 & 0 & 0 \\ 0 & 0 & 0 & 1 \end{pmatrix}, \quad \text{and} \quad A_5^6 = \begin{pmatrix} c_6 & 0 & s_6 & a_6 c_6 \\ s_6 & 0 & -c_6 & a_6 s_6 \\ 0 & 1 & 0 & 0 \\ 0 & 0 & 0 & 1 \end{pmatrix},$$

where and hereafter $c_i = \cos\theta_i$ and $s_i = \sin\theta_i$ for $i = 4, \cdots, 47$.

Likewise, the reader can write all the remaining one-step homogeneous transformation matrices by yourself without major difficulty. However, we are emphasizing the following transition matrix at each splitting point from the common trunk to the four limbs as well as the neck of a digital mannequin:

$$A_{11}^{12} = \begin{pmatrix} c_{12} & -s_{12} & 0 & -a_{12l}c_{12} \\ s_{12} & c_{12} & 0 & -a_{12l}s_{12} \\ 0 & 0 & 1 & -d_{12} \\ 0 & 0 & 0 & 1 \end{pmatrix}, \quad A_{11}^{21} = \begin{pmatrix} c_{12} & s_{12} & 0 & a_{12r}c_{12} \\ s_{12} & -c_{12} & 0 & a_{12r}s_{12} \\ 0 & 0 & -1 & -d_{12} \\ 0 & 0 & 0 & 1 \end{pmatrix},$$

where A_{11}^{12} is going to the left arm and A_{11}^{21} is going to the right arm. Similarly,

$$A_{11}^{n} = \begin{pmatrix} c_{12} & s_{12} & 0 & 0 \\ s_{12} & -c_{12} & 0 & 0 \\ 0 & 0 & -1 & -d_n \\ 0 & 0 & 0 & 1 \end{pmatrix}$$

is to transit from frame #11 to the neck frame towards the head. The transition from the H-triangle to the two legs is using a common constant transformation:

$$A_6^h = \begin{pmatrix} c_h & s_h & 0 & -a_h c_h \\ s_h & -c_h & 0 & -a_h s_h \\ 0 & 0 & -1 & 0 \\ 0 & 0 & 0 & 1 \end{pmatrix} = \begin{pmatrix} 1 & 0 & 0 & -a_h \\ 0 & -1 & 0 & 0 \\ 0 & 0 & -1 & 0 \\ 0 & 0 & 0 & 1 \end{pmatrix}$$

due to $\theta_h = 0$, the constant zero.

After the common transition A_6^h, the left leg and the right leg will start with

$$A_h^{31} = \begin{pmatrix} c_{31} & 0 & s_{31} & 0 \\ s_{31} & 0 & -c_{31} & 0 \\ 0 & 1 & 0 & -d_{31} \\ 0 & 0 & 0 & 1 \end{pmatrix} \quad \text{and} \quad A_h^{38} = \begin{pmatrix} c_{38} & 0 & s_{38} & 0 \\ s_{38} & 0 & -c_{38} & 0 \\ 0 & 1 & 0 & d_{38} \\ 0 & 0 & 0 & 1 \end{pmatrix},$$

respectively.

Once the complete 47-joints kinematic model for a digital human is developed by exactly following the D-H convention, we now turn our attention to driving the digital human model for motion. In general, we can have the following practical approaches to generating a desired motion:

1. **Local joint motion:** to record a desired initial posture and a desired final posture by manually adjusting the related joint values for a digital mannequin, and then to drive the mannequin to move via the joint motion algorithm given by equation (4.7) in Chapter 4. Since such a joint-by-joint adjustment is intuitively time-consuming and tedious, and may not guarantee to produce the motion realistically, it may only be used to maneuver the mannequin for a local or partial movement.
2. **Global joint motion:** to define an independent joint trajectory $q_i(t)$ for each of the 47 joints listed in the joint position vector:

$$q = (d_1 \; d_2 \; d_3 \; \theta_4 \; \cdots \; \theta_{47})^T$$

   to generate a mannequin motion; or like the teaching-by-doing method commonly used in most industrial robots with a teaching pendent to store all the joint values at each desired posture monitoring on the computer screen, and then reproduce a motion through an independent joint linear interpolation, i.e., the joint motion algorithm given by equation (4.7) from Chapter 4. This approach may possibly generate mannequin motions in a more global fashion, but adjusting manually each joint position while monitoring the mannequin's posture on the screen at each time of recording is a very lengthy and even painful process.
3. **Local Cartesian motion:** using a partial I-K (inverse kinematics) solution to drive part of the digital human body for a local segment movement.

A typical example that has been built in a number of major commercial digital human simulation packages, such as Safework^{TM}, is to drag either one or both the position and orientation of a hand frame through the cursor move for a single arm movement with respect to the relatively fixed shoulder frame, or drag a foot to move the leg locally with respect to the fixed hip frame. There are two major approaches to do so: (1) determine a set of I-K solutions symbolically first, and then program them to solve each involved joint value in terms of a desired targeting Cartesian position or orientation, or both, and (2) purely using an equation solving function, such as the internal function fsolve($\cdots$) or fzero($\cdots$) in the MATLAB^{TM} optimization toolbox, to resolve the Cartesian motion numerically.

4. **Global Cartesian motion:** there are also two major approaches: (1) to start with a motion capture by installing sensors with one at each marker of a real human body (a flog of birds), and record a motion trajectory of each sensor when the real human is making a desired motion, or by a single or multi-camera system to take a video for all the marker trajectories. Then, applying a set of local I-K algorithms to solve all the joint values at each sampling point in terms of every marker trajectory recorded. Since a sensor, such as a 3D accelerometer, on each marker of the real human body provides an output of 3D position in (x, y, z) format at each sampling point, to resolve up to 47 joint values, one has to install such sensors on at least $47/3 \approx 16$ markers over the entire real human body for recording; and (2) to path-plan a set of desired motion trajectories for a few extremities of the mannequin, such as one or two hands, and/or one or two feet, and then applying linear or nonlinear programming algorithms to solve a global kinematic equation under multiple optimization objectives in a real-time numerical manner. Typical way is the optimization-based digital human modeling to set one or more objective functions to be minimized or maximized under a number of constraints. In terms of the algorithms of problem-solving, it also belongs to the Cartesian motion category, and global or local solution depends on how many joint angles to be solved for a particular motion under optimization. The MATLAB^{TM} optimization toolbox has a number of built-in functions: fmincon($\cdots$) and fgoalattain($\cdots$), etc., available for such a real-time numerical solution purpose.
5. **Global differential motion:** to develop a global Jacobian matrix J and to update each joint position by an inverse solution of the Jacobian equation $J\dot{q} = V$ with every desired path planned in tangent space, as given by equation (4.21) in Chapter 4. Since a digital human model is redundant in kinematics, i.e., the number of joints is greater than the total d.o.f., it offers a big "room" in nature to carry out a number of subtask optimization algorithms simultaneously, see Chapter 5 for details. For instance, if two hands and two feet plus the H-point of the digital human are modeled to be the five "end-effectors", then, they produce up to $6 \times 5 = 30$ d.o.f. totally. Suppose that only 44 joints by excluding the neck/head are to be

updated by an inverse solution of the Jacobian equation. Then, it is obvious that a global Jacobian matrix J to be developed will be as large as 30 by 44 with at least $44 - 30 = 14$ degrees of redundancy.

It is commonly recognizable that to develop an I-K algorithm even for a local area of a digital human would be a very hard and even highly-skilled job if you wish to be done symbolically. Of course, you can also solve a kinematic equation in Cartesian motion numerically by an existing algorithm. However, since either a partial or a whole digital human body contains a large number of revolute joints, such a numerical solution is often not controllable or determinable due to the multi-configuration phenomenon. This multi-configuration issue that has been addressed in the previous chapters is not only because of the redundancy, but also caused by the multiple solution nature in a trigonometric equation. Unless one or more constraints could be imposed on the kinematic equation to narrow down the possibility of multi-configurations in a numerical programming, a symbolical I-K algorithm may still possess a better controllable advantage over the numerical solution.

For example, to derive an I-K for the left arm starting from the shoulder frame 14 and ending to the hand frame $21la$, one has to first compute either symbolically or numerically the following multiplication:

$$A_{14}^{21la} = A_{14}^{15} A_{15}^{16} A_{16}^{17} A_{17}^{18} A_{18}^{19} A_{19}^{20} A_{20}^{21la}, \tag{9.2}$$

where each matrix on the right-hand side is a one-step homogeneous transformation. After the product A_{14}^{21la} is achieved, one then compares both its 4-th column and the upper-left 3 by 3 block with a desired position vector and orientation matrix in numerical form between frame 14 and frame $21la$. Through the comparison, one finally has to deal with every required trigonometric equation towards up to seven $\text{atan2}(\cdot\,,\,\cdot)$ forms to resolve the seven joint angles θ_{15} through θ_{21} involved in the left arm if a symbolical solution is desired. Or, set a 3 by 4 matrix equation for the above comparison and let an existing numerical equation-solving algorithm give us a solution of the seven joint angles. To illustrate both the symbolical derivation and numerical solution, let us continue our discussion and development in detail below.

Actually, you don't have to calculate the entire A_{14}^{21la} symbolically to find the I-K solutions. Like most 6-joint industrial manipulators, the first three joints often contribute a motion to meet a desired end-effector position, and the last three joints that are squeezed within the wrist produce a desired orientation. For the 7-joint left arm of the digital human, the first three joints $\theta_{15} - \theta_{17}$ are all concentrated within the shoulder center, which can be modeled as a universal RRR type of joint, while the fourth one θ_{18} is the elbow joint, as shown in Figure 9.4. If we now decompose the one-step homogeneous transformation A_{18}^{19} into two matrix-factors: one is only for translation and the other one is only for rotation:

$$A_{18}^{19} = \begin{pmatrix} c_{19} & 0 & s_{19} & 0 \\ s_{19} & 0 & -c_{19} & 0 \\ 0 & 1 & 0 & d_{19} \\ 0 & 0 & 0 & 1 \end{pmatrix} = \begin{pmatrix} 1 & 0 & 0 & 0 \\ 0 & 1 & 0 & 0 \\ 0 & 0 & 1 & d_{19} \\ 0 & 0 & 0 & 1 \end{pmatrix} \begin{pmatrix} c_{19} & 0 & s_{19} & 0 \\ s_{19} & 0 & -c_{19} & 0 \\ 0 & 1 & 0 & 0 \\ 0 & 0 & 0 & 1 \end{pmatrix},$$

and let the first translation one to merge with the previous transformation A_{17}^{18} so that

$$A_{17}^{18d} = A_{17}^{18} \begin{pmatrix} 1 & 0 & 0 & 0 \\ 0 & 1 & 0 & 0 \\ 0 & 0 & 1 & d_{19} \\ 0 & 0 & 0 & 1 \end{pmatrix} = \begin{pmatrix} c_{18} & 0 & s_{18} & d_{19}s_{18} \\ s_{18} & 0 & -c_{18} & -d_{19}c_{18} \\ 0 & 1 & 0 & 0 \\ 0 & 0 & 0 & 1 \end{pmatrix}.$$

Then, the first four joints can have the following homogeneous transformation:

$$A_{14}^{18d} = A_{14}^{15}A_{15}^{16}A_{16}^{17}A_{17}^{18d} = \begin{pmatrix} R_{14}^{18} & p_{14}^{19} \\ 0_3 & 1 \end{pmatrix}, \tag{9.3}$$

where

$$R_{14}^{18} = \begin{pmatrix} b_1c_{18} - c_{15}s_{16}s_{18} & -c_{15}c_{16}s_{17} + s_{15}c_{17} & b_1s_{18} + c_{15}s_{16}c_{18} \\ b_2c_{18} - s_{15}s_{16}s_{18} & -s_{15}c_{16}s_{17} - c_{15}c_{17} & b_2s_{18} + s_{15}s_{16}c_{18} \\ s_{16}c_{17}c_{18} + c_{16}s_{18} & -s_{16}s_{17} & s_{16}c_{17}s_{18} - c_{16}c_{18} \end{pmatrix},$$

and

$$p_{14}^{19} = \begin{pmatrix} d_{19}(b_1s_{18} + c_{15}s_{16}c_{18}) + d_{17}c_{15}s_{16} \\ d_{19}(b_2s_{18} + s_{15}s_{16}c_{18}) + d_{17}s_{15}s_{16} \\ d_{19}(s_{16}c_{17}s_{18} - c_{16}c_{18}) - d_{17}c_{16} - d_{15} \end{pmatrix},$$

and $b_1 = c_{15}c_{16}c_{17} + s_{15}s_{17}$ and $b_2 = s_{15}c_{16}c_{17} - c_{15}s_{17}$.

Now, let the position vector p_{14}^{19} starting from the left clavicle elevation/depression joint point and ending to the wrist joint center be equal to a given desired numerical position vector:

$$p_{14}^{19} = \begin{pmatrix} x \\ y \\ z \end{pmatrix}.$$

Since the left-hand side of the above equation has four variable angles $\theta_{15} - \theta_{18}$ to be solved, while the right-hand side has only three known numbers x, y and z, we have to deduct one variable from the four. It is reasonable to let the shoulder medial/lateral rotation angle θ_{17} be fixed at its home position, i.e., $\theta_{17} = 90^0$. Then, the above position equation can be reduced to

$$p_{14}^{19} = \begin{pmatrix} d_{19}(s_{15}s_{18} + c_{15}s_{16}c_{18}) + d_{17}c_{15}s_{16} \\ d_{19}(-c_{15}s_{18} + s_{15}s_{16}c_{18}) + d_{17}s_{15}s_{16} \\ -d_{19}c_{16}c_{18} - d_{17}c_{16} - d_{15} \end{pmatrix} = \begin{pmatrix} x \\ y \\ z \end{pmatrix}, \tag{9.4}$$

along with the reduced $b_1 = s_{15}$ and $b_2 = -c_{15}$.

To solve this trigonometric equation, let us first square every element of the column vector p_{14}^{19} and then add them together to yield

$$d_{17}^2 + d_{19}^2 + 2d_{17}d_{19}c_{18} = x^2 + y^2 + (z + d_{15})^2.$$

This is exactly consistent to the law of cosine for the triangle formed by the shoulder, elbow and the wrist center points, as expected. Thus,

$$c_{18} = \frac{x^2 + y^2 + (z + d_{15})^2 - d_{17}^2 - d_{19}^2}{2d_{17}d_{19}}.$$

Since the desired motion range of the elbow joint angle θ_{18} is $[0,\ 180^0]$, $s_{18} \geq 0$ so that $s_{18} = \sqrt{1 - c_{18}^2}$. We can readily solve

$$\theta_{18} = \mathrm{atan2}(s_{18},\ c_{18}) = \mathrm{atan2}(\sqrt{1 - c_{18}^2},\ c_{18}).$$

After θ_{18} is solved, using the z-component of equation (9.4), we have

$$c_{16} = -\frac{z + d_{15}}{d_{17} + d_{19}c_{18}}.$$

We now encounter a two-configuration situation, i.e., the shoulder abduction/ adduction angle θ_{16} can be either within $(-180^0,\ 0]$ as an adducted and abducted below the horizontal level case, or $\theta_{16} > 0$ as the left arm abducted above the horizontal level. In the first case, $s_{16} = -\sqrt{1 - c_{16}^2}$, while in the second case, $s_{16} = \sqrt{1 - c_{16}^2}$. Thus,

$$\theta_{16} = \mathrm{atan2}(s_{16},\ c_{16}) = \mathrm{atan2}(\pm\sqrt{1 - c_{16}^2},\ c_{16}).$$

Once both θ_{16} and θ_{18} are solved, the shoulder flexion/extension angle θ_{15} can be resolved by using the first two components of equation (9.4). Namely,

$$\begin{cases} d_{19}(s_{15}s_{18} + c_{15}s_{16}c_{18}) + d_{17}c_{15}s_{16} = x \\ d_{19}(-c_{15}s_{18} + s_{15}s_{16}c_{18}) + d_{17}s_{15}s_{16} = y \end{cases}$$

so that

$$\begin{pmatrix} d_{19}s_{18} & d_{17}s_{16} + d_{19}s_{16}c_{18} \\ d_{17}s_{16} + d_{19}s_{16}c_{18} & -d_{19}s_{18} \end{pmatrix} \begin{pmatrix} s_{15} \\ c_{15} \end{pmatrix} = \begin{pmatrix} x \\ y \end{pmatrix}.$$

As soon as solving both s_{15} and c_{15}, the shoulder flexion/extension angle θ_{15} can be determined by

$$\theta_{15} = \mathrm{atan2}(s_{15},\ c_{15}).$$

Because θ_{16} can have two possible joint values as a typical multi-configuration phenomenon, the angle value θ_{15} will be affected to have two distinct solutions as well.

After the first four angles $\theta_{15} - \theta_{18}$, including the fixed $\theta_{17} = 90^0$, are found in terms of the given position coordinates x, y and z, by substituting them into equation (9.3), the orientation R_{14}^{18} can be determined numerically. If the desired orientation for the left hand frame $21la$ with respect to frame 14 is given by R_{14}^{21la}, then we can find what additional rotation should be done by the wrist in order to meet this desired R_{14}^{21la}. Namely, the additional rotation can be found numerically as follows:

$$R_{18}^{21la} = R_{18}^{14} R_{14}^{21la} = (R_{14}^{18})^T R_{14}^{21la} = \begin{pmatrix} r_{11} & r_{12} & r_{13} \\ r_{21} & r_{22} & r_{23} \\ r_{31} & r_{32} & r_{33} \end{pmatrix}.$$

With a further derivation, we achieve a symbolical form of the required additional rotation:

$$R_{18}^{21la} = R_{18}^{19} R_{19}^{20} R_{20}^{21la} = \begin{pmatrix} c_{19}c_{20}c_{21} - s_{19}s_{21} & c_{19}s_{20} & -c_{19}c_{20}s_{21} - s_{19}c_{21} \\ s_{19}c_{20}c_{21} + c_{19}s_{21} & s_{19}s_{20} & -s_{19}c_{20}s_{21} + c_{19}c_{21} \\ s_{20}c_{21} & -c_{20} & -s_{20}s_{21} \end{pmatrix}.$$

Let the above symbolical and numerical R_{18}^{21la} be equal and compared to each other. We can solve all the three joint angles involved in the left wrist rotation. First, $c_{20} = -r_{32}$, and notice that the wrist flexion/extension angle θ_{20} rotates between 0 and 180^0. Thus, $s_{20} = \sqrt{1 - c_{20}^2} = \sqrt{1 - r_{32}^2}$ so that

$$\theta_{20} = \operatorname{atan2}(\sqrt{1 - r_{32}^2},\ -r_{32}).$$

Now, since $s_{20} > 0$ (we assume that θ_{20} will not reach the limited angle 0 or 180^0),

$$\theta_{19} = \operatorname{atan2}(r_{22},\ r_{12}) \quad \text{and} \quad \theta_{21} = \operatorname{atan2}(-r_{33},\ r_{31}).$$

Of course, the values of the three wrist joint angles θ_{19}, θ_{20} and θ_{21} will also be in a multi-solution case due to the two possible solutions of θ_{16}. If we take off the condition of $0 \le \theta_{20} \le 180^0$ for the wrist flexion/extension, then two more configurations will occur within the wrist joint rotations. Furthermore, the above I-K solutions are derived under the fixed shoulder medial/lateral rotation angle $\theta_{17} = 90^0$. If the condition is released, then the number of the I-K solutions will become infinity. The reason is clear that the left arm has 7 joints in total and the hand motion is 6 d.o.f. so that it is a redundant arm.

To verify this local I-K algorithm, a MATLAB^{TM} main program is created. Let the position of frame $21la$ referred to frame 14 be set by $x = 0$, $y = -0.5$ and $z = -0.4$ in meters, and the orientation be set as

$$R_{14}^{21la} = \begin{pmatrix} -1 & 0 & 0 \\ 0 & 1 & 0 \\ 0 & 0 & -1 \end{pmatrix}.$$

This setting is supposed to be done by cursor-dragging. The main program is given as follows:

```
d15=0.16;  d17=0.3;  d19=0.2935;
                     % Left Arm Parameters in Meters
th_body=[0; 0; 0; 0; -90; 0];
th_trunk=[0; -90; -90; -90; -90; 0];
th_larm=[0; -90; 0; -80; 90; 10; -90; 90; -90];
th_rarm=[0; -90; 0; 100; 90; 10; -90; 90; 90];
th_lleg=[0; -85; 90; 10; 0; 10; 90];
th_rleg=[180; 95; 90; 10; 0; 10; 90];
th_head=[0; -90; 10];
q=[th_body; th_trunk; th_larm; th_rarm; th_lleg;...
                          th_rleg; th_head]*pi/180;
     % Initialize All the 47 Joint Values at Home

x=0; y=-0.5; z=-0.4;
          % The Desired Hand Position from Frame 14
R21=[-1 0 0;0 1 0;0 0 -1];
     % The Desired Orientation Referred to Frame 14

for i=15:21
    s(i)=sin(q(i));  c(i)=cos(q(i));
end
c(18)=(x^2+y^2+(z+d15)^2-d17^2-d19^2)/(2*d17*d19);
s(18)=sqrt(1-c(18)^2);
q(18)=atan2(s(18),c(18));

c(16)=-(z+d15)/(d17+d19*c(18));
s(16)=-sqrt(1-c(16)^2);   % Or s(16)=sqrt(1-c(16)^2);
q(16)=atan2(s(16),c(16));

A=[d19*s(18) d17*s(16)+d19*s(16)*c(18)
   d17*s(16)+d19*s(16)*c(18) -d19*s(18)];
b=A\[x; y];
q(15)=atan2(b(1),b(2));

R18=[b(1)*c(18)-b(2)*s(16)*s(18) -b(2)*c(16) ...
                          b(1)*s(18)+b(2)*s(16)*c(18)
    -b(2)*c(18)-b(1)*s(16)*s(18) -b(1)*c(16) ...
                         -b(2)*s(18)+b(1)*s(16)*c(18)
    c(16)*s(18) -s(16) -c(16)*c(18)];

Rw=R18'*R21;

q(20)=atan2(sqrt(1-Rw(3,2)^2),-Rw(3,2));
q(19)=atan2(Rw(2,2),Rw(1,2));
q(21)=atan2(-Rw(3,3),Rw(3,1));
```

```
f_ag6(q);        % Call a Subroutine to Draw the Mannequin
axis([-0.5 2 -1.5 1.5 0 2]), daspect([1 1 1]),
zoom(2.0), view([1 -0.3 0.2]),


>> q(15:21)'     % Joint Angles at Home in Degrees

   0   -80    90    10   -90    90   -90

>> q(15:21)'     % Joint Angles at Desired Point When s(16) < 0

  72.3  -59.8  90.0  31.2  -70.8  30.5  -122.0

>> q(15:21)'     % Joint Angles at Desired Point When s(16) > 0

 -67.0  62.5  90.0  41.7  59.4  29.6  -44.3

>>
```

The above MATLAB^{TM} code is to calculate all the joint angles involved in the left arm, and then to call a subroutine f_ag6(q) for the mannequin drawing, the input of which is a 47 by 1 joint position vector q. The result is shown in Figure 9.5, where the left posture is under $\sin\theta_{16} < 0$, and the right one is in $\sin\theta_{16} > 0$ case.

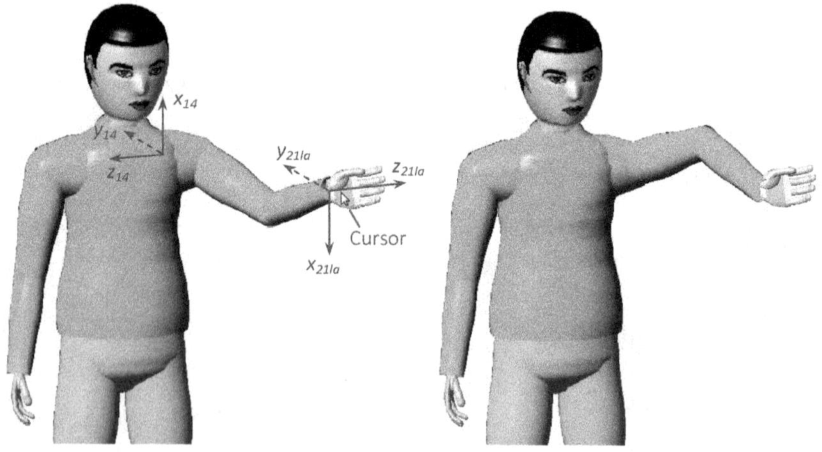

Fig. 9.5 The left arm of a digital mannequin is manually maneuvered by a local I-K algorithm with at least two distinct configurations

In addition, the last three rows of the program print the output data of the seven joint angles along the left arm in both cases before dragging the hand and after moving the hand to the desired position and orientation. In other words, the first set of joint angles in the print is the initial values at the home position, while the second set of joint angles is the I-K solution in degrees when $\sin\theta_{16} < 0$, and the last set of joint angles is in the case of $\sin\theta_{16} > 0$, i.e., the left shoulder is abducted up beyond the horizontal level.

If we now directly solve the kinematic equation (9.2) for the left arm numerically by the internal function fsolve($\cdots$) in MATLAB^{TM}, i.e., solve the following matrix equation:

$$F(q) \triangleq A_{14}^{21la}(1:3,1:4) - \begin{pmatrix} -1 & 0 & 0 & 0 \\ 0 & 1 & 0 & -5 \\ 0 & 0 & -1 & -4 \end{pmatrix} = 0,$$

then a new program and its result are given below:

```
d15=0.16;  d17=0.3;  d19=0.2935;
                    % Left Arm Parameters in Meter
th_body=[0; 0; 0; 0; -90; 0];
th_trunk=[0; -90; -90; -90; -90; 0];
th_larm=[0; -90; 0; -80; 90; 10; -90; 90; -90];
th_rarm=[0; -90; 0; 100; 90; 10; -90; 90; 90];
th_lleg=[0; -85; 90; 10; 0; 10; 90];
th_rleg=[180; 95; 90; 10; 0; 10; 90];
th_head=[0; -90; 10];

q0=[th_body; th_trunk; th_larm; th_rarm; th_lleg;...
                             th_rleg; th_head]*pi/180;
     % Initialize All the 47 Joint Values at Home

q = fsolve(@f_pos,q0);   % Call the Equation Solver

f_ag6(q);      % Call a Subroutine to Draw Mannequin
axis([-0.5 2 -1.5 1.5 0 2]), daspect([1 1 1]),
zoom(2.0), view([1 -0.3 0.2]),

% -------------------- %

function F = f_pos(q)      % Equation To Be Solved

    d15=0.16;  d17=0.3;  d19=0.2935;

    q(17)=pi/2;        % Force q(17) To Be Fixed

    for i=15:21
        s(i)=sin(q(i));  c(i)=cos(q(i));
```

```
    end

    A15=[c(15) 0 s(15) 0; s(15) 0 -c(15) 0
                       0 1 0 -d15; 0 0 0 1];
    A16=[c(16) 0 s(16) 0; s(16) 0 -c(16) 0
                       0 1 0 0; 0 0 0 1];
    A17=[c(17) 0 -s(17) 0; s(17) 0 c(17) 0
                       0 -1 0 d17; 0 0 0 1];
    A18=[c(18) 0 s(18) 0; s(18) 0 -c(18) 0
                       0 1 0 0; 0 0 0 1];
    A19=[c(19) 0 s(19) 0; s(19) 0 -c(19) 0
                       0 1 0 d19; 0 0 0 1];
    A20=[c(20) 0 -s(20) 0; s(20) 0 c(20) 0
                       0 -1 0 0; 0 0 0 1];
    A21=[c(21) 0 -s(21) 0; s(21) 0 c(21) 0
                       0 -1 0 0; 0 0 0 1];

    A=A15*A16*A17*A18*A19*A20*A21;

    F=A(1:3,1:4)-[-1 0 0 0; 0 1 0 -5; 0 0 -1 -4];
              % Make the Two 3 By 4 Matrices Equal

end


>> q(15:21)'   % Joint Angles at Desired Point

 -67.0  62.5  90.0  41.7  -120.6  -29.6  135.7

>>
```

The digital mannequin in the resulting posture is nearly the same as the right picture in Figure 9.5, but microscopically the last three wrist joint values are different from the symbolical solution, though the first four angles are exactly same between the numerical solution and the second case $s_{16} > 0$ of the symbolical solution. The reason is due to the different sign for θ_{20}, as mentioned before. Therefore, the left arm alone has already had four distinct configurations, let alone for the entire digital human I-K solutions in a global Cartesian motion.

Nevertheless, any two distinct configurations are not close to each other. Because of this reality, if the current joint position $q(t_i)$ at $t = t_i$ is placed in the internal function fsolve(@f_pos, q0) as an initial condition q0 to update for a new joint value $q(t_{i+1})$ at the next sampling point $t = t_{i+1}$, and both are close to each other, then, the solution will be smooth without any multi-configuration problem, unless it touches a singular point. Therefore, both the symbolical I-K algorithm and the internal function are feasible and can be implemented to locally generate a desired continuous Cartesian motion.

Moreover, if we can drag the left wrist of a digital human by the computer mouse on a MATLAB^{TM} working window to a desired position and/or a desired orientation, a possible data-cursor function may immediately tell the above local I-K algorithm the instantaneous position and orientation as input data for the left wrist to move with respect to the left shoulder frame 14 that is relatively fixed during this particular cursor-dragging operation. In fact, a number of commercial digital human simulation packages, such as Safework^{TM}, have implemented the similar local I-K algorithm to allow the user to drag the cursor and maneuver the arm around.

In contrast to the local I-K algorithm development effort, the same operation can also be done by using a local differential motion with a local Jacobian matrix J_{la}. In the above case of dragging the left hand to maneuver the left arm, based on equation (4.14) in Chapter 4, the local Jacobian matrix can be determined by

$$J_{la} = \begin{pmatrix} p_{21la}^{14} \times r_{21la}^{14} & \cdots & p_{21la}^{20} \times r_{21la}^{20} \\ r_{21la}^{14} & \cdots & r_{21la}^{20} \end{pmatrix},$$

where p_{21la}^{i} and r_{21la}^{i} are the 4th and 3rd columns of the homogeneous transformation matrix A_{21la}^{i}, respectively, for $i = 14, \cdots, 20$. This local Jacobian can be further projected onto frame #14 by

$$J_{la(14)} = \begin{pmatrix} R_{14}^{21la} & O \\ O & R_{14}^{21la} \end{pmatrix} J_{la},$$

and both J_{la} and $J_{la(14)}$ can be prepared numerically in a MATLAB^{TM} program without any symbolical derivation needed.

Once one drags the left hand frame of the digital human, the data-cursor function (if it could exist) would generate a Cartesian velocity vector augmented by a linear velocity v and an angular velocity ω, both are 3-dimensional in tangent space. Then, with the pseudo-inverse of $J_{la(14)}$, each joint angle involved in the left arm motion can be updated by

$$q(t + \Delta t) \approx q(t) + \dot{q}\Delta t = q(t) + J_{la(14)}^{+} \begin{pmatrix} v \\ \omega \end{pmatrix} \Delta t, \tag{9.5}$$

where the minimum-norm solution to the Jacobian equation is adopted, due to the redundancy, i.e., 7 joints versus 6 d.o.f., and $J_{la(14)}^{+}$ is the pseudo-inverse of the local Jacobian $J_{la(14)}$. In MATLAB^{TM}, the built-in internal function $\text{pinv}(J) = J^{+}$ can directly calculate a numerical pseudo-inverse for any non-square matrix J.

Because of the numerical advantage, the local differential motion can be easily extended to the global differential motion with a global Jacobian matrix J to uniquely represent the instantaneous posture of a digital mannequin. Since calculating a global Jacobian matrix J and further taking its pseudo-inverse can be done numerically by MATLAB^{TM}, no symbolical derivation is necessary. Therefore, the global differential motion will be better, more potential and suitable to generating digital human motions, although it may cause more computational error than an I-K algorithm of Cartesian motion due to the first-order approximation in the updating procedure, as seen in equation (4.21) in Chapter 4 and the above equation (9.5). However, such a cumulative computational error can always be remedied in a certain extent.

Furthermore, a differential motion based I-K algorithm to drive the digital human in motion requires a set of velocity trajectories, instead of the position trajectories in a Cartesian motion based I-K algorithm. The DC-bias, or any constant component of each position trajectory will be automatically removed after taking time-derivatives. In other words, the set of trajectories in tangent space will no longer contain any DC-bias, and this will definitely make the digital human motion generation be less sensitive to the variation of anthropometric data, especially to a percentile change.

9.2 Local and Global Jacobian Matrices in a Five-Point Model

As an open serial-parallel hybrid chain structure, a global Jacobian matrix for a digital human model can be constructed by a certain augmentation of several local Jacobian matrices to match the list of all the joint positions in the vector q under a certain order. Similar to dividing the entire D-H table into several D-H subtables with one for each of the major segment groups over the entire mannequin body, a global Jacobian matrix J can also be developed by finding a local Jacobian for each corresponding segment group first. However, every local Jacobian matrix formation must start from the human base, i.e., frame 0 without exception.

Let us now begin the construction from the local Jacobian J_H of the H-triangle to the local Jacobians J_{lh} and J_{rh} of the two hands, and then it will be followed by the local Jacobians J_{lf} and J_{rf} of the two feet. We exclude the neck/head group, because the three joint angles: head turn, nod and tilt will not participate in the differential motion updating process.

Based on equation (4.14) and the related equations in Chapter 4, and noticing that the first three joints of the digital human model are prismatic, the H-triangle Jacobian matrix can be formed by

$$J_H = \begin{pmatrix} r_6^0 & r_6^1 & r_6^2 & p_6^3 \times r_6^3 & p_6^4 \times r_6^4 & p_6^5 \times r_6^5 \\ 0 & 0 & 0 & r_6^3 & r_6^4 & r_6^5 \end{pmatrix}, \tag{9.6}$$

where, as aforementioned, each p_6^i and each r_6^i are the 4th and 3rd columns of A_6^i for $i = 0, \cdots, 5$, respectively, and each 0 inside the Jacobian matrix is the 3 by 1 zero column. Therefore, one must prepare all the A_6^i's by multiplying the corresponding one-step homogeneous transformation matrices and inverting them in order to determine J_H, and all such multiplications and inversions can be done numerically in MATLAB^{TM}. Clearly, this H-triangle Jacobian matrix J_H is 6 by 6 in dimension.

The formation of J_{lh} is much more lengthy than the above J_H, because the hand is far more distal from the base than the H-triangle. Its general formula is given as follows:

$$J_{lh} = \begin{pmatrix} r^0_{21la} & r^1_{21la} & r^2_{21la} & p^3_{21la} \times r^3_{21la} & \cdots & p^{20}_{21la} \times r^{20}_{21la} \\ 0 & 0 & 0 & r^3_{21la} & \cdots & r^{20}_{21la} \end{pmatrix}, \tag{9.7}$$

which is a 6 by 21 matrix and must also start from the human base frame 0.

Likewise, for the right arm,

$$J_{rh} = \begin{pmatrix} r^0_{30ra} & r^1_{30ra} & r^2_{30ra} & p^3_{30ra} \times r^3_{30ra} & \cdots & p^{29}_{30ra} \times r^{29}_{30ra} \\ 0 & 0 & 0 & r^3_{30ra} & \cdots & r^{29}_{30ra} \end{pmatrix} \tag{9.8}$$

with a frame number jump between 11 and 21 after the 12th column, i.e.,

$$J_{rh} = \begin{pmatrix} \cdots & p^{11}_{30ra} \times r^{11}_{30ra} & p^{21}_{30ra} \times r^{21}_{30ra} & \cdots \\ \cdots & r^{11}_{30ra} & r^{21}_{30ra} & \cdots \end{pmatrix}.$$

This Jacobian matrix J_{rh} is also 6 by 21 dimensional.

The left foot has the following local Jacobian matrix:

$$J_{lf} = \begin{pmatrix} r^0_{37ll} & r^1_{37ll} & r^2_{37ll} & p^3_{37ll} \times r^3_{37ll} & \cdots & p^{36}_{37ll} \times r^{36}_{37ll} \\ 0 & 0 & 0 & r^3_{37ll} & \cdots & r^{36}_{37ll} \end{pmatrix} \tag{9.9}$$

with a frame number jump between 5 and 30 after the 6th column, i.e.,

$$J_{lf} = \begin{pmatrix} \cdots & p^{5}_{37ll} \times r^{5}_{37ll} & p^{30}_{37ll} \times r^{30}_{37ll} & \cdots \\ \cdots & r^{5}_{37ll} & r^{30}_{37ll} & \cdots \end{pmatrix}.$$

Clearly, this is a 6 by $6 + 7 = 13$ matrix.

Similarly, the local Jacobian for the right foot is obtained as follows:

$$J_{rf} = \begin{pmatrix} r^0_{44rl} & r^1_{44rl} & r^2_{44rl} & p^3_{44rl} \times r^3_{44rl} & \cdots & p^{43}_{44rl} \times r^{43}_{44rl} \\ 0 & 0 & 0 & r^3_{44rl} & \cdots & r^{43}_{44rl} \end{pmatrix} \tag{9.10}$$

with a frame number jump between 5 and 37 after the 6th column, i.e.,

$$J_{rf} = \begin{pmatrix} \cdots & p^{5}_{44rl} \times r^{5}_{44rl} & p^{37}_{44rl} \times r^{37}_{44rl} & \cdots \\ \cdots & r^{5}_{44rl} & r^{37}_{44rl} & \cdots \end{pmatrix},$$

which is also 6 by 13 in dimension.

All of the above five local Jacobian matrices J_H, J_{lh}, J_{rh}, J_{lf} and J_{rf} start from the human base towards each individual "end-effector". However, based on their formations, each of them is projected on its own last frame as default, instead of the human base. For the future applications to maneuvering the digital mannequin and generating motions, it is better to re-project each local

Jacobian matrix to the human base, frame 0, or even onto the world base, frame b, to make them consistent to the global path planning and workcell creation.

According to equation (4.15) in Chapter 4, each of the five local Jacobian matrices to be projected onto the common world base, frame b, can be done by

$$J_{H(b)} = \begin{pmatrix} R_b^0 R_0^6 & O \\ O & R_b^0 R_0^6 \end{pmatrix} J_H = \begin{pmatrix} R_b^6 & O \\ O & R_b^6 \end{pmatrix} J_H, \tag{9.11}$$

where R_b^0 is the rotation matrix to convert the projection between the human base and the world base, as given in (9.1), and R_0^6 can be found from the numerical computation of A_0^6. Likewise,

$$J_{lh(b)} = \begin{pmatrix} R_b^{21la} & O \\ O & R_b^{21la} \end{pmatrix} J_{lh},$$

$$J_{rh(b)} = \begin{pmatrix} R_b^{30ra} & O \\ O & R_b^{30ra} \end{pmatrix} J_{rh},$$

$$J_{lf(b)} = \begin{pmatrix} R_b^{37ll} & O \\ O & R_b^{37ll} \end{pmatrix} J_{lf},$$

and

$$J_{rf(b)} = \begin{pmatrix} R_b^{44rl} & O \\ O & R_b^{44rl} \end{pmatrix} J_{rf}, \tag{9.12}$$

where each O is the 3 by 3 zero matrix.

Once all the local Jacobian matrices projected on the common world base are ready, we now augment them together to form a global Jacobian matrix $J_{(b)}$ to cover the entire digital human body, but excluding the head motion and hand fingers. Since the targeting global Jacobian matrix must eventually be post-multiplied by the 44 by 1 joint velocity vector:

$$\dot{q} = (\dot{d}_1 \;\; \dot{d}_2 \;\; \dot{d}_3 \;\; \dot{\theta}_4 \;\; \cdots \;\; \dot{\theta}_{44})^T,$$

the augmentation should make itself compatible and consistent to the rule of matrix multiplication. By noticing the frame number jumps in the last three of the five local Jacobian matrices: J_{rh}, J_{lf} and J_{rf} in equations (9.8), (9.9) and (9.10), we finally achieve a complete global Jacobian construction: $J =$

$$\begin{pmatrix} J_H & O_{6\times 6} & O_{6\times 9} & O_{6\times 9} & O_{6\times 7} & O_{6\times 7} \\ J_{lh}(:,1\!:\!6) & J_{lh}(:,7\!:\!12) & J_{lh}(:,13\!:\!21) & O_{6\times 9} & O_{6\times 7} & O_{6\times 7} \\ J_{rh}(:,1\!:\!6) & J_{rh}(:,7\!:\!12) & O_{6\times 9} & J_{rh}(:,13\!:\!21) & O_{6\times 7} & O_{6\times 7} \\ J_{lf}(:,1\!:\!6) & O_{6\times 6} & O_{6\times 9} & O_{6\times 9} & J_{lf}(:,7\!:\!13) & O_{6\times 7} \\ J_{rf}(:,1\!:\!6) & O_{6\times 6} & O_{6\times 9} & O_{6\times 9} & O_{6\times 7} & J_{rf}(:,7\!:\!13) \end{pmatrix}, \tag{9.13}$$

where $J_{lh}(:,1\!:\!6)$ is a typical notation in MATLAB^{TM}, which means to take every rows but only take columns between 1 and 6 from J_{lh}, and similar notations for the other local Jacobian blocks. In addition, $O_{6\times 9}$ in (9.13) is the 6 by 9 zero matrix and similar for the others.

The above global Jacobian matrix J is 30 by 44 in dimension and is an original version, i.e., it has not been projected onto the world base yet. Once each local Jacobian block inside (9.13) is projected to the world base by (9.11) and (9.12), it reaches to a final version of the global Jacobian matrix $J_{(b)}$ projected onto the world base.

Let a linear velocity and an angular velocity of each "end-effector" be defined as v_b^j and ω_b^j, respectively, with each as a 3 by 1 vector and both projected on the world base. Then, a 6 by 1 Cartesian velocity vector is defined by $V_b^j = \begin{pmatrix} v_b^j \\ \omega_b^j \end{pmatrix}$ for each of the five end-effectors, and all the five individual Cartesian velocities are augmented to form a 30 by 1 global Cartesian velocity vector $V_b \in \mathbb{R}^{30}$. Then, the Jacobian equation for a digital human has been unified to be the same form as that of an industrial robot manipulator but in a much larger dimension:

$$V_b = J_{(b)}\dot{q}, \tag{9.14}$$

where both V and J must be projected on the same reference frame, and here, of course, on the world base.

Since the entire digital human kinematics is a redundant robotic model, according to equation (5.7) in Chapter 5, a general solution of the above Jacobian equation (9.14) should be

$$\dot{q} = J_{(b)}^{+} V_b + (I - J_{(b)}^{+} J_{(b)})z \tag{9.15}$$

for an arbitrary vector $z \in \mathbb{R}^{44}$, where $J_{(b)}^{+} = J_{(b)}^{T}(J_{(b)} J_{(b)}^{T})^{-1}$ is the pseudo-inverse of $J_{(b)}$, and I is the 44 by 44 identity. The arbitrary vector z offers a wide-opened door and allows us to fill one or more optimization algorithms to improve the I-K (inverse kinematics) solutions under the differential motion.

It is conceivable that the above modeling process has proposed a concept of five "end-effectors" to attempt an explicit representation of digital human postures and motions, and the global Jacobian matrix J formed by (9.13) can completely and uniquely represent each instantaneous posture of the digital human model. If each end-effector requires up to 6 d.o.f. motion, then the digital human can possess up to a total number of independent d.o.f. equal to $5 \times 6 = 30$. The development has actually arrived at a **five-point model** along with a 30 by 44 **global Jacobian matrix** J or $J_{(b)}$.

To continue the development, we first give a comparative study in Table 9.1 among the three major approaches to generating digital human motions: the proposed five-point model, the local/global Cartesian motion based model and the motion capture that records the motion trajectories by more than 16 sensors distributed over a real human body, usually one on each body marker.

In Table 9.1, the last two approaches rely on their respective mathematical models to automatically generate digital human motions. While the quality

Table 9.1 A comparative study among the three major methods of digital human motion generation

Method and Model	**Motion Capture Method**	**Five-Point Model with Global J**	**Cartesian Motion Model**
Number of Traj's	≥ 16	≤ 5	> 5
Type of Traj's	Position Profiles	Velocity Profiles	Position Profiles
Algorithms	Local I-K Solutions	Global Jacobian Equation	Local/Global I-K Equation Solvers
Solutions	Exact without Room	Multi-Potential Function Opt.	Carry Multi-Opt. Objectives
Motion Update or Change	Need Re-Capture Off-line Again	Only Modify ≤ 5 V-Trajectories	Modify > 5 Pos./Orient. Paths
Requirement	A Real Human with Sensors	No Real Human & Devices Needed	No Real Human & Devices Needed
Realistic Motion	Best	Better	Better
Traj-Tracking Accuracy	N/A	Good	Better
Sensitivity to the Variation of Anthropometric Data	N/A	Lower	Higher
Joint Torque Analysis	N/A	Yes with Global J	Unable to Do It

of motion generated, in the current status, may not be compared with the real human motion capture method, the two mathematical model-based approaches possess a number of unique advantages. With more extensive studies on human body structures and motion patterns ahead, both the differential motion based and Cartesian motion based models have a great potential to promise a further improvement of their digital human motion quality and performance towards more realistic motion results.

Moreover, between the two mathematical model-based approaches, since there is a scale limitation of the numerical equation solvers in nonlinear programming, the Cartesian motion model may face a stability challenge, especially to carry a number of optimization objectives as a set of constraints on the equation solver. In contrast, the differential motion based model has developed a global Jacobian matrix that covers the entire digital human body except the head and fingers. The solution to the global Jacobian equation along with a nearly unlimited number of potential functions to be optimized is very stable and robust, unless it touches a singular point. However, the singularity is a common issue in I-K for both the two mathematical model-based approaches, and it can be avoided as one of the optimization objectives.

Therefore, how to mathematically create a set of coordinated position and velocity profiles with one for each end-effector and how to define a single or multiple optimization criteria to be implemented into the general solution become two key steps towards a realistic digital human motion generation without any motion capture. Figure 9.6 summarizes a block diagram of the five-point model with a global Jacobian matrix to aim the goal in a purely mathematical fashion.

While the five-point model along with the formation of the global Jacobian matrix have been accomplished thus far by borrowing the major kinematic modeling tools from robotics, we have to recognize the common fact that a digital human realistic motion is far beyond the robotic kinematics. Even for the real human basic motions, such as walking, running and jumping, in addition to obeying the natural laws of kinematics and dynamics, many sophisticated factors, such as the personality, mood, age, gender, infirmity and physical fitness or health, have also to be taken into serious consideration. In other words, to generate a realistic motion for a digital human, a complete and accurate kinematic model is just a foundation as a necessary condition, and many other issues from the biomechanics and kinesiology perspectives must be incorporated to enhance the quality of digital human motions [3, 4, 5].

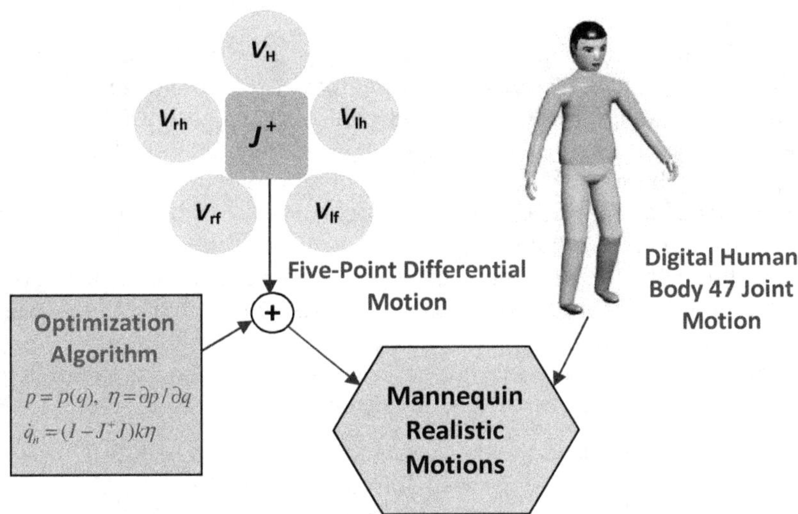

Fig. 9.6 A block diagram of the five-point model

9.3 The Range of Motion (ROM) and the Range of Strength (ROS)

9.3.1 Basic Concepts of the Human Structural System

The human structure is constituted by a skeleton and a number of muscles, which are collectively called the human musculoskeletal system. The human skeleton is a framework that consists of more than 200 bones, among which 126 are forming the so-called appendicular skeleton and comprise the bones of the upper and lower extremities [10]–[17]. The bones of upper extremity include the scapula, clavicle, humerus, ulna and radius and carpal bones, etc. The lower extremity has the pelvis, femur, tibia, fibula, tarsal bones, as well as the metatarsals and phalanges. The remaining bones are forming the axial skeleton, such as the skull, spinal column, sternum and ribs. Bones articulate with the other bones to produce joints, and they are the locations of movement. Arthrology is the special area of studying joints and their articulations, which often begins with the classification of joints:

1. The gliding joint (arthrodia), such as tarsal and carpal bones on a wrist or ankle, and on a clavicle;
2. The hinge joint (ginglymus) that is uniaxial, such as the elbow and knee in flexion and extension;
3. The ball-and-socket joint (enarthrosis) that is triaxial, such as the shoulder with a spherical head of one bone fitting into the hollowed concave surface of the other bone, forming a tri-axial joint of many actions;
4. The condyloid joint (ovoid or ellipsoid joint) that is biaxial, such as the articulations between the carpal bones of the wrist and the metacarpals of the fingers, similar to the ball-and-socket type, but only in the sagittal and frontal planes without rotation;
5. The saddle joint (sellar joint) that is also biaxial as a modification of the condyloid joint, such as the carpal-metacarpal joint of the thumb with a concave-convex surface;
6. The pivot joint (trochoid joint) that is also uniaxial, such as the neck with only rotary movement about the longitudinal axis of the bone.

The above classification has been previewed in Figure 9.1. All the joints listed above are of the diarthrodial joints category. Namely, they possess an articular cavity and have a ligamentous capsule that encases the joint and the synovial fluid. The articular surfaces of these joints are smooth and covered with cartilage. In addition to the diarthrodial joints, two other types of joints: synarthrodial and amphiarthrodial, have no separation or articular cavity, and thus they are almost negligible without movement.

Although the types of movements are predetermined by the structure and type of the joints, the range of motion (ROM) is determined not only by the mechanical structure, but also by many human factors, such as the use, disease, injury, extendability of muscles, tendons and ligaments, etc. In order to drive a joint to move, force and torque are required. Muscle contraction plays a main role in producing and delivering the force/torque on the joints and between the body segments at a biomechanical standpoint.

Regarding the human muscles, there are three kinds of muscles: cardiac, smooth and skeletal (striated) that also vary in accordance with their functions. The cardiac and smooth muscles have similar functions and both surround hollow organs. Only the skeletal muscles are usually attached to bone and cartilage with fibers to produce a driving force by the contractions. Moreover, there are three types of muscle contraction: concentric, eccentric and static or isometric. Concentric or shortening contraction occurs when a muscle develops sufficient tension to overcome a resistance and shortens to move the body segment from one attachment to the other. Eccentric or lengthening contraction occurs when the resistance cannot be overcome, but the muscle develops tension and lengthens during the action against the moving force of gravity or the other external forces. Static or isometric contraction occurs when a muscle that develops tension is unable to move the load but does not change the length.

Moreover, human living on the earth adapts a constant and uniform influence by gravity. Intuitively, the comfort zone of each joint, which must be a subset of the corresponding ROM, is different between the environments with gravity and without gravity. In addition to gravity, many other objective and subjective factors may also vary the joint comfort zones, and even the ROM itself. Hence, they are not uniquely fixed individual by individual.

The above biomedical and biomechanical perspectives are, indeed, the foundations to underlie our digital human modeling and realistic motion generation. However, facing such a huge challenging reality, we have to mimic the human structure and mechanism of motion step by step towards a realistic digital human model.

9.3.2 An Overview of the Human Movement System

In a human body, energy derived from muscular contraction is converted to the bones, resulting in movement of the body segments. These segments will further transmit energy to external objects even more advantageously. The activity of muscle can not only cause an effective and purposeful movement of the human body, but can also make a coordinative action of relatively large number of muscles for multi-axis coordinated motion. Although the muscle contraction is the only source to produce body movements, those bones that are not engaged in the movements can provide attachment for one end of such muscles to offer a stable base for their contractions [10]–[17].

The most typical example of the movement subsystem is the shoulder complex that consists of the shoulder (glenohumeral), acromioclavicular, and sternoclavicular joints, which link together the humerus, scapula, clavicle and sternum in a chain. The shoulder complex, in association with the muscles that link it to the axial skeleton, provides the upper extremity with a range of motion (ROM) exceeding that of any other joint or joint complex. Such a large ROM is due to four sources of relative motion in the shoulder complex: the shoulder, acromioclavicular and sternoclavicular joints, as well as the scapulothoracic gliding mechanism.

On the other hand, due to a variety of restrictions on the movement over the joints of the shoulder complex, most upper extremity movements that involve a large abduction of the arm with respect to the trunk are affected by a combination of movement in two or more of the four sources of movement in the shoulder complex. Such a muscle that is coupled with the combination of movement is referred to as *two-joint muscles* or *bi-articular muscles*, as illustrated in Figure 9.7. Therefore, lack of flexibility in the shoulder complex and lack of extendability in the muscles that control the scapulothoracic gliding mechanism may impair abduction of the upper extremity and result in shoulder impingement.

Just like the movements involving the shoulder joints that are associated with movements of the shoulder girdle, movements of the hip joints are also associated with movements of the vertebral column. For example, abduction of the hip is usually associated with lateral flexion of the trunk, as in standing on one leg with the free leg abducted. Similarly, hip flexion is usually associated with trunk flexion as in a sit-up exercise posture, as shown in Figure 9.8. Therefore, reduced flexibility in the hip is partially compensated by movement in the vertebral column. To this extent, the hip joints and joints

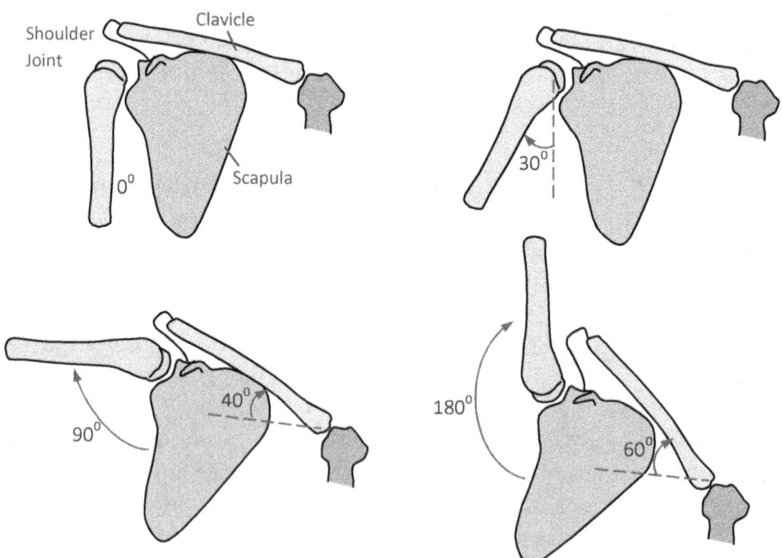

Fig. 9.7 Shoulder abduction and its clavicle joint combination effect

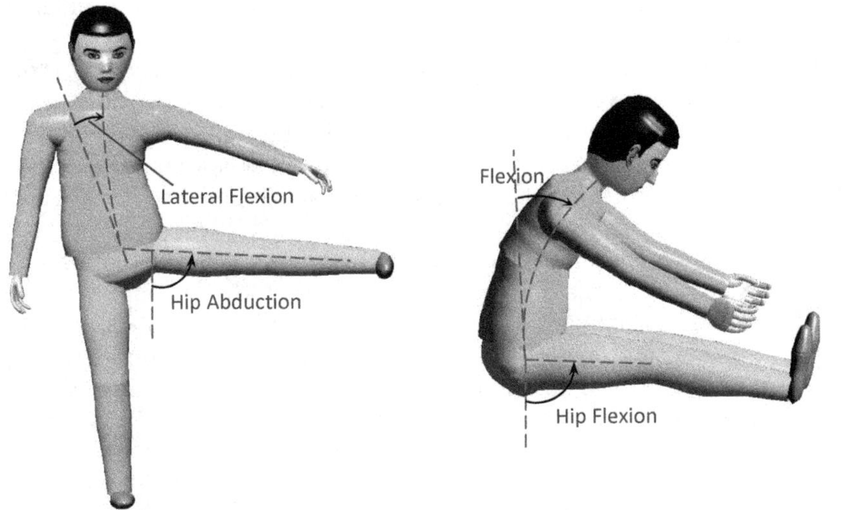

Fig. 9.8 Hip flexion and abduction with joint combination effects to the trunk flexion and lateral flexion

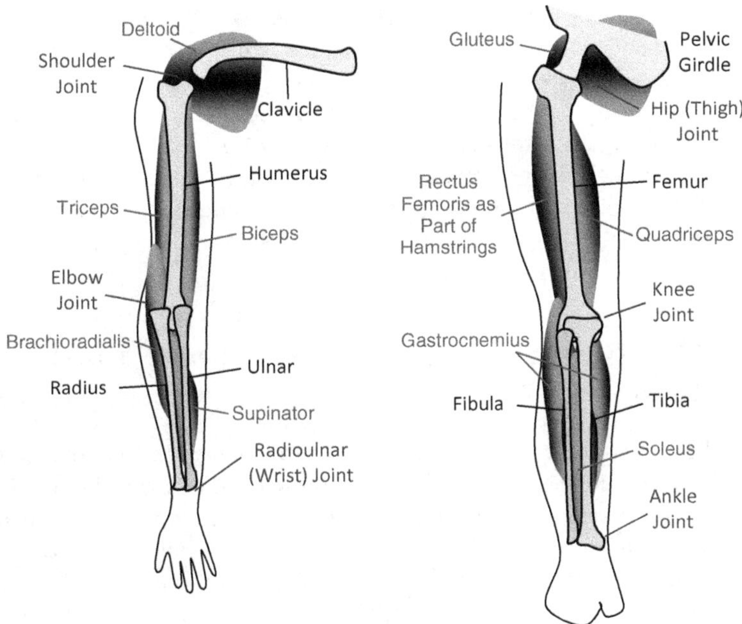

Fig. 9.9 Two-joint muscles on the arm and leg

of the vertebral column may be regarded as a functional group similar to the shoulder complex.

In fact, the joint combination phenomenon due to the skeletal structure coordination is equivalent to the change of the reference coordinates for the next joint by the previous joint movement. The movement of the clavicle joint will lead to a new neutral position of the shoulder abduction, and the hip flexion or abduction will result in a new orientation of the H-triangle of the human body.

The action of two-joint muscles on the levers is similar to that of a pulley. For example, the rectus femoris causes flexion at the hip and extension at the knee; the gastrocnemius helps flex the lower leg and plantar flexes the foot; the hamstrings cause flexion at the knee and extension at the hip, as depicted in Figure 9.9. In addition, the flexors and extensors of the fingers might be called multi-joint muscles, since they pass over the wrist and at least two joints of the fingers.

One outstanding characteristic of these muscles is that they are not long enough to permit a complete range of action simultaneously in the joints involved because of their location on two joints, either because the antagonist muscles prevent full range of action on these joints, or because antagonistic action occurs in the two joints, such as flexion in one and extension in the other. Furthermore, muscles can contribute to a limitation in joint mobility because of their inherent length-tension properties. When the muscle is extremely shortened, it cannot further produce a useful amount of tension. If gravity or tissue compression causes resistance to extreme rotation of a body segment, then joint motion can be limited by the diminished force capability of the muscles. When a muscle acts about only one joint, such limitation is not great enough. However, many joints are spanned by two-joint muscles, and thus the mobility at one joint depends in some degree on the position of an adjacent joint. Those factors directly cause a limited range of strength (ROS) in each human body joint.

9.3.3 The Range of Motion (ROM) and Joint Comfort Zones

The ROM for each joint over a human body vary with respect to age, gender, body build, level of activity, health condition and many other factors [10, 12]. With such a large deviation, we have to take an average for a regular person. The comfort zone of each joint, on the other hand, should be a subset of the corresponding joint ROM. Tables 9.2–9.5 list those average joint ROM's and the comfort zones just as a reference, where the upper or lower limit of each comfort zone is calculated by $0.35\times$ the upper or lower limit of the corresponding joint ROM. Tables 9.2–9.5 also consider the two-joint muscle effect so that many joint ROM's are conditional under the adjacent joint positions.

Table 9.2 The average joint ROM's and the joint comfort zones along the vertebral column

Joint Mobility	ROM in Degree	Comfort Zone in Degree	Conditions When
Cervical Flex/Ext	40 / -75	14 / -26.25	N/A
Cervical Lat Flex	35 / -35	12.25 / -12.25	N/A
Cervical Rotation	50 / -50	17.5 / -17.5	N/A
Thoracic Flex/Ext	45 / -25	15.75 / -8.75	Hip Neutral 0^0
	30 / -25	10.5 / -8.75	Two Hip Flex 90^0
Thoracic Lat Flex	20 / -20	7 / -7	Hip Neutral 0^0
	20 / -10	7 / -3.5	One Hip Abducts 90^0
Thoracic Rotation	35 / -35	12.25 / -12.25	Hip Neutral 0^0
	35 / -15	12.25 / -5.25	One Hip Abducts 90^0
Lumbar Flex/Ext	60 / -35	21 / -12.25	Hip Neutral 0^0
	40 / -35	14 / -12.25	Two Hip Flex 90^0
Lumbar Lat Flex	20 / -20	7 / -7	Hip Neutral 0^0
	20 / -10	7 / -3.5	One Hip Abducts 90^0
Lumbar Rotation	5 / -5	2 / -2	Hip Neutral 0^0
	5 / -2	2 / -1	One Hip Abducts 90^0

Thus, according to the ROM table, the total flexion of the thoracic/lumbar part of the human vertebral column is $45 + 60 = 105^0$, the total extension of the same part is $25 + 35 = 60^0$, the total lateral flexion is $20 + 20 = 40^0$ and the total axial rotation is $35 + 5 = 40^0$, as the sums of the ROM's of the thoracic and lumbar regions at the neutral hip position without a hip flexion or abduction [11]. With the hip flexion or abduction up to 90^0, those sums will be reduced due to the two-joint muscles combination. As mentioned above, most comfort zones are determined as 35% of the ROM angles.

All the ROM data in the above tables are referred to the literature of biomechanics and kinesiology as average ranges. The comfort zone of each joint is basically determined by 35% of the ROM values. We can then find the comfort center θ_i^c for each joint angle θ_i by the following formula:

$$\theta_i^c = \frac{\theta_{iCZ}^{up} - \theta_{iCZ}^{low}}{2} + \theta_i^h, \tag{9.16}$$

Table 9.3 The average joint ROM's and the joint comfort zones for the shoulder

Joint Mobility	ROM in Degree	Comfort Zone in Degree	Conditions When
Shoulder Flex/Ext	188 / -61	65.8 / -21.35	Elbow Neutral 0^0
	170 / -30	59.5 / -10.5	Elbow Flexes 90^0
Shoulder Abd/Add	198 / -50	69.3 / -17.5	Elbow 0^0 & Shoulder Neutral (Vertical)
	134 / -48	46.9 / -16.8	Elbow 0^0 & Shoulder Flexes 90^0 (Horizontal)
	130 / 0	45.5 / 0	Elbow 90^0 & Shoulder Neutral 0^0
	120 / -30	42 / -10.5	Elbow 90^0 & Shoulder Flexes 90^0
Shoulder Med/Lat Rotation	90 / -45	31.5 / -15.75	Elbow 90^0 & Shoulder Neutral 0^0
	97 / -34	34 / -11.9	Elbow 90^0 & Shoulder Flexes 90^0
	90 / -90	31.5 / -31.5	Elbow 90^0 & Shoulder Abducts 90^0

Table 9.4 The average joint ROM's and the joint comfort zones for the forearm

Joint Mobility	ROM in Degree	Comfort Zone in Degree	Conditions When
Elbow Flexion	142	49.7	Shoulder Neutral 0^0
	< 142	< 49.7	Shoulder Flexes 90^0
Elbow Supin/Pron	113 / -77	39.55 / -27	Elbow Neutral 0^0
	< 113 / -77	< 39.55 / -27	Elbow Flexes 90^0
Wrist Flex/Ext	90 / -81	45 / -40.5	N/A
Wrist Rad/Uln Deviation	27 / -47	13.5 / -23.5	N/A

where θ_{iCZ}^{up} and θ_{iCZ}^{low} are the first and second angles of the comfort zone, respectively, for the corresponding joint i, and θ_i^h is the i-th joint home position. For instance, θ_{16} is the joint angle of the left shoulder abduction/adduction with its home position at $\theta_{16}^h = -90^0$. According to the ROM table, $\theta_{16CZ}^{up} = 69.3^0$ and $\theta_{16CZ}^{low} = -17.5^0$. Thus, $\theta_{16}^c = (69.3 + 17.5)/2 - 90 = -46.6^0$. This is the comfort center for joint 16, but under no gravity condition. If taking the gravity effect into consideration, this particular comfort center θ_{16}^c may

Table 9.5 The average joint ROM's and the joint comfort zones for the leg

Joint Mobility	ROM in Degree	Comfort Zone in Degree	Conditions When
Hip Flex/Ext	113 / -45	39.5 / -15.75	Knee Neutral 0^0
	90 / -30	31.5 / -10.5	Knee Flexes 90^0
Hip Abd/Add	45 / -40	15.75 / -14	Knee 0^0 & Hip Neutral (Vertical)
	53 / -31	18.55 / -10.8	Knee 0^0 & Hip Flexes 90^0 (Horizontal)
Hip Med/Lat Rotation	39 / -34	13.65 / -11.9	Knee 90^0 & Hip Neutral 0^0
	31 / -30	10.8 / -10.5	Knee 90^0 & Hip Flexes 90^0
Knee Flexion	113 (Stand) 125 (Prone) 159 (Kneel)	39.5 43.75 55.65	Hip Neutral 0^0
	80 (Stand)	28	Hip Flexes 90^0
Knee Med/Lat Rotation	43 / -35	15 / -12.25	Knee Neutral 0^0
	< 43 / -35	< 15 / -12.25	Knee Flexes 90^0
Ankle Plan/Dorsi	38 / -35	19 / -17.5	Knee Neutral 0^0
	36 / -33	18 / -16.5	Knee Flexes 90^0
Foot Inversion /Eversion	24 / -23	12 / -11.5	N/A
Toe Flex/Ext	35 / -20	17.5 / -10	N/A

have to be modified to about -65^0. The joint comfort center θ_i^c will be useful to set up a joint-comfort optimization criterion in digital human optimal posture and motion generations.

9.3.4 The Joint Range of Strength (ROS)

National Institute of Occupational Safety and Health (NIOSH) had estimated before 1981 that approximate 33% of the U.S. workforce was required to exert significant force as part of their jobs. The heavy exertion can lead to external, internal and cumulative back injuries. National Safety Commission (NSC) had also reported since 1975 that 27% of all the compensible injuries were due to overexertion, of which 70% were due to lifting and 20% due to pushing and pulling. Therefore, a guideline on lifting are critical to reduce all kinds of the overexertion injuries [12].

NIOSH then developed a comprehensive guideline for manual material handling (MMH), and published in "Work Practices Guide for Manual Lifting" that covered many aspects of MMH in 8 chapters and 150 pages. This guideline for lifting was based on the considerations of epidemiology, biomechanical models, physiological limits and psychophysical data [18, 19]. Under the following assumptions:

1. Limited to symmetric (two-handed) lifting;
2. Smooth lifting (i.e., no sudden accelerations);
3. Moderate width objects (the two hands separation is less than 75 cm.);
4. Unrestricted lifting postures (no bracing);
5. Good couplings (i.e., hands hold are secure, and shoe/floor slip potential is low); and
6. Favorable temperature conditions;

NIOSH had developed a lifting equation in 1981, which is given below:

$$AL = 40(15/H)(1 - (0.004|V - 75|))(0.7 + 7.5/D)(1 - F/F_{max}) \quad \text{in Kg.},$$

and

$$MPL = 3AL,$$

where AL is the action limit, MPL is the maximum permissible limit, H is a horizontal distance in centimeter from the load center of gravity (CG) at the origin of lift to the midpoint between the ankles with the minimum 15 and maximum 80 cm., $|V|$ is an absolute vertical distance in cm. to the load CG at the origin of lift measured from the floor with the maximum 175 cm., D is a vertical distance traveled with the minimum 25 to the maximum 200 cm., F is an average frequency of lifting with the minimum 0.2 lifts/minute, and F_{max} is the maximum value determined by the period of lifting and body position.

This AL equation was further revised by NIOSH and published a new equation in 1991 to calculate a recommended weight limit (RWL) that can also be applied for an asymmetric lifting case:

$$RWL = LC \cdot HM \cdot VM \cdot DM \cdot AM \cdot FM \cdot CM \quad \text{in Kg.},$$

where $LC = 23$ Kg. is the load constant, $HM = 25/H$ is a horizontal multiplier, $VM = 1 - 0.003|V - 75|$ is a vertical multiplier, $DM = 0.82 + 4.5/D$ is a distance multiplier, $AM = 1 - 0.0032A$ is an asymmetric multiplier with A the angle of asymmetry in degree, FM is the frequency multiplier and CM is the coupling multiplier. The new equation has no maximum permissible limit.

For example, a group of workers is assigned to lift boxes by two hands and then deposit each of them on another platform. If the vertical distance of lift is $D = 25$ cm., starting the height $V = 50$ cm. The average horizontal distance is $H = 40$ cm. in front of the mid-point between the two ankles to lift the box and then turn 60^0 to deposit it on the platform. Suppose that the box-lifting task is once every two minutes. Based on the table of NIOSH, with less than 1 hour duration of the work, $FM = 0.97$ and $CM = 1.0$ for a good coupling condition. Then, the recommended weight limit in this task case can be calculated as follows:

$$RWL = 23 \cdot \frac{25}{40} \cdot (1 - 0.003 \cdot 25) \cdot \left(0.82 + \frac{4.5}{25}\right) \cdot (1 - 0.0032 \cdot 60) \cdot 0.97 \cdot 1$$

$$= 23 \cdot 0.625 \cdot 0.925 \cdot 1 \cdot 0.808 \cdot 0.97 \cdot 1 = 10.42 \text{ Kg.} = 102.24 \text{ N.}$$

This means that for such a particular lifting-depositing task, the recommended weight is at most 10.42 Kg.

The above guideline was developed based on a variety of aspects, especially the L5/S1 vertebral disc compressive force will not exceed 3400 N. at the biomechanical standpoint, and the energy expenditure will not exceed 2200-4700 cal/min. at the physiological point of view. However, the result of computation from the RWL equation is just a recommendation of external load to be lifted, instead of the internal exertion. For a detailed estimation about human joint strengths, we have to review more specifically the quantitative analysis and discussion from the research of biomechanics.

The researchers in biomechanics have also made extensive investigations on human internal force exertion due to gravity and external load [12, 13, 14]. A common recognizable viewpoint is that one cannot accurately calculate muscle force by any existing model, but can calculate more precisely the muscle moment exerted at each human body joint. The following equations have been well developed to predict major human joint torques (moments of strength) in the sagittal plane with their mean values and standard variations [12] all in N-m.:

1. Elbow flexion:

$$S_{Ef} = (336.29 + 1.544\alpha_E - 0.0085\alpha_E^2 - 0.5\alpha_S)G,$$

   where G is a gender adjustment, see the next table in detail, and all the angles are in degrees and defined in Figure 9.10;
2. Elbow extension:

$$S_{Ee} = (264.153 - 0.575\alpha_E - 0.425\alpha_S)G,$$

3. Shoulder flexion:

$$S_{Sf} = (227.338 + 0.525\alpha_E - 0.296\alpha_S)G,$$

4. Shoulder extension:

$$S_{Se} = (204.562 - 0.099\alpha_S)G,$$

5. Seated torso flexion:

$$S_{Tf} = (141.179 + 3.694\alpha_T)G,$$

6. Seated torso extension:

$$S_{Te} = (3365.123 - 23.947\alpha_T)G,$$

7. Standing torso flexion:

$$S_{Tf} = (17.17\alpha_T - 0.0796\alpha_T^2)G,$$

8. Standing torso extension:

$$S_{Te} = (3894 - 13.9\alpha_T)G,$$

9. Hip flexion:

$$S_{Hf} = (-820.21 + 34.29\alpha_H - 0.11426\alpha_H^2)G,$$

10. Hip extension:

$$S_{He} = (3338.1 - 15.711\alpha_H + 0.04626\alpha_H^2)G,$$

11. Knee flexion:

$$S_{Kf} = (-94.437 + 6.3672\alpha_K)G,$$

12. Knee extension:

$$S_{Ke} = (1091.9 - 0.0996\alpha_K + 0.17308\alpha_K^2 - 0.00097\alpha_K^3)G,$$

13. Ankle extension:

$$S_{Ae} = (3356.8 - 18.4\alpha_A)G.$$

The gender adjustment G and the coefficient of variation $\sigma = SD/\bar{S}$ in the above equations are different from each other. According to [12], they are given in Table 9.6.

For example, in an elbow flexion case, if let $\alpha_S = 0$ and find the maximum for S_{Ef} with respect to α_E, i.e., to solve $\frac{\partial S_{Ef}}{\partial \alpha_E} = (1.544 - 0.017\alpha_E)G = 0$, then, $\alpha_E = 90.8235^0$ and $S_{Ef} = 77.7454$ N-m. for male ($G = 0.1913$), and $S_{Ef} = 40.8438$ N-m. for female ($G = 0.1005$). These results are the mean values of the elbow joint torque when the forearm is lifting up a load at the elbow angle $\alpha_E = 90.8235^0$ and the shoulder angle $\alpha_S = 0$.

To find the maximum value over all the population, we need to determine the standard deviation SD from the coefficient of variation σ, i.e., $SD = \sigma S_{Ef}$ and then $S_{Efmax} = S_{Ef} + SD = (1+\sigma)S_{Ef}$ in each case. Based on the above table, we find that $S_{Efmax} = 96.8552$ N-m. for male and $S_{Efmax} = 51.5816$ N-m. for female.

Table 9.6 The table of gender adjustment and coefficient of variation

Human Body Major Joints	G = Gender Adjust		Coeff. of Variation	
	Male	Female	Male	Female
Elbow flexion	0.1913	0.1005	0.2458	0.2629
Elbow extension	0.2126	0.1153	0.2013	0.3227
Shoulder flexion	0.2845	0.1495	0.2311	0.2634
Shoulder extension	0.4957	0.2485	0.3132	0.3820
Seated torso flex.	0.2796	0.1654	0.2932	0.3965
Seated torso ext.	0.3381	0.1890	0.3152	0.3455
Stand torso flex.	0.2146	0.1129	0.2939	0.3962
Stand torso ext.	0.1559	0.1016	0.3152	0.3455
Hip flexion	0.1304	0.0871	0.2729	0.3364
Hip extension	0.0977	0.0516	0.4016	0.3779
Knee flexion	0.1429	0.0851	0.2934	0.3212
Knee extension	0.0898	0.0603	0.3503	0.3466
Ankle extension	0.0816	0.0489	0.3307	0.2745

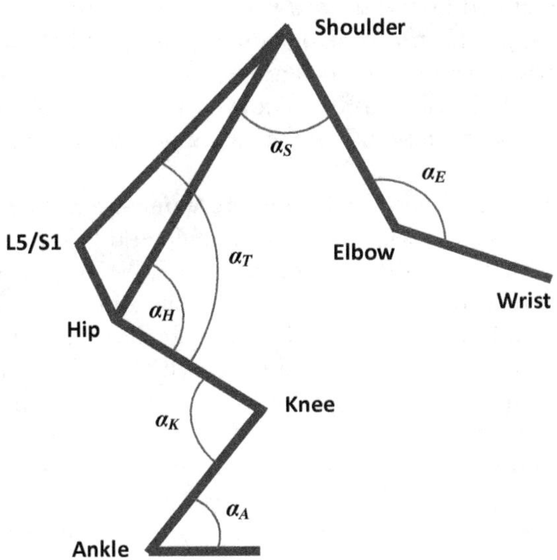

Fig. 9.10 The angles of human posture in sagittal plane for a joint strength prediction

Likewise, in an elbow extension case, if the elbow and shoulder angles are the same when the forearm is pushing an object down, then, $S_{Ee} = 45.0562$ N-m. for male and $S_{Ee} = 24.4355$ N-m. for female. Using the σ value from Table 9.6, we can further calculate the maximum joint torques in the elbow extension case, and $S_{Eemax} = (1+\sigma)S_{Ee} = 54.1260$ N-m. for male and $S_{Eemax} = 32.3208$ N-m. for female.

The above experimental-based empirical equations, indeed, cover a large population with different genders and percentiles. All the estimated values from the equations along with Table 9.6 will be the basis to design a set of weights, or a weighting matrix W to predict the optimal joint torque distribution for a digital human in operating a material-handling task. However, the above equations that have been published thus far provide only 6 to 7 major joints with their moments of strength, which can be called the range of strength (ROS) for each major human body joint. The application will be mentioned again in the next section.

If we plot both the ROM and ROS for each joint in a chart of joint torque versus joint angle, then a closed loop will be shown up, which represents a boundary of the maximum motion range and maximum torque range. In other words, any joint activity cannot exceed this closed boundary. Otherwise, either this particular joint moves beyond its ROM or beyond its ROS. For example, Figure 9.11 shows a shoulder flexion/extension joint chart. Let the elbow angle $\alpha_E = 90^0$ in the third equation of the ROS. Also, let the shoulder angle α_S varies from the lower bound -30^0 to the upper bound 170^0 of its ROM under $\alpha_E = 90^0$, see Table 9.3. Then, the left and right bounders of the closed boundary are just the lower and upper bounds of the ROM, respectively. The top and bottom bounders of the closed boundary in Figure 9.11 are both curved that are the maximum value of S_{Sf} and the minimum value of $-S_{Se}$ in the ROS equations 3 and 4, respectively, where G and σ are picked from Table 9.6 on the shoulder flexion and extension lines for a male.

Now, a shoulder joint activity is running along a curve of the joint torque versus joint angle, which starts at 0 joint angle and 0 joint torque, and then moves forward and exceeds the upper bound of the ROS. Although it comes back into the closed loop, it finally exceeds the upper bound of the ROM. Therefore, this kind of activity for the shoulder flexion and extension is not allowed. If every joint can plot such a chart with its own closed boundary in terms of the ROM and ROS, or even in terms of the comfort zones of the ROM and ROS, we will be able to monitor all the joint activities to see if any violation occurs for a realistic real-time ergonomic assessment.

Another advantage of such a joint torque versus joint angle chart is that the area under each activity curve is the energy dissipated by this joint in activity, because of the fact that since the mechanical power is always $P = \tau\dot{\theta}$, the energy increment is $dE = Pdt = \tau d\theta$ so that

$$E = \int_{\theta_0}^{\theta_1} \tau \, d\theta.$$

Therefore, the chart may provide an opportunity to optimize each joint activity curve in order to minimize the joint effort.

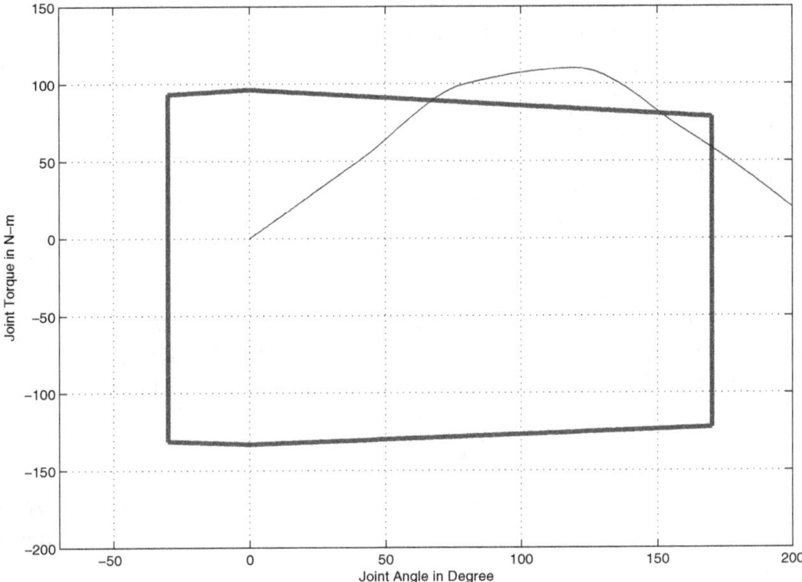

Fig. 9.11 A closed boundary for the shoulder ROM and ROS in a chart of joint torque vs. joint angle

However, only 6 to 7 major joint ROS's have been thus far published, which do not cover the majority of human joints. Even if we just monitor those major joints to compare with the available ROS's, we can easily trace each joint angle at any time instant, but how can we acquire the data of instantaneous joint torques? Since we have developed a global Jacobian matrix for a digital human model, to find an instantaneous joint torque for each joint quantitatively will no longer be a difficult job.

9.4 Digital Human Statics

9.4.1 Joint Torque Distribution and the Law of Balance

Recalling the discussion on robotic statics and equation (4.30) in Chapter 4, we now intend to apply the principle of statics that is a duality counterpart of the kinematics to a digital human model in order to determine its instantaneous joint torque distribution when the two hands are operating a material-handling task in either a standing, or a sitting or even a kneeling posture. The general form of the statics equation is repeated here again:

$$\tau = J_{(k)}^T F_{(k)} = J_{(b)}^T F_{(b)}, \tag{9.17}$$

where both the Jacobian matrix $J_{(k)}$ and Cartesian force or wrench vector $F_{(k)}$ must be projected onto the common frame k, no matter which k is. For the most convenient task description, both are often projected onto the base frame b.

In this statics equation, the Jacobian matrix of the digital mannequin can either be global or be local (partial). For instance, if the left hand of the mannequin is lifting up a load with a vertical down force $f_{lh} = mg$ when the mannequin is standing, where m is the mass of the load and $g = 9.81$ m/sec.^{2} is the gravitational acceleration, then the 6 by 1 Cartesian force (wrench) vector can be written as $F_{lh(b)} = (0 \;\; 0 \;\; -mg \;\; 0 \;\; 0 \;\; 0)^T$ that is projected onto the world base if no other forces or moments (torques) exerted on the left hand. Based on equations (9.7) and (9.12), all the 21 joints are exerted with torques to form a joint torque distribution given by

$$\tau_{lh} = J_{lh(b)}^T F_{lh(b)},$$

where τ_{lh} is a 21 by 1 column vector that lists all the joint torques from the last joint at frame $21la$ of the left hand to the first joint of the mannequin as the entire body forward-backward translation.

Similarly, the 21 by 1 joint torque vector from the last joint #30 of the right hand backup to the first joint of the digital mannequin can be determined in the same way. However, since the joint number in the list has a jump from 12 to 22 between the torso and right arm, it is better to augment both $J_{lh(b)}$ and $J_{rh(b)}$ together to form a single two-hand Jacobian matrix $J_{2h(b)}$. According to the global Jacobian augmentation rule in (9.14),

$$J_{2h(b)} = \begin{pmatrix} J_{lh(b)}(:, 1:12) & J_{lh(b)}(:, 13:21) & O_{6\times 9} \\ J_{rh(b)}(:, 1:12) & O_{6\times 9} & J_{rh(b)}(:, 13:21) \end{pmatrix},$$

which is a 12 by 30 matrix.

If the right hand has the similar or same force $F_{rh(b)} = (0\; 0\; -mg\; 0\; 0\; 0)^T$ exerted as a two-hand lifting task, then the augmented Cartesian force for the two hands becomes

$$F_{2h(b)} = \begin{pmatrix} F_{lh(b)} \\ F_{rh(b)} \end{pmatrix}$$

that is a 12 by 1 column vector. The joint torque distribution for all the 30 joints involved in the torso and two arms can be determined by

$$\tau_{2h} = J_{2h(b)}^T F_{2h(b)}. \tag{9.18}$$

As a numerical example, let a standing digital human lift an object of $mg = 10$ N. as the total weight by two hands symmetrically so that the lifting force on each hand is 5 N. The 12-dimensional Cartesian force vector becomes

$$F_{2h(b)} = (0 \;\; 0 \;\; -5 \;\; 0 \;\; 0 \;\; 0 \;\; 0 \;\; 0 \;\; -5 \;\; 0 \;\; 0 \;\; 0)^T.$$

Figure 9.12 shows the instantaneous posture of the digital mannequin when the two hand are lifting the load. Then, all the 30 joint forces/torques listed in

$$\tau_{2h} = (f_1 \; f_2 \; f_3 \; \tau_4 \; \cdots \; \tau_{21} \; \tau_{22} \; \cdots \; \tau_{30})^T$$

can be immediately calculated by (9.18) and plotted in MATLAB^{TM} as a bar chart in Figure 9.13.

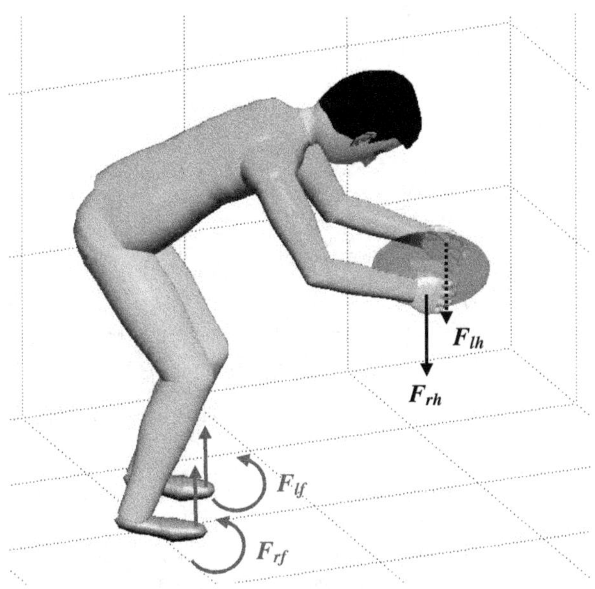

Fig. 9.12 Analysis of mannequin force balance in standing posture

Note that the resultant bar chart of joint torque distribution in Figure 9.13 is only due to the 10 N. lifting load and excluding the joint torques due to gravity of each body segment. We will make a detailed analysis of the joint forces/torques due to gravity in the later discussions.

It can be further observed that at any moment of steady-state as a digital human is situated in either a standing or a sitting or kneeling posture, the entire body should be balanced as motionless. Since the first six joints of our digital human model are defined as the x, y and z translations plus the body spin, fall over and fall aside rotations, the motionless fact implies **the Law of Balance**, i.e.,

The first six joints of the digital human model must have all zero joint forces/torques at a steady-state moment.

However, the first six joint forces/torques in Figure 9.13 currently do not vanish yet. The reason should be understandable that the current joint torque distribution is considering the joint torques only due to the lifting load on the

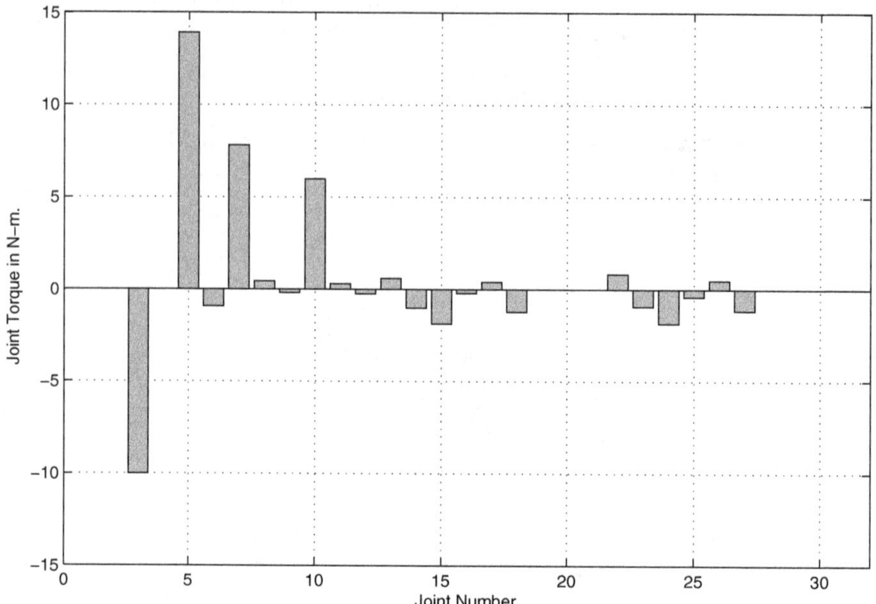

Fig. 9.13 Two arms and torso joint torque distribution in standing posture

two hands. In fact, once the digital mannequin is lifting the load, the floor will immediately react to push up the mannequin through its two feet for a balance. Taking the push-up reacting forces from the floor into consideration, based on the Law of Balance, the final combined result of the first six joint forces/torques of the mannequin must be all zero.

Let the first 6 columns of the 12 by 30 two-hand Jacobian matrix $J_{2h(b)}$ form a 12 by 6 partial Jacobian matrix $J^6_{2h(b)}$ for the hands. Likewise, let the first 6 columns of the following 12 by 20 two-foot Jacobian matrix:

$$J_{2f(b)} = \begin{pmatrix} J_{lf(b)}(:,1:6) & J_{lf(b)}(:,7:13) & O_{6\times 7} \\ J_{rf(b)}(:,1:6) & O_{6\times 7} & J_{rf(b)}(:,7:13) \end{pmatrix}$$

form another 12 by 6 partial Jacobian matrix $J^6_{2f(b)}$ for the feet. For an arbitrary two-hand load $F_{2h(b)}$, the reaction forces/torques $F_{2f(b)}$ from the floor to the feet in order to balance the entire body of the mannequin must satisfy the following equation:

$$(J^6_{2h(b)})^T F_{2h(b)} + (J^6_{2f(b)})^T F_{2f(b)} = 0. \tag{9.19}$$

Now, returning back to the above load-lifting example, the first six joint forces/torques due to the weight-lifting by the two hands at the moment of standing posture given in Figure 9.12 can be found directly from the bar chart in Figure 9.13:

$$\tau(1{:}6) = (J^6_{2h(b)})^T F_{2h(b)} = (0 \;\; 0 \;\; -10 \;\; 0 \;\; 13.9128 \;\; -0.9209)^T.$$

Hence, based on (9.19),

$$(J^6_{2f(b)})^T F_{2f(b)} = -\tau(1{:}6).$$

Since $(J^6_{2f(b)})^T$ is 6 by 12, one cannot uniquely solve the above equation for the 12 by 1 two-foot reaction force vector $F_{2f(b)}$. This non-uniqueness of solution actually reflects a reality that a real human can always adjust his/her two feet pressure distribution on the floor in order to balance his/her entire body in relatively different ways. However, at a mathematical standpoint, one can readily achieve one of the most meaningful solutions for the above balance equation, which is known as the minimum-norm solution (see also in Chapter 5):

$$F_{2f(b)} = -\left(J^{6T}_{2f(b)}\right)^+ \cdot \tau(1{:}6), \tag{9.20}$$

where $(J^{6T}_{2f(b)})^+$ is the pseudo-inverse of the 6 by 12 partial Jacobian matrix transpose $(J^6_{2f(b)})^T$. The numerical minimum-norm solution has been programmed in MATLAB^{TM} and calculated as follows:

```
>> F_2f(b) = -pinv(J_2f(b)') * tau(1:6)

F_2f(b) =

   0.0001    0.0002    5.0323    2.9541   -1.1454    0.0000
  -0.0001   -0.0002    4.9677    2.9541   -1.1454    0.0000

>>
```

The first 6 elements of the resulting 12 by 1 Cartesian force vector are the reaction forces/torques from the floor to the left foot, and the rest 6 elements are the reaction on the right foot. Comparing them with the standard definition of a Cartesian force vector given from equation (4.29) in robotic statics (see Chapter 4):

$$F = (f_x \;\; f_y \;\; f_z \;\; m_x \;\; m_y \;\; m_z)^T,$$

it can be seen that the third element 5.0323 is the vertical-up reaction force in N. from the floor to the left foot, while the 9th element 4.9677 in N. is the vertical-up reaction force from the floor to the right foot, and the sum of both two is exactly equal to 10 N. to balance the total weight of the lifting load.

In order to completely balance the mannequin entire body, additional nonzero Cartesian moments (Cartesian torques) may also be required, such as the nonzero 4th to 6th elements for the left foot and the nonzero 10th to 12th elements for the right foot. In fact, the reaction forces from the floor to the real human two feet for a balance are often not concentrated at a single point. Instead, as everyone can be self-experienced, there must be a multi-reacting-point distribution on the bottom of each foot in order to completely and comfortably balance the load-lifting with the hands in a standing posture.

Once all the action and reaction forces and Cartesian moments are either given or determined, a complete joint torque distribution after the balance in a standing posture, as shown in Figure 9.12, can be computed in MATLAB^{TM} and displayed as a bar chart in Figure 9.14. The numerical computation is based on the global statics in (9.17), where the Cartesian force $F_{(b)}$ is 30 by 1 that is augmented by the 6 by 1 $F_H = 0$ for the H-triangle in this particular standing posture, the 12 by 1 $F_{2h(b)}$ for the two hands and the 12 by 1 $F_{2f(b)}$ for the two feet. One can clearly see that the first six joint forces/torques now all vanish, and this exactly reflects that the mannequin is standing and lifting a load by the two hands at a steady-state moment. One may compare the final balanced joint torque distribution bar chart in Figure 9.14 with the unbalanced one in Figure 9.13 to see the difference.

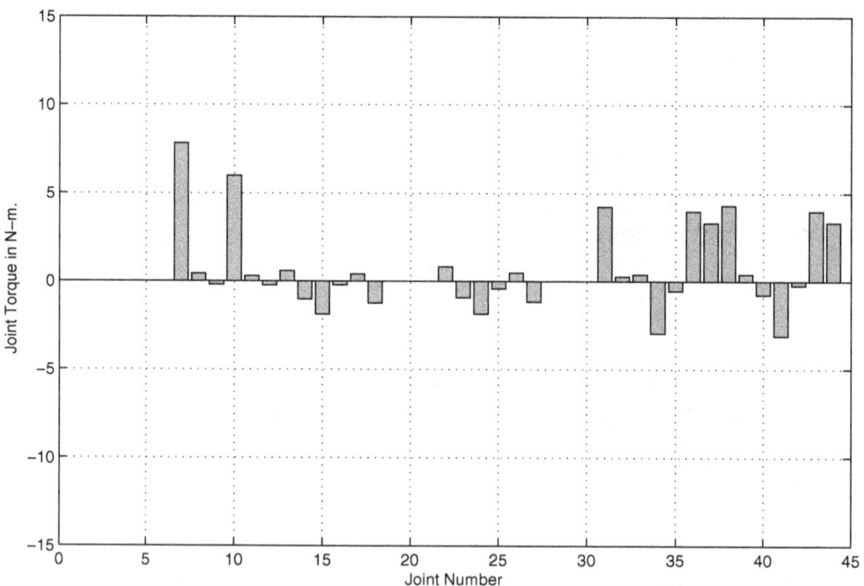

Fig. 9.14 A complete joint torque distribution in standing posture

Using the same approach to the standing posture, we can also determine an instantaneous joint torque distribution for the mannequin in sitting or kneeling posture and holding a weight by two hands symmetrically, as shown in Figures 9.15 and 9.16.

Unlike the standing case, the H-triangle is no longer force-free in a sitting case. Instead, the H-triangle of the mannequin is exerted by the majority of

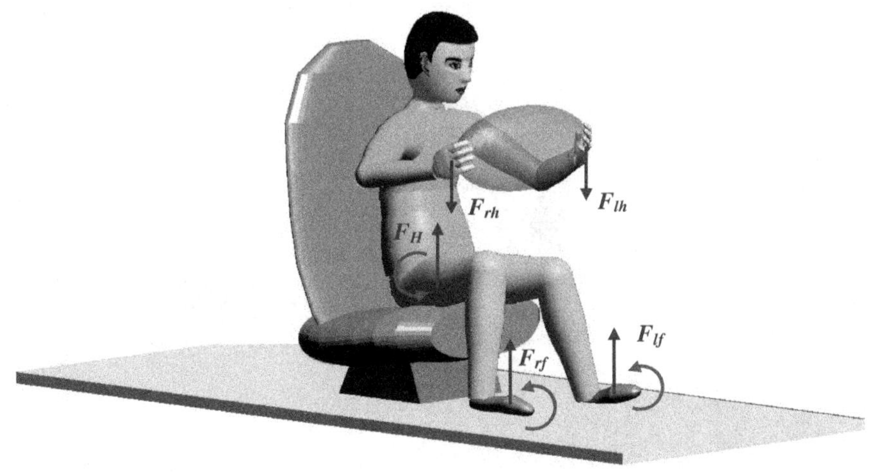

Fig. 9.15 Analysis of mannequin force balance in sitting posture

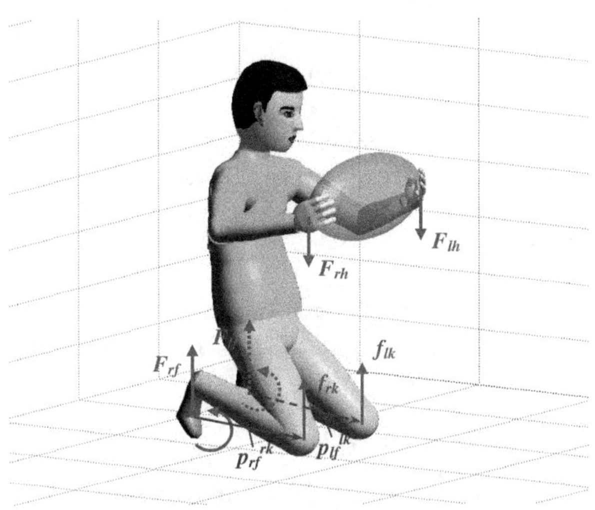

Fig. 9.16 Analysis of mannequin force balance in kneeling posture

reaction from the seat in order to balance the hands weight-lifting, as shown in Figure 9.15. Under the sitting posture, we can solve the same equation in (9.18) with the same Cartesian force $F_{2h(b)}$ and the same Jacobian $J_{2h(b)}$ for the two hands but totally different joint angles between the standing and sitting postures. The joint torque distribution from joint 1 to joint 30 under the condition that all the joint torques along the two legs are zero for the sitting posture is given in Figure 9.17.

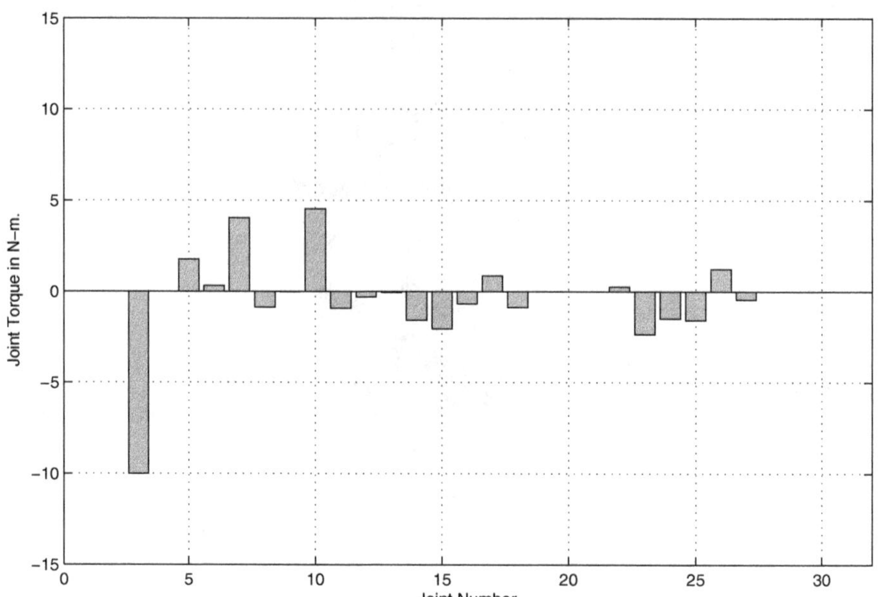

Fig. 9.17 The joint torque distribution over two arms and torso in sitting posture

As we can see, the first six joint torques are non-zero and required to be balanced by a Cartesian force $F_{H(b)}$ exerted at the H-triangle through the buttock if no additional Cartesian forces are imposed by the two feet from the floor. Based on the Law of Balance, equation (9.19) should be changed to

$$(J_{2h(b)}^{6})^{T} F_{2h(b)} + J_{H(b)}^{T} F_{H(b)} = 0 \tag{9.21}$$

in the sitting case. From the bar chart in Figure 9.17, the first six joint forces/torques are

$$\tau_{2h}(1:6) = (J_{2h(b)}^{6})^{T} F_{2h(b)} = (0 \;\; 0 \;\; -10 \;\; 0 \;\; 1.7636 \;\; 0.3064)^{T}.$$

Since $J_{H(b)}$ is a 6 by 6 square matrix, it makes easier to solve equation (9.21) and results in a Cartesian force $F_{H(b)}$ exerted at the H-triangle for balance:

$$F_{H(b)} = (0 \;\; 0 \;\; 10 \;\; -0.1652 \;\; -3.8516 \;\; 0)^{T}.$$

A complete joint torque distribution in the sitting posture case after the balance is shown in Figure 9.18. Likewise, the first six joints have all zero forces/torques to meet the Law of Balance. Since we don't consider any reaction from the floor to the two feet, all the joint torques along the two legs are zero. Although two thighs of the mannequin may be pushed up by the seat cushion, those pushing forces can be equivalent to the part of the Cartesian force $F_{H(b)}$ acting on the H-triangle based on equation (4.35) and its discussion in Chapter 4.

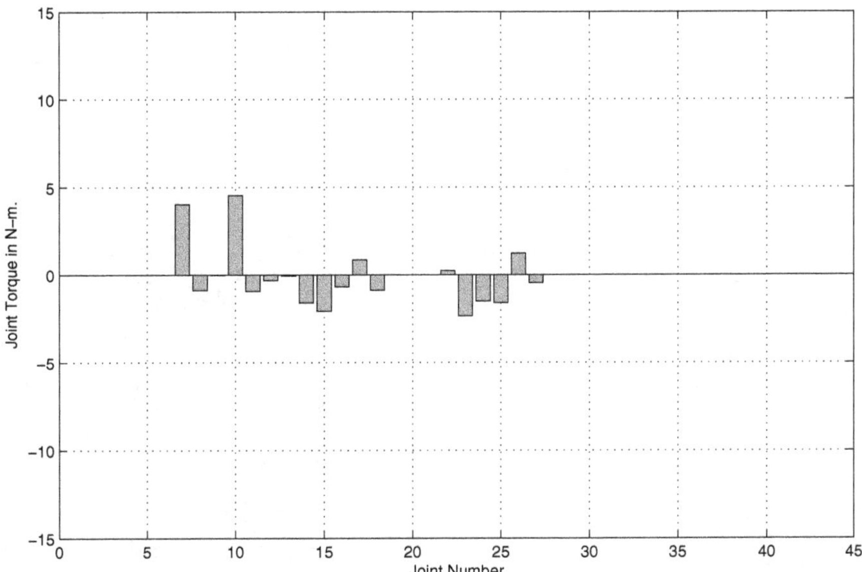

Fig. 9.18 A complete joint torque distribution in sitting posture

In the kneeling posture case, as shown in Figure 9.16, two knees and two feet receive reaction forces and moments from the floor if the mannequin is doing the same weight-lifting task by the two hands. Based on the discussion in Chapter 4 about a force applied at a point C other than the end-effector of a robot, specifically with equation (4.35) to find an equivalent Cartesian force, we can equivalently convert the knee reaction forces to be part of the Cartesian force $F_{2f(b)}$ at the two feet. Then, the formulation and calculation become the same as in the previous standing posture case, but under the different joint angles in the kneeling posture.

The joint torque distribution bar chart for the two hands and torso in the kneeling case is shown in Figure 9.19, where the first six joint forces/torques have not been balanced yet. Using the same equation in (9.20), we calculate the Cartesian force acting on the two feet in order to meet the Law of Balance in MATLAB^{TM} by the following program:

```
>> F_2f(b) = -pinv(J_2f(b)') * tau(1:6)

F_2f(b) =

   -0.0241    0.0000    5.0388   -0.0092   -2.7379   -0.0038
    0.0241   -0.0000    4.9612   -0.0092   -2.7379   -0.0038

>>
```

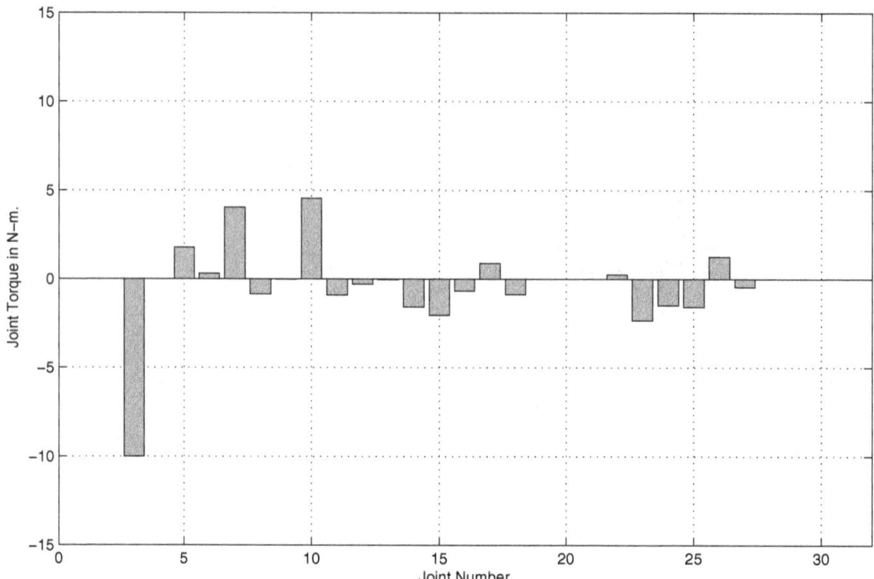

Fig. 9.19 The joint torque distribution over two arms and torso in kneeling posture

Finally, a complete joint torque distribution bar chart in the kneeling case after the balance is plotted in Figure 9.20.

Wrapping up the discussions on finding the joint torque distributions for the entire body of a digital human model in standing, sitting and kneeling postures, we make the following remarks:

1. A global Jacobian matrix $J_{(b)}$ formulated in equation (9.13) is exclusively useful not only in representing a unique instantaneous posture for a digital mannequin, but also in performing a unique transformation between Cartesian and joint tangent spaces as well as in determining a joint torque distribution at each time instant as the mannequin is doing a weight-related work;
2. Because of being able to uniquely represent a human posture, the global Jacobian matrix will also be the key to unveil the secret of posture optimization;
3. A Jacobian matrix also depends on the digital human kinematic parameters, which are part of the anthropometric data. A human in different percentile may have a different joint torque distribution from the others even if he or she is performing the same task. It is a complicated issue. However, the variation only causes a small deviation in each numerical computation, but it does not alter any property and formulation;
4. All the discussions thus far have not taken the joint torques due to gravity into account yet. We are considering this issue in the following analysis.

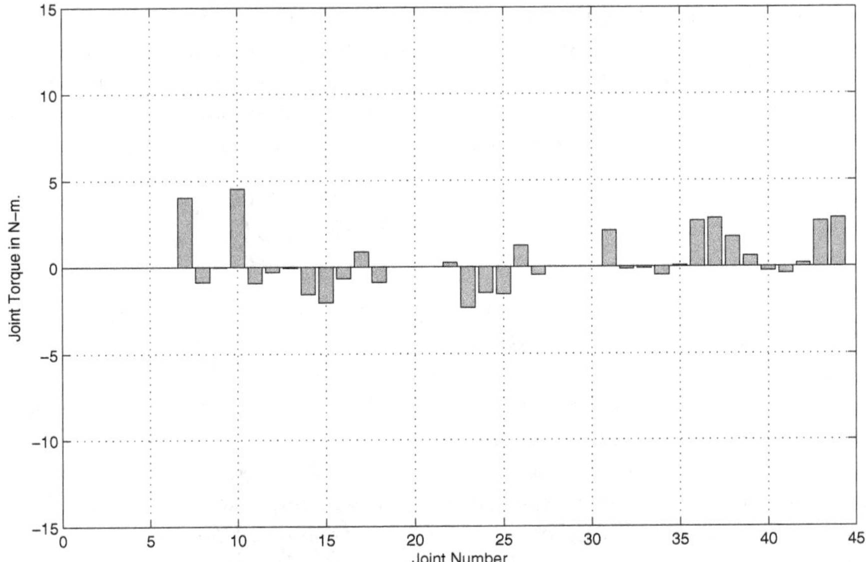

Fig. 9.20 A complete joint torque distribution in kneeling posture

9.4.2 Joint Torque Distribution due to Gravity

In order to find a joint torque distribution due to gravity, we must first recognize the fact that gravity is acting at every point of the entire human body, instead of some isolated points. In other words, the action of gravity is distributive, not centralized in nature. Even if we can determine the center of gravity CG for a digital mannequin at a time instant without major difficulty, provided that the mass and the coordinates of the mass center of each segment are known, the single CG location along with the total weight Mg of the mannequin are still insufficient to determine the joint torque distribution τ_g due to gravity.

Suppose that the center of gravity CG is located near the H-triangle of the mannequin at a time instant. We may equivalently transfer the gravity force Mg from the CG point to the H-triangle with a force $f = Mg$ plus a Cartesian moment $m = p_H^{CG} \times f$ added according to equation (4.34) in Chapter 4. However, in such a way to make equivalence, only the first six joint torques below the H-triangle are calculated by $\tau_H = J_{H(b)}^T F_{H(b)}$, while the two arms and two legs will be missed to cover their joint torques, which is not true. This intuition tells us that the center of gravity is virtually a model used to check and determine the body balance, but is not a reality. Therefore, we must seek a different approach to representing the distributive nature of gravity, or at least to decentralizing the global CG model down to the segment level.

Since the gravitation is a typical conservative force field, we may first find a total gravitational potential energy P for a digital human and then determine the joint torque vector τ_g due to gravity. Let h_j be the height of the mass center of segment j for $j = 1, \cdots, n$ over the digital human body with respect to the base. Then, it is well-known and was also discussed in Chapter 7 that

$$\tau_g = -\frac{\partial P}{\partial q} = -\frac{\partial}{\partial q}\sum_{j=1}^{n} m_j g h_j = -\sum_{j=1}^{n} m_j g \frac{\partial h_j}{\partial q},$$

where m_j is the mass of segment j, $g = 9.81$ m./sec^2 is the gravitational acceleration constant on the earth, and $q \in \mathbb{R}^n$ is the joint position vector of the digital mannequin. Here we must point out that in mathematics, each time to take a partial derivative for a scalar field with respect to a column vector, the result must be a row vector, called the gradient vector of the scalar field. Under this convention, τ_g in the above equation is supposed to be a row vector. However, in engineering tradition, it will still be considered as a column vector in order to save the transpose notation. We may keep this engineering tradition now and hereafter just for a statement convenience, and will indicate otherwise in case any confusion occurs.

Recalling equation (4.19) in Chapter 4, the partial derivative of a position vector p_0^n with respect to q can always be equivalent to the Jacobian matrix J_v related to the linear velocity v, as long as J_v is projected onto the base. Because each height h_j is the third component (z component) of the mass center vector that is tailed at the digital human base origin and arrow-pointing to the mass center of segment j, $\frac{\partial h_j}{\partial q}$ is just the third row of the related Jacobian matrix, i.e.,

$$\frac{\partial h_j}{\partial q} = J_{cj(b)}(3,:),$$

where $J_{cj(b)}$ is the sub-Jacobian matrix at the mass center of segment j according to equation (7.31) and its discussion in Chapter 7, which must be projected onto the base frame b:

$$J_{cj(b)} =$$

$$\begin{pmatrix} R_b^{cj} & O_{3\times3} \\ O_{3\times3} & R_b^{cj} \end{pmatrix} \begin{pmatrix} r_j^0 & r_j^1 & r_j^2 & p_{cj}^3 \times r_j^3 & \cdots & p_{cj}^{j-1} \times r_j^{j-1} & O_{3\times(n-j)} \\ 0 & 0 & 0 & r_j^3 & \cdots & r_j^{j-1} & O_{3\times(n-j)} \end{pmatrix}, \tag{9.22}$$

where each $p_{cj}^{i-1} = p_j^{i-1} - c_j$ with c_j the mass center vector, whose tail is at the origin of frame j and arrow is at the mass center of segment j, and $R_b^{cj} = R_b^j$ due to the parallel-shift between frames cj and j, as mentioned in Chapter 7 and illustrated in Figure 9.21. As we can also recall, each p_j^{i-1} and r_j^{i-1} in the sub-Jacobian are the 4th and 3rd columns of the corresponding homogeneous transform matrix A_j^{i-1}, which need to be prepared before constructing such a sub-Jacobian by (9.22).

Therefore, once all the sub-Jacobian matrices of mass center formulated in (9.22) are computed for $0 < j \leq n$, the joint torque distribution τ_g of the mannequin due to gravity can be readily found based on the general statics equation in (9.17) with a Cartesian force $F_{cj(b)}$ for segment j at its mass center defined by

$$F_{cj(b)} = (0 \;\; 0 \;\; -m_j g \;\; 0 \;\; 0 \;\; 0)^T.$$

Namely,

$$\tau_g = -\sum_{j=1}^{n} m_j g \frac{\partial h_j}{\partial q} = \sum_{j=1}^{n} J_{cj(b)}^T F_{cj(b)} = -\sum_{j=1}^{n} m_j g J_{cj(b)}^T(3,:). \tag{9.23}$$

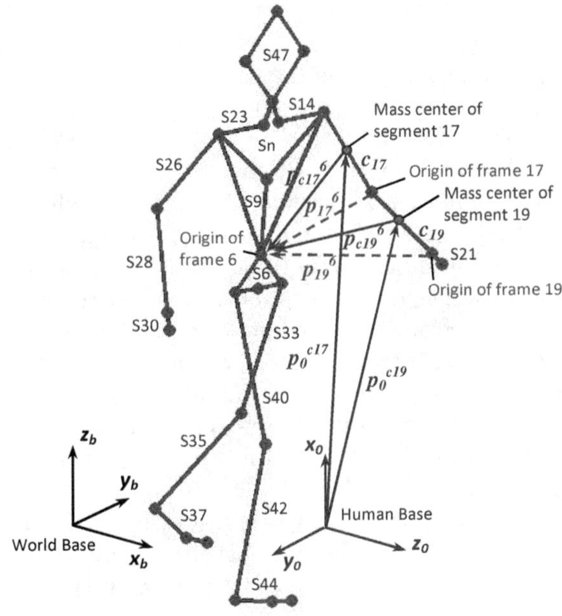

Fig. 9.21 A digital human skeleton model with segment numbering

Equation (9.23) becomes a foundation to evaluate an instantaneous joint torque distribution due to gravity for a digital human in any posture. To apply the equation, it should be emphasized that

1. Since many joints share a common joint center in our digital human model, such as the three joints at each shoulder, wrist and hip, as well as at the lumbar and thorax, the number of segments to be considered for the sub-Jacobian determination will be much less than $n = 47$.

We may select about 18 segments, each of which has a nonzero link length and/or a nonzero joint offset, to be ready for deriving their sub-Jacobian matrices, and they are all numbered in Figure 9.21. Each segment number matches the frame number. For instance, S26 is the segment of the right upper arm, whose coordinate system is frame 26. S47 is the head and its coordinate system is frame 47. Sn is the segment from the cervical column to the neck with frame n. Therefore, the summation in (9.23) can be reduced to add only 18 sub-statics terms, instead of $n = 47$ terms. In other words, the subscript index j has 18 values: $j = 6, 9, 14, 17, 19, 21, 23, 26, 28, 30, 33, 35, 37, 40, 42, 44, n$ and 47, as shown in Figure 9.21, but the superscript index i must scan every joint from joint 1 up to joint j, excluding the joint number skips along each passage of the right arm, two legs and head;

2. It can also be seen that all the sub-Jacobian matrices defined by (9.22) have the same dimension, 6 by n, but each sub-Jacobian $J_{cj(b)}$ has only j columns nonzero and the rest $n - j$ columns are all zero. This means that the weight of segment j will only contribute to the joint torques on and below joint j along each passage of joint assignments and will not affect the joints beyond joint j, which is consistent to the reality;
3. When one implements and programs equations (9.22) and (9.23) into MATLAB^{TM}, it should be more patient that the formulations of up to 18 sub-Jacobian matrices are very lengthy, each of which requires a large number of multiplications of many one-step homogeneous transformation matrices. It should also be careful to deal with the joint number jumps for most passages of joint assignments, such as from the thorax to the right arm and head, and from the H-triangle to the two legs. In other words, the summation in (9.23) requires correct joint torque numbers to be added together, and each term inside the summation as well as the final sum τ_g must all be the 47 by 1 column vectors. The reader can refer to the global Jacobian matrix formation in (9.13) for all the joint number assignments and allocations;
4. As aforementioned in Chapter 7, when we develop a dynamic model for either a robot or a digital human, we must first compute the inertial matrix W by equation (7.32), which requires preparing every 6 by n sub-Jacobian matrix. Thus, to determine the joint torque distribution τ_g by (9.23) is virtually a byproduct of the dynamic modeling process with no extra work needed. We will further discuss this issue in Chapter 11. Because the calculation of (9.23) requires only the third row of each sub-Jacobian matrix, it becomes a small subset of the digital human dynamic modeling effort;
5. Once τ_g is determined at a time instant, one needs to check its first six joint forces/torques again. They are not all zero at this time. One must apply equation (9.20), if the current posture is standing in a steady state, to find the Cartesian reaction force from the floor to the two feet in order to meet the Law of Balance, as illustrated in the previous weight-lifting examples.

As a numerical illustration, let us roughly (it may not be anthropometrically correct) assign a mass to each of the 18 selected segments as follows:

$$m_6 = 5,\ m_9 = 5,\ m_{14} = 2,\ m_{17} = 5,\ m_{19} = 4,\ m_{21} = 3,$$

$$m_{23} = 2,\ m_{26} = 5,\ m_{28} = 4,\ m_{30} = 3,\ m_{33} = 7,\ m_{35} = 6,$$

$$m_{37} = 3,\ m_{40} = 7,\ m_{42} = 6,\ m_{44} = 3,\ m_n = 4,\ m_{47} = 4,$$

and all the masses are in Kg. By adding all of them together, the total weight of the mannequin is 78 Kg. = 765.18 N. Now, using the above mass data as well as the mannequin kinematic parameters, we can program the 18 sub-Jacobian matrices based on (9.22) plus equation (9.23) into MATLAB^{TM} to create a function subroutine. Of course, all the one-step homogeneous transformation matrices and their products must be written into the program, too, before formulating the sub-Jacobian matrices in the subroutine. The input of this subroutine is the 47 joint positions in q vector. Therefore, it will print out a bar chart of the joint torque distribution due to gravity for any given posture in terms of q. The program is, indeed, quite lengthy, but is worthwhile for a permanent use. Let us look at the following two examples of calling this function subroutine.

First, for a mannequin standing in a neutral posture, as shown in Figure 9.22, the joint torque distribution for all the 47 joints, including the head/neck, due to gravity can immediately be computed by the function subroutine once the joint position vector q is entered. The output bar chart is plotted in Figure 9.23, which clearly shows the exact amount of the weight force equal to 765.18 N. acted on joint 3 that is the global body up-down translation, while the other joints are exerted only with very minor torques. By taking a closer look, among all the joints except joint 3, only the two joints at both the left and right clavicles may have a little higher torques than the others to overcome the weight of the two arms, as expected.

Secondly, if we now input the joint position vector q in the weight-lift standing posture that is same as Figure 9.12 but without holding any load into the function subroutine, the joint torque distribution only due to gravity for all the 47 joints is plotted in Figure 9.24. Likewise, we can follow the same procedure to calculate the Cartesian reaction forces/moments from the floor to the two feet as follows:

```
>> F_2f(b) = -pinv(J_2f(b)') * tau_g(1:6)

F_2f(b) =

    0.0004    0.0033  385.1235   51.8619   -7.4913    0.0005
   -0.0004   -0.0033  380.0565   51.8619   -7.4913    0.0005

>>
```

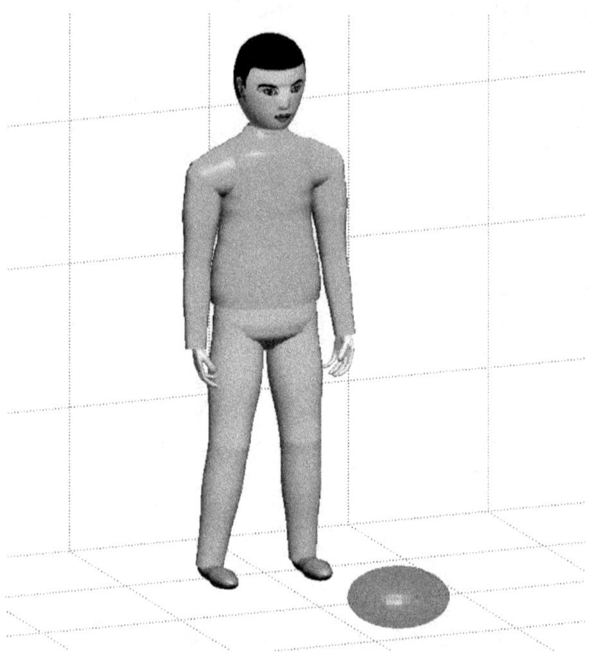

Fig. 9.22 A mannequin is in neutral standing posture and ready to pick an object

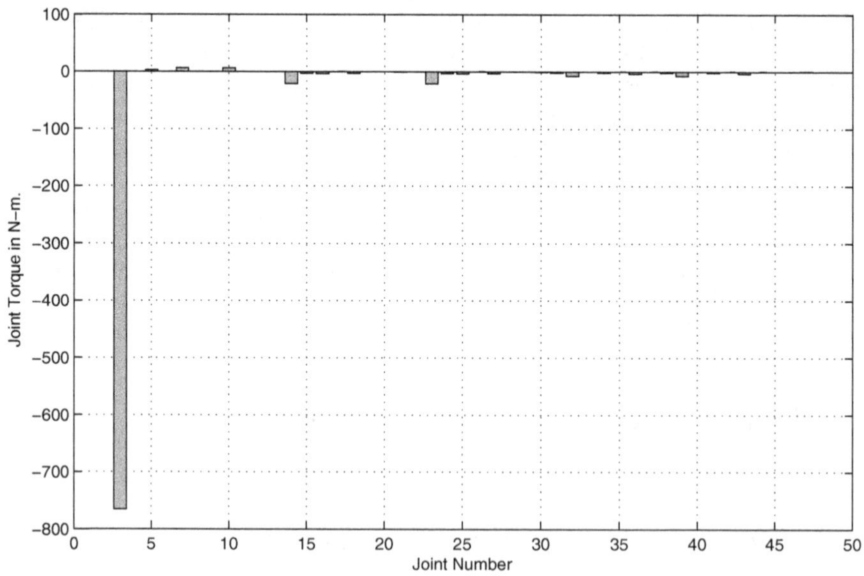

Fig. 9.23 A 47-joint torque distribution due to gravity in neutral standing posture

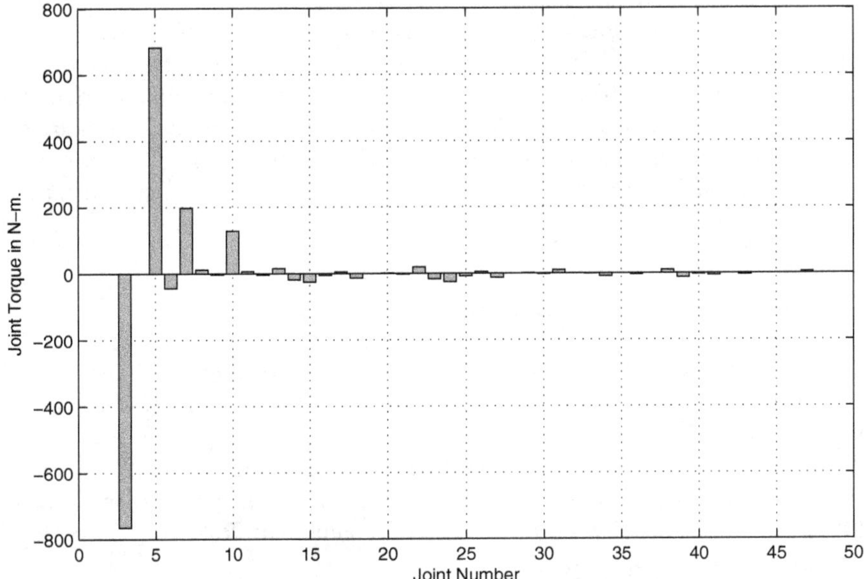

Fig. 9.24 A 47-joint torque distribution due to gravity in standing posture before the balance

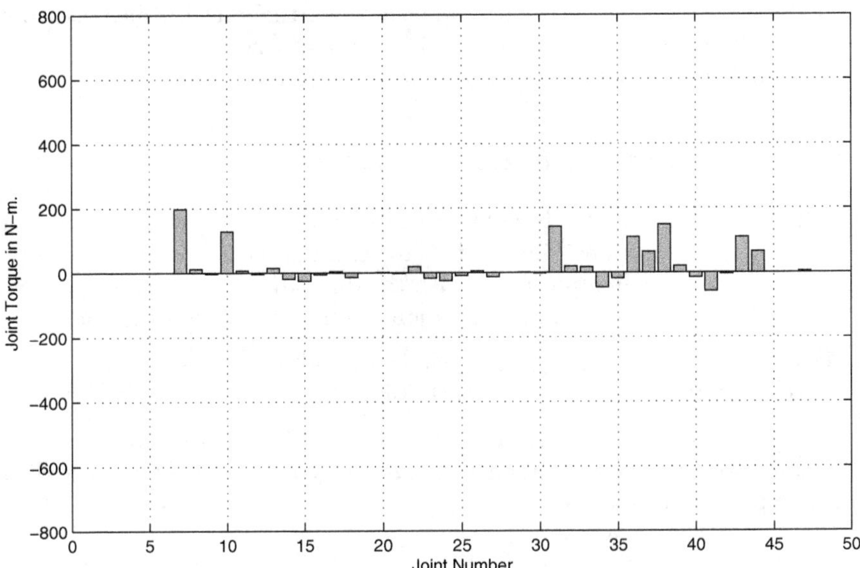

Fig. 9.25 A 47-joint torque distribution due to gravity after balancing the reaction forces

After taking care of the balance in the second standing case, the final joint torque distribution is shown in Figure 9.25. It can be clearly observed that while the first six joint forces/torques are balanced, the two legs from joint 31 to joint 44 will be responsible of exerting higher joint torques for a total balance against the reaction. Here we purposely make the same scale of the vertical joint torque axis in both Figures 9.24 and 9.25 in order to easily compare to each other between the two bar charts before and after the balance.

9.5 Posture Optimization Criteria

Human body is a typical open serial/parallel hybrid-chain structure with high degree of redundancy at the kinematics point of view. If one wishes to generate a realistic digital human motion, in addition to determining a set of motion trajectory-patterns, a certain number of criteria and/or objective functions must be developed for an overall or global posture optimization [20]–[24]. More specifically, it is desired to have the following three fundamental criteria for optimization:

1. Make each joint angle be falling into its comfort zone, or as close to its comfort center as possible;
2. In operating a material-handling task, or a force related task, the entire body of a digital human is desired to have all the joint torques distributed as evenly or uniformly as possible in each movement;
3. Control each dynamic movement with a minimum effort.

9.5.1 The Joint Comfort Criterion

The first criterion requires to define a scalar potential function in joint space, as discussed in equations (5.9) and (5.10) of Chapter 5, which can be monotonically decreasing in association with the differential motion equation given in (5.8). The key steps to create this joint comfort criterion include

1. Define a comfort center q_i^c for each joint; and
2. Define a certain weight w_i for each joint.

In other words, the criterion from Chapter 5 is now extended from a redundant robot to digital human application by the following weighted joint comfort potential function:

$$p(q) = \frac{1}{2}(q - q^c)^T W (q - q^c) = \frac{1}{2}\sum_{i=1}^{n} w_i (q_i - q_i^c)^2, \tag{9.24}$$

where the n by n weighting matrix W is diagonal with each diagonal element w_i as a weight on the corresponding joint position q_i. Because this potential function is defined in a joint-by-joint form, its gradient (column) vector becomes very simple:

$$\eta = W(q - q^c) = \begin{pmatrix} w_1(q_1 - q_1^c) \\ \vdots \\ w_n(q_n - q_n^c) \end{pmatrix}.$$

When it is implemented into equation (9.15), the gain constant $k < 0$ in order to drive the value of $p(q)$ in (9.24) towards the minimum.

Since every joint in a digital human kinematic model created by the D-H convention has "equal opportunity" to each other at a mathematical standpoint, the joint comfort potential function for a digital human model must be imposed by an individual weight on each joint to either enhance or weaken the chance and range of the movement in order to match with its range of motion (ROM).

For instance, the two joints on each side of the clavicles may have to be heavily weighted in order to reduce their relative chance and range of rotations. In contrast, the first three body translation joints should be zero-weighted to make them widely open for free motion. Furthermore, in order to avoid singularity, both the elbow on each arm and knee on each leg may need a little more weights w_i, and their comfort centers q_i^c's should be defined away from their zero home positions accordingly. Because such a weighted joint comfort potential function defined in (9.24) will be very effective to adjust each digital human instantaneous posture after its gradient η is implemented into equation (9.15), we should be more careful to specify and fine-tune each weight w_i and comfort center q_i^c on joint i for $i = 1, \cdots, n$.

Regarding the determination of each joint comfort center q_i^c, equation (9.16) may offer a suggestion that is referred to the tables of ROM and comfort zone in the preceding section. Since that equation as well as the ROM tables fully rely on the experimental measurements and statistics, each suggestion of the comfort center q_i^c or θ_i^c from (9.16) may need an additional adjustment. It is a fact that the ROM and comfort zone for each joint are, indeed, very complicated and diversified, and also subject to person by person and individual health condition. Therefore, it will have no clear winning solution to the comfort center determination.

9.5.2 The Criterion of Even Joint Torque Distribution

The second optimization objective is quite significant and effective for a digital human to operate a material handling or force related task/activity [23]. As we have introduced and justified equation (5.12) in Chapter 5, the following scalar function:

$$p(q) = \mathrm{tr}(JWJ^T) \tag{9.25}$$

is a meaningful and also effective potential function to make the digital human weighted joint torque distribution be as even, or as uniform as possible. Even if there is no weighting matrix W sandwiched by the Jacobian matrix J and J^T, i.e.,

$$p(q) = \mathrm{tr}(JJ^T), \tag{9.26}$$

such a potential function can still play an effective role in distributing every joint torque as evenly as possible.

Because the global Jacobian matrix J of a digital human model can, not only uniquely represent an instantaneous posture in the global sense, but also play a key role in solving both differential motion and statics equations as a unique transformation between Cartesian and joint spaces, the best posture for a mannequin, intuitively, should be directly related to the Jacobian matrix. The even joint torque distribution potential function defined in (9.25) or (9.26) just reflects this intuition.

Let us now revisit the mathematical justification plus equation (5.11) from Chapter 5 again. It is well-known that for n positive real numbers: $a_1, \cdots, a_n$ with each $a_i > 0$, their arithmetic mean value is always greater than or equal to their geometric mean value, i.e.,

$$\frac{a_1 + \cdots + a_n}{n} \geq \sqrt[n]{a_1 \cdots a_n},$$

and it becomes equal if and only if all the positive numbers a_i's are equal to each other, $a_1 = \cdots = a_n$.

Now, let a weighted joint torque norm square for a digital human body be defined by

$$\tau^T W \tau = w_1 \tau_1^2 + \cdots + w_n \tau_n^2,$$

where the joint torque vector is $\tau \in \mathbb{R}^n$ and the weighting matrix W is an n by n diagonal matrix with each diagonal element $w_i > 0$. Therefore, it is true that

$$\frac{\tau^T W \tau}{n} \geq \sqrt[n]{w_1 \tau_1^2 \cdots w_n \tau_n^2},$$

and likewise, both sides are equal if and only if the weighted joint torques are uniformly distributed, i.e.,

$$w_1 \tau_1^2 = \cdots = w_n \tau_n^2.$$

On the other hand, based on the statics, the joint torque vector $\tau = J^T F$, where J is the global Jacobian matrix of the digital human model, and $F \in \mathbb{R}^m$ is a Cartesian force vector representing a load imposed on the digital human at one or more end-effectors. Then,

$$\tau^T W \tau = F^T J W J^T F.$$

This means that the weighted joint torque norm square is dependent of the external Cartesian forces/moments.

However, based on the Rayleigh Quotient Theorem from linear algebra,

$$\lambda_{min} \leq \frac{F^T J W J^T F}{F^T F} \leq \lambda_{max},$$

where λ_{min} and λ_{max} are the minimum and maximum eigenvalues of the positive-definite matrix JWJ^T for any vector F. Without loss of generality, let the load Cartesian force vector be normalized, i.e., $F^T F = 1$ to test the joint torque distribution, then the above equation is reduced to

$$\lambda_{min} \leq F^T J W J^T F = \tau^T W \tau \leq \lambda_{max}.$$

This implies that the weighted joint torque norm square is always upper-bounded by λ_{max} and lower-bounded by λ_{min} of the m by m positive-definite weighted Jacobian matrix JWJ^T, no matter what the unity external Cartesian force F is.

Now, because each eigenvalue λ_i for the positive-definite weighted Jacobian matrix JWJ^T is a positive number as long as J is full-ranked, they also obey

$$\frac{\lambda_1 + \cdots + \lambda_m}{m} \geq \sqrt[m]{\lambda_1 \cdots \lambda_m},$$

and it becomes equal if and only if all the m eigenvalues are equal to each other, i.e., $\lambda_1 = \cdots = \lambda_m$. Therefore, the even distribution of eigenvalues of JWJ^T seen in Cartesian space is equivalent to the even distribution of the weighted joint torques seen in joint space.

Furthermore, according to linear algebra,

$$\det(JWJ^T) = \lambda_1 \cdots \lambda_m \quad \text{and} \quad \text{tr}(JWJ^T) = \lambda_1 + \cdots + \lambda_m,$$

so that

$$\frac{\text{tr}(JWJ^T)}{m} \geq \sqrt[m]{\det(JWJ^T)}.$$

Therefore, minimizing $\text{tr}(JWJ^T)$ can squeeze all the eigenvalues of JWJ^T to be equal to each other, and thus, every weighted joint torque will approach to being evenly distributed as well. This is the rationality why (9.25) or (9.26) can be defined as a potential function to be minimized for an even joint torque distribution.

Also, it is quite interesting that $\det(JWJ^T)$ is the (weighted) manipulability, while $\text{tr}(JWJ^T)$ is the potential function for even joint torque distribution. If every eigenvalue could be equal to each other, i.e., $\lambda_1 = \cdots = \lambda_m$, then, the potential function for even joint torque distribution would reach the minimum, and at the same time, the manipulability would reach the maximum. Therefore, an instantaneous posture with all the eigenvalues of JWJ^T equal to each other is obviously the best and most ideal posture for a digital human. However, such an ideal posture may never be reachable in the reality, but at least it can be approached to as closely as possible.

In addition, because for any two matrices A and B in dimensions m by n and n by m, respectively, $\mathrm{tr}(AB) = \mathrm{tr}(BA)$ based on linear algebra, the Jacobian matrix J and J^T in (9.25) or (9.26) can be projected onto any coordinate frame without affecting the trace value. In other words, the trace of JJ^T or JWJ^T is invariant of the Jacobian matrix projection.

It can be easily justified that due to $J_{(k)} = R^j_k J_{(j)}$ with an augmented rotation matrix R^j_k in matching the dimension of $J_{(j)}$, as similar to equation (4.15) in Chapter 4,

$$\mathrm{tr}(J_{(k)} J^T_{(k)}) = \mathrm{tr}(R^j_k J_{(j)} J^T_{(j)} R^k_j) = \mathrm{tr}(J_{(j)} J^T_{(j)} R^k_j R^j_k) = \mathrm{tr}(J_{(j)} J^T_{(j)}).$$

This invariant property will bring some convenience to our future algorithm development and programming.

Since we wish to minimize $\mathrm{tr}(JJ^T)$, the gain constant k must be negative, i.e., $k < 0$, whenever equation (5.8) from Chapter 5 or equation (9.15) is being implemented along with the gradient vector

$$\eta = \frac{\partial p(q)}{\partial q} = \frac{\partial \, \mathrm{tr}(JJ^T)}{\partial q}. \tag{9.27}$$

However, to take partial derivatives for $\mathrm{tr}(JJ^T)$ symbolically is extremely difficult and tedious, and even impossible by hand, because the global Jacobian matrix J for a digital human is too large to perform symbolical derivations.

On the other hand, *the accuracy of the gradient vector computation is critical for the effectiveness of posture optimization.* One can use the first-order approximation via a one-step memory to simplify its numerical computation algorithm. Namely, if $q(k)$ and $q(k-1)$ are the current and the last (one-step before) joint angle vectors, respectively, and also $\mathrm{tr}(J(k)J(k)^T) = p(k)$ and $\mathrm{tr}(J(k-1)J(k-1)^T) = p(k-1)$ are the current and the last trace values of JJ^T, respectively, then the i-th element of the first-order approximated gradient vector can be written as

$$\eta_i \approx \frac{p(k) - p(k-1)}{q_i(k) - q_i(k-1)} \quad \text{for } i = 1, \cdots, n.$$

Of course, this first-order approximation for the gradient vector is low quality, which may degrade the effectiveness of posture optimization.

In fact, MATLAB^{TM} offers a number of built-in optimization functions for numerical solutions. One of the most typical functions is

fmincon(fun,x0,A,b,Aeq,beq,lb,ub,nonlcon,options)

to solve the minimum of a problem specified by

$$\min_x f(x) \quad \text{such that} \quad \begin{cases} Ax \le b \\ A_{eq}x = b_{eq} \\ lb \le x \le ub \\ c(x) \le 0 \\ c_{eq}(x) = 0. \end{cases}$$

This internal function of constrained minimization can be employed for a digital human in global Cartesian motion planning and generation. However, the function program is often asking the user to specify a gradient formula for the objective function $f(x)$. Otherwise, fmincon will simply estimate the gradient using a finite difference, like the first-order approximation, as a default. Therefore, no matter which approach is adopted between the differential motion and Cartesian motion generations, all such constrained optimization algorithms commonly require to know the gradient vector of an objective function, and *the quality of the gradient vector determines how effective and successful the optimization algorithm will be.*

To accurately find the gradient vector η for the even joint torque distribution potential function, we recommend here the following two approaches to taking derivatives for the trace of the very large global Jacobian matrix as numerical solutions:

1. As we can recall, when formulating a Jacobian matrix, each column requires a cross product: $p_n^{i-1} \times r_n^{i-1}$, and each p_n^{i-1} and each r_n^{i-1} are the 4th and 3rd columns of the homogeneous transformation matrix A_n^{i-1}, respectively. Since such a homogeneous transformation matrix is also a product of many factors:

$$A_n^{i-1} = A_n^{n-1} A_{n-1}^{n-2} \cdots A_j^{j-1} \cdots A_i^{i-1},$$

each factor, such as A_j^{j-1}, is the inverse of the one-step homogeneous transformation A_{j-1}^{j} that is the only factor depending on θ_j and all the other factors are independent of θ_j. Therefore, one can easily find the derivative of A_n^{i-1} with respect to θ_j as follows:

$$\frac{\partial A_n^{i-1}}{\partial \theta_j} = \begin{cases} A_n^{n-1} \cdots \frac{\partial A_j^{j-1}}{\partial \theta_j} \cdots A_i^{i-1} & \text{for } i \le j \le n \\ 0 \quad \text{if } 0 < j < i. \end{cases} \tag{9.28}$$

On the other hand, taking a derivative for each one-step homogeneous transformation and its inverse symbolically is an easy job. For instance, the inverse of the body fall-aside homogeneous transformation A_5^6 can be derived based on equation (3.5) from Chapter 3 by

$$A_6^5 = \begin{pmatrix} c_6 & 0 & s_6 & a_6 c_6 \\ s_6 & 0 & -c_6 & a_6 s_6 \\ 0 & 1 & 0 & 0 \\ 0 & 0 & 0 & 1 \end{pmatrix}^{-1} = \begin{pmatrix} c_6 & s_6 & 0 & -a_6 \\ 0 & 0 & 1 & 0 \\ s_6 & -c_6 & 0 & 0 \\ 0 & 0 & 0 & 1 \end{pmatrix}$$

so that

$$\frac{\partial A_6^5}{\partial \theta_6} = \begin{pmatrix} -s_6 & c_6 & 0 & 0 \\ 0 & 0 & 0 & 0 \\ c_6 & s_6 & 0 & 0 \\ 0 & 0 & 0 & 0 \end{pmatrix}.$$

If we now replace every A_6^5 involved as a factor wherever in the Jacobian matrix J by the above derivative, then the trace of the resultant matrix times J^T is half of the 6th element η_6 of the gradient vector (9.27). More precisely, the j-th element of η is calculated by

$$\eta_j = \frac{\partial \mathrm{tr}(JJ^T)}{\partial \theta_j} = \mathrm{tr}\left(\frac{\partial J}{\partial \theta_j} J^T\right) + \mathrm{tr}\left(J \frac{\partial J^T}{\partial \theta_j}\right) = 2\,\mathrm{tr}\left(\frac{\partial J}{\partial \theta_j} J^T\right).$$

Although the algorithm is quite lengthy, the resultant numerical gradient vector η is exactly accurate.

2. Using the dual-number calculus, let the i-th joint angle θ_i appeared in the Jacobian matrix J be replaced by its dual angle $\hat{\theta}_i = \theta_i + \epsilon 1$ and no more replacements for the joint angles other than θ_i. Then, the dual part of the trace result is the i-th element η_i of the gradient vector (9.27). If MATLAB^{TM} would build-in such a dual-number calculus, it would significantly simplify the algorithm to find an accurate gradient vector numerically, see also equation (2.19) and its discussion in Chapter 2.

Moreover, if one wishes to adopt the potential function (9.25) by inserting a weighting matrix W between J and J^T, then each weight element w_i for joint i is supposed to be determined based on the ROS (range of strength) of joint i, as introduced in the preceding section. In general, the smaller the ROS maximum value is, the higher weight the joint should be assigned to. However, only a few major joints over a real human body have been studied and reported with their ROS data thus far in the biomechanics literature, and thus, it lacks a basis to define every weight reasonably for all the 47 joints of the digital human model.

Let us now verify the effectiveness of (9.26) without a weighting matrix W by the following two simulation case studies. The first one is the same as what we adopted for a standing posture joint torque distribution analysis in Figure 9.12. This posture is actually a mid-way as the mannequin started from a neutral standing posture, as shown in Figure 9.22, and ready to bow the trunk down and pick up the load.

We now compare two different resulting postures between the cases without the potential function of even joint torque distribution given in (9.26) and with such potential function. However, the main tasks of the mannequin in both cases are exactly the same as to pick up the load and then move it to a destination. The mid-way postures of carrying the load in the two cases are depicted in Figure 9.26, and the values of $\mathrm{tr}(JJ^T)$ can be immediately evaluated as $\mathrm{tr}(J_bJ_b^T) = 101.3124$ in the case without optimization versus $\mathrm{tr}(J_bJ_b^T) = 96.7105$ with optimization.

Furthermore, if the load is defined to have a weight of 50 N. and each hand shares 25 N. force down as a symmetric material-handling task, a double-bar chart of the joint torque distribution over the 44 joints (excluding the neck/head) in only considering the weight-lift after balancing the floor-feet reaction is plotted in Figure 9.27. Also, a double-bar chart for all the 47 joint torques in considering both the weight-lift and gravity is shown in Figure 9.28.

In both two double-bar charts, each dark blue bar is the joint torque without optimization, while the light green bar represents the joint torque with optimization. It is evident enough that after the even joint torque distribution potential function is implemented, every higher joint torque is pressed down, while some lower ones are raised up to a little bit higher torques to automatically compensate with each other towards an overall uniform distribution. Since every reaction force balance between the floor and feet is considered, the first six joint torques all vanish in both the two double-bar charts, but the two legs have to be responsible to exert more joint torques for the balance.

The second example is in the moment of nearly final posture for the mannequin carrying the same load and placing it on an overhead shelf, also in the two cases without and with optimization, as shown in Figure 9.29. At this time instant, the value of the Jacobian trace is calculated to be $\mathrm{tr}(J_bJ_b^T) = 109.2005$ in the case without optimization and $\mathrm{tr}(J_bJ_b^T) = 105.3465$ in the case with optimization. Likewise, their double-bar charts are plotted in Figure 9.30 for only considering the weight-lift and in Figure 9.31 for considering

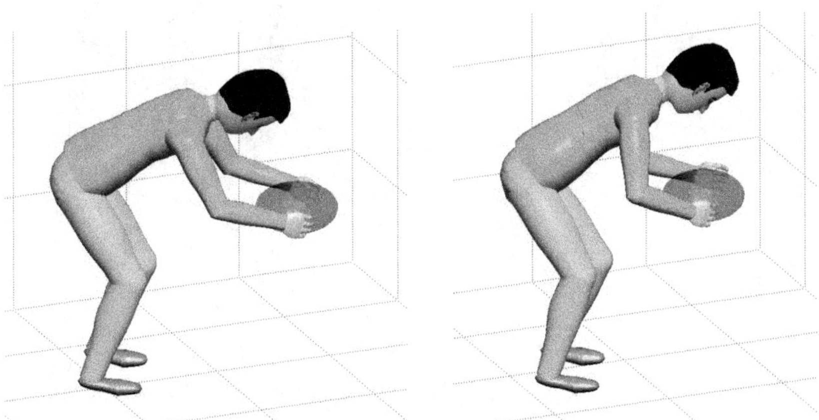

Fig. 9.26 Mannequin postures in picking up a load without and with optimization

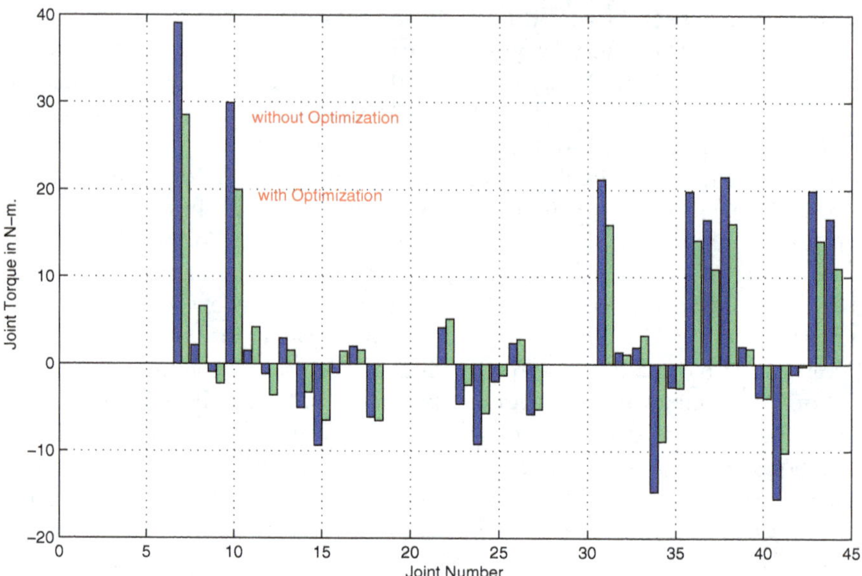

Fig. 9.27 A joint torque distribution due to weight-lift without and with optimization

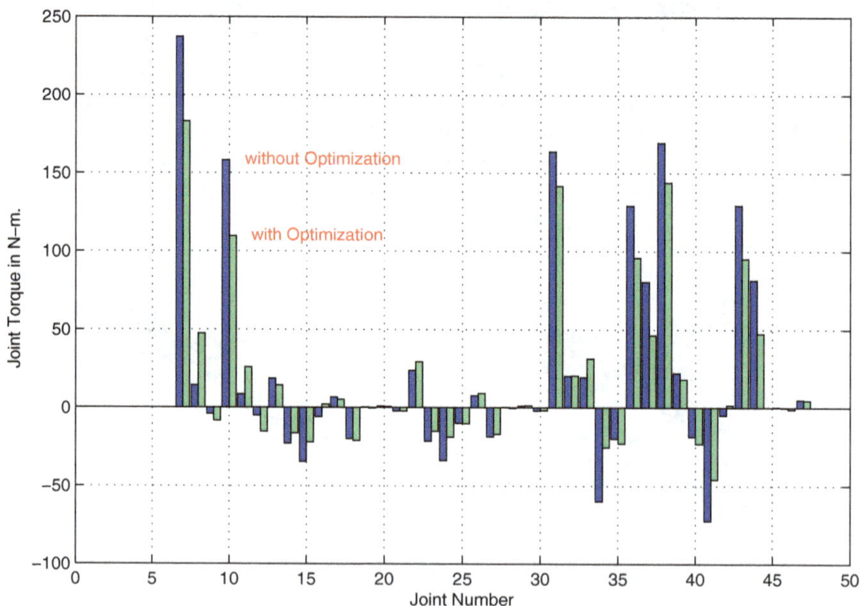

Fig. 9.28 A complete joint torque distribution with and without optimization

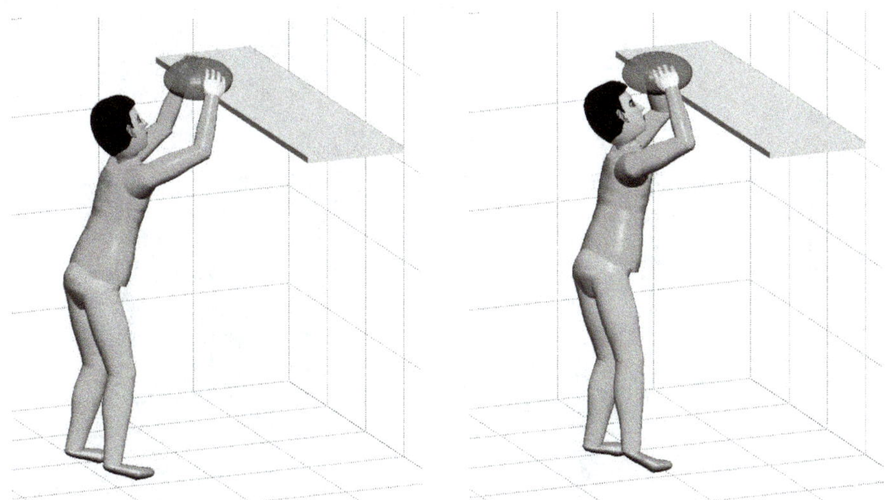

Fig. 9.29 The mannequin postures in placing a load on the overhead shelf without and with optimization

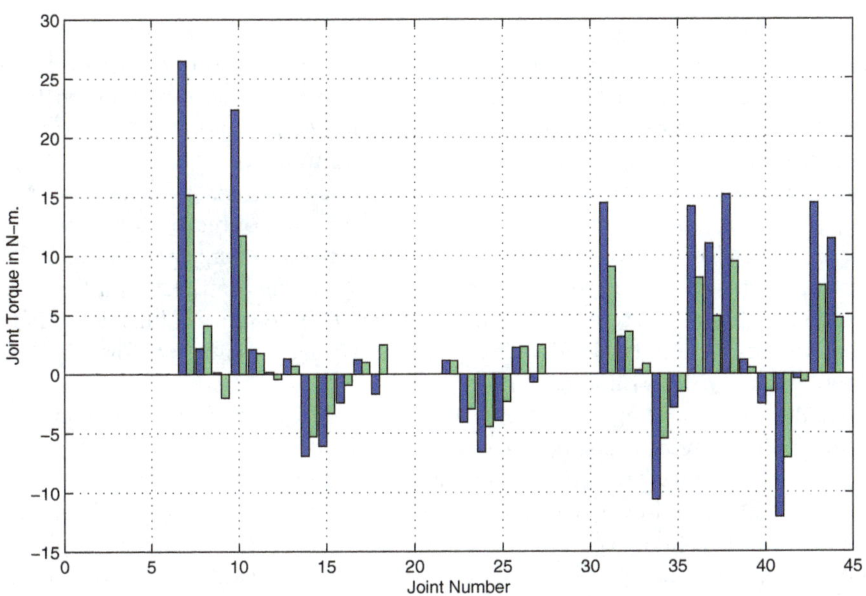

Fig. 9.30 A joint torque distribution in placing a load with and without optimization

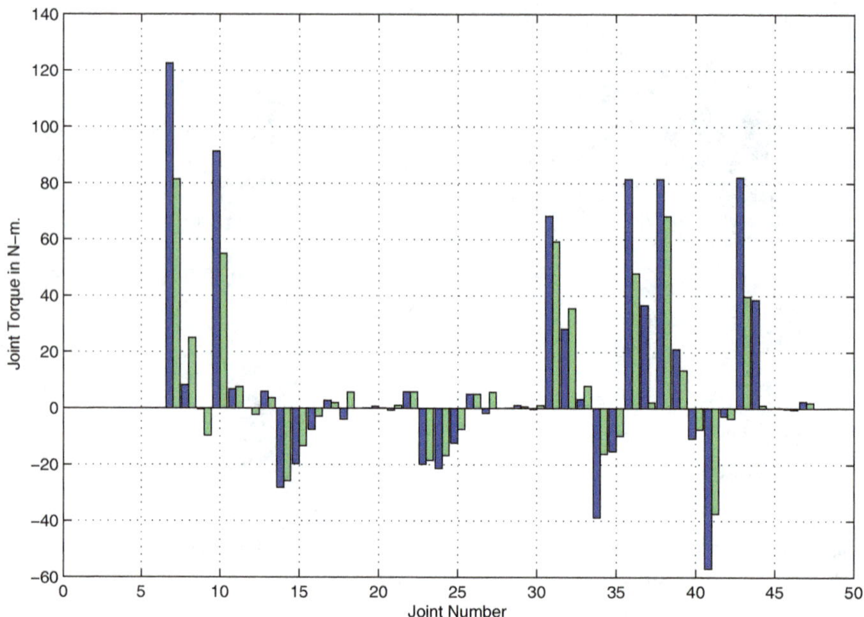

Fig. 9.31 A complete joint torque distribution with and without optimization

both the weight-lift and gravity. They all showcase the effect of even joint torque distribution after the potential function (9.26) is implemented by using the first accurate approach, as given in (9.28), to numerically determine the gradient vector η in MATLAB^{TM} programming.

If we can appropriately define a weighting matrix W and adopt the weighted potential function in equation (9.25), the double-bar chart will look better and more uniform effect. However, it may not improve very much the optimization effectiveness. Instead, more critical factor is the quality of the gradient vector. Therefore, we may keep using the potential function without the weighting matrix W given in (9.26) until each weight w_i for joint i can be reasonably determined.

In summary, the optimization criterion of even joint torque distribution is, indeed, quite significant and meaningful both conceptually and practically. A worker gets injury often due to an excessive load that hurts one or few individual joint(s), not everywhere over the human body. If the worker is handling a load that can be distributed evenly over his/her entire body, the injury can be avoided even if the task is exactly the same. How can one control the joint torque distribution uniformly? This is just a posture optimization issue. In other words, a human worker can do a material-handling work without injury if he/she is in a good posture. With the same load, he/she will be locally injury if the posture is awkward. In most practical cases, a real human can always adjust his/her posture to adapt a given task automatically in nature.

However, how can a digital human model do the same? Therefore, the potential function (9.25) or (9.26) along with its gradient vector (9.27) plays a vital role in automatically adjusting the posture towards its optimization in a digital environment. More significantly, the potential function defined in either (9.25) or (9.26) is independent of the external Cartesian force and solely depends on the global Jacobian matrix that is the unique representation of digital human instantaneous postures.

9.5.3 On the Minimum Effort Objective

Regarding the attempt of defining a potential function for minimum effort, we have to comment that it is beyond the scope of kinematics and statics. Since effort is a measure of cumulative energy, such a targeting potential function may have to be developed in the modeling phase of digital human dynamics [24]. One of the greatest mathematician Lagrange defined a scalar function that is known as the Lagrangian $L(q, \dot{q}) = K(q, \dot{q}) - P(q)$, where K is the kinetic energy and P is the potential energy for a mechanical system. An effort was further defined as a cumulative amount of the Lagrangian between time instants t_1 and t_2, i.e.,

$$I = \int_{t_1}^{t_2} L(q(t), \dot{q}(t))dt.$$

In order to minimize such an effort, by applying the calculus of variation, a general solution obeys the well-known Lagrange equation:

$$\frac{d}{dt}\left(\frac{\partial L}{\partial \dot{q}}\right) - \frac{\partial L}{\partial q} = 0,$$

see also the same formulation in (7.14) along with the discussion in Chapter 7.

The solution of this homogeneous force-free Lagrange equation is a trajectory $q(t)$ with the minimum effort. The Lagrange equation is also exactly consistent to the *geodesic equation* at the differential geometry point of view in mathematics, the solution of which is a trajectory with the shortest arc length, see equation (7.20) and the discussion in Chapter 7. Therefore, in order to make every movement for a digital human meet the minimum effort objective, the entire body must be governed by the Lagrange equation, not just by the inverse kinematics. In other words, the digital human dynamics has to be involved in every motion generation. It is getting more recognizable that only the kinematics will not be sufficient enough to create a realistic digital human motion, unless a sophisticated motion capture is applied. More detailed dynamic modeling and motion generation will be discussed in Chapter 11.

9.6 Exercises of the Chapter

1. Following the procedure started from equation (9.2) for the I-K of the left arm, develop an I-K symbolical algorithm for the left leg with its homogeneous transformation A_{30}^{37ll}, i.e., from frame 30 as a temporary base to the left toe frame $37ll$. It also has totally 7 joints and you may fix the left hip medial/lateral joint $\theta_{33} = 90^0$ at the home position. Namely, find θ_{31} through θ_{37} except θ_{33} in terms of a given position and orientation of frame $37ll$ with respect to frame 30, i.e., p_{30}^{37ll} and R_{30}^{37ll}. After the symbolical I-K algorithm is derived, verify it by a numerical I-K solution through the internal function fsolve$(\cdots)$ in MATLAB^{TM}.
2. Write a MATLAB^{TM} program to determine a 6 by 7 local Jacobian matrix J_{ll} for the left leg with seven joints numerically. Then, take away the 3rd column of J_{ll} due to the fixed joint angle $\theta_{33} = 90^0$ to make a 6 by 6 square Jacobian J_{ll}^s. By comparing the differential motion-based I-K algorithm $\dot{q} = (J_{ll}^s)^{-1}V$ with the above Cartesian motion-based I-K solution, discuss the common results and their differences.
3. Following the D-H tables and coordinate frames assignment, as shown in Figure 9.4, try to create a 3D skeletal mannequin in MATLAB^{TM} and using equation (9.13) to program and update the global Jacobian matrix $J_{(b)}$ in each sampling interval. Then, generating simple trajectories for each of the four points: two hands and two feet, and taking time-derivatives in the first-order approximation to determine the trajectory of each velocity ready for updating the joint values by the minimum-norm I-K solution $\dot{q} = J_{(b)}^+ V$. When the four sets of trajectories are generated, double check and remove any potential singularity, such as the distance between any one of the two hands and one of the two feet would never reach beyond the normal range at any time instant. If successful, you will see the skeletal mannequin moving.
4. A short matrix A of dimension m by n with $m < n$ is said to be α-orthogonal if each row a_i of A for $i = 1, \cdots, m$ satisfies

$$a_i a_j^T = \begin{cases} 0 & \text{if } i \neq j \\ \alpha > 0 & \text{if } i = j. \end{cases}$$

a. Prove that if A is a short α-orthogonal matrix, then,

$$\frac{\mathrm{tr}(AA^T)}{m} = \sqrt[m]{\det(AA^T)} = \alpha;$$

b. Also, justify that if A is not α-orthogonal, then,

$$\frac{\mathrm{tr}(AA^T)}{m} > \sqrt[m]{\det(AA^T)}.$$

References

1. Badler, N., Phillips, C., Webber, B.: Simulating Humans: Computer Graphics Animation and Control. Oxford University Press, New York (1993)
2. Chaffin, D.: On Simulating Human Reach Motions for Ergonomics Analysis. Human Factors and Ergonomics in Manufacturing 12(3), 235–247 (2002)
3. Chaffin, D.: Digital Human Modeling for Workspace Design. In: Reviews of Human Factors and Ergonomics, vol. 4, p. 41. Sage (2008), doi:10.1518/155723408X342844
4. Moes, N.: Digital Human Models: An Overview of Development and Applications in Product and Workplace Design. In: Proceedings of Tools and Methods of Competitive Engineering (TMCE) 2010 Symposium, Ancona, Italy, April 12-16, pp. 73–84 (2010)
5. Duffy, V. (ed.): Handbook of Digital Human Modeling, Research for Applied Ergonomics and Human factors Engineering. CRC Press (2008)
6. Abdel-Malek, K., et al.: Santos: A Digital Human In the Making. In: IASTED International Conference on Applied Simulation and Modeling, Corfu, Greece, ADA542025 (June 2008)
7. Abdel-Malek, K., Yang, J., et al.: Towards a New Generation of Virtual Humans. International Journal of Human Factors Modelling and Simulation 1(1), 2–39 (2006)
8. Abdel-Malek, K., et al.: Santos: a Physics-Based Digital Human Simulation Environment. In: The 50th Annual Meeting of the Human Factors and Ergonomics Society, San Francisco, CA (October 2006)
9. Yang, J., et al.: Santos^{TM}: A New Generation of Virtual Humans. In: 2005 SAE World Congress, Detroit, Michigan, USA (2005)
10. Hamilton, N., Luttgens, K.: Kinesiology Scientific Basis of Human Motion, 10th edn. McGraw Hill (2002)
11. Watkins, J.: Structure and Function of the Musculoskeletal System. Human Kinetics (1999)
12. Chaffin, D., Andersson, G.: Occupational Biomechanics, 2nd edn. John Wiley & Sons (1991)
13. Adrian, M., Cooper, J.: Biomechanics of Human Movement. WCB McGraw-Hill, Boston (1989)
14. Robertson, D., Caldwell, G., Hamill, J., Kamen, G., Whittlesey, S.: Research Methods in Biomechanics. Human Kinetics (2004)
15. Chapman, A.: Biomechanical Analysis of Fundamental Human Movements. Human Kinetics (2008)
16. Griffiths, I.: Principles of Biomechanics and Motion Analysis. Lippincott Williams & Wilkins (2006)
17. Wilson, A.: Effective Management of Musculoskeletal Injury, A Clinical Ergonomics Approach to Prevention, Treatment, and Rehab. Churchill Livingstone (2001)
18. Mital, N., Ayoub, A.: Guide to Manual Material Handling. Taylor & Francis, London (1993)
19. Snook, S., Ciriello, V.: The Design of Manual Handling Tasks: Revised Tables of Maximum Acceptable Weights and Forces. Ergonomics 34(9), 1197–1213 (1991)

20. Kim, H., Horn, E., Arora, J., Abdel-Malek, K.: An Optimization-Based Methodology to Predict Digital Human Gait Motion. In: Proceedings of 2005 SAE Digital Human Modeling for Design and Engineering, Iowa City, IA, USA, June 14-16 (2005)
21. Kim, J., Abdel-Malek, K., Yang, J., Marler, T.: Prediction and Analysis of Human Motion Dynamics Performing Various Tasks. International Journal of Human Factors Modeling and Simulation 1(1), 69–94 (2006)
22. Yang, J., Pitarch, E., Kim, J., Abdel-Malek, K.: Posture Prediction and Force/Torque Analysis for Human Hands. In: Proceedings of 2006 SAE Digital Human Modeling for Design and Engineering, Lyon, France, July 4-6 (2006)
23. Gu, E., Teng, Y., Oriet, L.: A Study of Human Joint-Torque Distribution for Optimal Posture Auto-Generation. In: Proceedings of the 2002 SAE International Conference on Digital Human Modeling for Design and Engineering, Munich, Germany, pp. 421–430 (June 2002)
24. Gu, E., Teng, Y., Oriet, L.: A Minimum-Effort Motion Algorithm for Digital Human Models. In: Proceedings of the 2003 SAE International Conference on Digital Human Modeling for Design and Engineering, Montreal, Canada, June 17-20, Paper number: 2003-01-2228 (2003)

Chapter 10
Digital Human Modeling: 3D Mock-Up and Motion Generation

10.1 Create a Mannequin in MATLAB^{TM}

In Chapter 6, we have extensively studied how to draw a robot arm in MATLAB^{TM} and how to plan a motion and then animate it in a computer. To create a digital mannequin, the major procedures of generating every part and assembling them together are nearly the same as those for a robot drawing in principle. However, since a digital human consists of much more joints and segments along with more complex body surfaces, it will definitely be a more challenging job than creating a robot arm.

By browsing the existing major commercial digital human modeling and simulation packages, it is quite noticeable that most of their mannequins have made a significant breakthrough in terms of the realistic appearance. One of the most representative efforts is the virtual soldier Santos that was initiated and developed by the University of Iowa [1]–[4], as well as the Jack, RAMSIS, Safework and HUMOS mannequins. While the achievement of realistic looking has widely been recognized and appreciated by the users, modeling and generating digital human realistic motions and their environmental interactions may still be a number of pressing challenges in front of the digital human modeling research community to expect a further substantial improvement [5]–[8].

In this chapter, we intend to introduce a basic modeling procedure to mock up a digital human and drive it for motion in MATLAB^{TM}. The appearance is not our major focus and concern. Instead, starting from the basic modeling procedure, the reader may be able to spend more time and effort to microscopically sculpture or locally deform each high-resolution surface to further improve the digital mannequin creation for better looking. Let us now specifically describe how to develop a digital human model in MATLAB^{TM}.

An entire human body in appearance consists of a large number of parts to form a symmetric serial-parallel hybrid-chain structure. Basically, the major parts include a head, a neck, a hip/abdomen, a torso, and two shoulders, two arms and two hands, as well as two legs and two feet. The left and right

E.Y.L. Gu, *A Journey from Robot to Digital Human*,
Modeling and Optimization in Science and Technologies 1,
DOI: 10.1007/978-3-642-39047-0_10,

shoulders, arms, legs, hands and feet are typically symmetric as a mirror image to each other. Furthermore, since each part of a real human body is muscular-skeleton based and covered by skin, it is deformable. One can develop a deformable model for more realistic looking. However, it is only superficial and does not substantially help for generating more realistic motion. Therefore, each body part and their assembly to be introduced here for creating a digital human model are based only on non-deformable rigid body towards a rigid motion.

The first part to be created is the head of a digital mannequin. Through a fine sculpturing and texture-mapping process, the final result in MATLAB^{TM} is given in Figure 10.1.

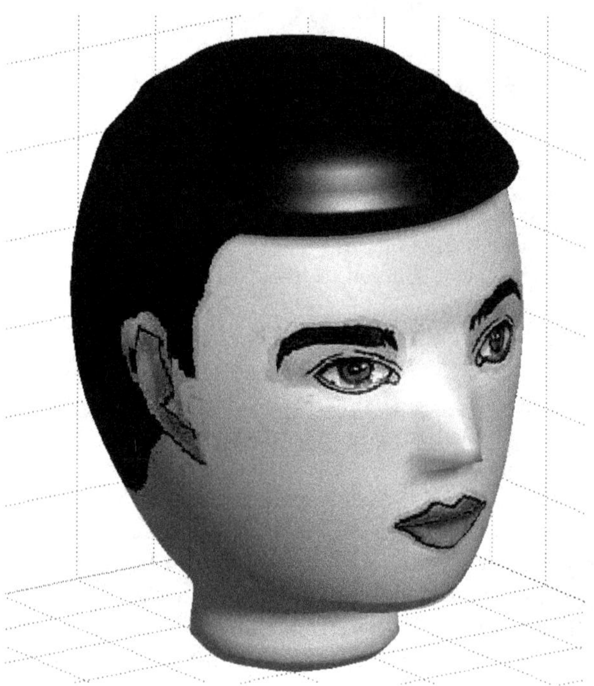

Fig. 10.1 A digital human head model

In overall, the entire head-creating process is outlined as follows:

1. Generate a high resolution cylinder;
2. Mathematically sculpture this cylindrical surface;
3. Prepare a human face picture to be texture-mapped onto the sculptured cylindrical surface.

In order to sculpture every major detail for a realistic face creation, the cylinder must be generated in a relatively higher resolution. Then, using a number of simple functions, such as a linear function, sine and cosine functions and

Fig. 10.2 A face picture for texture-mapping onto the surface of a digital human head model

their squares to mathematically sculpture the cylindrical surface for desired local deformations, such as the convex areas for nose, lips, ears and the front hair, and the concave areas for two eyes and mouth. After the necessary sculptures, we then prepare a human face picture in jpg file, as typically shown in Figure 10.2, ready to paste it on the entire sculptured cylindrical surface by a texture-mapping procedure. The MATLAB^{TM} program for the above three steps of the head creation is given as follows:

```
Head=[0 0.25 0.49 0.56 0.55 0.54 0.54 0.6 0.68 0.74 0.8 ...
      0.85 0.89 0.93 0.97 1.00 1.04 1.07 1.09 1.12 1.14 ...
      1.16 1.17 1.1852 1.1934 1.1984 1.20 1.1984 1.1934 ...
      1.1852 1.1738 1.1591 1.1413 1.1203 1.0963 1.0692 ...
      1.0392 0.9708 0.9 0.8485 0.7053 0.6 0];  % In Decimeters

rgb=imread('face.jpg');       % Retrieve the Face-Pic File

nh=size(Head);    [xh,yh,zh]=cylinder(Head,60);

     % Start a Mathematical Sculpture Process
for i=9:34
   for j=10:50
      xh(i,j)=(1+0.55*(34-i)^2/25^2*sin((j-10)*pi/40))*xh(i,j);
   end
end
for i=20:27
   for j=30:33
      xh(i,j)=(1+0.23*(27-i)/7*sin((j-30)*pi/3))*xh(i,j);
   end
end               % For Nose Bulging Out
```

```
for i=25:28
   for j=25:38
      xh(i,j)=xh(i,j)+0.04;     % Eyes In
   end
end

for i=18:28
   for j=17:19
      yh(i,j)=yh(i,j)-0.004*(j-19)*(i-18);   % Ear Out
   end
   for j=45:47
      yh(i,j)=yh(i,j)+0.004*(45-j)*(i-18);   % Ear Out
   end
end
for i=9:20
   xh(i,:)=xh(i,:)+0.01*(i-20);     % Chin Out
end

xh(8,:)=xh(8,:)-0.06;
xh(14,31:32)=xh(14,31:32)-0.04;
xh(15,31:32)=xh(15,31:32)+0.05;
xh(16,30:33)=xh(16,30:33)-0.05;     % Mouth In/Out

for i=36:42
   for j=19:41
      xh(i,j)=xh(i,j)-0.2;        % Front Hair Out
   end
end

A01=eye(4);  % Define a Homogeneous Transformation to Send
             % the Head to a Destination, Here is the Identity

x=[];  y=[];  z=[];  % Initialize the Cylinder Coordinate Arrays

for ii=1:nh(2)
   cylin=A01*[xh(ii,:); yh(ii,:); zh(ii,:)*3.5; ones(1,61)];
   x=[x;cylin(1,:)];
   y=[y;cylin(2,:)];
   z=[z;cylin(3,:)];
end

surf(x,y,z,'CData',flipud(rgb),'FaceColor','texturemap', ...
     'FaceLighting','Phong','EdgeColor','None', ...
     'AmbientStrength',0.5), grid, axis([-2 2 -2 2 0 4]),
daspect([1 1 1]), zoom(2), grid, view([-1 1 0.4]),
light('Position',[-1 0.2 1],'Style','infinite'),
```

In the program, since the data structure of each coordinate array xh, yh or zh is that the first row corresponds to the bottom layer of the cylindrical surface, while the image matrix rgb after calling the function imread(face.jpg) is counting the top line of the picture as the first row, we must use a flipping upside down command to reverse the order of the rows, i.e., flipud(rgb). Alternatively, if the face picture has purposely been rotated by 180^0 before saving it to the file face.jpg, then you can just input rgb into 'CData', instead of flipud(rgb).

Moreover, since the generated cylindrical surface counts the longitudinal columns starting from the x-axis of the base, as shown in Figure 6.1 in Chapter 6, to make a correct texture mapping, the x-axis of the reference frame must be at the middle point of the head back and outward normal to the head back surface. Thus, the number of columns for each of the xh, yh and zh arrays is counted starting from the middle point of the head back to turn counterclockwise about the z-axis of the base and returning to the same point with totally 61 columns, as generated by the cylinder(Head,60) function in this particular program.

Of course, one can define even more columns or more rows to enhance the resolution for better face sculpturing as desired. In the above program, the total number of rows for each of the xh, yh and zh arrays is $nh = 43$ that is the size of the Head array. As mentioned before, counting the latitudinal rows is from bottom to top on the cylindrical surface. Each of the local deformations to bulge a nose, two ears and upper and lower lips as well as to carve a concave area around the two eyes in the above program utilizes a small range of the latitudes with index i and longitudes with index j. Finally, the 4 by 4 homogeneous transformation matrix A_0^1 in the program sends the head to a required destination with a desired orientation, but here is temporarily to set the identity.

We now turn our attention to discussing how the other major parts of the digital human body can be created. The entire human trunk along the vertebral column was modeled to have two universal-joint areas: one is around the lumbar and the other one is around the thorax with three axes each in a roll-pitch-yaw configuration, as described in Figure 9.2 in Chapter 9. In other words, the first universal-joint area consists of θ_7, θ_8 and θ_9 three revolute joint angles, and the second one consists of θ_{10}, θ_{11} and θ_{12} three more revolute joint angles, as depicted in Figure 9.3 of Chapter 9. Therefore, an entire trunk has to be divided into three segments of cylindrical surface: an abdomen/hip segment and two segments for the torso, as shown in Figures 10.3 and 10.4, respectively. Similar to the head, each segment of the trunk is sculptured by a few local deformations on a cylindrical surface, and the MATLAB^{TM} program of the trunk creation is illustrated below:

```
Trunk1=[0 0.35 0.5 0.7 0.85 1.0 1.1 1.13 1.16 1.2 1.2 1.195 ...
        1.19 1.18 1.175 1.172 1.17 1.165 1.162 1.16 1.15 ...
        1.14 1.13 1.12 1.1 1.1 1.0 0.8 0.6 0.4 0.2];
Trunk2=[1.25 1.26 1.27 1.28 1.28 1.28 1.27 1.26 1.25 1.24 ...
        1.23 1.22 1.2 1.18 1.16 1.14 1.12 1.1 1.1 1.09 1.08 ...
        1.07 1.06 1.05 1.05];
```

```
Trunk3=[1.05 1.05 1.05 1.06 1.06 1.08 1.1 1.11 1.11 1.14 ...
       1.17 1.18 1.2 1.2 1.21 1.21 1.2 1.15 1.1 1.05 ...
       0.9 0.7 0.55 0.5 0.4];         % In Decimeter

[xt1,yt1,zt1]=cylinder(Trunk1,60);    % Abdomen/Hip
[xt2,yt2,zt2]=cylinder(Trunk2,60);    % Lower Torso
[xt3,yt3,zt3]=cylinder(Trunk3,60);    % Upper Torso

nt1=size(Trunk1);  nt2=size(Trunk2);  nt3=size(Trunk3);

A12=eye(4);
A23=[eye(3) [0; 0; 2]; 0 0 0 1];   % Send to Destination

for it=4:24
   for jt=13:47
      xt1(it,jt)=(1+0.16*sin((jt-13)*pi/34)^2* ...
                 sin((it-4)*pi/20)^2)*xt1(it,jt);
   end
end          % Sculpture To Bulge the Hip

x1=[];  y1=[];  z1=[];
for ii=1:nt1(2)
   cylin=A12*[xt1(ii,:); yt1(ii,:)*1.5; zt1(ii,:)*3.8
             ones(1,61)];
   x1=[x1;cylin(1,:)];
   y1=[y1;cylin(2,:)];
   z1=[z1;cylin(3,:)];
end

for it=7:25
   for jt=18:42
      xt3(it,jt)=(1-0.16*sin((jt-18)*pi/24)^2* ...
                 sin((it-7)*pi/18)^2)*xt3(it,jt);
   end         % Sculpture the Back of the Upper Torso
   for jt=48:61
      xt3(it,jt)=(1-0.14*sin((jt-48)*pi/24)^2* ...
                 sin((it-7)*pi/18)^2)*xt3(it,jt);
   end
   for jt=1:12
      xt3(it,jt)=(1-0.14*sin((12-jt)*pi/24)^2* ...
                 sin((it-7)*pi/18)^2)*xt3(it,jt);
   end        % Sculpture the Front of the Upper Torso
end

x2=[];  y2=[];  z2=[];
for ii=1:nt2(2)
   cylin=A12*[xt2(ii,:)*1.1; yt2(ii,:)*1.5; zt2(ii,:)*3-0.7
          ones(1,61)];
   x2=[x2;cylin(1,:)];
   y2=[y2;cylin(2,:)];
   z2=[z2;cylin(3,:)];
```

```
end
for ii=1:nt3(2)
   cylin=A23*[xt3(ii,:)*1.1; yt3(ii,:)*1.5; zt3(ii,:)*3.6-0.4
               ones(1,61)];
   x2=[x2;cylin(1,:)];
   y2=[y2;cylin(2,:)];
   z2=[z2;cylin(3,:)];
end                 % Connect the Lower and Upper Torso Together

surf(x1,y1,z1,'FaceColor','c','FaceLighting','Phong', ...
     'EdgeColor','None','AmbientStrength',0.4), grid,
axis([-4 4 -4 4 -1 6]), daspect([1 1 1]), view([1 -1 0.4]),
light('Position',[1 -1 1],'Style','infinite'), hold on,

surf(x2,y2,z2,'FaceColor','g','FaceLighting','Phong', ...
     'EdgeColor','None','AmbientStrength',0.4),
```

In the above program, both the lower and upper torso segments are interfaced with each other by sharing the same coordinate arrays x_2, y_2 and z_2. In other words, first, fill the lower torso segment coordinates into the arrays, and then concatenate them by the upper torso segment coordinates. The interfacing zone is adjusted to have a zero gap in a neutral torso posture and can also effectively cover any opening gap caused by a twisting rotation between the two segments. Moreover, each local deformation in the program is illustrated to use a number of simple functions, such as a linear function, or a sine or its square function. The reader can also use different functions and methods to sculpture each local area on the cylindrical surfaces for more desirable deformations. To make each surface smoother and better looking, each surf function in the program sets a 'Phong' to the 'FaceLighing'. The reader can refer to the MATLAB^{TM} online help for more detailed explanations.

To digitally generate and sculpture the coordinates data for two arms and two legs drawing, a cylindrical surface is initiated for each segment and ready for necessary local deformations. The pictures of the segments and their connections are shown in Figure 10.5 for an arm and Figure 10.6 for a leg. The corresponding MATLAB^{TM} program for the arm is given as follows:

```
[xa,ya,za]=cylinder(0.63:-0.009:0.45,40);   % Upper Arm
[xm,ym,zm]=cylinder(0.45:-0.005:0.35,40);   % Forearm
[xs,ys,zs]=sphere(40);    % Spheres for Joint Connection
                          % Length Unit in Decimeter

ph1=135*pi/180;   ph2=100*pi/180;
A0a=[cos(ph1) 0 sin(ph1) -2; 0 1 0 0
     -sin(ph1) 0 cos(ph1) 7; 0 0 0 1];
A0m=[cos(ph2) 0 sin(ph2) 1.3; 0 1 0 0
     -sin(ph2) 0 cos(ph2) 4.3; 0 0 0 1];
Asa=[eye(3) [-2.4; 0; 7.4]; 0 0 0 1];
```

```
Asm=[eye(3) [0.7; 0; 4.5]; 0 0 0 1];  % Send to Destinations

for it=2:16
   for jt=1:41
      xa(it,jt)=(1+0.12*sin((it-2)*pi/14)^2)*xa(it,jt);
      ya(it,jt)=(1+0.12*sin((it-2)*pi/14)^2)*ya(it,jt);
   end
end          % The Upper Arm Local Deformation

x=[];  y=[];  z=[];
for ii=1:21
   cylin=A0a*[xa(ii,:); ya(ii,:); za(ii,:)*3.2; ones(1,41)];
   x=[x;cylin(1,:)];
   y=[y;cylin(2,:)];
   z=[z;cylin(3,:)];
end

surf(x,y,z,'FaceColor','g','FaceLighting','Phong', ...
     'EdgeColor','None','AmbientStrength',0.4), grid,
axis([-4 4 -4 4 1 9]), daspect([1 1 1]), grid, view([1 -1 0.4]),
light('Position',[1 -1 1],'Style','infinite'), hold on,

x=[];  y=[];  z=[];
for ii=1:40
    cylin=Asa*[xs(ii,:)*0.62; ys(ii,:)*0.62; zs(ii,:)*0.62
               ones(1,41)];
    x=[x;cylin(1,:)];
    y=[y;cylin(2,:)];
    z=[z;cylin(3,:)];
end         % Add a Spherical Ball to the Shoulder Joint

surf(x,y,z,'FaceColor','g','FaceLighting','Phong', ...
     'EdgeColor','None','AmbientStrength',0.4),

for it=2:16
   for jt=1:41
      xm(it,jt)=(1+0.1*sin((it-2)*pi/14)^2)*xm(it,jt);
      ym(it,jt)=(1+0.1*sin((it-2)*pi/14)^2)*ym(it,jt);
   end
end          % The Forearm Local Deformation

x=[];  y=[];  z=[];
for ii=1:21
   cylin=A0m*[xm(ii,:); ym(ii,:); zm(ii,:)*2.85; ones(1,41)];
   x=[x;cylin(1,:)];
   y=[y;cylin(2,:)];
   z=[z;cylin(3,:)];
end

surf(x,y,z,'FaceColor','g','FaceLighting','Phong', ...
     'EdgeColor','None','AmbientStrength',0.4),
```

```
x=[];  y=[];  z=[];
for ii=1:40
    cylin=Asm*[xs(ii,:)*0.45; ys(ii,:)*0.45; zs(ii,:)*0.45
               ones(1,41)];
    x=[x;cylin(1,:)];
    y=[y;cylin(2,:)];
    z=[z;cylin(3,:)];
end         % Add a Spherical Ball to the Elbow Joint

surf(x,y,z,'FaceColor','g','FaceLighting','Phong', ...
     'EdgeColor','None','AmbientStrength',0.4),
```

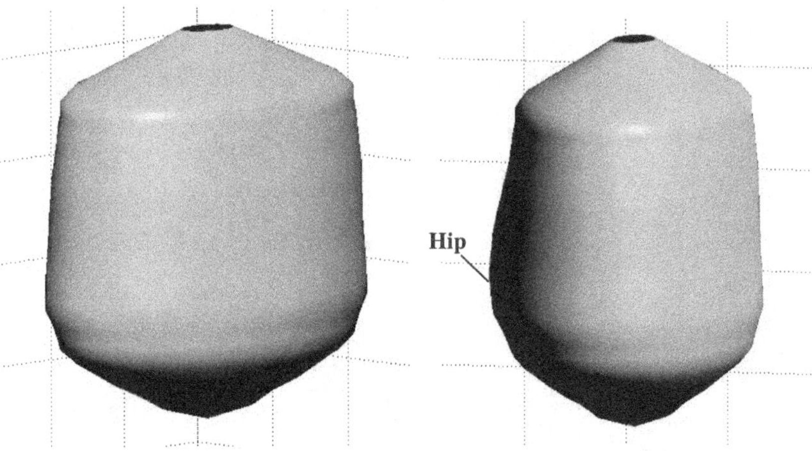

Fig. 10.3 A digital human abdomen/hip model

Unlike the torso connection, the upper arm to the shoulder and to the forearm are interfaced by a spherical ball at each, instead of the data concatenation in their x, y and z coordinate arrays. The reason is that both the shoulder and elbow rotations have much larger ROM (range of motion) than the universal joint between the lower and upper torso segments. Likewise, the program for the leg drawing is given below:

```
[xt,yt,zt]=cylinder(0.9:-0.01:0.7,40);      % The Thigh
[xl,yl,zl]=cylinder(0.7:-0.012:0.46,40);  % The Leg
[xs,ys,zs]=sphere(40);    % Sphere for Knee Joint Connection

ph1=110*pi/180;  ph2=150*pi/180;
```

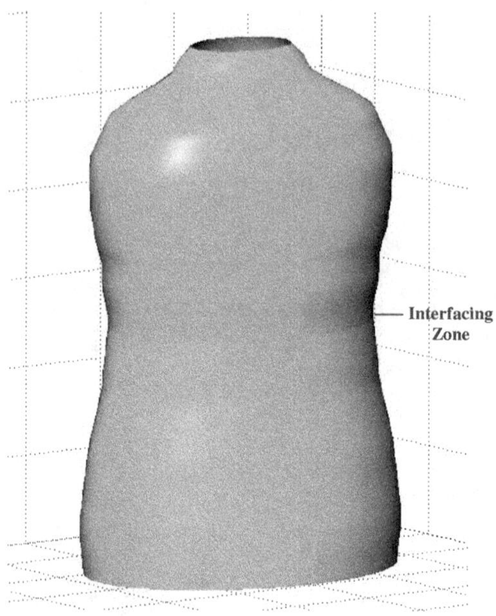

Fig. 10.4 A digital human torso model

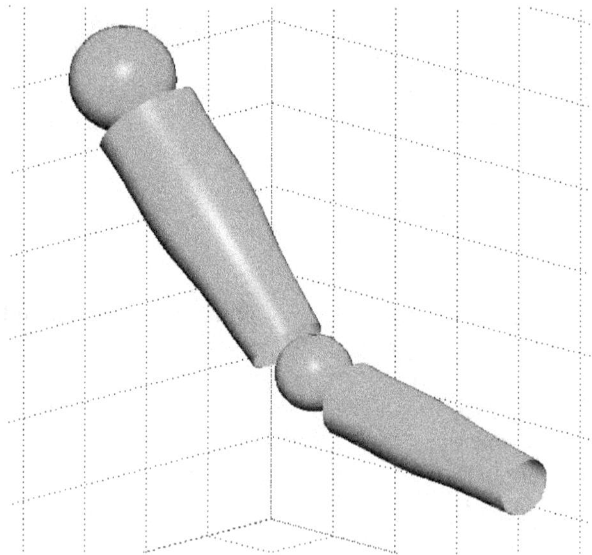

Fig. 10.5 A digital human upper arm/forearm model

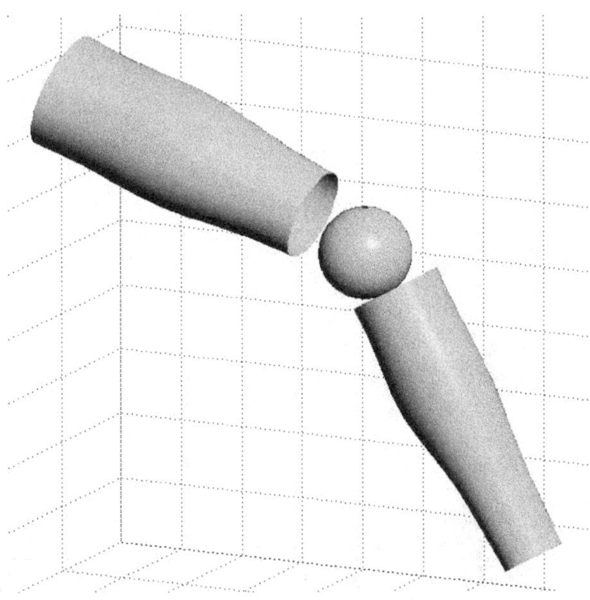

Fig. 10.6 A digital human thigh/leg model

```
A0t=[cos(ph1) 0 sin(ph1) -3; 0 1 0 0
     -sin(ph1) 0 cos(ph1) 8; 0 0 0 1];
A01=[cos(ph2) 0 sin(ph2) 2.5; 0 1 0 0
     -sin(ph2) 0 cos(ph2) 5.5; 0 0 0 1];
A12=[-1 0 0 0; 0 -1 0 0; 0 0 1 0; 0 0 0 1];
Asl=[eye(3) [2; 0; 6]; 0 0 0 1]; % Send to the Destinations

for it=3:19
   for jt=1:41
      xt(it,jt)=(1+0.14*sin((it-3)*pi/16)^2)*xt(it,jt);
      yt(it,jt)=(1+0.14*sin((it-3)*pi/16)^2)*yt(it,jt);
   end
end
for it=1:12
   for jt=20:40
      yt(it,jt)=(1+0.14*sin((jt-20)*pi/20)^2* ...
                 cos((it-1)*pi/22)^2)*yt(it,jt);
   end
end       % Sculpture the Thigh

x=[];  y=[];  z=[];
for ii=1:21
    cylin=A0t*A12*[xt(ii,:); yt(ii,:); zt(ii,:)*4.4
                   ones(1,41)];
```

```
      x=[x;cylin(1,:)];
      y=[y;cylin(2,:)];
      z=[z;cylin(3,:)];
end

surf(x,y,z,'FaceColor','c','FaceLighting','Phong', ...
     'EdgeColor','None','AmbientStrength',0.4), grid,
axis([-4 6 -4 4 0 10]), daspect([1 1 1]), view([1 -1 0.4]),
light('Position',[1 -1 1],'Style','infinite'), hold on,

for it=1:16
   for jt=6:34
      xl(it,jt)=(1+0.2*sin((jt-6)*pi/28)^2* ...
                  sin((it-1)*pi/15)^2)*xl(it,jt);
   end
end
for it=2:16
   for jt=1:41
      xl(it,jt)=(1+0.1*sin((it-2)*pi/14)^2)*xl(it,jt);
      yl(it,jt)=(1+0.1*sin((it-2)*pi/14)^2)*yl(it,jt);
   end
end       % Sculpture the Leg

x=[];  y=[];  z=[];
for ii=1:21
   cylin=A01*A12*[xl(ii,:); yl(ii,:); zl(ii,:)*4.55
                  ones(1,41)];
   x=[x;cylin(1,:)];
   y=[y;cylin(2,:)];
   z=[z;cylin(3,:)];
end
surf(x,y,z,'FaceColor','c','FaceLighting','Phong', ...
     'EdgeColor','None','AmbientStrength',0.4),

x=[];  y=[];  z=[];
for ii=1:40
   cylin=Asl*[xs(ii,:)*0.7; ys(ii,:)*0.7; zs(ii,:)*0.7
              ones(1,41)];
   x=[x;cylin(1,:)];
   y=[y;cylin(2,:)];
   z=[z;cylin(3,:)];
end         % Add a Spherical Ball to the Knee Joint

surf(x,y,z,'FaceColor','c','FaceLighting','Phong', ...
     'EdgeColor','None','AmbientStrength',0.4),
```

Table 10.1 Segment drawing and sending information for mocking up a digital mannequin

Segment Name	Part Surface surf(···)	Sent By A_b^j	Data Adjustments x	y	z
Abdomen/Hip	cylinder($Trunk1$, 60)	A_b^6	$z * 3.8 - 2.5$	x	$y * 1.5$
Lower Torso	cylinder($Trunk2$, 60)	$A_b^8 A_8^t$	$-y * 1.5$	$-x * 1.1$	$0.7 - z * 3$
Upper Torso	cylinder($Trunk3$, 60)	A_b^n	$-y * 1.5$	$x * 1.1$	$z * 3.6 - 0.4$
Head	f_head	A_b^{47}	$z - a_{47}$	$-x$	$-y$
Neck	cylinder([0.68 0.56], 60)	A_b^n	x	y	$z * 0.3$
Left Shoulder	cylinder([0.95 0.9 0.6], 40)	A_b^{14}	x	$y * 0.95$	$-z * 1.9$
	sphere(40)	A_b^{16}	$x * 0.6$	$y * 0.6$	$z * 0.6 + 0.1$
L. Upper Arm	cylinder($uarm$, 40)	A_b^{16}	x	y	$z * 3.2$
L. Elbow	sphere(40)	A_b^{18}	$x * 0.45$	$y * 0.45$	$z * 0.45$
L. Forearm	cylinder($farm$, 40)	A_b^{18}	x	y	$z * 2.85$
L. Hand	f_hand(···)	A_b^{21}	x	y	z
Right Shoulder	cylinder([0.95 0.9 0.6], 40)	A_b^{23}	x	$y * 0.95$	$z * 1.9$
	sphere(40)	A_b^{25}	$x * 0.6$	$y * 0.6$	$z * 0.6 - 0.1$
R. Upper Arm	cylinder($uarm$, 40)	A_b^{25}	x	y	$-z * 3.2$
R. Elbow	sphere(40)	A_b^{27}	$x * 0.45$	$y * 0.45$	$z * 0.45$
R. Forearm	cylinder($farm$, 40)	A_b^{27}	x	y	$-z * 2.85$
R. Hand	f_hand(···)	A_b^{30}	x	y	z
L. Thigh	cylinder($thigh$, 40)	A_b^{32}	x	y	$z * 4.4$
L. Knee	sphere(40)	A_b^{34}	$x * 0.7$	$y * 0.7$	$z * 0.7$
L. Leg	cylinder(leg, 40)	A_b^{34}	x	y	$z * 4.55$
L. Foot	cylinder($foot$, 20)	A_b^{36}	x	$y * 1.4$	$z * 1.8$
L. Toe	cylinder(toe, 20)	A_b^{37}	x	$y * 1.4$	$z - 0.52$
R. Thigh	cylinder($thigh$, 40)	A_b^{39}	x	y	$-z * 4.4$
R. Knee	sphere(40)	A_b^{41}	$x * 0.7$	$y * 0.7$	$z * 0.7$
R. Leg	cylinder(leg, 40)	A_b^{41}	x	y	$-z * 4.55$
R. Foot	cylinder($foot$, 20)	A_b^{43}	x	$y * 1.4$	$z * 1.8$
R. Toe	cylinder(toe, 20)	A_b^{44}	x	$y * 1.4$	$z - 0.52$

In this program, the thigh and leg segments are also connected at the knee joint with a spherical ball for better looking in such a larger motion range case. Also, the homogeneous transformation matrix A_1^2 is used here to reverse both the thigh and leg due to the asymmetric shapes between their front and back.

Once every digital human body segment is prepared, the key important step to mock up a digital mannequin is to send each part to its right place by the corresponding homogeneous transformation $A_b^j = A_b^0 A_0^j$. Similar to Table 6.1 in Chapter 6 for assembling the 7-joint redundant robot, how to

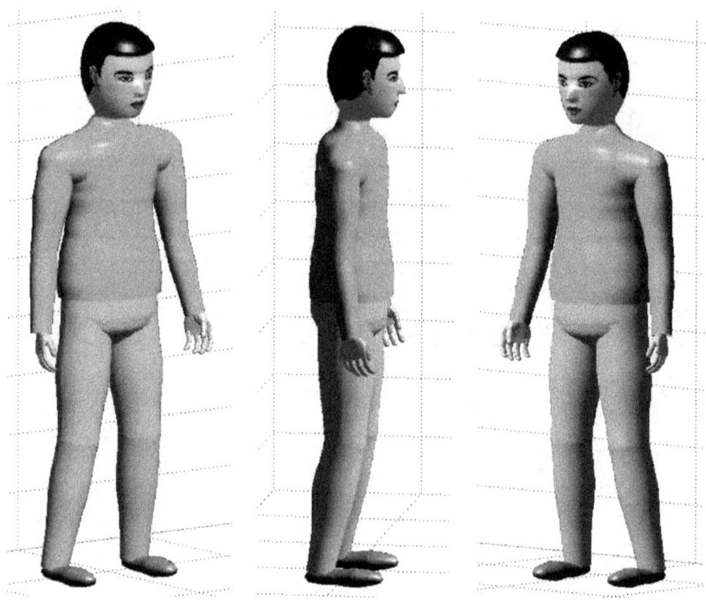

Fig. 10.7 Three different views of the finally assembled digital human model

send each part of the digital mannequin to its destination is outlined in Table 10.1. Every homogeneous transformation appeared in Table 10.1 is based on the digital human model under the D-H convention in Figure 9.4 of Chapter 9. With all the segments created and transformed, the final graphical result of the digital mannequin after assembling all the parts together in three different view angles is shown in Figure 10.7.

In Table 10.1, the homogeneous transformation matrix A_8^t in the second row for the lower torso segment is defined in the following program:

```
x=[];  y=[];  z=[];
for ii=1:nt2(2)
   th9=q(9)+ii*q(12)/nt2(2);  % To Make a Linearly Increasing
                              % Twist Angle Across the Layers
   A8t=[cos(th9) -sin(th9) 0 0;sin(th9) cos(th9) 0 0;
        0 0 1 0; 0 0 0 1];
   Abt=Ab8*A8t;

   cylin=Abt*[-yt2(ii,:)*1.5; -xt2(ii,:)*1.1; 0.7-zt2(ii,:)*3
             ones(1,61)];
   x=[x;cylin(1,:)];
   y=[y;cylin(2,:)];
   z=[z;cylin(3,:)];
end
```

As aforementioned, the x, y and z coordinate arrays of the lower and upper torso segments are concatenated to each other to avoid any opening gap. However, when the upper torso is twisted by a larger joint angle θ_{12}, while the lower torso spinning joint angle θ_9 is smaller or zero, the initial smooth interface between the two adjacent segments will become rough due to their relative twist if the concatenation is very simple. Now, let each latitudinal layer of the lower torso surface be gradually and linearly twisted from the bottom to the top in order to smoothly interface itself to the upper torso surface. This is why an additional matrix A_8^t is inserted into the homogeneous transformation when sending the lower torso segment to the right place.

More specific explanations for Table 10.1 are given below:

1. The pre-defined arrays $Trunk1$, $Trunk2$ and $Trunk3$ in the first three rows for the entire trunk surface creation have been specified in the previous MATLAB^{TM} program;
2. The function subroutine f_head is based on the head generation program, as given earlier;
3. The arrays $uarm$ and $farm$ for creating the upper arm and forearm surfaces as well as the $thigh$ and leg arrays for the thigh and leg segments have also been defined in the preceding MATLAB^{TM} programs;
4. The function f_hand($\cdots$) is to generate either a right hand or a left hand, the modeling procedure of which will be introduced and detailed in the next section;
5. Finally, the $foot$ and toe arrays inside the cylinder function for generating a foot model and a toe model can be defined in many ways as desired, but one of the simplest ways is given here as an example:

```
ts=0:0.1:1;      % Define an Index Sequence

[xf,yf,zf]= ...
   cylinder((1-ts).^0.2*0.34.*(0.95+0.05*cos(3*pi*ts)),20);
                        % To Generate the Foot Data
[xo,yo,zo]=cylinder((1-ts).^0.3*0.34,20);
                        % To Generate the Toe Data
```

Since each homogeneous transformation A_b^j that sends the corresponding segment to the right place is a function of the joint positions/angles, the digital mannequin created can be effectively animated to generate and visualize a pre-planned motion if every computation is included inside a for-loop or a while-loop. Of course, the resulting digital human motion could be slow due to a large amount of surface generations in comparison with the robotic animation that was illustrated in Chapter 6.

Unless your computer CPU and the graphics card are fast enough, a skeletal mannequin may be a solution to significantly reduce the computational time for a real-time realistic digital human motion on the screen. To this end, by keeping the same digital human mathematical model with the same transformations and the same global Jacobian matrix, one can only draw a thick line for each segment, instead of a 3D surface, as seen in Figure 9.1 from Chapter 9. The result in terms of the human motion and every joint movement function will be exactly same and much faster, but just does not look like the same appearance as a full-surfaced model.

We now describe in detail how to draw a skeletal mannequin based on the same digital human mathematical model. First, let us display such a skeletal digital mannequin in dancing in Figure 10.8. This mannequin possesses the same number of joints, i.e., 47 joints in total that obey the same D-H tables in Chapter 9. In terms of drawing, it has up to 27 nodes and 31 connecting thick lines. Because all the 47 required homogeneous transformation matrices A_b^j for $j = 1, \cdots, 47$ must be numerically prepared in a MATLAB^{TM} program, the position vector of each node with respect to the base can be directly found from the 4th column of the corresponding homogeneous transformation.

For example, by referring to the coordinate frame assignment for a digital human model in Figure 9.4 of Chapter 9, the left and right clavicle nodes are p_b^{12} and p_b^{21}, and the left and right shoulder nodes are p_b^{16} and p_b^{25}, respectively. Of course, one may also use p_b^{15} and p_b^{24} to represent the left and right shoulder nodes, because the origins of frame 15 and frame 16 are at the same point, and so are the origins of frame 24 and frame 25. Also, the left and right hip nodes are located by p_b^{32} and p_b^{39}, respectively, and the middle node between them is p_b^{30} that may be included or excluded as the only option for the skeletal mannequin drawing.

Regarding the four nodes of the head, the top and the neck ones are obviously p_b^{47} and p_b^n, respectively, with a_{47} being the total height of the head. For the left and right ear nodes, one may first find the middle point between the top and neck nodes: $p_b^{mid} = (p_b^n + p_b^{47})/2$. If the width of the head is w, then starting from the middle point along the direction of z_{47}-axis to determine the left and right ear nodes by $p_b^{le} = p_b^{mid} + \frac{w}{2} r_b^{47}$ and $p_b^{re} = p_b^{mid} - \frac{w}{2} r_b^{47}$, respectively, where r_b^{47} is the 3rd column of the homogeneous transformation A_b^{47} to represent the z_{47}-axis.

After all the position vectors of up to 27 nodes are prepared numerically at each sampling point in the for-loop or while-loop, using the internal function plot3$(\cdots)$ to connect the 27 nodes together in a certain order to create a complete skeletal mannequin, as depicted in Figure 10.8 as the mannequin is spinning for dance.

10.2 Hand Models and Digital Sensing

A human hand, no matter it is a left or right hand, consists of a palm and five fingers: thumb, index, middle, ring and pinky fingers. The entire palm can also be divided into the lower and upper palms. The lower palm is the house for the five metacarpals, while the upper palm partially covers the proximal phalanges of the index, middle, ring and pinky four fingers. Each of the five fingers has four revolute joints, and makes a total of 20 joints for each hand.

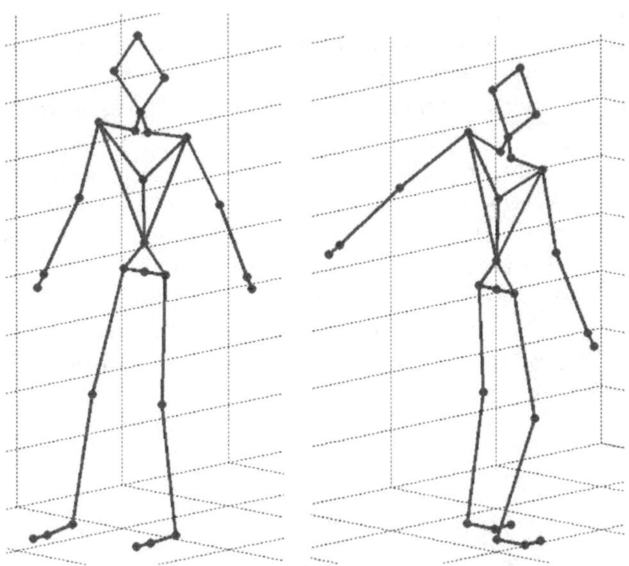

Fig. 10.8 A skeletal digital mannequin in dancing

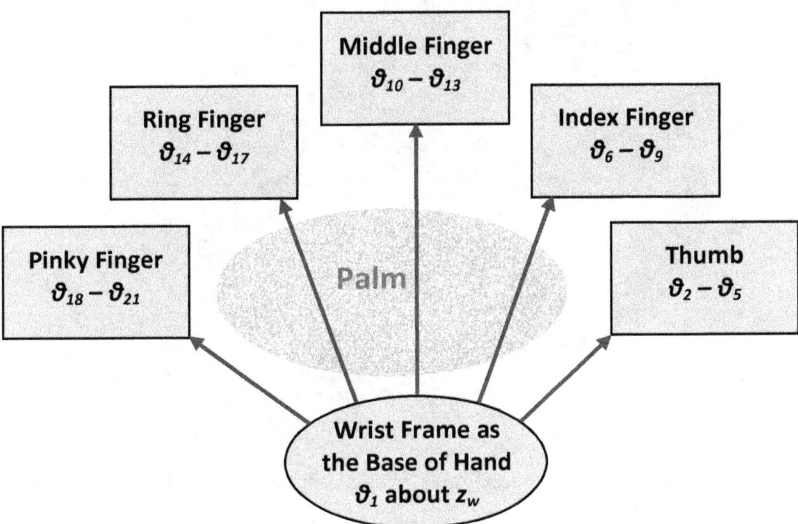

Fig. 10.9 A block diagram for the right hand modeling and reversing the order for the left hand

Since the carpal bone is directly connected to the wrist, the hand rotation about the wrist may be counted as the first revolute joint θ_1 about the axis z_w of the wrist, as shown in Figure 10.9.

Using the Denavit-Hartenberg (D-H) convention, as introduced and discussed in Chapter 4, we can readily model a hand by first assigning all the 21 coordinate frames, including the wrist frame (x_w, y_w, z_w) as the base of the hand. Then, the D-H table for each of the five fingers can be determined. In the thumb table, Table 10.2, θ_1 is a rotation about z_w and is -90^0 at the home position because x_1 has to be defined as a common normal vector to both the z_w and z_1 axes according to the D-H convention. It is interesting that the twist angle α_1 between z_w and z_1 about x_1, see Figure 10.10, may be slightly different from person to person. In our digital hand model and simulation study, we set $\alpha_1 = -36^0$.

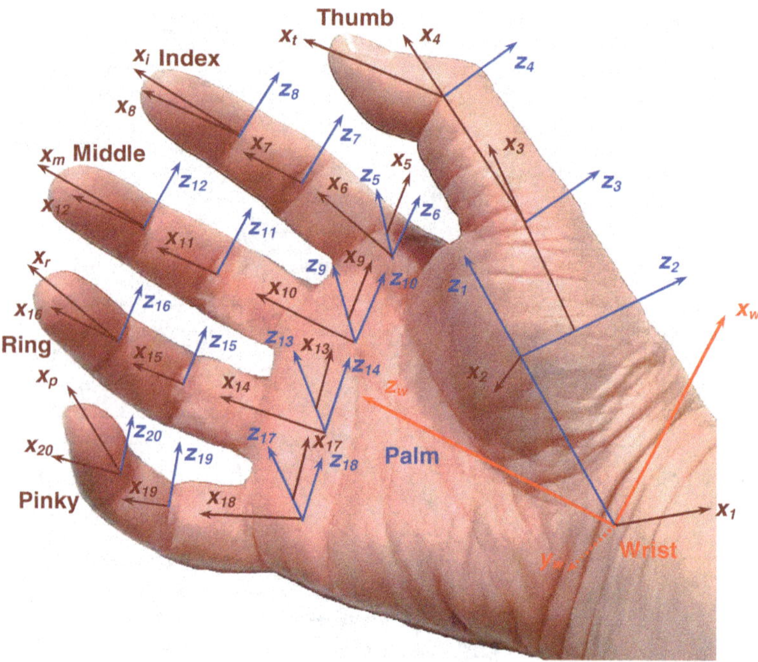

Fig. 10.10 The joint/link coordinate frame assignment for hand modeling based on the D-H convention

Each of the D-H tables has also included a column of the joint range of motion (ROM) for each of the 20 joints over a hand. It can be seen that most joints of the hand are very limited in motion due to the special structure of hand bone. Also, some of the joints may have the effect of two-joint muscle, as addressed and discussed in Chapter 9. The ROM's in the above D-H tables do not take this special effect into account and all the ranges are just estimated for an average human hand.

Table 10.2 The D-H table for the thumb

Finger	θ_i	d_i	α_i	a_i	$\theta_i(home)$	ROM
	θ_1	0	α_1	0	-90^0	
	θ_2	d_2	-90^0	a_2	θ_2^h	$0 \sim 45^0$
Thumb	θ_3	d_3	-90^0	a_3	-90^0	$-120 \sim -30^0$
	θ_4	0	0	a_4	0	$-10 \sim 60^0$
	θ_5	0	0	a_5	0	$0 \sim 90^0$

Table 10.3 The D-H table for the index finger

Finger	θ_i	d_i	α_i	a_i	$\theta_i(home)$	ROM
	θ_{f1}	d_f	90^0	a_{f1}	0	
	θ_6	$-d_6$	90^0	0	90^0	$60 \sim 120^0$
Index	θ_7	0	0	a_7	0	$-10 \sim 90^0$
	θ_8	0	0	a_8	0	$0 \sim 110^0$
	θ_9	0	0	a_9	0	$0 \sim 70^0$

Table 10.4 The D-H table for the middle finger

Finger	θ_i	d_i	α_i	a_i	$\theta_i(home)$	ROM
	θ_{f2}	d_f	90^0	a_{f2}	0	
	θ_{10}	$-d_{10}$	90^0	0	90^0	$70 \sim 110^0$
Middle	θ_{11}	0	0	a_{11}	0	$-10 \sim 90^0$
	θ_{12}	0	0	a_{12}	0	$0 \sim 110^0$
	θ_{13}	0	0	a_{13}	0	$0 \sim 70^0$

Table 10.5 The D-H table for the ring finger

Finger	θ_i	d_i	α_i	a_i	$\theta_i(home)$	ROM
	θ_{f3}	d_f	90^0	a_{f3}	0	
	θ_{14}	$-d_{14}$	90^0	0	90^0	$70 \sim 110^0$
Ring	θ_{15}	0	0	a_{15}	0	$-10 \sim 90^0$
	θ_{16}	0	0	a_{16}	0	$0 \sim 120^0$
	θ_{17}	0	0	a_{17}	0	$0 \sim 70^0$

Table 10.6 The D-H table for the pinky finger

Finger	θ_i	d_i	α_i	a_i	$\theta_i(home)$	ROM
	θ_{f4}	d_f	90^0	a_{f4}	0	
	θ_{18}	$-d_{18}$	90^0	0	90^0	$70 \sim 110^0$
Pinky	θ_{19}	0	0	a_{19}	0	$-10 \sim 90^0$
	θ_{20}	0	0	a_{20}	0	$0 \sim 120^0$
	θ_{21}	0	0	a_{21}	0	$0 \sim 70^0$

In the digital hand model, all the five fingers are in parallel connected to the common wrist base frame through the palm. In other words, each of the starting z-axes: z_1 for the thumb, z_5 for the index finger, z_9 for the middle finger, z_{13} for the ring finger and z_{17} for the pinky finger, is directly rotated from z_w of the wrist base by each individual twist angle: $\alpha_1 = -36^0$ about x_1, α_{f1} about x_5, α_{f2} about x_9, α_{f3} about x_{13} and α_{f4} about x_{17}, see Figure 10.10. In the digital model, each starting twist angle of the last four fingers is set to be $\alpha_{fi} = 90^0$ for $i = 1, \cdots, 4$.

Because the z_w-axis is intentionally defined at the middle point between frames 9 and 13 of the middle and ring fingers, we may define the link lengths $a_{f3} = -a_{f2}$ and $a_{f4} = -a_{f1}$ as symmetric to z_w. In the simulation study, $a_{f1} = 3$ cm. and $a_{f2} = 1$ cm. such that $a_{f3} = -1$ and $a_{f4} = -3$ cm. Such a symmetric treatment has a unique advantage that can minimize the number of parameters to be changed in order to switch the same hand model from the right hand to the left hand. Moreover, all the joint offsets along z_w from x_w to x_5, x_9, x_{13} and x_{17} are commonly defined as $d_f = 4.9$ cm. as the length of the lower palm. This is also beneficial to making the right-left hand models exchangeable.

In fact, with the same hand model and drawing program, using the following small number of changes to select and switch the hand model from the right to the left without need to rewrite a new hand drawing program:

```
a_f1 = 3;   a_f2 = 1;    % All in centimeter.

rl=input('Right Hand=1, Left Hand=2, Your Choice = ');
          % Choose Which Hand You Want to Create

if rl == 1     % To Draw the Right Hand

    alpha_1 = -36*pi/180;   % The 1st Thumb Twist Angle
    alpha_2 = -pi/2;        % The 2nd Thumb Twist Angle

  else         % To Draw the Left Hand

    alpha_1 = 36*pi/180;
    alpha_2 = pi/2;
    a_f1 = -a_f1;
    a_f2 = -a_f2;   % Mirror the Four-Finger Positions

    q(3) = -q(3);        % Only the Five Joint Angles
    q(6) = pi-q(6);      % Need To Be Adjusted
    q(10) = pi-q(10);    % When Switching from
    q(14) = pi-q(14);    % the Right to the Left Hand
    q(18) = pi-q(18);
end
```

In the above switch program, we keep the same palm unchanged, and mirror the positions of the four fingers by minus the link lengths a_{f1} and a_{f2}, and also minus the first two twist angles α_1 and α_2 for the thumb. Due to the mirror and sign change, only the five related joint angles, one for each finger, need to be adjusted accordingly. Therefore, one single program with the above switch routine can generate and draw both the right and left hand models.

Furthermore, with the D-H tables determined, we can readily find each one-step homogeneous transformation A_{i-1}^i for all the 21 frames. For example, the first five one-step homogeneous transformation matrices for the thumb are given as follows:

$$A_w^1 = \begin{pmatrix} c_1 & -s_1 & 0 & 0 \\ s_1 & c_1 & 0 & 0 \\ 0 & 0 & 1 & 0 \\ 0 & 0 & 0 & 1 \end{pmatrix} \begin{pmatrix} 1 & 0 & 0 & 0 \\ 0 & c\alpha_1 & -s\alpha_1 & 0 \\ 0 & s\alpha_1 & c\alpha_1 & 0 \\ 0 & 0 & 0 & 1 \end{pmatrix},$$

$$A_1^2 = \begin{pmatrix} c_2 & 0 & -s_1 & a_2c_2 \\ s_2 & 0 & c_2 & a_2s_2 \\ 0 & -1 & 0 & d_2 \\ 0 & 0 & 0 & 1 \end{pmatrix}, \quad A_2^3 = \begin{pmatrix} c_3 & 0 & -s_3 & a_3c_3 \\ s_3 & 0 & c_3 & a_3s_3 \\ 0 & -1 & 0 & d_3 \\ 0 & 0 & 0 & 1 \end{pmatrix},$$

and

$$A_3^4 = \begin{pmatrix} c_4 & -s_4 & 0 & a_4c_4 \\ s_4 & c_4 & 0 & a_4s_4 \\ 0 & 0 & 1 & 0 \\ 0 & 0 & 0 & 1 \end{pmatrix}, \quad A_4^5 = \begin{pmatrix} c_5 & -s_5 & 0 & a_5c_5 \\ s_5 & c_5 & 0 & a_5s_5 \\ 0 & 0 & 1 & 0 \\ 0 & 0 & 0 & 1 \end{pmatrix},$$

where each $s_i = \sin\theta_i$, $c_i = \cos\theta_i$ as well as $c\alpha_i = \cos\alpha_i$ and $s\alpha_i = \sin\alpha_i$ as a traditional short notation in robotics.

After programming all the one-step homogeneous transformation matrices, every required transformation for both the 3D drawing and hand motion can be numerically calculated in MATLAB^{TM}. As a result, Figures 10.11 and 10.12 show the right hand model and the left hand model, respectively, in MATLAB^{TM} along with each in a gesture of ball-grasping.

To generate the right hand model in a relax gesture, as shown on the left of Figure 10.11, all the 20 joint angles have the following values:
For the thumb,

$$\theta_2 = 0^0,\ \theta_3 = -60^0,\ \theta_4 = 0^0,\ \theta_5 = 10^0;$$

For the index finger,

$$\theta_6 = 80^0,\ \theta_7 = 10^0,\ \theta_8 = 10^0,\ \theta_9 = 10^0;$$

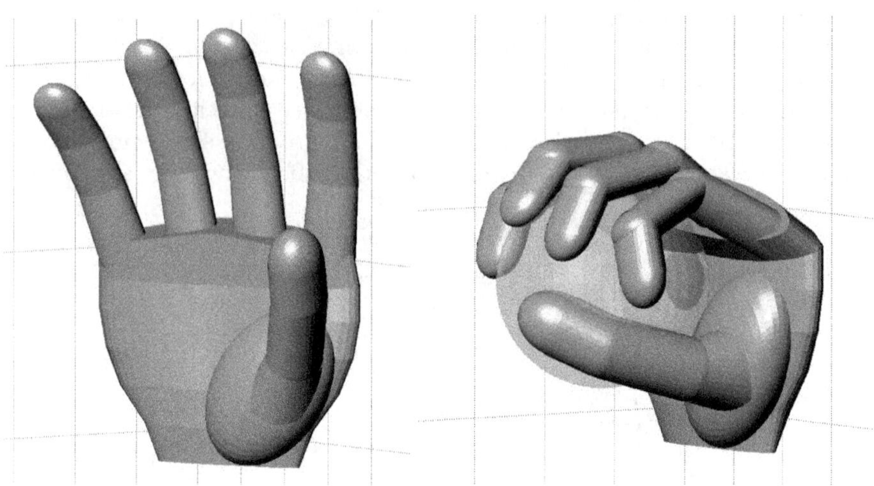

Fig. 10.11 The right hand digital model with a ball-grasping gesture

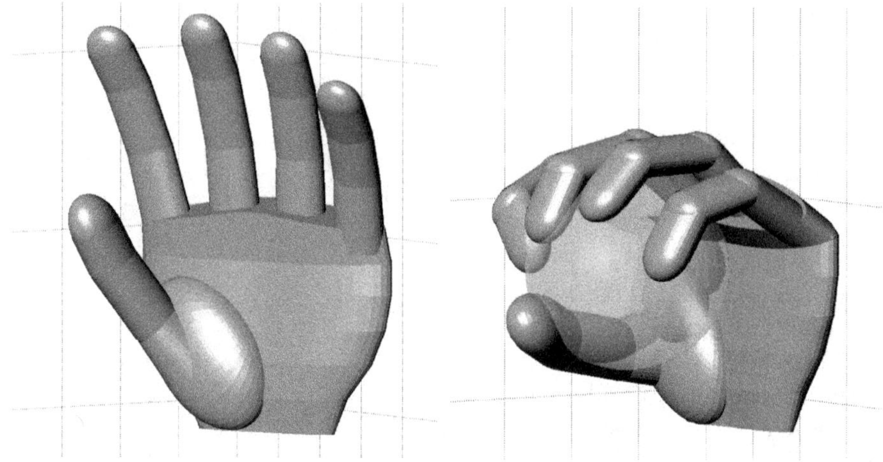

Fig. 10.12 The left hand digital model with a ball-grasping gesture

For the middle finger,

$$\theta_{10} = 90^0,\ \theta_{11} = 10^0,\ \theta_{12} = 10^0,\ \theta_{13} = 10^0;$$

For the ring finger,

$$\theta_{14} = 95^0,\ \theta_{15} = 10^0,\ \theta_{16} = 10^0,\ \theta_{17} = 10^0;$$

For the pinky finger,

$$\theta_{18} = 100^0,\ \theta_{19} = 20^0,\ \theta_{20} = 10^0,\ \theta_{21} = 10^0.$$

When the right hand is grasping a ball of radius 2.5 cm., as shown on the right of Figure 10.11, the joint angles reach the following positions:
For the thumb,

$$\theta_2 = -4.81^0,\ \theta_3 = -2.25^0,\ \theta_4 = 9.63^0,\ \theta_5 = 19.63^0;$$

For the index finger,

$$\theta_6 = 113.69^0,\ \theta_7 = 58.13^0,\ \theta_8 = 58.13^0,\ \theta_9 = 58.13^0;$$

For the middle finger,

$$\theta_{10} = 90.00^0,\ \theta_{11} = 50.91^0,\ \theta_{12} = 50.91^0,\ \theta_{13} = 50.91^0;$$

For the ring finger,

$$\theta_{14} = 95^0,\ \theta_{15} = 46.10^0,\ \theta_{16} = 46.10^0,\ \theta_{17} = 46.10^0;$$

For the pinky finger,

$$\theta_{18} = 100.00^0,\ \theta_{19} = 58.50^0,\ \theta_{20} = 48.50^0,\ \theta_{21} = 48.50^0.$$

In terms of the 3D graphical drawing, each hand model consists of a variety of basic 3D components, such as cylinders, cones or wedged cylinders and spheres or ellipsoids. Figure 10.13 demonstrates that the right hand model is making a big "V" gesture. The figure also depicts all the drawing components necessary to create each joint and segment for the right hand model, and similar for the left hand model. In each joint of the five fingers, a sphere or an ellipsoid is inserted to fill the gap due to any possible opening between the adjacent segments that are created by two cylinders when they are rotating. Although, adding so many spheres or ellipsoids into the 3D drawing may cause more time-consuming, it is a quite effective way to make each finger and its movement more realistic.

In contrast to the relax gesture, to create a big "V" gesture for the right hand model in Figure 10.13, all the joint angles are set as follows:
For the thumb,

$$\theta_2 = -20^0,\ \theta_3 = -20^0,\ \theta_4 = 20^0,\ \theta_5 = 60^0;$$

For the index finger,

$$\theta_6 = 80^0,\ \theta_7 = 5^0,\ \theta_8 = 5^0,\ \theta_9 = 5^0;$$

For the middle finger,

$$\theta_{10} = 100^0,\ \theta_{11} = 5^0,\ \theta_{12} = 5^0,\ \theta_{13} = 5^0;$$

For the ring finger,

$$\theta_{14} = 95^0,\ \theta_{15} = 30^0,\ \theta_{16} = 120^0,\ \theta_{17} = 30^0;$$

For the pinky finger,

$$\theta_{18} = 100^0,\ \theta_{19} = 30^0,\ \theta_{20} = 120^0,\ \theta_{21} = 30^0.$$

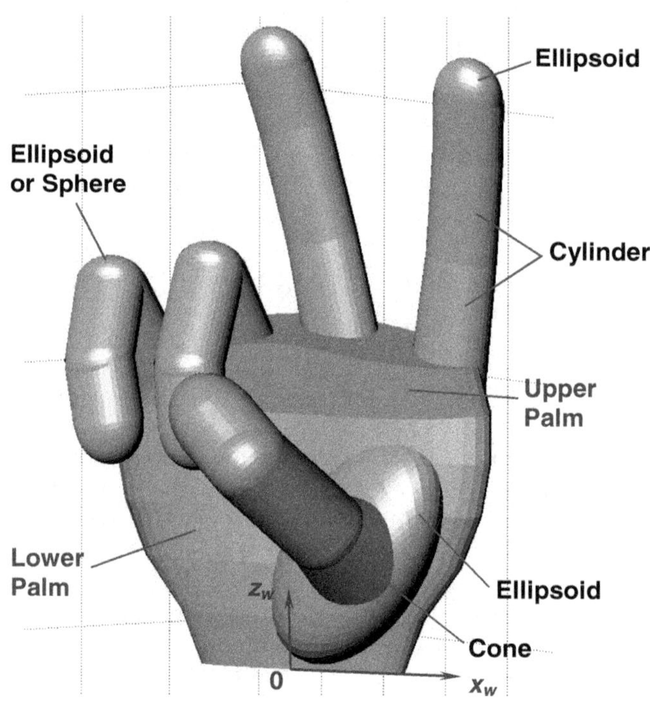

Fig. 10.13 A digital hand model consists of various drawing components

Basically, two hands are usually driven by a joint motion algorithm. Since the position of each frame origin over a hand model can always be found at any time instant in the hand drawing and motion generation program, we may take advantage of the position data to create a motion of hand grasping or any other hand operations. For instance, when the right hand at the ball-grasping gesture on the right of Figure 10.11, the last column of every homogeneous transformation matrix A_w^j for $j = 2, \cdots, 21$ is the position of the j-th frame origin, and can be retrieved as follows:

$$p_w^{thumb} = (0.96 \quad -5.26 \ \ 2.86)^T,$$
$$p_w^{index} = (3.19 \quad -4.42 \ \ 4.45)^T,$$
$$p_w^{middle} = (0.99 \quad -4.99 \ \ 5.40)^T,$$
$$p_w^{ring} = (-1.10 \quad -4.76 \ \ 6.00)^T,$$
$$p_w^{pinky} = (-2.97 \quad -4.20 \ \ 4.70)^T,$$

where all the numbers in centimeters and each 3D position vector is tailed at the origin of the wrist frame (the base of the hand) and arrow-pointing to the tip point of each finger. At the same gesture, the origin positions of the other frames can also be found immediately from the hand drawing program, such as

$$p_w^{3} = (1.88 \quad -2.13 \ \ 2.04)^T,$$
$$p_w^{7} = (2.36 \quad -2.51 \ \ 6.35)^T,$$
$$p_w^{11} = (1.00 \quad -2.28 \ \ 6.92)^T,$$
$$p_w^{15} = (-1.18 \quad -1.96 \ \ 6.97)^T,$$
$$p_w^{19} = (-3.24 \quad -2.17 \ \ 6.24)^T,$$

and

$$p_w^{6} = (3.00 \ \ 0.04 \ \ 4.90)^T,$$
$$p_w^{10} = (1.00 \ \ 0.20 \ \ 4.90)^T,$$
$$p_w^{14} = (-1.00 \ \ 0.20 \ \ 4.90)^T,$$
$$p_w^{18} = (-3.00 \ \ 0.04 \ \ 4.90)^T.$$

We now look at the right hand model to grasp a spherical ball of radius $R = 3.2$ cm., as shown in Figure 10.14. The center of this ball is sitting at $c_{ball} = (0 \quad -3.6 \ \ 4)^T$ that is referred to the wrist (base) frame. Then, the motion constraint of each point on the five fingers must obey the following inequality:

$$\text{norm}(p_w^j - c_{ball}) \geq R + b, \tag{10.1}$$

where R is the ball radius, b is an additional bias to reflect the half-thickness of each finger, and norm is the built-in function in MATLAB^{TM} to calculate the distance between a point and the ball center.

A partial MATLAB^{TM} program to sense the motion constraints for this ball-grasping example is given below:

```
q_thumb = [0 0 -60 0 10]';      % q(1) - q(5)
q_index = [80 10 10 10]';       % q(6) - q(9)
q_mid = [90 10 10 10]';         % q(10) - q(13)
q_ring = [95 10 10 10]';        % q(14) - q(17)
q_pinky = [100 20 10 10]';      % q(18) - q(21)
```

```
q=[q_thumb; q_index; q_mid; q_ring; q_pinky]*pi/180;
   % The Initial Joint Angles at a Relax Gesture

rate1 = 0.04;  rate2 = 0.04;  rate3 = 0.04;
rate4 = 0.04;  rate5 = 0.04;
   % Initiate a Non-zero Rate for Each Finger

R=3.2;  c_ball=[0; -R-0.4; 4];
   % Specify the Ball Radius and Center

for j = 0:N

    % ---------------------- %
    % Numerical Calculation of All
    % the Homogeneous Transformation
    % Matrices and the Program for
    % Hand Drawing and Animation
    % ---------------------- %

    q(2) = q(2) - rate1;
    q(3:5) = q(3:5) + rate1;
    q(6:9) = q(6:9) + rate2;
    q(11:13) = q(11:13) + rate3;
    q(15:17) = q(15:17) + rate4;
    q(19:21) = q(19:21) + rate5;
     % Joint Motion for Five Fingers
     % to Grasp a Ball

    if norm(Aw5(1:3,4)-c_ball) < R+0.5
        rate1 = 0;
    end
    if norm(Aw9(1:3,4)-c_ball) < R+0.4
        rate2 = 0;
    end
    if norm(Aw13(1:3,4)-c_ball) < R+0.4
        rate3 = 0;
    end
    if norm(Aw17(1:3,4)-c_ball) < R+0.4
        rate4 = 0;
    end
    if norm(Aw21(1:3,4)-c_ball) < R+0.4
        rate5 = 0;
    end
    % Once Exceed the Constraint for
    % Each Finger, Stop Its Motion
    % Respectively.

end
```

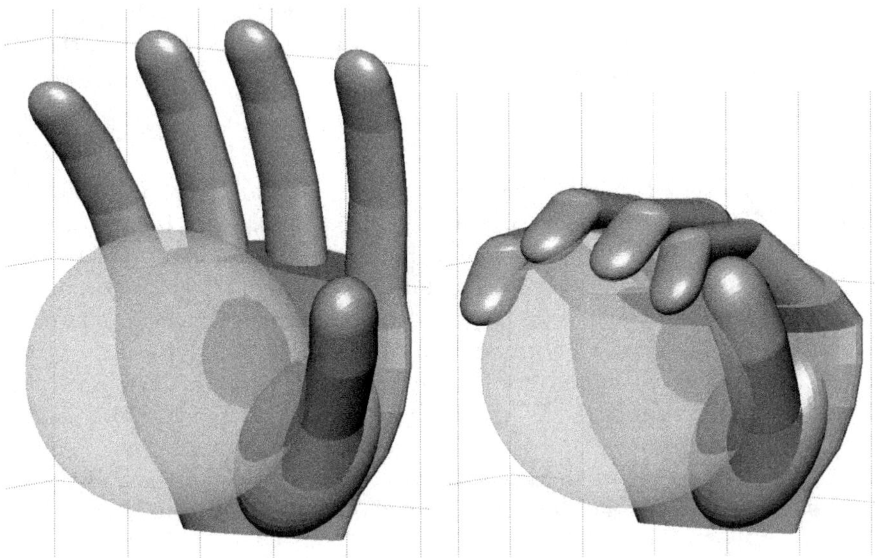

Fig. 10.14 The right hand is going to grasp a big ball

This partial program is to test each finger tip point individually. If one of them exceeds the corresponding motion constraint given by the inequality in (10.1), then, that finger will stop the grasping motion respectively. The hand grasping process will sustain until all the fingers stop their motion. This is called a **digital sensing**. Unless a hand statics or even dynamics subroutine is included, in such a joint motion based kinematics-alone algorithm, a touching force or friction will not be able to help the kinematic program, as an input, to control the movement for each finger. Therefore, this kind of digital sensing has to fully rely on detecting every point of the fingers kinematically to see if anyone violates the finger motion constraint.

While the above ball-grasping program with the digital sensing algorithm is quite effective, the object to be grasped is limited thus far to a spherical ball, or relatively simple shapes, such as an ellipsoid, a cylinder or a cube. The main difficulty causing the limitation is that there is no closed form of equation available to represent the boundary for a grasping object with a relatively complex surface. If possible, a complex object may be circumscribed by a minimum spherical or ellipsoidal envelope, as a motion constraint, before running the digital sensing algorithm. Such a circumscription approach may have a potential to extend the digital sensing algorithm to more general hand grasping and manipulation model developments.

In order to perform a static force/torque analysis when one or two hands are grasping an object in a manipulation task, it becomes necessary to formulate a Jacobian matrix J for each of the hands either locally or globally. If the k-th finger tip is acted with a 3-dimensional linear force f_k for $k = 1, \cdots, 5$, then the Cartesian force vector of all the five finger tips can be augmented by the five f_k's towards a $3 \times 5 = 15$ dimensional vector F. Thus, the joint torque distribution over a single hand is given by

$$\tau = J^T F,$$

where both J and F must be referred to the same frame and J is 15 by 21 if it is global and only a linear force acting on each finger tip is considered.

By following equation (4.15) in Chapter 4, we can readily find either the global Jacobian matrix J, or a local Jacobian matrix J_k for finger k ($k = 1, \cdots 5$) under all the corresponding homogeneous transformation matrices prepared. Furthermore, to determine the hand dynamics, with all the sub-Jacobian calculated, we can directly apply equation (7.30) or (7.32) from Chapter 7 to find the 21 by 21 hand inertial matrix W numerically in MATLAB^{TM}.

As a detail, based on the hand joint/link frame assignment and the D-H tables, the local Jacobian matrix for the thumb ($k = 1$) is given by

$$J_t = \begin{pmatrix} p_t^w \times r_t^w & p_t^1 \times r_t^1 & p_t^2 \times r_t^2 & p_t^3 \times r_t^3 & p_t^4 \times r_t^4 \\ r_t^w & r_t^1 & r_t^2 & r_t^3 & r_t^4 \end{pmatrix},$$

and the local Jacobian matrices for the index, middle, ring and pinky fingers can also be formulated, respectively, as follows:

$$J_i = \begin{pmatrix} p_i^w \times r_i^w & p_i^5 \times r_i^5 & p_i^6 \times r_i^6 & p_i^7 \times r_i^7 & p_i^8 \times r_i^8 \\ r_i^w & r_i^5 & r_i^6 & r_i^7 & r_i^8 \end{pmatrix},$$

$$J_m = \begin{pmatrix} p_m^w \times r_m^w & p_m^9 \times r_m^9 & p_m^{10} \times r_m^{10} & p_m^{11} \times r_m^{11} & p_m^{12} \times r_m^{12} \\ r_m^w & r_m^9 & r_m^{10} & r_m^{11} & r_m^{12} \end{pmatrix},$$

$$J_r = \begin{pmatrix} p_r^w \times r_r^w & p_r^{13} \times r_r^{13} & p_r^{14} \times r_r^{14} & p_r^{15} \times r_r^{15} & p_r^{16} \times r_r^{16} \\ r_r^w & r_r^{13} & r_r^{14} & r_r^{15} & r_r^{16} \end{pmatrix},$$

and

$$J_p = \begin{pmatrix} p_p^w \times r_p^w & p_p^{17} \times r_p^{17} & p_p^{18} \times r_p^{18} & p_p^{19} \times r_p^{19} & p_p^{20} \times r_p^{20} \\ r_p^w & r_p^{17} & r_p^{18} & r_p^{19} & r_p^{20} \end{pmatrix},$$

where each p_k^j and each r_k^j are the 4th and 3rd columns of the corresponding homogeneous transformation A_k^j. For instance, p_t^2 and r_t^2 of the thumb are the 4th and 3rd columns of $A_t^2 = (A_2^t)^{-1}$.

Because all the five fingers share the first joint rotation by θ_1 about the z_w-axis of the wrist, the complete 30 by 21 global Jacobian matrix J for either one of the two hands can be augmented by

$$J = \begin{pmatrix} J_t(:,1) & J_t(:,2:5) & O & O & O & O \\ J_i(:,1) & O & J_i(:,2:5) & O & O & O \\ J_m(:,1) & O & O & J_m(:,2:5) & O & O \\ J_r(:,1) & O & O & O & J_r(:,2:5) & O \\ J_p(:,1) & O & O & O & O & J_p(:,2:5) \end{pmatrix},$$

where each O is the 6 by 4 zero matrix.

Let the 21 by 1 joint position vector be defined by $q = (\theta_1\ \theta_2\ \cdots\ \theta_{21})^T \in \mathbb{R}^{21}$. Then, the global Jacobian equation can be written as

$$V = \begin{pmatrix} V_t \\ V_i \\ V_m \\ V_r \\ V_p \end{pmatrix} = J\dot{q},$$

where both the $6 \times 5 = 30$ dimensional Cartesian velocity V and the 30 by 21 global Jacobian matrix J must be projected onto a common frame. However, it can be immediately discovered that the above complete global Jacobian J is a tall matrix so that the Jacobian equation cannot have a compatible solution in general.

In fact, each finger tip motion may be limited only to translation without rotation in many grasping cases. If so, each local Jacobian matrix can be reduced to 3 by 5 by taking only the top 3 rows from the above 6 by 5 matrix form so that the global Jacobian matrix J will be reduced to 15 by 21, which is now a short matrix and is thus a kinematically redundant system. In this case, each local Jacobian matrix J_k can also be determined by directly taking partial derivatives of the finger tip position vector p_w^k with respect to the local joint position vector q_k. Namely,

$$J_{k(w)} = \frac{\partial p_w^k}{\partial q_k},$$

which is projected on the wrist frame w, i.e., the base of the hand. For instance, the 3 by 5 local Jacobian $J_{m(w)}$ for the middle finger, if only the tip linear velocity is considered, can be found by

$$J_{m(w)} = \frac{\partial p_w^m}{\partial q_m},$$

where $q_m = (\theta_1\ \theta_{10}\ \theta_{11}\ \theta_{12}\ \theta_{13})^T \in \mathbb{R}^5$ is the local joint position vector for the middle finger, and the middle finger tip position vector $p_w^m \in \mathbb{R}^3$ is the last column of the homogeneous transformation A_w^m.

10.3 Motion Planning and Formatting

In Chapter 9, a *five-point model* has been developed and discussed. The theoretical foundation is based on the Jacobian equation (9.14) with a global Jacobian matrix given in (9.13). The general solution is shown in equation (9.15) along with a vector z to carry multiple potential functions for optimization, as illustrated in Figure 9.6. In order to successfully generate a realistic digital human motion, two major things must be prepared before implementing the five-point model, and they are (1) find up to five Cartesian velocities that are augmented in V_b, and (2) define a number of potential functions for optimization.

To specify a Cartesian velocity V_i for each of the five points: the left hand, right hand, H-triangle, left foot, and right foot over a digital human body, we first have to plan a motion trajectory at each of the five points, such as a set of trajectories when the digital human in walking, running or jumping. Often, it suffices to prepare three trajectories: $x(t)$, $y(t)$ and $z(t)$ coordinates versus time to represent each point movement. The orientation changes, on the other hand, can be added only if they are necessary. Then, utilizing a lower-order approximation numerical algorithm to differentiate each trajectory with respect to time towards a Cartesian velocity V_i for each point.

In general cases, to plan a basic motion or a motion with small posture changes, such as walking or running, four sets of trajectories for two hands and two feet will be sufficient enough. Often, the trajectory of the H-triangle can be determined by taking average of the two feet paths. For instance, if V_{lf} and V_{rf} are the linear velocities of the left foot and right foot, respectively, then, the linear velocity of the H-point can be found by

$$V_H = \frac{V_{lf} + V_{rf}}{2}.$$

However, this averaging formula is only for linear velocity, not for angular velocity, and also, only for motions with a small posture variation.

In order to create all the desired trajectories for the four extremity points, let us first investigate what kind of each point movement a real person can be by taking a video on the real human in some basic motion activity, such as walking, and then capture frame by frame the coordinates of each of the four points from the video clip. To do so, a regular digital camera was utilized in the laboratory, which can generate a video clip in an avi file format with 24 frames per second (fps) and 1280×720 pixel resolution for each frame. Each video was taken about 6 to 8 seconds to record 144 to 192 frames in total. It may not have to capture the coordinates of the four points from every frame. Instead, just capture the data from each skipped frame, for instance, every even (or odd) number of frames, to reduce the data process. There are two approaches that can be adopted for such a motion capture and process in MATLAB^{TM} after a video clip is saved:

1. Directly call an internal function

```
implay('walk.avi')
```

   if the video file name is walk.avi. Once the video file is open in a MATLAB^{TM} window, using the "pixel region" tool to manually move the small pixel region window and place it to each of the four points, and then to copy and paste the pixel coordinates into a word processor file, such as Microsoft Word. Repeating the pixel region move, copy and paste steps again for every even (or odd) number of frames to record a sequence of coordinates for all the four points: the left and right hands and the left and right feet.
2. Let each of the four extremity points of the real human body be stuck with a high-contrast color spot, such as a small red sticker, and make sure that no any other place has the red color used over the entire background before taking a video. Then, we run the following program in MATLAB^{TM} to automatically retrieve all the coordinates on each video frame captured:

```
obj = VideoReader('walk.avi');
        % To Read the Video Data
nf = get(obj,'numberofframes');
        % Count the Number of Frames

P=[];   % Initiate a Coordinates Array

for ii=2:2:nf   % Recording Every Even Frame
    rgb = read(obj,ii);   % Get a 3D Array
    [M,N,k] = size(rgb);
    for i = 1:M
        for j = 1:N
            if rgb(i,j,1) > 220 && ...
               rgb(i,j,2) < 80 && ...
               rgb(i,j,3) < 80

                P=[P [i;j;ii]];
               % To Capture the Red Spots

            end
        end
    end
end
```

The resulting 3 by $2 \cdot nf$ array P in the above program contains all the useful data: the first two elements of each column of P are the x and

z-coordinates in numbers of pixels as a unit, and the third one of each column is the corresponding frame number.

Note that the value of the 3D array rgb(i, j, k) in the program represents the level of color k at the pixel of the i-th row and j-th column of frame ii, where $k = 1, 2, 3$ indicates the red, green and blue, respectively, and the level of each prime color is digitized from 0 to 248. Clearly, the above small program can effectively discriminate and retrieve the red spots frame by frame through a contrast threshold defined for each prime color.

Since only a single camera was used in the laboratory to capture the $x - z$ image on each video frame without the depth dimension y, such a simple but very convenient method is limited to a 2D motion capture. Due to the limitation, if the camera only shoots the side view of a real human in walking, one of the two hands will be blocked by the human body when the two hands swing alternatively. Therefore, it may only detect three red spots on some frames so that a necessary data interpolation may be required.

In addition to the 2D limitation, when reading an image file by MATLAB^{TM}, the pixel origin $(1, 1)$ is defined at the upper-left corner, because the program treats each image as a 720 by 1280 matrix. Thus, the output z-coordinates have to be converted from the the second element p_2 of each column of the 2D array P to $z = 720 - p_2$, while the x-coordinates remain unchanged.

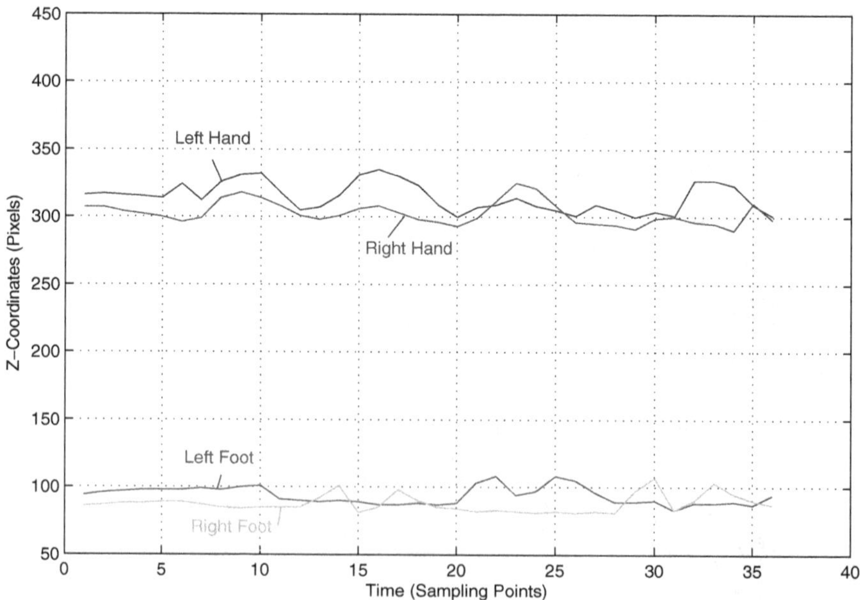

Fig. 10.15 A walking z-coordinates profile for the hands and feet from a motion capture

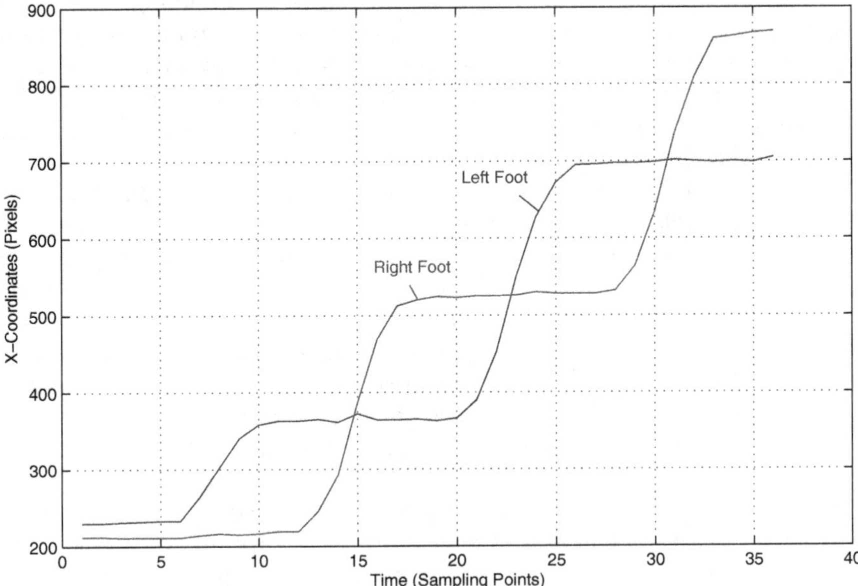

Fig. 10.16 A walking x-coordinates profile for the feet from a motion capture

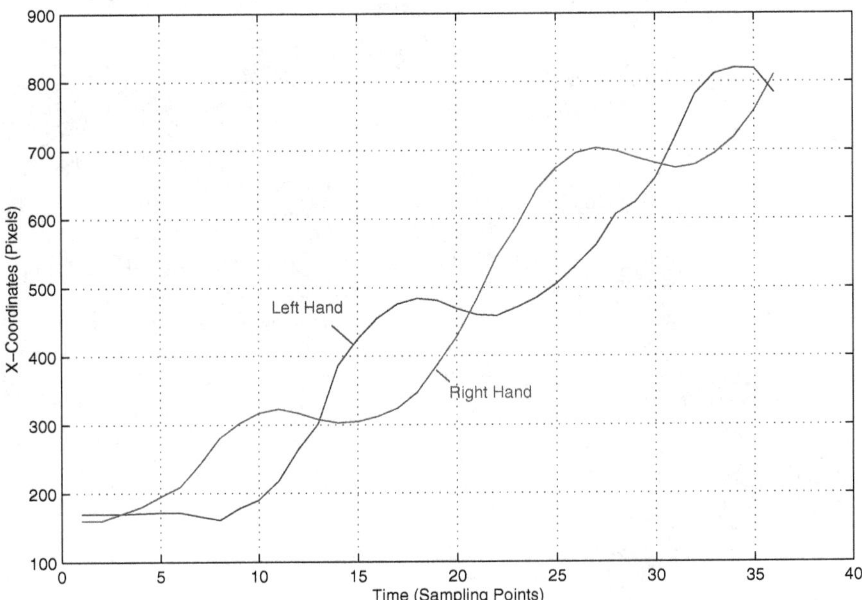

Fig. 10.17 A walking x-coordinates profile for the hands from a motion capture

Once the captured and recorded coordinates of each moving point for the real human in walking are connected respectively, they will generate in total four x-trajectories and four z-trajectories versus time in numbers of frames as a unit, which are plotted in Figures 10.15, 10.16 and 10.17. The vertical z-trajectories for the left and right hands and feet are shown in Figure 10.15. It is quite interesting that the horizontal x-trajectories for the left and right feet in Figure 10.16 are alternatively crossing to each other periodically, and each of them looks like a flat stair with a period about 16 sampling points (frames), and is referred to as an *F-stair*. While the horizontal x-trajectories for the left and right hands are also alternatively crossing to each other with the same period. However, each "stair" on the x-trajectories of the two hands emerges an N shape with a little overshooting and oscillation, and is called an *N-stair* due to each hand swinging motion during the walk. In terms of the synchronization, if the left foot strides first, then the right hand is the first one to swing forward, as can be observed and compared between Figure 10.16 and Figure 10.17.

Inspired by the motion capture trajectories recorded from a real human in walking, we now try a following simple numerical algorithm to mathematically generate a pair of the x and z-trajectories for each of the four "end-effector" points and make them ready to drive a digital mannequin to walk:

```
t = 0:19; tt = 0:0.5:19;  % The Original Time Scales

x_r=[0 0 0 1 2 2 2 3 4 4 4 5 6 6 6 7 8 8 8 9];
x_l=[0 0 1 1 1 2 3 3 3 4 5 5 5 6 7 7 7 8 9 9];

zm = 6;   zp = 0.6;

z_r1=[5 5 6 5 6 5 6 5 6 5 6 5 6 5 6 5 6 5 6 5]/zm;
z_r2=[0 0 zp 1 0 0 0 0 0 0 zp 1 0 0 0 0 0 0 zp ...
      1 0 0 0 0 0 0 zp 1 0 0 0 0 0 0 zp 1 0 0 0]/zm;
z_l2=[0 0 0 0 0 0 zp 1 0 0 0 0 0 0 zp 1 0 0 0 0 ...
      0 0 zp 1 0 0 0 0 0 0 zp 1 0 0 0 0 0 0 zp]/zm;

ti = 0:0.1:19;   % The New Time Scale

x_rf = pchip(t,x_l,ti);
x_lf = pchip(t,x_r,ti);   % Two Feet x-Trajectories

x_rh = spline(t,x_r,ti);
x_lh = spline(t,x_l,ti);  % Two Hands x-Trajectories

z_rh = pchip(t,z_r1,ti);
z_lh = pchip(t,z_r1,ti);
z_rf = pchip(tt,z_r2,ti);
z_lf = pchip(tt,z_l2,ti);  % All the z-Trajectories
```

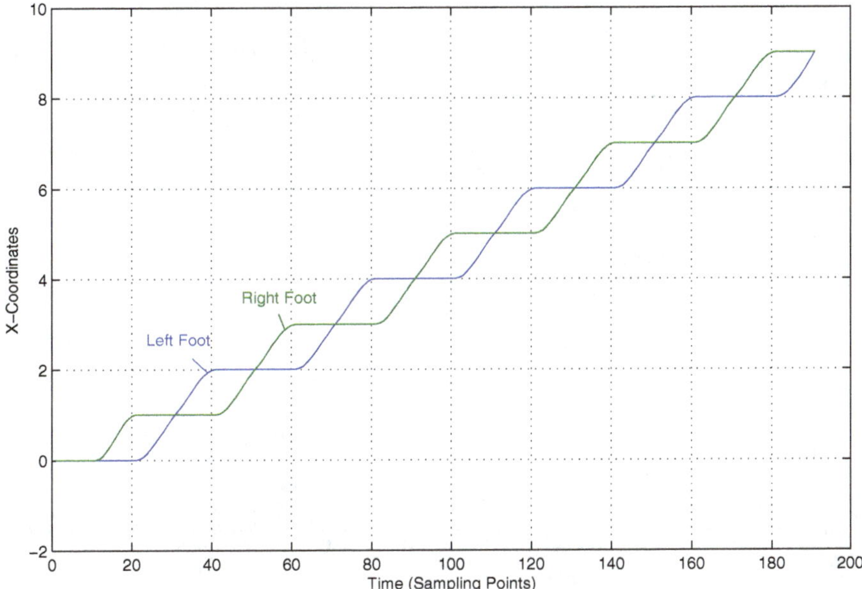

Fig. 10.18 A walking x-coordinates profile for the feet created by a numerical algorithm

With such a numerical algorithm based motion planning, all the coordinate profiles are plotted in Figures 10.18, 10.19 and 10.20. In the above program, we first define a numerical array for each trajectory, and then apply two internal functions: pchip($\cdots$) and spline($\cdots$) in MATLAB^{TM} to interpolate each coordinate array for more details. The function pchip(t, x, ti) is a piecewise cubic Hermite interpolating polynomial that can generate a more detailed and higher resolution data array sampled by ti from the original x sampled by t with the sampling time interval $\Delta ti < \Delta t$. The new trajectory by such a special piecewise interpolation preserves the shape of the original trajectory with less oscillation and no overshooting, which is well-suitable to emulate the F-stair x-trajectories for the two feet as well as the z-trajectories for both the two hands and two feet, as depicted in Figures 10.18 and 10.20.

In contrast, to numerically generate a shape similar to the N-stair x-trajectories for the two hands, as given in Figure 10.17, we may adopt another function: spline(t, x, ti) that can smooth every corner of turning segments but may create some overshooting and oscillation, resulting in Figure 10.19, which is, however, perfect for emulating the N-stair x-trajectories of the two hands.

By more closely inspecting the z-trajectories for the left and right two feet in Figure 10.20, we can further observe that while both the two feet move up and down alternatively and periodically, they never have any overlap between their "up" phases. This shows that both the two feet never leave from the ground

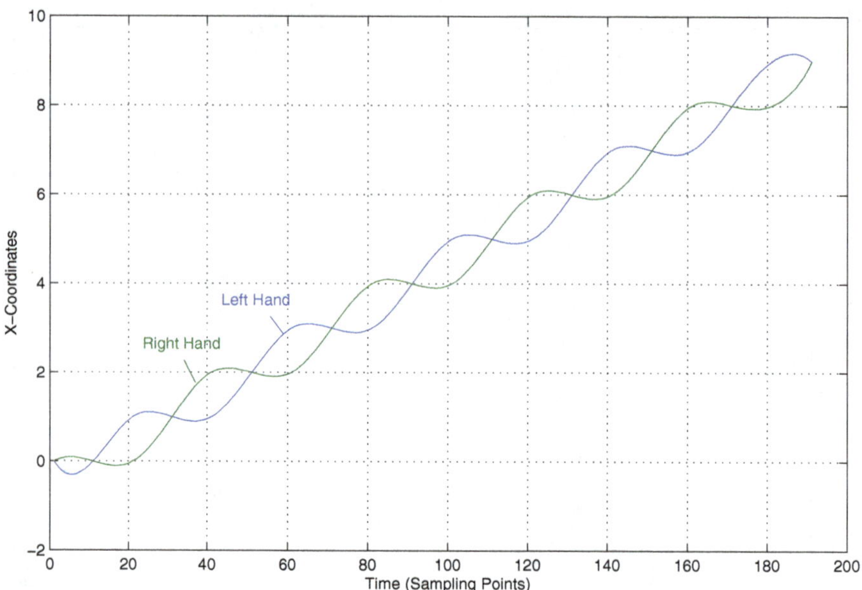

Fig. 10.19 A walking x-coordinates profile for the hands created by a numerical algorithm

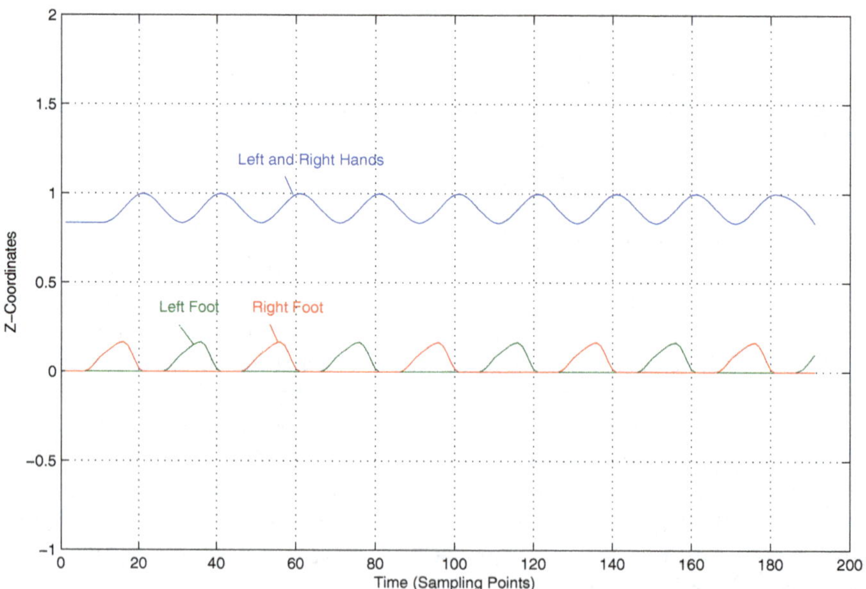

Fig. 10.20 A walking z-coordinates profile for both the feet and hands created by a numerical algorithm

simultaneously for a real human in walking. In contrast, there must be a short time interval in each period, where both the left and right feet move up and leave from the ground simultaneously when a real human is running. Based on this reality, we may keep the similar patterns of all the x-trajectories for walking, and only remodel the z-trajectories for the two feet to make each of them have an overlap in each period, as shown in Figure 10.21. Such a simple modification can immediately extend a walking activity to running by adding a little more joint and Cartesian accents or decorations.

For instance, the initial values of the two elbow joint angles θ_{18} and θ_{27} may be increased from 10^0 to 60^0 for running. Also, a linearly increasing angular velocity ω_y is added into the 6-dimensional H-triangle Cartesian velocity vector V_H, i.e., $V_H(5) = \omega_y$, in order to allow the mannequin body to gradually lean forward during the first a few periods. Of course, the entire $5 \times 6 = 30$-dimensional augmented velocity vector V_b must be magnified in value as well in the running case.

As a result, some snapshots of the mannequin in walking are shown in Figure 10.22, while the mannequin in running is displayed in Figure 10.23. Note that in the latter case, the mannequin is running along a circular track, instead of straight forward. To do so, the easiest way is to impose a rotation on joint 4 that is the global body spin axis, see the D-H tables for a digital human model in Chapter 9. More accents and decorations in both joint and Cartesian spaces may be added in order to make the digital human in walking or running be more realistic, and they will be further addressed in the next section.

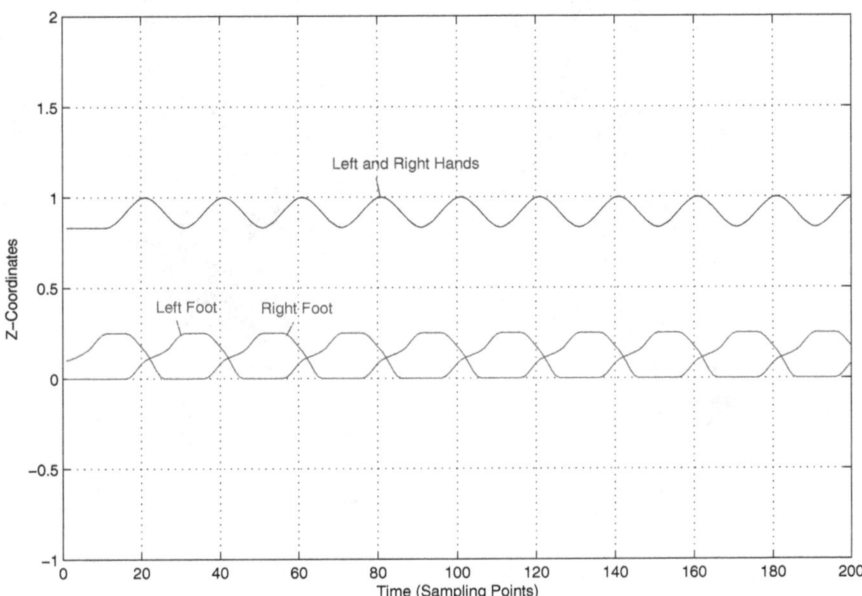

Fig. 10.21 z-trajectories in a running case for the feet and hands created by a numerical model

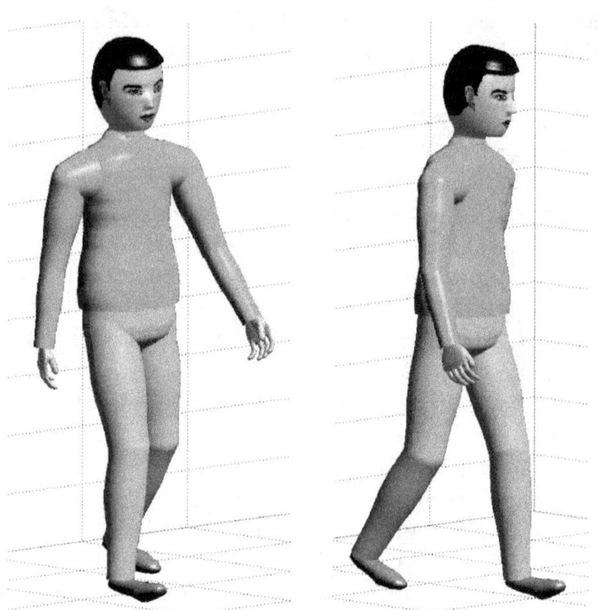

Fig. 10.22 A digital human in walking

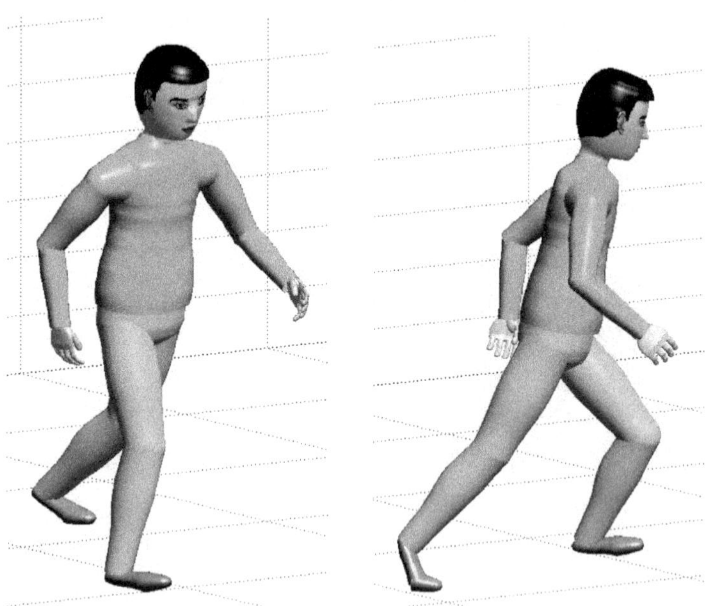

Fig. 10.23 A digital human in running

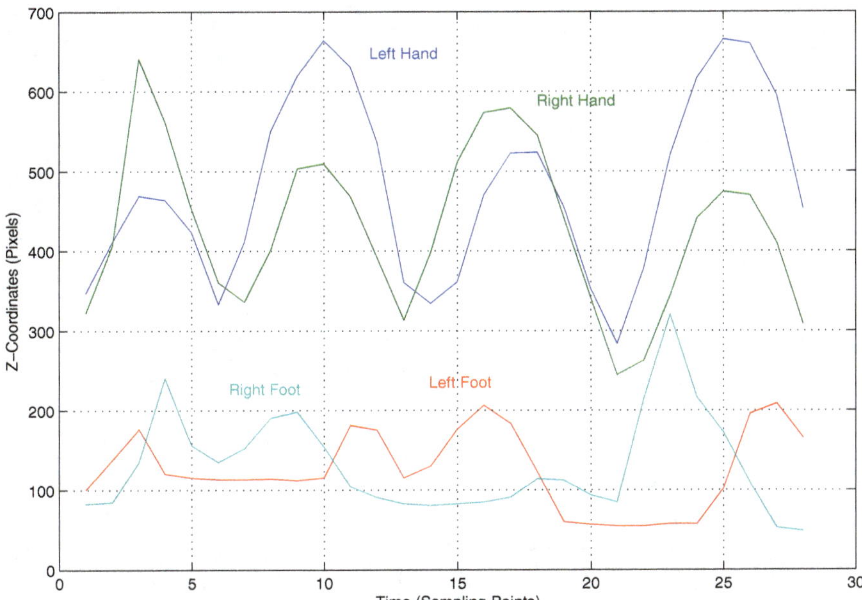

Fig. 10.24 z-trajectories in a jumping case for the feet and hands by a motion capture

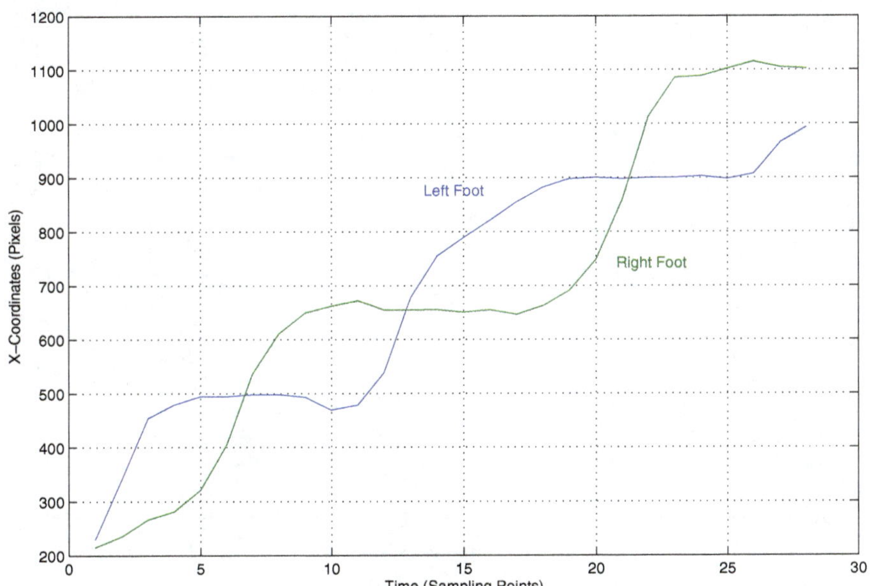

Fig. 10.25 x-trajectories in a jumping case for the two feet by a motion capture

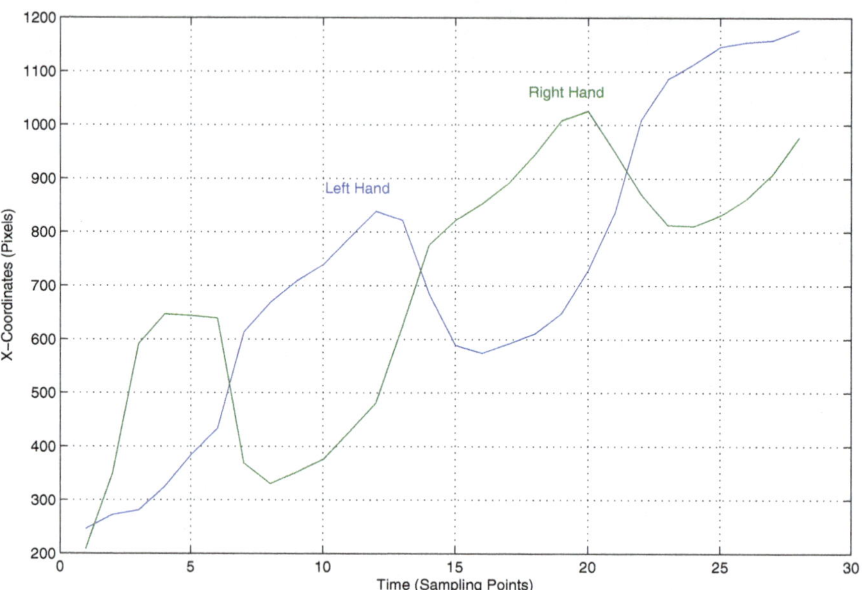

Fig. 10.26 x-trajectories in a jumping case for the two hands by a motion capture

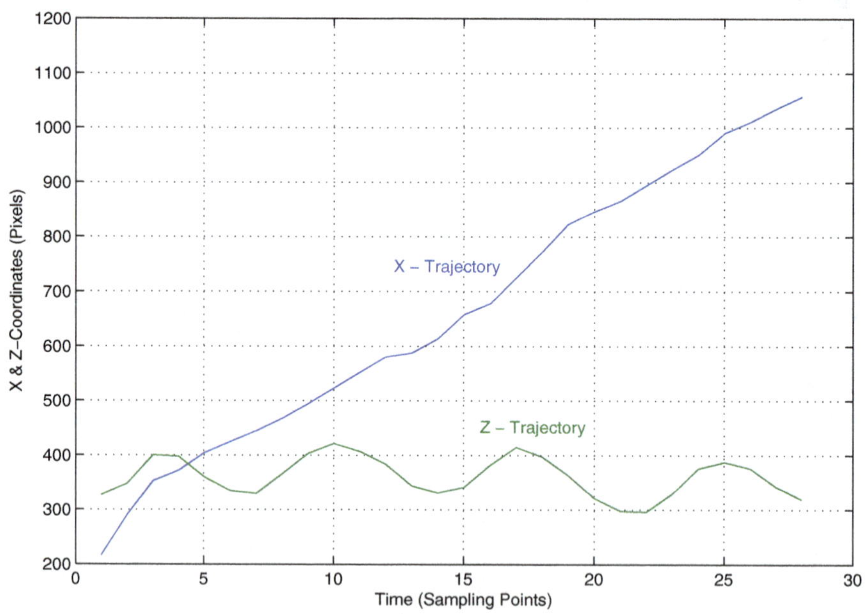

Fig. 10.27 x and z-trajectories in a jumping case for the H-triangle by a motion capture

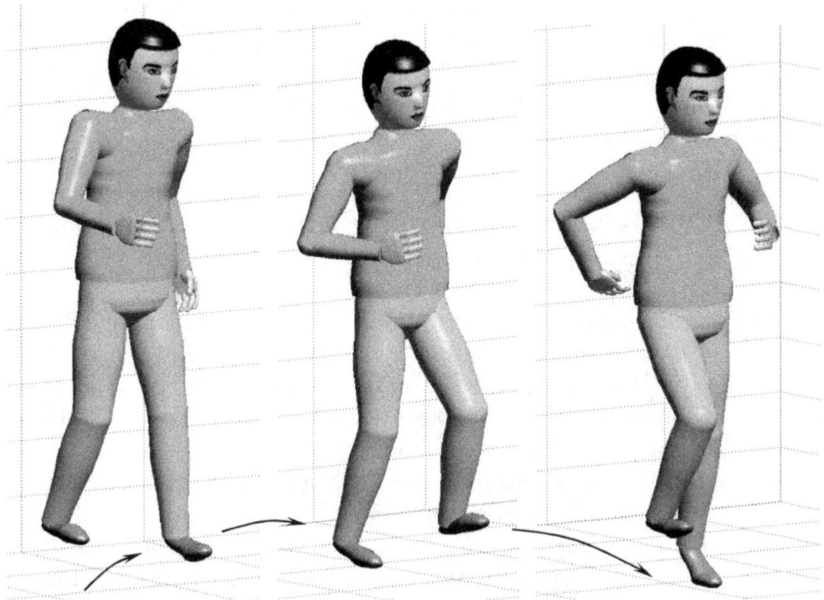

Fig. 10.28 A digital human in jumping

Furthermore, in order to generate a digital human in jumping, both the two feet must have even longer time interval to leave the ground simultaneously and every trajectory may have much higher magnitude than the running or walking. Therefore, it is necessary to capture or numerically create the coordinate trajectories for all the five moving points, including the H-triangle in a jumping case. In the laboratory, a real person in jumping has also been captured by the same camera and data-retrieving process, and both the z and x-trajectories for all the five points are plotted in Figures 10.24, 10.25, 10.26 and 10.27. Once all the real human motion capture trajectories are implemented into the five-point model to drive the digital human in motion (the implementation procedures will be discussed in the later section), a realistic jumping motion can immediately be generated for a digital mannequin in computer, and three frames are saved and demonstrated in Figure 10.28.

It is obvious that a jumping motion can be in many different ways, and the above particular jumping motion capture with its snapshots is just a kind of typical periodic jumping case. No matter what kind of jumping activity is simulated, unlike the walking and running cases, it is often required to find all five sets of trajectories to meet the five-point model. As we can clearly see in Figure 10.27, the H-triangle trajectories in both x and z directions do not exactly match with the average of the two feet trajectories in Figures 10.24 and 10.25 due to the variation and complication of the H-triangle movement when jumping. In fact, an instantaneous position and velocity of the H-triangle is a major or even dominant factor to affect each instantaneous posture for a real human in jumping or in other high-speed and high-magnitude movements, such as in most sport activities. An analysis of the three basic human motions: walking, running and jumping with their comparative study will be detailed in the next section.

10.4 Analysis of Basic Human Motions: Walking, Running and Jumping

The three basic human motions: walking, running and jumping have been introduced, modeled and illustrated by a number of simulation studies in the last section. It is commonly recognizable that these three basic motion activities have a large number of variations, subject to the age, gender, anthropometric size and weight, personality, mood and physical condition. Indeed, women walk, run or jump differently from men. A taller man (woman) often has a distinct gait from a short man (woman). An elderly walks also differently in comparison with a young person in terms of their gaits and speeds [9]. The variety of different gaits and speeds in walking, running and jumping apparently reflect different kinematics, but they are all root-caused by different dynamics, and even by different physiology and psychology. In a large volume of the literature on kinesiology and biomechanics, researchers constantly emphasize those variations and complications. In other words, all the research results, conclusions, comments and suggestions intend to justify a fact that there is no machine or mathematical model made by the best engineering scientist that could replace the natural movement of a real human machine [10]–[13]. It could be true, but would never discourage our efforts to model and mimic a real human movement.

Facing so many challenging issues, variations and complications, we will continue to aim at a general mathematical model development that can mimic real human motions with less sensitivity to the variation of anthropometric data and more flexibility to adapt for different ages and genders. To pursue the goal, we now start with a detailed analysis of the three basic human motions. Let us first look at the following comparison in Table 10.7 among the three basic motions in terms of their kinematic modeling and motion-planning methodologies as well as the necessary joint and Cartesian accents and decorations:

At first glance, Table 10.7 seems to be a comparison only in mathematical modeling aspects. Actually, with a small number of accents and decorations added into the five-point model, the resulting digital human basic motions will be better to match the real human movements. In other words, in addition to specifying all the required five-point trajectories as a kinematic foundation, adding more accents and decorations at both the joint and Cartesian levels can mimic the basic motions more realistically. For instance, the clavicle joints: θ_{13} and θ_{14} on the left side, and θ_{22} and θ_{23} on the right side, may

Table 10.7 A comparative study among the three basic motion generations

Comparison	Walking	Running	Jumping
Motion Model	Four-Point	Four-Point	Five-Point
x-Traj for Feet	F-Stair	F-Stair	High F-Stair
x-Traj for Hands	N-Stair	N-Stair	High N-Stair
x-Traj for H-Point	Average of Feet	Average of Feet	More Dynamic Move
z-Traj for H/F	Waves	Waves	High Waves
z-Traj for H-Point	Average of Feet	Average of Feet	More Dynamic Move
Clavicle Up/Down	Mild	Moderate	Higher
Elbow Joint Angles	$\sim 10^0$	$\sim 60^0$	$\sim 10 - 120^0$
H-Swinging ω_x	Small	Medium	High
H-Rocking ω_y	No	Small	Less or No
H-Rolling ω_z	Lower	Higher	Lower
Ankle Rotation ω_y	Small	Medium	High

be activated more sensitively for a digital human in the three basic motions, as an accent at the joint level beyond the five sets of trajectories, especially in the jumping case. This can be done by reducing the weights for those joints in the joint comfort potential function defined by equation (9.24) in Chapter 9.

Another example is to add a decoration for a small variation of the H-triangle or pelvis orientation. It is commonly observable that a woman walks often with a more swinging pelvis than a man. In contrast, a man walks with more rolling pelvis than a woman. More precisely, in the motion planning, the 6-dimensional H-triangle Cartesian velocity V_H should not shut off the last three angular velocity components and only consider the first three linear velocity components. Instead, we should define

$$V_H = (v_x \;\; v_y \;\; v_z \;\; \omega_x \;\; 0 \;\; \omega_z)^T, \tag{10.2}$$

and make the two non-zero angular velocities ω_x and ω_z be synchronized with the periodic movements of the two feet. Specifically, if the linear velocities of the left and right feet in the x-direction are v_{xlf} and v_{xrf}, respectively, then, let $dw = v_{xrf} - v_{xlf}$, and let both ω_x and ω_z be defined to be proportional to dw but with two different proportional coefficients, i.e., $\omega_x = k_1 dw$ and $\omega_z = -k_2 dw$, where $k_1 > 0$ and $k_2 > 0$ are two positive constants.

Therefore, in mathematical language, women often have higher amplitude of the periodic angular velocity ω_x for the H-triangle than men, while men have higher amplitude of ω_z than women when they are walking or running. By setting different k_1 and k_2, the generated gaits on the screen would be distinguishable between the two genders.

A small ankle rotation is also a noticeable decoration to enhance the realistic basic motion generation. This can be done by adding a $\omega_y \neq 0$ into each 6-dimensional Cartesian velocity vector of the two feet. Namely, add $\omega_y = kdw$ to the left foot Cartesian velocity V_{lf} and add $\omega_y = -kdw$ to the right foot V_{rf} with a constant scalar $k > 0$. Such a small periodic adjustment for the two ankle rotations can also improve the walking and running motions effectively.

A gaze control by dynamically adjusting the three neck/head angles θ_{45}, θ_{46} and θ_{47} will also help the digital human motion for more natural looking. A gaze-target trajectory associated with a possible cognitive interaction model is required to control the orientation of digital human vision via the head joint motion.

Moreover, in the entire motion planning process, either to generate every trajectory for the five moving points, or to decorate the digital human motion by adding more accents, a common key issue is to clarify what frame is the reference to define all of the quantities. Clearly, it is desirable and also preferable that all the five sets of trajectories as well as every accent can be defined with respect to a frame that always follows the digital human in motion, instead of the fixed world base frame b or the digital human base frame 0.

For instance, when a mannequin is running along a circular track, the entire body has to spin gradually. If one stands on the world base to look at the mannequin in gradually spinning, all the trajectories that are originally planned have to be re-projected onto the fixed world base. Then, the initial x or z-trajectory will no longer be the same x or z-profile, while the y-trajectory will no longer be zero. If now a new reference frame can follow the spinning digital human body, all the initial trajectories will have no need to change. This will definitely bring a lot of convenience to planning and controlling the digital human motions.

By revisiting the digital human modeling with the assignment of all the coordinate frames based on the D-H convention, as seen in Figure 9.4 from Chapter 9 as well as the D-H tables, we can immediately discover that the first three joints are all prismatic so that they will never change the orientation of frame 3 with respect to the digital human base frame 0 or the world base frame b. Fortunately, frame 3 always has exactly the same orientation as the world base. Therefore, frame 4 that is the frame right after the body spin by the joint angle θ_4 with respect to frame 3 has the following orientation directly referred to frame b:

$$R_b^4 = R_3^4 = \begin{pmatrix} c_4 & 0 & -s_4 \\ s_4 & 0 & c_4 \\ 0 & -1 & 0 \end{pmatrix}.$$

Thus, if we define a frame that has the same orientation at $\theta_4 = 0$ and is rotated by $\theta_4 \neq 0$ about the z_b-axis of frame b to be a **human centered frame**, then, this frame has the following simple orientation with respect to the world base frame b:

$$R_b^{hc} = R_3^{hc} = \begin{pmatrix} c_4 & -s_4 & 0 \\ s_4 & c_4 & 0 \\ 0 & 0 & 1 \end{pmatrix}. \tag{10.3}$$

With such a new motion-planning reference frame defined, every trajectory that was planned by the previous procedures for walking and running can just be pre-multiplied by R_b^{hc} to refer them to the world base if a digital human body is gradually spinning. The human centered frame will also offer the easiest way to define and add any accent or decoration in Cartesian space. For instance, the aforementioned ω_x, ω_y and ω_z for either the H-triangle or the two feet of the digital human can now be referred to the human centered frame. In the case of a mannequin running along a circular track, each Cartesian velocity vector, including V_H in (10.2), can keep using the same one as the mannequin running linearly straightforward, but pre-multiplying them by the rotation matrix R_b^{hc} given in (10.3), i.e.,

$$V_H = \begin{pmatrix} R_b^{hc} \begin{pmatrix} v_x \\ v_y \\ v_z \end{pmatrix} \\ R_b^{hc} \begin{pmatrix} \omega_x \\ 0 \\ \omega_z \end{pmatrix} \end{pmatrix}.$$

This pre-multiplication should also be applied to all the other Cartesian velocities: V_{lh}, V_{rh}, V_{lf} and V_{rf} by a spinning angle ϕ that is equivalent to

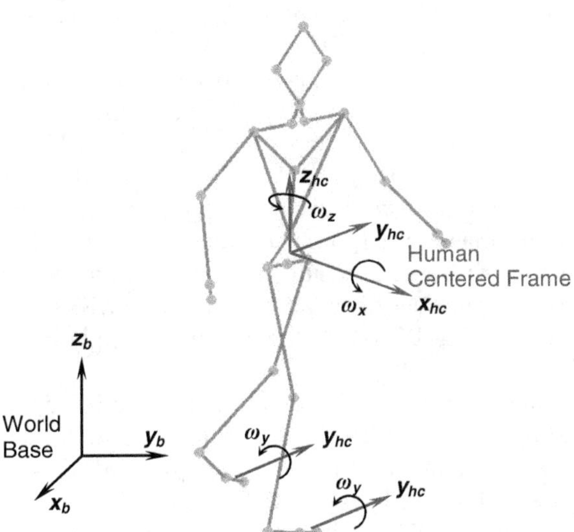

Fig. 10.29 A relation diagram between the human centered frame and the world base

θ_4 in a velocity-generating subroutine. Note that the spinning angle ϕ in the subroutine is represented in Cartesian space, while θ_4 in the main program is a joint value. If using the subroutine with ϕ, the joint angle θ_4 in the main program should not be rotated again, and just leave θ_4 as is. The definition of the human centered frame and the decorations ω_x, ω_y and ω_z as well as the world base frame b are all depicted in Figure 10.29.

10.5 Generation of Digital Human Realistic Motions

After the analysis of the three basic human motions: walking, running and jumping, we discover that to mimic a real human motion in a wide spectrum: from slow to fast motion and from less active to more active motion, in addition to planning a set of trajectories to meet the five-point model requirements, a number of appropriate accents and decorations are also necessary. Therefore, an entire motion planning and implementation process for simulating a human activity should consist of three major steps: (1) trajectory determination; (2) accent and decoration identification; and (3) five-point model implementation. More specifically,

Step 1: Trajectory planning, motion partitioning, sequential motion spline and refinement, and numerical differentiation.

Step 2: Joint accent identification, posture alteration, motion trimming and decorating.

Step 3: Local and global singularity avoidance, subtask optimization, and Cartesian velocity on/off control.

Step 1 paves the road for a journey of digital human realistic motion generation as an infrastructural task. In most cases, modeling a set of trajectories for the five-point differential motion based model is often inspired by a motion capture result from a real human. Although the basic motions, such as walking and running, are getting matured without need of any more motion capture inspiration, they are just a small fraction of the various kinds of human motion. Therefore, we wish to collect more different human motion patterns into a database to help for creating a more variety of digital human motion trajectories with less dependence on the real human motion captures. Let us now browse a number of typical human activities that have been modeled and animated in MATLAB^{TM} before further explaining the three major steps.

The first animation study is a mannequin going to run and throw a ball forward, as exhibited in Figures 10.30, 10.31 and 10.32. The second one is a mannequin climbing up a stair and then jumping down to the floor, as shown in Figures 10.33, 10.34 and 10.35. The third animation is a digital human in springboard diving, a few frames of which are saved and shown in Figures 10.36, 10.37, 10.38 and 10.39.

One of the most challenging digital human motion generations is to create a mannequin in a sequential ingress motion: walking and getting into a

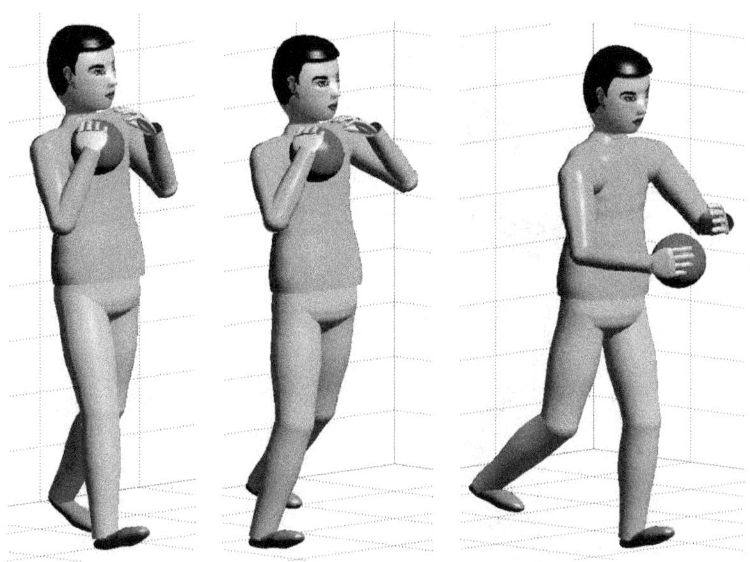

Fig. 10.30 A digital human in running and ball-throwing

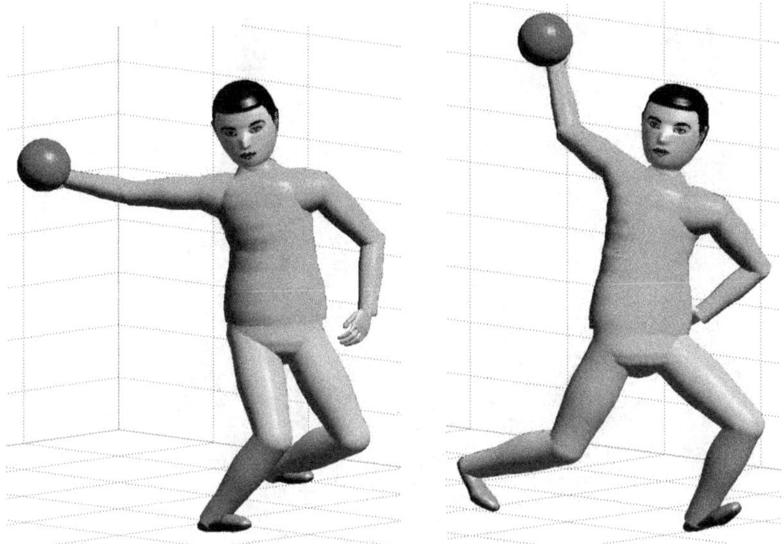

Fig. 10.31 A digital human in ball-throwing

car and then to seat, under a number of car interior physical constraints. Figures 10.40, 10.41 and 10.42 exhibit a few snapshots in the sequence of the ingress activity.

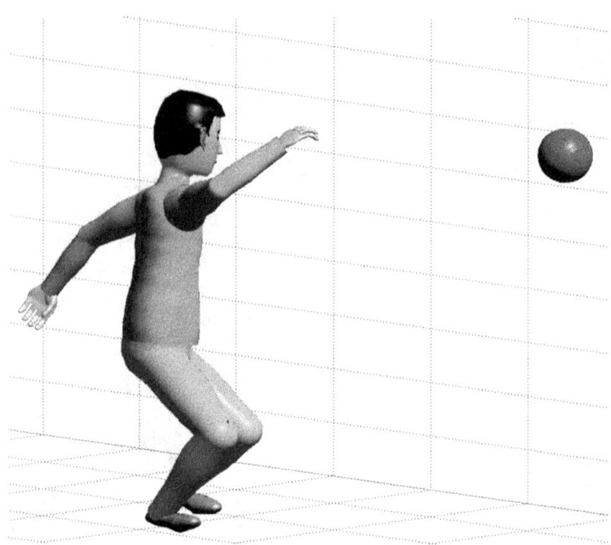

Fig. 10.32 A digital human in ball-throwing

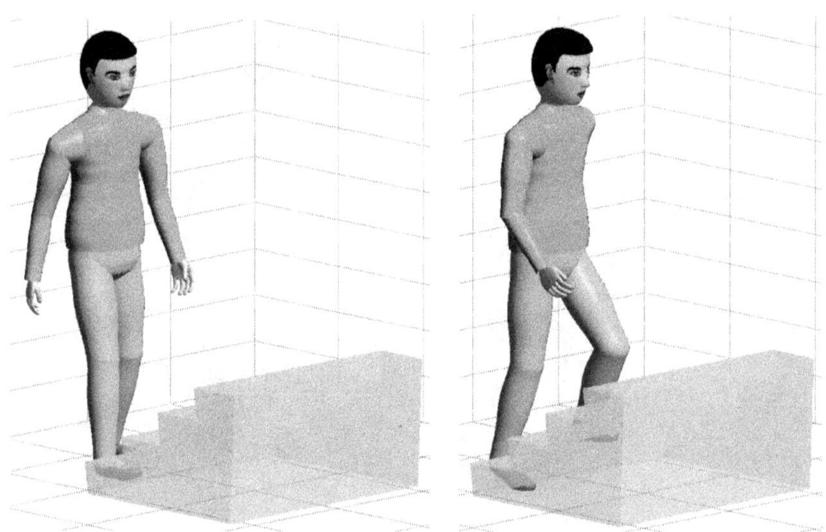

Fig. 10.33 A digital human is climbing up a stair

All the trajectories in the first running and ball-throwing case were motion captured by a real person in such a typical sport activity, see Figures 10.43, 10.44, 10.45 and 10.46.

It is quite interesting in Figure 10.43 that the two hands have nearly the same z-trajectories when both the left and right hands are carrying and hold-

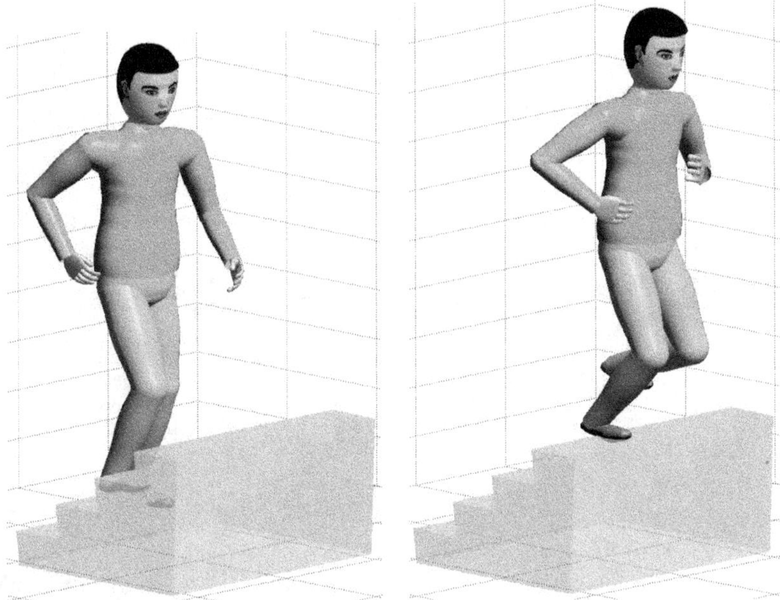

Fig. 10.34 A digital human is climbing up a stair and then jumping down

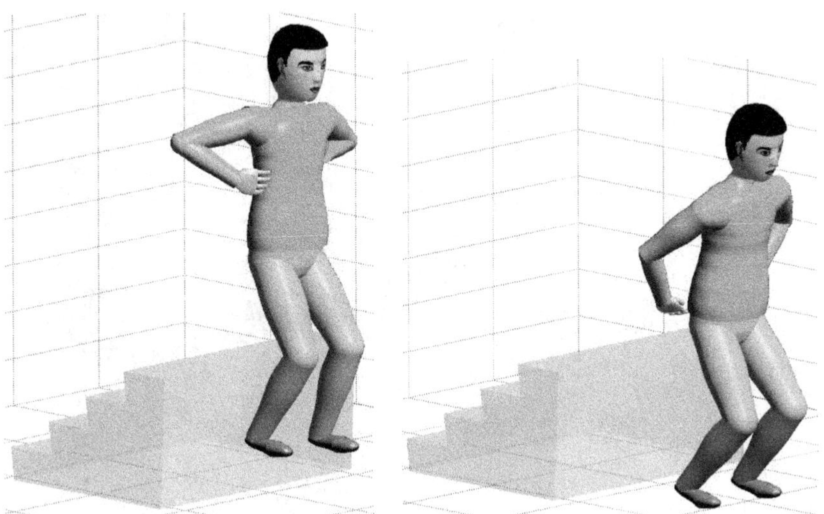

Fig. 10.35 A digital human is jumping down from the stair

ing the ball on the chest before the 18th sampling point. After that, the two hands start departing and the right hand is moving down first to make it ready to turn and lift the ball up after the 22nd sampling point, while the left hand is just moving down to balance the right hand, as seen on the third

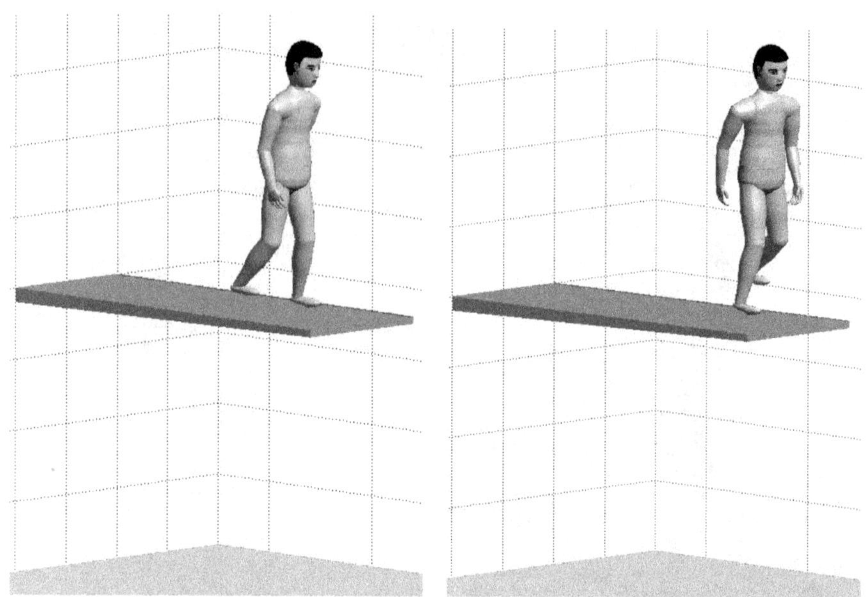

Fig. 10.36 A digital human in springboard diving

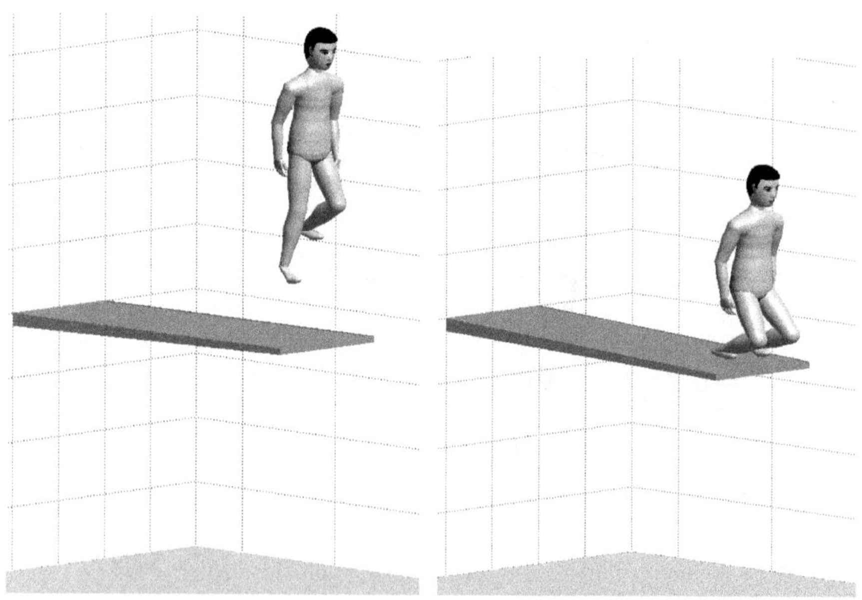

Fig. 10.37 A digital human in springboard diving

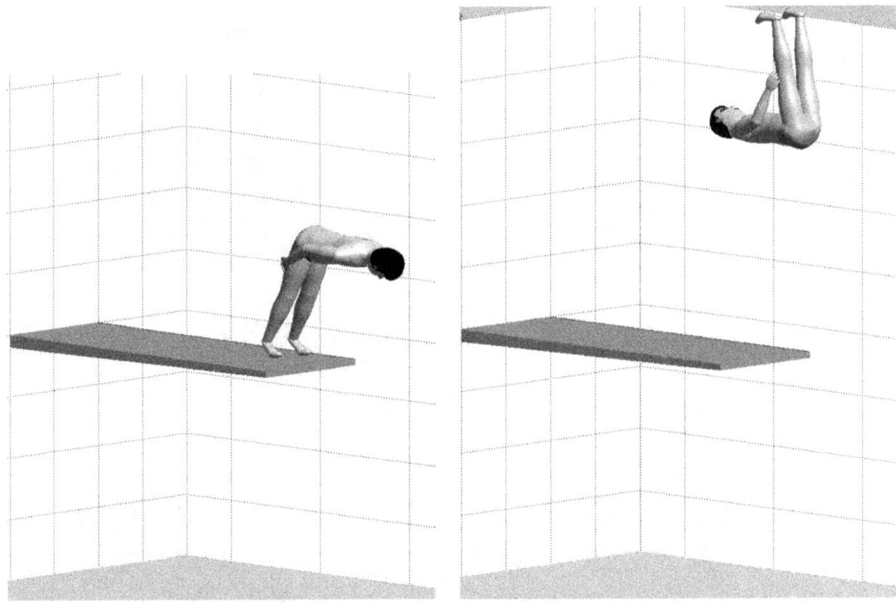

Fig. 10.38 A digital human in springboard diving

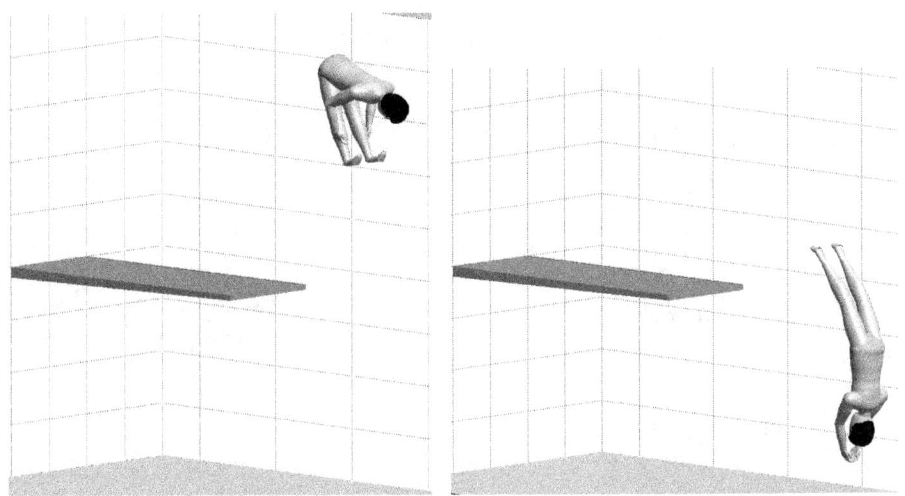

Fig. 10.39 A digital human in springboard diving

(rightmost) snapshot in Figure 10.30. Once the right hand reaches near the top level at the 30th sampling point, it becomes an instantaneous posture to throw the ball forward. The x-trajectories of the two feet in Figure 10.44

Fig. 10.40 A digital human is walking and getting into a car

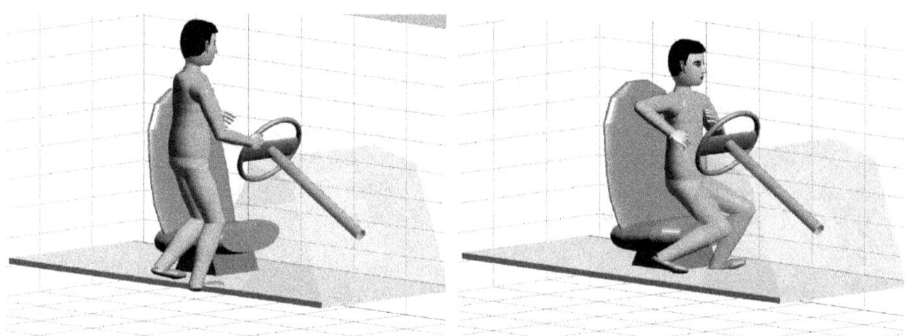

Fig. 10.41 A digital human is getting into the car

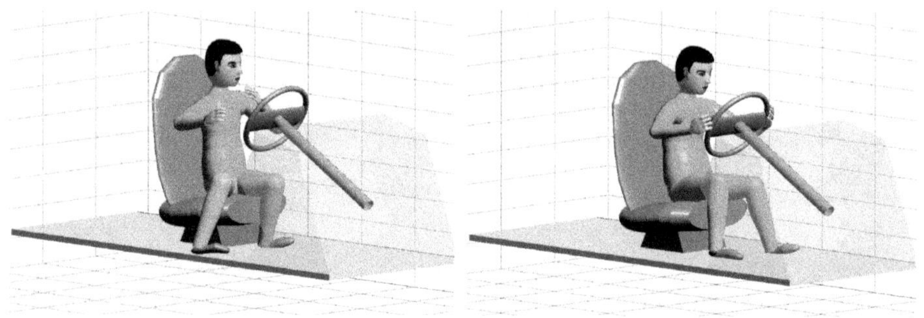

Fig. 10.42 A digital human is getting and seating into the car

look very similar to the walking case. However, a little deviation of the synchronization could lead to a very different instantaneous posture.

The two feet have a relatively longer standstill moment with the left foot behind the right one between the 10th and 13th sampling points, while the H-point keeps moving forward in the x-direction, see in Figure 10.46, resulting in a leaning forward instantaneous posture, as shown on the middle (second) snapshot in Figure 10.30.

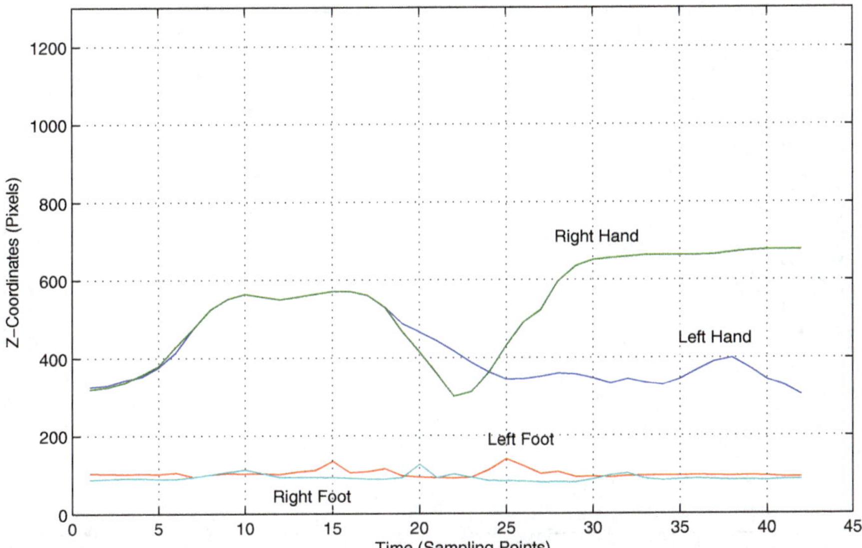

Fig. 10.43 z-trajectories in the ball-throwing case for the feet and hands by the motion capture

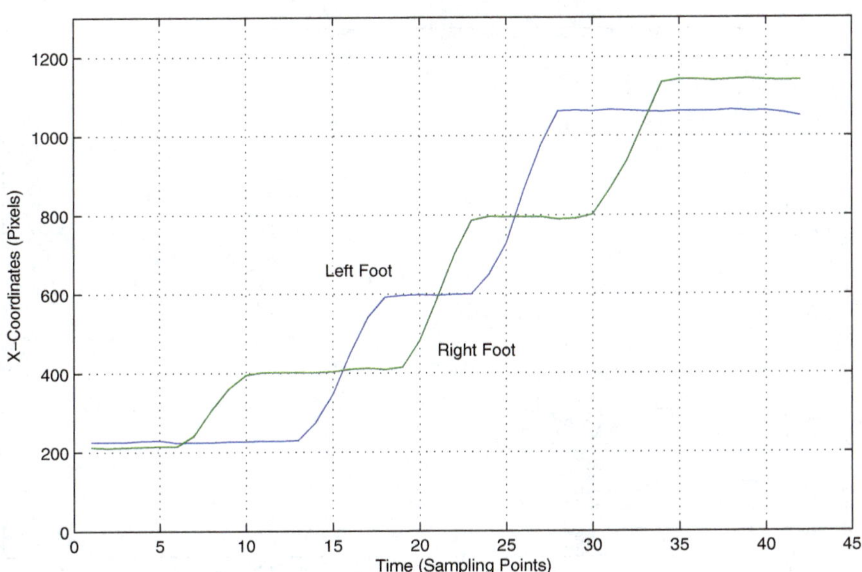

Fig. 10.44 x-trajectories in the ball-throwing case for the two feet by the motion capture

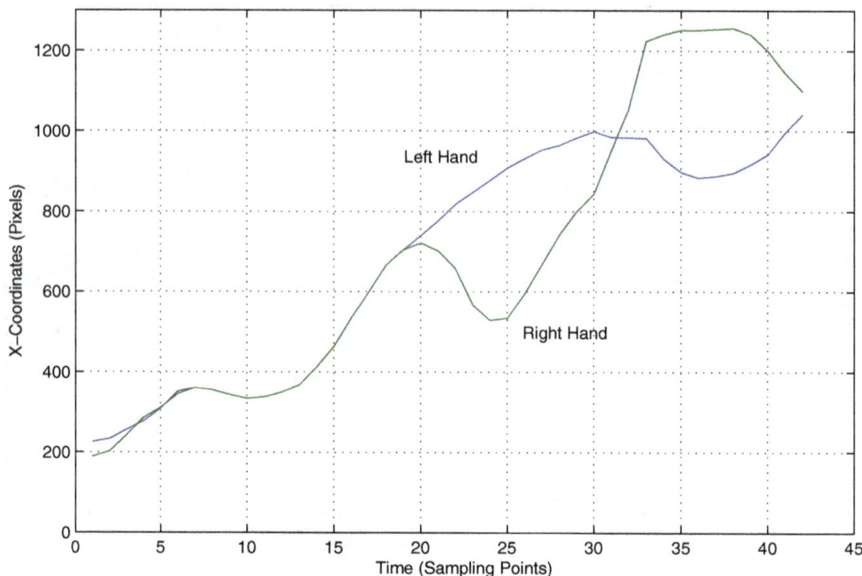

Fig. 10.45 x-trajectories in the ball-throwing case for the two hands by the motion capture

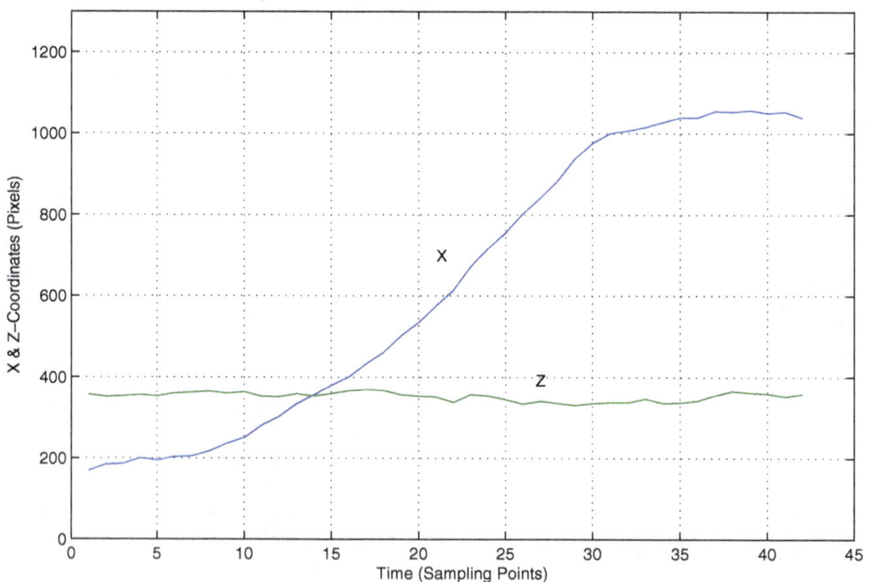

Fig. 10.46 x and z-trajectories in the ball-throwing case for the H-triangle by the motion capture

There is another standstill moment of the two feet between the 27th and 30th sampling points in Figure 10.44, which is the phase of driving the right hand along a circle clockwise in order to gain a momentum to speed up the ball for final ball-throwing, as can be seen from the two instantaneous postures in Figure 10.31. Also, Figure 10.45 demonstrates that the right hand, as a ball-carrier, is going backwards first and then turning forward between the 21st and 30th sampling points, which is actually a circular motion in clockwise. After the 30th sampling point, both the right hand and right foot are moving forward in the x-direction until the ball is thrown out at the 33rd sampling point.

Likewise, we are also interested in interpreting the trajectories recorded by a real human motion capture in the case of climbing up a stair and then jumping down to the floor, as animated in Figures 10.33, 10.34 and 10.35. All the trajectories are plotted in Figures 10.47, 10.48, 10.49 and 10.50.

It can be clearly seen that the z-trajectories for both the two hands and two feet in Figure 10.47 are nearly the same as the walking case, except that they are imposed with a bias of linearly stepping up and then steeply dropping down. In addition, on each step up, there is a small overshooting for the two feet z-trajectories to reflect a fact that the real human in climbing stair has to carefully move up each foot a little more inches before pressing down on each step of the stair.

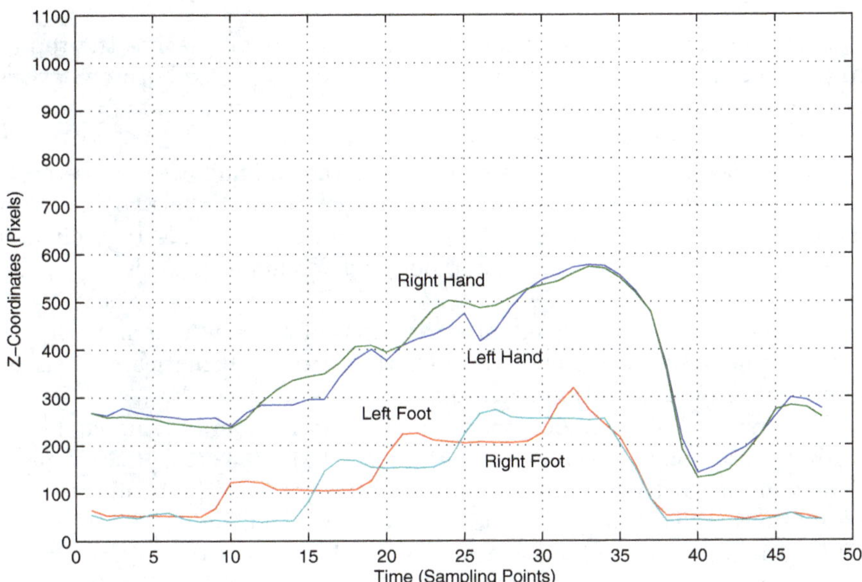

Fig. 10.47 z-trajectories in the stair-climbing/jumping case for the feet and hands by the motion capture

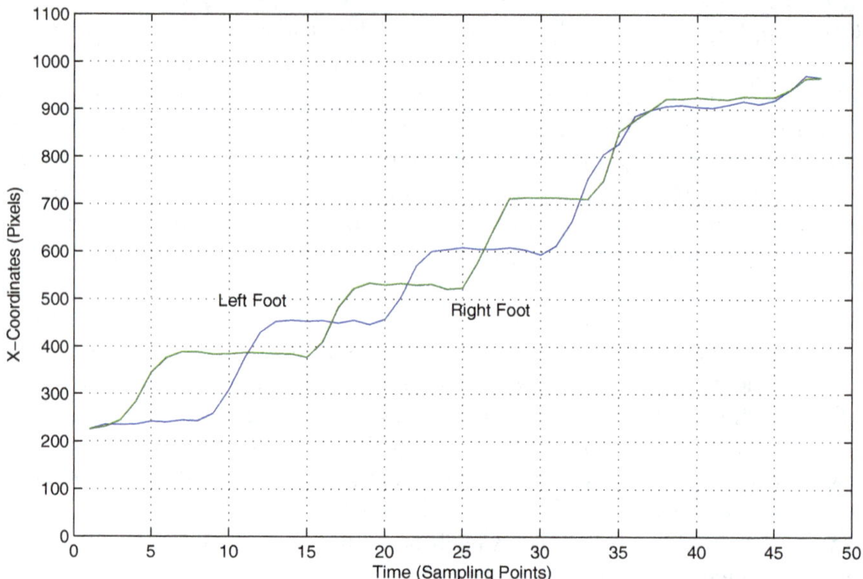

Fig. 10.48 x-trajectories in the stair-climbing/jumping case for the two feet by the motion capture

It is also natural enough that after the real human reaches the top step to be jumping down to the floor, all the z-trajectories for both the two feet and hands as well as the z-trajectory of the H-point commonly emerge a dropping-down shape, as exhibited in Figures 10.47 and 10.50. While the x-trajectories of the two feet are similar to the walking case before getting flat in the jumping-down phase, the x-trajectories of the two hands look more complicated than the walking and running cases due to their active movements as part of balancing the body when climbing up the stair, see in Figure 10.34.

Through the above detailed interpretations of the real human motion capture data, we now try to create a set of trajectories mathematically to animate more digital human motion activities.

The first trial is to mimic a mannequin for springboard diving, and the resulting animation is shown in Figures 10.36, 10.37, 10.38 and 10.39. This particular motion activity is partitioned into two phases: (1) running/jumping on the springboard, and (2) rolling in the sky and diving into the water. Phase 1 requires a set of trajectories to match the five-point differential motion model, and then a complete joint motion algorithm will take over the differential motion in Phase 2. Therefore, only the trajectories used in Phase 1 were created numerically by inspiring and learning from the motion capture data interpretations and are plotted in Figures 10.51, 10.52 and 10.53.

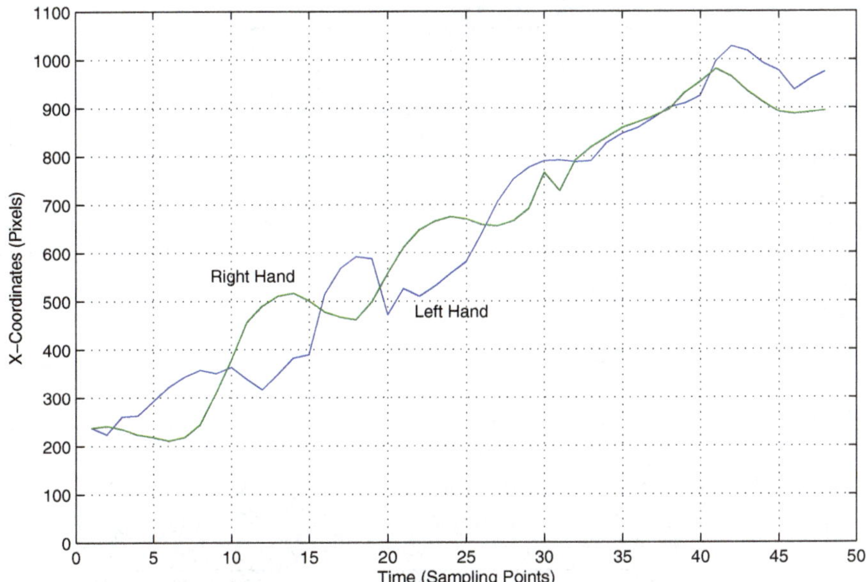

Fig. 10.49 x-trajectories in the stair-climbing/jumping case for the two hands by the motion capture

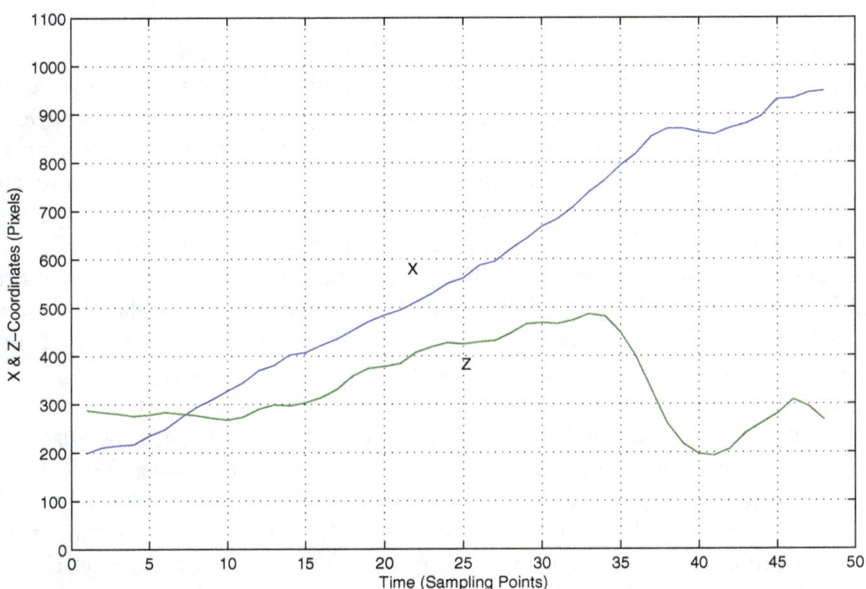

Fig. 10.50 x and z-trajectories in the stair-climbing/jumping case for the H-triangle by the motion capture

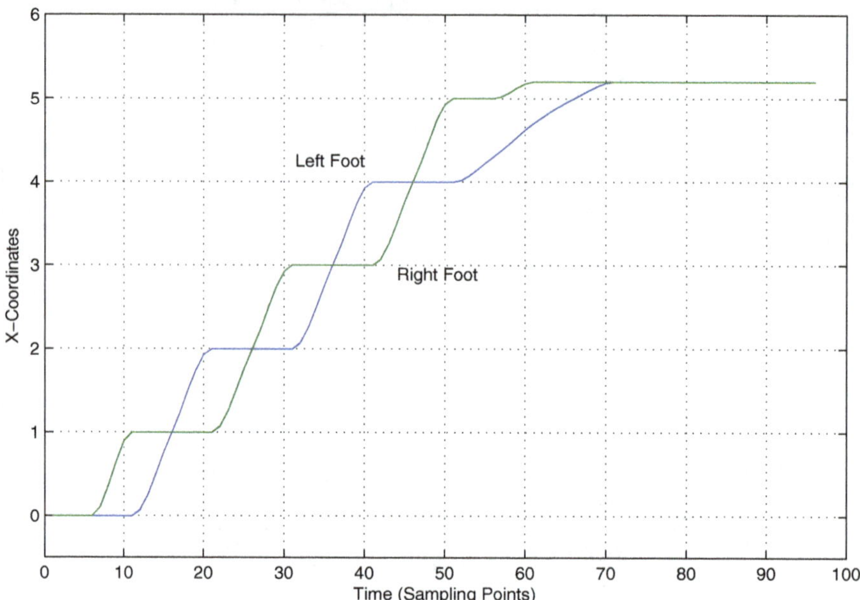

Fig. 10.51 x-trajectories in the springboard diving case for the two feet by a math model

As can be observed in Figure 10.53, before the 50th sampling point, every trajectory looks like a regular walking movement. After the 50th sampling point, the right foot starts pressing down the springboard firmly in order to gain a high rebound, and then both the two feet are jumping up with the left leg kneeling up, as shown in Figures 10.36 and 10.37. Once it reaches the 90th sampling point, the two feet are landed nearly simultaneously on the springboard and ready to make a final jump and then to take off for sky-rolling and diving. In fact, after the 70th sampling point, the x-trajectories of both the feet and hands have already been saturated to be flat, indicating no further movement in the forward direction, and only an up/down jumping motion, as depicted in Figures 10.51 and 10.52.

In the second phase of diving, a joint motion algorithm takes over the control of almost all the diving movements after leaving the springboard. The joint angles involved in this particular joint motion includes the shoulder, elbow, hip and knee flexions as well as the ankle dorsiflexions on both the left and right sides. However, to more effectively control the entire mannequin body rolling in the sky, a differential motion algorithm of the single H-point model that rotates about the y-axis of the human centered frame is employed.

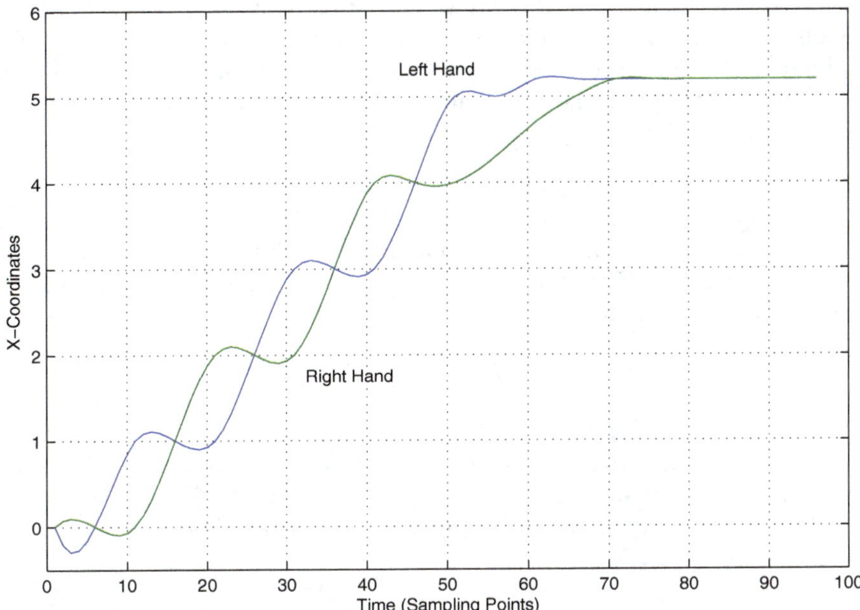

Fig. 10.52 x-trajectories in the springboard diving case for the two hands by a math model

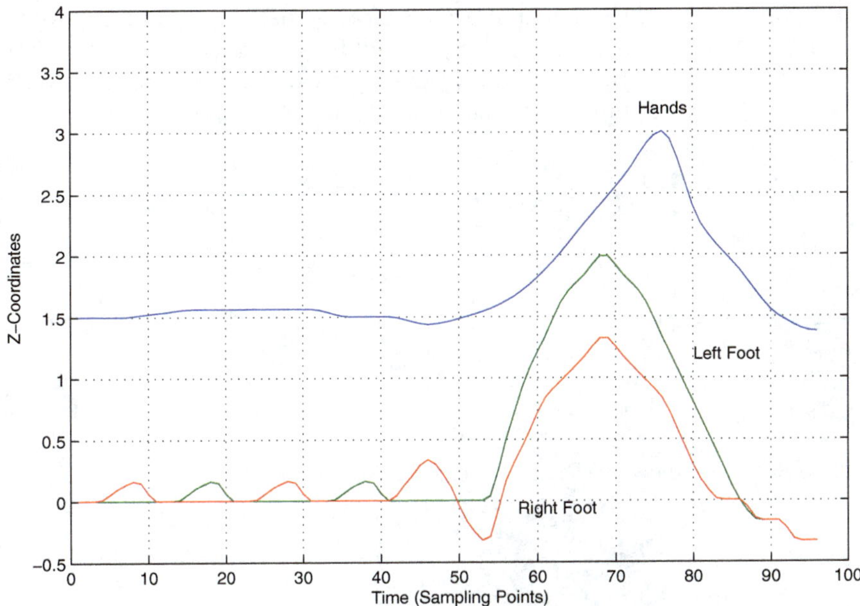

Fig. 10.53 z-trajectories in the springboard diving case for the two feet and two hands by a math model

One of the unique features for the differential motion is easy to switch on/off any one or more components of the Cartesian velocity for each of the five moving points. In fact, such a velocity switch is virtually to turn on/off the degrees of freedom (d.o.f.) at each of the five moving points. For example, let a Cartesian velocity be

$$V = (v_x \ v_y \ v_z \ \omega_x \ \omega_y \ \omega_z)^T.$$

Thus, the corresponding Jacobian matrix J is 6 by n for $n > 6$ joints of a redundant system. If we set $v_y = 0$ and $\omega_x = 0$, and no more changes for both V and J, then the solution

$$\dot{q} = J^+V + (I - J^+J)z$$

with an arbitrary vector $z \in \mathbb{R}^n$ controls the system motion without any linear translation along the y-axis and without any rotation about the x-axis. In contrast, if we shut off both v_y and ω_x, i.e., to take out the second and the fourth components of V, the new Cartesian velocity is reduced to 4 by 1:

$$V = (v_x \ v_z \ \omega_y \ \omega_z)^T,$$

and the corresponding Jacobian matrix has to be reduced to 4 by n as well by removing the second and fourth rows. Then, the resultant solution will make the motion be totally different from the former $v_y = 0$ and $\omega_x = 0$ case. Namely, both the linear movement along the y-axis and the rotation about the x-axis become all free, or gain full freedom without any restriction. Such an on/off switch in the above example can be easily programmed in MATLAB^{TM} by defining a switching index array as follows:

```
m = 6;          % The Total Number of Components in V
j = 2;   k = 4;   % The Components to Be Switch Off

index = [1:(j-1) (j+1):(k-1) (k+1):m];

V = V(index);
J = J(index,:);   % The V and J after Off

% Define an n by 1 Vector z

PI_J = pinv(J);   % The Pseudo-Inverse of J
dot_q = PI_J*V+(eye(n)-PI_J*J)*z

% ........... %
```

Since the augmented Cartesian velocity vector V_b in the differential motion based five-point model is up to 30 by 1, it is desired to define a switchable index array to represent which components of V_b should stay and the rest should be out at each sampling point. At the same time, the same index array will be used to switch on or off the corresponding number of rows of the global Jacobian matrix J_b, too. In fact, such an index-switchable approach associated with the differential motion algorithm is quite effective to control every digital mannequin instantaneous posture in more active and high-speed single or sequential motions, such as a sport activity.

One of more challenging digital human simulation studies is the ingress/egress motion. Let us look at a following example, where the digital mannequin is walking towards a car (no ceiling in this example) and then stepping up on the car floor to get into the car and seat. It is quite obvious that every possible collision must be avoided with a natural and realistic posture at each time instant. To ease the motion planning, the entire ingress motion activity is decomposed into three phases: walking towards the car, making ready to step up, and getting into the car to seat. In this special case, all the three directions x, y and z of trajectories should be generated, instead of only the x and z-trajectories in the previous cases. Furthermore, all the x, y and z-trajectories are developed with respect to the human centered frame due to the global body spin in the last two phases, and they are shown in Figures 10.54 through 10.58.

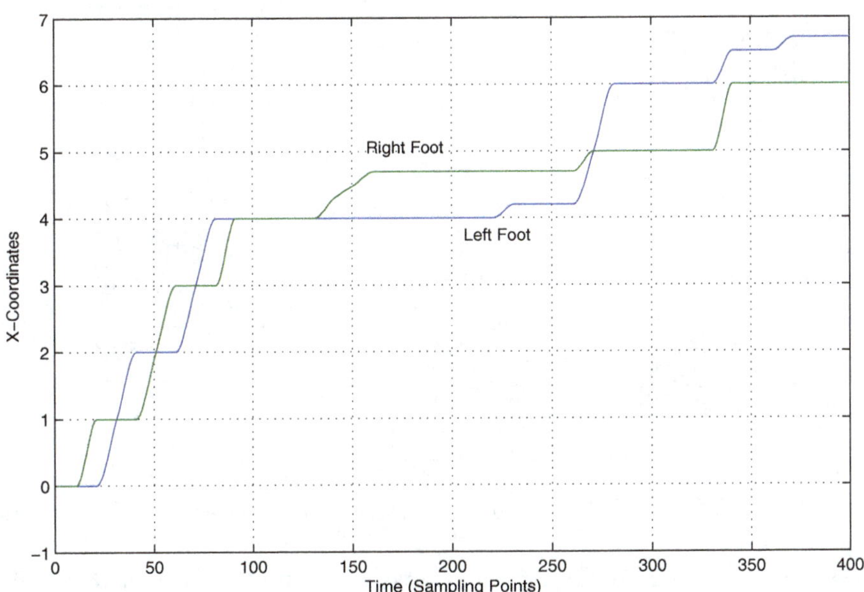

Fig. 10.54 x-trajectories in the ingress case for the two feet by a math model

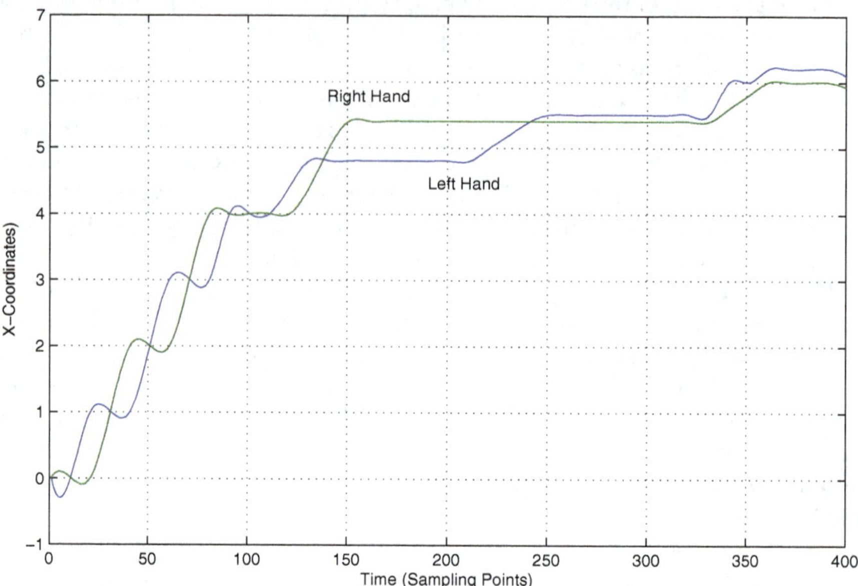

Fig. 10.55 x-trajectories in the ingress case for the two hands by a math model

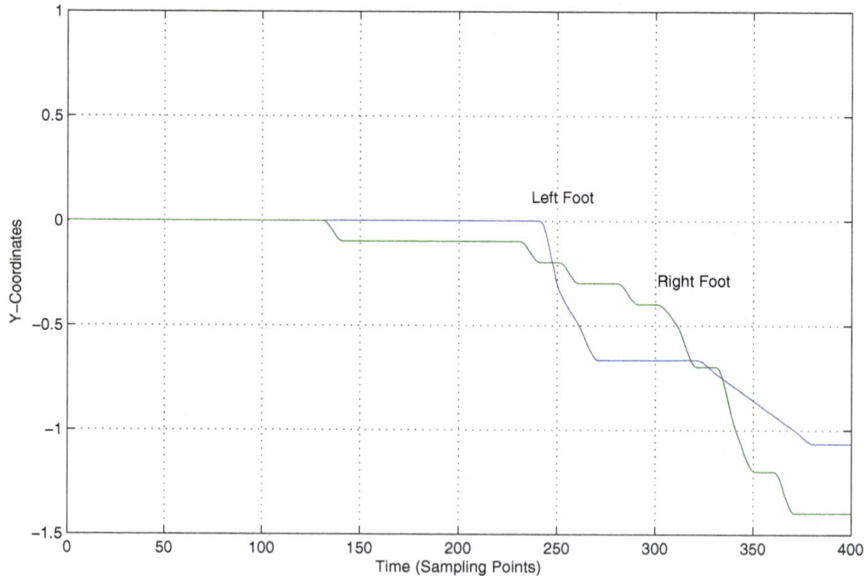

Fig. 10.56 y-trajectories in the ingress case for the two feet by a math model

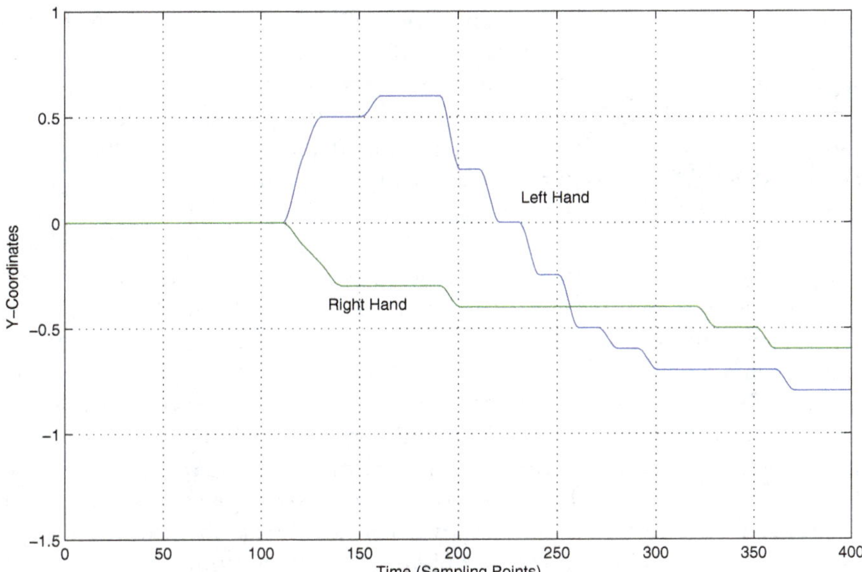

Fig. 10.57 y-trajectories in the ingress case for the two hands by a math model

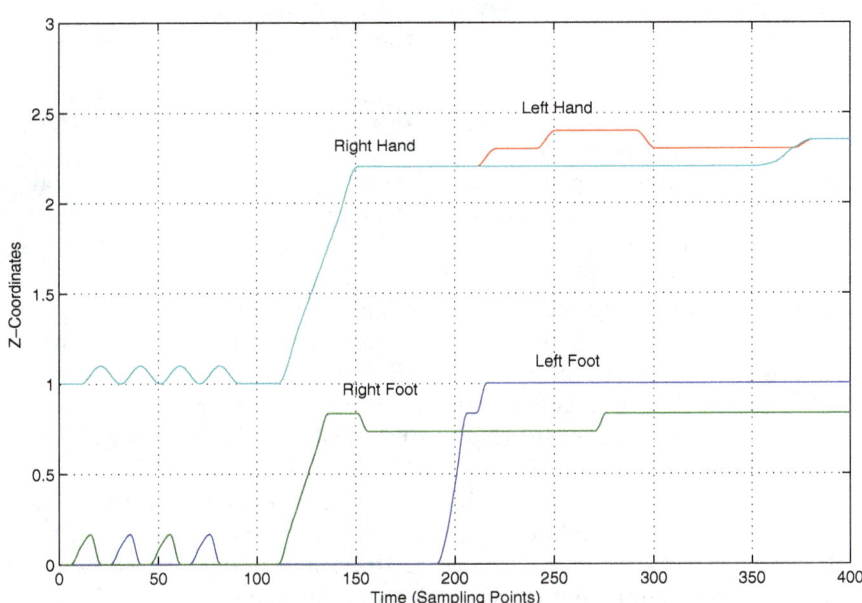

Fig. 10.58 z-trajectories in the ingress case for the two feet and two hands by a math model

It is also observable that among the 400 sampling points, the first 100 points are spent as Phase 1: walking towards the car, as seen on the left picture in Figure 10.40. The time interval between the 100th and 200th sampling points is Phase 2 to move the two feet up, one at a time, on the car floor to make ready for the next ingress, as depicted on the right picture of Figure 10.40. After about the 200th sampling point, the most complicated ingress process of Phase 3 begins until the mannequin is seated, as shown in Figures 10.41 and 10.42.

By inspecting all the numerically created x, y and z-trajectories, we can see that the first phase is almost the same as the regular walking motion but along a circular track about a total of 90^0 body spinning counterclockwise. Since each trajectory is referred to the human centered frame, the body spinning does not affect all the shapes of the x and z-trajectories, while the y-trajectories of both the feet and hands maintain zeros in Phase 1, as observed in Figures 10.56 and 10.57. In contrast, both the y and z-trajectories in Phase 2 become more active while the x-trajectories keep nearly flat because the mannequin is stopping to move forward in the front direction and only to step up on the car floor. In the last phase, the digital mannequin is gradually spinning back about 90^0 clockwise while the entire body is getting into the car under a collision avoidance until being finally seated.

After looking over all the above digital mannequin motion generation examples, we can be further convinced by the following differential motion based five-point model advantages:

1. It requires only five sets of trajectories to cover most human motion activities to be digitally simulated;
2. Since the motion is planned in tangent space, i.e., to generate all the velocity profiles by differentiating every position trajectory to meet the five-point kinematic model requirement, any constant bias or shift in each trajectory does not alter the final movement;
3. Since each component of the augmented Cartesian velocity V_b represents a d.o.f. at one of the five points, it is easy to switch on/off to control the digital human motion in a more natural and realistic fashion;
4. Because a digital human kinematic model is highly redundant, it features a high-dimensional null space that is able to carry a large number of subtasks for optimization;
5. It offers a lot of convenience in motion partitioning, sequential motion planning, spline and refinement;
6. It is also easy to add any accent or decoration in both Cartesian space and joint space to improve digital mannequin movements.

On the other hand, the downsides of the differential motion based model are outlined as follows:

1. Since each joint position updating is often based on the first-order approximation, any cumulative error will cause a deviation for the digital human motion. It is good for simulating a large number of general human motion

activities, but it needs a certain compensation when the motion requires high accuracy, such as a digital mannequin is assembling and installing parts into a car body in an automotive manufacturing operation;
2. Singularity in both global and local areas is always the biggest common enemy for motion generations in any kinematic model. Although a local singularity may not stop the entire motion program, it will cause an unwanted big jump with high speed to mess up the rest trajectory-tracking process. Unfortunately, most cases of singularity are unpredictable. Therefore, it must be avoided by sacrificing some degree of trajectory-tracking accuracy before it occurs.

10.6 Exercises of the Chapter

1. First, create a high-resolution cylindrical surface with 100 latitudinal and 100 longitudinal grids in MATLAB^{TM} by

```
[x,y,z] = cylinder(ones(100),100);
surf(x,y,z*2,'FaceColor','c','FaceLighting','Phong', ...
     'EdgeColor','None','AmbientStrength',0.4), grid,
     axis([-2 2 -2 2 0 2]), daspect([1 1 1]), grid,
     view([1 1 0.5]),
     light('Position',[1 0.5 1],'Style','infinite'),
     xlabel('X'), ylabel('Y'), zlabel('Z')
```

   Then, design a periodic function to sculpture a local area of the cylindrical side surface in your choice. Where are the starting coordinates of the surface? If the amplitude of the periodic function is replaced by another periodic function of time t, either sine or cosine or their squares, then, display a dynamic deformation effect around the local area of the cylindrical surface in MATLAB^{TM}.
2. Following the coordinate frames assignment in Figure 10.10 and D-H tables, draw a right hand in MATLAB^{TM}. Then, make each finger move joint-by-joint, and further move all the five fingers simultaneously to operate a simple grasping task with the digital sensing algorithm.
3. Develop a set of x, y and/or z-trajectories for each of the five points: two hands, two feet and the H-point for a mannequin in climbing up a stair and then walking down the stair.
4. Develop a complete set of trajectories for the five points mathematically to generate a digital athlete in a high-jumping sport activity. How to predict a singularity during the trajectory creation at each sampling point of the entire process of high-jumping? Does any joint of the digital athlete need an accent or decoration to enhance the realism of this particular sport activity?

5. Referring to the same environment in the car ingress motion case study, as demonstrated in Figures 10.40 through 10.42 as well as in Figures 10.54 through 10.58, develop a complete set of trajectories for the five points mathematically to generate a process of egress motion, starting from a sitting posture and ending to step down from the car floor to the ground.
6. If a digital mannequin is slipped and fell down by accident from an original standing posture to sitting on the ground, can you create a complete set of trajectories in the five-point model to simulate the entire process of such an accidental motion?

References

1. Abdel-Malek, K., et al.: Santos: A Digital Human In the Making. In: IASTED International Conference on Applied Simulation and Modeling, Corfu, Greece, ADA542025 (June 2008)
2. Abdel-Malek, K., Yang, J., et al.: Towards a New Generation of Virtual Humans. International Journal of Human Factors Modelling and Simulation 1(1), 2–39 (2006)
3. Abdel-Malek, K., et al.: Santos: a Physics-Based Digital Human Simulation Environment. In: The 50th Annual Meeting of the Human Factors and Ergonomics Society, San Francisco, CA (October 2006)
4. Yang, J., et al.: Santos^{TM}: A New Generation of Virtual Humans. In: 2005 SAE World Congress, Detroit, Michigan, USA (2005)
5. Chaffin, D.: On Simulating Human Reach Motions for Ergonomics Analysis. Human Factors and Ergonomics in Manufacturing 12(3), 235–247 (2002)
6. Chaffin, D.: Digital Human Modeling for Workspace Design. In: Reviews of Human Factors and Ergonomics, vol. 4, p. 41. Sage (2008), doi:10.1518/155723408X342844
7. Moes, N.: Digital Human Models: An Overview of Development and Applications in Product and Workplace Design. In: Proceedings of Tools and Methods of Competitive Engineering (TMCE) 2010 Symposium, Ancona, Italy, April 12-16, pp. 73–84 (2010)
8. Duffy, V. (ed.): Handbook of Digital Human Modeling, Research for Applied Ergonomics and Human factors Engineering. CRC Press (2008)
9. Whittle, M.: Gait Analysis, an Introduction, 4th edn. Butterworth Heinemann, Edinburgh (2007)
10. Hamilton, N., Luttgens, K.: Kinesiology Scientific Basis of Human Motion, 10th edn. McGraw Hill (2002)
11. Adrian, M., Cooper, J.: Biomechanics of Human Movement. WCB McGraw-Hill, Boston (1989)
12. Chapman, A.: Biomechanical Analysis of Fundamental Human Movements. Human Kinetics (2008)
13. Griffiths, I.: Principles of Biomechanics and Motion Analysis. Lippincott Williams & Wilkins (2006)

Chapter 11
Digital Human Modeling: Dynamics and Interactive Control

11.1 Dynamic Models, Algorithms and Implementation

Dynamic modeling, formulation and algorithm development for robotic systems have been investigated and discussed in fair details in Chapter 7. The theories and approaches can now be borrowed to benefit for digital human dynamic modeling and computations. The only difference is the modeling scale between a robot and a digital human. In order to implement a **global dynamics algorithm** developed based on the Lagrange equation into a digital human model, it is required to prepare the following three key components:

1. The inertial matrix W ready for inertial force/torque evaluation;
2. The inertial derivative matrix W_d used to find all the centrifugal and Coriolis forces/torques; and
3. The joint torque τ_g due to gravity, and/or any additional conservative force, such as a spring force.

Equation (7.30) or (7.32) in Chapter 7 developed for a robot is also well-suitable for a digital human model to find its inertial matrix W numerically, but the dimension of W is jumped to 47 by 47, or 44 by 44 if excluding the neck/head joints. For a convenient cross-reference, we repeat the two equations here:

$$W = \sum_{j=1}^{n} J_j^T U_j J_j,$$

or

$$W = \sum_{j=1}^{n} J_{cj}^T U_{cj} J_{cj}.$$

As described in Chapter 7, the first equation requires that each sub-Jacobian matrix J_j be referred to the origin of the corresponding segment frame j, while in the second one, J_{cj} is referred to the mass center frame of segment j. The total number of terms in each of the summations is only 18 by skipping over the segments with zero masses among the 47 segments in total for a

E.Y.L. Gu, *A Journey from Robot to Digital Human*,
Modeling and Optimization in Science and Technologies 1,
DOI: 10.1007/978-3-642-39047-0_11,

digital human dynamic model. In other words, only 18 segments have non-zero masses, see the discussion in Chapter 7. The dynamic parameter matrix U_j or U_{cj} defined in equations (7.26) and (7.33) for each segment can be specified by referring to the anthropometric data of dynamics, such as the mass, mass center coordinates, and the moments of inertia for each human body segment in biomechanics [2]–[5].

However, to determine the inertial derivative matrix W_d, we have to first develop a numerical algorithm to compute the formula defined in equation (7.17), i.e.,

$$W_d = \begin{pmatrix} \dot{q}^T \frac{\partial W}{\partial q_1} \\ \vdots \\ \dot{q}^T \frac{\partial W}{\partial q_n} \end{pmatrix}.$$

Then, all the centrifugal and Coriolis forces/torques can be calculated by

$$C(q, \dot{q})\dot{q} = c(q, \dot{q}) = \left(W_d^T - \frac{1}{2} W_d \right) \dot{q}. \tag{11.1}$$

Clearly, it raises a similar challenging question: how to take partial derivatives for such a large-scale inertial matrix W with respect to each joint position q_i numerically for $i = 1, \cdots, n$? Namely, we wish each derivative $\partial W / \partial q_i$ could be programmed and computed in MATLAB^{TM} either accurately without error or with an acceptable approximation.

Similar to the accurate approaches to calculating the gradient vector for a potential scalar function in Chapter 9, we may implement either the factorization approach to taking the derivative on each one-step homogeneous transformation or its inverse, as given by equation (9.28) with the explanation in Chapter 9, or the dual-number calculus to find a numerical solution of W_d. Only in the cases if the acceleration of a digital human motion is overwhelmingly high, such as in a car crash accident or in a moment of response to an IED (Improvised Explosive Device) explosion, those centrifugal and Coriolis forces/torques given in (11.1) may be negligible in comparison with the inertial force/torque term $W\ddot{q}$.

However, no matter in which special case, the joint torque τ_g due to gravity should not be neglected. Fortunately, to determine τ_g, according to the discussion around equation (9.23) in Chapter 9, i.e.,

$$\tau_g = -\sum_{j=1}^{n} m_j g \frac{\partial h_j}{\partial q} = \sum_{j=1}^{n} J_{cj(b)}^T F_{cj(b)} = -\sum_{j=1}^{n} m_j g J_{cj(b)}^T(3, :),$$

it is a "byproduct" of the numerical computation for the inertial matrix W. Specifically, if we adopt equation (7.32), instead of (7.30), then the 3rd row of each sub-Jacobian matrix $J_{cj(b)}$ for segment j of the digital human body can be utilized to find part of the joint torque distribution due to the gravity force $m_j g$ acting on segment j through the above equation. Therefore, the complete dynamic algorithm implementation for a digital human model is reduced to an only two-step program: (1) calculate every sub-Jacobian matrix towards the inertial matrix W determination, and (2) take derivatives on the sub-Jacobian matrices with respect to each q_i towards a solution of the inertial derivative matrix W_d. The joint torque vector τ_g due to gravity will be determined as a "byproduct" accordingly without extra effort.

The two-step program will be written into a subroutine f_dyn.m file to represent the following digital human state-space equation:

$$\dot{x} = f(x, u).$$

If the state vector is defined by augmenting both the joint position q and joint velocity vector $\dot{q}$ together, i.e., $x = \begin{pmatrix} q \\ \dot{q} \end{pmatrix} \in \mathbb{R}^{2n}$, and $u \in \mathbb{R}^n$ is a joint torque vector as an input to the digital human dynamic system, then the above tangent vector field $f(x, u)$ can be expressed in terms of the three key components W, W_d and τ_g as follows:

$$f(x, u) = \begin{pmatrix} \dot{q} \\ -W^{-1}[(W_d^T - \frac{1}{2}W_d)\dot{q} - \tau_g - u] \end{pmatrix}. \tag{11.2}$$

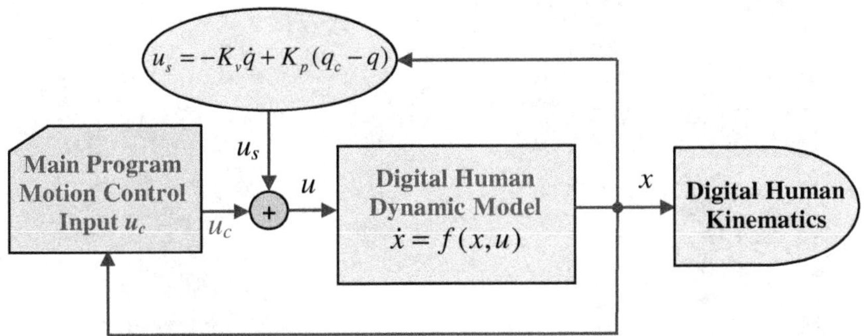

Fig. 11.1 A structure of digital human dynamic model and motion drive

The structure of implementing the digital human dynamics is described in Figure 11.1. As mentioned above, the global dynamic equation (11.2) is placed as a function subroutine in the f_dyn.m file, and let a main program in MATLAB^{TM} call it each time as needed. The input variables of f_dyn.m are the state vector x and the input u, and it returns a corresponding $2n$ by 1 function value of f_dyn(x, u). In general, the input u is a sum of two control sources. The first one u_s is a "self-defense" input defined by a typical PD (Proportional-Derivative) control law:

$$u_s = -K_v\dot{q} + K_p(q_c - q), \tag{11.3}$$

where q_c can be either a comfort center vector, or an initial joint position vector for the digital human, and $K_v > 0$ and $K_p > 0$ are either two positive-definite gain constant matrices, or just two positive scalar constants. This PD control law is the same as the set-point control law in robotic systems given by equation (8.67) in Chapter 8, and reflects a human "self-defense" resilient nature that makes each joint as close to its comfort center or initial position as possible. This is also the same reason why a dummy model in a real car crash test installs a spring at each joint.

The second input u_c is a motion control law to be generated in the main program. The state output x will directly go to another function subroutine to draw and drive the digital human in motion.

The main program includes a differential equation solver to numerically solve the state equation $\dot{x} = f(x, u)$ at each sampling point starting from an initial state $x(0)$. We can either call the MATLAB^{TM} built-in function ode45$(\cdots)$, or use the following 4th order Runge-Kutta algorithm to numerically integrate the state equation for updating the state value x at each sampling point:

```
for i = 0 : N

    u = -Kv*dq + Kp*(qc - q);

    .....

    x = [q; dq];

    r1 = f_dyn(x,u)*dt;
    r2 = f_dyn(x+r1/2,u)*dt;
    r3 = f_dyn(x+r2/2,u)*dt;
    r4 = f_dyn(x+r3,u)*dt;

    x = x + (r1+2*r2+2*r3+r4)/6;

end
```

In the above program, dt is the sampling interval. By placing this Runge-Kutta algorithm into a for-loop or a while-loop inside the MATLAB^{TM} main program, the joint position q, as part of the state x, will be updated and directly forwarded as an input to the digital human drawing subroutine f_dhuman(q) to display and maneuver the digital human in motion. However, the 4th order Runge-Kutta algorithm requires to call the function subroutine f_dyn.m four times in each sampling interval, which could be extremely time-consuming. It is also a clear fact that the function subroutine f_dyn.m is a very large program that covers all the computations of the sub-Jacobian matrices $J_{cj(b)}$'s at up to 18 segment mass centers, the 47 by 47 inertial matrix W and its derivative matrix W_d as well as the 47 by 1 joint torque vector τ_g due to gravity. If the computer processor and graphics card are not fast enough, the digital human dynamic motion on the screen will be very slow. Therefore, the lower order differential equation solver may be necessary, such as the first order one that only needs to call the function subroutine f_dyn.m once per sampling interval.

In order to drive a digital mannequin and effectively generate a desired motion by the global dynamics algorithm, the following two questions must be answered:

1. How to define a motion control input vector u_c to excite the global dynamics algorithm?
2. How to automatically generate a reaction against the environment, such as a reacting force from the floor if standing, or from a seat if sitting, or from an obstacle if colliding?

In the digital human dynamic model, the motion control input port for u_c is widely open, which allows us to define a control law to drive the digital human body for global motion. According to Chapter 8, one of the most typical control algorithms is the state-feedback linearization based trajectory-tracking control, as described in equation (8.76) and developed in great details in Chapter 8. Suppose that we want the digital human to walk, and all the kinematic profiles of desired trajectories for the two hands plus two feet are known. Then, those desired trajectories can be augmented to form a desired output vector $y_d(t)$.

More specifically, if each of the two hands and two feet follows a 3D trajectory that is positional only, then the dimension of $y_d(t)$ is $3 \times 4 = 12$ for both the two hands and two feet. If the H-point of the digital human is also included to follow its own desired trajectory, then the dimension of $y_d(t)$ will be extended to 15 by 1. However, in comparison with the total number of joints for the digital human, this is a highly redundant case. Thus, it is presumable that in the state-feedback control law given by

$$u = \alpha(x) + \beta(x)v,$$

both $\alpha(x)$ and $\beta(x)$ will contain the pseudo-inverse of the global Jacobian matrix J for the digital human if a minimum-norm solution is adopted, i.e.,

$$\alpha(x) = -WJ^{+}\dot{J}\dot{q} + C(q,\dot{q})\dot{q} - \tau_g \quad \text{and} \quad \beta(x) = WJ^{+},$$

along with a new input for tracking the desired trajectory:

$$v = \ddot{y}_d + K_2 \Delta\dot{y} + K_1 \Delta y,$$

where $\Delta y = y_d(t) - y(t)$ is the trajectory error vector.

Regarding the question how to automatically generate a reaction in response to the environment interaction, it is necessary to know every obstacle location, such as the coordinates of the floor, wall, and any object that will possibly be hit by the digital human body with respect to the world base. If the digital human is standing on the floor, then it can be balanced by either a solution of the statics equation or a PD control, or both. At each sampling point, the joint torque τ_g due to gravity is calculated by the digital human dynamics subroutine f_dyn.m, and the first six components of τ_g have not vanished yet at the beginning. Once a statics algorithm that can implement the Law of Balance (see Chapter 9) is programmed, the feet reaction from the floor will automatically produce additional joint torques along the two legs to cancel the first six components of τ_g. Then, the digital human will be dynamically balanced on the floor at any time instant.

As a detail, according to equations (9.19) and (9.20) in Chapter 9, the Cartesian reaction force $F_{2f(b)}$ from the floor to the two feet in order to balance the entire body gravity can be determined based on the Law of Balance:

$$F_{2fg(b)} = -\left(J_{2f(b)}^{6T}\right)^{+} \cdot \tau_g(1:6).$$

If now an additional Cartesian force $F_{2fc(b)}$ that includes reacting forces and/or moments other than balancing the body gravity is applied on both feet of the digital human in order to walk or run, then the total Cartesian force (or wrench) acting on the two feet should be

$$F_{2f(b)} = F_{2fg(b)} + F_{2fc(b)}.$$

This 12-dimensional Cartesian force vector will become an additional part of the 30-dimensional global Cartesian force vector $F_{(b)}$ in terms of the five-point model in digital human kinematics and statics. If the global Jacobian matrix of the digital human is $J_{(b)}$ that is 30 by 47 in dimension, then the total joint control input u_{rc} due to the environment interaction can be determined by

$$u_{rc} = J_{(b)}^T F_{(b)}.$$

If one adds a PD control law to further cancel out the joint force $u(3)$ that is a prismatic joint perpendicular to the floor, the digital human body will be even more effectively balanced in the vertical direction. Suppose that p_b^{37} and p_b^{44} are the position vectors of the left and right feet, respectively, which can be extracted directly from the dynamics subroutine f_dyn.m at each sampling point. If both of them are referred to the world base, their third components are just the coordinates in the vertical up-down direction that can be directly used to determine the PD control law for the input $u(3)$ of the digital human dynamic model, as illustrated in the following program:

```
if min(p37(3),p44(3)) > 0.01
    u(3)=u(3)-Kv*dq(3)+Kp*(-min(p37(3),p44(3)));
end
```

```
if min(p37(3),p44(3)) < -0.01
    u(3)=u(3)-Kv*dq(3)+Kp*(-min(p37(3),p44(3)));
end
```

where $dq(3)$ is the 3rd joint velocity $\dot{q}(3)$, and both K_p and K_v are the position and velocity gain constants. The threshold above or below the zero vertical position at the floor level is set to be 0.01 m. = 1 cm. in the above PD control program.

Utilizing a similar PD control law but in Cartesian space of the form

$$f = -K_2 v + K_1(p_{ob} - p), \tag{11.4}$$

the dynamics program can also automatically generate every reaction from touching or hitting an obstacle, a wall or other objects. In the PD control law (11.4), $K_1 > 0$ and $K_2 > 0$ are two gain scalar constants or matrices, $p \in \mathbb{R}^3$ is an actual position vector of the hitting point on the digital human body and can be determined readily by the forward kinematics as a function of the instantaneous joint angles q, p_{ob} is the hitting position that is fixed on the obstacle, while v is a linear velocity of the digital human at the hitting point and can be found by using a local Jacobian or sub-Jacobian matrix J under the equation $v = J\dot{q}$.

For instance, if the head of the digital human is hitting to the wall, then J is the first three rows of the sub-Jacobian matrix J_{47} from the digital human body segment #47 (head) down to the base, see Figure 9.21 in Chapter 9, and $\dot{q}$ is the corresponding joint velocity vector. The head position vector can also be found directly from the homogeneous transformation derivations in the digital human dynamics program. Once the 3D instantaneous force f is calculated by the PD control law (11.4), using the statics equation

$$\tau = J^T f \tag{11.5}$$

to find the additional joint torques in reacting to the force f, and then add the joint torque vector τ of (11.5) as a part into the motion control input u_c. Appropriately adjusting both the two gain constants K_1 and K_2 in equation (11.4) will make such a hitting-and-bouncing dynamic balance look very naturally.

Therefore, the digital human dynamics program f_dyn.m is not only essential to generate a desired dynamic motion, but is also a vast resource to supply all the useful data, such as the local sub-Jacobian matrices, rotation matrices and position vectors, necessary for calculating and determining every possible human-environment interaction.

In fact, beyond the virtual realistic motion generation, the research of the digital human dynamics and modeling can also benefit the control design and development for humanoid robots to improve and enhance their stability and realistic dynamic motion.

11.2 δ-Force Excitation and Gait Dynamics

In Chapter 10, we have discussed in detail and also effectively animated a digital human model in walking, running, jumping and beyond. All the motions planned and generated are thus far based only on the digital human kinematics. However, in the reality with a real human, there is no such a set of kinematic trajectories available to guide the hands and feet to walk, to run or to jump. Even the person himself/herself is never thinking about how each of his/her hands and feet can follow a desired trajectory in order to walk, run or jump. In fact, everything happens naturally and spontaneously, and obeys the natural law of dynamics. Under such an interpretation, the digital human basic motions and beyond would be created more realistically if the dynamics could be taken into consideration, especially to mimic a fast and wide range human motion, such as a sport activity [1]–[7].

Motivated by the above notion, let us launch a study on real human gait dynamics. Researchers in health science, physiology, kinesiology and biomechanics have made a comprehensive gait analysis in terms of various aspects: walking kinematics and kinetics, gaits in different ages and in different genders as well as the gait pathology in different health conditions. However, most of the research results lie in the analysis of gait cycles and energy consumption, statistical analysis, comparative studies and methodologies of measurement and clinical assessment [1]–[4]. Although the research of walk and gait had nearly a century long history, the study on gait dynamics can still be considered as a young research topic.

In the literature [6], a normal human gait cycle in terms of each single leg is divided into two major phases: a stance phase and a swing phase. The two consecutive phases in a gait cycle are alternatively exchanged between two legs. Namely, if the right foot at a moment presses the floor and receives a reacting force to push itself up/forward to leave the floor and starts its stance phase, the left leg is in its swing phase, and vice verso. Once the right leg completes a stride to touch down the floor again, its stance phase is over and immediately turns to a swing phase until the next stance phase starts. For a human in walking, unlike running or jumping, the stance phases of both the right and left legs have a short overlapping, and both two feet never leave the floor simultaneously in any time interval.

On the other hand, according to the digital human dynamic modeling structure in Figure 11.1, we have already prepared a global dynamic model as a real-plant subroutine to be called by the Runge-Kutta differential equation solver. The only thing we need to do is to design and determine a pattern of the motion control input u_c. Figure 11.2 shows a typical force/moment analysis in order to explore what kind of the motion control u_c will be in each of the two phases in a gait cycle.

From a detailed force/moment analysis, we discover that if the right leg is in its stance phase and the left leg is in its swing phase, then,

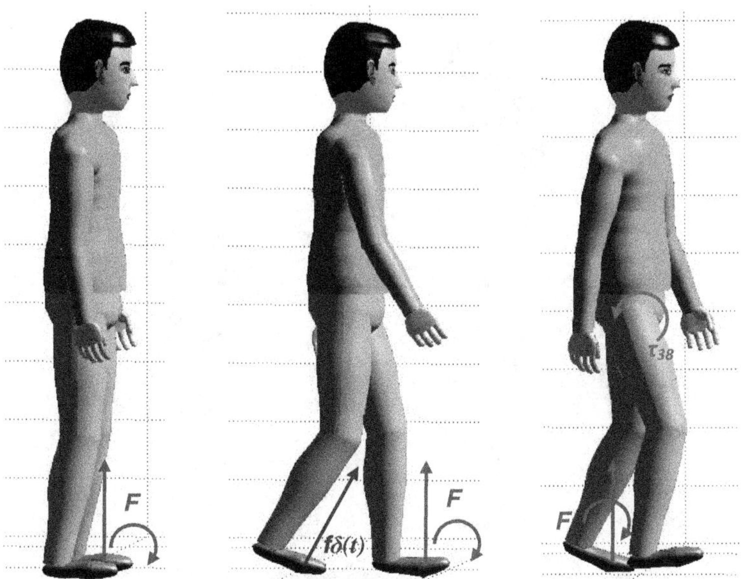

Fig. 11.2 Dynamic balance in standing case and δ-force excitation in a walking case

1. At the initial position before the walk starts, the digital human in a neutral standing posture must be balanced by additional joint torques along the two legs based on the Law of Balance, see Chapter 9.
2. There is a δ-force $f\delta(t)$ reacting on the right foot from the floor at the beginning of the stance phase, the direction of which is up and leaning forward. The leaning angle depends on the walking speed and other human factors. This δ-force is just a mathematical model, which means that the reacting force f excites the right foot only in one or two sampling interval(s).
3. At a few middle points of the stance phase, the right thigh flexion will be exerted with a joint torque τ_{38} produced by a local muscle contraction to keep and reinforce the momentum of the δ-force excitation. The sustain time may span a few sampling intervals.
4. The right knee joint in the middle of the stance phase is supposed to keep relax and allow the gravity to hang the right foreleg down naturally before the right foot touches down the floor to end the stance phase.
5. Since the left leg is standing on the floor in its swing phase, in addition to being fully responsible of balancing the entire body due to gravity, the left foot will also be required to interact with the floor and produce an additional moment (torque) to help the entire body to lean forward.

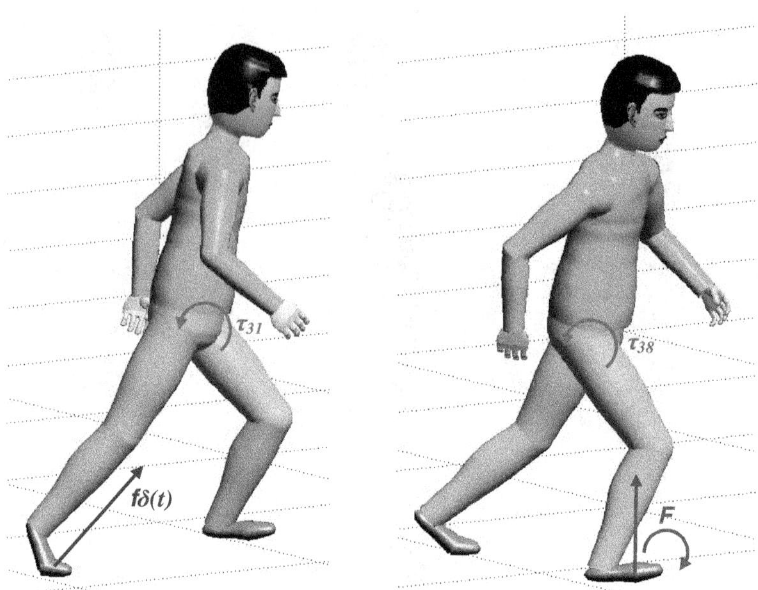

Fig. 11.3 Dynamic balance and δ-force excitation in a running case

6. The two arms should be fully relax without additional joint torques imposed. Their alternative swings during the walk are produced naturally by the entire body dynamic balance due to gravity as well as the dynamic coupling with the two legs alternative movements based on the principle of angular momentum conservation.

In contrast to the walking case, both the two feet of the digital mannequin in running must have a short time interval Δt leaving from the floor. When the right foot is bounced from the floor by a δ-force excitation at the beginning of its stance phase, the left foot has not been touching down the floor yet, see on the left snapshot in Figure 11.3. Therefore, this δ-force must be stronger than that in the walking case. At the same time, the left thigh also needs a joint torque τ_{31} by a local muscle contraction in order to make a larger forward pace.

Once the right foot touches down the floor at the end of its stance phase, see the right picture of Figure 11.3, since the left foot has already received a δ-force and departed from the floor, the right leg will be solely responsible of balancing the entire body, overcoming the reacting force F from the floor as well as producing more Cartesian moment to move the body leaning forward in order to keep running. The two elbow joint angles may be preset with an angle, such as 60^0 each. Similar to the walking case, the two-arm alternative swings can be produced naturally by the dynamic coupling effect based on the principle of angular momentum conservation.

Since the motion control input u_c into the global dynamics algorithm is at the joint level, every force and moment in Cartesian space must be converted to the corresponding joint torques by the digital human statics, as discussed in Chapter 9. Also, the "relax" status in the above analysis means that the self-defense torque u_s at each relevant joint should be either set to zero or weakened by reducing the PD gain constants.

Because in each sampling interval, there is a large amount of computations required to run according to the digital human dynamic motion generation structure in Figure 11.1, such as the 4th order Runge-Kutta algorithm to call four times the large-scale digital human dynamics subroutine in the real-plant, plus the large digital mannequin drawing subroutine with complete kinematics, the motion generated on a computer screen could be very slow. Therefore, the computer power and graphics capacity must be further upgraded in order to see a real-time realistic motion effect.

Once the digital human walking and running activities can be animated more realistically by the global dynamics algorithm than those driven only by the kinematic motion algorithm, it will be much easier to extend the basic motions to more varieties of dynamic motions, including the sport activities and working activities in a very restricted environment.

It is also conceivable that the global dynamics algorithm plus a certain pattern of the motion control input u_c will automatically generate a desired motion with the minimum effort for a digital human model, because each solution of the Lagrange equation is a trajectory with the minimum integral of the Lagrangian function, i.e.,

$$I = \int_{t_1}^{t_2} L(q(t), \dot{q}(t))dt.$$

In the next two sections, we will demonstrate two realistic applications of digital human dynamic modeling and their animations. The human-environment interactions, such as the reaction force and body balance between the digital human feet and floor, between the hip and seat, and between the head and steering wheel, etc., will also be effectively modeled and simulated based on equations (11.4) and (11.5).

11.3 Digital Human Dynamic Motion in Car Crash Simulations

The digital human global dynamics algorithm can be applied quite effectively to simulate a human driver's dynamic response to a car crash accident, where the car hits an obstacle in either frontal or side collision cases [8, 9, 10]. It is the fact that during a small fraction of a second after the collision, the acceleration of both the vehicle and human driver is huge, and the peak value can reach up to 80 g, where $g = 9.81$ m./sec^2 is the earth gravitational acceleration, if the vehicle hitting speed is at 45 mph., see Figure 11.4. Therefore, the human dynamic equation can be approximately reduced by ignoring all the centrifugal and Coriolis forces/torques given by the term (11.1) so that

$$W\ddot{q} - \tau_g \approx \tau. \tag{11.6}$$

Accordingly, the state-space equation (11.2) can also be reduced to

$$f(x,u) \approx \begin{pmatrix} \dot{q} \\ W^{-1}(\tau_g + u) \end{pmatrix} = \begin{pmatrix} \dot{q} \\ W^{-1}\tau_g \end{pmatrix} + \begin{pmatrix} O \\ W^{-1} \end{pmatrix} u, \tag{11.7}$$

where O is the zero matrix with the same dimension as the inertial matrix W, and $u = \tau$ is the system input vector.

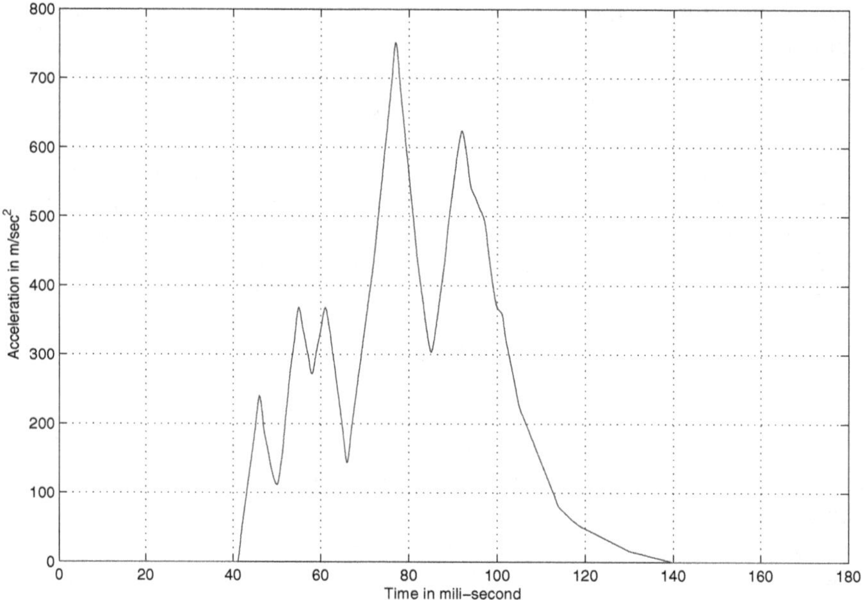

Fig. 11.4 A frontal collision acceleration profile as a vehicle speed at 45 mph

We can have the following two approaches to modeling the dynamics of the car-human system in a frontal crash accident:

1. Combine both the vehicle and digital human into a single cascaded model. In other words, let the vehicle/seat be link 1 and let its horizontal sliding movement be the first joint of the new cascaded model. Then, the entire 47-joint digital human model will follow to start as joint 2 through joint 48, which is similar to a 6-joint robot arm mounted on a linear track, as shown in Figure 5.2 from Chapter 5. Or, simply remodel the first three prismatic joints of the digital human to define the vehicle/seat as the first

link to slide along the original z_0-axis of the digital human model. It is then followed by the second prismatic joint defined for the digital human but along the tangent direction of the seat cushion surface. The third joint may keep the original d_3 along the z_2-axis. With such a remodeling treatment, the first element $w(1,1)$ of the new inertial matrix W will contain a very heavy mass M that represents the vehicle and seat. Accordingly, the first joint force/torque input $u(1)$ will be $-MG$, where G is the acceleration profile given in Figure 11.4.

2. Let y be the horizontal displacement of the vehicle/seat. Then, $\ddot{y} = -G$, where G is again the acceleration profile in Figure 11.4. If the state variables $z_1 = y$ and $z_2 = \dot{y}$, then,

$$\begin{pmatrix} \dot{z}_1 \\ \dot{z}_2 \end{pmatrix} = \begin{pmatrix} z_2 \\ -G \end{pmatrix}.$$

Let the above state-space equation be the function inside the Runge-Kutta algorithm in order to numerically integrate the equation and solve for the vehicle displacement $z_1 = y$ and speed $z_2 = \dot{y}$ at each sampling point. Then, let a PD control law be defined as

$$f = -K_2\dot{y} + K_1(y_c - y)$$

to couple the moving vehicle/seat with the digital human model together around the interacting area, such as the H-point of the digital human slid by a reaction force f from the seat cushion surface. In the above PD control law, y_c is the car seat displacement at the time instant of hitting the obstacle, and $K_1 > 0$ and $K_2 > 0$ are the PD gain constants.

In this section, we adopt the first remodeling approach for the car frontal crash simulation study, while in the next section of the dynamic response to an IED explosion, the second interacting approach will be applied.

In order to realistically simulate all the human-environment interactions in this car frontal collision case, in addition to applying the local PD control law and statics given by (11.4) and (11.5), we must define the PD gain constants K_1 and K_2 respectively in each of the following hitting cases:

1. The two feet of the digital human driver hit the car floor;
2. The buttock sits on the seat and hits the seat cushion;
3. The backs of the two thighs press and hit the seat cushion;
4. The shoulder/clavicle and waist are restrained by the upper and lower seat belts;
5. The chest and/or the head hit the activated airbag;
6. The chest and/or the head hit the steering wheel if no airbag is activated;
7. The head and/or the torso bounce back and hit the seat back.

In general, K_1 is proportional to the stiffness of the hit object, while K_2 is a damping coefficient that may absorb and reduce the vibration.

To explicitly demonstrate the dynamic motions before and after the frontal collision occurs, let us first look at the worst case scenario with a digital human driver who forgot to wear the upper seat belt and only buckled up the lower one at a driving speed 45 mph., as shown in Figure 11.5. Once the vehicle hits the obstacle, because no upper belt is restrained and also no airbag is activated, the chest of the driver moves quickly towards the steering wheel and severely hits it, see the snapshot in Figure 11.6. It is immediately followed by the head hitting the steering wheel, too, in Figure 11.7.

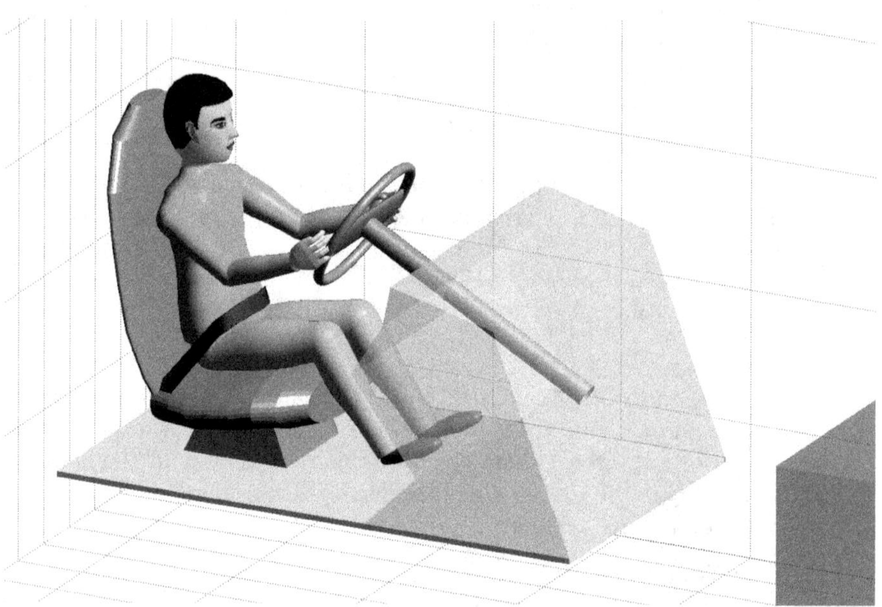

Fig. 11.5 The mannequin forgets wearing an upper seat belt before the vehicle crashes at 45 mph

Within a very short moment of hit, the driver's head is bouncing back from the steering wheel and going backwards to hit the seat back, as shown in Figures 11.8 and 11.9. The entire process is sustained just within a very small fraction of a second.

Now, this driver wears both the upper (shoulder) and lower (waist) seat belts as well as having a frontal airbag available in the second case of car crash accident, see Figure 11.10. After a new frontal collision, the driver is now protected by both the seat belts and airbag from hitting to the steering wheel with his chest and head, as depicted in Figures 11.11 and 11.12.

If both the upper and lower seat belts are controlled by an active restraint system that will be introduced and discussed in the later section, even though no frontal airbag is activated, the digital human driver will be much safer to adapt such a high-speed (45 mph) crash accident and will also significantly

Fig. 11.6 At the moment of collision, the mannequin's chest Hits the steering wheel

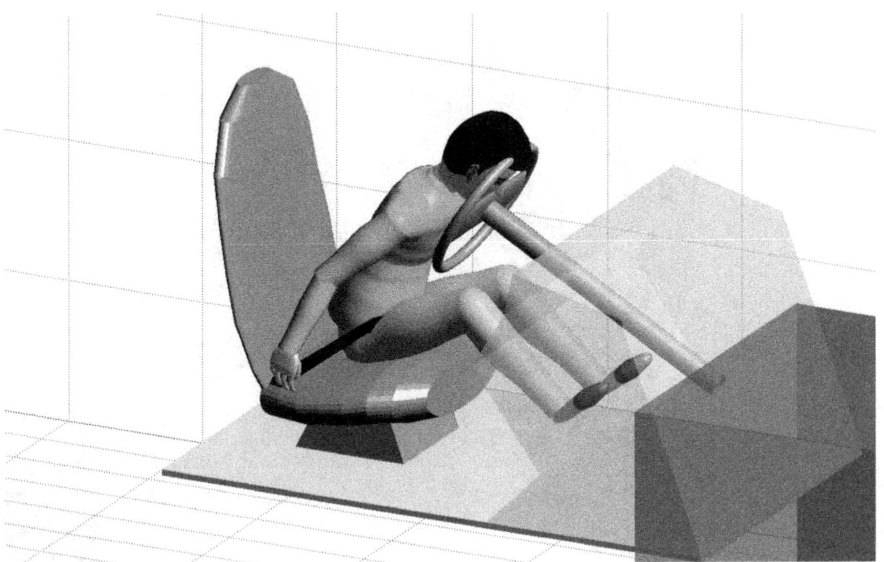

Fig. 11.7 After the chest impact, the head immediately follows to hit the steering wheel

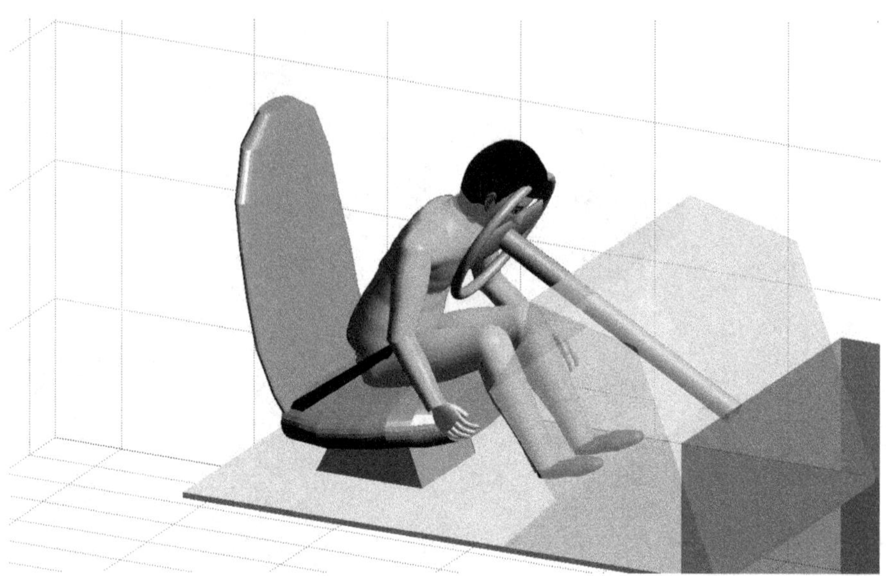

Fig. 11.8 The mannequin's head is bouncing back after hitting the steering wheel

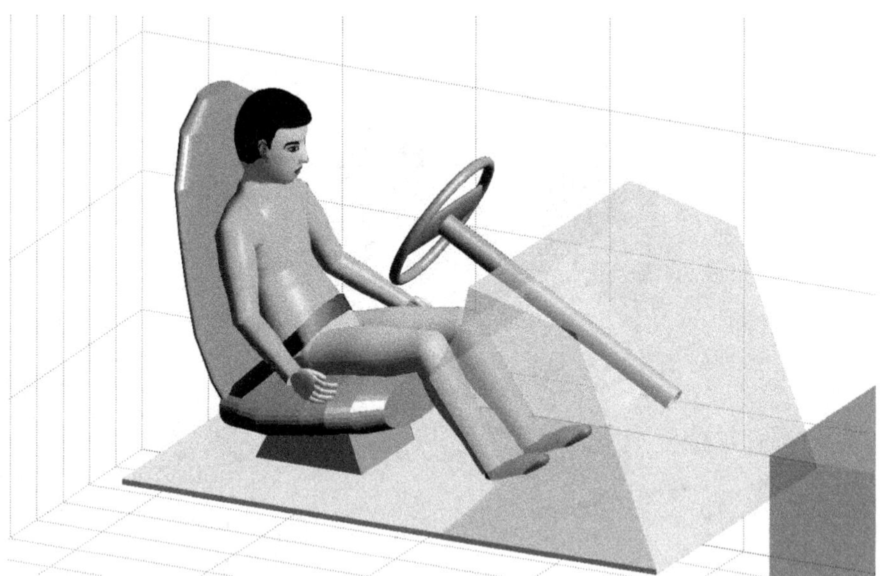

Fig. 11.9 With the momentum of bouncing back, the mannequin's head and back hit the car seat back

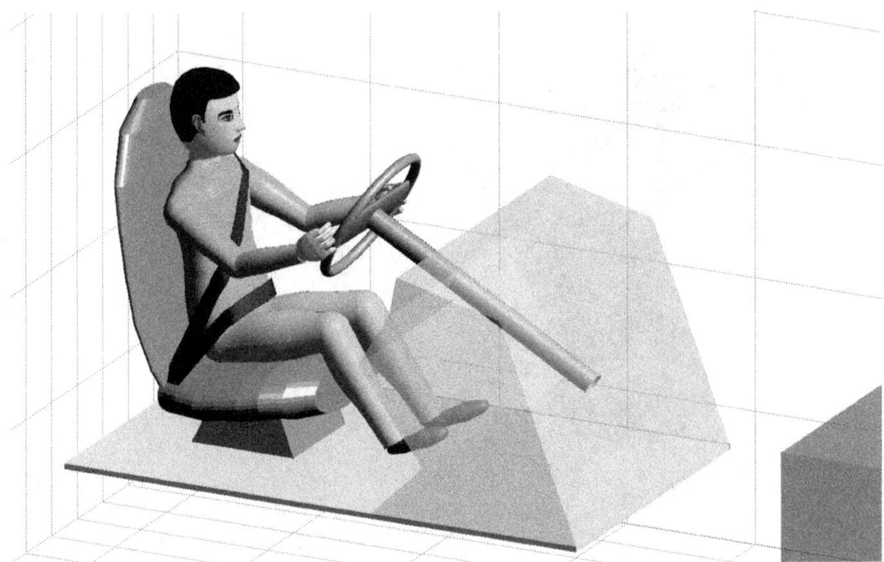

Fig. 11.10 The mannequin now wears both upper and lower seat belts and drives the car at 45 mph

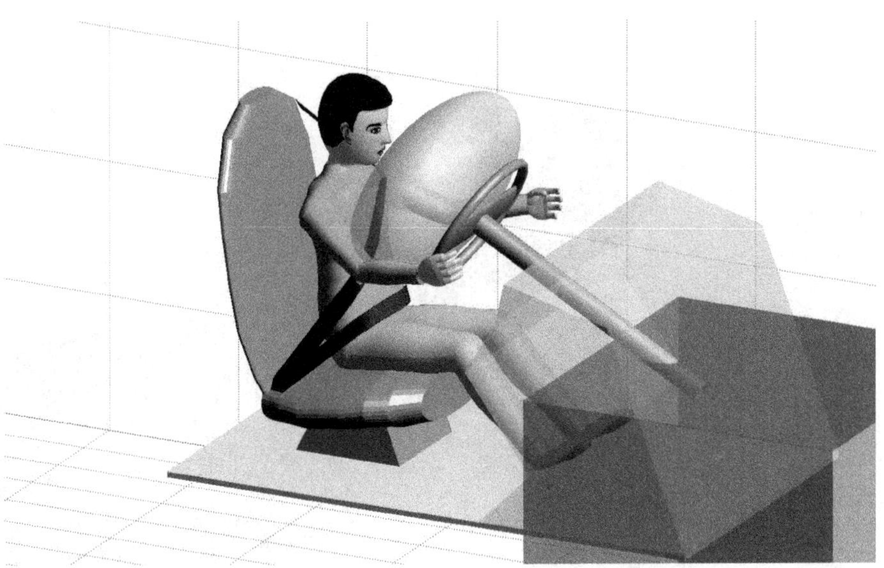

Fig. 11.11 After a frontal impact occurs, the mannequin's chest hits the activated frontal airbag

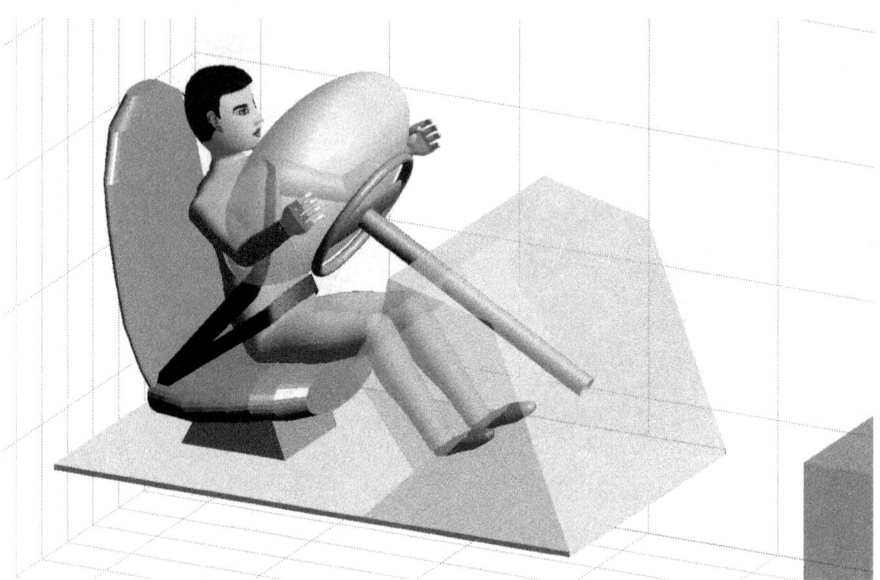

Fig. 11.12 With the airbag, the mannequin's chest and head are protected from the deadly hit

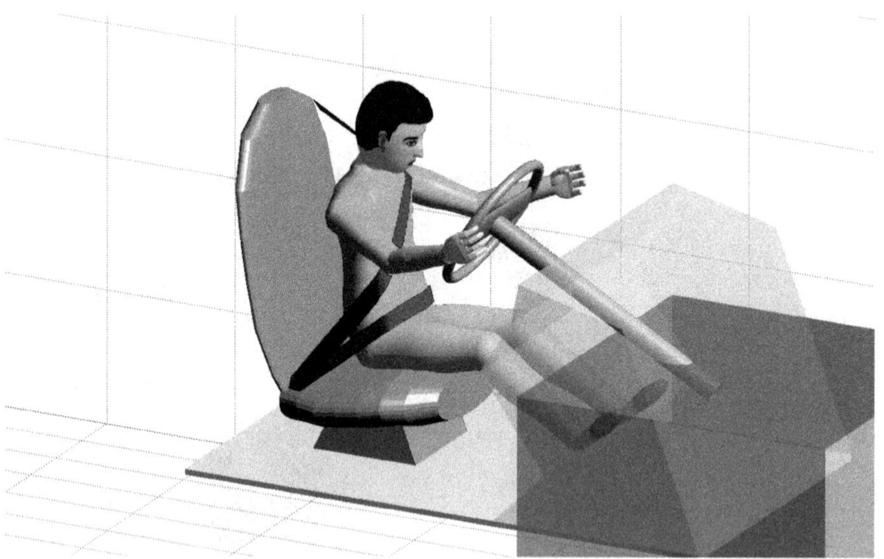

Fig. 11.13 Under an active restraint control, the mannequin is much safer in a crash accident

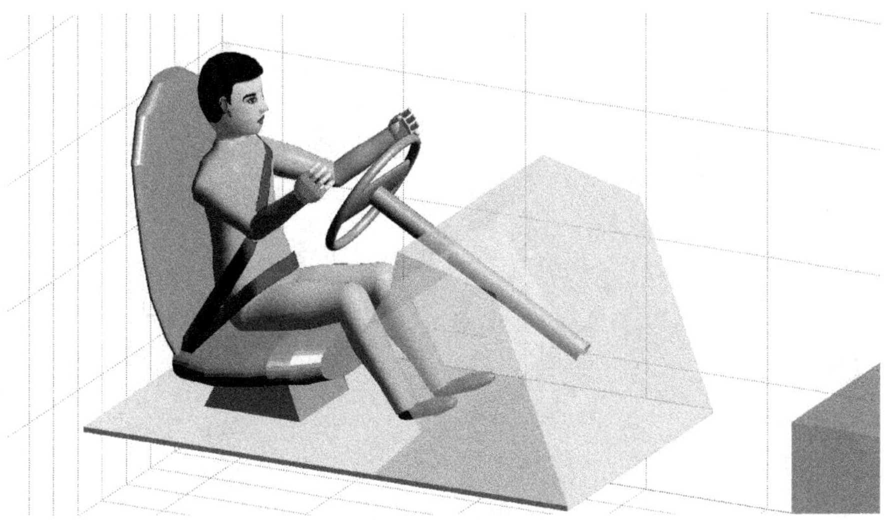

Fig. 11.14 With the active restraint control, severe bouncing back to hit the car seat back is also avoided

reduce the chance of bouncing back to hit the seat back, see Figures 11.13 and 11.14.

The dynamic animation study of car crash accident has a unique advantage in evaluating every joint acceleration over the digital driver's body at any time instant. Figures 11.15, 11.16 and 11.17 plot the acceleration profiles for the three major joints: the lumbar, thorax and the neck/head before and after a frontal collision with a vehicle speed at 45 mph in cases 1, 2 and 3, respectively. More precisely, the three plots display the accelerations in m./sec^2 for joints 7, 10 and 47, i.e., $\ddot{\theta}_7$, $\ddot{\theta}_{10}$ and $\ddot{\theta}_{47}$ versus time. Since in case 1, the driver did not wear the upper belt and also no airbag was activated, all the three joint accelerations are very high in overall with the neck joint acceleration boosting extremely high spikes at a few special time instants as the chest and head hit the steering wheel and also the head hits the seat back.

In contrast, all the three joint accelerations in case 2 are much lower due to the protection of both the upper and lower seat belts and the airbag activation. However, at the time instant the head is hitting the inflated airbag, the neck acceleration still emerges a spike.

In comparison with case 2, case 3 has even lower and more uniform accelerations that are attributed by the active restraint control. The required belt strain control forces versus time in case 3 are plotted in Figure 11.18, and will be discussed in detail in the later section. In fact, the acceleration of any joint other than the lumbar, thorax and neck for the digital human driver during the collision can also be plotted and evaluated readily without extra effort.

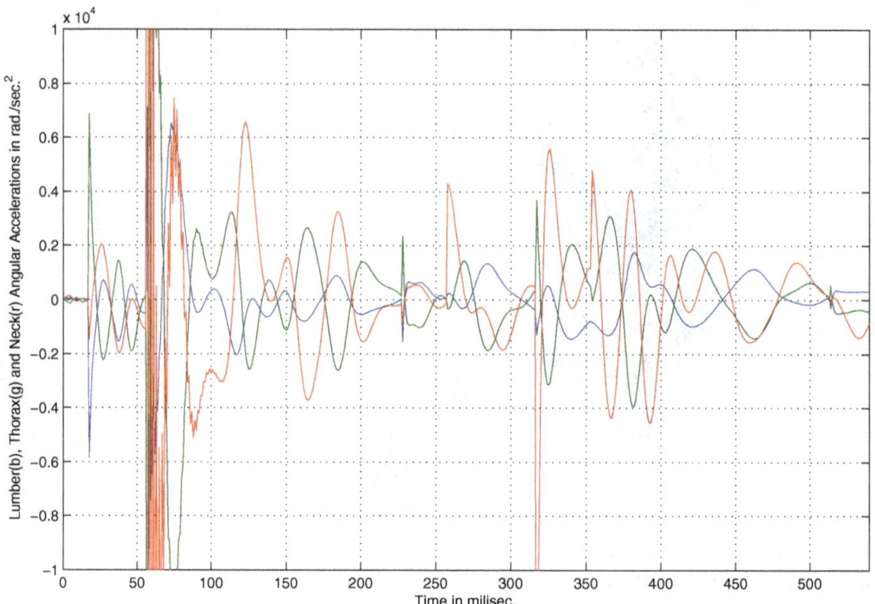

Fig. 11.15 The lumbar, thorax and head accelerations in Case 1

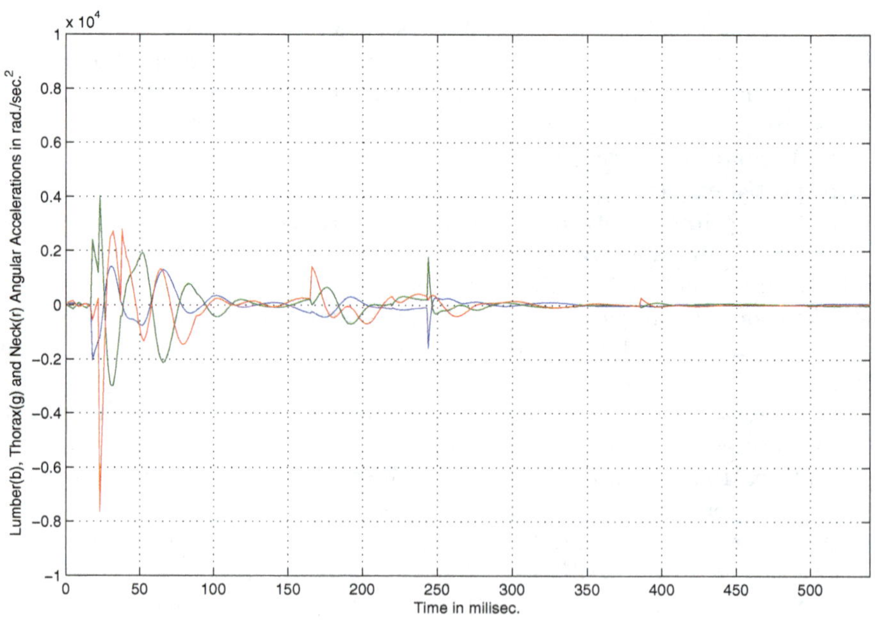

Fig. 11.16 The lumbar, thorax and head accelerations in Case 2

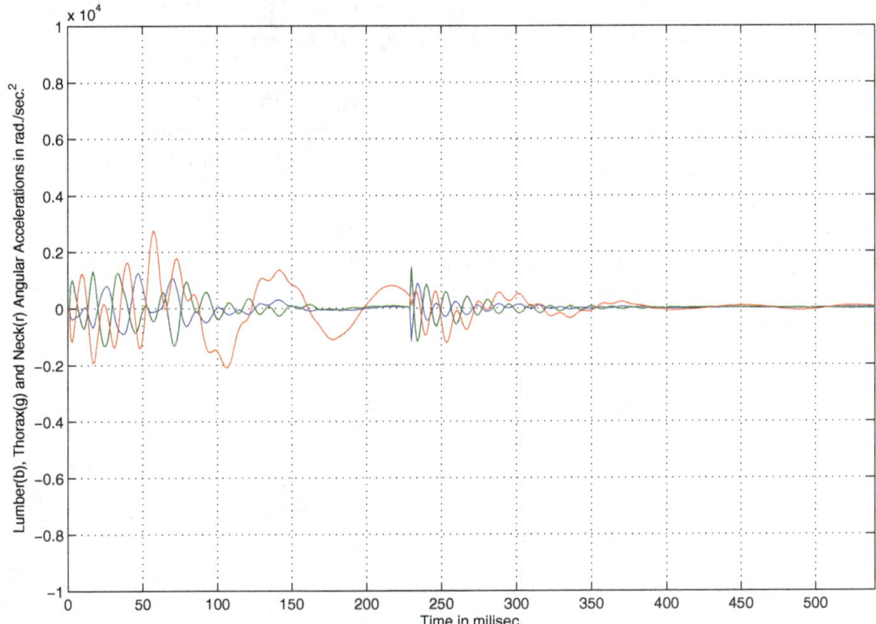

Fig. 11.17 The lumbar, thorax and head accelerations in the case with active restraint control

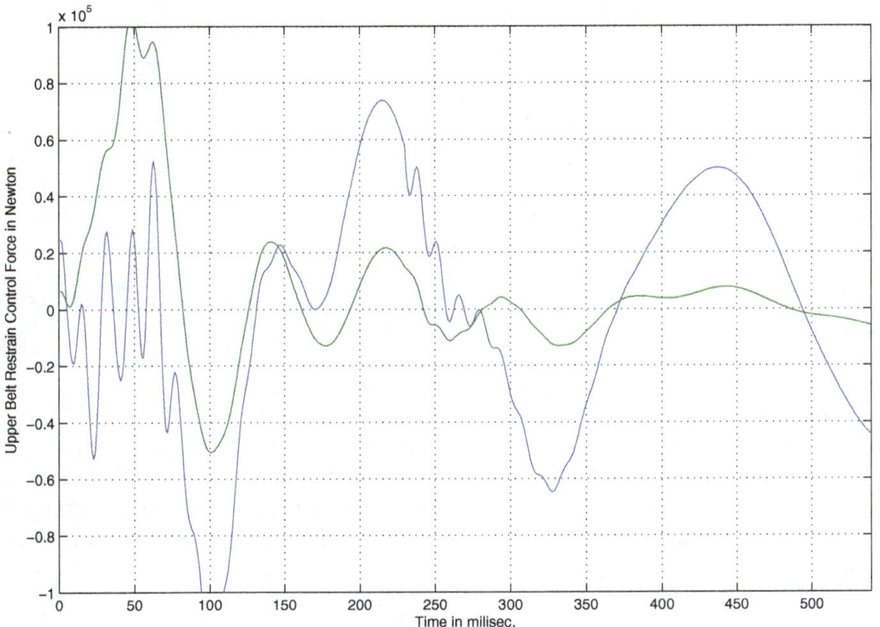

Fig. 11.18 The control inputs in the case with an active restraint system

11.4 Modeling and Analysis of Mannequin Dynamics in Response to an IED Explosion

In addition to the simulation study on the digital human dynamic motion in a car frontal collision, the same digital human dynamic model and algorithm can be employed for simulating a warfighter dynamic motion in response to an IED (Improvised Explosive Device) explosion underneath a military vehicle. In this case, the vehicle as well as the seat are acted with nearly an impulse of acceleration in a vertical direction. A typical profile of such an acceleration impulse versus time is depicted in Figure 11.19, and its peak value can exceed $20g$.

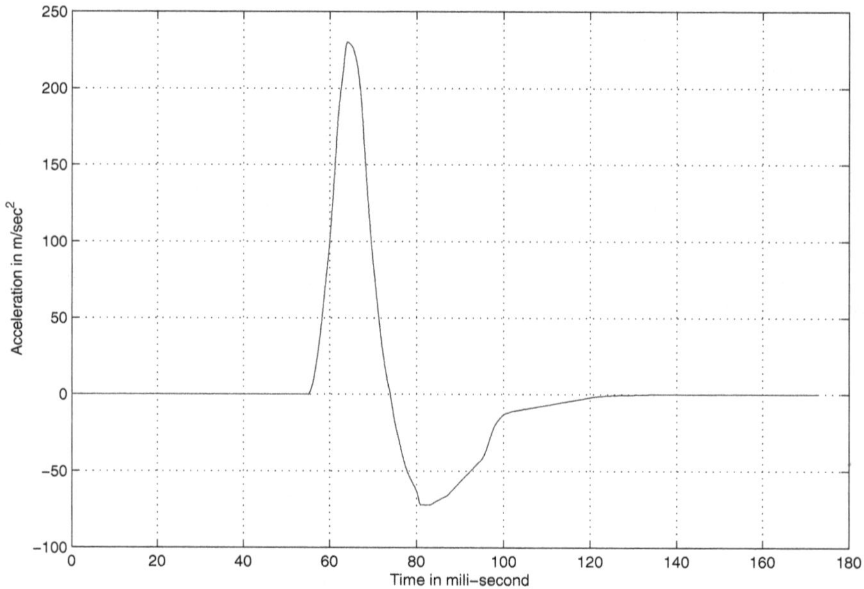

Fig. 11.19 The acceleration profile of an IED explosion underneath the vehicle seat

Due to the extremely high acceleration, the same dynamic equation as given in (11.6) and (11.7) can be adopted for the digital warfighter in response to the IED explosion. Similar to the car crash case study in the last section, the acceleration of the vehicle, including the seat but excluding the digital warfighter, before and after the explosion satisfies the following simple relation:

$$\ddot{y} = G - g,$$

where y is a vertical coordinate, G is the acceleration profile as plotted in Figure 11.19 and $g = 9.81$ m./sec^2 is the earth gravitational acceleration.

Let the state variables be $z_1 = y$ and $z_2 = \dot{y}$. Then, the equivalent state-space equation becomes

$$\begin{pmatrix} \dot{z}_1 \\ \dot{z}_2 \end{pmatrix} = \begin{pmatrix} z_2 \\ G - g \end{pmatrix}.$$

By running the 4th order Runge-Kutta algorithm, as shown earlier, the above state-space equation can be solved for $z_1 = y$ to find the vertical position of the vehicle/seat at each sampling point. Afterwards, using a spring-damper (PD control) model, we can determine the 3rd joint force $u(3)$ as an input to the digital warfighter dynamic model to simulate the seat cushion to bounce up the entire body through the digital human H-triangle. Then, the dynamic motion of the digital warfighter starts a response to the IED explosion from the bottom of the vehicle. A portion of the MATLAB^{TM} program for generating the above input $u(3)$ is given as follows:

```
for j = 0 : N

   u = -Kv*dq + Kp*(qc - q);    % Self-Defense

   if j > 3*n1 && spos <= 0 || j < n1
       v = 9.81;    z = [0; 0];
       else   v = G(j+1);
   end

   r1 = f_seat(z,v)*dt;
   r2 = f_seat(z+r1/2,v)*dt;
   r3 = f_seat(z+r2/2,v)*dt;
   r4 = f_seat(z+r3,v)*dt;

   z = z + (r1+2*r2+2*r3+r4)/6;

   spos = z(1);     % The Seat vertical Position

   if hip < b+spos
           % When the hip hits the Seat
       u(3) = u(3)-K2*dq(3)+K1*(b+spos-hip);
   end

   % Then, continue to call f_dyn(x,u)
   % to update the state x for digital human,
   % and call f_dhuman(q,spos) to show animation.

end


   % The Function Subroutine %
```

```
function f_s = f_seat(z,v)

    dz = z(2);   g = 9.81;
    f_s = [dz; v-g];

end
```

In the above partial MATLAB^{TM} program, $n1$ is a small time interval before and after the digital warfighter dynamic response, *hip* is a vertical displacement of the H-point that can directly be found from the digital human dynamics subroutine, b is a bias between the H-point and the hitting point of the buttock with the seat cushion , and $K1$ and $K2$ are the PD gain constants to be predetermined and fine-tuned.

Likewise, the first line inside the for-loop is a "self-defense" PD control law, as formulated in (11.3), to keep every joint angle in q towards its comfort center q_c or the initial value. Once the input u is updated in each sampling interval, both the state x and input u will go to the digital human dynamics Runge-Kutta algorithm to solve for new x by calling the f_dyn.m subroutine. The updated joint position vector q as well as the seat position *spos* will be the inputs into the digital human drawing subroutine f_dhuman.m.

By repeating the above updating process, we will be able to see a realistic animation of the digital warfighter dynamic motion in response to the IED explosion underneath the vehicle. Figure 11.20 shows the digital warfighter

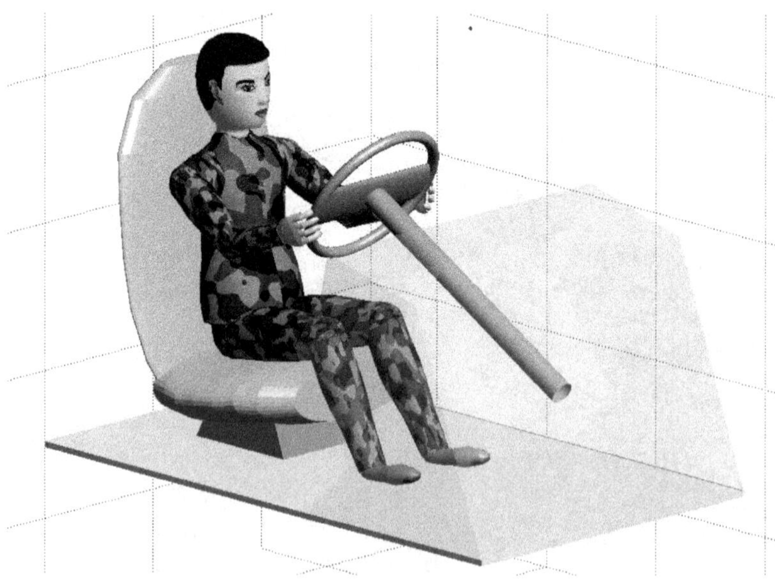

Fig. 11.20 A digital warfighter is sitting in a military vehicle with a normal posture

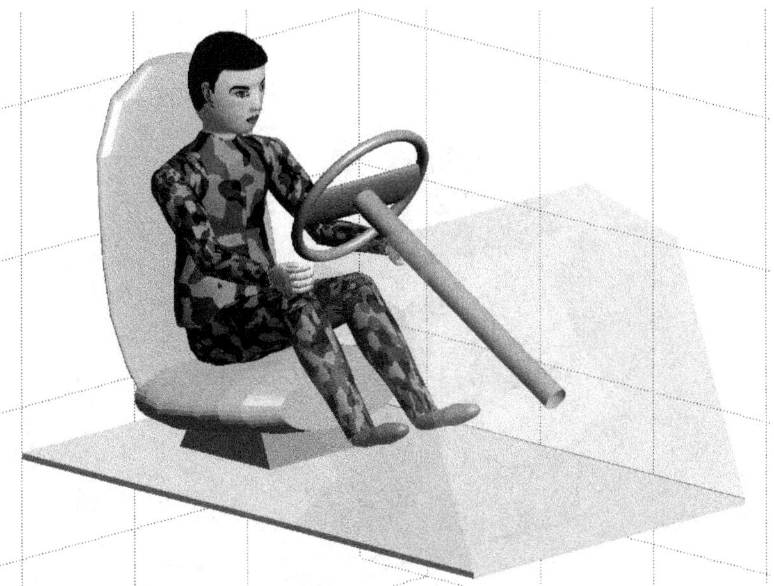

Fig. 11.21 An IED explosion blasts the vehicle and bounces up the mannequin

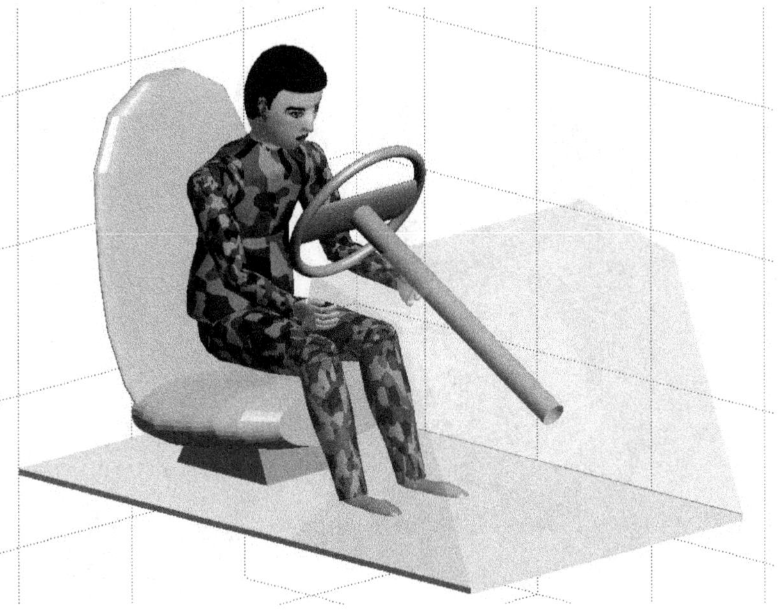

Fig. 11.22 The explosion makes the mannequin further jump up

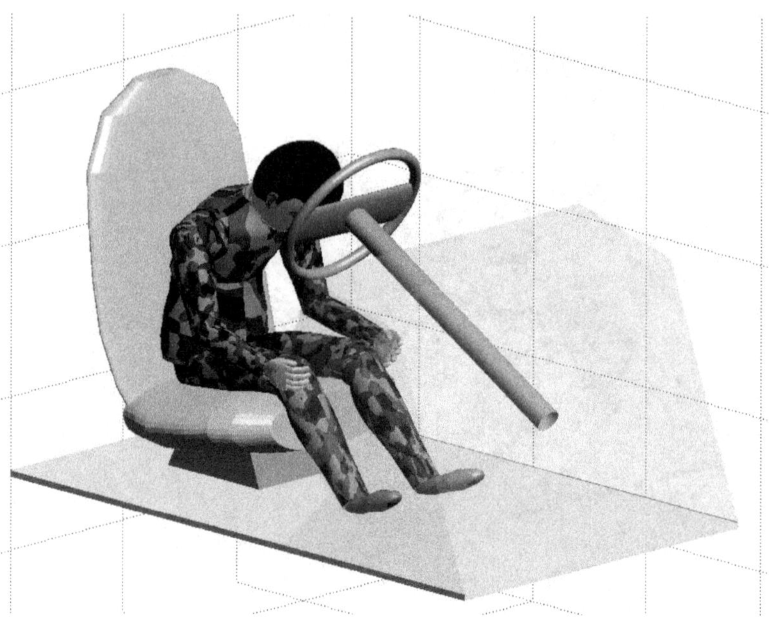

Fig. 11.23 The head would severely hit the steering wheel without any protection in response to the IED explosion

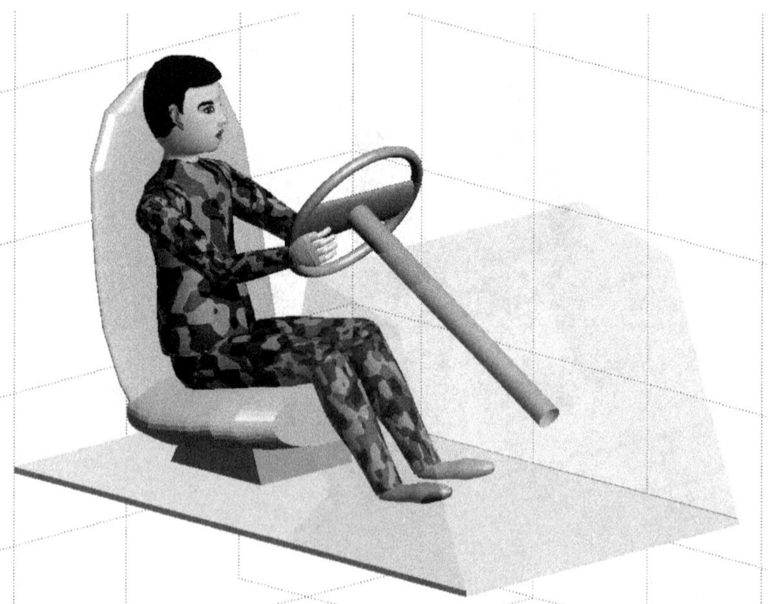

Fig. 11.24 The digital warfighter is sitting with a 20^0 turning angle before an IED explodes

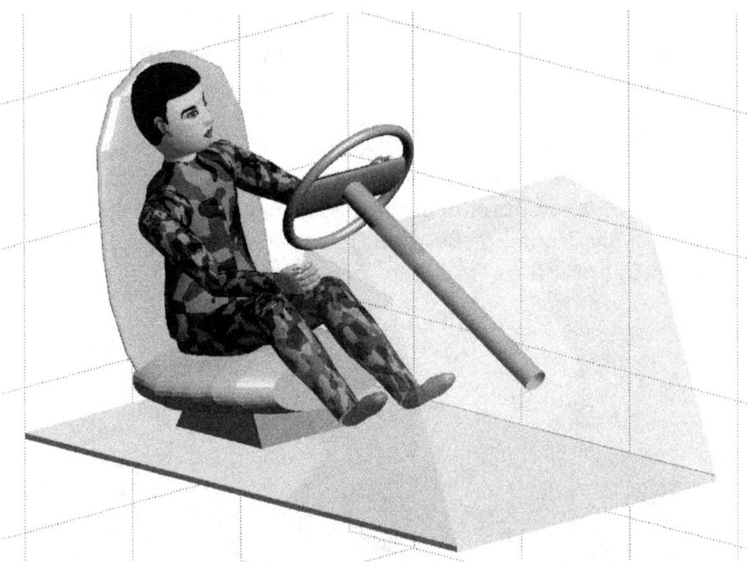

Fig. 11.25 The digital warfighter body is not only bouncing up, but also starting leaning off

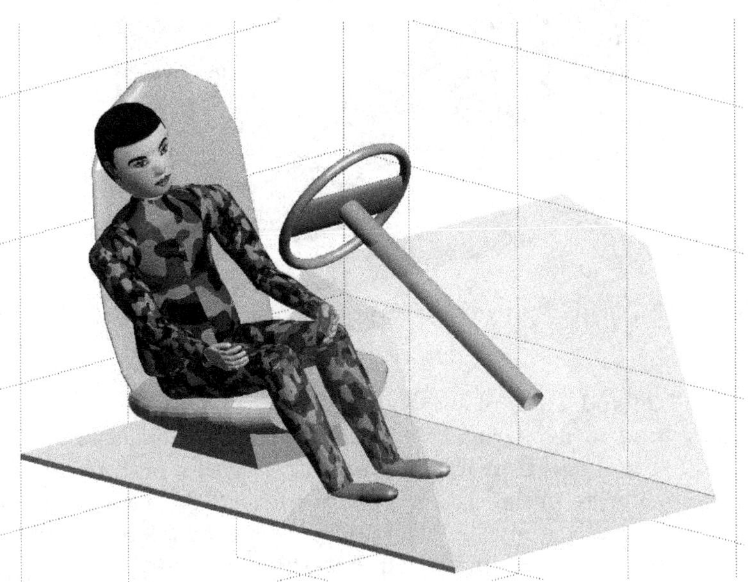

Fig. 11.26 The digital warfighter is further leaning away

is sitting on the military vehicle seat in a normal posture but does not wear any protecting seat belt as the worst case scenario just for a simulation purpose. After the explosion, the sequence of dynamic motion is visualized in Figures 11.21, 11.22 and 11.23.

If now the digital warfighter is sitting on the seat with a 20^0 turning angle before an IED explodes, as shown in Figure 11.24, to simulate a more general accident case, then the consequence will be very different from the neutral initial posture case. Namely, the digital warfighter will soon be leaning off and eventually falling down from the seat, as shown in Figures 11.25, 11.26 and 11.27.

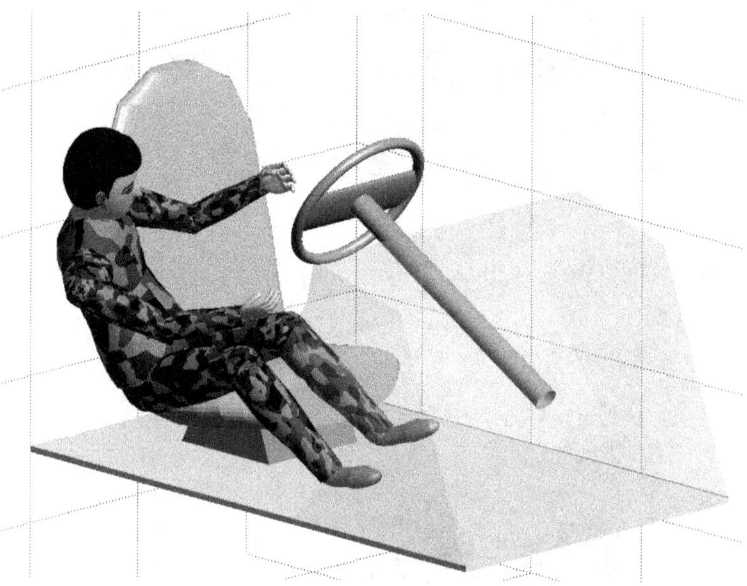

Fig. 11.27 The digital warfighter is struggling and finally falling down from the seat

Figures 11.28 and 11.29 plot the angular accelerations in rad./sec^2 of the three joints: head turn $\ddot{\theta}_{45}$, head tilt $\ddot{\theta}_{46}$ and head nod $\ddot{\theta}_{47}$ around the digital human neck area versus time in the two cases under an initial normal sitting posture and under an initial sitting posture with a 20^0 turning angle, respectively. In the first case, only the head nod acceleration $\ddot{\theta}_{47}$ is overwhelmingly high and the other two are very low due to the initial normal sitting posture without any turning or leaning angle. In contrast, due to the initial 20^0 turning angle in the second case, after the explosion, the digital warfighter was struggling for a balance at the beginning and is then leaning off and falling down from the seat. All the three joints of the neck are suffering from higher accelerations until the warfighter body drops down to the vehicle floor in about 3 seconds, see Figure 11.29.

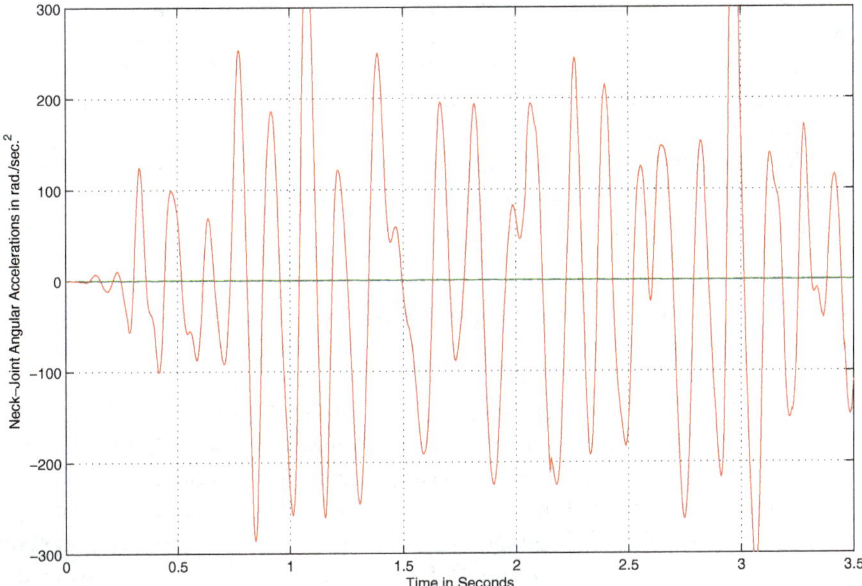

Fig. 11.28 Three joint accelerations of the neck vs. time under an initial normal posture

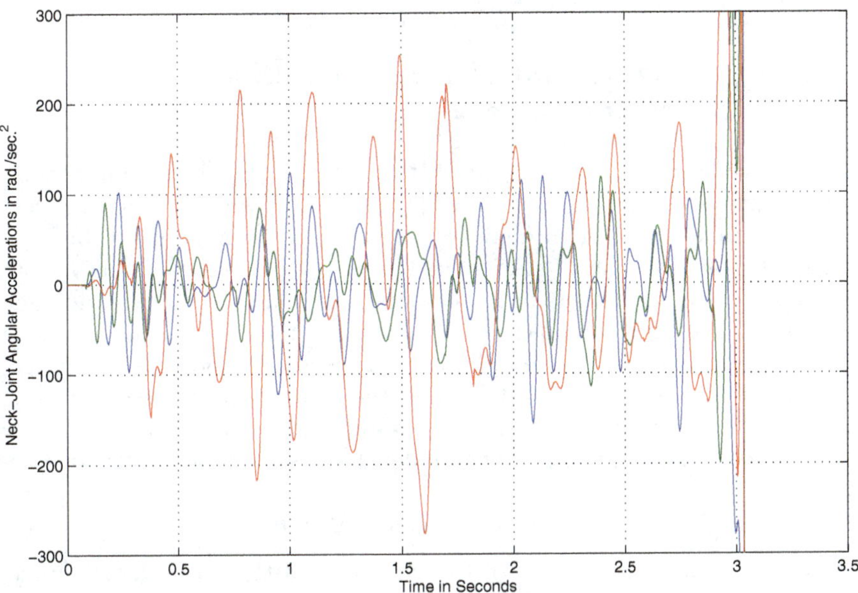

Fig. 11.29 Three joint accelerations of the neck vs. time under an initial posture with a 20^0 turning angle

Since every digital human joint velocity and acceleration can be evaluated directly from the above dynamic motion simulation study, it becomes a virtual but powerful tool to determine a transfer matrix, the input of which is the warfighter's initial posture, military vehicle restraint, and interior packaging condition, etc., and the output of which is the degree of injury for every body joint and/or segment. Clearly, for a more reliable and creditable prediction by the virtual simulation study, the human-environment interactions must be more carefully modeled and fine-tuned.

11.5 Dynamic Interactive Control of Vehicle Active Systems

In the previous two sections, we have demonstrated two typical simulation studies of digital human dynamics with a number of snapshots during the digital human dynamic response to the car frontal collision and the IED explosion underneath the vehicle. However, in most cases, the restraint and protection systems inside the vehicle are all passive. Only in the case, as shown in Figures 11.13 and 11.14, an active restraint control system is applied on the upper seat belt. Theoretically, such an active restraint control scheme can effectively absorb the impact and significantly reduce the responded accelerations over the entire human body [8, 9]. Let us now introduce how the active restraint control system is developed.

11.5.1 Modeling and Control of Active Vehicle Restraint Systems

A regular vehicle seat-belt restraint system often consists of an upper shoulder belt segment and a lower waist belt segment, both of which are interconnected in series from a retractor through two slip-rings to form a unified three-point belt system, as shown in Figure 11.30. The upper segment restrains one of the two shoulders and controls both the occupant's lumbar and thoracic joints in terms of the digital human model. Whereas the lower segment can only control the human H-triangle to limit its translational movement. Therefore, the two levels of seat-belts have to be modeled in slightly different ways.

The retractor offers a spring and a damping forces so that the two segments of the entire seat-belt can be modeled as a second-order PD-type linear system:

$$m\Delta\ddot{\xi} + k_v\Delta\dot{\xi} + k_p\Delta\xi = u,$$

where $\Delta\xi$ is the displacement of the belt beyond the initial length under a pretension, m is the mass of the belt, $k_v > 0$ and $k_p > 0$ are the damping and spring constants, respectively, and the input u is the tension force on the belt beyond the pretension and produced from the retractor.

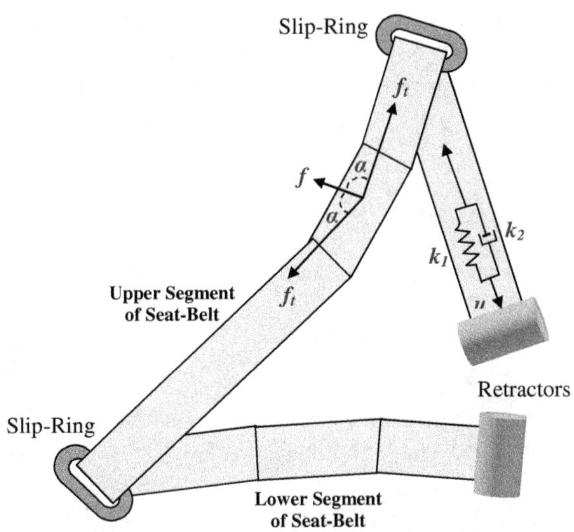

Fig. 11.30 A typical seat-belt restraint system

Since the human occupant's shoulder/chest area receives a distributive pressure due to the additional tension of the belt, the chest may be modeled as an operating point to be exerted with a pressing force $f = 2f_t \cos\alpha$ perpendicular to the chest area, where $f_t = k_2\Delta\dot{\xi} + k_1\Delta\xi$ is the strain force along the belt and α is the angle between the belt line and the normal vector to the chest area. Combining all the parameters together, by defining $\xi_1 = \Delta\xi$ and $\xi_2 = \Delta\dot{\xi}$, we finally have a model for the shoulder/chest segment of the seat-belt as follows:

$$\begin{cases} \dot{\xi}_1 = \xi_2 \\ \dot{\xi}_2 = -K_1\xi_1 - K_2\xi_2 + bu \end{cases} \tag{11.8}$$

with two positive constants $K_1 > 0$ and $K_2 > 0$.

Let a state vector be defined as $\xi = (\xi_1 \; \xi_2)^T$. Equation (11.8) can be converted to a linear state-space equation form:

$$\dot{\xi} = A\xi + Bu, \tag{11.9}$$

where

$$A = \begin{pmatrix} 0 & 1 \\ -K_1 & -K_2 \end{pmatrix}, \quad \text{and} \quad B = \begin{pmatrix} 0 \\ b \end{pmatrix}.$$

The pressing force on the human shoulder/chest can be defined as an output, i.e.,

$$f = \bar{K}_1\xi_1 + \bar{K}_2\xi_2 = C\xi = (\bar{K}_1 \; \bar{K}_2)\xi$$

with two constants $\bar{K}_1 > 0$ and $\bar{K}_2 > 0$.

Equation (11.9) along with the above output equation can be extended to a 4th-order linear system with the output dimension $r = 2$, provided that the restraint system has two independent displacements ξ_1 and ξ_2. Namely, $\xi = (\xi_1 \; \xi_2 \; \xi_3 \; \xi_4)^T$ with $\xi_3 = \dot{\xi}_1$ and $\xi_4 = \dot{\xi}_2$, and

$$A = \begin{pmatrix} O & I \\ -K_1 & -K_2 \end{pmatrix}, \quad \text{and} \quad B = \begin{pmatrix} O \\ b \end{pmatrix},$$

where O and I are the 2 by 2 zero and identity matrices, respectively, and

$$K_1 = \begin{pmatrix} k_{p1} & 0 \\ 0 & k_{p2} \end{pmatrix}, \quad K_2 = \begin{pmatrix} k_{v1} & 0 \\ 0 & k_{v2} \end{pmatrix}, \quad \text{and} \quad b = \begin{pmatrix} b_1 & 0 \\ 0 & b_2 \end{pmatrix}.$$

With such an extension, the system becomes "square", because both the input $u = (u_1 \; u_2)^T$ and the output $F = (f_1 \; f_2)^T = C\xi$ are 2-dimensional with

$$C = (\bar{K}_1 \; \bar{K}_2) = \begin{pmatrix} \bar{k}_{p1} & 0 & \bar{k}_{v1} & 0 \\ 0 & \bar{k}_{p2} & 0 & \bar{k}_{v2} \end{pmatrix},$$

where $\bar{k}_{pi} > 0$ and $\bar{k}_{vi} > 0$ are all constants for $i = 1, 2$.

The above output vector F is actually a Cartesian force acting on the shoulder/chest of the human body. Based on the robotic statics from Chapter 4, the corresponding joint torque is $\tau_s = J^T F$, where J is a local Jacobian matrix of the shoulder/chest with respect to the hip sitting on the seat. More specifically, if $F \in \mathbb{R}^2$ and $\tau_s \in \mathbb{R}^2$ that may represent a vector of the lumbar and thoracic flexion/extension joint torques:

$$\tau_s = \tau_{lt} = \begin{pmatrix} \tau_7 \\ \tau_{10} \end{pmatrix},$$

then this local Jacobian matrix J is 2 by 2 for the human chest area directly with respect to the hip. It can be seen that the local Jacobian matrix J must at least be a 2 by 2 square matrix in order to make itself invertible as well as to cover both the lumbar and thoracic two major joints on the human trunk for this particular frontal collision dynamic response case. This is also the main reason why equation (11.9) is extended to the above 4th-order system to create a two-input/two-output active restraint system. Therefore, the output of the entire seat-belt restraint system (11.9) becomes

$$y = \tau_s = J^T F = J^T C\xi. \tag{11.10}$$

By taking a time-derivative on the output (11.10), we obtain

$$\dot{y} = \dot{J}^T C\xi + J^T C\dot{\xi} = \dot{J}^T C\xi + J^T CA\xi + J^T CBu.$$

Since the decoupling matrix

$$CB = \begin{pmatrix} \bar{k}_{v1}b_1 & 0 \\ 0 & \bar{k}_{v2}b_2 \end{pmatrix},$$

it should be non-singular because both $\bar{k}_{v1}b_1 \neq 0$ and $\bar{k}_{v2}b_2 \neq 0$. Then, by defining a new input $v_\xi = \dot{y}$, and supposing that the local Jacobian matrix J is also non-singular, the input u can be resolved as

$$u = (CB)^{-1}J^{-T}(v_\xi - \dot{J}^T C\xi - J^T CA\xi).$$

This means that the system is invertible after taking the first time-derivative of the output y, and thus the system has a relative degree $r_i = 1$ at each output channel. If the dimension of y is $r = 2$, then the total relative degree is $r_d = r_1 + r_2 = 2 = r$. Here and thereafter all the concepts of the decoupling matrix, relative degree as well as the internal dynamics, zero dynamics and minimum-phase can be found in Chapter 8.

Because the decoupling matrix CB in this case is clearly related to the states ξ_3 and ξ_4, it implies that $\zeta = (\xi_1\ \xi_2)^T$ is a right definition for the state vector of the internal dynamics. Hence,

$$\dot{\zeta} = \begin{pmatrix} \dot{\xi}_1 \\ \dot{\xi}_2 \end{pmatrix} = \begin{pmatrix} \xi_3 \\ \xi_4 \end{pmatrix}.$$

According to the output definition (11.10),

$$y = J^T C\xi = J^T \bar{K}_1 \begin{pmatrix} \xi_1 \\ \xi_2 \end{pmatrix} + J^T \bar{K}_2 \begin{pmatrix} \xi_3 \\ \xi_4 \end{pmatrix} = J^T \bar{K}_1 \zeta + J^T \bar{K}_2 \begin{pmatrix} \xi_3 \\ \xi_4 \end{pmatrix}.$$

Therefore, the internal dynamics turns out to be

$$\dot{\zeta} = \begin{pmatrix} \xi_3 \\ \xi_4 \end{pmatrix} = \bar{K}_2^{-1} J^{-T} y - \bar{K}_2^{-1} \bar{K}_1 \zeta. \tag{11.11}$$

Its zero-dynamics that is obtained by setting $y \equiv 0$ becomes

$$\dot{\zeta} = -\bar{K}_2^{-1} \bar{K}_1 \zeta, \tag{11.12}$$

which obviously has an asymptotically stable equilibrium point at $\zeta = 0$ due to $\bar{K}_1 > 0$ and $\bar{K}_2 > 0$. This sufficiently shows that the entire seat-belt restraint system not only has relative degree $r_i = 1$ at each output channel for $i = 1, 2$, but also guarantees to be asymptotically minimum-phase.

To develop an active control scheme for the entire seat-belt restraint system interacting with the human driver/occupant, let us apply the cascaded dynamic model and backstepping control design procedure [12, 8, 9] that have been developed in Chapter 8. Suppose that there is an n-dimensional linear or nonlinear system, the input of which is cascaded by another m-dimensional linear or nonlinear system through its output. Namely,

$$\dot{x} = f(x) + G(x)y, \tag{11.13}$$

$$y = h(x, \xi), \tag{11.14}$$

and

$$\dot{\xi} = \eta(\xi) + \Gamma(\xi)u, \tag{11.15}$$

where $x \in \mathbb{R}^n$, $y \in \mathbb{R}^r$, $\xi \in \mathbb{R}^m$ and $u \in \mathbb{R}^r$. Since the entire cascaded system is, in general, an MIMO (Multi-Input and Multi-Output) system, the coefficient $G(x)$ for the input y in (11.13) is an n by r matrix, i.e., $G(x) = (g_1(x) \ \cdots \ g_r(x))$, where each column is an n-dimensional vector field. We intentionally define the input u of (11.15) to have the same dimension as its output y to ensure that it is a "square" MIMO system. Similarly, $\Gamma(\xi) = (\gamma_1(\xi) \ \cdots \ \gamma_r(\xi))$ in (11.15) is an m by r matrix. We now apply a backstepping procedure, as introduced and discussed in Chapter 8, to perform our control design for the global input u of the cascaded system in equations (11.13), (11.14) and (11.15).

Suppose that equation (11.13) can be asymptotically stabilized with respect to its equilibrium point $x = 0$ by a static state-feedback input $y = \sigma(x)$, called a *virtual control law*, under a Lyapunov candidate $V(x)$, i.e.,

$$\dot{V} = \frac{\partial V}{\partial x} f(x) + \frac{\partial V}{\partial x} G(x)\sigma(x) \le -\epsilon(x)$$

for a p.d.f. $\epsilon(x)$. Moreover, let equation (11.15) be an asymptotically minimum-phase system with a relative degree $r_i = 1$ at each output channel for $i = 1, \cdots, r$ so that the total relative degree $r_d = r_1 + \cdots + r_r = r$, which becomes a condition to be imposed on (11.15) in order to successfully carry out the backstepping control design procedure. The reader can refer to Chapter 8 for details.

Thus, based on the theory of state-feedback linearization from Chapter 8, a new input can be defined by

$$v = \dot{y} = \frac{\partial h}{\partial x} f(x) + \frac{\partial h}{\partial x} G(x)y + \frac{\partial h}{\partial \xi}\eta(\xi) + \left(\frac{\partial h}{\partial \xi}\Gamma(\xi)\right) u, \tag{11.16}$$

and the coefficient of the input u here is an r by r non-singular matrix, which is known as the decoupling matrix in a certain neighborhood of the equilibrium point $\xi = 0$ for the system (11.15) due to each output channel relative degree $r_i = 1$. Therefore,

$$u = \left(\frac{\partial h}{\partial \xi}\Gamma(\xi)\right)^{-1} \left[v - \frac{\partial h}{\partial x} f(x) - \frac{\partial h}{\partial x} G(x)y - \frac{\partial h}{\partial \xi}\eta(\xi)\right]. \tag{11.17}$$

Let an error function be defined as $e(x) = y - \sigma(x)$, which represents a "distance" between the output (11.14) of the system (11.15) and the asymptotically stable input of the system (11.13). Clearly, we wish $e(x)$ would go to zero asymptotically. Thus, a global Lyapunov candidate for the entire cascaded system can be defined as follows:

$$V_\sigma(x, \xi) = V(x) + \frac{1}{2} e^T e.$$

Its time-derivative becomes

$$\begin{aligned}
\dot{V}_\sigma &= \dot{V} + e^T \dot{e} = \frac{\partial V}{\partial x} f(x) + \frac{\partial V}{\partial x} G(x) y + e^T (\dot{y} - \dot{\sigma}) \\
&= \frac{\partial V}{\partial x} f(x) + \frac{\partial V}{\partial x} G(x)(\sigma + e) + e^T (\dot{y} - \dot{\sigma}) \\
&\leq -\epsilon(x) + \frac{\partial V}{\partial x} G(x) e + e^T (\dot{y} - \dot{\sigma}).
\end{aligned} \tag{11.18}$$

Now, let

$$v = \dot{y} = -ke + \dot{\sigma} - G^T \left(\frac{\partial V}{\partial x} \right)^T.$$

Substituting it into (11.17), the time-derivative of the global Lyapunov function in (11.18) becomes

$$\dot{V}_\sigma \leq -\epsilon(x) - k e^T e \leq 0$$

for a positive gain constant $k > 0$.

Finally, the control law of the entire cascaded system turns out to be

$$u = -\left(\frac{\partial h}{\partial \xi} \Gamma(\xi) \right)^{-1} \left[ke + \frac{\partial e}{\partial x} f(x) + \frac{\partial e}{\partial x} G(x) y + G^T \left(\frac{\partial V}{\partial x} \right)^T + \frac{\partial h}{\partial \xi} \eta(\xi) \right]. \tag{11.19}$$

Since the system (11.15) is assumed to be asymptotically minimum-phase with $r_d = r$, its $m - r_d = m - r$ dimensional internal dynamics in the form of

$$\dot{\zeta} = \phi(\zeta, z) + \psi(\zeta, z) v,$$

or by setting $z \equiv 0$ and $v = \dot{z} = 0$, its zero dynamics

$$\dot{\zeta} = \phi(\zeta, 0)$$

must be asymptotically stable. Here, $z = y$ and $v = \dot{y}$ are the new state and new input after the exact linearization, respectively, while $\zeta \in \mathbb{R}^{m-r}$ represents the states of the unlinearizable subsystem, i.e., the internal dynamics. This ensures that the internal dynamics will not interfere the overall stability.

Specifically, in our application to the control of active vehicle restraint systems, equation (11.13) is virtually a partial model of digital human dynamics, while equation (11.15) represents the seat-belt restraint system model, and the output equation (11.14) follows the model of robotic statics, i.e.,

$$y = J^T F = J^T C\xi,$$

as discussed earlier. All the conditions imposed by the above backstepping control design procedure can be met as well.

In summary, a complete block diagram of the active restraint system control associated with the digital human dynamic model is depicted in Figure 11.31.

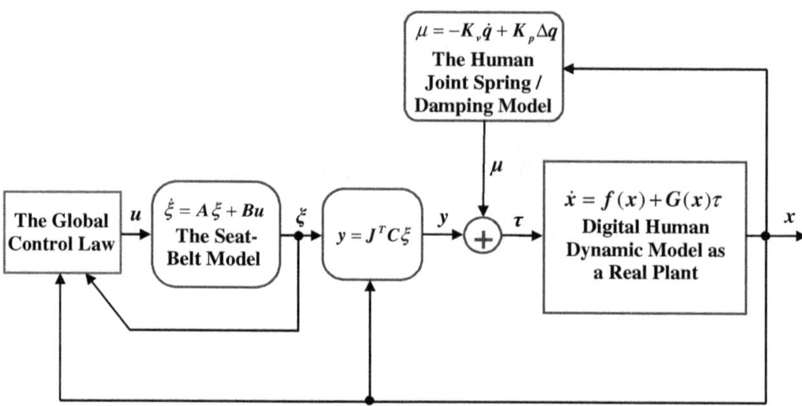

Fig. 11.31 A complete block diagram for the active restraint control system

Once again, the block containing $\mu = -K_v\dot{q} + K_p\Delta q$ is a "self-defense" joint torque that is the same as equation (11.3). In this car crash simulation study, since the upper seat belt can offer at most two-dimensional force to act on the area of the digital driver's shoulder/chest, the dynamic equation (11.13) used for developing a control law via the backstepping procedure has to be a 2-dimensional partial dynamic formulation:

$$W_{lt}\ddot{q}_{lt} - \tau_{glt} = \tau_{lt} \tag{11.20}$$

that corresponds to the flexion/extension joints of the lumbar and thorax, i.e.,

$$W_{lt} = \begin{pmatrix} W(7,7) & W(7,10) \\ W(10,7) & W(10,10) \end{pmatrix}, \quad q_{lt} = \begin{pmatrix} \theta_7 \\ \theta_{10} \end{pmatrix}, \quad \text{and } \tau_{lt} = \begin{pmatrix} \tau_7 \\ \tau_{10} \end{pmatrix},$$

while τ_{glt} is similar to the above joint torque τ_{lt} but is the joint torque due to gravity.

In the above 2 by 2 partial inertial matrix W_{lt}, each $W(i,j)$ is the i-th row and j-th column element of the 47 by 47 inertial matrix W in the digital human dynamic model, and this 2 by 2 W_{lt} just represents the two major joints to be controlled. Of course, the partial dynamic equation (11.20) is valid if every movement of the joints other than joint 7 and joint 10, such as the joints along the arms and legs, will not interfere the two joints. In other words, all but the 7th and 10th elements in each of the 7th and 10th rows of the inertial matrix W are zero. This is possible only if the digital human is in a sitting posture, and the car crash case belongs to this category.

However, while only two joints are controlled by the active restraint system, all the 47 joints of the digital human are involved in the entire dynamic model given by (11.6) and (11.7) with the full 47 by 47 inertial matrix W, as a real-plant in Figure 11.31. Therefore, due to the dynamic coupling, the entire digital human will still be able to make a realistic motion in response to the collision.

Two snapshots of the resulting 3D animation has been shown in Figures 11.13 and 11.14. They have realistically demonstrated a dynamic motion within a second after the frontal collision. Since the upper seat-belt touches and presses on the driver's shoulder/chest, the control force, intuitively, can interact directly to both the lumbar and thoracic flexion/extension joints as the driver is sitting on the seat.

In the simulation study, we adopt a joint-space PD control law, as developed in equation (8.67) from Chapter 8. Namely, the virtual control law is defined by

$$\sigma(x) = \tau_{lt} = -\tau_{glt} - K_v \dot{q}_{lt} + K_p \Delta q_{lt}. \tag{11.21}$$

The corresponding Lyapunov candidate is based on equation (8.74) in Chapter 8, but in a partial dynamic model form:

$$V = \frac{1}{2}\dot{q}_{lt}^T W_{lt} \dot{q}_{lt} + \frac{1}{2}\Delta q_{lt}^T K_p \Delta q_{lt}, \tag{11.22}$$

where $\Delta q_{lt} = q_{lt}^d - q_{lt}$ with a desired joint position vector $q_{lt}^d = \begin{pmatrix} \theta_7^d \\ \theta_{10}^d \end{pmatrix}$ for both the lumbar and thoracic flexion/extension joints.

To justify that (11.22) is a qualified Lyapunov function, first, it is sure that V is a p.d.f. in the state space spanned by θ_7, θ_{10} and their time-derivatives $\dot{\theta}_7$ and $\dot{\theta}_{10}$. Second, its time derivative is

$$\dot{V} = \dot{q}_{lt}^T W_{lt} \ddot{q}_{lt} + \frac{1}{2}\dot{q}_{lt}^T \dot{W}_{lt} \dot{q}_{lt} + \Delta \dot{q}_{lt}^T K_p \Delta q_{lt}.$$

We now assume that the second term is much smaller than the first one due to the overwhelmingly high accelerations in a car crash case so that the second term is negligible. Once both the partial dynamic equation (11.20) and the virtual control law (11.21) are substituted into the above $\dot{V}$ equation, as similar to equation (8.76) in Chapter 8, it results in

$$\begin{aligned}\dot{V} &= \dot{q}_{lt}^T(\tau_{lt} + \tau_{glt}) - \dot{q}_{lt}^T K_p \Delta q_{lt} \\ &= \dot{q}_{lt}^T(K_p \Delta q_{lt} - K_d \dot{q}_{lt}) - \dot{q}_{lt}^T K_p \Delta q_{lt} = -\dot{q}_{lt}^T K_d \dot{q}_{lt} \le 0.\end{aligned} \tag{11.23}$$

It can be further shown based on the La Salle's Theorem that the virtual control law (11.21) can asymptotically stabilize the system (11.20) at the equilibrium point $\Delta q_{lt} = 0$ and $\dot{q}_{lt} = 0$. Hence, by noticing that the state is $x = \begin{pmatrix} q_{lt} \\ \dot{q}_{lt} \end{pmatrix}$ and the state-space equation is

$$\dot{x} = f(x) + G(x)\tau_{lt} = \begin{pmatrix} \dot{q}_{lt} \\ W_{lt}^{-1}\tau_{glt} \end{pmatrix} + \begin{pmatrix} O \\ W_{lt}^{-1} \end{pmatrix} \tau_{lt}$$

in this partial dynamic model case, we have

$$\frac{\partial \sigma}{\partial x} = \begin{pmatrix} \frac{\partial \sigma}{\partial q_{lt}} & \frac{\partial \sigma}{\partial \dot{q}_{lt}} \end{pmatrix} = \begin{pmatrix} -K_p - \frac{\partial \tau_{glt}}{\partial q_{lt}} & -K_v \end{pmatrix},$$

so that

$$\frac{\partial \sigma}{\partial x} f(x) = -K_p \dot{q}_{lt} - \frac{\partial \tau_{glt}}{\partial q_{lt}} \dot{q}_{lt} - K_v W_{lt}^{-1} \tau_{glt}$$

if all the centrifugal and Coriolis terms are neglected, and

$$\frac{\partial \sigma}{\partial x} G(x) = -K_v W_{lt}^{-1}.$$

It can also be seen that in the specified virtual control law (11.21) with the Lyapunov function (11.22),

$$G^T \left(\frac{\partial V}{\partial x} \right)^T = W_{lt}^{-1} \frac{\partial V}{\partial \dot{q}_{lt}} = W_{lt}^{-1} W_{lt} \dot{q}_{lt} = \dot{q}_{lt}.$$

Furthermore, in the new input equation (11.16), the first two terms are equal to $\dot{J}^T C \xi$ in the simulation case due to $y = h(x, \xi) = J^T C \xi$, and $\partial h / \partial \xi = J^T C$ so that

$$\frac{\partial h}{\partial \xi} \eta(\xi) = J^T C A \xi \quad \text{and} \quad \frac{\partial h}{\partial \xi} \Gamma(\xi) = J^T C B.$$

All the above derivations are required by the global control law given in (11.19).

The control law (11.19) is, however, only to excite the driver's shoulder/chest through the upper segment of the active seat-belt, and then to interact both the lumbar and thoracic joints of the human trunk. In contrast, the real-plant in the simulation study is a full 47-joint dynamic model with each revolute joint of the driver excited by a PD control in the form:

$$\mu_i = -K_v \dot{\theta}_i + K_p(\theta_i^h - \theta_i),$$

where θ_i^h is the i-th joint angle in the driver's original sitting posture before the collision for $i = 4, \cdots, 47$. This PD excitation to each human revolute joint in the real-plant reflects the "self-defense" resilient nature to struggle for retaining the original posture. This is why a dummy model in a real car crash test is installed with a spring at each dummy's joint.

As a result, the accelerations of the three major joints: lumbar and thoracic flexion/extension and neck nod, have been plotted in Figure 11.17, and the required control input u is given in Figure 11.18.

In the literature and publication of National Highway Traffic Safety Administration (NHTSA) [11], the Head Injury Criteria *HIC* and the Chest Acceleration Limit *Ac* are two of the major injury criteria to assess the collision incidents. The *HIC* is defined by

$$HIC = \max \left[\frac{1}{t_2 - t_1} \int_{t_1}^{t_2} a(t)dt \right]^{2.5} (t_2 - t_1),$$

where t_1 and t_2 are two arbitrary distinct time instants during the period of acceleration pulse and $a(t)$ is the head acceleration in multiples of the gravitational acceleration g. NHTSA proposed to limit the testing time interval $t_2 - t_1 = 36$ miliseconds in 1986 as HIC_{36}, and then updated to 15 ms. in 2000, denoted as HIC_{15}.

Suppose that we roughly measure $d_c = 0.2$ m. from the H-point to the chest center for an average human body, $d_n = 0.2$ m. from the chest center to the neck joint and $d_h = 0.1$ m. from the neck to the center of the head. Based on the simulation result in the actively controlled case, at a 15 ms. time interval after a frontal collision with the vehicle speed at 45 mph, the lumbar joint acceleration is around $\beta_l = 500$ rad/sec^2, the thoracic joint acceleration is around $\beta_t = 800$ rad/sec^2, and the neck joint acceleration peak is about $\beta_n = 2000$ rad/sec^2. Then, the maximum head acceleration in the worst case becomes

$$a(t) = \beta_l(d_c + d_n + d_h) + \beta_t(d_n + d_h) + \beta_n d_h = 690 \approx 70g.$$

Using this as a constant head acceleration inside the above integral for 15 ms. in the worst case scenario, we can estimate $HIC_{15} = 615$, which is below the upper limit 700 for an adult according to the NHTSA standard. In fact, since the three major joint accelerations may not always be synchronized in the same direction, the value of HIC_{15} would be much lower.

If we estimate the chest acceleration, then $Ac = \beta_p d_c = 100 \approx 10g$, which is much smaller than the adult limit $60g$. More precisely, the linear velocity of the chest is given by $\dot{p}_{hc} = J\dot{q}_{hc}$, where J is the local Jacobian matrix of the chest with respect to the H-point and q_{hc} is the corresponding joint angle vector between the chest and the H-point. Therefore, the chest acceleration can be determined by

$$a_{hc} = \ddot{p}_{hc} = J\ddot{q}_{hc} + \dot{J}\dot{q}_{hc}.$$

This equation can also be used to find the head acceleration, provided that both J and q_{hc} are the local Jacobian matrix and joint angle vector between the head and the H-point [9].

11.5.2 An Active Suspension Model and Human-Machine Interactive Control

In many luxury cars, a hydraulic actuator is often equipped to act as an active suspension device between each wheel axle and the car chassis. Let the actuating length be z_a, the hydraulic flow be Q and the area of the piston in the actuator chamber be A. Then, $\dot{z}_a = Q/A$ and

$$\dot{Q} = f_q(Q) + g_q(Q)u$$

with some linear or nonlinear functions $f_q(\cdot)$ and $g_q(\cdot)$, and u is an actuator valve control input. A possible linear model can be given by $\dot{Q} = -c_1 Q + c_2 u$ with two constants $c_1 > 0$ and $c_2 > 0$. In a quarter-car model, let the height of the body chassis be z_b and the seat height be z_s, all referred to the ground, as shown in Figure 11.32.

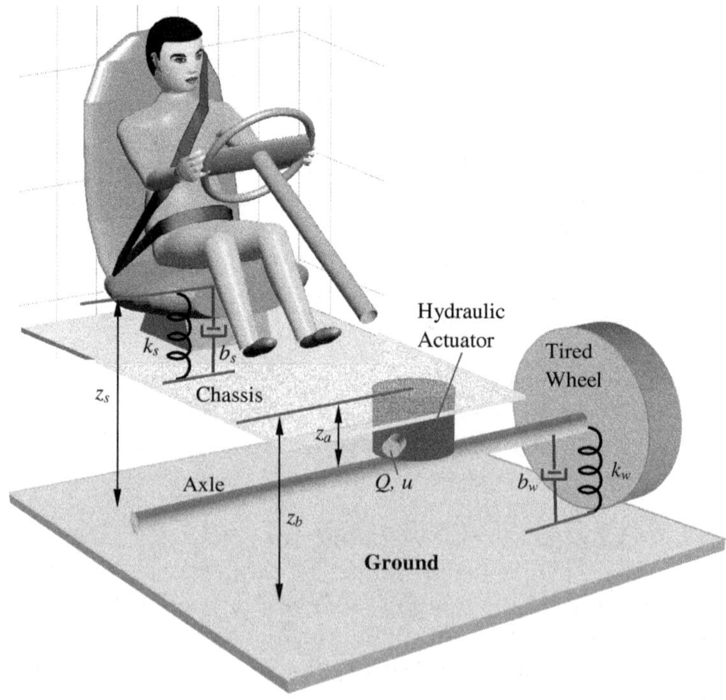

Fig. 11.32 A digital human drives a car with an active suspension system

Then, we can readily find the following linear dynamic equations:

$$M_b\ddot{z}_b + b_w(\dot{z}_b - \dot{z}_a) + k_w(z_b - z_a) + b_s(\dot{z}_b - \dot{z}_s) + k_s(z_b - z_s) = 0,$$

and

$$m_s\ddot{z}_s + b_s(\dot{z}_s - \dot{z}_b) + k_s(z_s - z_b) = f_h(z_s, \dot{z}_s),$$

where, b_w, k_w and b_s, k_s are the positive damping coefficients and spring constants for the tired wheel and the car seat, respectively, M_b and m_s are the effective masses for the car body and seat, respectively, and $f_h(z_s, \dot{z}_s)$ is the reacting force from the seat to the human driver.

To study this quarter-car active suspension control system, we can adopt a k-cascaded model that was given by equation (8.102) and developed in Chapter 8. Let $\xi_{11} = z_s$, $\xi_{12} = \dot{z}_s$, $\xi_{21} = z_b$, $\xi_{22} = \dot{z}_b$, $\xi_{31} = z_a$ and $\xi_{32} = Q$. Then, the above dynamic models are directly converted into a set of standard state-space equations:

$$\dot{x} = f(x) + G(x)y_1$$
$$y_1 = J^T(x)f_h(\xi_{11}, \xi_{12})$$

$$\begin{cases} \dot{\xi}_{11} = \xi_{12} \\ \dot{\xi}_{12} = -\frac{b_s}{m_s}\xi_{12} - \frac{k_s}{m_s}\xi_{11} + f_h(\xi_{11}, \xi_{12}) + y_2 \end{cases}$$

$$y_2 = \frac{b_s}{m_s}\xi_{22} + \frac{k_s}{m_s}\xi_{21}$$

$$\begin{cases} \dot{\xi}_{21} = \xi_{22} \\ \dot{\xi}_{22} = -\frac{b_w}{M_b}\xi_{22} - \frac{k_w}{M_b}\xi_{21} - \frac{b_s}{M_b}(\xi_{22} - \xi_{12}) - \frac{k_s}{M_b}(\xi_{21} - \xi_{11}) + y_3 \end{cases}$$

$$y_3 = \frac{b_w}{M_b A}\xi_{32} + \frac{k_w}{M_b}\xi_{31}$$

$$\begin{cases} \dot{\xi}_{31} = \frac{1}{A}\xi_{32} \\ \dot{\xi}_{32} = f_q(\xi_{32}) + g_q(\xi_{32})u \end{cases} \quad (11.24)$$

This evidently demonstrates that each quarter-car with its active suspension can be modeled as a k-cascaded dynamic system with $k = 3$, and the human dynamic model is the top of equation (11.24). Theoretically, it can also be extended to a full-car model with multiple active suspensions. By applying a backstepping procedure, as developed in Chapter 8 and also discussed earlier in this section, we will be able to resolve the hydraulic actuator valve control u to minimize the overall joint accelerations of the human body for achieving a better ride quality [12, 8].

Since we have completely developed a digital human dynamic formulation, the k-cascaded systems modeling and control design associated with the

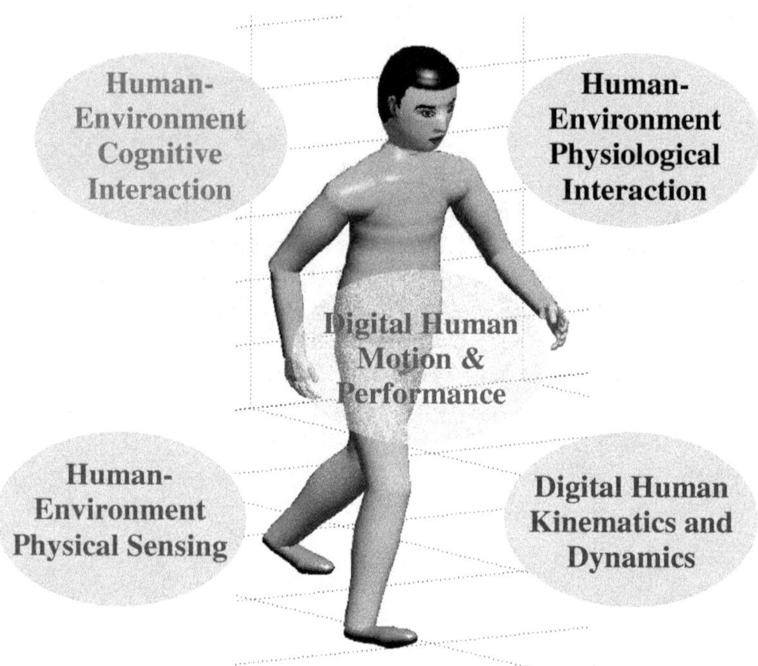

Fig. 11.33 A future integration in research and development of digital human modeling

digital human dynamics will be very useful in a wide range of applications, as long as the dynamic interactive control systems are related to human operators, such as a driver, passenger and warfighter.

11.6 Future Perspectives of Digital Human Modeling

What are we expecting the future of digital human modeling? The literature [13]–[16] has provided a comprehensive survey and review. Let us first celebrate the recent achievements and then look forward to expecting and predicting the future goals and challenges.

Realistic Appearance: In the past two decades, digital human appearance has made a revolutionary breakthrough. Not only a digital mannequin's overall look is now much more realistic and aesthetically pleased, but also the details of dynamically deformable muscles and skins plus the deformable cloths wearing are well achieved. One of the most representative mannequins is Santos^{TM}, a virtual soldier, created and developed by the University of Iowa [17, 18, 19] during the 2000's.

Realistic Motions: Regarding the digital human motion generation, it has also accomplished a big leap in comparison with the status more than a decade ago. However, there is still a long way to go in terms of, not only the posture and motion realism, but also the motion continuum and smoothness, especially in a sequential or high-speed motion. To further improve the major aspects of the realistic posture and motion generation, the accuracy and fidelity of digital human physical models must be the first consideration to take into. While the digital human dynamic models and motion algorithms have been well-developed and ready for numerical implementations, they still encounter a major challenge of how to design a control input into the dynamic model towards a more realistic and natural motion result.

Human-Environment Physical Interaction: This is another turn-key to achieve a new breakthrough in the study of realistic posture and motion generation. In a digital human kinematic model, the environmental limits and constraints can be described by a number of boundary positions and/or orientations so that a digital mannequin will kinematically understand to never go beyond the boundaries. In contrast, if a dynamics algorithm is implemented, like a real human moving in a restricted space, the digital human must be reacted with a force or torque from the environmental boundaries before the motion slows down and stops at the limit. Almost every convex point over the entire digital human physical body should be able to sense such an environmental reaction. Therefore, it seems to be a very natural reality, but is a huge challenge to modeling the digital human-environment physical interaction in a more effective and smarter manner.

Anthropometric Adaptation: The diversity of anthropometric data in mannequin creations is always one of the most challenging issues. The complication is caused by different genders, races, percentiles as well as the personal and environmental conditions. In the current status of kinematics-based digital human modeling and motion generation, each time of changing kinematic parameters, every "end-effector" trajectory plus the necessary accents and decorations have to be modified accordingly. Otherwise, the motion could possibly be distorted. If a complete digital human dynamic control model with a dynamic motion algorithm would be implemented in near future by taking advantage of the adaptive control strategy, every parameter could be automatically adapted to generate a natural posture and realistic motion at any time with any initial anthropometric data input. We wish an adaptation law associated with a digital human dynamic control model would ease the anthropometric complication in the future.

Physiological Interaction: Beyond the physical or geometrical interaction to the environment, we wish a digital human would also be able to adapt the environment with major human factors, such as the strength, effort, comfort and tiredness.

Cognitive Interaction: A desired cognitive interaction includes psychological and emotional actions and reactions to the environment. More specifically, the real human abilities of feeling, preference, emotion, intention, problem solving, memory recalling and decision making could also be made capable for a digital human model. The visual, sound, humidity and temperature sensing as well as the work load and pressure are the major signals as an input of the cognitive interaction. The facial expression, gaze, posture and gesture are the major physical representations as an output of the cognitive interaction. The existing cognitive architectures, cognitive models and psychological models are the best tools to help for developing feasible algorithms to be integrated into the digital human physical modeling and motion generation.

Integration: Every aspect and modeling consideration must eventually be integrated to enhance the final performance of digital human models. This is a multi-disciplinary joint effort to integrate knowledge and expertise from engineering, biomechanics, ergonomics, kinesiology, physiology, psychology and cognitive science. Since we are modeling a human, not just a robotic machine, every aspect must be taken into serious consideration in the modeling integration before a satisfactory digital human performance can be achieved.

11.7 Exercises of the Chapter

1. Write a program in MATLAB^{TM} to numerically calculate the sub-Jacobian matrix J_{33} of the digital human left thigh segment. Define your own dynamic parameters for the left thigh to determine the parameter matrix U_{33} and then calculate the partial inertial matrix $W_{33} = J_{33}^T U_{33} J_{33}$. Through necessary conversions, also find the partial joint torque τ_{g33} due to the left thigh segment gravity based on equation (9.23) in Chapter 9.
2. If the steering wheel of a car has a 60^0 angle from the horizontal plane, how can you model an interaction force if the head of a digital driver hits the steering wheel in an accident? In other words, how can you define every variable in equation (11.4)? Furthermore, what is the local Jacobian matrix J that can represent the head in order to convert the Cartesian reacting force f to the joint torques τ by equation (11.5)?
3. How do you plan to model the head of a digital driver hitting an inflated side airbag in a side collision accident? What are the differences between the side and frontal collision cases?
4. If an IED explosion impulse is blasting to the vehicle not in a vertical direction, but with an angle from the vertical axis, how will the original interaction model between the military vehicle seat and warfighter be modified? If the acceleration profile G due to the IED explosion with an angle is given, write a state-space equation to represent the vehicle seat dynamics.

References

1. Hamilton, N., Luttgens, K.: Kinesiology Scientific Basis of Human Motion, 10th edn. McGraw Hill (2002)
2. Adrian, M., Cooper, J.: Biomechanics of Human Movement. WCB McGraw-Hill, Boston (1989)
3. Chapman, A.: Biomechanical Analysis of Fundamental Human Movements. Human Kinetics (2008)
4. Griffiths, I.: Principles of Biomechanics and Motion Analysis. Lippincott Williams & Wilkins (2006)
5. Chaffin, D., Andersson, G.: Occupational Biomechanics, 2nd edn. John Wiley & Sons (1991)
6. Whittle, M.: Gait Analysis, an Introduction, 4th edn. Butterworth Heinemann, Edinburgh (2007)
7. Chaffin, D.: Digital Human Modeling for Workspace Design. In: Reviews of Human Factors and Ergonomics, vol. 4, p. 41. Sage (2008), doi:10.1518/155723408X342844
8. Gu, E.: Modeling of Human-Vehicle Dynamic Interactions and Control of Vehicle Active Systems. International Journal on Vehicle Autonomous Systems 10(4), 297–314 (2012)
9. Gu, E., Das, M.: Backstepping Control Design for Vehicle Active Restraint Systems. Transactions: ASME Journal of Dynamic Systems, Measurement and Control 135(1), 011012 (2013)
10. TNO, MADYMO Human Models manual, Release 7.2, Tass HQ, The Netherlands (January 2010)
11. Eppinger, R., Sun, E., Bandak, F., et al.: Development of Improved Injury Criteria for Assessment of Advanced Automotive Restraint Systems - II (1999), `http://www.nhtsa.gov/DOT/NHTSA/NRD/Multimedia/PDFs/Crashworthiness/Air%20Bags/rev_criteria.pdf`
12. Kristic, M., Kanellakopoulos, I., Kokotovic, P.: Nonlinear and Adaptive Control Design. John Wikey & Sons, New York (1995)
13. Moes, N.: Digital Human Models: An Overview of Development and Applications in Product and Workplace Design. In: Proceedings of Tools and Methods of Competitive Engineering (TMCE) 2010 Symposium, Ancona, Italy, April 12-16, pp. 73–84 (2010)
14. Chaffin, D.: Digital Human Modeling for Workspace Design. In: Reviews of Human Factors and Ergonomics, vol. 4, p. 41 (2008), doi:10.1518/155723408X342844
15. Duffy, V. (ed.): Handbook of Digital Human Modeling, Research for Applied Ergonomics and Human factors Engineering. CRC Press (2008)
16. Cai, Y. (ed.): Digital Human Modeling. LNCS (LNAI), vol. 4650. Springer, Heidelberg (2008)
17. Abdel-Malek, K., Yang, J., et al.: Towards a New Generation of Virtual Humans. International Journal of Human Factors Modelling and Simulation 1(1), 2–39 (2006)
18. Abdel-Malek, K., et al.: Santos: a Physics-Based Digital Human Simulation Environment. In: The 50th Annual Meeting of the Human Factors and Ergonomics Society, San Francisco, CA (October 2006)
19. Abdel-Malek, K., et al.: Santos: A Digital Human In the Making. In: IASTED International Conference on Applied Simulation and Modeling, Corfu, Greece, ADA542025 (June 2008)

Index

R

S

T

CPSIA information can be obtained
at www.ICGtesting.com
Printed in the USA
LVHW011616160619
621375LV00002B/29/P

9 783642 390463